The Ray Society

INSTITUTED 1844

This volume is No. 150 of the series.

LONDON

1975

1. Heracleum mantegazzianum Somm. & Lev. Bank of river Brent, Ealing

All photographs by T. G. Collet

frontispiece

THE HISTORICAL FLORA OF MIDDLESEX

An account of the wild plants found in the Watsonian vice-county 21 from 1548 to the present time

by

DOUGLAS H. KENT

LONDON
THE RAY SOCIETY
1975

MADE AND PRINTED IN GREAT BRITAIN
AT THE UNIVERSITY PRINTING HOUSE, CAMBRIDGE
(EUAN PHILLIPS, UNIVERSITY PRINTER)

DESIGNED FOR THE RAY SOCIETY BY
RUARI MCLEAN

ISBN NO. 0 903874 03 2

Sold by
The Ray Society
c/o The British Museum (Natural History)
Cromwell Road, London SW7 5BD
England

CONTENTS

FOREWORD

This is an account of changes which have taken place in the flora of an area much of which up to 50 years ago was semi-rural, but almost the whole of which is now highly urbanised. Population growth, pollution, and other factors have resulted in the rapid decrease of certain plant species but also in the establishment and spread of others.

Changes similar to those described in this book are taking place in many other areas in the British Isles and elsewhere, but the close proximity to London has resulted in what may be a longer and more detailed botanical investigation of the area than any other of a similar nature.

I am indebted to the many friends and correspondents who have contributed records and accompanied me on field excursions, and in particular to Gerald Collett, who has also kindly provided the photographs which appear in the book, Margaret Kennedy and Barbara Welch. Grateful thanks are also due to the many specialists who have identified plant material, to Dr W. T. Stearn who has provided much useful advice on the layout of the work, to Edgar Milne-Redhead and Peter Hunt for kindly bringing my manuscript to the notice of the Ray Society, and to that Society for undertaking the publication of the book.

Finally I must extend my deep gratitude to Doreen and Peter Hunt for much assistance in proof-reading.

Douglas H. Kent
Ealing, Dec. 1974

FOREWORD

INTRODUCTION

Middlesex (vice-county 21), containing the Cities of London and Westminster, and the greater part of Central London, is situated in the south-east corner of England, and despite having a tidal river for part of its southern boundary is wholly inland.

With the exception of Rutland (c. 152 sq. miles), Middlesex is the smallest of the English counties, and according to the Agricultural Returns for 1873[1] its total area is 181,317 statute acres, or 283·31 sq. miles. The Local Government Act of 1888, the outcome of which was the formation of the County of London, resulted in the transfer of approximately 51 sq. miles of the eastern part of Middlesex to the newly formed administrative county in January 1889. In the following year certain parts of Hadley and East Barnet were transferred to Hertfordshire. On 1 April 1965 Potters Bar and South Mimms were transferred to Hertfordshire, Staines, Yeoveney, Poyle, Stanwell, Ashford, Laleham, Littleton, Charlton, Shepperton, Halliford and Sunbury were transferred to Surrey, and the remainder of Middlesex placed under the auspices of the Greater London Council. In this work, which is based on the Watsonian vice-county, these changes are ignored.

The shape of the vice-county is that of a very irregular parallelogram which in the extreme north-west thrusts a small spur into Hertfordshire, and in the north-east is penetrated by a spur of Hertfordshire. In the south-west it forms a prominent bulge into Surrey. The greatest length from east to west is c. 19 miles, and from north to south c. 18 miles.

Middlesex is bounded in the south by the river Thames which divides it from Surrey (vice-county 17) and West Kent (vice-county 16), while in the east the old course of the river Lee, now passing through a chain of reservoirs, separates the vice-county from South Essex (vice-county 18). In the west the river Colne and its subsidiary streams divide Middlesex from Buckinghamshire (vice-county 24) and from Hertfordshire (vice-county 20). The northern boundary which is entirely artificial separates the vice-county in its whole length from

[1] This date is chosen in order to assess as accurately as possible the area of H. C. Watson's vice-county 21. The first edition of his *Topographical Botany* in which the vice-county system was first fully used was published in 1873.

Hertfordshire. This boundary is extremely irregular in outline, and mainly follows a series of ancient earthworks, the most important of which is Grimsdyke, along the high ground.

The general surface of Middlesex is a plain sloping gradually southwards to the Thames. The northern border, therefore, contains the highest ground, varying from 400 to just over 500 feet above sea level. The surface is usually undulating but the area lying south of the M4 Motorway and Bath Road is very flat and uniform, being nowhere more than 30 feet above the Thames level at Staines.

Apart from the Thames the vice-county possesses no large rivers but the principal tributaries of that river flowing down from the northern heights are the Colne, Crane, Brent and Lee.

Large areas of Middlesex were originally covered by forest, especially in the north where the subsoil is clay. The whole of the north-east corner was occupied by Enfield Chase, and extensive tracts of woodland covered parts of Hendon, Harrow, Hornsey, Highgate and Hampstead. In the north-west a great oak–hornbeam forest covered most of Harefield, Ruislip and Northwood. At the time of the Norman invasion more than fifty per cent of the vice-county was wooded, but by the mid-eighteenth century this had dwindled considerably. John Roque's map of Middlesex, published in 1754, shows a large area of woodland remaining at Enfield Chase, with smaller tracts at South Mimms, Highgate, Hampstead, Harefield, Northwood, Harrow and Greenford. In 1777 an Act of Parliament was passed for disafforesting Enfield Chase, and about 1800 much of the woodland about Muswell Hill and Hornsey was felled and the areas laid down to grass. By 1888 the total area of woodland remaining in the vice-county amounted to 2,545 acres. The planting of new woodlands, however, helped to improve the position and by 1905 the acreage had risen to 3,968. This improvement was unfortunately only temporary, for between 1914 and 1920 Bishop's and Turner's Woods at Hampstead, the Alder's Wood near Whetstone and Winchmore Hill Wood were largely cleared and built over.

In 1924 the Forestry Commission carried out a survey of woodlands, and the total acreage then remaining was estimated at 1,897 acres.

Today the area of woodland remains at c. 1,900 acres, and the following list summarises the most important woods extant in the vice-county:

*Bayhurst Wood, near Harefield	98 acres
*Coldfall Wood, Highgate	90
†Copse Wood, Ruislip–Northwood	276

* Bayhurst Wood has now been scheduled as a Nature Reserve by the Hertfordshire and Middlesex Naturalists' Trust. The other woods prefixed by an asterisk are also maintained either by the Greater London Council or the appropriate London Borough.

† A large portion of Copse Wood has now been cleared and built over.

*Hadley Wood	c. 50
*Highgate Wood	96
*Ken Wood, and North Wood, Hampstead	c. 130
*Mad Bess Wood, Ruislip	138
Mimmshall Wood, South Mimms	c. 150
*†Old Park Wood, Ruislip	39
*Park Wood, Ruislip	284
*Perivale Wood	27
*Queen's Wood, Highgate	55
†Scratch Wood, Boys Hill Wood, Thistle Wood and Froghall Wood, Edgwarebury	95
*Whitewebbs Park Wood, Enfield	c. 140
Small woods, coppices etc., under 25 acres in area	c. 250
	c. 1,918 acres

While the clay areas of the northern heights were wooded, the sandy tops of the hills and the extensive flat areas of the south-west consisted of barren heaths. At the beginning of the sixteenth century the area of common and waste land probably exceeded 45,000 acres. Continual enclosures were, however, being made, and by 1760 the area had probably decreased to c. 30,000 acres. Towards the end of the eighteenth century an Act of Parliament was passed ordering the inclosure and turning over to cultivation of common and waste lands; thus between 1800 and 1825 over 29,000 acres of common, heath and common fields were inclosed. In 1865 a survey of the commons and open spaces within a radius of twenty-five miles of London was undertaken, and the area of common land for Middlesex was estimated at 2,383 acres. This figure, however, included village greens, roadside wastes, etc.

Little typical heath remains in the vice-county today largely due to the lowering of the water table and the invasion of the areas by birch; this is particularly significant at Hampstead Heath, and Harrow Weald and Stanmore Commons. The most important heaths and commons which are left are tabulated overleaf:

* Part of Old Park Wood has now been scheduled as a Nature Reserve by the Hertfordshire and Middlesex Naturalists' Trust, and Perivale Wood is maintained as a Nature Reserve and Bird Sanctuary by the Selborne Society. The other woods prefixed by an asterisk are also preserved and maintained either by the Greater London Council or the appropriate London Borough.

† A small area of Old Park Wood has now been cleared. A portion of Scratch Wood has also been cleared in connection with the M1 Motorway Extension.

*Hadley Common	c. 180 acres
*Hampstead Heath	293·7
*Harrow Weald Common	44·8
†Hounslow Heath	c. 200
Ickenham Marsh	18·4
*Poor's Field, Ruislip	40
*Shortwood Common, Staines	20·8
†Staines Moor	289·4
*Stanmore Common	120·7
	1,207·8 acres

Many other commons, e.g. Uxbridge Common (14·5 acres), Ickenham Green (21·7 acres), Whitchurch Common, Stanmore (10·4 acres), Ealing Common (47 acres), Eel Brook Common, Parsons Green (14 acres) and Wormwood Scrubs (205 acres) still exist but have been extensively tidied up and are now little more than recreation grounds or playing fields. Finchley Common (c. 15 acres) has been largely spoiled by the tipping of household rubbish in its marshy hollows.

Apart from its heaths and commons Middlesex possesses a number of fine private estates and parks of considerable botanical interest to which the public are allowed freedom of access, e.g. Bushy Park (994 acres), Hampton Court Home Park (752 acres), Osterley Park (c. 250 acres) (all three are possibly remnants of the sixteenth-century Hounslow Heath), and closer to London, Hyde Park and Kensington Gardens (636 acres).

In ancient times a vast area of swamp, up to three miles wide in places, bordered the Thames from Brentford to the Isle of Dogs; large tracts also bordered the river in its higher reaches about Hampton, Sunbury, Shepperton and Staines, and extended along the valleys of the Colne and Lee. During the seventeenth century much of this marshland was reclaimed, and by 1754 most of that remaining was in the Isle of Dogs, and in the valleys of the Colne and Lee. In 1798 Middleton estimated that there were 1,000 acres of marsh in the Isle of Dogs; this was, however, drained during the early and mid-nineteenth century, and by 1880 almost the whole of the 'island' had become covered with docks, port installations, warehouses, factories and houses.

The only areas of marsh remaining in Middlesex today are narrow

* These heaths and commons are preserved and maintained by the Greater London Council or the appropriate London Borough.

† Hounslow Heath has now been extensively devastated by gravel-digging operations and the tipping of rubbish, and a large area of it has been scheduled to be built on. Part of the northern area of Staines Moor may possibly be used for gravel digging while extensive grazing by cattle over the whole area has resulted in the disappearance or decline of many interesting species, and has also resulted in the introduction and establishment of many ruderal species.

isolated tracts bordering the Colne and its tributaries and the Lee, and small patches on various heaths and commons and around the verges of some ponds and lakes.

Sphagnum bog formerly existed in a number of localities, principally on heaths and commons at Harefield, Ruislip, Stanmore, Harrow Weald, Hampstead, and at Ken Wood. Drainage schemes have now almost obliterated the bogs of the vice-county, and the only remaining site of any importance is at West Meadow, Ken Wood, with smaller areas on West Heath, Hampstead and at Stanmore Common.

Middlesex is fortunate in possessing a number of large artificial lakes, among which may be mentioned Ruislip and Brent Reservoirs, Highgate and Hampstead Ponds, Long Water, Hampton Court and Shortwood Common lake.* In addition there are many large gravel pits in the valleys of the Lee and Colne, and in various areas in the south-west. These all provide habitats for aquatic and lacustral species. It is unfortunate, however, that on completion of the gravel-digging operations many of the pits are filled in with household rubbish, soil and other debris, and the reclaimed ground often eventually built on.

Probably no other vice-county in the British Isles has lost so much of its natural area in proportion to its size as Middlesex. The most important factor that led to this was the close proximity of London, and the enormous increase in population, from c. 847,000 in 1801 to 3,560,000 in 1901 and c. 5,000,000 in 1961. This resulted in the greater part of the vice-county being virtually obliterated by a huge torrent of bricks and mortar and the creation of vast new suburbs which engulfed most of the old Middlesex villages and almost eradicated the London Clay grass plain. Large factory estates have been created at Uxbridge, Hayes, Southall, in various parts of the Lee valley and along the Great West, North Circular, and other modern arterial roads resulting in the loss of much valuable land. The construction of London Airport at Heathrow, and the building of a series of large reservoirs from Littleton to Staines and Laleham, has also blotted out some of the finest agricultural land in south-eastern England.

Enormous quantities of gravel, brickearth and sand, and smaller amounts of chalk, have been removed from various parts of the vice-county, the pits created subsequently being filled with household and other rubbish from the London suburbs.

The rapid spread of suburbia has so reduced the agricultural areas of the vice-county that today only a minimum remains. In 1869 over 145,000 acres were either arable or under crops, but by 1939 this had fallen below 21,000 acres, and today is considerably less. The bulk of the arable land remaining is situated in the flat market gardening plain of the south-west where very small quantities of wheat and barley are

* Shortwood Common lake has now been severely damaged by radical excavation works carried out on the common.

grown, together with fodder, cabbages, some root crops and a few cut flowers. The only other arable areas of any size are concentrated in the area between Harrow Weald, Potters Bar and South Mimms.

The Middlesex orchards, always small in size, are also mainly confined to the south-western areas. In 1910 the acreage of orchards was 5,345 but by 1944 had dwindled to 792, and is now undoubtedly considerably less. The produce consists largely of apples and pears, with very small quantities of cherries and plums.

The largest area of permanent grassland was formerly found on the London Clay plain stretching northwards from Uxbridge to Ruislip and Northwood, and eastwards through Hillingdon and Ickenham to Northolt, Greenford, Perivale and Sudbury to Wembley, Kingsbury, Harrow, Kenton, Stanmore and Edgware. During the last fifty years, however, many housing estates have arisen in these areas, though substantial tracts of grassland still survive within the precincts of the 'green belt'. Other areas of permanent grass include tracts in the valleys of the Colne, and, to a lesser extent, the Lee. The acreage of permanent grassland in the vice-county declined from 76,703 in 1869 to 19,890 in 1939, but the latter figure includes parks, recreation grounds, golf courses, etc.

REFERENCES

Andrews, W. (Editor). 1899. *Bygone Middlesex*. London.

Besant, Sir Walter. 1902. *London in the Eighteenth Century*. London.

Briggs, M. 1934. *Middlesex Old and New*. London.

Forestry Commission. 1928. *Report on the census of woodlands in 1924*. London.

Lawson, W. 1872. *Geography of the County of Middlesex*. London.

Middleton, J. 1798. *View of the Agriculture of Middlesex*. London.

Ministry of Agriculture and Fisheries. 1869–1949. *Agricultural Returns and Statistics*. London.

Page, W. (Editor). 1911. *The Victoria History of the County of Middlesex*. vol. 2. London.

Thorne, J. 1876. *Handbook to the Environs of London*. 2 vols. London.

Trimen, H. & Dyer, W. T. Thiselton. 1869. *Flora of Middlesex*. London.

Willatts, E. C. 1937. *Middlesex and the London Region: The Land of Britain: The Report of the Land Utilisation Survey of Britain*. Part 79. London.

SUPERFICIAL GEOLOGY OF MIDDLESEX

I am indebted to S. W. Hester of the Geological Survey and Museum, London, for revising and correcting the following brief account of the superficial geology of the vice-county.

The formations exposed at the surface in Middlesex belong to the Cretaceous, Eocene, Pleistocene and Holocene systems. They are summarised in the following table:

System	Formation	System	Formation
Holocene	Alluvium	Eocene	Bagshot Beds
Pleistocene	Brickearth		Claygate Beds
	Terrace Gravels		London Clay
	Glacial Gravels		Reading Beds
	Boulder Clay	Cretaceous	Chalk
	Pebble Gravels		

The geology of the area is comparatively simple, the vice-county forming part of the structure referred to as the London Basin. The various geological features each have their points of interest in providing useful correlations between geographical and physical features and between the soils and subsoils.

CHALK. The outcrop of chalk occupies a very small area and can be seen dipping below the overlying Eocene deposits at Harefield and Springwell. It is also found at the bottom of the small valley of the Mimmshall Brook at South Mimms.

READING BEDS. This valuable series, consisting of impervious clay in the upper half and sand and loam in the lower, outcrops on the east side of the Colne valley, and along the Pinn through Ickenham and Ruislip to Northwood and Pinner. It occurs also in the valley of the Mimmshall Brook at South Mimms.

LONDON CLAY. A stiff clay which covers a very extensive area in the vice-county and forms heavy land.

CLAYGATE BEDS. Transitional sandy beds at the top of the London Clay, found at Hampstead Heath, Highgate, Stanmore, Elstree, Mill Hill, Harrow and Hanger Hill, Ealing.

BAGSHOT BEDS. The lower Bagshot Sands cap the hills at Hampstead, Highgate and Harrow.

PEBBLE GRAVELS. This deposit occurs in small patches overlying the London Clay or Bagshot Sands on the high ground at Stanmore Common, Elstree, Mill Hill, Potters Bar and Hampstead Heath.

BOULDER CLAY and GLACIAL GRAVEL. A triangular area of high ground with Whetstone at the apex and the base extending from Hendon to Muswell Hill is covered with blue-grey clay, with many stones and gravel. Other areas occur around Belmont, near Stanmore, at Chase Side, Enfield, Harefield and Hillingdon.

TERRACE GRAVELS. The gravels associated with the Thames and

its tributaries originally formed some of the best land in Middlesex. Three terraces can be recognised by their constituent material and approximate elevation.

1. The BOYN HILL TERRACE GRAVEL lies at about 100 ft. O.D., and consists of coarse material. An area roughly two miles by one mile occurs south of Hillingdon and smaller patches are found at Islington and on the west side of the Lee valley.

2. The TAPLOW TERRACE GRAVEL and overlying brickearth have a wide and almost continuous outcrop at about 50 ft. O.D., from Yiewsley through Hayes and Southall, and south to Hounslow and Hampton, Acton, Shepherds Bush, Bayswater, Kensington, Hyde Park, the City of London, Stepney, Clapton, Tottenham, Edmonton, Ponders End and Enfield.

3. The FLOOD PLAIN TERRACE, consisting of fine gravel and sand, forms wide spreads bordering the Thames at Staines, Hampton Court, Teddington, Chiswick, Fulham, Chelsea and Limehouse.

ALLUVIUM. A deposit consisting of silt clay and peat occupies tracts in the Isle of Dogs and northwards along the Lee valley. It is also to be found along the Colne valley, and fringes the Thames and some of its tributaries. The City of Westminster is built upon alluvial soil, and was once an 'island' (Thorney Island).

THE CLIMATE OF MIDDLESEX

RAINFALL

Rainfall may be considered the most essential of climatic conditions; it is also the most variable both in place and occurrence, two localities but a short distance apart receiving extremely variable precipitation at the same time.

The average annual rainfall in England is 33 inches, and that for Middlesex between the years 1916 and 1950 varied from 27·72 inches at Potters Bar, on the high ground in the north-east, to 23·44 inches on the low ground at St James's Park, Westminster. Further west along the Thames valley averages of around 22·5 inches occurred. Variation in rainfall is, however, completely incalculable and annual falls may be as much as 60 per cent to 140 per cent of the average.

The greatest precipitation over a short period on record in the vice-county occurred at Camden Square, St Pancras, on 23 June 1878 when 1·04 inches fell in 10 minutes, 1·43 inches fell in 15 minutes, 1·82 inches fell in 20 minutes and 2·30 inches fell in 28 minutes. Another noteworthy downpour occurred at Campden Hill, Kensington, on 6 June 1917 when 4·65 inches of rain fell in two and a half hours. Amongst the highest actual monthly records of rainfall for Middlesex may be mentioned 7·13 inches at Hampstead in October 1865.

On an average falls of 0·01 inches or more occur on 139 days, and falls of 0·04 inches or more on 138 days during the year.

TEMPERATURE

The outstanding feature of Middlesex temperature records is their general similarity, although the averages are slightly higher in built-up areas than in the more rural districts. This is due to the retention of heat by the bricks and stonework of buildings combined with the artificial heating of the atmosphere by domestic and industrial appliances.

Over a long period at Kew, Surrey (just over the vice-county boundary), the average mean temperature was found to be 50·4 °F., while in the closer built-up areas of London the average mean temperature of St James's Park, Westminster, proved to be 51·4 °F., and at Camden Square, St Pancras, 51·2 °F. In the north-east Enfield produced a mean average of 50·1 °F., while Tottenham had a mean average of 51·1 °F.

The parts of Middlesex situated near the Thames appear to have a slightly higher temperature than the remainder of the vice-county, and in winter there is a very slight increase in temperature from east to west; this is reversed in the summer months when the eastern areas enjoy a slightly higher temperature than the western.

Extreme temperatures are rare and usually of short duration, the lowest temperature on record for the British Isles being −17 °F. at Braemar, Aberdeenshire, on the night of 11 February 1895. The lowest temperature on record for Middlesex is apparently that quoted by H. C. Watson in his *Cybele Britannica* (−11 °F. at the Royal Society, London, on the night of 19 January 1838); he also gives the following records for the same night: −4·5 °F. at Chiswick and −4·0 °F. at Hampstead Heath and Kensington. Howard records a temperature of −6·5 °F. at Edmonton on the night of 24 December 1796, and −5 °F. was recorded at Tottenham on 9 February 1816.

The highest temperature on record for the British Isles is 100·5 °F. at Tonbridge, Kent, on 22 July 1868, but readings exceeding 90 °F. are very rare even in warm summers. The highest temperature recorded in Middlesex ('north of London') was 97·0 °F. on 18 July 1825. A temperature of 94·4 °F. was recorded at Chiswick in 1836, and 94·1 °F. at Hampstead Heath on 9 August 1911.

SUNSHINE

The mean daily average of bright sunshine between 1921 and 1950 varied from 3·96 hours at Enfield (448 ft. O.D.) and 4·03 hours at Hampstead (450 ft. O.D.) in the north of the vice-county to 3·54 hours at Bunhill Row (80 ft. O.D.) in the heavily built-up area of

Central London. Regents Park (129 ft. O.D.) had an average of 3·66 hours, while in the north-east, Tottenham (51 ft. O.D.) had an average of 3·87 hours.

About fifty per cent of the winter sunshine of the City and inner London was previously cut off by smoke, and in December 1890 the aggregate duration for the month at Bunhill Row was only 0·1 hour, and every day except one was sunless. This depressing situation has doubtless been greatly improved by the enforcement of the Clean Air Act.

WIND

The prevailing winds blowing across Middlesex come from the south and south-west, though winds from other directions are not infrequent. Gales and extreme winds are uncommon.

ATMOSPHERIC POLLUTION

In common with other built-up areas atmospheric pollution fluctuates considerably in the vice-county, being relatively high at certain times during the autumn and winter months, and less concentrated at other times of the year. The types of air pollution are diverse and are controlled by a number of factors including wind direction, which determines the areas to which the pollution is directed, and rainfall which assists in the deposition of the pollutants.

The most frequent air pollutants are coal smoke containing quantities of soot and sulphur dioxide, and carbon monoxide and diesel oil fumes emitted from the exhausts of motor vehicles. Domestic sources were reputed to be responsible for an estimated 84 per cent of the smoke emission in the old County of London, but the Clean Air Act has undoubtedly greatly reduced that figure.

Bleasdale has pointed out that 'the environment for plant growth is affected in three ways by the presence of coal smoke: (1) by a reduction in the amount of light available to the plant, (2) by an alteration in soil conditions, and (3) by the contamination of the air by foreign gases'. Cohen and Rushton showed by a number of experiments that rainfall in polluted areas reduced the number of bacteria available in the soil, and Bleasdale has noted that 'this in turn could result in a reduction in the availability of plant nutrients'. It seems likely that atmospheric pollution is responsible for the loss or decrease of many species of liverworts and mosses in the vice-county, and to a lesser extent for the decrease of a number of phanerogams.

While atmospheric pollution is the likely cause of the disappearance and reduction in quantity of various species, Bowen has pointed out that certain other species thrive and spread in areas where the soil is heavily polluted; thus the sulphur dioxide precipitated from the poisoned atmosphere by rain into the soil appears to be an important factor in the spread of a number of weedy species, particularly in the Cruciferae and Compositae.

WATER POLLUTION

In recent years the discharge of effluents from factory wastes into the streams and canals of Middlesex has poisoned the waters to such a degree as to greatly reduce the numbers of many of the aquatic species growing in them. Major sources of pollution appear to be the large quantities of sodium hydroxide and soluble phosphates derived from household detergents which now reach the sewage farms in enormous quantities. While many species are adversely affected by the poisoning of the water there are, on the other hand, a few which thrive and increase under conditions of high pollution.

REFERENCES

Bilham, E. G. 1938. *The Climate of the British Isles*. London.

Bleasdale, J. K. A. 1959. The effects of air pollution on plant growth, in Yapp, W. B. *The Effects of Pollution on Living Material*. London.

Bowen, H. J. M. 1965. Sulphur and the distribution of British plants. *Watsonia* 6: 114–119.

Chandler, T. J. 1964. Climate and the built-up area, in Coppock, J. T. & Prince, H. C. (Editors). *Greater London*: 42–51. London.

Howard, L. 1818. *The Climate of London*. 2 vols. London.

Lamb, H. 1964. *The English Climate*. Edition 2. London.

Marshall, W. A. L. 1952. *A Century of London Weather*. M.O. 508. London.

Meteorological Office. 1953. *Averages of Temperature for Great Britain and Northern Ireland, 1921–1950*. M.O. 571. London.

—— 1953. *Averages of Bright Sunshine for Great Britain and Northern Ireland, 1921–1950*. M.O. 572. London. Reprinted 1959.

—— 1958. *Averages of Rainfall for Great Britain and Northern Ireland, 1916–1950*. M.O. 635. London.

—— 1865–1957. *British Rainfall*. London.

Watson, H. C. 1847–59. *Cybele Britannica*. 4 vols. London.

THE BOTANICAL EXPLORATION OF MIDDLESEX

SIXTEENTH CENTURY

The first author to record detailed localities of Middlesex plants was William TURNER (c. 1508–68), physician and churchman, whose religious beliefs caused him to spend much time on the Continent in enforced exile. In 1548, while living at Syon House, Brentford, then the residence of the Duke of Somerset, he published his *Names of Herbes, etc.* (republished by the Ray Society in 1965). This work contains references to a few species growing in the vicinity of Syon. In a *Newe Herball*, printed in three parts between 1551 and 1562, he recorded further Middlesex plants, chiefly from London, Brentford and Hounslow Heath.

Pierre PENA (c. 1530–1605) and Matthias de L'OBEL (1538–1616) came to England from France in 1566, and in 1570–71 published a volume entitled *Stirpium Adversaria Nova, etc.*; this provided a few references to aquatic species noted in the Thames about London.

Pena returned to France in 1572, but l'Obel remained in England and took up residence at Highgate. In 1576 his *Stirpium Observationes* was printed in Antwerp. This book included records of a number of species new to the vice-county from Hampstead, Highgate and the areas near the Thames. His final work, *Stirpium Illustrationes, etc.*, was published posthumously in 1655 under the editorship of his son-in-law, William How.

During 1597, one of the most famous British herbals made its appearance – *The Herball or generall Historie of Plantes, etc.*, by John GERARDE (or Gerard) (1545–1612), Barber-Surgeon of London. This provided descriptions and localities of many species new to Middlesex mainly from Hampstead Heath, Highgate and the northern suburbs, but it also contained a high percentage of dubious records.

SEVENTEENTH CENTURY

Jules Charles de L'ESCLUSE (or Carolus Clusius) (1526–1609), a Fleming, who in 1593 became Professor of Botany at Leyden, visited England on a number of occasions. He knew both Pena and l'Obel and made a number of excursions to various parts of England in their company. In 1601 he published his *Rariorum plantarum historia, etc.*, which contained records of a few plants noted in London.

The earliest known printed account of organised botanical excursions in the British Isles appeared in 1629 under the title *Iter Plantarum Investigationis Ergo Susceptem A decem sociis, in Agrum Cantianum. Anno Dom. 1629 Julii 13*, with an appendix of three pages headed *Ericetum Hamstedianum Sive Plantarum ibi crescentium observatio habita, Anno eodem* [*1629*] *1 Augusti*. This provided details of botanical excursions in Kent, and to Hampstead Heath, by the author, Thomas JOHNSON (c. 1604–44), an apothecary of London, and some of his friends. Data on further excursions to these places were published in 1632 under the heading *Descriptio Itineris Plantarum Investigationis ergo suscepti, in Agrum Cantianum*, with an appendix of seven pages entitled *Enumeratio Plantarum in Ericeto Hampstediano locisque vicinis Crescentium, etc.* The two works contained many records of species new to Hampstead Heath. During 1633 Johnson edited a new and very much enlarged and amended edition of Gerarde's *The Herball, etc.* This included records of a large number of species new to the vice-county, chiefly from Hampstead and Highgate. It was reprinted in 1636. In 1638 Johnson compiled a 'Catalogus plantarum juxta Tottenham lectarum'. This most useful catalogue of the plants noted growing about Tottenham contained numerous new records for Middlesex. It was unfortunately never published but a copy of the manuscript (the whereabouts of the original is not known) is preserved in the Library of the Royal Botanic Gardens, Kew. Johnson's association with Tottenham remains obscure, but the comprehensive and accurate list of plants given in his manuscript suggests that he knew the district well, and may possibly have had a country house there. In his various writings he added well over 250 species to the vice-county list.

Among the early British horticultural treatises one of the most renowned is *Paradisi in Sole. Paradisus Terrestris: or, A Garden of all sorts of Pleasant Flowers, etc.*, published in 1629. The author of this interesting work, in which *Scilla autumnalis* is recorded from beside the Thames at Chelsea, was John PARKINSON (1567–1650), a London apothecary. In 1640, his *Theatrum Botanicum: The Theater of Plants; or an Herball of large extent, etc.*, was published, and although somewhat overshadowed by the appearance of Johnson's excellent revision of Gerarde's *Herball* a few years previously, it nevertheless added a few more species, chiefly from London and Kentish Town, to the Middlesex list.

John GOODYER (1592–1664), of Mapledurham, Hampshire, visited Middlesex on a few occasions and noted *Sisymbrium irio* and *Geranium columbinum* near Whitechapel, and *Damasonium alisma* on Hounslow Heath.

William HOW (c. 1619–56), a London physician, published a work entitled *Phytologia Britannica, etc.*, in 1650. This included records of plants from London, Hampstead, Highgate, Hampton Court and Uxbridge.

Pinax Rerum Naturalium Britannicarum, by Christopher MERRETT (1614–95), a London physician, was printed in 1666. This interesting book may perhaps be regarded as one of the earliest attempts to produce a systematic work. Most of the copies, stored in the City of London, were destroyed in the Great Fire of 1666, and an unauthorised 'pirated' reprint appeared in 1667. It seems probable that Merrett considered producing a second edition of the book, for a copy containing many additions and corrections in his handwriting is preserved in the Library of the British Museum. A new edition was unfortunately not printed. Merrett apparently travelled widely in Middlesex, and recorded plants from Teddington, Hampton Court, Brentford, Hampstead and London.

John RAY (1627–1705), the celebrated naturalist and churchman, published a number of important botanical works, including *Catalogus Plantarum Angliae, etc.* (1670), *Fasciculus Stirpium Britannicarum, etc.* (1688), *Historia Plantarum, etc.* (1686–1704), and *Synopsis Methodica Stirpium Britannicarum, etc.* (1690), the latter book achieving three editions. Ray had an extremely critical knowledge of native British plants, and although his visits to Middlesex appear to have been infrequent he was the first to record a number of rushes and grasses from Hampstead Heath and London.

Between 1677 and 1679 Christopher Love MORLEY (fl. 1646–1702), a physician of Hampstead, compiled a 'Catalogus planta villâ nostrâ Hamstiadensis'. This catalogue of plants, both wild and cultivated, noted growing about Hampstead, included records of a few species new to Middlesex. The work remained unpublished but the manuscript is preserved at the British Museum.

In Edmund Gibson's edition of Camden's *Britannia* published in 1695 is a list of the 'More rare plants growing in Middlesex'. The compiler of this list was James PETIVER (c. 1658–1718), a London apothecary. Petiver was undoubtedly one of the most knowledgeable and competent botanists of his time, and made extensive collections of specimens in many branches of the biological sciences. His literary output was considerable for the period, and his botanical works included *Museum Peteverianum, etc.* (1695–1703), *Botanicum Londinense, etc.* (1709–10), *Herbarij Britannici clariss. D. Raii Catalogus cum Iconibus, etc.* (1713–15) and *Graminum, Muscorum, Fungorum, Submarinorum, etc.* (1716). He collected extensively about London, Chelsea, Hampstead and Highgate, and his large herbarium is now incorporated in the Sloane Herbarium.

Samuel DOODY (1656–1706), another London apothecary, contributed records of Middlesex plants to Ray's *Synopsis* (1690). He was a close friend of Petiver, and provided a number of records to the former's 'More rare plants growing in Middlesex'. Although much of Doody's botanising was confined to London and Hampstead, he appears to have had some connection with Isleworth, where he observed *Ranunculus parviflorus* and *Tordylium maximum*, while from the nearby Hounslow Heath he recorded *Pilularia globulifera* and *Utricularia minor*.

Adam BUDDLE (c. 1650–1715), a churchman, was a close friend of both Petiver and Doody, and spent much time in their company searching for plants. About 1708 he compiled his 'Hortus Siccus Buddleanus sive Methodus nova Stirpium Britannicarum', which consisted of a reasonably accurate account of the British flora as known at the time. The work was extensive and remained unpublished, though the manuscript is now preserved in the Sloane Collection at the British Museum, London. Buddle also accumulated a large herbarium, which is now incorporated in the Sloane Herbarium. Most of the species new to the vice-county found by Buddle were from the Hampstead area.

Leonard PLUKENET (1641–1706), a London physician, published a number of works on British botany at his own expense. His botanical activities appear to have been confined to the close proximity of London, though he paid particular attention to the interesting and varied flora of Tothill Fields, Westminster, where he was the first to record *Trifolium subterraneum*.

William STONESTREET (c. 1650–1716), Rector of St Stephen's, Walbrook, London, from 1689, knew Petiver, Doody and Plukenet, and botanised in the vicinity of London, particularly about Chelsea, where he noted a number of species new to Middlesex.

James NEWTON (c. 1639–1718), a medical man, botanised in the Hampstead–Highgate area towards the end of the seventeenth century, and observed a few species new to the vice-county.

Samuel DALE (1659–1739), an apothecary and physician of Braintree, Essex, visited Middlesex on a number of occasions and noted several species new to the vice-county list.

William SHERARD (1659–1728) contributed records of Middlesex plants to Ray's *Synopsis* (1690). He was consul at Smyrna from 1703 to 1715, and on his return to England took up residence on Tower Hill, London. In 1721 he was responsible for bringing Dillenius to England. He founded the Sherardian Chair of Botany at Oxford, and bequeathed his herbarium, library and manuscripts to the University. Sherard's botanical activities in Middlesex appear to have been chiefly confined to London and its immediate vicinity.

EIGHTEENTH CENTURY

The third edition of Ray's *Synopsis* appeared in 1724, under the editorship of Johann Jakob DILLENIUS (1684–1747), a German doctor who was brought to England by William Sherard in 1721. Dillenius spent much time at Sherard's home on Tower Hill, and visited many parts of Middlesex in search of plants. On the death of Sherard in 1728, Dillenius was appointed first Sherardian Professor of Botany at Oxford. Dillenius travelled widely, visiting many remote parts of Britain, while collecting data towards a proposed fourth edition of Ray's *Synopsis*. This project, however, remained unfulfilled. Most of the species new to Middlesex recorded or noted by Dillenius were from Hampstead and Hounslow Heaths. His large herbarium of British plants is now preserved at the University of Oxford.

Isaac RAND (d. 1743), a London apothecary, contributed records of Middlesex plants to the Dillenian *Synopsis* (1724), and compiled a herbarium which is now incorporated in the Herbarium of the British

Museum (Natural History). His records were mainly from London, Hampstead and Hounslow Heath.

John MARTYN (1699–1768), the son of a Hamburg merchant, was responsible for founding in 1721 one of the earliest botanical societies in England. The headquarters was in London, and meetings took place each Saturday evening at the Rainbow Coffee House in Watling Street, and after a time at the homes of various members. Dillenius was secured as President, and the members included Thomas Dale (1699?–1750), a medical man, John Wilmer (1697–1769), a London apothecary, Robert Fysher (1698–1747), a physician, John Chandler (1700–80), a London apothecary, and Joseph Harris (1704–?), a surgeon-apothecary, all of whom contributed in various ways to the knowledge of Middlesex botany. The Society flourished for a few years but was apparently dissolved some time after 1726. From 1727 Martyn practised medicine in London, and in 1733 he was elected Professor of Botany at Cambridge, a position which he held until 1761. He published a number of botanical works, of which the best known is probably his translation and adaptation of Tournefort's 'History of Plants growing about Paris' to the plants found in Britain, which appeared in 1732. Martyn appears to have been the first botanist to pay attention to the Isle of Dogs.

One of the most important books on Middlesex botany – *Fasciculus Plantarum circa Harefield sponte nascentium, etc.* – was published in 1737. The author of the work was John BLACKSTONE (1712–53), a London apothecary. His maternal grandfather, Francis Ashby, owned property at Harefield, and young Blackstone, who suffered indifferent health, spent much time there in the course of which he compiled his careful and accurate account of the plants growing in the area, often with details of habitat preference and frequency. A sketch of the topography and superficial geology of the district was also provided. These data were skilfully brought together in the *Fasciculus*, which may be regarded as one of the most competent early local Floras. Blackstone corresponded with Dillenius, Richard Richardson, Sir Hans Sloane and many other eminent botanists of the period, and in 1746 published *Specimen Botanicum quo Plantarum plurium rariorum Angliae indigenarum loci natales illustrantur*. This provided details of localities for rare and local species in many parts of England, but contained also most of the records previously published in the *Fasciculus*. Blackstone added well over 100 species new to the flora of Middlesex, and provided the earliest evidence of the occurrence of calcicole species in the vice-county.

John WILMER (1697–1769), an apothecary of London, knew Blackstone, and provided a number of records which were included in the latter's *Specimen Botanicum* (1746). He appears to have been well acquainted with the flora of the Cowley and Hackney districts.

John HILL (c. 1716–75), a London apothecary, medical man, and would-be actor, was the author of a number of botanical works containing Middlesex records. The most important of these were *The British Herbal, etc.* (1756–57), *The Vegetable System; etc.* (1759–95), and *Flora Britannica* [*sic*] *: sive, Synopsis Methodica Stirpium Britanicarum* [*sic*], *etc.* (1760). His botanising appears mainly to have been in the vicinity of London, Hampstead and Uxbridge.

William WATSON (1715–87), a London apothecary and physician, knew Blackstone and Hill, and contributed Middlesex records to their various publications. Watson carried out many experiments with electricity, and was knighted in 1786. He appears to have visited many parts of the vice-county, including Hampstead, Hounslow Heath, Hillingdon and Uxbridge, in search of plants.

Robert NICHOLLS (d. 1751), an apothecary of London, also knew Blackstone and Hill and provided records of Middlesex plants which were included in their various publications. Nicholls appears to have been one of the first botanists to pay attention to the Staines area.

In 1762, a small, slim volume entitled *Miscellaneous Tracts on Natural History, etc.* was published. This contained an appendix headed 'Observations on Grasses'. Its author was Benjamin STILLINGFLEET (c. 1702–71), one-time Barrack-master at Kensington. This book is of some interest as it was the first containing data relating solely to the Gramineae.

William HUDSON (1730–93), a London apothecary, was the author of *Flora Anglica, exhibens Plantas per regnum Angliae sponte crescentes, distributas secundum Systema Sexuale, etc.*, which appeared in 1762. This was a most important work, based on the binomial terminology of Linnaeus, and it was immediately adopted as the standard manual of British botany. A second, enlarged edition of the book, with additions and corrections, was published in 1778. Hudson visited many parts of the vice-county, and recorded species from the inner London suburbs, Colney Hatch, Enfield and Staines.

Stanesby or Stainsby ALCHORNE (1727–1800), a London apothecary, and one-time Assay Master at the Royal Mint, knew Hudson and contributed to *Flora Anglica* (1762). His botanical expeditions appear to have been concentrated to the more northern parts of the vice-county, and particularly the Stoke Newington, Hornsey and Southgate districts.

Peter COLLINSON (1694–1768), renowned for his famous garden containing many exotic trees and shrubs at Mill Hill, contributed a number of interesting records. His botanising was often undertaken in the company of his son Michael COLLINSON (c. 1727–95). Father and son were both very interested in British orchids, and they confirmed a number of Blackstone's records from Harefield, as well as providing new data from Mill Hill and Enfield.

Thomas MARTYN (1735–1825), a churchman, and the son of John

Martyn (1699–1768), was the author of a number of publications containing records of Middlesex species. His works included *Plantae Cantabrigiensis . . ., etc.* (1763), which included lists of species noted in other counties, and *Flora Rustica, etc.* (1792–95). Martyn also edited the ninth edition of Philip Miller's *The Gardeners' and Botanists' Dictionary, etc.*, which appeared in 1807.

William CURTIS (1746–99), a London apothecary, was the author of one of the finest and most famous botanical works of all time – *Flora Londinensis: or, Plates and Descriptions of such Plants as grow wild in the Environs of London, etc.* This was printed in numbers, commencing in 1775. Each number contained six plates with descriptions in folio. Twelve numbers formed a fasciculus, of which six were issued. The plates were of the highest quality and are still regarded by many authorities as the finest illustrations of British plants ever published. The work progressed very slowly and was still uncompleted at Curtis's death. Curtis was an extremely careful and accurate observer and provided a number of records of plants new to the vice-county, chiefly from the vicinity of London.

John LIGHTFOOT (1735–88), a churchman of Uxbridge, knew Hudson and Curtis, and contributed records to the second edition of the former's *Flora Anglica* (1778) and the latter's *Flora Londinensis* (1775–98). Lightfoot was a very competent botanist, and knew the flora of Hounslow Heath, Colnbrook and Uxbridge very well. He compiled a large herbarium, the remnants of which are now preserved at the Royal Botanic Gardens, Kew.

In Gough's edition of Camden's *Britannia*, published in 1789, there appeared a list of 'Rare plants found in Middlesex'. The compiler of this list was Edward FORSTER (1765–1849), a banker, who was a keen amateur botanist. Forster knew most of the leading botanists of his day and contributed Middlesex records to a number of their publications. He botanised extensively in the eastern half of the vice-county and provided many new records from Enfield, Edmonton, Stoke Newington and Hackney. The large herbarium which he compiled during his long life is now incorporated in the herbarium of the British Museum (Natural History). His brother, Thomas Furly FORSTER (1761–1825), contributed a few Middlesex records to various botanical works, but his botanical activities were mainly confined to Essex.

Colin MILNE (c. 1743–1815), a churchman, and Alexander GORDON (fl. 1770–1845), a Reader in Botany, were the joint authors of *Indigenous Botany, etc.*, published in 1793. This contains the results of a number of botanical excursions made in Kent, Middlesex and the adjacent counties between 1790 and 1792. It contains a few new Middlesex records but is blemished by a large number of inaccuracies.

In 1790 the publication of a new and very important British Flora commenced – its title was *English Botany; or, Coloured Figures of British Plants, etc.* The text was written by James Edward SMITH (afterwards Sir James Edward Smith) (1759–1828), and the figures were drawn by James SOWERBY (1757–1822). This excellent work was not completed until 1814. During 1794 Smith began his *Flora Britannica*, which was published in three volumes in 1804. His final work, *The English Flora*, appeared in four volumes between 1824 and 1828. Smith was one of the first British botanists to seriously study the critical genus *Salix*, and his treatment of many of the sections is still valid today. Most of his new vice-county records were from London, though a few were from the Harrow district.

Samuel GOODENOUGH (1743–1827), Bishop of Carlisle, made a careful study of British plants, and was one of the earliest British botanists to undertake a critical study of the large and complex genus *Carex*. He knew Curtis, Smith, Turner and Dillwyn, and contributed Middlesex records to their various publications. At one time resident in Ealing, most of his records were made there and in the adjacent area of Hanwell. He did, however, contribute a number of notes from Hounslow Heath.

Joseph BANKS (afterwards Sir Joseph Banks) (1743–1820), a man of considerable means, and famous for his expedition in the 'Endeavour' with Captain Cook, gathered many interesting species about Brompton and Chelsea at the end of the eighteenth century. He also visited the Harefield area where he discovered *Orchis militaris* at 'Gulch-well' (= Garett Wood, Springwell), a station rediscovered by Benbow nearly a century later. His specimens are now incorporated in the Herbarium of the British Museum (Natural History).

Aylmer Bourke LAMBERT (1761–1842), another early specialist on the genus *Salix*, contributed records of Middlesex plants to Smith's *English Botany* (1790–1814), and to Turner and Dillwyn's *Botanist's Guide* (1805).

NINETEENTH CENTURY

The Botanist's Guide through England and Wales, by Dawson TURNER (1775–1858) and Lewis Weston DILLWYN (1778–1855), was published in 1805. It contained a few records new to Middlesex, and although neither author appears to have spent much time botanising in the vice-county Dillwyn at least visited the Isle of Dogs.

Joseph WOODS (1776–1864), an architect, who was an enthusiastic amateur botanist, made a critical study of the British roses. He contributed records to Turner and Dillwyn's *Botanist's Guide*, and to the works of other authors of the period. Woods botanised chiefly in the north-eastern parts of Middlesex, especially about Potters Bar, Muswell Hill and Stoke Newington.

Edward HUNTER (fl. 1790–1824), steward at Caen (= Ken) Wood

Towers, provided a number of notes and records from Ken Wood to the *Topography and Natural History of Hampstead* by John James Park (1795–1833), which was published in 1814. Although many of Hunter's records were dubious, and some certainly erroneous, he was the first to note *Maianthemum bifolium* at Ken Wood.

The Spanish botanist, Mariano LA GASCA Y SEGURA (1776–1839), came to England in 1822 and stayed until 1831. Between 1826 and 1827, with the aid of his son, he collected a number of plants, including many from the Hampstead area. These were made into sets, entitled 'Hortus Siccus Londinensis, etc.', and were apparently sold. It is regrettable that no complete set exists in Britain, but an incomplete set is preserved in the herbarium of the University of Oxford.

Nathaniel John WINCH (1768–1838), a Newcastle botanist, provided records for Watson's *New Botanist's Guide* (1835–37). Born at Hampton, he appears to have removed to Newcastle at an early age, and to have made infrequent visits to his native county. His few records originated from Hampton and Hounslow Heath.

John Joseph BENNETT (1801–76), Keeper of Botany at the British Museum from 1858, botanised in the northern suburbs and compiled a herbarium which is now incorporated in the Herbarium of the British Museum (Natural History). His brother, Edward Turner BENNETT (1797–1836), a surgeon, who was a keen amateur zoologist, was also interested in botany and collected a number of interesting species in the vice-county. These were apparently passed to J. J. Bennett.

Between 1825 and 1834 a 'Catalogue of over 600 plants found within a two mile radius of Hampstead Heath' was compiled by Alexander IRVINE (1793–1873), a schoolmaster. In 1838 Irvine published *The London Flora; containing a concise description of the Phaenogamous British Plants which grow spontaneously in the vicinity of the Metropolis, etc.*; this included records of plants not only from London but also from many other parts of Britain. He contributed notes and records to *Loudon's Magazine of Natural History*, *The Phytologist*, *Science Gossip*, and various other botanical periodicals. In 1855 he became editor of *The Phytologist*, and in 1858 produced his *Illustrated Handbook of the British Plants*. He also contributed many records to Trimen and Dyer's *Flora of Middlesex* (1869). Irvine travelled widely in Middlesex and his records were made from localities as widely separated as Teddington and Enfield; the bulk, however, were from Hampstead and Chelsea.

Ezekiel George VARENNE (1811–87), a physician and surgeon, provided many records for Trimen and Dyer's *Flora of Middlesex* (1869). Although best known as an Essex botanist he spent part of his youth in residence at Marylebone and Stanmore.

Henry KINGSLEY (fl. 1830–64), a doctor and surgeon of Uxbridge, and a leading member of the Botanical Society of London, botanised extensively in the Uxbridge and Harefield districts where he collected many interesting species. His herbarium was acquired by J. C. Melvill and is now incorporated in the Melvill Herbarium at the Butler Museum, Harrow School.

The first volume of *The New Botanist's Guide to the Localities of the Rarer Plants of Britain*, by Hewett Cottrell WATSON (1804–81), was printed in 1835; the second volume appeared in 1837. This work listed the localities of some of the rare and local plants of Britain under their respective counties. In 1847 Watson published the first volume of his magnum opus *Cybele Britannica; or, British Plants and their Geographical Relations*. Further volumes followed in 1848, 1852 and 1859, and a supplement in 1860. Between 1868 and 1870 he produced the *Compendium of the Cybele Britannica*. He contributed a number of records, mainly from the area of the Thames between Sunbury and Twickenham, to Trimen and Dyer's *Flora of Middlesex* (1869), and in 1873 published the first volume of another important work, *Topographical Botany, etc.*, the second volume following in 1874. Watson compiled a large herbarium which is now preserved at the Royal Botanic Gardens, Kew.

William PAMPLIN (1806–99), a botanical bookseller and publisher of Soho, London, was friendly with Alexander Irvine, and in his younger days botanised about Hampstead, Highgate and Chelsea, often in the company of the latter. He contributed records to Cooper's *Flora Metropolitana* (1836), and to Trimen and Dyer's *Flora of Middlesex* (1869), and published many notes in *The Phytologist* and other botanical periodicals.

Flora Metropolitana; or Botanical Rambles within thirty miles of London, by Daniel COOPER (c. 1817–42), a leading member, and Curator, 1837–38, of the Botanical Society of London, was published in 1836. This was the product of many botanical excursions made within thirty miles of London between 1833 and 1835, though the author regretted (in the preface) that he had been able to give little attention to Middlesex. A Supplement to the Flora was printed in 1837. Cooper added a few new records to the vice-county, chiefly from areas near the Thames and from Hampstead Heath.

Daniel Chambers MACREIGHT (1799–1868), a medical man residing in London, published a *Manual of British Botany* in 1837. This included two new vice-county records – *Crataegus laevigata* from Tottenham and *Orobanche minor* from Twickenham.

Between 1837 and 1842, Edward BALLARD, a medical student of Islington, who later became an authority on public health, compiled 'A catalogue of plants indigenous to Islington and its immediate vicinity'. This work was not published in the author's lifetime but was printed in the *Journal of Botany* during 1928 under the editorship of John Ardagh. It provided a valuable account of the flora of the Islington and Hornsey areas in the early nineteenth century. Ballard also compiled a herbarium of plants from the Islington area, and this is now incorporated in the Herbarium of the British Museum (Natural History).

Thomas TWINING (1806–95), an authority on technical education, resided in Twickenham in the early part of the nineteenth century and made a study of the flora of the district. Although he failed to publish an account of his work he corresponded with H. C. Watson, and many of his specimens are in Herb. Watson at Kew.

Thomas MOORE (1821–87), a gardener and horticulturist, assisted in the laying out of the Regents Park gardens (formerly Marylebone Fields) between 1844 and 1847. In the course of this work he collected and preserved many specimens of the 'weeds' that he encountered. A keen pteridologist, he published in 1848 *A Handbook of British Ferns, etc.*, which was followed in 1851 by *A Popular History of British Ferns, etc.* During 1859–60 he produced a further book – *The Octavo Nature-printed British Ferns* – followed in 1862 by *The Field Botanist's Companion, etc.* Moore's botanising appears to have been mostly confined to the immediate vicinity of London, and in particular Regents Park and Hampstead Heath. Many of his specimens are in the Herbarium of the British Museum (Natural History).

The publication of the third edition of Smith's *English Botany* was commenced in 1863, under the editorship of John Thomas Irvine Boswell-Syme né SYME (1822–88), the work being finally completed in 1872. Syme lived at Hampstead for many years and contributed records from that district to Trimen and Dyer's *Flora of Middlesex* (1869). He compiled a large herbarium which is now preserved at the British Museum (Natural History).

William Marsden HIND (1815–89) was appointed to the curacy of Pinner in 1859. He took a keen interest in the local flora, publishing the results of his observations under the title of 'The Flora of Harrow and its Vicinity' in the *Harrow Gazette* during January and February 1860. He greatly assisted J. C. Melvill in the preparation of *The Flora of Harrow* which appeared in 1864, and contributed numerous records from Pinner, Harrow and Greenford to Trimen and Dyer's *Flora of Middlesex* (1869). In 1871 he published notes on a number of Middlesex species in the *Journal of Botany*, and in 1876 edited the second edition of Melvill's *Flora of Harrow*.

James BRITTEN (1846–1924), a professional botanist, and editor of the *Journal of Botany* from 1880 until his death, contributed many notes on London plants to the *Botanist's Chronicle*, *The Naturalist*, *Journal of Botany*, and other periodicals. He also donated many records from Chelsea, Brompton and Brentford, where he resided, to Trimen and Dyer's *Flora of Middlesex* (1869).

Edwin LEES (1800–87), better known as a Worcestershire botanist, contributed records to Melvill's *Flora of Harrow* (1864), and to Trimen and Dyer's *Flora of Middlesex* (1869). Lees appears to have been well acquainted with the flora of the Perivale district.

Frederic William FARRAR (1831–1903), Dean of Canterbury from 1895, was for sixteen years a house master at Harrow School, and provided records to Melvill's *Flora of Harrow* (1864), and to Trimen and Dyer's *Flora of Middlesex* (1869).

James Cosmo MELVILL (1844–1929) was educated at Harrow School, and during his residence there made a collection of the plants that he found growing in the district. In this venture he was greatly assisted by the Rev. W. M. Hind and the Rev. F. W. Farrar. His interest in the local flora culminated in the publication of *The Flora of Harrow* in 1864. His British herbarium is now at the Butler Museum, Harrow School.

Henry TRIMEN (1843–96) was a medical man, who in 1869 revoked his profession to enter the Department of Botany of the British Museum as an assistant. Interested in the flora of Middlesex from as early as 1859 he made numerous excursions to many parts of the vice-county in search of plants. In 1866 he discovered *Wolffia arrhiza*, a species hitherto unknown in the British Isles, at Staines, and in 1869 in collaboration with William Turner Thiselton DYER (afterwards Sir William Turner Thiselton Dyer) (1843–1928), Director of the Royal Botanic Gardens, Kew, from 1885 to 1905, he published the *Flora of Middlesex*. This excellent work, which received the highest praise from the botanical world, immediately became the model for many of the county floras that appeared at the end of the nineteenth century. Both Trimen and Dyer collected many specimens of Middlesex plants, and these are now in the Herbarium of the British Museum (Natural History).

William Williamson NEWBOULD (1819–86), a churchman, contributed numerous records to Trimen and Dyer's *Flora of Middlesex* (1869) and to many other county floras. Of a scholarly nature, he was much interested in the early history of British botany, and was probably largely responsible for much of the data on early Middlesex botanists that appeared in Trimen and Dyer's work. He resided at various times at Turnham Green and Kew, and spent much of his leisure botanising in many parts of the vice-county.

John Byrne Leicester WARREN, afterwards the Third Lord de Tabley (1835–95), poet and keen amateur botanist, greatly assisted Trimen and Dyer with their flora and provided many records for the work. He was particularly interested in the genus *Rumex*, and contributed notes on this and on the botany of the London area to the *Journal of Botany*. In 1871 he compiled a careful account of the flora of Hyde Park and Kensington Gardens. This was printed in the *Journal of Botany*. His *Flora of Cheshire* was published posthumously in 1899.

Alfred FRENCH (1839–79), a baker, was interested in botany from an early age, and compiled a herbarium which is now incorporated in the Herbarium of the British Museum (Natural History). His botanising was mainly confined to the northern suburbs.

In 1873 an excellent account of the plants found in the Enfield area appeared in *A History of Enfield* by George Henry Hodson and Edward Ford. The compiler of the account was Edward FORD.

A small octavo volume entitled *A New London Flora; or, Handbook to the Botanical Localities of the Metropolitan Districts*, appeared in 1877. Its author was Eyre Champion DE CRESPIGNY (1821–95), a medical man, and one-time Acting Conservator of Forests and Superintendent of Botanical Gardens in the Bombay Presidency. De Crespigny's book provided lists of plants found growing about London, arranged alphabetically under their locations, together with an alphabetical list of species and details of the localities in which they were to be found. Although many localities are listed for Middlesex species the work is marred by a number of errors. De Crespigny formed a herbarium which is now at the University of Manchester.

John BENBOW (1821–1908), a man of independent means, of Uxbridge Common, became interested in British plants, and particularly ferns, about 1860. In 1884 he began a detailed and careful survey of the flora of Middlesex, travelling many miles on foot in search of interesting plants. The partial results of his investigations were given in a series of papers in the *Journal of Botany* between 1884 and 1901. A fuller account of the plants noted in his explorations was given in his MS. 'Flora of Uxbridge and district, etc.' which he compiled about 1900. The latter was an extremely competent and accurate work based entirely on his own observations. It was unfortunately never published but the manuscript is now at the Department of Botany Library, British Museum (Natural History), and an identical copy is in the Library of the Botany School, University of Cambridge. Benbow was a most able botanist and knew the characters of British plants extremely well. He even undertook the study of such critical genera as *Rubus*, *Rosa* and *Salix*, and many of his determinations have been confirmed by modern specialists. He accumulated a very large herbarium of Middlesex plants which is now incorporated in the Herbarium of the British Museum (Natural History).

Theodore Dru Alison COCKERELL (1866–1948), one-time Professor of Zoology at the University of Colorado, lived at Chiswick as a youth, and in 1887 contributed a paper entitled 'The Flora of Bedford Park, Chiswick' to the *Journal of Botany*. This provided a useful account of the species to be found in that district in the late nineteenth century.

Henry Thomas WHARTON (fl. 1886–1900), a surgeon, provided a list of the plants to be found about Hampstead in J. Logan Lobley's *Hampstead Hill*, published in 1889.

James Eddowes COOPER (1864–1952), a tea-broker, and eminent amateur conchologist, was also interested in British botany, and made an extensive study of the casual adventive species found on waste ground and rubbish-tips, particularly at Yiewsley and Hackney Marshes. In 1887, and again in 1898, he contributed short notes on casual plants found in north London to *Science Gossip*, and in 1914 he published a comprehensive list of his discoveries in the *Journal of Botany*. Many of his specimens are preserved in the Herbaria of the British Museum (Natural History) and the Royal Botanic Gardens, Kew.

W. D. COCHRANE (fl. 1890) was responsible for an account of the flora of Hampstead which appeared in F. E. Baines' *Records of the Manor, Parish and Borough of Hampstead to December 31st 1889*, which was published in 1890.

Constance GARLICK (d. 1934) was a teacher in botany at University College School, London, and a leading member of the Hampstead Scientific Society. She provided a chapter on the wild flowers and trees of Hampstead in P. E. Vizard's *Guide to Hampstead* which appeared in 1890, and in 1912 contributed some general remarks on the botany of Hampstead to Thomas Walter Barratt's *Annals of Hampstead*.

At the end of the nineteenth century the flora of the Staines area was studied by Edward Fergusson SHEPHEARD (fl. 1890–1906). He contributed notes and records to the *Journal of Botany* and other periodicals.

TWENTIETH CENTURY

James Edward WHITING (c. 1850–1927), a taxidermist of Hampstead, contributed notes on the botany of Hampstead to the *Hampstead Annual* for 1901, and in 1912 provided an excellent account of the flora of Hampstead to Thomas Walter Barratt's *Annals of Hampstead*. A lifelong resident of Hampstead, Whiting possessed a unique knowledge of the flora of Hampstead Heath and Ken Wood.

During the early part of the twentieth century Alfred LOYDELL (1849–1910) visited many parts of Middlesex, often in the company of his friends Charles Bayliss Green (c. 1850–1918) and Albert Bruce Jackson (1876–1947), in search of plants. He compiled a herbarium which is now incorporated in the Herbarium of the Botany School, University of Oxford.

Charles Baylis GREEN (c. 1850–1918), a railway employee, was greatly interested in the flora of the vice-county, and accumulated a large herbarium of Middlesex plants which is now incorporated in the Herbarium of the South London Botanical Institute.

George Claridge DRUCE (1850–1932), a chemist, several times Mayor of Oxford, and the author of Floras of Berkshire, Buckinghamshire, Oxfordshire, Northamptonshire and Wester Ross, made a number of visits to the large rubbish-tips at Yiewsley, West Drayton and Hackney Marshes in search of exotic adventive species. In 1910 he published a paper on Middlesex plants in the *Journal of Botany*. This consisted largely of records made by Green, Jackson, Loydell and others, but did contain some original observations.

Archibald Sim MONTGOMERY (1844–1922), a timber merchant of Brentford, was a keen amateur botanist and collected many specimens in the Brentford, Hounslow and Twickenham areas. He retired to Cheltenham in 1909, and his herbarium is now preserved at Gloucester Museum.

James Chapman SHENSTONE (1855–1935), perhaps best known as an Essex botanist, was interested in the flora of building sites and waste ground in Central London, and in 1910 published in the *Selborne Magazine* a paper entitled 'A wild flower garden in the City of London'. This listed 28 species of vascular plants that he had noted on a waste site in Faringdon Street. In 1912 he produced a second paper entitled 'The Flora of London Building Sites'. This appeared in the *Journal of Botany* and dealt with the flora of building sites in the City of London and at Bloomsbury. During 1913 he contributed an account of the flora of the Brent Valley Bird Sanctuary (= Perivale Wood) to the *Selborne Magazine*.

About 1910, Frederic Newton WILLIAMS (1862–1933), a medical practitioner of Brentford, commenced the compilation of a new edition of the 'Flora of Middlesex'. Although he received assistance from C. B. Green, A. B. Jackson, A. Loydell and other botanists the project failed to proceed very far and was very incomplete at the time of his death. The fragmentary manuscript is now in the Department of Botany Library, British Museum (Natural History).

Williams was an authority on the Caryophyllaceae and published revisions of *Dianthus*, *Arenaria*, *Cerastium*, *Sagina*, *Herniaria*, and other genera in various British and foreign botanical periodicals.

Thomas Alfred DYMES (1866–1944) made special studies of seed germination and of the British Orchidaceae. He lived at West Drayton for a number of years and provided interesting notes on the flora of the district.

Albert Bruce JACKSON (1876–1947), an arboriculturist, compiled a *Catalogue of Hardy Trees and Shrubs growing in the grounds of Syon House, Brentford*. This was published in 1910. He was interested in the flora of Middlesex, and in the early years of the century visited many parts of the vice-county in search of plants often with his friends A. Loydell and C. B. Green.

A small booklet entitled *A Flower List of the Hampstead District* was printed in 1910. Its compiler was Mrs May CHAMPNEYS, née Drummond (c. 1860–1935). The work consisted of an annotated list of species with the localities in which they were to be found, and covered Hampstead, Golders Green and Hendon.

Between 1915 and 1921, Sir Christopher Howard ANDREWES, the eminent authority on the common cold, compiled a 'Flora of Highgate'. Through the kindness of the writer I have been able to examine the manuscript, which provides an interesting account of the flora of the Highgate area nearly fifty years ago.

Paul Westmacott RICHARDS, Professor of Botany, University College of North Wales, resided in Middlesex in his youth, and between 1921 and 1927 made a number of interesting records from Hampstead Heath. During the same period he made a careful survey of the bryophytes of the vice-county, the results of which were published in the *Journal of Ecology* for 1928 as a paper entitled 'Ecological Notes on the Bryophytes of Middlesex'.

A further account of the adventive flora of the London area appeared in the *Report of the Botanical Society and Exchange Club* for 1927 (1928). The authors of the paper, which dealt with the exotic species found on rubbish-tips at Yiewsley and Hackney Marshes, were Ronald MELVILLE, a professional botanist, and Royston Leslie SMITH (1892–1973).

William Charles Richard WATSON (1885–1954) carried out a critical study of the brambles of Middlesex from 1932 onwards, and visited many areas in the vice-county, sometimes in the company of his friend, and fellow batologist, Charles AVERY (1880–1960). The results of their work was published in the *London Naturalist* in 1947, and in Kent and Lousley's *Hand List of the Plants of the London Area*, part 2 (1952).

My own studies on the flora of Middlesex began as a schoolboy, about 1932, and over the last forty years I have visited every parish in Middlesex on numerous occasions at different times of the year. The results of some of these investigations have appeared in various issues of the *London Naturalist* from 1946 onwards.

Job Edward LOUSLEY, a leading British amateur botanist, and a specialist on the genus *Rumex*, investigated the flora of bombed areas in the City and Central London between 1942 and 1948, and published a number of papers on his observations. He has also visited many parts of the vice-county, often in company with myself.

Between 1944 and 1952, Mrs Barbara WELCH, née Gullick, visited many parts of Middlesex, often in company with myself, and added many interesting records to the vice-county lists.

From about 1949 onwards Ronald Archie BONIFACE provided numerous interesting notes and records from many parts of Middlesex, particularly of bryophytes, some of which were new vice-county records.

John George DONY, author of the Floras of Bedfordshire and Hertfordshire, visited the East Bedfont area in the late 1940s, where he studied the complex series of Junci, including the Australian *Juncus pallidus*, which grew in gravel pits there. In more recent years in the company of his wife, Mrs Christina Mayne DONY, née Goodman, he paid some attention to the northern areas of the vice-county adjacent to Hertfordshire. On these visits it was sometimes my good fortune to join them. Dr Dony may be credited with a somewhat unusual trio of new vice-county records from railway tracks in the Mill Hill–Hadley Wood areas – viz. *Cochlearia danica*, *Cerastium diffusum* and *C. pumilum*. The last-mentioned species he found in the company of Edgar Milne-Redhead.

In the early 1950s Thomas Gerald COLLETT commenced excursions to various parts of the vice-county, often in the company of the author, and photographed many rare and local species; he was later joined by his son Michael George COLLETT, and both father and son have added interesting records from many areas of Middlesex.

During 1950 an accurate and detailed survey of the flora of bombed sites in the London Borough of Holborn was undertaken by James Edward WHITTAKER (1901–57), and although he failed to publish the results of his studies the numerous plants that he collected are now in the Herbarium of the British Museum (Natural History).

Between 1952 and 1955 the flora of bombed areas in the City of London were studied by Alfred W. JONES, the results of his observations appearing in a paper in the *London Naturalist* in 1958.

A small book entitled *Flowers and Ferns of Harrow*, by Raymond Mervyn HARLEY, was published in 1953. This was virtually a modern version of Melvill's *Flora of Harrow*, though the area studied in the new work was considerably smaller than that covered by Melvill. A supplement containing additions and corrections was printed in the *Proceedings of the Botanical Society of the British Isles* in 1960.

Dion MURRAY became interested in the flora of Chiswick about 1955, and in 1956 published a list of the plants that he had observed in the district. Revised editions of the list were printed annually until 1960, but since that time revisions have been published at less frequent intervals.

From the mid-1950s Christopher HOLME carried out a detailed investigation of the wild flora of Regents Park, and provided many interesting and unusual records from the area. During the same period Brian PICKESS and Ian JOHNSON contributed numerous notes and records from the Uxbridge, Denham, Ruislip and Harefield areas, and the former was the first to note *Epipactis phyllanthes* in the vice-county.

Between 1958 and 1962 the wild flora of Kensington Gardens and Hyde Park was subjected to an extremely detailed study by David Elliston ALLEN. His work produced many interesting records, including two species new to Middlesex – *Rumex tenuifolius* and *Plantago maritima*. The results of the survey finally appeared as a paper in the *Proceedings of the Botanical Society of the British Isles* during 1965.

In 1959, Eric William GROVES contributed a paper entitled 'Notes on the flora of the Brent Valley Bird Sanctuary (= Perivale Wood)' to the *Selborne Magazine*, and during 1962 contributed a much more detailed account of the flora of the wood to Thomas L. Bartlett's *Bird Sanctuary*. These two papers provided interesting comparisons to J. C. Shenstone's account of the flora of the wood written nearly fifty years earlier. Further studies on this interesting area are now being carried out by Peter John EDWARDS.

A survey of the wild and naturalised vascular plants and bryophytes growing in the grounds of Buckingham Palace was carried out by David MCCLINTOCK between 1960 and 1963. Over 260 species of vascular plants were noted and a full account of the study was printed in the *Proceedings and Transactions of the South London Entomological Society* in 1964. This study continues.

From the early 1960s John A. MOORE has contributed many notes and records from the Northwood area, while Mrs Patricia MOXEY, née Firrell, has carried out studies on the flora of the Ruislip and District Natural History Society's Nature Reserve. The results of many of her observations have been printed as a series of papers in the *Journal of the Ruislip and District Natural History Society*.

In the 1960s Lady Anne BREWIS made a detailed investigation of the species present in various parts of Central London, and succeeded in discovering many rare and interesting plants in most unusual habitats.

During the 1960s and later many parts of Middlesex were explored by a trio of enthusiastic and highly competent young amateur botanists – Eric CLEMENT, John MASON and Brian WURZELL. Lists of their interesting discoveries, especially of casual adventives, reached me annually.

In conclusion it has, since 1965, given me great pleasure to revisit the more outlying areas of Middlesex in company with my friend Margaret Elizabeth KENNEDY. Alone, in connection with the London Natural History Society's Plant Distribution Mapping Scheme, Miss Kennedy has compiled records of the species found growing about Tottenham, Enfield and Potters Bar, and has succeeded in refinding a number of rare plants which had not been reported from these areas for many years.

REFERENCES

Allen, D. E. 1967. John Martyn's Botanical Society: A biographical analysis of the membership. *Proc. Bot. Soc. Brit. Isles* 6: 305–324.

Apothecaries Company of London. 1693–1752. *A catalogue of several members of the Society of Apothecaries, London: Living in and about the City of London*. London.

Barnhart, J. H. (Compiler). 1965. *Biographical Notes on Botanists.* 3 vols. New York.

Barrett, C. R. B. 1965. *The History of the Society of Apothecaries.* London.

Botanical Society [and Exchange Club] of the British Isles, etc. 1880–1948. *Report.* Vols. 1–13. Oxford, Manchester, etc.

—— 1949–69. *Watsonia. Journal of the Botanical Society of the British Isles.* Vols. 1–7. London.

—— 1954–69. *Proceedings.* Vols. 1–7. London.

Brett-James, N. G. 1926. *The Life of Peter Collinson.* London.

Britten, J. and Boulger, G. S. 1931. *A Biographical Index of Deceased British and Irish Botanists.* Edition 2. Edited by A. B. Rendle. London.

Clokie, H. N. 1965. *An Account of the Herbaria of the Department of Botany in the University of Oxford.* Oxford.

Copeman, W. S. C. 1967. *The Worshipful Society of Apothecaries of London. A History. 1617–1967.* Oxford.

Dandy, J. E. (Editor). 1958. *The Sloane Herbarium: an annotated List of the Horti Sicci comprising it; with biographical accounts of the principal contributors. Based on records compiled by the late James Britten. Revised and edited by J. E. Dandy.* London.

Druce, G. C. 1908. La Gasca and his 'Hortus Siccus Londinensis'. *J. Bot. (London)* 46: 163–170.

—— 1917. John Goodyer of Mapledurham, Hampshire. *Rep. Bot. Soc. & E.C.* 5: 532–550.

—— 1926. *Flora of Buckinghamshire.* Arbroath.

Druce, G. C. and Vines, S. H. 1907. *The Dillenian Herbaria, an account of the Dillenian collections in the herbarium of the University of Oxford.* Oxford.

Edwards, P. I. 1963. The Botanical Society (of London), 1721–1726. *Proc. Bot. Soc. Brit. Isles* 5: 117–118.

Foster, C. W. and Green, J. J. 1888. *History of the Wilmer Family.* Leeds.

Gorham, G. C. 1830. *Memoirs of John Martyn, F.R.S. and of Thomas Martyn, B.D., F.R.S., F.L.S.* London.

Gunther, R. T. 1922. *Early British Botanists and their Gardens, etc.* Oxford.

Kent, D. H. 1949. John Blackstone, Apothecary and Botanist (1712–53). *Watsonia* 1: 141–148.

—— 1950. Tothill Fields, Westminster: A lost botanical area. *Lond. Nat.* 29: 3–6.

Kew, H. W. and Powell, H. E. 1932. *Thomas Johnson, Botanist and Royalist.* London.

Lee, S., *et al.* 1885 →. *Dictionary of National Biography.* London.

Legré, L. 1899. *Pierre Pena et Matthias de Lobel.* Paris.

McVaugh, R. 1968. Rare old publications in Michigan herbaria. *Michigan Botanist* 7: 3–13. [La Gasca].

Pulteney, R. 1790. *Historical and Biographical Sketches of the Progress of Botany in England.* 2 vols. London.

Raven, C. E. 1942. *John Ray, Naturalist, his Life and Works.* Cambridge.
—— 1947. *English Naturalists from Neckam to Ray.* Cambridge.
Stearns, R. P. 1953. James Petiver, promoter of natural science, c. 1663–1718. *Proc. Amer. Antiq. Soc. new ser.* 62: 243–365.
Wall, C., Cameron, H. C. and Underwood, E. A. 1963. *A History of the Worshipful Society of Apothecaries of London.* Vol. 1. 1617–1815. London.
Weber, W. A. 1965. Theodore Dru Alison Cockerell, 1866–1948. *Univ. Colorado Studies. Series Bibliography* 1.
Young, S. 1890. *The Annals of the Barber-Surgeons of London.* London.

ABBREVIATIONS OF BOOKS, MSS. AND PAPERS CITED

Alch. MSS. Notes written (c. 1750–60) by Stanesby Alchorne in a copy of Blackstone's *Specimen Botanicum, etc.* See *Phyt.* 3: 166–170, 189–190 (1848).

Allen Fl. The Flora of Hyde Park and Kensington Gardens, 1958–1962. By D. E. Allen. *Proc. Bot. Soc. Brit. Isles* 6: 1–20 (1965).

Andrewes MSS. Flora of Highgate. By [Sir] Christopher Andrewes. MSS. compiled c. 1915–21. In the possession of the compiler.

Ann. & Mag. Nat. Hist. *Annals and Magazine of Natural History, including Zoology, Botany and Geology.* Vols. 6–20. London (1841–47): Ser. II. Vols. 1–20. (1848–57): ser. III. Vols. 1–20. (1858–67): ser. IV. Vols. 1–20. (1868–77).

Ann. Bot. (Oxford) *Annals of Botany.* New series. Vol. 1 →. London and Oxford. (1937 →).

Ann. Rep. & Proc. Bristol Nat. Soc. *Annual Report and Proceedings of the Bristol Naturalists' Society.* Vol. 1 →. Bristol. (1863 →).

Bab. Manual *Manual of British Botany, etc.* By C. C. Babington. London. (1843): Edition 2. (1847): Edition 3. (1851): Edition 4. (1856): Edition 5. (1862): Edition 6. (1867): Edition 7. (1874): Edition 8. (1881): Edition 9, edited by H. and J. Groves. (1904): Edition 10, edited by A. J. Wilmott. (1922).

Ball. MSS. Catalogue of plants indigenous to Islington and its immediate vicinity. By Edward Ballard. MSS. (compiled c. 1837–42). Printed under the editorship of J. Ardagh, *J.B. (London)* 66: 185–194. (1928).

Baxt.	*British Phaenogamous Botany, etc.* By William Baxter. 6 vols. Oxford. (1834–43).
Baylis	*A New and Compleat Body of Practical Botanic Physics, etc.* London. (1791 [–93]).
Benbow MSS.	The Flora of Uxbridge and district (in Middlesex) with some additions to the *Flora of Middlesex* (arranged after Babington's *Manual of British Botany* and in districts after Trimen and Dyer's *Flora of Middlesex*). By John Benbow. MSS. written c. 1900. Now at Department of Botany Library, British Museum (Natural History), London. A second identical copy is at the Botany School Library, University of Cambridge.
Benn. MSS.	Notes written (c. 1880–1900) by A. W. Bennett *et al.*, in a copy of Trimen and Dyer's *Flora of Middlesex*. Formerly in the possession of the late L. James of Ickenham. Present whereabouts unknown.
B.G.	*The Botanist's Guide through England and Wales.* By Dawson Turner and Lewis Weston Dillwyn. 2 vols. London. (1805).
Biol. J. Linn. Soc.	*Biological Journal of the Linnean Society.* Vol. 1 →. London. (1969 →).
Blackst. Fasc.	*Fasciculus Plantarum circa Harefield sponte nascentium, cum appendice ad loci historiam spectante.* By John Blackstone. London. (1737).
Blackst. litt.	Letters written by John Blackstone to Sir Hans Sloane *et al.* British Museum (Sloane MSS.).
Blackst. MSS.	Notes written (c. 1733–45) by John Blackstone in a copy of Johnson's *Mercurius Botanicus, etc.*, and in a copy of Ray's *Synopsis, etc.*, edition 2. The whereabouts of the first volume, cited by Trimen and Dyer [*Flora of Middlesex* (1869)], is unknown, but the second volume is in the Library of the British Museum, London.
Blackst. Spec.	*Specimen Botanicum quo Plantarum plurium rariorum Angliae indigenarum loci natales illustrantur.* By John Blackstone. London. (1746).
Blair Pharm. Bot.	*Pharmaco-Botanologia: or, an alphabetical and classical dissertation on all the British indigenous and garden plants, etc.* By Patrick Blair. London. (1723–28).
Borr. MSS.	Notes written (c. 1830) by William Borrer in a copy of Turner and Dillwyn's *Botanist's Guide, etc.* Department of Botany Library, British Museum (Natural History), London.

Bot. Chron. — *The Botanist's Chronicle.* [Edited by Alexander Irvine]. *Nos.* 1–17. London. (1863–65).

Bot. Gaz. — *The Botanical Gazette, A Journal of the progress of British botany and the contemporary literature of science.* Edited by A. Henfrey. 3 vols. London. (1849–51).

Brewer MSS. — Adversariorum hodoeporicum. By Samuel Brewer. 1691. 'Transcribed from the original MSS. in the possession of Miss Currer'. [A transcript of the author's account of his journey from Yorkshire to London in 1691]. Library, Royal Botanic Gardens, Kew, Surrey. Present whereabouts of original MSS. unknown.

Brit. Fern Gaz. — *British Fern Gazette.* Vol. 1 → Kendal, etc. (1909 →).

Brittain — *South Mymmes, the Story of a Parish.* By F. Brittain. Cambridge. (1931).

Brown Chron. — *The Chronicles of Greenford Parva; or, Perivale past and present.* By J. Allen Brown. London. (1890).

BSBI Abstr. — *BSBI Abstracts. Abstracts from Literature relating to the Vascular Plants of the British Isles.* Parts 1 →. London. 1971 →.

Budd. MSS. — Hortus Siccus Buddleanus sive Methodus nova Stirpium Britannicarum. [By Adam Buddle]. MSS. written (c. 1700–15). British Museum, London. [Sloane MSS. 2970–2980].

Bull. Brit. Mus. (Bot.) — *Bulletin of the British Museum (Natural History). Botany.* Vol. 1 →. London. (1951 →).

Burn. Med. Bot. — *Medical Botany, etc.* By John Stephenson and James M. Churchill. 3 vols. London. (1828–31). New edition, edited by Gilbert T. Burnett. 3 vols. London. (1834–36).

Butch. MSS. — Notes written (c. 1925) by R. W. Butcher in a proof copy of Butcher and Strudwick's *Further Illustrations of British Plants.* Department of Botany Library, British Museum (Natural History), London.

Champ. List. — *A Flower List of the Hampstead Neighbourhood.* By May Champneys. Hampstead. (1914).

Clus. Rar. Pl. — *Rariorum Plantarum Historia, etc.* By J. C. Clusius. Antwerp. (1601).

Cochrane Fl. — The Flora of Hampstead. By W. D. Cochrane, in F. E. Baines' *Records of the Manor, Parish and Borough of Hampstead to December 31st 1889.* Pp. 117–123. London. (1890).

Cockerell Fl. — The Flora of Bedford Park, Chiswick. By T. D. A. Cockerell. *J.B. (London)* 25: 107–110. (1887).

Cockfield Cat. — *The Botanist's Guide: A Catalogue of Scarce Plants*

	Found in the Neighbourhood of London. [By Joseph Cockfield]. London. (1813).
Coles	*Adam in Eden; or, Natures Paradise. The History of Plants, Fruits, Herbs and Flowers, etc.* By William Coles. London. (1657).
Coll. MSS.	Notes written (c. 1765–90) by Michael and Peter Collinson in a copy of Blackstone's *Fasciculus Plantarum circa Harefield, etc.* See *Phyt. N.S.* 5: 171–176. (1861). Present whereabouts of volume unknown.
Coop. Cas.	Casual plants in Middlesex. By J. E. Cooper. *J.B. (London)* 52: 127–131. (1914).
Coop. Fl.	*Flora Metropolitana; or, Botanical Rambles within Thirty miles of London: being the result of excursions mainly in 1833–35, etc.* By Daniel Cooper. London. (1836): Edition 2. (1837).
Coop. Fl. MSS.	Notes written in an unknown hand in a copy of Cooper's *Flora Metropolitana, etc.* Now in the possession of J. E. Lousley, Streatham, London, S.W.16.
Coop. Suppl.	*Supplement to Flora Metropolitana, etc.* By Daniel Cooper. London. (1837).
Corn. Nat.	*The Naturalist on the Thames.* By J. C. Cornish. London. (1902).
Corn. Surv.	Surviving London wild flowers. Letter from J. C. Cornish printed in *The Times*, 17 Oct. 1903.
Cottam	No. 8 Whitehall. Letter from Arthur Cottam printed in *The Standard*, 8 Sept. 1891.
Country-side	*Country-side. The official publication of the British Naturalists' Association. New series.* Vol. 1 →. Kingston-upon-Thames, etc. (1920 →).
Cowper	*Descriptive Historical Account of Millwall commonly called the Isle of Dogs.* By B. H. Cowper. London. (1853).
Crowe MSS.	Notes written (c. 1780) by James Crowe in a copy of Hudson's *Flora Anglica, etc.*, edition 2. Library of the Linnean Society of London, Burlington House, Piccadilly, London, W.1.
Cullum Fl.	*Flora Anglicae Specimen, imperfectum et ineditum, etc.* By Sir Thomas Gery Cullum. London. (1774).
Cund. Guide	*London: A Guide for the Visitor, Sportsman and Naturalist.* By J. W. Cundall. London. (1898).
Curt. Brit. Ent.	*British Entomology, etc.* By John Curtis. 16 vols. London. (1824–39).
Curt. F.L.	*Flora Londinensis: or, Plates and Descriptions of such Plants as grow wild in the Environs of London, etc.* By William Curtis. 6 Fasciculi. London. (1775–98).

De Cresp. Fl. *A New London Flora; or, Handbook to the Botanical Localities of the Metropolitan Districts, etc.* By Eyre Champion de Crespigny. London. (1877).

De Ves. MSS. Catalogue of a collection of dried plants gathered between 1875 and 1884. By Roland Ellis [de Vesian]. MSS. now in the possession of the author.

Dickens *Dictionary of the Thames, etc.* By Charles Dickens Junr. [Botanical notes contributed by James Britten]. Edition 3. London. (1888).

Dicks. Hort. Sicc. *Hortus Siccus Britannicus; being a collection of dried British plants.* By James Dickson. 19 Fasciculi. London. (1793–1802).

Dill. MSS. Notes written (c. 1728–45) by Johann Jakob Dillenius, *et al.*, in a copy of Ray's *Synopsis, etc.*, edition 3. Recently in the possession of the late N. D. Simpson, Bournemouth, Hampshire.

Dillwyn Hort. Coll. *Hortus Collinsonianus, an account of the plants cultivated by the late P. Collinson, etc.* By L. W. Dillwyn. Swansea. (1843).

Don Gen. Hist. *A General History of the Dichlamydeous Plants, etc.* By George Don. 4 vols. London. (1831–38).

Doody MSS. Notes written (c. 1695–1705) by Samuel Doody, in a copy of Ray's *Synopsis, etc.*, edition 2. Library of the British Museum, London.

Druce C.F. *The Comital Flora of the British Isles, etc.* By George Claridge Druce. Arbroath. (1932).

Druce Fl. *Flora of Buckinghamshire, etc.* By George Claridge Druce. Arbroath. (1926).

Druce Notes Notes on the flora of Middlesex. By George Claridge Druce. *J.B.* (*London*) 48: 269–278. (1910).

Druett *Pinner Through the Ages.* By Walter W. Druett. Harrow. (1937).

Dunn Fl. *Alien Flora of Britain.* By S. T. Dunn. London. (1905).

Dunn MSS. Notes written (c. 1906 →) by S. T. Dunn, *et al.*, in a copy of the *Alien Flora of Britain.* Library, Royal Botanic Gardens, Kew, Surrey.

Dyer List Plants found on the site of the Exhibition of 1862, South Kensington. By William T. Thiselton Dyer, in a *Key to the London International Exhibition for June 17, 1872.* London. (1872).

E.B. *English Botany; or Coloured Figures of British Plants, with their essential characters, synonyms, and places of growth, etc.* By James Edward Smith; the figures by James Sowerby. 36 vols. London. (1790–1814): Edition 2. 12 vols. (1832–46). [reissued as a spurious edition 3. (1848–54)]: Edition 3, edited by John T. Boswell Syme. 12 vols. (1863–89).

E.B. Suppl. — *Supplement to English Botany, etc.* By W. J. Hooker, *et al.* Vol. 1. London. (1831): Vol. 2. (1834): Vol. 3. (1843): Vol. 4. (1849): Vol. 5. (1863–66). *Supplement* to edition 3. By N. E. Brown. London. (1891–92).

Essex Nat. — *The Essex Naturalist; being the Journal [, Transactions and Proceedings] of the Essex Field Club.* Vol. 1 →. Stratford, etc. (1887 →).

Evans Hist. — *History of Hendon, etc.* By T. Evans. London. (1898).

Faulkn. Hist. — *Historical and Topographical Account of Fulham, etc.* By Thomas Faulkner. London. (1813).

Find. MSS. — Notes written (c. 1880–1910) by Charles J. B. Findon, in a copy of Trimen and Dyer's *Flora of Middlesex*. Now in the possession of the author.

Fl. B.P.G. — Natural History of the Garden of Buckingham Palace: wild and naturalized vascular plants. By D. McClintock. *Proc. & Trans. S. Lond. Ent. & N.H.S.* 1963 (2): 14–25. (1964): bryophytes and fungi. Co-ordinated by D. McClintock. *loc. cit.* 1963 (2): 36–38. (1964).

Fl. Herts — *Flora of Hertfordshire: The wild plants of the County of Hertford and adjoining areas included in the Watsonian Vice County 20.* By John G. Dony. Hitchin. (1967).

Foley — *Our Lanes and Meadow Paths: Rambles in Rural Middlesex.* By H. J. Foley. London. [c. 1887].

Forst. Midd. — Rare plants found in Middlesex. [By Edward Forster], in vol. 2 of Richard Gough's edition of Camden's *Britannia, etc.* London. (1789).

Forst. MSS. — Edward Forster's botanical notebooks for 1791–94. Library of the Department of Botany, British Museum (Natural History), London.

Francis — *An Analysis of British Ferns and their Allies, etc.* By G. W. Francis. London. (1837): Edition 2. (1842): Edition 3. (1847): Edition 4. (1850): Edition 5, revised and enlarged by A. Henfrey. (1855).

Galp. Syn. — *A Synoptical Compendium of British Botany, etc.* By John Galpine. Salisbury. (1806): Edition 2. London. (1820): Edition 3. (1829): Edition 4. (1834).

Gard. News — *Garden News.* Peterborough. 1958 →.

Garlick W.F. — The wild flowers of Hampstead. By C. Garlick, in P. E. A. Vizard's *Guide to Hampstead.* Pp. 47–49. London. (1890): Edition 2. (1898).

Garry Notes — Notes on the drawings for English Botany. By F. N. A. Garry. Supplement to *J.B. (London)* 41–42. (1903–4).

Gentleman's Mag. *The Gentleman's Magazine: or, Monthly Intelligencer.* Vols. 1–303. London. (1731–1907).

Ger. Hb. *The Herball, or Generall Historie of Plantes, etc.* Gathered by John Gerarde. London. (1597).

Gray Nat. Arr. *A Natural Arrangement of British Plants according to their Relations to each other, etc.* By Samuel Frederick Gray. 2 vols. London. (1821).

Groves Veg. Vegetation of the sanctuary. By E. W. Groves, in T. L. Bartlett's *Bird Sanctuary.* Pp. 39–57. Selborne Society. London. (1962).

Gunther *Early British Botanists and their Gardens, based on unpublished writings of Goodyer, Tradescant and others.* By R. T. Gunther. Oxford. (1922).

H. & F. *A History of Enfield, etc.* By George Henry Hodson and Edward Ford. [Botany by Edward Ford. Pp. 147–150]. London. (1873).

Hall Nat. *Nature Rambles in London.* By K. M. Hall. London. (1908).

Hall Tham. *The Book of the Thames from its Rise to its Fall.* By Samuel Carter Hall and Anna Maria Hall. London. (1859): Edition 2. (1877).

Hampst. *Hampstead Heath, its Geology and Natural History. Hampstead Scientific Society.* London. (1913).

Hampst. Sci. Soc. Rep. *Hampstead Scientific Society. Annual Report.* Hampstead. (1901–39).

Harley Add. Additions to the Flora of Harrow. By Raymond Harley. *Proc. Bot. Soc. Brit. Isles* 3: 380–383. (1960).

Harley Fl. *Flowers and Ferns of Harrow.* By Raymond Harley. London. (1953).

Henrey *The King of Brentford.* By Robert Henrey. London. (1946).

Hill Brit. Herb. *The British Herbal: An History of Plants and Trees, Natives of Britain, cultivated for use, or raised for beauty.* By John Hill. London. (1756–57). [See W. T. Stearn, *Taxon* 16: 494–498. (1967)].

Hill Fl. Brit. *Flora Britanica [sic]: sive, Synopsis Methodica Stirpium Britanicarum [sic], etc.* By John Hill. London. (1760).

Hill MSS. Notes written (c. 1755–70) by John Hill in a copy of Ray's *Synopsis*, edition 2. Library of the Botany School, University of Oxford.

Hill Veg. Syst. *The Vegetable System; etc.* by John Hill. 26 vols. London. (1759–95).

Hind Fl. The Flora of Harrow and its vicinity. By W. M. Hind. *Harrow Gazette*, 16 Jan. and 3 Feb. 1860. Reprinted in part in *Phyt. N.S.* 4: 107–119. (1861).

Horton Rep. Report on wild flowers, 1947. By W. R. G. Horton. *Ann. Rep. Harrow School N.H.S.* 3(2): 3–18. (1947).

How MSS. Notes written (c. 1650–56) by William How, in a copy of his *Phytologia Britannica, etc.* Library of Magdalen College, Oxford.

How Phyt. *Phytologia Britannica, natales exhibens indigenarum Stirpium sponte emergentium.* By William How. London. (1650).

Huds. Fl. Angl. *Flora Anglica, exhibens Plantas per regnum Angliae sponte crescentes, distributas Secundum Systema Sexuale, etc.* By William Hudson. London. (1762): Edition 2. 2 vols. (1778): Reprinted in one vol. (1798).

Hull Brit. Fl. *The British Flora, etc.* By John Hull. London. (1799): Edition 2. (1808).

Irv. Ill. Handb. *The Illustrated Handbook of the British Plants.* By Alexander Irvine. London. (1858).

Irv. Lond. Fl. *The London Flora; containing a concise description of the Phaenogamous British Plants, which grow spontaneously in the vicinity of the Metropolis, etc.* By Alexander Irvine. London. (1838).

Irv. MSS. A list of plants growing within a two-mile radius of Hampstead Heath. By Alexander Irvine. MSS. written c. 1825–34. Present whereabouts of MSS. unknown and all references used have been taken from Trimen and Dyer's *Flora of Middlesex* (1869).

Jacks. Ann. *The Annals of Ealing from the 12th century to the present time.* By Edith Jackson. London. (1898).

Jacks. Cat. *Catalogue of Hardy Trees and Shrubs growing in the grounds of Syon House, Brentford.* By A. Bruce Jackson. London. (1910).

Jacks. MSS. Notes written (c. 1908–19) by Albert Bruce Jackson in a copy of Trimen and Dyer's *Flora of Middlesex.* Now in the possession of Dr C. E. Hubbard, Hampton, Middlesex.

J.B. (London) *Journal of Botany, British and Foreign.* Vols. 1–80. London. (1863–1942).

J. Ecol. *Journal of Ecology.* British Ecological Society. Vol. 1 →. Cambridge and Oxford. (1913 →).

Jenk. Gen. Spec. *A Generic and Specific description of British Plants, translated from the Genera et Species Plantarum of Linnaeus, etc.* By James Jenkinson. Kendal. (1775).

Jesse Glean. *Gleanings in Natural History, etc.* By Edward Jesse. London. (1832): Second series, etc. (1834): Third and last series, etc. (1835).

J. Linn. Soc. (Bot.) *Journal of the Linnean Society of London (Botany).* Vols. 1–61. London. (1857–1968).

Johns. Cat. Catalogus plantarum juxta Tottenham lectarum anno dom. 1638. By Thomas Johnson. 1638. Copy of an original MSS. Library, Royal Botanic Gardens, Kew, Surrey. The present whereabouts of the original MSS. is unknown.

Johns. Enum. *Enumeratio Plantarum in Ericeto Hampstediano locisque vicinis Crescentium, etc.* By Thomas Johnson. London. (1632).

Johns. Eric. *Ericetum Hamstedianum Sive Plantarum ibi crescentium observatio habita, Anno eodem* [1629] *1 Augusti.* By Thomas Johnson. London. (1629).

Johns. Ger. *The Herball, or Generall Historie of Plantes, etc. . . . Gathered by John Gerarde . . . very much enlarged and amended.* By Thomas Johnson. London. (1633). Reprinted in 1636.

Johns. MSS. Notes written (c. 1640) by Thomas Johnson in a copy of his *Descriptio itineris Plantarum, etc.* At one time the property of William How, and now in the Library of Magdalen College, Oxford.

Johns. Nat. *The Nature World of London: 1. Trees and Plants.* By Walter Johnson. London. (1924).

Jones Fl. The Flora of City of London bombed sites. By A. W. Jones. *L.N.* 37: 189–210. (1958).

Jones & Turrill *British Knapweeds. A Study in Synthetic Taxonomy.* By E. M. Marsden-Jones and W. B. Turrill. Ray Society. London. (1954).

J. Roy. Hort. Soc. *Journal of the Royal Horticultural Society.* New series. Vol. 1 →. London. (1866 →).

J. Ruisl. & Distr. N.H.S. *Journal of the Ruislip and District Natural History Society.* Vol. 1 →. Ruislip. (1952 →).

Kalm *En resa til Norra America, etc.* By Pehr Kalm. 3 vols. Stockholm. (1753–61). Partly translated into English by Joseph Lucas. London. (1893).

Kew Bull. *Bulletin of Miscellaneous Information.* Royal Botanic Gardens, Kew. London. (1887–1942). Continued as *Kew Bulletin.* Vol. 1 →. London. (1946 →).

K. & L. *A Hand List of the Plants of the London Area.* By Douglas H. Kent and J. Edward Lousley. Supplement to *L.N.* 30–36. London. (1951–57).

Knapp Gram. Brit. *Gramina Britannica: or representations of the British Grasses, etc.* By J. L. Knapp. London. (1804): Edition 2. (1842).

La Gasca Hort. Sicc. Hortus Siccus Londinensis or, a Collection of

Dried Specimens of Plants growing wild within twenty miles round London named on the authority of the Banksian Herbarium and other original Collections. By D. Mariano La Gasca, etc. 4 Fasciculi. London. 1826–27. [See G. C. Druce, *J.B.* (*London*) 46: 163–170 (1908) and R. McVaugh, *Michigan Botanist* 7: 3–13 (1968).]

Lamb. MSS. Notes written (c. 1800–30) by Aylmer Bourke Lambert, in a copy of Hudson's *Flora Anglica, etc.*, edition 2. Library of the British Museum, London.

Lawson MSS. Thomas Lawson's botanical note-book. MSS. written c. 1670–75. Library of the Linnean Society of London, Burlington House, Piccadilly, London, W.1. [See C. E. Raven, *Proc. Linn. Soc.* 160: 3–12. (1949)].

Lightf. MSS. Notes written (c. 1762–88) by John Lightfoot in copies of Hudson's *Flora Anglica, etc.*, and Ray's *Synopsis, etc.*, edition 3. Library of the Botany School, University of Oxford.

L.N. *The London Naturalist. Journal of the London Natural History Society*. Vol. 1 →. London. (1921 →).

Lob. Stirp. Ill. *Stirpium Illustrationes . . ., etc.* By Matthias de l'Obel. Edited by William How. London. (1655).

Lob. Stirp. Obs. *Plantarum seu Stirpium Historia . . . Stirpium Observationes.* By Matthias de l'Obel. Antwerp. (1576).

Lond. Nat. MSS. Botanical records compiled by members of the London Natural History Society. c. 1900 →. In the care of the Botanical Recorder, J. Edward Lousley, Streatham, London, S.W.16.

Loud. Arb. *Arboretum et Fruticetum Britannicum; or the Trees and Shrubs of Britain, native and foreign, etc.* 8 vols. By J. C. Loudon. London. (1835–38).

Lousley Fl. The Flora of bombed sites in the City of London in 1944. By J. Edward Lousley. *Rep. Bot. Soc. & E.C.* 12: 875–883. (1946).

Lousley Pion. Fl. The pioneer flora of bombed sites in central London. By J. Edward Lousley. *Rep. Bot. Soc. & E.C.* 12: 528–531. (1944).

Macreight Man. *Manual of British Botany, etc.* By D. C. Macreight. London. (1837).

Mag. Nat. Hist. *The Magazine of Natural History and Journal of Zoology, Botany, Mineralogy, Geology and Meteorology.* Vols. 1–9. London. ([1828] 1829–36): new ser. Vol. 1. (1837). Continued as *Magazine of Natural History*, new ser. Vols. 2–4. (1838–40).

Mart. Fl. Rust. *Flora Rustica: exhibiting accurate figures of such plants*

	as are either useful or injurious in Husbandry, etc. By Thomas Martyn. 4 vols. London. (1792–95).
Mart. Mill. Dict.	*The Gardener's and Botanist's Dictionary . . . by the late Philip Miller . . . to which are now first added, A complete Enumeration and Description of all Plants hitherto known.* By Thomas Martyn. 2 vols. London. (1807). [The ninth edition of Miller's *Gardener's Dictionary.*]
Mart. Pl. Cant.	*Plantae Cantabrigiensis: or, a Catalogue of the Plants which grow wild in the county of Cambridge . . . Herbationes Cantabrigiensis: or, directions to the places where they may be found. . . . To which are now added lists of the more rare Plants growing in many parts of England and Wales.* By Thomas Martyn. London. (1763).
Mart. Tourn.	*Tournefort's History of Plants growing about Paris. . . . Translated into English, . . . and accommodated to the Plants growing in Great Britain.* By John Martyn. 2 vols. London. (1732).
Melv. Fl.	*The Flora of Harrow.* By J. C. Melvill. London. (1864): Edition 2, edited by W. M. Hind. (1876).
Merr. MSS.	Notes written (c. 1670) by Christopher Merrett in a copy of his *Pinax, etc.* Library of the British Museum, London.
Merr. Pin.	*Pinax Rerum Naturalium Britannicarum, continens Vegetabilia, Animalia et Fossilia in hac Insula reperta inchoatus, etc.* By Christopher Merrett. London. (1666): spuriously reprinted in 1667.
M. & G.	*Indigenous Botany; or habitations of English Plants: containing the results of several botanical excursions chiefly in Kent, Middlesex and the adjacent counties in 1790, 1791 and 1792.* By Colin Milne and Alexander Gordon. London. (1793).
Mill. Bot. Off.	*Botanicum Officinale; or a compendious Herbal, etc.* By Joseph Miller. London. (1722).
Monckt. Fl.	*The Flora of the district of the Thames Valley Drift between Maidenhead and London.* By H. W. Monckton. London. (1919).
Montg. MSS.	Notes written (c. 1897–1908) by Archibald Sim Montgomrey, in a copy of Trimen and Dyer's *Flora of Middlesex.* Formerly in the possession of J. W. Haines, Hucclecote, Gloucestershire. Present whereabouts unknown.
Moore	*The Octavo Nature-printed British Ferns.* By Thomas Moore. 2 vols. London. (1859–60).

Moring MSS.	Records of Middlesex plants. By Percy Moring. MSS. written c. 1920–28. Library of the Department of Botany, British Museum (Natural History), London.
Moris. Hist.	*Plantarum Historia Universalis Oxoniensis, etc.* By Robert Morison. Pars. 2. Oxford. (1680): Pars. 3, edited by Jacob Bobart. (1699): Pars. 1, on trees and shrubs, was never published but exists in MS. at the Bodleian Library, Oxford.
Moris. Pl. Umb.	*Plantarum Umbelliferarum distributio nova, per tabulas cognationis et affinitatis ex libro naturae observata & detecta.* By Robert Morison. Oxford. (1672).
Moris. Prael. Bot.	*Hortus Regius Blesensis auctus, cum notulis durationis & charactismis Plantarum . . . additarum . . . Item Plantarum . . . nemini hucusque scriptarum . . . delineatio. Quibus accessere observationes generaliores . . . & cognitu perutiles. Praeludiorum Botanicorum pars prior, etc.* By Robert Morison. London. (1669).
Morley MSS.	Catalogus planta villâ nostrâ Hamstiadensis. By Christopher Love Morley. MSS. written c. 1677–79. British Museum, London. (Sloane MSS).
Moxey Br.	The bryophytes of the Ruislip Local Nature Reserve. By P. Moxey. *J. Ruisl. & Distr. N.H.S.* 14: 33–41. (1965).
Moxey List	Systematic list of the plants of Ruislip Local Nature Reserve. By P. Moxey. *J. Ruisl. & Distr. N.H.S.* 13: 18–20. (1964).
M. & S.	Adventive flora of the Metropolitan area. By R. Melville and R. L. Smith. *Rep. Bot. Soc. & E.C.* 8: 444–454. (1928).
Murray List	*The Flora of Chiswick as observed by Dion Murray.* London. (1956, *et seq.*).
Nat.	*The Naturalist. A Quarterly Journal, principally for the North of England.* Hull, London, etc. (1864 →).
Nat. Camb.	*Nature in Cambridgeshire.* Vol. 1 →. Cambridge. (1958 →).
Nature	*Nature.* Vol. 1 →. (London. 1869 →).
Nature Notes	*Nature Notes. The Magazine of the Selborne Society.* Vols. 1–19. London. (1890–1908).
Newb. MSS.	Notes written (c. 1870–83) by William Williamson Newbould, in a copy of Trimen and Dyer's *Flora of Middlesex.* Now in the possession of the author.
New Phyt.	*The New Phytologist.* Vol. 1 →. Cambridge and Oxford. (1902 →).
Newt. MSS.	Notes written (c. 1680) by James Newton, in a copy

of Ray's *Catalogus Plantarum Angliae*. Present whereabouts of volume unknown, and all references used have been taken from Trimen and Dyer's *Flora of Middlesex*. (1869).

Notes Roy. Bot. Gard. Edinb. *Notes from the Royal Botanic Garden, Edinburgh*. Vol. 1 →. Edinburgh, etc. (1900 →).

Pamplin MSS. Notes written (c. 1869–70) by William Pamplin, in a copy of Trimen and Dyer's *Flora of Middlesex*. Library, Royal Botanic Gardens, Kew, Surrey.

Park Hampst. *Topography and Natural History of Hampstead*. By J. J. Park. London. (1814): Edition 2. (1818).

Park. Par. *Paradisi in Sole. Paradisus Terrestris: or, A Garden of all sorts of Pleasant Flowers which our English ayre will permitt to be noursed up, etc.* By John Parkinson. London. 1629; Edition 2. (1656).

Park. Theat. *Theatrum Botanicum: The Theater of Plants; or an Herball of large extent, etc.* By John Parkinson. London. (1640).

Pena & Lob. Stirp. Adv. *Stirpium Adversaria Nova, perfacilis vestigatio, luculentaque accessio ad priscorum, presertim Dioscoridis & recentiorum, materiam medicam, etc.* By Pierre Pena and Matthias de l'Obel. London. (1570 [–71]).

Pet. Bot. Lond. Botanicum Londinense; or, The London Herbal, etc. By James Petiver. *Monthly Miscellany; or, Memoirs for the Curious 3*. London. (1709–10).

Pet. Gram. Conc. *Graminum, Muscorum, Fungorum, Submarinorum, etc. Britannicorum Concordia. A methodical concordance of British grasses, etc.* By James Petiver. London. (1716).

Pet. H.B.C. *Herbarij Britannici clariss. D. Raii Catalogus cum iconibus ad vivum delineatis, etc.* By James Petiver. Tab. 1–50. London. (1713): Tab. 51–72. (1715).

Pet. Midd. *More rare plants growing wild in Middlesex, communicated by Mr. James Petiver*, in Edmund Gibson's translation of Camden's *Britannia, etc.* London. (1695) [and later editions].

Pet. MSS. Notes written (c. 1715) by James Petiver, in a copy of Ray's *Synopsis, etc.*, edition 2. Library of the British Museum, London.

Pet. Mus. *Musei Petiveriani centuria prima (secunda-decima), Rariora Naturae continens: viz. Animalia, Fossilia, Plantas, ex variis mundi plagis advecta, ordine digesta, et nominibus propriis signata, etc.* By James Petiver. *Centuriae x*. London. (1695 [–1703]).

Phyt. *The Phytologist; a Popular Botanical Miscellany*. Conducted by George Luxford. 5 vols. London. (1841–54).

Phyt. N.S. *The Phytologist; a Botanical Journal.* Edited by Alexander Irvine. [A new series]. 6 vols. London. (1855–63).

Pluk. Alm. *Almagestum Botanicum, sive Phytographiae Pluc'netianae onomasticon methodo synthetica digestum . . . Adjiciuntur & aliquot novarum Plantarum icones, etc.* [By Leonard Plukenet]. London. (1696).

Pluk. Amalth. *L. Plukenetii Amaltheum Botanicum* (i.e.) *Stirpium indicarum alterum copiae cornu, etc.* London. (1705).

Pluk. Mant. *Almagest; Botanici Mantissa novissime detectarum . . . complectens . . . Cum Indice. totius operis ad calcem adjecto.* [By Leonard Plukenet]. London. (1700).

Pluk. MSS. Notes written (c. 1690–1705) by Leonard Plukenet, in a copy of Ray's *Catalogus Plantarum Angliae, etc.* Library of the British Museum, London.

Pluk. Phyt. *L. Plukenetij Phytographia, sive Stirpium illustriorum & minus cognitarum icones, etc.* Partes 1 et 2. London. (1691): Pars. 3. (1692): Pars. 4. (1694).

Proc. Bot. Soc. Brit. Isles *Proceedings of the Botanical Society of the British Isles.* Vols. 1–7. London. (1954–69).

Proc. Croyd. N.H. & Sci. Soc. *Proceedings of the Croydon Natural History and Scientific Society.* Vol. 1 →. Croydon. (1901 →).

Proc. Linn. Soc. *Proceedings of the Linnean Society of London.* Vol. 1 →. London. ([1838] 1839 →).

Proc. & Trans. S. Lond. Ent. & N.H.S. *Proceedings and Transactions of the South London Entomological and Natural History Society.* 1897 →. (London. 1898 →).

Pryor Fl. *A Flora of Hertfordshire, etc.* By the late Alfred Reginald Pryor. [Edited by Benjamin Daydon Jackson]. London. (1887).

Ray Cat. *Catalogus Plantarum Angliae, et insularum adjacentium . . . una cum. observationibus & experimentis novis medicis & physicis.* By John Ray. London. (1670): Edition 2. (1677).

Ray Cat. MSS. Notes written (c. 1680–90) by John Ray, in a copy of his *Catalogus Plantarum Angliae, etc.*, edition 2. Library of the British Museum, London.

Ray Fasc. *Fasciculus Stirpium Britannicarum, post editum Plantarum Angliae Catalogum orum, etc.* [By John Ray]. London. (1688).

Ray Hist. *Historia Plantarum, species huctenus editas aliasque insuper multas noviter inventas et descriptas complectens, etc.* By John Ray. Vol. 1. London. (1686): Vol. 2. (1688): Vol. 3 [supplementary]. (1704).

Ray Syn. *Synopsis Methodica Stirpium Britannicarum, in qua tum notae generum characteristicae traduntur, tum species singulae breviter describuntur, etc.* By John Ray. London. (1690): Edition 2. (1696): Edition 3 [Edited by J. J. Dillenius]. (1724).

Ray Syn. MSS. Notes written (c. 1733–40) in an unknown hand, in a copy of Ray's *Synopsis, etc.*, edition 3. Library of the British Museum, London.

Rep. Bot. Loc. Rec. Club *Report of the Botanical Locality Record Club, 1873–78.* London. (1874–79). Continued as *Report of the Botanical Record Club,* 1879–86. Manchester. (1880–87).

Rep. Bot. Soc. & E.C. *Report. London Botanical Exchange Club.* London. (1866–68). Continued as *Report. Botanical Exchange Club of the British Isles.* Manchester. (1880–1901). Continued as *Report Botanical Exchange Club and Society of the British Isles.* Vols. 2–3. Manchester. (1901–14). Continued as *Report. Botanical Society and Exchange Club of the British Isles.* Vols. 4–13. Arbroath. (1914–48).

Rep. Brit. Bryol. Soc. *Report of the British Bryological Society.* Vols. 1–4. Cambridge. (1923–46). Continued as *Trans. Brit. Bryol. Soc.*

Rep. Moss Exch. Club *Report of the Moss Exchange Club.* London. (1896–1922). Continued as *Rep. Brit. Bryol. Soc.*

Rep. Watson B.E.C. *Report of the Watson Botanical Exchange Club.* London. (1884–1934).

Richards Ec. Ecological notes on the bryophytes of Middlesex. By P. W. Richards. *J. Ecol.* 16: 239–300. (1928).

Richards Tarax. Fl. The *Taraxacum* flora of the British Isles. By A. J. Richards. Suppl. *Watsonia* 9. Pp. 141. (1972).

Robson Fl. *The British Flora, etc.* By Stephen Robson. York. (1776).

R. & P. *England Displayed: Being a new complete and accurate survey and description of the Kingdom of England and Principality of Wales.* By P. Russell and Owen Price. 2 vols. London. (1769).

School Nat. Stud. J. *School Nature Study Journal.* Vol. 1 →. London. (1903 →).

Sci. Goss. *Hardwicke's Science Gossip, etc.* Vols. 1–29. London. (1865–93). Continued as *Science Gossip, etc.*, new series. Vols. 1–9. (1894–1902).

Selb. Mag. *Selborne Magazine [and Nature Notes], etc.* Selborne Society. Vol. 1 →. London. (1888 →).

Shenst. Fl. The flora of the Brent valley Bird Sanctuary. By

J. C. Shenstone. *Selb. Mag.* 24: 26–27, 50–51, 105–106, 146–148, 169–171, 181–185 and 214–216. (1913).

Shenst. Fl. Lond. The flora of London building sites. By J. C. Shenstone. *J.B. (London)* 50: 117–124. (1912).

Shove Fl. The flora of a derelict site in the Zoological Gardens, Regents Park. By R. F. Shove. *School Nat. Stud. J.* 40: 58–60. (1945).

Sibth. MSS. Notes written (c. 1780) by John Sibthorp. Present whereabouts of MSS. unknown, and all references used have been taken from G. C. Druce's *Flora of Buckinghamshire* (1926).

Smith Engl. Fl. *The English Flora.* By Sir James Edward Smith. 4 vols. London. (1824–28): Edition 2. 4 vols. (1828–30).

Smith Fl. Brit. *Flora Britannica, etc.* By J. E. Smith. 3 vols. London. (1800–04).

Smith Linn. Corr. *A selection of the Correspondence of Linnaeus and other naturalists, etc.* Edited by Sir James Edward Smith. 2 vols. London. (1821).

Smith MSS. Notes written by Sir James Edward Smith on the original drawings for *English Botany, etc.* Library of the Department of British Museum, British Museum (Natural History), London.

Sole Menth. Brit. *Menthae Britannicae: being a new botanical arrangement of all the British Mints . . . with several new species, etc.* By William Sole. Bath. (1798).

Spencer *The Complete British Traveller, etc.* By Nathaniel Spencer. London. (1771).

Still. Misc. Tracts. *Miscellaneous Tracts relating to Natural History, etc.* By Benjamin Stillingfleet. Stratton, Norfolk. (1759): Edition 2 . . *augmented with additional notes, etc.* London. (1762): Edition 3. (1775): Edition 4. (1791).

T. & D. *Flora of Middlesex: A topographical and historical account of the plants found in the county; with sketches of its physical geography and climate, and of the progress of Middlesex botany during the last three centuries.* By Henry Trimen and William T. Thiselton Dyer. London. (1869).

Trans. Brit. Bryol. Soc. *Transactions of the British Bryological Society.* Vols. 1–6. London. (1947–71).

Trans. Herts N.H.S. *Transactions of the Hertfordshire Natural History Society and Field Club.* Vol. 1 →. London. (1879 →).

Trans. Hort. Soc. *Transactions of the Horticultural Society of London.* Vols. 1–7. London. (1812–30): Ser. 2. Vols. 1–3. (1831–48).

Trans. Linn. Soc. *Transactions of the Linnean Society [of London].* Vols. 1–30. London. (1791–1875). Continued as *Trans-*

actions of the Linnean Society of London. Botany. Ser. 2. Vols. 1–9. (1875–1922). Continued as *Transactions of the Linnean Society of London.* Ser. 3. Vol. 1 →. London. (1939 →).

Trans. L.N.H.S. *Transactions of the London Natural History Society.* London. (1914–20). Continued as *L.N.*

Trans. Norf. & Norw. N.H.S. *Transactions of the Norfolk and Norwich Naturalists' Society.* Vol. 1 →. Norwich. (1869 →).

Trim. MSS. Notes written (c. 1869–1910) by James Britten, Henry Trimen, *et al.*, in Henry Trimen's interleaved copy of *Flora of Middlesex, etc.* Library of the Department of Botany, British Museum (Natural History), London.

Turn. Bot. *Botanologia. The British Physician: or, the Nature and Vertues of English Plants, etc.* By Robert Turner. London. (1664).

Turn. Hb. *A Newe Herball, etc.* By William Turner. Pars. 1. London. (1551): Pars. 2. Cologne. (1562): Pars. 3. Cologne. (1566): reprinted in one volume. Cologne. (1568).

Turn. Names *The Names of Herbes in Greke, Latin, Englishe, Duche and Frenche, wyth the commune names that Herbaries and Apotecaries use.* Gathered by William Turner. London. [1548]. Ray Society facsimile. 1965.

Walker Ramb. *Saturday Afternoon Rambles round London: Rural and Geological.* By Henry Walker. London. (1871).

Waring A letter from Richard Hill Waring ... On some Plants found in Several Parts of England. *Philosophical Transactions of the Royal Society* 61: 359. (1770).

Warren Fl. The Flora of Hyde Park and Kensington Gardens. By J. L. Warren. *J.B.* (*London*) 9: 227–238. (1871).

Wats. MSS. 'Plants observed between Staines and Twickenham'. A copy of *The London Catalogue of British Plants,* edition 1 (1844) marked by H. C. Watson. Library, Royal Botanic Gardens, Kew, Surrey.

Wats. New Bot. *The New Botanist's Guide to the Localities of the Rarer Plants of Britain, etc.* By H. C. Watson, Vol. 1. London. (1835): Vol. 2. (1837).

Wats. New Bot. Suppl. *Supplement to The New Botanist's Guide to the Localities of the Rarer Plants of Britain, etc.* By H. C. Watson. London. (1837).

Watsonia *Watsonia. Journal of the Botanical Society of the British Isles.* Vols. 1–7. London. (1949–69): *Watsonia. Journal and Proceedings of the Botanical Society of the British Isles.* Vols. 8 →. London. (1970 →).

Webb Br. Vall. *The Brent Valley Bird Sanctuary.* By Wilfred Mark Webb. Selborne Society. London. (1911).

Webb & Colem. Fl. *Flora Hertfordiensis; or a Catalogue of plants found in the County of Hertford, etc.* By R. H. Webb and W. H. Coleman. London. (1849).

Webb & Colem. Suppl. *Supplement to Flora Hertfordiensis, etc.* By R. H. Webb and W. H. Coleman. London. (1851): *Second Supplement.* (1851).

Webst. Reg. *The Regents Park and Primrose Hill History and Antiquities.* By A. D. Webster. London. (1911).

Wedg. Cat. *The Wedgwood Herbarium at Marlborough College.* By M. L. Wedgwood. Arbroath. (1945).

Whale Fl. *Egham Wild Flowers.* By William Whale. Egham. [c. 1875].

Wharton Fl. The Flora of Hampstead. By Henry T. Wharton, in J. Logan Lobley's *Hampstead Hill.* Pp. 73–80. London. (1889).

White Hampst. *Sweet Hampstead and its Associations.* [A fragment of the flora of Hampstead. Pp. 362–368]. By Caroline A. White. London. (1900).

Whiting Fl. The Flora of Hampstead. By J. E. Whiting, in W. H. Barratt's *Annals of Hampstead*, vol. 3. pp. 199–234. London. (1912).

Whiting Notes Some notes on the flora of Hampstead. By J. E. Whiting. *Hampstead Annual* 1901: 116–124. (1901).

Whitwell MSS. Notes written (c. 1880–90) by William Whitwell and B. B. Le Tall, in a copy of Trimen and Dyer's *Flora of Middlesex.* Now in the possession of the author.

Wild Fl. Mag. *The Wild Flower Magazine. Journal of the Wild Flower Society.* Tunbridge Wells, etc. (1921 →).

Williams MSS. Flora of Middlesex, etc. [*Campanula* to *Rubus*]. By Frederic Newton Williams. MSS. written c. 1910. Library of the Department of Botany, British Museum (Natural History), London.

Wilson Syn. *A Synopsis of British Plants in Mr Ray's method, etc.* By John Wilson. Newcastle upon Tyne. (1744).

Winch MSS. Notes written (c. 1800–30) by Nathaniel Winch, in copies of Turner and Dillwyn's *Botanist's Guide, etc.* and Smith's *Flora Britannica, etc.* Library of the Linnean Society of London, Burlington House, Piccadilly, London, W.1.

With. Bot. Arr. *A Botanical Arrangement of all the Vegetables naturally growing in Great Britain, etc.* By William Withering. 2 vols. Birmingham. (1776): Edition 2, as *A Botanical Arrangement of British Plants, etc.* Edited by J. Stokes.

	3 vols. Birmingham and London. (1787–92): Edition 3. 4 vols. (1796): Edition 4, as *A Systematic Arrangement of British Plants, etc.* By William Withering Junr. (1801): Edition 5. (1812): Edition 6, as *An Arrangement of British Plants, etc.* (1818): Edition 7. (1830).
Wood Ramb.	*Rambles in the Home Counties.* By C. A. Wood. London. (1914).
Wright MSS.	Notes written (c. 1872–92) by C. A. Wright, in a copy of Trimen and Dyer's *Flora of Middlesex.* Now in the possession of T. G. Collett, Ealing, London, W.13.

ABBREVIATIONS FOR HERBARIA CONSULTED

BM	British Museum (Natural History), London.
BO	Herbarium of William Borrer. Now at the Royal Botanic Gardens, Kew, Surrey.
CGE	Botany School, University of Cambridge.
CYN	Croydon Natural History and Scientific Society. Not seen, but details of Middlesex specimens kindly provided by the late Dr D. P. Young.
D	Herbarium of George Claridge Druce. Now at the Botany School, University of Oxford.
DB	Herbarium of Charles Du Bois. Now at the Botany School, University of Oxford. Not thoroughly searched and some localities given are those cited by G. C. Druce. (The Du Bois Herbarium, *Rep. Bot. Soc. & E.C.* 8: 463–493 (1928)).
DILL	Herbarium of Johann Jakob Dillenius. Now at the Botany School, University of Oxford. Not thoroughly searched, and some localities given are those cited by G. C. Druce and S. H. Vines (*The Dillenian Herbaria, etc.* Oxford. (1907)).
G & R	Herbarium of W. F. Goodger and R. Rozea. Not seen, and present whereabouts unknown. Localities given are those cited by Trimen and Dyer (*Flora of Middlesex.* London. (1869)).
HE	Herbarium of Robert Hardwicke. Not seen, and apparently sent to Bombay, though present whereabouts unknown. Localities given are those cited by Trimen and Dyer (*Flora of Middlesex.* London. (1869)).

HPD	Hampstead Scientific Society. Small herbarium, examined in 1948. Since destroyed owing to insect damage.
HY	Herbarium of Frederick Janson Hanbury. Now at the British Museum (Natural History), London.
K	Royal Botanic Gardens, Kew, Surrey.
L	Private herbarium of J. Edward Lousley, Streatham, London, S.W. 16. Not examined, but details of some Middlesex specimens kindly provided by the owner.
LINN	British herbarium of the Linnean Society of London, Burlington House, Piccadilly, London, W.1. Part transferred to the British Museum (Natural History), London, in November 1963.
LNHS	London Natural History Society. Part transferred to the British Museum (Natural History), London, c. 1950–53. Remainder at present at the South London Botanical Institute, 323 Norwood Road, London, S.E.24.
LT	Herbarium of John Lightfoot. Now at the Royal Botanic Gardens, Kew, Surrey.
MANCH	University of Manchester, Manchester Museum. Not seen, but details of some Middlesex specimens kindly provided by various correspondents.
ME	Herbarium of James Cosmo Melvill. Now at the Butler Museum, Harrow School, Harrow, Middlesex.
MY	Herbarium of Archibald Sim Montgomery. Now at City Museum, Brunswick Road, Gloucester. Not seen, but details of the more important Middlesex specimens kindly provided by J. W. Haines.
OXF	Botany School, University of Oxford.
QMC	Queen Mary College, University of London, Mile End Road, London, E.1.
S	Herbarium of Jonathan Salt. Now at City Museum, Weston Park, Sheffield. Not seen, but details of Middlesex plants extracted from *List of plants collected . . . by Jonathan Salt, etc.* Sheffield. 1889.
SBY	Salisbury, South Wilts. and Blackmore Museum, St Ann Street, Salisbury, Wilts. Not seen, but details of various Middlesex specimens kindly provided by Mrs B. Welch. Part of the collection was transferred to the British Museum (Natural History), London, in 1963.
SH	Herbarium of William Sherard. Now at the Botany School, Oxford. Not thoroughly searched.

SLBI	South London Botanical Institute, 323 Norwood Road, London, S.E.24.
SLO	The Sloane Herbarium. Now at the British Museum (Natural History), London. Not thoroughly searched.
STR	Passmore Edwards Museum, Romford Road, Stratford, Essex. Not seen, but details of various Middlesex specimens kindly provided by B. T. Ward.
SY	Herbarium of John Thomas Irvine Boswell-Syme. Now at the British Museum (Natural History), London.
TLS	Borough of Tunbridge Wells Museum, Mount Pleasant, Tunbridge Wells, Kent. Not thoroughly searched.
W	Herbarium of Hewett Cottrell Watson. Now at the Royal Botanic Gardens, Kew, Surrey.
WD	Herbarium of Samuel Pickworth Woodward. Now at Royal Agricultural College, Cirencester, Glos. Not seen. Localities given are those cited by Trimen and Dyer (*Flora of Middlesex* (1869)).
WE	Herbarium of Allen Wedgwood and Mrs Maria Louisa Wedgwood at Marlborough College, Marlborough, Wilts. Not seen, but details of Middlesex specimens extracted from M. L. Wedgwood's *The Wedgwood Herbarium*. Arbroath (1945).

CRITICAL GENERA, ETC.

The following specialists have kindly assisted in naming material in the various genera, etc., listed below.

CHARACEAE G. O. Allen; Mrs S. P. Phillips.
BRYOPHYTA R. A. Boniface; A. H. Norkett; J. H. G. Peterken; F. Rose; T. D. V. Swinscow; C. C. Townsend; E. C. Wallace; E. F. Warburg.
EQUISETUM P. Taylor.
RANUNCULUS Sect. BATRACHIUM R. W. Butcher.
FUMARIA H. W. Pugsley; N. Y. Sandwith.
CARDAMINE D. E. Allen.
BARBAREA A. B. Jackson.
HYPERICUM N. K. B. Robson.
CERASTIUM J. K. Morton.
MONTIA S. M. Walters.
AMARANTHUS J. P. M. Brenan; I. A. W. Kloos.
CHENOPODIUM P. Aellen; J. P. M. Brenan.

OXALIS D. P. Young.
RUBUS W. C. R. Watson; J. E. Woodhead.
ROSA N. Y. Sandwith.
CRATAEGUS S. Batko.
EPILOBIUM G. M. Ash; C. D. Pennington.
RUMEX J. E. Lousley; K. H. Rechinger.
SALIX R. D. Meikle.
CALYSTEGIA R. K. Brummitt.
VERONICA E. B. Bangerter; J. H. Burnett.
EUPHRASIA H. W. Pugsley; E. F. Warburg.
MENTHA R. A. Graham; R. M. Harley.
THYMUS C. D. Pigott.
HIERACIUM H. W. Pugsley; P. D. Sell; C. West.
CREPIS J. B. Marshall.
POTAMOGETON J. E. Dandy; Sir George Taylor.
EPIPACTIS D. P. Young.
DACTYLORHIZA H. W. Pugsley.
CYPERUS S. S. Hooper.
CAREX E. Nelmes.
GRAMINEAE N. L. Bor; C. E. Hubbard; A. Melderis.

DIVISIONS OF MIDDLESEX FOR BOTANICAL PURPOSES

The following divisions are based on river drainage, and apart from being outlined in greater detail conform closely with the divisions set out by Trimen and Dyer (*Flora of Middlesex*, xxxvii–xli (1869)).

1. UPPER COLNE. **N.** The boundary of the county from its western extremity to the railway tunnel at Deacon's Hill, near Elstree. **W.** The boundary of the county from its northern extremity to the ford near Little Britain Lake, Cowley Peachey. **S.** From the point last mentioned along Packet Boat Lane, thence passing through Pield Heath and Colham Green (via High Road, Cowley Peachey, Peachey Lane, Pield Heath Road and Harlington Road) to the Uxbridge Road at Hillingdon. **E.** From the point last mentioned in an easterly direction along the Uxbridge Road, thence in a northerly direction to Ickenham and Ruislip (via Long Lane, High Road, Ickenham, Ickenham Road and Kings End), thence through Pinner to Harrow (via High Street, Ruislip, Eastcote Road, Cheney Street, Bridle Road, Eastcote Road, Pinner, Marsh Road and Harrow Road), and continuing through Harrow Weald and Bushey Heath to Brockley Hill (via Headstone Lane, Headstone Avenue, Uxbridge Road, Brookshill, Common Road, London Road, Stanmore Hill and Wood Lane),

thence through Elstree to the railway tunnel at Deacon's Hill (via Elstree Road and Barnet Lane). There is also an outlying portion of this district, the most northerly part of the vice-county, which as it were locks into Hertfordshire. The vice-county boundary limits this on all sides except the east, where it is bounded by the Great North Road from Hadley Green through Potters Bar.

2. LOWER COLNE. **N.** The south boundary of district 1. **W.** The vice-county boundary from the ford near Little Britain Lake to its southern extremity. **S.** The vice-county boundary from its western extremity to the eastern end of Eelpie Island, Twickenham. **E.** From the point last mentioned westwards through Fulwell, towards Hanworth (via Lebanon Park, Richmond Road, York Street, The Green, Twickenham, Staines Road and Twickenham Road), thence in a northerly direction through Hatton and Harlington to Hayes (via Hampton Road, Uxbridge Road, Harlington Road, Faggs Road, Hatton Road, High Street, Harlington, Station Road and Coldharbour Lane), and so into the Uxbridge Road which it follows westward to join the northern boundary.

3. CRANE. **W.** The eastern boundary of district 2, and the eastern boundary of district 1 as far north as Harrow Weald. **S.** The vice-county boundary from the eastern end of Eelpie Island, Twickenham, to the western boundary of Syon House grounds, Isleworth. **E.** From the point last mentioned through Spring Grove and Lampton (via Park Road, London Road and High Street, Hounslow), thence northwards through Lampton, Heston and Norwood Green to Southall (via Lampton Road, Heston Road, Norwood Road, King Street, The Green and South Road), thence crossing the Uxbridge Road and continuing in a northerly direction through Northolt, Wood End and Harrow to Harrow Weald (via Lady Margaret Road, Ealing Road, Wood End Lane, Wood End Gardens, Wood End Road, South Vale, Sudbury Hill, London Road, High Street, Harrow, Peterborough Road, Tyburn Lane, Lowlands Road and Pinner Road), where it joins the northern extremity of the western boundary. The district is thus pointed at its northern extremity.

4. UPPER BRENT. **N.** The vice-county boundary from the railway tunnel at Deacon's Hill to the Great North Road north of Whetstone. **W.** The eastern boundary of district 1 as far south as Harrow Weald, and the eastern boundary of district 3 as far south as Harrow. **S.** The road from Harrow through Sudbury and Wembley (Kenton Road, Watford Road, Harrow Road and High Road, Wembley) to the Midland Region railway line at Wembley station, thence following the line south-eastwards to Kensal Green station. **E.** From Kensal Green station in a north-easterly direction along the railway line to Kilburn station, thence through Temple Fortune to

the southern end of the West Heath at Hampstead (via Kilburn High Road, West End Lane, Platts Lane, Telegraph Hill, West Heath Road, Branch Hill, Frognal Rise, Holly Hill and Heath Street), and across to 'The Spaniard's' Inn (via Spaniard's Road), and so to Highgate and Finchley (via Hampstead Lane, High Street, Highgate, Hornsey Lane, Archway Road, Muswell Hill Road, Queen's Avenue, Fortis Green and High Road, East Finchley), and continuing through North Finchley and Whetstone (via the Great North Road) to the northern boundary of the vice-county.

5. LOWER BRENT. **N.** The southern boundary of district 4. **W.** The eastern boundary of district 3 as far north as Harrow. **S.** The vice-county boundary from the western end of Syon House grounds to Hammersmith Bridge. **E.** From the point last mentioned to Kensal Green railway station (via Hammersmith Bridge Road, Brook Green Road, Shepherds Bush Road, The Lawn, Wood Lane, Scrubs Lane and Harrow Road).

6. LEE. **N.** The vice-county boundary from the Great North Road at Potters Bar to its eastern extremity. **W.** The eastern boundary of the outlying part of district 1, the very irregular vice-county boundary from Chipping Barnet to north of Whetstone, and the eastern boundary of district 4 as far south as Highgate. **S.** From the point last mentioned to Hornsey (via Hornsey Lane, Crouch End Hill, Crouch End Broadway, Tottenham Lane and Church Lane), thence to Tottenham High Cross and the vice-county boundary (via Turnpike Lane, West Green Road, Broad Lane and Ferry Lane). **E.** The vice-county boundary from its northern extremity to Tottenham Hale (Ferry Lane).

7. METROPOLITAN. **N.** The eastern boundary of district 4 from Kensal Green to Highgate, and the southern boundary of district 6 from Highgate to the vice-county boundary. **W.** The eastern boundary of district 5. **S.** The vice-county boundary from Hammersmith Bridge to its eastern extremity. **E.** The vice-county boundary from Tottenham Hale (Ferry Lane) southwards.

THE PLAN OF THE FLORA

The sequence and nomenclature of families, genera and species are, with certain exceptions, based on the following publications: Charophyta – Allen, G. O. *British Stoneworts (Charophyta)*. Haslemere, Surrey. 1950. Hepaticae – Paton, J. A. *Census Catalogue of British Hepatics*. Edition 4. Ipswich. 1965. Bryophyta – Richards, P. W. and Wallace, E. C. An annotated list of British Mosses. *Trans. Brit. Bryol. Soc.* 1, part 4, supplement, i–xxi. 1950. Pteridophyta and Spermatophyta – Clapham, A. R., Tutin, T. G. and Warburg, E. F. *Flora of the British Isles*. Edition 2. Cambridge. 1962.

The Latin name of the species is given at the top left-hand side of each account, followed, where appropriate, in the vascular plants, by the English name or names.

Synonyms in general use, together with certain archaic names used in Trimen and Dyer's *Flora of Middlesex*, are placed below the supposedly correct Latin name of the species.

Below the Latin name, or synonym(s), literary references to taxonomic accounts of the species, or notes upon them, are provided in a number of instances. Where the account is in a foreign journal, reference is made to an abstract in English. These are prefixed by the abbreviation Lit., and in the case of abstracts (*Abstr.*).

The supposed status of the species in Middlesex is then given, and the terms used are defined as follows:

Native. A species indigenous to a particular habitat, or to particular habitats in the vice-county, and believed to have been present in pre-Roman times.

Denizen. A species which has the appearance of being native but which for various reasons is suspect of having been accidentally or deliberately introduced prior to the mid-sixteenth century.

[1]*Colonist*. A weed of cultivated and waste ground believed to have been introduced by the activities of man.

Alien. A species of known foreign origin which is now established, or was formerly established, in the vice-county.

Introduced. A species which though possibly native in parts of the British Isles is not considered so in Middlesex.

Details of the habitat, or habitats, in which the species may be found are then given, and this is followed by an indication of its frequency within the vice-county. The categories of frequency which apply at the present time are defined as follows:

Very common. Generally distributed in numerous localities in all the districts.

Common. Occurring in all the districts, but less frequently in some than in others.

Rather common. Widely distributed, but in disjunct localities, and not necessarily present in all the districts.

Local. Occurring in from 13 to 24 localities, and showing a marked distribution pattern which may be correlated with soil, habitat, etc.

Rather rare. Occurring in from 13 to 24 localities, but apparently failing to show a marked distribution pattern which can be correlated with soil, habitat, etc.

[1] In the section on bryophytes the term 'colonist' refers to species which have become established in natural or semi-natural habitats, though originating from areas outside the vice-county.

Rare. Occurring in from 7 to 12 localities.
Very rare. Occurring in from 1 to 6 localities.
Extinct. No longer occurring in the vice-county.

On the next line is given, where known, the name of the person who is believed to have first recorded, or collected the species in Middlesex, together with the appropriate date. This information is not supplied for Charophytes and Bryophytes. Where a species is assumed to be extinct the date of the last known record and the recorder's name is provided. If the date of the record, or records, is based on herbarium material it is referred to as 'first evidence' or 'last evidence'.

The localities in which the species has been recorded then follow, and modern spelling of place names, etc., is used except in certain circumstances. These are arranged under the seven districts outlined on pp. 52–54. No localities are provided for 'very common' species, but for common species records are usually given for district 7 – the metropolitan area. Records not followed by a reference or a recorder's name were made by the author. When several different records are separated by colons or semi-colons the authority, or herbarium abbreviation, cited after the last is applicable to all. The occurrence of a species in a particular locality over a period of time is indicated by the separation of the dates by colons or semi-colons. Abbreviations of herbaria, a list of which is given on pp. 49–51, are enclosed in round brackets and follow Kent, D. H. *British Herbaria*. London (1958). A list of the abbreviations used for books, periodicals, papers, MSS., etc., is provided on pp. 31–49.

Details of subspecies, varieties, forms and hybrids are placed beneath the account of the species to which they are relative.

The following symbols are used:

! Following a locality indicates that the plant has been seen there by the author.

! Following a recorder's name, or recorders' names, indicates that the record was made in the company of the author.

† Following a locality indicates that the plant is presumed to be extinct there. The symbol is not used for localities of species believed to be extinct in the vice-county.

LIST OF RECORDERS, ETC.

The following persons are known to have recorded or collected Middlesex plants. The list is acknowledgedly incomplete and for omissions I offer my apologies.

Absolon, E. M. →
Aitken, Mrs, fl. 1923
'A.J.', fl. 1856–60
Alchorne, S., 1727–1800

Allen, D. E. →
Alletson, J. →
Alletson, S. G., 1901-66
Alston, A. H. G., 1902–58
Alston, R. →
Alyland, Mrs D. →
Ambrose, F. →
Anderson, J. →
Andrewes, Sir Christopher H. →
Andrews, C., fl. 1831–83
Andrews, C. R. P., 1870–1951
Andrews, J., fl. 1710–62
Andrews, R. G., fl. 1874–83
Arnold, M., 1822–88
Ash, G. M., 1900–59
Ashby, F., 1660–1743
Atkins, Mrs A., 1797–1871
August, V. E., d. 1955
Avery, C., 1880–1960
Ayland, W. →

Babington, C. C., 1808–95
Bacon, G., fl. 1913
Bacon, Miss, fl. 1925
Bacot, A. W., 1866–1922
Bagnall, J. E., 1830–1918
Bain, P. C. →
Baker, E. C. →
Baker, E. G., 1864–1949
Baker, J. G., 1834–1920
Balfour, J. H., 1808–84
Ballard, E., 1820–97
Ballard, F. →
Bangerter, E. B. →
Banker, J., fl. 1850–52
Banks, H. →
Banks, Sir Joseph, 1743–1820
Barham, F., fl. 1845–1900
Barlow, J. W. →
Bartlett, T. L. d. 1973
Barton, W. C., d. 1955
Batko, S. →
Battley, A. U., d. 1905
Baxter, W., 1787–1871
Baylis, E., fl. 1790–94
Bayliss, B. E. →
Bayliss, H. A. →
Baynes, J., fl. 1791
Bedford, J. →
Bedford, R., fl. 1909
Beeby, W. H., 1849–1910
Beevis, J., fl. 1837
Bell, P. →
Bell, W., 1862–1925
Bell, W., c. 1840–1920
Benbow, J., 1821–1908
Bence, T. A. →
Bendle, L. S. →
Bennett, A., 1843–1929
Bennett, A. W., 1833–1902
Bennett, Mrs D. →
Bennett, E. A. →
Bennett, E. T., 1797–1836
Bennett, J. J., 1801–76
Bennett, R. L. →
Benoit, P. M. →
'B.G.', fl. 1836
Bicheno, J. E., 1785–1851
Bickerstaff, Mrs M. S. →
Biden, W., fl. 1845
Bigg-Wither, K. →
Binstead, C. H., 1862–1941
Bird, G., 1814–54
Bishop, E. B., 1864–1947
Blackstone, J., 1712–53
Blair, P., 1666–1728
Blake, W. J., 1805–75
Bliss, J., fl. 1800–26
Bloss, H. R., fl. 1710
Blow, T. B., 1854–1941
Bloxam, A., 1801–78
Blundell, J., fl. 1900–23
Boniface, R. A. →
Boodle, L. A., 1865–1941
Boott, F., 1792–1863
Borrer, W., 1781–1862
Bostock, M. W., d. 1959
Bosworth, G. F., fl. 1900–38
Boucher, F. P. D. →
Boulger, G. E. S., 1853–1922

Bourne, Mrs, fl. 1871
Bowman, J. E., 1785–1841
Bradley, A. E., 1873–1944
Braithwaite, J. O., d. 1937
Braithwaite, R., 1824–1917
Braybon, A. →
Bredwell, fl. 1597
Bree, W. T., 1787–1863
Brenan, J. P. M. →
Brewer, E., d. 1922
Brewer, J. A., 1818–86
Brewer, S., 1670–1743
Brewis, Lady Anne →
Briggs, Mrs E. A. →
Brittain, F. →
Britten, J., 1846–1924
Britton, C. E., 1872–1944
Britton, H. →
Broadway, W. E., d. 1935
Brocas, F. Y., fl. 1840–59
Bromwich, H., 1828–1907
Broome, C. E., 1812–86
Brown, G. C., d. 1969
Brown, J. A., 1831–1903
Brown, L., 1699–1749
Brown, N. E., 1849–1934
Brown, R., 1773–1858
Brown, W. H., fl. 1891–1912
Browner, G., fl. 1875
Browning, F. R. →
Brummitt, R. K. →
Buckle, W. F. →
Buddle, A., c. 1650–1715
Bull, K. E. →
Burchell, W. J., 1781–1863
Burges, R. C. L., 1901–59
Burkill, H. J., 1871–1956
Burnett, G. T., 1800–35
Burton, R. M. →
Butcher, R. W., 1897–1971
Button, E. H., fl. 1819–60
Bywater, W. M., fl. 1863–64

Caley, G., 1770–1829
Cameron, Mrs M. →
Campbell, A. →
Cannon, J. F. M. →
Cargill, J., fl. 1600–03
Carrington, J. A., fl. 1889–1926
Carter, —, fl. 1836–40
Casey, Mrs J. →
Castell, C. P., 1907–72
Castles, R., fl. 1835–40
Catcheside, D. G. →
Cattley, M. E. →
Chambers, R., 1784–1858
Champneys, Mrs M., fl. 1860–1935
Chandler, S. E., 1880–1957
Chapple, J. F. G. →
Chatterley, W. M., fl. 1835–39
Cherry, J. W., 1846–1935
Children, J. G., 1777–1852
Christy, W., c. 1807–39
Church, Sir Arthur H., 1834–1915
Clarke, C. B., 1832–1906
Clarke, F. →
Clarke, J., fl. 1871–91
Clarkson-Birch, W., fl. 1903
Clement, E. →
Clements, R., fl. 1760
Clusius, J. C., 1526–1609
Coates, N. H. →
Cobbe, A. B., d. 1952
Cobbe, M., d. 1936
Cochrane, W. D., fl. 1890
Cockerell, T. D. A., 1866–1948
Cockfield, J., 1740?–1816
Codrington, J. C. →
Coker, P. D. →
Cole, A. B., fl. 1860–69
Cole, M. →
Coleman, W. H., 1816?–63
Coles, J., fl. 1597
Coles, W., 1626–62
Collenette, C. L., 1888–1959
Collett, G. W., d. 1964
Collett, Mrs M. →
Collett, M. G. →
Collett, T. G. →

Collins, S., fl. 1720–30
Collinson, M., 1728?–95
Collinson, P., 1694–1768
Colman, H. M., fl. 1840
Congreve, C. R. T., 1876–1952
Cooke, M. C., 1825–1914
Cooke, P. H., 1859–1950
Cooper, C. A., 1871–1944
Cooper, D., 1817?–42
Cooper, J. E., 1864–1952
Corbett, Mrs L. →
Corke, H. →
Corner, J. H. →
Cornish, J. C., 1858–1906
Cottam, A., 1838–1912
Cotton, A. D., 1879–1962
Coules, —, fl. 1920
Coward, R. E. →
Cowper, B. H., fl. 1848–80
Crabbe, J. A. →
Craig, J., fl. 1866
Cramp, S. →
Crawford, A. →
Creasey, H., fl. 1910
Crooks, S. E. →
Crowe, J., 1750–1807
Cruttwell, N. E. G. →
Cubitt, M., fl. 1845
Cullen, W. H., fl. 1836–37
Cullum, Sir Thomas G., 1741–1831
Cundall, J. H., 1808–84
Curtis, W., 1746–99

Dale, E. M., fl. 1899–1913
Dale, J., d. 1662
Dale, S., 1659–1739
Dale, T., 1699?–1750
Dandridge, (T.) J., 1664–1746
Dandy, J. E. →
David, R. W. →
Davidge, G. J. →
Davies, H., 1739–1821
Davies, Mrs H. R., d. c. 1960
Davies, T., fl. 1868
Davies, W., 1814–91
Davy, Lady, 1865–1955
Day, F. M., 1890–1962
Day, G., fl. 1888–90
'D.E.', fl. 1890
De Crespigny, E. C., 1821–95
Denison, J., fl. 1834
Denner, G. B. →
Dennes, G. E., fl. 1817–60
De Vesian, R. E., fl. 1870–90
Dickson, J., 1738–1822
Dillenius, J. J., 1684–1747
Dillwyn, L. W., 1778–1855
Dixon, H. N., 1861–1944
Dodsworth, M., 1654–97
Donald, J., 1815–72
Donovan, J. E. →
Dony, Mrs C. M. →
Dony, J. G. →
Doody, S., 1656–1706
Drabble, E. F., 1877–1933
Drabble, Mrs H., d. 1965
Druce, F., 1873–1941
Druce, G. C., 1850–1932
Druett, W. W., fl. 1919–38
Drummond, Mrs, fl. 1912
Du Bois, c., 1656–1740
Dunn, S. T., 1869–1938
Dupree, T. W. J. D. →
Dussan, Mme., fl. 1916
Dyer, Sir William T. T., 1843–1928
Dymes, T. A., 1865–1944

Eagles, T. R., d. c. 1970
Eastwood, Mrs D. M. →
Edwards, E., 1812–86
Edwards, P. J. →
Edwards, R. K. →
'E.I.', fl. 1855–56
'E.K.'—See Kent, E.
Eland, T., fl. 1908
Eliot, Lady Alethea →
Ellis, J. →
Ettlinger, D. M. T. →

Evans, A., fl. 1854
Evans, J. B. →
Evans, T., fl. 1898
Every, A., fl. 1897
Evetts, Mrs E. F., d. 1947

Farend, J. →
Farenden, W. →
Farrar, F. W., 1831–1903
Faulkner, T., 1777–1855
Field, J., fl. 1713–52
Finch, Louisa, Countess of Aylesford, née Thynne, 1760–1832
Findlay, R., fl. 1869–1927
Findon, C. J. B., fl. 1880–1913
Firminger, T. A. C., 1812–84
Firrell, P. →
Fisher, H., 1860–1935
Fitter, R. S. R. →
Flippance, F. →
Flower, T. B., 1817–99
Foley, H. J., fl. 1887–1900
Ford, E., fl. 1873
Forster, E., 1765–1849
Forster, T. F., 1761–1825
Forsyth, W., 1772–1835
Fowle, H., fl. 1836–63
Fox, H. E., 1841–1926
Francis, G. W., 1800–65
Franklin, H. →
Franklin, J. →
Franks, H. →
Fraser, J., 1854–1935
Freeman, J., 1813–1907
French, A., 1839–79
Fritsch, F. E., 1879–1954
Frost, A. →
'F.W.', fl. 1848–62
Fysher, R., 1698–1747

'G.A.', fl. 1890–1902
Gade, G. K., fl. 1886–87
Gale, J. E., fl. 1898
Galpine, J., 1769?–1806
Gardiner, J. C. →
Gardiner, Mrs M. W. →
Gardner, J. E., fl. 1906–30
Gardner, R. C. B., fl. 1930
Garlick, C., d. 1934
Garry, F. N. A., 1861–1940
Gawler, J. B., fl. 1835–37
George, J. B., d. 1871
Gepp, A. K. D., 1862–1955
Gérard, F., d. 1840
Gerarde, J., 1545–1612
Gerrans, M. B. →
Gibbs, J., d. c. 1829
Gibson, Mrs G. M., d. 1965
Gifford, B., fl. 1864
Gilbert, E. G., 1840–1915
Gilbert, J. L. →
Gilmour, J. S. L. →
Girard
Gleadow, P. R. →
Goater, B. →
Goodenough, S., 1743–1827
Goodger, W. F., fl. 1815-20
Goodyer, J., 1592–1664
Goom, N. →
Gordon, A., fl. 1770–1845
Gowan, T. Y., fl. 1835–40
Graham, R. A., 1915–58
Gray, J. E., 1800–75
Gray, P., fl. 1843–73
Gray, S. F., 1766-1828
Gray, S. O., 1828–1902
Green, C. B., d. 1918
Greenwood, E. F. →
Greenwood, P. R. →
Griffin, W. H., 1841–1921
Griffith, A. W., 1866–69
Grigg, H. C. →
Grizet, H. E., fl. 1892
Groult, P., fl. 1800–04
Groves, E. W. →
Groves, H., 1855–1912
Groves, J., 1858–1933
Grubbe, J., fl. 1879
Grugeon, A., 1826–1913

Gunn, S. J., fl. 1876
Gunning, M., fl. 1883
Gush, G. H. →

Hales, A. F. →
Hall, Mrs A. M., 1800–81
Hall, A. R. →
Hall, Mrs J. F. →
Hall, K. M., fl. 1908–13
Hall, L. B., 1878–1945
Hall, P. C. →
Hall, S. C., 1800–89
Halligey, P. →
Hamilton, W., fl. 1829–42
Hanbury, D., 1825–75
Hanbury, F. A., fl. 1858–63
Hanbury, F. J., 1951–1938
Hankey, J. A., 1810?–82
Hanson, P. J., d. 1948
Hardwicke, R., 1824–76
Harley, R. M. →
Harrington, Mrs E., fl. 1918
Harris, H. C. →
Harris, J., fl. 1704–29
Harris, J., 1791–1873
Harris, K. C. →
Harris, S. →
Harrison, D. N. →
Harting, J. E., 1841–1928
Hartog, M. M., 1851–1924
Hasler, J. K. →
Haussknecht, H. C., 1838–1903
Haworth, A. H., 1768–1833
Haywood, J. A. →
Heath, F. G., 1843–1913
Heathfield, R., 1802–48
Helsby, I. A., d. c. 1940
Hemsley, W. B., 1843–1924
Henfrey, A., 1819–59
Henrey, R. →
Henson, A. →
Hepper, F. N. →
Hepworth, Mrs →
Higgins, Mrs D. M., 1856–1920
Hill, J., 1716–75
Hill, M. O. →
Hill, R. S., d. 1872
Hillaby, J. D. →
Hind, W. M., 1815–94
Hinds, A. J., fl. 1914
Hinson, D. J. →
Hockaday, Mrs M. C. →
Hodson, J. →
Holland, P. C. →
Holme, H. C. →
Hooper, Dr, fl. 1820
Horder, H. →
Horsman, S., 1698–1751
Horsnell, G., 1625?–97
Horton, W. R. G. →
Horwood, A. R., 1879–1937
Horwood, C. M. B. →
How, W., 1620–56
Hubbard, C. E. →
Hudson, W., 1730–93
Hudson, W. H., d. 1922
Hull, J., 1761–1843
Hunt, P. F. →
Hunter, E., fl. 1790–1824
Hurlock, S., d. 1748
Hutchinson, J., 1884–1972
Hutchinson, R. R., 1870–1951
Hyde, H., fl. 1869

I'Anson, J., 1784–1821
Ing, B. →
Irons, E., fl. 1863
Irvine, A., 1793–1873
Irvine, D. E. G. →
Irvine, J., fl. 1864
Irvine, Mrs L. →
Isaac, M., 1878–1964
Isherwood, E. M. C. →

'J.A.', fl. 1856–63
Jackson, A. B., 1876–1947
Jackson, B. D., 1846–1927
Jackson, E., fl. 1879–1907
James, L., d. 1967
James, Mrs W., fl. 1840–50

Jameson, Mrs H., fl. 1840
Jefferey, H. J., 1885–1950
Jeffkins, A. C. →
Jefford, G. →
Jenkinson, G., 1739?–1808
Jermy, A. C. →
Jesse, E., 1780–1868
Jewitt, T. O. S., 1799–1869
'J.F.G.', fl. 1856
'J.G.', fl. 1881
'J.L.G.', fl. 1840
Johns, L. J., d. 1970
Johnson, G., fl. 1869
Johnson, I. G. →
Johnson, T., d. 1644
Johnson, W., d. c. 1925
Johnston, Sir Harry H., 1858–1927
Jones, A. W. →
Jones, E. M. M., 1887–1960
Jones, Mrs J. A. W. →
Jones, —, fl. 1776
'J.W.', fl. 1875–81
'J.W.T.', fl. 1856

Kalm, P., 1715–79
Kemble, W., fl. 1817–51
Kennedy, M. E. →
Kent, D. H. →
Kent, E., fl. 1820–30
Kent, Mrs I. M., née Austen →
Kent, M. P. →
Kiddle, Mrs E. →
King, Mrs D. →
King, J. M. B. →
King, P. S., fl. 1883–84
King, S., fl. 1897–1934
Kingsley, H., fl. 1830–64
Kippist, R., 1812–82
Kirkpatrick, R., fl. 1922
Kite, E. C. →
Knapp, J. L., 1767–1845
Knight, F. P. →
Knight, R. J., 1822–99
Knightley, —, fl. 1910
Knipe, P. R. →
Knowlton, T., 1692–1782
Knox, M., d. 1952

Laflin, T., 1914–72
La Gasca y Segura, M., 1776–1839
Lake, K. E. →
Lambert, A. B., 1761–1842
Lambert, T., fl. 1839
Lankester, Sir Edwin R., 1847–1929
Lansbury, I. G. →
Laundon, J. R. →
Lawfield, W. N., 1913–62
Lawrence, J. W., fl. 1843–54
Lawson, M. A., 1840–96
Lawson, T., 1630–91
'L.B.', fl. 1929
Lees, E., 1800–87
Leete, A. C. →
Le Gros, A. →
Leith, R. F. C., 1854–1936
Lemman, C. M., 1806–52
Lester-Garland, L. V., 1860–1944
Le Tall, B. B., 1858–1906
Lewis, J. →
Lewisham, Third Viscount, 1731–1801
Libbey, R. P. →
Lightfoot, J., 1735–88
Ligo, —, fl. 1735
Lillford, G. →
Lindley, J., fl. 1775
Lindley, J., 1799–1865
Lindsay, G., fl. 1904
Linton, E. F., 1848–1928
Linton, W. R., 1850–1908
Little, J. E., 1861–1935
Lloyd, A. K. →
Lloyd, G., 1804–89
Lloyd, J., 1810–96
Lloyd, J. →
Lloyd, W. H., fl. 1850
l'Obel, M. de, 1538–1616

Lobley, J. L., 1868–1914
Lockett, G. H. →
Longman, F. W., fl. 1846–64
Lording, T. A. →
Lorkin, R., fl. 1629–34
Loudon, J. C., 1783–1843
Lousley, J. E. →
Lovell, M. A. G., fl. 1906
Lovis, J. D. →
Lowe, Mrs, fl. 1836
Lowne, B. T., 1878–1956
Loydell, A., 1849–1910
'L.S.', fl. 1839
Luxford, G., 1807–54

Mackay, J. B., fl. 1869–87
Macreight, D. C., 1799–1868
Maher, D., fl. 1847
Mann, R. J., 1817–86
Markham, J. →
Marks, C. E. →
Marks, K. M., d. 1971
Marsden, P., fl. 1900–10
Marshall, E. S., 1858–1919
Martin, F. E., fl. 1899–1915
Martin, L. →
Martin, N. H. →
Martin, W., fl. 1597
Martyn, J., 1699–1768
Martyn, S. →
Martyn, T., 1735–1825
Mason, J. →
Masterman, N., d. 1938
Masters, M. T., 1833–1907
Maude, A. H., 1850–1933
Maude, J. →
Maurice, —, fl. 1823
'M.A.W.', fl. 1861–63
Mawer, W., fl. 1833–83
Maxwell, Mrs M. Y., fl. 1899
McClintock, D. →
McIvor, W., d. 1876
McLean, Mrs J. K. →
M'Creight, —, fl. 1830–70
Meehan, T., 1826–1901
Meikle, R. D. →
Meinertzhagen, R., 1878–1967
Melderis, A. →
Melvill, J. C., 1845–1929
Melville, R. →
Merison, S., 1901–73
Merrett, C., 1614–95
Metcalfe, E., fl. 1829–40
Metcalfe, W., fl. 1829–40
Middleton, R. M., 1846–1909
Miles, B., 1937–70
Mill, J. S., 1806–73
Miller, C., 1739–1817
Miller, J., d. 1748
Miller, P., 1691–1771
Mills, W. H., 1873–1959
Milne, C., 1743?–1815
Milner, C. A. →
Milne-Redhead, E. →
Milsom, F. E., 1886–1945
Minnion, W. E. →
Missen, Mrs C. E. →
Mitchell, J., fl. 1835–36
Mitchell, N. S. P. →
Mitten, W., 1819–1906
M'Nab, J., fl. 1874
Moggeridge, J. T., 1824–74
Molesworth, C., 1764–1872
Monckton, H. W., 1857–1931
Montgomrey, A. S., 1844–1922
Moody, Mrs, fl. 1947
Moore, J. A. →
Moore, T., 1821–87
Morgan, B. M. C. →
Morgan, E., c. 1619–89
Morgan, H., fl. 1540–88
Morgan, J. H., fl. 1894
Moring, P., 1865–1930
Morison, R., 1620–83
Morley, C. L., fl. 1646–1702
Morris, A., fl. 1845–69
Morris, D. →
Morris, J., 1810–86
Morris, L. E. →
Morris, W. H., fl. 1897

Morton, J. K. →
Mountford, J. O. →
Moxey, Mrs P., née Firrell →
Moxey, P. A. →
Moxon, J. E., fl. 1837–54
Mullin, J. M. →
Munby, G., 1813-76
Murison, —, fl. 1912
Murray, D. →

Naylor, F., 1811–82
Naylor, S. A., fl. 1868
Nelmes, E., 1895–1959
Newbould, W. W., 1819–86
Newey, P. M. →
Newman, E., 1801–76
Newman, H., fl. 1840
Newton, J., 1639–1718
Newton, S. →
Nicholls, R., fl. 1713–50
Nicholson, C. S., d. 1918
Nicholson, G., 1847–1908
Norman, P. R. →
Norton, M., fl. 1888

O'Byrne, J. Kennedy →
O'Dell, J. W., fl. 1872–84
Ord, J., fl. 1864
Ormerod, G. E., fl. 1874–87

Page, W. T., fl. 1908
Palmer, E., fl. 1843–54
Palmer, J. R. →
Palmer, S., fl. 1830
Palmer, W. J. S., fl. 1934
Pamplin, W., 1806–99
Park, J. J., 1795–1833
Parker, Mrs, fl. 1845
Parker, F. W., fl. 1930
Parkinson, J., 1567–1650
Parnell, R., 1810–82
Parr, Mrs, fl. 1871
Parsons, H. F., 1846–1914
Paterson, J. H. →
Paulson, R., 1875–1935
Payne, F. W., 1852–1927
Payne, L. G., 1893–1949
Payne, R. →
Payne, R. M. →
Payton, H. W. →
Pearce, W. A., fl. 1883
Pegler, L. W. H., 1852–1927
Pena, P., fl. 1530–1605
Perkins, H. →
Perring, F. H. →
Peterken, J. F. G., 1893–1973
Petiver, J., c. 1658–1718
Pettet, A. →
Phelp, S. →
Philcox, D. →
Philipson, W. R. →
Phillips, J. R. →
Phillips, S. P. →
Pickess, B. P. →
Pierce, E., fl. 1909
Pierson, H., 1852–1915
Pigott, C. D. →
Planchon, J. E., 1823–88
Platten, Mrs →
Plukenet, L., 1641–1706
Pocock, R., 1760–1830
Polunin, O. V. →
Potter, J. J. →
Potts, A. J., d. 1950
Powell, J. T., 1833–1904
Powles, E. M., fl. 1839–46
Powles, R. C., fl. 1869
Preedy, W. P., fl. 1837
Price, O., fl. 1769
Proskauer, J. →
Pryor, A. R., 1839–81
Pryor, R., fl. 1830–46
Pugsley, H. W., 1868–1947
Pulteney, R., 1730–1801

Quekett, E. J., 1808–47

'R', fl. 1860
Rake, —, fl. 1916
Ralfs, J., 1807–90

Ralph, T. S., 1813–91
Rand, I., d. 1743
Raven, P. H. →
Ray, J., 1627–1705
Rayer, J., 1735–97
Redgrove, H. S., 1887–1943
Reeley, A.
Reeves, W. W., 1819–92
Reid, D. M., d. 1955
Renson, J. →
Reynolds, B., fl. 1915–36
Reynolds, J., 1792–1868
'R.G.', fl. 1890
Richards, P. W. →
Richardson, R., 1663–1741
Riddelsdell, H. J., 1866–1941
Ridley, H. N., 1855–1956
Robbins, J. C., 1905–32
Robbins, R. W., 1871–1941
Roberts, P., fl. 1710
Robinson, Mrs, d. 1847
Robinson, E. K., 1857–1928
Robinson, T., fl. 1684–1748
Robinson, W. K. →
Robson, S., 1741–79
Roebuck, M. →
Roffey, J., 1860–1927
Rogers, W. M., 1835–1920
Rolfe, R. A., 1855–1921
Rönaasen, R. →
Rose, F. →
Roseway, Sir David, 1890–1969
Rowbotham, F. J., fl. 1882
Royle, W. →
Rozea, R., fl. 1815–30
'R.T.', fl. 1870
Rudge, S., 1728–1817
Rudiger, A. W. →
Russell, Mrs, B. H. S. →
'Rusticus' = Newman, E.
Ryall, R. →
Ryves, T. B. →

Sabine, J., 1770–1837
Sabine, L. A. C., fl. 1906–15
Salisbury, Sir Edward J. →
Salisbury, R. A., 1761–1829
Salt, J., 1759–1810
Salter, T. B., 1814–58
Sanders, R., fl. 1771
Sandford, H. A. →
Sandwith, N. Y., 1901–65
Savage, C., fl. 1863–64
Sayle, M. →
Schmolle, G. E. →
Scholey, M. A. R. S., d. 1970
Scott-White, J., fl. 1872
Seagrott, E., fl. 1886
Seeley, W. J., fl. 1841
Sennitt, B. F. C. →
Sewell, P., 1865–1928
Shaw, H. K. Airy →
Shenstone, J. C., 1855–1925
Shepheard, E. F., fl. 1896–1906
Sherard, J., 1666–1737
Sherard, W., 1659–1728
Sherrin, W. R., 1871–1955
Shillito, J. F. →
Short, V. B. →
Shove, R. F., d. c. 1960
Shute, Mrs, d. 1866
Shuttleworth, R. J., 1810–74
Sibthorp, H., 1712–97
Sibthorp, J., 1758–96
'Sigma', fl. 1861
Sim, J., fl. 1876–1911
Simmonds, J. →
Simpson, N. D., 1890–1974
Sladen, W. J. L. →
Slatter, J. W., 1829–96
Small, J. →
Small, Mrs L. M. P. →
Small, W., d. 1974
Smith, G. E., 1804–81
Smith, J., fl. 1867
Smith, Mrs J. E. →
Smith, Sir James E., 1759–1828
Smith, R. L., 1892–1973
Smith, W. G., 1835–1917
Sneyd, R., fl. 1781–1820

Sole, W., 1741–1802
Soper, L. C. →
Souster, H. B. →
Sowerby, J., 1757–1822
Sowerby, J. de C., 1787–1871
Spencer, N.—Pseudonym of Sanders, R.
Spooner, H. →
Sprague, T. A., 1877–1958
Spreadbury, W. H. →
Springett, K. E. →
Stearn, W. T. →
Stebbing, T. R. R., 1835–1926
Stevens, W., fl. 1846–97
Stillingfleet, B., 1702–71
Stocks, J. E., 1822–54
Stonestreet, W., d. 1716
Stott, P. A. →
Stratton, F., 1840–1916
Street, J. →
Streeter, D. J. →
Stuart, Mrs G. W., fl. 1864
Studley, F. C. →
Summerhayes, V. S., 1897–1974
Sutton, A., 1859–1932
Sweet, R., 1783–1835
Sworder, R. I. →
Syme, J. T. B., 1822–88

Talbot, Rt. Hon. Sir George J., 1861–1938
Talbot, W. H. F., 1800–77
Tansley, Sir Arthur G., 1871–1955
Taylor, Sir George →
Taylor, H., fl. 1822–49
Taylor, J., fl. 1812–36
Taylor, M., fl. 1870–1945
Taylor, P. →
Teagle, W. G. →
Teesdale, R., d. 1804
Tempere, J., fl. 1873–75
Temple, H. W. →
Tester, J. A. →
Thompson, B. B., fl. 1905
Thompson, F. P., fl. 1891
Todd, W. A., fl. 1910–22
Tooke, W. A., d. 1884
Tooke, Mrs W. A., fl. 1864–72
Tooke, Miss —, fl. 1872
Townsend, C. C. →
Tremayne, L. J., 1873–1959
Trimen, H., 1843–96
Trimmer, K., 1804–87
Tuck, G., fl. 1831–42
Tuck, G. →
Tucker, R., 1832–1905
Tuff, J., fl. 1858
Turner, D., 1775–1858
Turner, D. N. →
Turner, R., fl. 1626–87
Turner, W., c. 1508–1568
Turrill, W. B., 1890–1961
Twining, T., 1806–95
Tyrrell, A. R. W. →

Uvedale, R., 1642–1722

Vallance, A. H., fl. 1893
Vardy, J., d. 1675
Varenne, E. G., 1811–87
Vaughan, A. →
Vaughan, J., 1855–1922
Veál, Mrs M. A. G. →
Vernon, P. F. →
Vivian, C., d. c. 1957

Waddell, C. H., 1858–1919
Walker, F., 1829–89
Walker, H., fl. 1871–82
Walker, V. E., fl. 1828–58
Walkin, W. A., fl. 1912
Wall, G., 1821?–94
Wallace, E. C. →
Waller, J., fl. 1889
Wallis, A., fl. 1810
Warburg, E. F., 1908–66
Ward, B. T. →
Ward, F. H., fl. 1838–84
Warde, A. →
Ware, A. H., fl. 1902

Waring, R. H., 1720?–94
Warmington, E. F. →
Warner, Mrs, fl. 1841–43
Warren, D. S. →
Warren, J. B. L., 1835–95
Watson, H. C., 1804–81
Watson, Sir William, 1715–87
Watson, W. C. R., 1885–1954
Watt, H. B., 1858–1941
Watts, G. A. R., 1873–1949
'W.D.', fl. 1841
Webb, R. H., 1806?–80
Webb, W. M., 1868–1952
Webster, A. D., fl. 1893–1925
Webster, M. McC. →
Wedgwood, Mrs M. L., 1854–1953
Welch, Mrs B., née Gullick →
Wells, R. J., fl. 1874
Wernham, H. F., 1879–?
West, C. →
West, W., d. 1961
Westcombe, T., 1815–92
Westcott, B. F., 1825–1901
Westrup, A. W., d. 1964
Whale, W., fl. 1870–76
Wharton, H. J., fl. 1864–76
Wharton, H. T., 1846–95
Wheeler, T., 1754–1847
Whellan, J. A. →
Whichelmore, P., fl. 1901
Whidborne, B. S. →
White, Mrs. A. →
White, C. A., 1811–1912
White, C. F., fl. 1878
White, E. C., fl. 1866–96
White, Mrs H. T., fl. 1915
White, J. W., 1846–1932
Whiting, J. E., fl. 1850–1927
Whittaker, J., 1901–57
Whitton, W. A., fl. 1912
Whitwell, W., 1839–1920
Widgery, J. P. →
Wilkinson, T., fl. 1864
Williams, F. N., 1862–1923
Williams, I. A., 1890–1962
Williams, W. S. →
Williamson, A., 1819–70
Willisel, T., d. 1675?
Willmott, E. A., 1858–1934
Wilmer, J., 1697–1769
Wilmott, A. J., 1888–1950
Wilson, W., 1799–1871
Winch, N. J., 1768–1838
Wing, W. E., 1827–55
Winston, —, fl. 1885
Withering, W., 1741–99
Wollaston, G. B., 1814–99
Wolley-Dod, A. H., 1861–1948
Wood, A., fl. 1869–1914
Wood, C. A., fl. 1914
Wood, W., fl. 1800–05
Woodhead, J. E., 1883–1967
Woods, J., 1776–1864
Woodward, S. P., 1821–65
Woodward, T. J., 1745–1820
Worsdell, W. C., 1867–1957
Wright, C. A., 1834–1907
Wrighton, F. E. →
Wroughton, R. C., fl. 1907–9
'W.S.', fl. 1846
Wurzell, B. →

Yeo, P. F. →
Young, D. P., 1917–72
Young, E. →
Young, J. F., 1796–1860

ANALYSIS OF THE FLORA OF MIDDLESEX

The table below shows the supposed status of the 1,109 species of vascular plants established, or formerly established, in Middlesex with relative data taken from Trimen and Dyer's *Flora of Middlesex* for comparison.

	PRESENT WORK	TRIMEN AND DYER
Native	756	768
Denizen	37	
Colonist	22	
[1]Introduced	49	91
Alien		
Europe	148	
N. Africa	4	
S. Africa	3	
Asia	32	
N. America	38	
S. America	7	
Australasia	3	
Cosmopolitan	4	
Hortal origin	6	
	1,109	859

In addition to the above there are over 600 species, mostly of foreign origin, which have occurred as 'casuals' in the vice-county and which are not included in the above analysis. The microspecies of *Capsella*, *Rubus*, *Hieracium* and *Taraxacum* are also excluded.

Trimen and Dyer list 57 species as probably extinct, but no less than 34 of these have been rediscovered, though 12 of that total are now almost certainly lost. A further three species given in their list as probably extinct were almost certainly ephemerals.

It seems probable that no less than 129 species are now lost to Middlesex, and in view of the precarious state of many others this total is likely to be added to annually.

The supposed extinct species in Middlesex are *Lycopodiella inundata*, *Lycopodium clavatum*, *Thelypteris limbosperma*, *Cystopteris fragilis*, *Pilularia globulifera*, *Juniperus communis*, *Helleborus viridis*, *Anemone apennina*,

[1] It should be noted that Trimen and Dyer's application of the term 'Introduced' differs from that of the present author.

Ranunculus sardous, *R. parviflorus*, *R. aquatilis*, *Myosurus minimus*, *Papaver lecoqii*, *P. hybridum*, *P. atlanticum*, *Corydalis bulbosa*, *Fumaria purpurea*, *Teesdalia nudicaulis*, *Cochlearia anglica*, *Cardamine impatiens*, *Arabis glabra*, *Viola lactea*, *V. palustris*, *Hypericum elodes*, *Helianthemum nummularium*, *Cucubalus baccifer*, *Petrorhagia nanteulii*, *Sagina subulata*, *S. nodosa*, *Beta vulgaris* subsp. *maritima*, *Althaea officinalis*, *Radiola linoides*, *Geranium versicolor*, *Trifolium ornithopodioides*, *T. glomeratum*, *T. scabrum*, *T. squamosum*, *Vicia lathyroides*, *Potentilla palustris*, *P. argentea*, *Agrimonia procera*, *Rosa pimpinellifolia*, *R. stylosa*,* *R. mollis*, *R. micrantha*, *R. agrestis*, *Chrysosplenium alternifolium*, *Parnassia palustris*, *Lythrum hyssopifolia*, *Bupleurum rotundifolium*, *B. tenuissimum*, *Petroselinum segetum*, *Cicuta virosa*, *Oenanthe pimpinelloides*, *O. silaifolia*, *O. lachenalii*, *Tordylium maximum*, *Asarum europaeum*, *Euphorbia platyphyllos*, *Rumex maritimus*, *Myrica gale*, *Pyrola minor*, *Anagallis tenella*, *A. minima*, *Samolus valerandi*, *Centaurium pulchellum*, *Gentiana pneumonanthe*, *Cynoglossum germanicum*, *Myosotis secunda*, *Lithospermum purpurocaeruleum*, *L. officinale*, *Solanum pygmaeum*, *Scrophularia umbrosa*, *Limosella aquatica*, *Pedicularis palustris*, *Melampyrum arvense*, *Euphrasia confusa*, *E. borealis*, *Orobanche rapum-genistae*, *O. elatior*, *Utricularia minor*, *Mentha pulegium*, *M. suaveolens*, *Littorella uniflora*, *Campanula rapunculus*, *Legousia hybrida*, *Jasione montana*, *Galium parisiense* subsp. *anglicum*, *Centranthus calcitrapa*, *Lonicera xylosteum*, *Scabiosa columbaria*, *Pulicaria vulgaris*, *Filago pyramidata*, *Carlina vulgaris*, *Carduus tenuiflorus*, *Cirsium dissectum*, *Centaurea jacea*, *Arnoseris minima*, *Hypochoeris glabra*, *Lactuca saligna*, *Sonchus palustris*, *Crepis foetida*, *Baldellia ranunculoides*, *Damasonium alisma*, *Potamogeton polygonifolius*, *P. alpinus*, *P. compressus*, *Allium oleraceum*, *A. triquetrum*, *Juncus gerardii*, *Leucojum aestivum*, *Crocus purpureus*, *Spiranthes spiralis*, *Coeloglossum viride*, *Platanthera chlorantha*, *P. bifolia*, *Ophrys insectifera*, *Orchis militaris*, *O. ustulata*, *Dactylorhiza incarnata*, *Trichophorum cespitosum*, *Blysmus compressus*, *Schoenoplectus triquetrus*, *Carex laevigata*, *C. rostrata*, *C. appropinquata*, *C. divisa*, *C. dioica* and *Poa chaixii*. In addition to these a number of species have become extinct as natives or denizens but have been either accidentally or deliberately reintroduced, e.g. *Osmunda regalis*, *Aquilegia vulgaris*, *Drosera intermedia*, *D. rotundifolia*, *Cuscuta epithymum* and *Maianthemum bifolium*.

The table on p. 70 shows the relative frequency of the various species of phanerogams and vascular cryptogams found in the vice-county with data taken from Trimen and Dyer's *Flora of Middlesex* for comparison.

* Now known to be still present in the Potters Bar area.

	PRESENT WORK		TRIMEN AND DYER	
Very common	101	365	129	377
Common	244		126	
Rather common	20		122	
Local	142	615	—	425
Rather rare	29		132	
Rare	108		159	
Very rare	336		134	
	980		802	
Possibly extinct	129		57	
	1,109		859	

A HISTORICAL SURVEY OF THE FLORA OF MIDDLESEX, 1869–1968

A study of the table given on p. 68 will show that despite the vast, and many, changes which have taken place in the vice-county, particularly the loss of much land to building, since the publication of Trimen and Dyer's *Flora of Middlesex* in 1869, the number of vascular plant species believed to be growing in the area has increased by nearly 22 per cent. If species which are now presumed to be extinct are taken into consideration the increase becomes nearly 35 per cent. It will be seen also that while the proportion of the more common species has remained fairly stable, that of the rarer species has increased by over 44 per cent and that of the supposedly extinct species has risen by just over 100 per cent.

The reasons for a substantial increase in the number of species found in the vice-county over the last hundred years, and for the even greater figure if species presumed now to be extinct are included, are complex and involve a number of factors which are reviewed below.

TRIMEN AND DYER'S FLORA

EARLY RECORDS. Despite the very detailed research into the early history of Middlesex botany, much of which was carried out by the indefatigable W. W. Newbould, a few species which had become extinct before 1869 were overlooked. These were *Trifolium squamosum*, gathered by the Thames by Joseph Andrews and John Field prior to 1721: *Chrysosplenium alternifolium*, collected at Harefield by Henry Kingsley in 1839: *Drosera intermedia*, recorded from Hampstead Heath by John Blackstone in 1746: *Hypochoeris glabra*, gathered at Teddington by John Vaughan in 1843: *Allium triquetrum* collected from the Isle of Dogs, where it had been confused with *Leucojum aestivum*, by J.

Banker in 1852, and *Carex rostrata* and *C. dioica*, noted at Teddington by Edward Forster in 1792.

TAXONOMY. The taxonomic views of the present time differ greatly from those of a century ago, especially in critical genera, and the table on pp. 71–75 illustrates the varying viewpoints on the conception of species and subordinate taxa as accepted by Trimen and Dyer and the present author.

CHANGES IN TAXONOMICAL CONCEPTION, 1869–1968

The table below shows the differing viewpoints on the conception of species as accepted by Trimen and Dyer and as accepted by the present author.

TRIMEN AND DYER, 1869	PRESENT WORK
Ranunculus peltatus Fries	*Ranunculus aquatilis* L.
	R. peltatus Schrank
Fumaria capreolata L.	*Fumaria purpurea* Pugsl.
1. *F. pallidiflora* Jord.	*F. muralis* Sond. ex Koch subsp.
2. *F. boraei* Jord.	*boraei* (Jord.) Pugsl.
3. *F. muralis* Sond.	
Nasturtium officinale R.Br.	*Nasturtium officinale* R.Br.
	N. microphyllum (Boenn.) Reichb.
Viola canina L.	*Viola canina* L.
	V. lactea Sm.
V. tricolor L.	*V. tricolor* L.
Var. β. *V. arvensis* Murr.	*V. arvensis* Murr.
Sagina apetala L.	*Sagina apetala* L. subsp. *erecta*
S. ciliata Fries	(Hornemann) F. Hermann
	subsp. *apetala*
Lotus corniculatus L.	*Lotus corniculatus* L.
Var. β. *L. tenuis* Kit.	*L. tenuis* Waldst. & Kit. ex Willd.
Alchemilla arvensis Scop.	*Aphanes arvensis* L.
	A. microcarpa (Boiss. & Reut.)
	Rothm.
Rosa canina L.	*Rosa canina* L.
	R. corymbifera Bork.
Prunus communis Huds.	*Prunus spinosa* L.
1. *P. spinosa* L.	*P. domestica* L. subsp. *insititia* (L.)
2. *P. insititia* L.	C. K. Schneid.
3. *P. domestica* L.	subsp. *domestica*
	P. cerasifera Ehrh.
Pyrus malus L.	*Malus sylvestris* (L.) Mill.
Var. β. *tomentosa* Koch	*M. domestica* Borkh.

Epilobium tetragonum L.	*Epilobium tetragonum* L. subsp. *tetragonum*
	subsp. *lamyi* (F. W. Schultz) Nyman
Oenothera biennis L.	*Oenothera biennis* L.
	O. erythrosepala Borbás
Apium graveolens L.	*Apium graveolens* L. subsp. *graveolens*
	subsp. *dulce* (Mill.) Lemke & Rothm.
Rumex acetosella L.	*Rumex tenuifolius* (Wallr.) A. Löve
	R. acetosella L.
R. obtusifolius L.	*R. obtusifolius* L. subsp. *obtusifolius*
	subsp. *transiens* (Simonk.) Reching. f.
R. pratensis Mert. & Koch	*R.* × *acutus* L.
Populus nigra L.	*Populus nigra* L.
	P. × *canadensis* Moench
Salix rubra Huds.	*Salix* × *rubra* Huds.
S. undulata Ehrh.	*S.* × *mollissima* Ehrh.
S. stipularis Sm.	*S.* × *stipularis* Sm.
S. smithiana Willd.	*S.* × *smithiana* Willd.
Calystegia sepium L.	*Calystegia sepium* (L.) R.Br.
	C. × *lucana* (Ten.) G. Don
	C. pulchra Brummitt & Heywood
	C. silvatica (Waldst.) Griseb.
Cuscuta epithymum Murr.	*C. epithymum* (L.) L.
C. trifolii Bab.	
Veronica anagallis L.	*Veronica anagallis–aquatica* L.
	V. catenata Pennell
Euphrasia officinalis L.	*Euphrasia nemorosa* (Pers.) Mart.
	E. confusa Pugsl.
	E. borealis (Towns.) Wettst.
	E. anglica Pugsl.
E. odontites L.	*Odontites verna* (Bellardi) Dumort. subsp. *verna*
	subsp. *serotina* Corbiére
Utricularia vulgaris L.	*Utricularia vulgaris* L.
	U. australis R.Br.
Mentha gentilis L.	*Mentha* × *gentilis* L.
M. gracilis Sm.	
M. sativa L.	*M.* × *verticillata* L.
M. rubra Sm.	*M.* × *smithiana* R. A. Grah.

M. sylvestris L.	*M. spicata* L.
M. viridis L.	
M. piperita Huds.	*M.* × *piperita* L.
Thymus serpyllum L.	*T. pulegioides* L.
1. *T. eu-serpyllum* Syme	
2. *T. chamaedrys* Fries	
Stachys sylvatica L.	*Stachys sylvatica* L.
Var. β. *S. ambigua* Sm.	*S.* × *ambigua* Sm.
Galeopsis tetrahit L.	*Galeopsis tetrahit* L.
Var. β. *bifida* Bönn.	*G. bifida* Boenn.
Galium palustre L.	*Galium palustre* L.
Var. α. *elongatum* Syme	subsp. *elongatum* (C. Presl) Lange
Var. β. *genuinum* Syme	subsp. *palustre*
Var. γ. *witheringii*. Syme E. B.	
Dipsacus sylvestris L.	*Dipsacus fullonum* L. subsp. *fullonum*
	subsp. *sativus* (L.) Thell.
Centaurea nigra L.	*Centaurea nigra* L. subsp. *nigra*
	subsp. *nemoralis* (Jord.) Gugler
Tragopogon pratensis L.	*Tragopogon pratensis* L.
Var. α. *genuinus* Syme	subsp. *pratensis*
Var. β. *T. minor* Fries	subsp. *minor* (Mill.) J. C. Hartman
Hieracium pilosella L.	*Pilosella officinarum* C. H. & F. W. Schultz
	subsp. *officinarum*
	subsp. *concinnata* (F. J. Hanb.) Sell & West
Crepis biennis L.	*Crepis biennis* L.
	C. vesicaria L. subsp. *taraxacifolia* (Thuill.) Thell.
Leontodon taraxacum L.	*Taraxacum officinale* Weber
Var. β. *T. erythrospermum* DC.	*T. laevigatum* (Willd.) DC.
Alisma plantago L.	*Alisma plantago-aquatica* L.
Var. β. *A. lanceolata* With.	*A. lanceolatum* With.
	A. lanceolatum × *plantago-aquatica*
Potamogeton pusillus L.	*Potamogeton pusillus* L.
	P. berchtoldii Fieb.
Narcissus pseudo-narcissus L.	*Narcissus pseudonarcissus* L.
	N. hispanicus Gouan
? *Epipactis media* Fr.	*Epipactis helleborine* (L.) Crantz
	E. purpurata Sm.
Orchis incarnata L.	*Dactylorhiza incarnata* (L.) Soó subsp. *incarnata*
	D. praetermissa (Druce) Soó

O. maculata L.	*D. maculata* (L.) Soó
	subsp. *ericetorum* (E. F. Linton) P. F. Hunt & Summerhayes
	D. fuchsii (Druce) Soó subsp. *fuchsii*
Sparganium ramosum Huds.	*Sparganium erectum* L. subsp. *erectum*
	subsp. *neglectum* (Beeby) Schinz & Thell.
Eleocharis palustris R.Br.	*Eleocharis palustris* (L.) Roem. & Schult.
	subsp. *palustris*
	subsp. *vulgaris* Walters
Scirpus carinatus Sm.	*Schoenoplectus* × *carinatus* (Sm.) Palla
Carex paniculata L.	*Carex paniculata* L.
	C. appropinquata Schumach.
C. axillaris Good.	*C.* × *pseudoaxillaris* K. Richt.
C. muricata L.	*C. muricata* L.
	C. spicata Huds.
Glyceria plicata Fr.	*Glyceria plicata* Fr.
Var. β. *G. pedicellata* (Towns.)	*G.* × *pedicellata* Townsend
	G. declinata Bréb.
Festuca duriuscula L.	*Festuca rubra* L. subsp. *rubra*
	subsp. *commutata* Gaudin
	F. longifolia Thuill.
F. pratensis Huds.	*F. pratensis* Huds.
Var. β. *F. loliacea* Huds.	× *Festulolium loliaceum* (Huds.) P. Fourn.
F. ovina L.	*Festuca ovina* L.
	F. tenuifolia Sibth.
Poa pratensis L.	*Poa pratensis* L.
Var. β. *subcaerulea*	*P. angustifolia* L.
	P. subcaerulea Sm.
Serrafalcus mollis Parl.	*Bromus hordeaceus* L. subsp. *hordeaceus*
	B. × *pseudothominii* P. Smith
	B. lepidus Holmberg
S. racemosus Parl.	*B. racemosus* L.
Var. β. *S. commutatus* Bab.	*B. commutatus* Schrad.
Hordeum murinum L.	*Hordeum murinum* L. subsp. *murinum*
	subsp. *leporinum* (Link) Aschers. & Graebn.
Aira caespitosa L.	*Deschampsia cespitosa* (L.) Beauv.
	subsp. *cespitosa*
	subsp. *parviflora* (Thuill.) Richter

Agrostis canina L.	*Agrostis canina* L.
	A. stricta J. F. Gmel.
A. vulgaris With.	*A. tenuis* Sibth.
Var. β. *pumila* Lightf.	*A. gigantea* Roth.
Phleum pratense L.	*P. bertolonii* DC.
	P. pratense L.

UNDER RECORDING. Trimen and Dyer and their botanical friends undoubtedly visited many parts of the vice-county collecting records for the *Flora of Middlesex*, mainly between the years 1861 to 1868. Nevertheless several areas, e.g. the Colne valley from Cowley to Springwell, and the Staines district, were inadequately explored. This is well confirmed by John Benbow's series of papers published in the *Journal of Botany* between 1884 and 1887, which showed that many species unrecorded or stated to be rare about Uxbridge and Harefield by the authors of the *Flora* were in fact locally abundant in those areas, which indeed in the case of certain species is still so today. In fairness, however, it must be appreciated that a journey from London to Harefield in the mid-nineteenth century would not have been easily accomplished, while Benbow lived near the area.

SPECIES PRESUMED TO HAVE BECOME EXTINCT SINCE 1869

It seems probable that 78 native and naturalised species known to Trimen and Dyer have become extinct during the last century. The most typical habitats in which these species were found are listed below in order of frequency.

Bogs, marshes, watery places, etc.	17
Heaths and moors	15
Cornfields, arable ground, etc.	8
Grassy places, bare ground, etc., on sandy and gravelly soils	8
Woods, thickets, etc.	5
Meadows, etc.	5
Banks, etc.	5
Hedgerows, etc.	4
Thames side, in semi-maritime situations	4
Grassy places, scrub, etc., on calcareous soils	3
Waste ground, etc.	4
	78

The periods in which the supposedly extinct species were last recorded are shown below arranged under their most typical habitats.

	1869–90	1891–1910	1911–30	1931–50	1951–70	Total
Bogs, marshes, watery places, etc.	7	3	3	4	—	17
Heaths and moors	11	2	1	1	—	15
Cornfields, arable ground, etc.	3	3	2	—	—	8
Grassy places, bare ground, etc., on sandy and gravelly soils	4	3	—	1	—	8
Woods, thickets, etc.	2	3	—	—	—	5
Meadows, etc.	1	1	1	2	—	5
Banks, etc.	3	—	1	1	—	5
Hedgerows, etc.	3	1	—	—	—	4
Thames side, in semi-maritime situations	4	—	—	—	—	4
Grassy places, scrub, etc., on calcareous soils	—	2	1	—	—	3
Waste ground, etc.	2	—	1	—	1	4
	40	18	10	9	1	78

BOGS, MARSHES AND WATERY PLACES. In common with the rest of southern England the greatest loss has been in species of wet habitats. This has been due to drainage, the filling in of ponds, and the general lowering of the water table. Species involved include *Viola palustris*, *Parnassia palustris*, *Oenanthe lachenalii*, *Rumex maritimus*, *Anagallis tenella*, *Limosella aquatica*, *Pedicularis palustris*, *Mentha pulegium*, *Littorella uniflora*, *Pulicaria vulgaris*, *Baldellia ranunculoides*, *Damasonium alisma*, *Potamogeton polygonifolius*, *P. compressus*, and *Dactylorhiza incarnata*.

HEATHS AND MOORS. Most of the heaths of the vice-county have been invaded by *Betula verrucosa*, which has resulted in the lowering of the water table and the creation of shade, which is intolerant to certain heath species. Among the species which have vanished from their former habitats are *Lycopodiella inundata*, *Lycopodium clavatum*, *Thelypteris limbosperma*, *Viola lactea*, *Sagina nodosa*, *Radiola linoides*, *Rosa pimpinellifolia*, *Orobanche rapum-genistae*, and *Jasione montana*.

CORNFIELDS, ARABLE GROUND, ETC. Modern grain cleaning methods and the greatly reduced area of cultivated ground has resulted in the elimination of a number of species, including *Myosurus minimus*, *Papaver hybridum*, *Bupleurum rotundifolium*, *Euphorbia platyphyllos* and *Legousia hybrida*.

GRASSY PLACES, BARE GROUND, ETC., ON SANDY AND GRAVELLY SOILS. The sandy and gravelly areas near the Thames, particularly in the south-western parts of the vice-county, have been extensively disturbed by the removal of sand and gravel. These operations have probably been responsible for the eradication of the following species: *Teesdalia nudicaulis*, *Trifolium ornithopodioides*, *T. glomeratum*, *T. scabrum*, *Vicia lathyroides*, *Potentilla argentea*, *Filago pyramidata* and *Crepis foetida*. Apparently suitable habitats for all but the last two, however, still remain in the Hampton Court area, and it is possible that some of them may yet be rediscovered.

WOODS, THICKETS, ETC. Despite the considerable reduction of woodland in Middlesex, few species have been rendered extinct. The five that have – *Anemone apennina*, *Cardamine impatiens*, *Platanthera chlorantha*, *P. bifolia* and *Orchis militaris* – were at all times very rare, and with the exception of *Cardamine impatiens*, attractive ornamental plants. Collecting was almost certainly responsible for the eradication of *Orchis militaris*, and possibly also for the two species of *Platanthera*.

MEADOWS, ETC. The area of meadowland in Middlesex has decreased considerably over the last century, and among the species formerly found in this habitat but which are now lost are *Centaurea jacea*, *Campanula rapunculus*, *Crocus purpureus* and *Coeloglossum viride*.

BANKS, ETC. Species favouring well-drained banks, usually on light soils, which are now extinct include *Ranunculus parviflorus*, *Arabis glabra*, *Petroselinum segetum* and *Spiranthes spiralis*.

HEDGEROWS, ETC. The lost species, always very rare, from the Middlesex hedgerows are *Rosa stylosa*,* *R. mollis*, *R. micrantha* and *Lonicera xylosteum*.

THAMES SIDE, IN SEMI-MARITIME SITUATIONS. Prior to the building of the Thames embankment in the mid-nineteenth century a number of semi-maritime species grew in mud and adjacent brackish ditches. These had dwindled considerably by 1869, and since that time the following have become extinct: *Cochlearia anglica*, *Samolus valerandi*, *Schoenoplectus triquetrus* and *Carex divisa*.

GRASSY PLACES, SCRUB, ETC., ON CALCAREOUS SOILS. The flora of calcareous areas in the vice-county has remained remarkably stable, despite considerable disruption by chalk-digging operations, the tipping of rubbish, and building. Three species – *Lithospermum officinale*, *Carlina vulgaris* and *Ophrys insectifera* – have, however, become extinct. The latter is known to have been destroyed by chalk digging.

WASTE GROUND, ETC. The grassy waste areas of Middlesex have possibly increased during the last century, nevertheless, four species – *Ranunculus sardous*, *Bupleurum tenuissimum*, *Solanum pygmaeum* and *Carduus tenuiflorus* – formerly found in such habitats are now lost.

* Now known to be still present in the Potters Bar area.

SPECIES WHICH HAVE DECREASED IN FREQUENCY SINCE 1869

The greatest decline in frequency has been in the species of bogs, marshes, ponds and riversides. These include *Caltha palustris*, *Hypericum tetrapterum*, *Lychnis flos-cuculi*, *Trifolium fragiferum*, *Lythrum salicaria*, *Oenanthe fistulosa*, *Hottonia palustris*, *Menyanthes trifoliata*, *Veronica beccabunga*, *Valeriana officinalis*, *Sagittaria sagittifolia*, *Butomus umbellatus*, *Hydrocharis morsus-ranae*, *Potamogeton perfoliatus*, *Groenlandia densa*, *Scirpus sylvaticus*, *Carex demissa*, *C. panicea* and *C. nigra*. The decline is also very marked on heaths and commons, where the following species have become much less frequent: *Blechnum spicant*, *Ranunculus hederaceus*, *Calluna vulgaris*, *Erica tetralix*, *E. cinerea*, *Odontites verna* and *Chamaemelum nobile*. In meadows and fields the pattern is similar and a reduction in frequency is apparent in *Lepidium campestre*, *Primula veris*, *Centaurium erythraea*, *Rhinanthus minor* and *Orchis morio*. In the sandy areas of south-west Middlesex *Lepidium heterophyllum* and *Cerastium semidecandrum* have become scarce, while *Primula vulgaris*, *Ajuga reptans* and *Adoxa moschatellina* are much less common in the woods of the vice-county than formerly.

Other species of varying habitats which have decreased in frequency are *Polypodium vulgare*, *Erophila verna*, *Agrostemma githago*, *Potentilla sterilis*, *Saxifraga tridactylites*, *Verbascum thapsus*, *Sherardia arvensis*, *Chrysanthemum segetum* and *Catapodium rigidum*.

The decrease of many of these species may be attributed to a variety of factors, including drainage, gravel digging, building, atmospheric pollution and possibly also to microclimatic changes.

SPECIES THOUGHT TO BE PROBABLY EXTINCT IN 1869 WHICH HAVE SINCE BEEN 'REDISCOVERED'

As mentioned previously, Trimen and Dyer listed 57 species as probably extinct. Some of these were still extant at the time, while others have since been rediscovered, sometimes only to be lost again. Details of these are given below.

OSMUNDA REGALIS. Certainly long extinct as a native, but now known as an established introduction in at least four localities.

CETERACH OFFICINARUM. Almost certainly still present in Harmondsworth Churchyard when the *Flora* was published, and survived there until at least 1920. Discovered in 1881 in Perivale Churchyard and in 1904 in Stanwell Churchyard. Still known to survive in at least two localities in the vice-county.

AQUILEGIA VULGARIS. Long extinct as a native species, but still present as an established escape from cultivation.

SISYMBRIUM IRIO. Rediscovered near the Tower of London by Mrs E. F. Evetts in 1945, and in a second locality at Regents Park by H. C. Holme in 1956. In both localities it was an introduction, but has spread and was still present in good quantity in both places in 1970.

DESCURAINEA SOPHIA. In Middlesex always of casual occurrence. It should have been excluded from Trimen and Dyer's list, as on page 421 of the *Flora* it is recorded from Cricklewood, 1869, as a grain introduction, by Warren. It has since occurred in the vice-county on a number of occasions, the most recent date on record being 1965.

DIANTHUS ARMERIA. Gathered at Hayes by John Benbow in 1862, and at East Finchley by J. E. Cooper in 1883. Collected again by Benbow between Uxbridge and West Drayton in 1886–90, and more recently noted near Highgate by H. C. Harris in 1948.

HERNIARIA GLABRA. Recorded from Harefield by G. Browner in 1875, and in more recent times gathered at Kenton by A. J. Potts in 1943. It persists in small quantity in the latter locality but is obviously an introduction.

H. HIRSUTA. A mere casual which should have been excluded from Trimen and Dyer's list. It was, however, collected at Uxbridge by C. B. Green in 1907.

TRIFOLIUM ORNITHOPODIOIDES. Collected at Hillingdon by Warren in 1871, and on Uxbridge Common in 1872 and 1884 by John Benbow. In 1885 Benbow gathered it near Hounslow Heath. It has not been recorded since 1885 and is presumed extinct, though it is possible that it may be rediscovered in the Hampton Court area where it was recorded by William Borrer over a century ago.

T. GLOMERATUM. A solitary plant was found in Hyde Park by Newbould in 1871, and the species not having been recorded since is regarded as extinct. It may yet be refound in the Hampton Court district.

T. SCABRUM. Collected at Hillingdon Heath by John Benbow in 1872, and at Hampton Court by C. A. Wright in 1892. Although unrecorded since 1892 it may eventually be rediscovered in the Hampton Court area.

T. OCHROLEUCON. Recorded from Southall in 1903, and from East Acton by G. C. Druce, but undoubtedly as casual adventives.

LATHYRUS APHACA. Collected, or recorded, as a casual adventive from a number of localities between 1907 and 1965, but in 1966 discovered by Miss M. E. Kennedy and myself in a rough field near Enfield where it had the appearance of being native.

MESPILUS GERMANICA. Discovered near Staines Moor by Mrs B. Welch in 1947, but the trees have since been felled. Found by myself in another locality near Heston in 1965.

LYTHRUM HYSSOPIFOLIA. Recorded as casual at Yiewsley by

R. Melville and R. L. Smith in 1928, but the adventive species *L. junceum* may have been seen.

BUPLEURUM ROTUNDIFOLIUM. Recorded from cultivated ground between Hanwell and Ealing by E. C. White in 1871, from Whetstone in 1906 by J. E. Cooper, and from Hampstead in 1918. The two twentieth century records may refer to the closely allied European adventive *B. lancifolium*. The species is now considered extinct in Middlesex.

RUMEX MARITIMUS. A single plant was noted at Hampstead by Newbould, and the species was detected in five other localities in different parts of the vice-county between 1877 and 1924, when it was last recorded. It is now presumed to be extinct.

MYRICA GALE. A single bush was rediscovered on Hampstead Heath by John Benbow in 1887. Last recorded there by J. E. Whiting in 1901, it is now considered extinct.

SALIX PENTANDRA. Always an introduction in Middlesex, it still persists on the North-west Heath, Hampstead, Highgate Ponds and Brent Reservoir.

ANAGALLIS TENELLA. Recorded from Enfield by E. Ford in 1873. Apparently not seen since and almost certainly extinct.

CUSCUTA EUROPAEA. Rediscovered near the Thames at Shepperton by C. E. Britton about 1915; recorded from Penton Hook, 1917–19, by J. E. Cooper; found near Chertsey Lock by J. E. Lousley in 1942. It persisted in the latter locality until about 1960, and although not recently seen is probably still present by the Thames between Staines and Shepperton.

LATHRAEA SQUAMARIA. Still present in Blackstone's locality near Harefield when the *Flora* was published, as indeed it is today.

UTRICULARIA AUSTRALIS. Rediscovered in Long Water, Hampton Court, by G. C. Druce. It still persists there but rarely flowers. In 1947, L. G. Payne refound it on Hounslow Heath, where it was first recorded (as *U. vulgaris*) by William Hudson in 1778. It still survives there in very small quantity. Known also from Shortwood Common, Staines, though never seen in flower, and more recently from Little Common, Stanmore.

MENTHA PULEGIUM. Rediscovered by a pond on Pinner Hill by W. M. Hind in 1871, Not recorded since, and almost certainly now extinct.

M. × GENTILIS. Rediscovered at Chelsea by R. A. Graham in 1947, and since then reported from at least ten different localities.

CAMPANULA RAPUNCULUS. Rediscovered at Harrow Weald Common by C. B. Green in 1908. Not recorded since, and presumed to be extinct.

XANTHIUM STRUMARIUM. A rare casual which did not warrant a place in Trimen and Dyer's list. It was collected at Acton Green by

A. Loydell in 1900, at Uxbridge by Mrs M. L. Wedgwood in 1912, and was recorded from Hackney Marshes by R. Melville and R. L. Smith in 1927.

GYMNADENIA CONOPSEA. Rediscovered near Harefield by John Benbow in 1889, and seen in the area at intervals until 1946. A single plant was found in a new locality at Northwood by J. A. Moore in 1965.

OPHRYS APIFERA. Rediscovered in the Harefield area by John Benbow in 1884. As recently as 1946 it was known from four stations in the district, but now appears to be reduced to one, where it is still present.

ORCHIS MILITARIS. Rediscovered near Harefield by John Benbow in 1885. It survived until 1900 but is now lost.

ANACAMPTIS PYRAMIDALIS. Rediscovered in a single locality at Harefield by John Benbow in 1885. It still occurs there almost annually but in varying quantity.

SCHOENOPLECTUS TABERNAEMONTANI. Discovered in a new locality at Syon Park by N. Y. Sandwith in 1955.

CYPERUS FUSCUS. Discovered in a new locality near Staines by T. W. J. Dupree in 1957.

HELICTOTRICHON PUBESCENS. Found near Denham by Trimen, and near Hillingdon by Warren in 1871. It still occurs about Harefield and South Mimms, and by the Thames between Hampton Court and Kingston Bridge, and at Syon Park.

ADDITIONS TO THE FLORA OF MIDDLESEX SINCE 1869

NATIVE SPECIES. Eighteen presumed native species have been added to the vice-county list since the publication of *Flora of Middlesex*; of these eleven may be regarded as critical and would have been 'lumped' under a collective species by Trimen and Dyer.

HELIANTHEMUM NUMMULARIUM was discovered on a bank at Pinner Hill by W. A. Tooke in 1871, and a second locality was found in a chalkpit under Garett Wood, Springwell, by John Benbow in 1901. The first locality was presumably eradicated by road widening and the second by chalk-digging operations.

POTAMOGETON FRIESII was collected in the Lee Navigation Canal near Tottenham by E. C. de Crespigny in 1871. It is now known from several localities in the vice-county.

CAREX APPROPINQUATA was gathered by the canal near West Drayton by J. B. L. Warren in 1873. It was at first confused with *C. paniculata*, but Benbow showed that the species was locally abundant in marshes between Harefield and Spingwell with odd tussocks about the canal at Denham, Uxbridge, Cowley and West Drayton. The species survived until shortly before the Second World War but has now been rendered extinct by gravel-digging operations.

CALLITRICHE OBTUSANGULA was noted near Sunbury by H. and J. Groves in 1877. It is now known from a few isolated localities in Middlesex but is a very rare species.

POTAMOGETON ALPINUS was discovered in ditches near Harefield by E. C. de Crespigny. Last recorded there in 1886, it is now considered to be extinct.

EUPHRASIA BOREALIS was gathered on Harefield Moor by John Benbow in 1884. This is the sole evidence of its occurrence in the vice-county, and it is regarded as extinct.

POTAMOGETON TRICHOIDES was gathered by John Benbow at Eastcote and Southall in 1884. It is now known from a few scattered localities in the vice-county. A critical species which could easily have been overlooked as a slender form of *P. pusillus*.

EUPHRASIA CONFUSA was collected by John Benbow at Pield Heath, Hillingdon, in 1887. There is no other Middlesex record, and the species is considered to be extinct.

BETA VULGARIS subsp. MARITIMA was gathered in the Isle of Dogs by John Benbow in 1887. There are no other records and the species is assumed to be extinct.

FUMARIA PURPUREA was found in two places near Uxbridge by John Benbow in 1892. It has not since been recorded and is thought to be extinct.

GENTIANELLA GERMANICA was discovered at Springwell by John Benbow in 1892. Although possibly still present it is very rare and almost eradicated as a result of chalk-digging operations.

BRACHYPODIUM PINNATUM was found near Northwood by John Benbow in 1892. It survives in quantity, and in 1968 was found established on a railway bank at North Acton by Mrs J. McLean.

OROBANCHE ELATIOR was gathered in a chalkpit at Harefield by C. B. Green in 1902. This is the sole Middlesex record, and the species is regarded as extinct.

OENANTHE PIMPINELLOIDES was collected in a field near Hendon by Miss H. T. White in 1915. There is no other record from the vice-county, and the species is considered extinct.

CAREX FILIFORMIS was discovered between Shepperton and Chertsey by I. A. Williams in 1928. The area has been much disturbed by gravel-digging operations but the species may survive.

POTAMOGETON OBTUSIFOLIUS was discovered by the author in a pond at Little Common, Stanmore, in 1947. In 1956 it was collected in the Colne north of Harefield by C. A. Milner. The pond at Stanmore was cleaned out about 1963 and the species apparently eradicated.

RUMEX TENUIFOLIUS was found in Hyde Park by D. E. Allen in 1960. It is now known in a number of localities in Middlesex. A critical species, easily passed over as a starved form of *R. acetosella*, it may be more widespread than the records suggest.

EPIPACTIS PHYLLANTHES was discovered near Harefield by B. P. Pickess in 1965; a second locality, on Harefield Moor, was found by S. E. Crooks in 1966.

INTRODUCED SPECIES. Twenty-one species that are native in various parts of the British Isles, but which are not so in Middlesex, have been discovered in the vice-county during the last century.

ANTHYLLIS VULNERARIA was found on the Thames bank near Kingston Bridge by H. C. Watson in 1873, having presumably been washed down from a calcareous area in the upper regions of the river. It is now known from at least four localities in the vice-county.

STRATIOTES ALOIDES was recorded from Enfield by E. Ford in 1873, and was rediscovered there, growing in quantity in a lake near Whitewebbs Park in 1965 by Miss M. E. Kennedy and the author. It has also been long naturalised in lakes at Stanmore Common.

ASPARAGUS OFFICINALIS was first recorded from a railway bank at Gunnersbury by A. R. Pryor in 1874. It is now common and widespread on waste ground, etc., as an escape from cultivation.

PYROLA MINOR was gathered at The Grove, Stanmore Common, by W. M. Hind in 1874, and was also seen there by C. B. Green in 1908. It is now regarded as extinct.

LINARIA REPENS was discovered in railway sidings at Acton by G. C. Druce in 1902. It is now known in a number of localities, chiefly about railway premises, in the vice-county.

ASARUM EUROPAEUM was collected at The Grove, Stanmore Common, by C. B. Green in 1906. There is no other Middlesex record and the species is considered to be extinct.

LEPIDIUM LATIFOLIUM was first noted on the banks of the Colne near Drayton Ford by the author in 1939. Since that time it has spread southwards in profusion, and is now abundant by the Colne, Grand Union Canal; and in gravel pits from Drayton Ford and Springwell to Cowley, it has been noted also in a rough field between Hayes and West Drayton, and by the canal at Frogmore Green, near Southall. In 1961 E. Clement noted it on railway banks at Feltham.

EPILOBIUM LANCEOLATUM was first recorded from near Shadwell Basin by J. E. Lousley in 1945. In 1961 it was reported from Buckingham Palace grounds by D. McClintock, in 1965 from Holloway by D. E. G. Irvine, and in 1967 it was gathered at Tottenham by Miss M. E. Kennedy. During 1967 Mrs J. McLean noted it as an abundant weed of flower-beds in St Paul's Churchyard, Covent Garden, and it was still present in 1968. It is likely that the Middlesex plants have originated from wind-borne seed from colonies in north Kent where the species is native and locally abundant in woods, etc.

ASTRANTIA MAJOR was first recorded at Hampton Court by Mrs B. Welch in 1946. In 1953 it was found in Regents Park by H. C. Holme, and in 1966 was seen in two localities at Eastcote.

TILIA CORDATA was noted regenerating at Hanger Hill, Ealing, by F. P. D. Boucher in 1946. Since that time self-sown seedlings have been seen at Bushy Park and Osterley Park.

ADIANTUM CAPILLUS-VENERIS was first seen on a bombed site near Holborn by James Whittaker. It is now known to be established on old walls and other brickwork in a few localities. The plants originate from spores from cultivated plants.

COCHLEARIA DANICA was found growing in quantity between the metals of the railway main line by Scratch Wood sidings by Dr J. G. Dony in 1950.

GERANIUM SANGUINEUM was first gathered at Uxbridge by Mrs L. M. P. Small in 1950. It is now known to be naturalised in grassy places in a few localities. The plants are always of garden origin.

MYRRHIS ODORATA was first recorded from Southgate by Miss M. A. R. S. Scholey in 1950. It persisted in the locality until 1953. In 1963 it was found in quantity on railway embankments at Hampton Wick by T. B. Ryves. The plants are almost certainly of cultivated origin.

CERASTIUM DIFFUSUM subsp. DIFFUSUM was first found between the metals of the railway main line by Scratch Wood sidings by Dr J. G. Dony in 1951. In the same year it was noted on railway tracks at Brentford by the author, and in 1965 was seen on railway banks at Potters Bar by Dr and Mrs Dony.

CERASTIUM PUMILUM was discovered on railway tracks at Mill Hill by Dr J. G. Dony and E. Milne-Redhead in 1951.

NARDURUS MARITIMUS was found growing on railway tracks between Uxbridge and Denham by the author in 1951. More recently B. Wurzell discovered a large colony on a bank by the railway near Denham.

GYMNOCARPIUM ROBERTIANUM was discovered in quantity on a wall by the Thames at Staines by R. M. Payne in 1964. The colony is thriving, and increasing.

TILIA PLATYPHYLLOS was noted to be regenerating at Hampton Court by the author in 1964.

HYPERICUM MACULATUM was collected from a railway bank at West Acton by Mrs J. McLean in 1965.

OROBANCHE HEDERAE was discovered growing in Highgate Cemetery by K. E. Bull in 1972.

ALIEN SPECIES. Over 100 species of adventive origin apparently unknown to Trimen and Dyer have been recorded or collected in Middlesex during the last century. No less than 85 of these originated as escapes or outcasts from gardens, or as self-sown seedlings of planted trees and shrubs. The two species which have shown the most rapid spread in a comparatively short period are *Buddleja davidii* and *Veronica filiformis*. The first mentioned, introduced into Britain as recently as 1896, and at one time considered to be possibly half-hardy was first recorded as a naturalised plant in Middlesex at Yiewsley by R. Melville and R. L. Smith in 1928. It produces large quantities of very light, viable seed which is readily wind borne, and the species is now frequent throughout the vice-county in a variety of habitats including waste ground, walls, chalk pits, railway banks and woods. *Veronica filiformis*, on the other hand, rarely sets seed in Britain and its spread is entirely by vegetative means. Although it was introduced long before *Buddleja davidii*, it was not until the early part of the present century that it became popular as a rock garden plant. It was first noted as an established escape at Hounslow by H. Banks in 1942. In recent years it has become a serious pest of lawns in many parts of Middlesex, and near the Thames and Colne it is locally abundant and completely naturalised in turf. Other species which have increased rapidly and readily compete with natural vegetation include *Galega officinalis*, *Oenothera erythrosepala*, *Solidago altissima*, *S. gigantea*, *Aster novi-belgii* and *A. lanceolatus*.

The most rapid spread of an adventive species of non-garden origin has been achieved by the North American *Epilobium adenocaulon*. Long known as a naturalised species in Surrey, where it was thought to have been introduced with fodder in the First World War, it was first noted on bombed sites in the City of London in 1945 by J. E. Lousley. Since that time it has spread to a great variety of natural and artificial habitats, ranging from marshes, woods and fields to roadsides, old walls, and rubbish-tips, to become the most common species of willow herb in the vice-county. Two other species which appear to have spread considerably but which were probably present but overlooked a century ago are *Artemisia verlotorum* and *Bromus lepidus*. The earliest evidence for the former, a native of China, is a specimen collected on Hounslow Heath (where it is still to be found) by C. B. Green in 1908. It is now known to be locally abundant, especially near the Thames, and it seems probable that it was formerly overlooked as a form of *A. vulgaris*. The earliest evidence of the occurrence of *Bromus lepidus* in Middlesex is a specimen collected in Finsbury Park by A. French in 1871. Probably originally introduced with grass seed, it is now widespread on roadside verges and on cultivated and waste ground. It is almost certain that it was formerly confused with *B. hordeaceus*. Other weedy species which have spread widely on waste ground are *Sisymbrium loeselii*, *S. orientale* and *Galinsoga ciliata*, though the latter may in former times have been confused with *G. parviflora*. Possibly the most recent spread has been that of *Bromus carinatus*, a North American species. First noted as an escape from the Royal Botanic Gardens, Kew, about 1919 by Dr C. E. Hubbard, it was first seen in Middlesex at Brentford in 1945 by the author. It is now widespread near the Thames from Chiswick to Twickenham, and has been seen in other widely dispersed localities at Harmondsworth, Hounslow Heath, Ealing, Greenford and Regents Park.

SPECIES WHICH HAVE INCREASED IN FREQUENCY SINCE 1869

The greatest increase in frequency has been in the weedy species found on waste ground, etc., and many plants considered to be rare or very rare by Trimen and Dyer are now common, or relatively so. These include *Coronopus didymus*, *Lepidium ruderale*, *Sisymbrium altissimum*, *Medicago sativa* subsp. *sativa*, *Melilotus officinalis*, *M. alba*, *Trifolium hybridum*, *Galinsoga parviflora*, *Senecio squalidus*, *S. viscosus* and *Lactuca serriola*. Other species in this category which have increased their range but to a lesser extent are *Bunias orientalis*, *Sedum reflexum*, *Linaria purpurea* and *Puccinellia distans*. The increase is also very marked in a number of species occupying a variety of habitats ranging from heaths to cultivated and waste ground, roadsides and pond and river sides; among these may be mentioned *Equisetum telmateia*, *Cardaria draba*, *Epilobium angustifolium*, *Heracleum mantegazzianum*, *Polygonum cuspidatum*, *Conyza canadensis* and *Crepis vesicaria* subsp. *taraxacifolia*. A number of species found in grassy places ranging from meadows to railway banks have shown a considerable increase in range; these include *Silene vulgaris*, *Geranium pyrenaicum*, *Euphorbia esula*, *Pentaglottis sempervirens* and *Pilosella aurantiaca*. Among the weeds of cultivated ground, formerly considered rare, but which are now quite common, are *Chenopodium ficifolium*, *Mercurialis annua*, *Veronica persica* and *V. polita*. A few species of calcareous and sandy soils, originally local or rare, have adopted a new role and become colonists of railway tracks; these include *Reseda lutea*, *Chaenorhinum minus* and *Vulpia myuros*. On lake and pond verges and by rivers and canals, *Impatiens capensis* and *I. glandulifera* have spread considerably, while *Scirpus maritimus* is now naturalised in a number of localities away from the vicinity of the tidal Thames. In hedgerows *Calystegia silvatica* has become locally abundant, though it may well have been present a century ago and have been confused with *C. sepium*. *C. pulchra* is also locally frequent, especially near the Thames, and was indeed present in Trimen and Dyer's time when it was considered to be a pink-flowered form of *C. sepium*. Another hedgerow species which has certainly increased is *Lycium barbarum*. Few species of woodlands and shady places appear to have increased their range, an exception being *Circaea lutetiana*. On old walls *Phyllitis scolopendrium* is now common even in heavily built-up areas, while *Sedum album* is rapidly increasing, particularly on the river wall of the Thames.

THE FLORA OF MIDDLESEX

CHARACEAE

NITELLEAE

NITELLA Ag.

N. opaca Ag.
Formerly native in lakes, etc., but now extinct.
1. Ruislip Reservoir, 1884, *Benbow* (BM).

N. flexilis Ag.
Chara flexilis L.
Formerly native in ponds and lakes but now extinct.
1. Ruislip Reservoir, 1884, *Benbow* (BM). **4.** Hendon, *Quekett*, *T. & D.*, 405. **7.** Near Hornsey, *Huds. Fl. Angl.*, 399.

N. mucronata Miq.
Formerly native in ponds but long extinct.
3. Isleworth, 1719 (DILL).

N. translucens Ag.
Chara translucens Pers.
Formerly native in ponds and lakes but now extinct.
1. Stanmore Common, 1827–30, *Varenne*, *T. & D.*, 405. Ruislip Reservoir, 1884, *Benbow* (BM).

TOLYPELLA Leonh.

T. intricata Leonh.
Formerly native in lakes, but now extinct.
2. Shortwood Common, Staines, 1892, *H.* and *J. Groves*, *J.B.* (*London*) 33: 292.

T. glomerata Leonh.
Formerly native in lakes, etc., but now extinct.
2. Staines Moor, 1877, *H.* and *J. Groves*, *Rep. Bot. E.C.* 1: 228; 1882, *H.* and *J. Groves* (K).

CHAREAE

CHARA L.

C. vulgaris L.
C. foetida Br.

Native in ponds, lakes, ditches, etc., with sandy or gravelly bottoms. Rare. **1.** Woodridings, Pinner,† *Hind, T. & D.*, 405. Ditch near Ruislip Reservoir. Pond, Wrotham Park, Bentley Heath, 1965. **2.** Near Staines, 1927, *Findlay* (K). Shortwood Common, *Welch*!, det. *G. O. Allen, L.N.* 27: 35. Staines Moor, *Welch*!; Hampton Court Park!; Bushy Park!, all det. *G. O. Allen, K. & L.*, 333. **3.** Gravel pit west of Hounslow Heath!,† det. *G. O. Allen, L.N.* 27: 35. Roxeth,† *Hind, T. & D.*, 405. **5.** Alperton,† *Hind, T. & D.*, 405. **7.** Isle of Dogs;† Notting Hill,† *Quekett*; Kilburn,† *Warren, T. & D.*, 405. Stoke Newington reservoirs, 1964, *Wurzell*, det. *S. P. Phillips*.

Var. **longibracteata** Kütz.
2. Shortwood Common, 1883, *Groves* (K). Staines Moor, *Westrup*, det., *G. O. Allen, K. & L.*, 333.

Var. **papillata** Wallr.
2. Pool near Yeoveney, Staines Moor, 1946, *Welch*!, det. *G. O. Allen, L.N.* 27: 35.

Var. **crassicaulis** Kütz.
2. Staines Moor, 1949, *Westrup*, det. *G. O. Allen, K. & L.*, 333.

C. hispida L.
Formerly native in ponds and ditches, but now extinct.
1. Ruislip Common, *De Cresp. Fl.*, 80. **6.** Finchley Common, *Woods, B.G.*, 405. **7.** Kensal Green, 1864, *Parsons* (CYN).

C. contraria Kütz.
Native in lakes, etc. Very rare.
2. Long Water, Hampton Court Park, 1949!, det. *G. O. Allen, L.N.* 29: 12.

C. globularis Thuill.
C. fragilis Desv.
Native in lakes, ponds, slow streams, etc. Rare.
1. Ruislip Reservoir, 1860, *Hind* (SY). Pinner Hill; Eastcote, *T. & D.*, 405. New Years Green, 1953, *Davidge, Haywood* and *T. G. Collett*!; 1966. Pinner, *W. R. Linton, Rep. Bot. E.C.* 1: 119. Pond on Harefield Green, 1965!, *Pickess*, det. *S. P. Phillips*. **2.** Shortwood Common!, *W. R. Linton, Rep. Bot. E.C.* 1: 119; 1967. Staines Moor, 1966. **4.** Finchley, 1877, *Parsons* (CYN). **5.** Ealing, 1873,† *E. C. White* (SY).

Var. **capillacea** (Thuill.) Zanev.
2. Wyrardisbury river, Staines!, det. *G. O. Allen, K. & L.*, 333. **7.** Clapton, *C. S. Nicholson, K. & L.*, 333.

C. delicatula Ag.
Native in lakes, etc. Very rare.
2. Shortwood Common, 1964, *Wurzell*, det. *S. P. Phillips, L.N.* 44: 17.

2. Symphytum orientale L. Perivale

3. Senecio squalidus L.   A frequent Middlesex plant

4. Epipactis phyllanthes G.E.Sm. Old chalkpit near Harefield

5. Nardurus maritimus (L.) Murb. Railway side near Denham

BRYOPHYTA

There is, unfortunately, little modern literature on the bryophytes of Middlesex. A very inadequate list was included as an appendix in Trimen and Dyer's *Flora of Middlesex* (1869); this was followed in the late nineteenth century by a short series of papers by John Benbow in the *Journal of Botany*. These dealt mainly with the bryophytes of the Uxbridge, Ruislip, Harefield areas. In 1928 P. W. Richards's excellent paper 'Ecological notes on the bryophytes of Middlesex' appeared in the *Journal of Ecology*. In recent times only three important contributions on the mosses and liverworts of the vice-county have appeared, viz. R. A. Boniface's 'Hepatics of the London Area'. *L.N.* 29: 23–27 (1950), J. H. G. Peterken's 'Hand List of the Plants of the London Area: Bryophytes'. *L.N.* 40: 43–71 (1960), based on records made by members of the London Natural History Society, and Mrs P. Moxey's 'The bryophytes of the Ruislip Local Nature Reserve'. *J. Ruisl. & Distr. N.H.S.* 14: 33–41 (1965).

In the last thirty years there is little doubt that, apart from the investigations of R. A. Boniface, this group of plants has been seriously neglected in the vice-county by bryologists, who quite understandably prefer to visit more rural areas which yield a much richer and more varied bryophyte flora.

The account which follows is consequently very incomplete, and many species referred to as rare may prove to be locally common, while others which are thought to be possibly extinct are likely to be rediscovered. Despite, however, the inadequacies of the list it will be seen that many species have been recorded from the vice-county, and I hope that it may stimulate renewed interest in these plants in their metropolitan habitats.

HEPATICAE

ANTHOCEROTACEAE

ANTHOCEROS L.

A. punctatus L.
Formerly native on damp acid heaths, but now long extinct.
3. Hounslow Heath, *Huds. Fl. Angl.*, 519.

MARCHANTIACEAE

CONOCEPHALUM Wiggers

C. conicum (L.) Underw.
Native in moist shady places. Very rare.
6. Pond embankment, Town Park, Enfield, *Proskauer*, *L.N.* 29: 24.

LUNULARIA Adans.

L. cruciata (L.) Dum.
Native on stream banks, in ditches, etc., established at the bases of damp walls, on gravel paths, in greenhouses, etc. Rather common.
1. Uxbridge!; Harefield!, *Benbow*, *J.B.* (*London*) 32: 370. Ickenham. Ruislip. New Years Green. Hillingdon. Potters Bar. **2.** Banks of the Thames near Shepperton!, locally abundant, *Richards Ec.* Staines. Laleham. Sunbury. Hampton Court. Bushy Park. Teddington. Fulwell. **3.** Twickenham. Isleworth. St Margaret's. Hounslow. **4.** West Heath, Hampstead. Willesden. Scratch Wood. Mill Hill. Edgware. Harrow. **5.** Chiswick!; Brentford!, *Boniface*; locally common, 1968. Ealing. Hanwell. Perivale. Heston. Norwood Green. Osterley. Near Shepherds Bush. South Acton. Near Kew Bridge. **6.** Whitewebbs Park, Enfield. Trent Park, Southgate. North Finchley. Whetstone. **7.** On pots in greenhouses, Buckingham Palace grounds, *Rose*, *Fl. B.P.G.* Near the Thames, Hammersmith and Fulham. Holland House grounds, Kensington. Regents Park. Hampstead Ponds, East Heath, Hampstead. By the Lee, Hackney Marshes. Stonework by the Thames near Wapping and Isle of Dogs. Base of old wall near Chelsea Physic Garden. By the New river, Clissold Park, Stoke Newington.

MARCHANTIA Raddi

M. polymorpha L.
Native on wet banks, damp waste ground, etc., established at the bases of damp walls, in gardens, etc. Common.
7. Bombed sites, Cripplegate, abundant, *Wrighton*; Grays Inn Road, E.C., *E. G. Wallace*; Shepherds Bush, *Peterken*, *L.N.* 29: 24. Buckingham Palace grounds, *Rose*, *Fl. B.P.G.* Fulham Palace grounds.

RICCIACEAE

RICCIA L.

R. fluitans L.
Native. Floating in ponds, lakes, ditches, etc., on wet mud, etc. Local.

1. Ruislip Reservoir, abundant!, *Wrighton*; Stanmore Common!, *Boniface, L.N.* 29: 24. Pinner, *Hind, T. & D.*, 405. **2.** Hampton Court Park!, *Boniface, L.N.* 29: 24. Staines, 1965. **4.** Between Bishop's Wood and Finchley, 1866,† *W. Davies*; Harrow, *Hind, T. & D.*, 405. West Heath, Hampstead, 1960. **6.** Hadley Common, *Warren, T. & D.*, 405. Pond near East Barnet, 1965, *Bangerter* and *Kennedy*! Enfield, 1965, *Kennedy*!. **7.** Viaduct Pond, East Heath, Hampstead, abundant!, *H. C. Harris, L.N.* 29: 24.

RICCIOCARPUS Corda

R. natans (L.) Corda
Introduced or native. Floating in still water. Very rare.
2. Old canal, West Drayton, 1943, *H. Banks, L.N.* 29: 24.

RICCARDIACEAE

RICCARDIA Gray

R. multifida (L.) Gray
Native in bogs, on wet clay, etc. Very rare, and perhaps extinct.
4. Hampstead Heath,† *Coop. Fl.*, 103. Kingsbury,† *De Cresp. Fl.*, 89. Highgate golf course, c. 1914, *Andrewes, L.N.* 40: 46. **7.** Ken Wood,† *Doody, Huds. Fl. Angl.*, 437.

R. sinuata (Dicks.) Trev.
Native in damp places. Very rare, or overlooked.
1. Grassland, east of Harefield, *Stott, L.N.* 47: 15. **5.** Perivale Wood, 1968, *P. J. Edwards* and *A. R. Hall.*

Var. **major** Lindb.
4. Hendon, 1926, *Richards Ec.*

R. pinguis (L.) Gray
Native in bogs, wet places, etc. Very rare, or overlooked.
1. Stanmore Common, *Richards Ec.* **5.** Perivale Wood, in two places, 1968, *P. J. Edwards* and *A. R. Hall.* Floor of disused reservoir, Ealing, 1969, *P. J. Edwards*, det. *Townsend.*

PELLIACEAE

PELLIA Raddi

P. epiphylla (L.) Corda
Native on damp shady banks, in wet shady places, etc. Rather common, especially on acid soils.

1. Harrow Weald Common!, *Hind, T. & D.*, 405. Harefield!, *De Cresp. Fl.*, 89. Park Wood, Ruislip!, *Richards Ec.* Northwood. Uxbridge. Mimmshall Wood. **3.** Hatch End. **4.** Hampstead Heath!, *Coop. Fl.*, 103. Highgate golf course, *Andrewes MSS.* Scratch Wood. Mill Hill. **5.** Osterley Park. Walpole Park, Ealing. Gunnersbury Park, Acton. Perivale Wood, *P. J. Edwards* and *A. R. Hall.* **6.** Whitewebbs Park, Enfield. Trent Park. **7.** East Heath, Hampstead!, *Richards Ec.* Ken Wood!, *Boniface.*

P. endiviifolia Dicks.
P. fabbroniana Raddi
Native on shady wet banks, damp ground on basic soils, in ditches, by streamsides, etc. Very rare, or overlooked.
1. Old chalkpit, Springwell, *Boniface, L.N.* 29: 25. **2.** Banks of the Thames near Shepperton, locally abundant, *Richards Ec.* **4.** Harrow, *Hind, T. & D.*, 405.

METZGERIACEAE

METZGERIA Raddi

M. furcata (L.) Dum.
Native on moist tree trunks. Very rare, or overlooked.
1. Near South Mimms; near Batchworth Heath; near Harefield, *Richards Ec.* Harefield!, *Boniface, L.N.* 29: 25. **4.** Near Edgwarebury, *Richards Ec.* **7.** Near Ken Wood, *Budd. MSS.*

BLASIACEAE

BLASIA L.

B. pusilla L.
Formerly native on damp sandy heaths, but now long extinct.
3. Hounslow Heath, *Huds. Fl. Angl.* 2: 519.

FOSSOMBRONIACEAE

FOSSOMBRONIA Raddi

F. pusilla (L.) Dum.
Native on damp heaths, in woodland rides on acid soils, etc. Very rare, or overlooked.
1. Ruislip Common, *Harris, L.N.* 29: 25. Ruislip Woods, 1947, *Peterken, L.N.* 40: 47. **4.** Bishop's Wood, *T. Martyn, B.G.*, 413. Highgate golf course, 1916, *Andrewes, L.N.* 40: 47. **7.** Ken Wood, *Coop. Fl.*, 103.

PTILIDIACEAE

PTILIDIUM Nees

P. ciliare (L.) Hampe
Native on heaths, in damp woods on acid soils, etc. Very rare, and perhaps extinct.
1. Stanmore Common, *Richards Ec.* **6.** Coldfall Wood, Highgate,† *Dandridge, Ray Syn.* 3: 111.

TRICHOCOLEA Dum.

T. tomentella (Ehrh.) Dum.
Formerly native in damp woods, but long extinct, if ever it occurred.
4. Wood between Highgate and Hornsey, *Budd. MSS.*

LEPIDOZIACEAE

LEPIDOZIA (Dum.) Dum.

L. reptans (L.) Dum.
Native on humus or rotten wood in woods, on heaths, damp banks, etc. Local.
1. Stanmore Common!; Harrow Weald Common, rare!, *Richards Ec.* Mad Bess Wood, Ruislip, *Boniface* and *K. C. Harris, L.N.* 29: 27. **4.** Bishop's Wood, *Budd. MSS.* Hampstead, *T. Martyn, B.G.*, 413; Hampstead Woods, frequent, *Richards Ec.* **7.** Ken Wood, abundant!, *Boniface, L.N.* 29: 27.

L. trichoclados K. Müll. **1.** Mad Bess Wood, Ruislip, *K. C. Harris, L.N.* 29: 27. Probably an error, *Trans. Brit. Bryol. Soc.* 1: 495. Definitely an error, the plant seen was *L. reptans, Boniface.*

CALYPOGEIACEAE

CALYPOGEIA Raddi

C. trichomanis (L.) Corda. There are a number of old records for this species but Mrs Jean Paton considers that it is unlikely to have ever occurred in Middlesex, and suggests that the records are referable either to *C. fissa* or *C. muellerana.*

C. muellerana (Schiffn.) K. Müll.
C. trichomanis auct., non (L.) Corda

Native on heaths, etc. Probably local, but confused with *C. fissa*.
1. Stanmore Common, *Richards*, fide *Paton*.

C. fissa (L.) Raddi
Native on heaths, banks, in woods on acid soils, etc. Local.
1. Ruislip Woods, rare!; Stanmore Common; Harrow Weald Common, *Richards Ec.*; 1949, *Boniface*. **4.** West Heath, Hampstead, 1916, *Andrewes MSS.*; c. 1927, *Richards Ec.*; 1949, *Boniface, L.N.* 29: 26. **7.** East Heath, Hampstead, *Richards Ec.* Ken Wood, *Boniface, L.N.* 29: 26.
This species has been much confused with *C. muellerana* and some of the above records are probably in need of revision.

LOPHOZIACEAE

LOPHOZIA (Dum.) Dum.

L. ventricosa (Dicks.) Dum.
Native in woods on acid soils. Very rare.
1. Park Wood, Ruislip, *Boniface*; Harefield, *L.N.* 40: 49.

L. bicrenata (Schmid.) Dum.
Native on heaths. Very rare.
3. Hounslow Heath, *Boniface, L.N.* 29: 25; 1967, *Townsend.*

PLECTOCOLEA (Mitt.) Mitt.

P. hyalina (Lyell ex Hook.) Mitt.
Native on damp acid clay soil. Very rare.
4. By damp rill, Hampstead Heath, 1968, *Trans. Brit. Bryol. Soc.* 5: 871.

PLAGIOCHILACEAE

PLAGIOCHILA (Dum.) Dum.

P. asplenioides (L.) Dum.
Perhaps formerly native in shady places, but extinct if ever it occurred.
1. Pinner, *Hind, T. & D.*, 404. **6.** Woods at Southgate, abundant, *Dill. MSS.*

HARPANTHACEAE

LOPHOCOLEA (Dum.) Dum.

L. bidentata (L.) Dum.
Native in moist shady places on acid soils. Rare.

1. Pinner Wood, *Hind, T. & D.*, 404. Harefield, 1907, *Sherrin*. Near Harefield, abundant; Ruislip Woods!, frequent; Stanmore Common; Harrow Weald Common, occasional!, *Richards Ec.* **3.** Hounslow Heath, *Richards Ec.* **4.** West Heath, Hampstead, *W. Davies, T. & D.*, 404. Kingsbury, *De Cresp. Fl.*, 90. Highgate, *Andrewes MSS.* Hampstead Woods; Wood Lane, Brockley Hill; near Edgwarebury, occasional, *Richards Ec.* **5.** Clay banks, Osterley Park, *Boniface*.
This species has been much confused with *L. cuspidata*, and some of the records given above may be in need of revision, fide *Paton*.

L. cuspidata (Nees) Limpr.
Native in moist shady places, on tree stumps, bark, logs, etc. Rare.
1. Northwood, *Benbow, J.B.* (*London*) 32: 370. Ruislip Woods, occasional!, *Richards Ec.* **2.** Bushy Park, *Boniface, L.N.* 29: 26. **5.** Perivale Wood, *P. J. Edwards* and *A. R. Hall.* **6.** Queen's Wood, Highgate, 1916, *Andrewes, L.N.* 40: 50. **7.** Damp grassland by lake, Buckingham Palace grounds, *Rose, Fl. B.P.G.*
This species has been much confused with *L. bidentata*, and some of the records given above may be in need of revision, fide *Paton*.

L. heterophylla (Schrad.) Dum.
Native on tree stumps, logs and soil in damp woods, etc. Local, or overlooked.
1. Pinner Wood!, *Melv. Fl.* 2: 138. Ruislip Woods!, abundant; Stanmore Common; Harrow Weald Common, rare; near Harefield; South Mimms, *Richards Ec.* Old Park Wood, Harefield. **4.** Hampstead Heath, *De Cresp. Fl.*, 90. Hampstead Woods, very rare; Wood Lane, Brockley Hill; near Edgwarebury, *Richards Ec.* Stanmore, 1908, *Sherrin*. Mill Hill, 1922, *Moring MSS.* **5.** Perivale Wood!, *Lillford* and *Ing*, *Groves Veg.* **6.** Queen's and Highgate Woods, *Andrewes MSS.* **7.** Ken Wood. Buckingham Palace grounds, *Rose, Fl. B.P.G.*

CHILOSCYPHUS Corda

C. polyanthos (L.) Corda
Native in wet places in woods and shady places, etc. Very rare.
1. Side of water courses, Ruislip Woods, *Benbow, J.B.* (*London*) 32: 370; c. 1949, *Peterken, L.N.* 29: 26. **2.** Bushy Park, *Boniface, L.N.* 29: 26.

C. pallescens (Ehrh.) Dum.
Native in damp woods on heavy soils, etc. Very rare.
1. Old Park Wood, Harefield, 1950, *Boniface.* **6.** Queen's Wood, Highgate, 1916, *Andrewes, L.N.* 40: 50.

CEPHALOZIELLACEAE

CEPHALOZIELLA (Spruce) Schiffn.

C. starkei (Funck) Schiffn.
Native on heaths, in woods on acid soils, etc. Very rare.
1. Park Wood, Ickenham, *Benbow, J.B. (London)* 32: 270. Ruislip Woods, very rare, 1927, *Richards Ec.*; c. 1949, *K. C. Harris, L.N.* 29: 26. Stanmore Common; Harrow Weald Common, very rare; Poor's Field, Ruislip, *Richards Ec.* **4.** Hampstead Heath, *De Cresp. Fl.*, 90; c. 1949, *Rose, L.N.* 29: 26.

CEPHALOZIACEAE

CEPHALOZIA (Dum.) Dum.

C. bicuspidata (L.) Dum.
Native in wet places, damp woods, on rotten wood, etc. Common.
7. East Heath, Hampstead. Ken Wood. Holland House grounds, Kensington.

C. connivens (Dicks.) Lindb.
Formerly native on wet heaths, but now extinct if ever it occurred.
4. Hampstead Heath, *De Cresp. Fl.*, 90.

ODONTOSCHISMA (Dum.) Dum.

O. sphagni (Dicks.) Dum.
Formerly native in bogs but now long extinct.
4. Hampstead Heath, 1832 (CGE).

SCAPANIACEAE

DIPLOPHYLLUM Dum.

D. albicans (L.) Dum.
Native on heaths, moist banks, wet woods on acid soils, etc. Very rare.
1. Harrow Weald Common!; Stanmore Common, locally abundant!; Ruislip Woods!; near Harefield, *Richards Ec.* **4.** Hampstead, *Budd. MSS.* Hampstead Heath, *Coop. Fl.*, 903. **7.** Ken Wood!, *Boniface, L.N.* 29: 27.

SCAPANIA (Dum.) Dum.

S. curta (Mart.) Dum. **1.** Ruislip Woods, *Richards Ec.*; c. 1949, *K. C. Harris* and *Peterken*, *L.N.* 29: 27. *S. curta* is most unlikely to have occurred in Middlesex, fide *Paton*.

S. irrigua (Nees) Dum.
Native in wet woodland rides, etc. Very rare.
1. Ruislip Woods, locally abundant, *Richards Ec.*; Park Wood, Ruislip, *Boniface*, *L.N.* 29: 27. Harrow Weald Common, *Richards Ec.*

S. nemorea (L.) Grolle
S. nemorosa (L.) Dum.
Native on tree stumps in woods, etc. Very rare.
1. Uxbridge; Ruislip Woods, *Benbow*, *J.B.* (*London*) 32: 370; 1927, *Richards Ec.*; c. 1949, *K. C. Harris*, *L.N.* 29: 27; c. 1960, *Peterken*, *L.N.* 40: 51.

S. undulata (L.) Dum.
Native on heaths, in damp woods on acid soils, etc. Very rare.
1. Harrow Weald Common, *Hind*, *T. & D.*, 404; c. 1877, *De Cresp. Fl.*, 89. Stanmore Common, *Richards Ec.* Park Wood, Ruislip, *Boniface*, *L.N.* 29: 27. **4.** Hampstead Heath, *Budd. MSS.*; c. 1877,† *De Cresp. Fl.*, 89.

RADULACEAE

RADULA Dum.

R. complanata (L.) Dum.
Formerly native on tree trunks in shady places, but now extinct.
1. Pinner Wood, *Hind*, *T. & D.*, 405; c. 1877, *De Cresp. Fl.*, 90. **4.** Kingsbury, *De Cresp. Fl.*, 90.

PORELLACEAE

PORELLA L.

P. platyphylla (L.) Lindb.
Native on bare ground, at bases of trees, etc., established on walls, etc. Very rare, and mainly confined to calcareous soils.
1. Uxbridge!; Harefield!, *Benbow*, *J.B.* (*London*) 32: 370. South Mimms, *Richards Ec.* **4.** Near Edgwarebury, *Richards Ec.*

LEJEUNEACEAE

LEJEUNEA Lib.

L. cavifolia (Ehrh.) Lindb.
Perhaps formerly native in shady places on calcareous soils, but extinct if ever it occurred.
1. Lane by Jack's Lock, Harefield, *Benbow*, *J.B.* (*London*) 32: 370.

FRULLANIA Raddi

F. tamarisci (L.) Dum. **4.** Hampstead Heath, *Burn. Med. Bot.*, 105 and *Coop. Fl.*, 103. Almost certainly an error.

F. dilatata (L.) Dum.
Formerly native on sheltered tree trunks, etc., but now extinct.
1. Pinner Wood, *Melv. Fl.* 2: 38; c. 1877, *De Cresp. Fl.*, 90. **4.** Harrow, *T. & D.*, 405.

MUSCI

SPHAGNACEAE

SPHAGNUM L.

S. palustre L.
S. cymbifolium Ehrh.
Native in bogs. Very rare.
7. Bog west of Ken Wood!, 1949, *Boniface*.

S. compactum DC.
S. rigidum Schp.
Native in bogs, wet heathy places, etc. Very rare, and perhaps extinct.
1. A few tufts among *Calluna* in shade of birches, lower part of Harrow Weald Common, 1925,† *Richards*. Ruislip Woods, *Richards Ec.*

S. squarrosum Pers. ex Crome
Native in bogs. Very rare.
7. Bog west of Ken Wood!, 1949, *Boniface*; 1959, *Rose*; 1967.

S. recurvum P. Beauv.
S. intermedium Hoffm.
Formerly native in wet bogs, but now extinct.
4. West Heath, Hampstead, *Benbow*, *J.B.* (*London*) 32: 106; very

sparingly among *Molinia* tussocks in the bog, 1920; a few plants, 1921; two tufts, 1926, *Richards*.

S. cuspidatum Ehrh.
Native in bogs, wet heathy places, etc. Very rare, and perhaps extinct.
1. Harrow Weald Common, *Hind*, *T. & D.*, 401. **4.** West Heath, Hampstead, *Richards Ec.*

S. subsecundum Nees
Native in bogs, wet heathy places, etc. Very rare.

Var. **inundatum** (Russ.) C. Jens.
4. Bog, West Heath, Hampstead, *Richards Ec.*

Var. **auriculatum** (Schimp.) Lindb.
1. Harefield;† Harrow Weald Common; Stanmore Common; Park Wood, Ruislip, *Richards Ec.*

S. fimbriatum Wils.
Native in bogs. Very rare.
7. West Meadow Bog, Ken Wood!, 1949, *Boniface*; 1967.

S. capillaceum (Weiss) Schrank
S. nemoreum Scop., *S. acutifolium* Ehrh.
Native in bogs, wet heathy places in shade, etc. Very rare, and perhaps extinct.
4. West Heath, Hampstead, 1916, *Andrewes*, *L.N.* 40: 53.

S. plumulosum Röll
S. acutifolium Ehrh. var. *subnitens* (Russ. & Warnst.) Dix.
Native in bogs. Very rare.
7. West Meadow Bog, Ken Wood, 1959, *Rose*.

POLYTRICHACEAE

ATRICHUM P. Beauv.

A. undulatum (Hedw.) P. Beauv.
Catharinea undulata Web. & Mohr
Native in woods, on heaths, banks, etc. Common.
7. East Heath, Hampstead. Ken Wood!; Ken Wood Fields!, *Boniface*.

POLYTRICHUM Hedw.

P. nanum Hedw.
Native on banks, heaths, etc. Very rare, and mainly confined to sandy soils.

1. Harrow Weald Common, 1922, *Richards.* Poor's Field, Ruislip, *Richards Ec.* **2.** Staines, *R. Braithwaite, T. & D.*, 402.

P. aloides Hedw.
Formerly native on heaths, dry banks, etc., on sandy and clay soils, but now extinct.
1. Pinner, *Hind, T. & D.*, 402. **4.** Harrow Weald, *Hind*; Hampstead Heath, *R. Braithwaite, T. & D.*, 402.

P. urnigerum Hedw.
Formerly native on heaths, but now extinct.
1. Harrow Weald Common, 1907, *Sherrin.*

P. piliferum Hedw.
Native on heaths. Rare, and confined to sandy soils.
1. Stanmore Common; Poor's Field, Ruislip, *Richards Ec.* Copse Wood, Ruislip, a small patch, *Boniface.* **3.** Hounslow Heath, *Richards Ec.*; 1950, *Boniface.* **4.** Hampstead Heath,† *Syme* and *R. Braithwaite, T. & D.*, 402. North-west Heath, Hampstead, 1916,† *Andrewes, L.N.* 40: 54.

P. juniperinum Hedw.
Native in woods, on heaths, etc. Locally plentiful on sandy soils.
1. Stanmore Common!; Poor's Field, Ruislip, locally abundant!, *Richards Ec.* Ruislip Woods, abundant!, *Boniface.* **2.** Hampton Court Park!, *Boniface.* **3.** Hounslow Heath, *Richards Ec.* **4.** Hampstead Heath!, *E.B.*, 200; West and North-west Heaths, Hampstead, 1967. Whitchurch Common,† *Richards Ec.* **7.** East Heath, Hampstead. Ken Wood.

P. aurantiacum Sw.
P. gracile Sm.
Native in woods. Very rare, and confined to acid soils.
1. Mimmshall Wood, 1951, *Boniface.*

P. formosum Hedw.
Native in woods, on heaths, banks, etc. Locally plentiful on acid soils.
1. Uxbridge; Harefield!, *Benbow, J.B. (London)* 32: 106. Ruislip Woods, abundant!; Stanmore Common!; Harrow Weald Common!; Poor's Field, Ruislip, locally abundant!; near South Mimms, *Richards Ec.* Breakspeares; Northwood!, *Richards.* **3.** Hounslow Heath, 1927, *Richards.* **4.** Hampstead Woods, very rare; Wood Lane, Brockley Hill, occasional,† *Richards Ec.* **5.** Perivale Wood, rare, *P. J. Edwards* and *A. R. Hall.* **6.** Near Enfield!, 1925, *Richards*; 1965, *Kennedy*!. **7.** Ken Wood, 1926, *Richards.*

P. commune Hedw.
Native in bogs, on wet heaths, etc. Formerly local, but now very rare.
1. Harrow Weald Common! (ME). Stanmore Common!; Watt's

Common, Harefield,† *Richards Ec.* Pinner,† *Hind, T. & D.*, 402. Ruislip Common, *Richards*. Ruislip Woods, 1927, *Richards, L.N.* 40: 54. **4.** Hampstead Heath, *T. & D.*, 402; 1916, *Andrewes, L.N.* 40: 54. **6.** Near Enfield, 1925, *Richards*. **7.** Bog near Ken Wood, 1949, *Bangerter* and *H. C. Harris*!.

FISSIDENTACEAE

FISSIDENS Hedw.

F. viridulus Wahlenb.
Formerly native on banks, etc., but now extinct.
1. Harefield, *Benbow, J.B. (London)* 32: 107.

F. bryoides Hedw.
Native in woods, ditches, on banks, by river- and streamsides, etc., established on lawns, etc. Common, especially on clay soils.
7. Hampstead!, *Syme, T. & D.*, 403: East Heath, Hampstead. Ken Wood. Holland House grounds, Kensington.

F. incurvus Starke ex Web. & Mohr
Native on shady banks, etc. Very rare, and confined to calcareous and clay soils.
1. Harefield, *Benbow, J.B. (London)* 32: 107. Near Harefield, rare, *Richards*. **4.** Monk's Mere, Hendon, 1922,† *Richards, Rep. Brit. Bryol. Soc.* 1923, 19.

F. crassipes Wils. ex Bruch, Schimp. & Guemb.
Native on stones, etc., by water, sometimes submerged, established on river and canal walls, etc. Locally plentiful.
1. Stones by the canal, Harefield, *Richards Ec.* **2.** River wall between Shepperton and Penton Hook Lock, locally abundant!, *Richards Ec.* Stones by the Thames, Sunbury. River wall near Hampton Court. **7.** In an iron pipe by river Lee, Hackney, 1939, *Peterken, L.N.* 40: 54.

F. exilis Hedw.
Native in woodland rides, wet places, etc. Very rare, or overlooked.
1. Ruislip Nature Reserve, *Moxey Br.* **4.** Monk's Mere, Hendon, rare,† *Richards Ec.* **6.** Enfield Chase, *Dickson* (BM).

F. taxifolius Hedw.
Native in woods, ditches, wet grassy places, on cultivated ground, by river- and streamsides, etc. Common.
7. Kensington Gardens, *E.B.*, 426. Near Highgate, *R. Braithwaite, T. & D.*, 403. East Heath, Hampstead.

F. cristatus Wils.
F. decipiens De Not.
Native in grassy places. Very rare, and confined to calcareous soils.
1. Harefield, 1951, *Boniface.*

F. adianthoides Hedw.
Native in wet grassy places, marshes, etc., established in damp chalk-pits, etc. Very rare, or overlooked.
1. Stanmore Common; Ruislip Common, *Richards Ec.* Chalkpit near Harefield, *Richards.* **7.** Kensington Gardens,† *E.B.*, 426.

OCTODICERAS Brid.

O. fontanum (La Pyl.) Lindb.
O. julianum Brid.
Native on wet stones by rivers, etc., sometimes submerged. Very rare, and apparently confined to the Thames.
2. Staines, *Richards* and *Catcheside*, *Rep. Brit. Bryol. Soc.* 1926–27, 280.

ARCHIDIACEAE

ARCHIDIUM Brid.

A. alternifolium (Hedw.) Mitt.
Formerly native in bare places on poor sandy soils, but now extinct.
1. Ruislip Reservoir, *Benbow*, *J.B.* (*London*) 32: 106.

DICRANACEAE

PLEURIDIUM Brid.

P. acuminatum Lindb.
Native on banks, in short turf, woodland rides, etc. Very rare, and sporadic.
1. New Years Green, 1922, *Richards.* Ruislip Common, *Richards Ec.* Poor's Field, Ruislip, *Boniface.* **4.** Hampstead Heath, *Syme* and *R. Braithwaite*, *T. & D.*, 401; c. 1877, *De Cresp. Fl.*, 92. Highgate golf course, 1916, *Andrewes*, *L.N.* 40: 55. Hendon, 1924, *Richards.*

P. subulatum (Hedw.) Lindb.
Native on heaths, banks, in woodland rides, etc. Very local.
1. Northwood; between Ruislip Reservoir and Eastcote, *Benbow*, *J.B.* (*London*) 32: 369. Copse Wood, Northwood, abundant, *Richards Ec.* Ruislip Common and golf course, *Richards*; Poor's Field, Ruislip, *Boniface.*

CERATODON Brid.

C. purpureus (Hedw.) Brid.
Native on bare ground, etc., established on walls, etc. Very common, especially on burnt ground.
7. Abundant in the district.

Var. **brevifolius** Milde
1. Ruislip, 1904, *Sherrin*.

SELIGERIA Bruch, Schimp. & Guemb.

S. pusilla (Hedw.) Bruch, Schimp. & Guemb.
Formerly in shady places on calcareous soils, but extinct, if ever it occurred.
1. Old chalkpit, Harefield Park, *Benbow*, *J.B.* (*London*) 32: 369.

S. calcarea (Hedw.) Bruch, Schimp. & Guemb.
Native? Established in chalkpits. Very rare.
1. Harefield!, *Benbow*, *J.B.* (*London*) 32: 106; 1950.

PSEUDEPHEMERUM (Lindb.) Loeske

P. nitidum (Hedw.) Reim.
Pleuridium axillare (Dicks.) Lindb.
Native in ditches, bare places, woodland rides, etc. Very rare, and perhaps extinct.
3. Southall, 1904, *Sherrin*. **4.** Hampstead Heath, *Coop. Fl.*, 103; c. 1927; near Edgwarebury, *Richards Ec.* Wembley Park, 1898, *Sherrin* (SLBI). **6.** Muswell Hill, *Dickson*, *Burn. Med. Bot.*, 105. **7.** Kensington Gardens,† *Syme*, *E.B.* East Heath, Hampstead, *Richards Ec.*

DICRANELLA (C. Müll.) Schimp.

D. varia (Hedw.) Schimp.
Native in bare places, woodland rides, by river- and streamsides, etc., established in chalkpits, etc. Rare, or overlooked.
1. Harrow Weald Common, *Hind*, *T. & D.*, 401; c. 1877, *De Cresp. Fl.*, 85. Old chalkpit, Springwell!, 1922, *Richards*; very abundant, 1951, *Boniface*. Chalkpits, Harefield!, rare, *Richards Ec.* Ruislip Nature Reserve, *Moxey Br.* **2.** Thames bank near Shepperton, rare; bank by Penton Hook backwater, abundant, *Richards Ec.* **4.** Hampstead Heath, 1879, *De Crespigny* (MANCH).

D. staphylina Whitehouse
Colonist on arable ground on non-calcareous soils, etc. Possibly common but confused with immature *D. varia*.
7. Garden, Highgate, 1966, *M. O. Hill*, *Trans. Brit. Bryol. Soc.* 5: 765.

D. rufescens (With.) Schimp.
Formerly in old chalkpits, but extinct if ever it occurred.
1. Chalkpits, Harefield and Springwell, *Benbow, J.B.* (*London*) 32: 106.

D. crispa (Hedw.) Schimp.
Native on moist heaths, in woodland rides, etc. Very rare.
4. West Heath, Hampstead, 1959, *Jefford*, det. *Warburg*. **6.** Woods near Southgate, *With. Bot. Arr.* 2: 97.

D. cerviculata (Hedw.) Schimp.
Native on peaty banks on heaths, in woods, etc. Very rare.
4. Hampstead Heath, *Dickson, With. Bot. Arr.* 3: 813; c. 1877, *De Cresp. Fl.*, 85. **7.** Sparingly on a peaty bank by Ken Wood, 1950!, *Rose.*

D. heteromalla (Hedw.) Schimp.
Native on banks, in woods, ditches, grassy places, etc. Common on acid soils, but absent from the chalk.
7. Hampstead Heath!, *R. Braithwaite, T. & D.*, 401; East Heath, Hampstead. Ken Wood, *H. C. Harris*! Parliament Hill Fields. Kensington Gardens.

DICRANOWEISIA Lindb.

D. cirrata (Hedw.) Lindb.
Native on tree trunks established on fences, walls, etc. Formerly local, but now apparently very rare, or overlooked.
1. Ruislip Woods!; near South Mimms, *Richards Ec.* Fences at Ruislip and near Elstree, 1925, *Richards*. Harefield. **4.** Harrow; near Harrow Weald; Cricklewood, *De Cresp. Fl.*, 95. Near Edgwarebury, 1926, *Richards*. **7.** East Heath, Hampstead, *Syme, T. & D.*, 401. Ken Wood, 1950, *Boniface.*

DICRANUM Hedw.

D. montanum Hedw.
Native on trunks and branches of trees in woods. Very rare.
1. By Bayhurst Wood, near Harefield; Duck Wood, Northwood, abundant, *Benbow, J.B.* (*London*) 32: 107 and 37: 441. Copse and Park Woods, Ruislip, *Richards Ec.*; still plentiful in Ruislip Woods, 1961, *Boniface.*

D. flagellare Hedw.
Native on tree stumps, etc., in woods. Very rare, or overlooked.
1. Harefield; Ruislip Woods, *Benbow, J.B.* (*London*) 37: 441; locally abundant, 1927, *Richards Ec.*; 1960, *Peterken, L.N.* 40: 56; Park Wood,

Ruislip, locally abundant, *Boniface*. Harrow Weald Common, *Sherrin*. **4.** Scratch Wood, 1949, *Boniface*.

D. strictum Schleich. ex Schwaegr.
Native on tree trunks, etc. Very rare.
1. Ruislip, 1937, *Milsom*, *Rep. Brit. Bryol. Soc.* 1937, 24.

D. majus Sm.
Native in woods on acid soils. Very rare.
1. Pinner, *Hind*, *T. & D.*, 401. On ground, Park Wood Ruislip, 1949, *Rose* and *Boniface*, *Trans. Brit. Bryol. Soc.* 5: 880.

D. bonjeanii De Not.
Native in bogs, wet heathy places, etc. Very rare or overlooked.
1. Stanmore Common, 1922, *Richards*, *Rep. Moss Exch. Club* 1923, 18. Watt's Common, Harefield;† Poor's Field and Ruislip Common!, *Richards Ec.* Harrow Weald Common, 1922, *Richards*. East of Harefield, *Stott*, *L.N.* 47: 15.

Var. **rugifolium** Bosw.
1. Ruislip Common, *Richards Ec.*

D. scoparium Hedw.
Native on hedgebanks, tree trunks, in woodland rides, grassy places, etc. Formerly locally common, but now rare or overlooked.
1. Pinner, *Hind*, *T. & D.*, 401. Ruislip Woods, abundant!; Stanmore Common!; Harrow Weald Common, rare, *Richards Ec.*; 1949, *Boniface*. Harefield, *Stott*, *L.N.* 47: 14. **2.** Fulwell Park, 1969, *Townsend*. **3.** Hounslow Heath, 1927, *Richards*. **4.** Hampstead Heath, *Curt. F.L.*; c. 1877, *De Cresp. Fl.*, 85; West Heath, Hampstead, sparingly, 1921,† *Richards*; Brockley Hill; Whitchurch Common,† *Richards Ec.* Harrow Weald, *Melv. Fl.* 2: 130. **5.** Near Osterley, *Boniface*.

Var. **orthophyllum** Brid.
1. Poor's Field, Ruislip, locally abundant, *Richards Ec.* **3.** Hounslow Heath, *Richards Ec.*

CAMPYLOPUS Brid.

C. fragilis (Dicks.) Bruch, Schimp. & Guemb.
Native in dry heathy places. Very rare.
1. Stanmore Common, 1949, *Boniface*.

C. pyriformis (Schultz) Brid.
Native in open woods, on heaths, etc. Formerly local, but now very rare or overlooked.
1. Stanmore Common!; Harrow Weald Common, *Richards Ec.*; 1949; Ruislip, *Boniface*. **4.** Hampstead Heath, *R. Braithwaite*, *T. &*

D., 401; North-west Heath, Hampstead, very abundant, *Andrewes MSS.* **6.** Queen's Wood, Highgate, *Andrewes MSS.* **7.** Buckingham Palace grounds, *Fl. B.P.G.*

C. flexuosus (Hedw.) Brid.

Native in woods, on heaths, etc. Formerly local on acid soils, but now very rare, or overlooked.

1. Ruislip Woods, *Benbow*, *J.B.* (*London*) 32: 106; 1927; Stanmore Common!; Harrow Weald Common, *Richards Ec.* Pinner Hill, 1923; Mimmshall Wood, 1925, *Richards.* **4.** Hampstead Heath, *Hampst.* Probably an error; ? *C. pyriformis* intended, *Richards.* Seed-pans in a greenhouse, Highgate, 1916, *Andrewes MSS.*

C. introflexus (Hedw.) Brid.

Native in open woods on acid soils, etc. Very rare.

4. On peaty soil near bottom of south-facing slope, Scratch Wood, Mill Hill, 1962, *Goater*, *Trans. Brit. Bryol. Soc.* 4: 725.

LEUCOBRYUM Hampe

L. glaucum (Hedw.) Schimp.

Native in woods, on heaths, etc. Very rare, and mainly confined to dry acid soils.

1. Harefield, *F. H. Ward*, teste *R. Braithwaite*; Harrow Weald Common, *T. & D.*, 401; 1879, *De Crespigny* (MANCH). Pinner, *Melv. Fl.* 2: 131. Mad Bess Wood, Ruislip, *Coker*, *Trans. Brit. Bryol. Soc.* 5: 411.

ENCALYPTACEAE

ENCALYPTA Hedw.

E. vulgaris Hedw.

Native on banks, etc., established on the river-wall of the Thames, etc. Very rare, and mainly confined to calcareous and alluvial soils.

1. Harefield, *Newbould*, *T. & D.*, 402. **2.** River-wall between Shepperton and Penton Hook Lock, very rare, *Richards Ec.* **4.** Hendon, 1922,† *Richards*, *L.N.* 40: 57.

POTTIACEAE

TORTULA Hedw.

T. ruralis (Hedw.) Crome

Formerly a colonist on walls and roofs, but now apparently extinct.

4. Roof near 'The Spaniards', Hampstead, *Syme, T. & D.*, 422; persisted until 1922, *Richards.*

T. intermedia (Brid.) Berk.

Colonist on brickwork, stonework, etc. Very rare.

2. River wall below Penton Hook Lock, *Richards Ec.*; c. 1960, *Peterken, L.N.* 40: 57. **6.** Wall at Southgate, 1879,† *De Crespigny* (MANCH).

T. laevipila (Brid.) Schwaegr.

Colonist on tree trunks, walls, etc. Very rare.

1. Uxbridge; Harefield, *Benbow, J.B. (London)* 32: 106. On pollarded willows near Harefield, 1923, *Richards.* Harrow Weald Common, 1907, *Sherrin.* **2.** On alders and willows by the Thames between Shepperton and Penton Hook Lock, occasional!, *Richards Ec.*; 1952, *Boniface*; 1967. **4.** Colindeep Lane, Hendon,† *Richards Ec.*

Var. **laevipiliformis** (De Not.) Limpr.

1. Wall near Copper Mills Down, Harefield, *Coker, J. Ruisl. & Distr. N.H.S.* 16: 8.

T. papillosa Wils. ex Spruce

Formerly a colonist on tree trunks, but now extinct.

1. Frays Meadows, Uxbridge, *Benbow, J.B. (London)* 32: 106.

T. latifolia Hartm.

Native or colonist on banks, bases of trees, stones, etc., by water. Very rare.

1. Frays Meadows, Uxbridge, *Benbow, J.B. (London)* 32: 106. **2.** Staines, 1926, *Richards* and *Catcheside, Rep. Brit. Bryol. Soc.* 1926–27: 282. River-wall between Shepperton and Penton Hook Lock, rare! *Richards Ec.*; 1952, *Boniface*; c. 1960, *Peterken, L.N.* 40: 57; 1967.

T. subulata Hedw.

Native or colonist on banks, heaths, in woods, etc., chiefly on sandy soils. Very rare, or overlooked.

1. Pinner, *Hind, T. & D.*, 402. Near Garett Wood, Springwell, *Benbow, J.B. (London)* 34: 400. Ruislip Nature Reserve, *Moxey Br.* **4.** Hampstead, *Syme, T. & D.*, 402. Harrow Weald, *De Cresp. Fl.*, 94.

T. muralis Hedw.

Colonist on walls, stonework, in crevices between paving stones, etc. Very common.

7. Frequent in the district.

Var. **aestiva** Hedw.

1. Ruislip, 1904, *Sherrin.*

T. stanfordiensis Steere

Colonist on paths, etc., in shady places, etc., Very rare, or overlooked.

3. On shaded path under trees above flood level by river Crane near Heathrow Airport, 1968, *M. O. Hill, Trans. Brit. Bryol. Soc.* 5: 881.

ALOINA Kindb.

A. rigida (Hedw.) Limpr. Cited for Middlesex in the British Bryological Society's *Census Catalogue*, edition 1, but I have been unable to trace the source of the record.

A. ambigua (Bruch & Schimp.) Limpr.
Tortula ambigua Bruch & Schimp.
Native or colonist on banks, etc., established in chalkpits, etc. Very rare, and confined to calcareous soils.
1. Harefield, *Benbow*, *J.B.* (*London*) 32: 106. Disused chalkpit, Springwell!, 1923, *Richards*; plentiful, 1950!, *Boniface*.

A. aloides (Schultz) Kindb.
Tortula aloides (Schultz) De Not.
Native or colonist on banks, etc., established in chalkpits, etc. Very rare, and confined to calcareous soils.
1. Harefield, *Benbow*, *J.B.* (*London*) 32: 106. Old chalkpit, Springwell, 1950!, *Boniface*.

POTTIA Fürnr.

P. lanceolata (Hedw.) C. Müll.
Native on banks, etc., on calcareous and other dry soils, established on wall-tops, etc. Very rare, or overlooked.
1. Above chalkpits, Harefield, *Benbow*, *J.B.* (*London*) 34: 400. Ruislip Nature Reserve, *Moxey Br.* **4.** Harrow, 1907, *Sherrin*.

P. heimii (Hedw.) Fürnr.
Colonist on mud by the tidal Thames. Very rare.
5. Chiswick, 1960, *Townsend*, *Trans. Brit. Bryol. Soc.* 4: 164.

P. intermedia (Turn.) Fürnr.
Colonist on bare soil, etc. Very rare, and perhaps extinct.
1. Harefield, *Benbow*, *J.B.* (*London*) 32: 369.

P. truncata (Hedw.) Fürnr.
Colonist on bare ground, in woods, fallow-fields, cart-ruts, etc. Formerly local, but now very rare or overlooked.
1. Near Harefield, locally frequent; New Years Green; Ruislip Common, *Richards Ec.* Potters Bar, *Richards*. **2.** Gravel pits, Ashford, 1967, *Townsend*. **3.** Southall, 1903, *Sherrin*. **4.** Hampstead Heath, *Syme*; near Bishop's Wood, *T. & D.*, 402. Highgate golf course, *Andrewes MSS*. Monk's Mere, Hendon;† Hampstead Heath Extension; Edgwarebury, *Richards Ec.* Canons Park, Edgware, *Richards*. Harrow, *Melv. Fl.* 2: 131. **5.** Perivale Wood, *P. J. Edwards* and *A. R. Hall*.

P. davalliana (Sm.) C. Jens.
Colonist on bare ground, etc. Very rare, and perhaps extinct.
1. Ruislip Common, *Richards Ec.*

PHASCUM Hedw.

P. cuspidatum Hedw.
Colonist on banks, in fields, cart-ruts, etc. Very rare, or overlooked.
1. Near Harefield, locally frequent; New Years Green; Ruislip Common, *Richards Ec.* South Mimms, 1925, *Richards.* **2.** Gravel pits, Ashford, 1967, *Townsend.* **4.** Willesden, 1879,† *De Crespigny* (MANCH). Near Edgwarebury, very rare; Hendon, 1927, *Richards.*

ACAULON C. Müll.

A. muticum (Hedw.) C. Müll.
Colonist on tree trunks, recently disturbed soil, paths, etc. Very rare.
1. New Years Green, *Richards Ec.* **2.** On alders and willows by the Thames between Shepperton and Penton Hook Lock!, *Richards Ec.*; 1965. **4.** Hampstead Heath, *R. Braithwaite, T. & D.*, 402. Disused allotments, Hendon, 1922–23, *Richards Ec.*

CINCLIDOTUS P. Beauv.

C. fontinaloides (Hedw.) P. Beauv.
Native on submerged stones and shingle, and established on the river-wall, of the Thames. Very rare.
2. Between Shepperton and Penton Hook Lock!, *Richards Ec.*; 1952, *Boniface*; c. 1960, *Peterken, L.N.* 40: 58; 1965. **7.** Thames about London, frequent,† *Huds. Fl. Angl.*, 398; on stones among mud at low water in the bed of the river opposite Lambeth Palace,† *Mackay, T. & D.*, 402.

BARBULA Hedw.

B. convoluta Hedw.
Native, or colonist, in grassy places, bare ground, etc., established on paths, wall-tops, etc. Rare, or overlooked.
1. Harefield, *Benbow, J.B.* (*London*) 32: 106; 1952, *Boniface.* Springwell, common on bare soil, *Stott, L.N.* 47: 14. Uxbridge. Ruislip Nature Reserve, *Moxey Br.* **2.** Near Penton Hook Lock, *Boniface.* **4.** West Heath, Hampstead, *Boniface.* **4.** Ken Wood, 1922, *Sutton*, fide *Richards.*

var. **commutata** (Jur.) Husn.
1. Harefield, *Sherrin.*

B. unguiculata Hedw.
Native, or colonist, on cultivated and waste grounds, banks, etc., established on paths, wall-tops, etc. Common.
7. Near Highgate, *Hind*, *T. & D.*, 402. East Heath, Hampstead, *Richards*.

Var. **cuspidata** (Schultz) Brid.
4. Wembley Park, 1908, *Sherrin*, fide *Paton*, *Trans. Brit. Bryol. Soc.* 4: 726.

B. revoluta Brid.
Formerly native, or a colonist, on bare ground, established on walls, etc., but now extinct.
1. Ickenham; Ruislip; canal walls, Uxbridge, *Benbow*, *J.B.* (*London*) 32: 106.

B. hornschuchiana Schultz
Native, or colonist, on bare ground, etc., established on walls, etc. Very rare, or overlooked.
2. Laleham, *R. Braithwaite*, *T. & D.*, 402. West Drayton, 1926, *Sherrin*. Littleton Churchyard, 1971, *Townsend*.

B. fallax Hedw.
Native, or colonist, on bare ground, etc., established on paths and walls, etc. Rare, or overlooked.
1. Ruislip!, *Hind*, *T. & D.*, 402. Harrow Weald Common, *De Cresp. Fl.*, 94. Harefield!, *Lockett*, teste *Richards*, *J. Ecol.* 33: 207; 1950!, *L.N.* 40: 59; 1962. **2.** Shepperton, occasional, *Richards Ec.* **6.** Southgate, 1879, *De Crespigny* (MANCH).

Var. **brevifolia** Schultz
3. Southall, 1902, *Sherrin*, teste *Dixon*.

B. nicholsonii Culm.
Native on stones and shingle by the Thames, and established on the river-wall. Very rare.
2. Laleham, 1906, *Bishop* (SLBI); 1926, *Catcheside*, *Rep. Brit. Bryol. Soc.* 1926–27, 283; here and there on the river bank from Chertsey to Penton Hook Lock; very fine and abundant below Penton Hook Lock, *Richards Ec.*; 1947, *Peterken*, *L.N.* 40: 59.

B. trifaria (Hedw.) Mitt.
B. lurida (Hornsch.) Lindb.
Native, or colonist, on stones in woods, fields, etc., established on walls, etc. Very rare, and perhaps extinct.
1. Harefield, *Benbow*, *J.B.* (*London*) 32: 106. Harrow Weald Common, 1907, *Sherrin*.

B. tophacea (Brid.) Mitt.
Colonist on damp calcareous ground, damp walls, etc. Very rare, or overlooked.
1. Uxbridge; Harefield; Garett Wood,† *Benbow, J.B. (London)* 32: 106. **2.** Gravel pit, Shepperton, 1967, *Townsend.* **7.** Buckingham Palace grounds, *Fl. B.P.G.*

B. cylindrica (Tayl.) Schimp.
Formerly native, or a colonist, on banks, etc., on calcareous soils, tree bases and roots by water, established on walls, etc., but now extinct.
1. Uxbridge; Harefield, *Benbow, J.B. (London)* 32: 270. **4.** Harrow Weald, 1907, *Sherrin.*

B. vinealis Brid.
Colonist on walls, etc. Very rare, or overlooked.
1. Uxbridge; Harefield, *Benbow, J.B. (London)* 32: 370. Near Pinner Hill, *Richards.* **2.** River wall between Shepperton and Penton Hook Lock, *Richards Ec.* **3.** Near Hatch End, *Richards.* **4.** Colindeep Lane, Hendon, 1923,† *Richards.*

B. recurvirostra (Hedw.) Dix.
B. rubella (Hoffm.) Lindb.
Colonist on banks, tree bases, etc., particularly on calcareous soils. Very rare, or overlooked.
1. Near Harefield; South Mimms, *Richards Ec.* Ruislip Nature Reserve, *Moxey Br.* **2.** Shepperton, rare, *Richards Ec.* **4.** Harrow, *Hind, T. & D.*, 402. Hampstead, *Richards.* **6.** Edmonton, *De Cresp. Fl.*, 85. **7.** Tottenham, *De Cresp. Fl.*, 85.

GYROWEISIA Schimp.

G. tenuis (Hedw.) Schimp.
Formerly a colonist on canal walls, but now extinct.
1. Canal wall near Harefield Moor, *Benbow, J.B. (London)* 32: 106.

TRICHOSTOMUM Bruch

T. sinuosum (Mitt.) Lindb. ex Herzog
Barbula sinuosa Wils.
Native, or colonist, on tree roots, etc., on calcareous soils, and by the Thames; established on the river-wall of the Thames. Local.
1. Garett Wood, Springwell !,† *L.N.* 40: 59. **2.** Thames bank below Penton Hook Lock, *Richards Ec.* Thames bank near Sunbury, and at Laleham, 1965. Stone embankment of river Thames near Hampton Court; between Kingston and Hampton Court Bridges, 1967, *J. C. Gardiner, Trans. Brit. Bryol. Soc.* 5: 636.

WEISSIA Hedw.

W. controversa Hedw.
W. viridula Hedw.
Native, or colonist, on banks, cultivated ground, in fields, etc. Very rare, and mostly confined to calcareous and sandy soils.
1. Near Harefield, *Richards*. **4.** Hampstead, *Syme*, *T. & D.*, 401. **6.** Near Edmonton, 1879, *De Crespigny* (MANCH). **7.** Near Highgate, *R. Braithwaite*, *T. & D.*, 401. Sandy soil in pots in greenhouses, Buckingham Palace grounds, *Fl. B.P.G.*

W. microstoma (Hedw.) C. Müll.
Formerly native, or a colonist, on bare ground on sandy soils, but now extinct.
1. Near Ruislip Reservoir, *Benbow*, *J.B.* (*London*) 32: 106.

W. crispa (Hedw.) Mitt.
Native, or colonist, in grassy places on calcareous soils, chalkpits, etc. Very rare.
1. Harefield, *Benbow*, *J.B.* (*London*) 32: 106. Old chalkpit, Springwell, 1949, *Boniface*.

LEPTODONTIUM Hampe

L. flexifolium (Sm.) Hampe
Formerly native in woods on peaty soil, but now extinct.
1. Woods east of Ruislip Reservoir, *Benbow*, *J.B.* (*London*) 32: 369.

GRIMMIACEAE

GRIMMIA Hedw.

G. apocarpa Hedw.
Colonist on stones, etc., on calcareous soils, established on walls, etc. Very rare.
1. On flints in old chalkpit, Harefield, 1949, *Boniface*. **6.** Walls of Colney Hatch Asylum,† *De Cresp. Fl.* 86.

G. pulvinata (Hedw.) Sm.
Colonist on walls and stonework. Formerly local, but now very rare or overlooked.
1. Harrow Weald Common, *Hind*, *T. & D.*, 402. South Mimms; near Ruislip; near Harefield, *Richards Ec.* Pinner, *Melv. Fl.* 2: 132. Brickwork of sluice, Ruislip Nature Reserve, *Moxey Br.* **2.** Shepperton, very rare, *Richards Ec.* Sunbury, 1965. **4.** Hendon, 1922; Stanmore, 1925, *Richards*. Harrow Weald, *Melv. Fl.* 2: 132. **6.** Tottenham, *De Cresp. Fl.*, 86.

RHACOMITRIUM Brid.

R. canescens (Hedw.) Brid.

Formerly native on heaths, etc., mainly on sandy soils, but now extinct.

1. Uxbridge Common, *Benbow*, *J.B.* (*London*) 32: 369. **4.** Hampstead, *Dillenius*, *With. Bot. Arr.* 2: 95. **6.** Enfield Chase, *Dillenius*, *With. Bot. Arr.* 2: 95.

R. lanuginosum (Hedw.) Brid.

Formerly native on heaths, but now extinct.

4. Hampstead Heath, *Burn. Med. Bot.*, 105.

FUNARIACEAE

FUNARIA Hedw.

F. hygrometrica Hedw.

Colonist on disturbed or burnt ground, heaths, cinder paths, etc. Very common.

7. Abundant in the district.

F. attenuata (Dicks.) Lindb. Cited for Middlesex in the British Bryological Society's *Census Catalogue*, edition 1. I have been unable to trace the source of the record, and the species is an unlikely one to have occurred in the vice-county.

PHYSCOMITRIUM Brid.

P. pyriforme (Hedw.) Brid.

Colonist on bare soil in wet places. Very rare, or overlooked.

2. Thames bank, Penton Hook Lock, 1952, *Boniface*. **4.** Hampstead Heath Extension, rare, *Richards Ec.* **5.** By the Thames near Chiswick Bridge, *Boniface*. **7.** Near Highgate, *R. Braithwaite*, *T. & D.*, 403. Frequent by the Lee [Navigation] Canal, *De Cresp. Fl.*, 92. East Heath, Hampstead, *Richards Ec.* Marshy edge of lake, Buckingham Palace grounds, *Rose*, *Fl. B.P.G.*

PHYSCOMITRELLA Bruch, Schimp. & Guemb.

P. patens (Hedw.) Bruch, Schimp. & Guemb.

Colonist on wet mud, etc. Very rare, or overlooked.

1. Uxbridge, 1904, *Sherrin*. **2.** Staines, 1926, *Richards* and *Catcheside*, *Rep. Brit. Bryol. Soc.* 1926–27, 286. Mud by Penton Hook backwater, *Richards Ec.* **4.** Near Edgwarebury, *Richards Ec.*

EPHEMERACEAE

EPHEMERUM Hampe

E. serratum (Hedw.) Hampe
Formerly a colonist on bare ground, etc., but now extinct.
4. Wembley Park, 1919, *Blundell*, comm. *Sutton*, *Rep. Brit. Bryol. Soc.* 1925, 146. **6.** Muswell Hill, *Dickson*, *E.B.*, 460.

TETRAPHIDACEAE

TETRAPHIS Hedw.

T. pellucida Hedw.
Colonist on tree stumps, peaty banks, etc., on heaths, in woods, etc. Formerly local, but now very rare.
1. Ruislip Woods, locally abundant!; near Batchworth Heath; Breakspeares, *Richards Ec.* Harrow Weald Common; Garett Wood, 1955,† *Boniface*. **2.** Staines, *R. Braithwaite*, *T. & D.*, 402. **4.** Bishop's Wood; Turner's Wood, *Andrewes*, *L.N.* 40: 61. Wood Lane, Brockley Hill,† *Richards Ec.* **5.** Osterley Park, rare, 1949, *Boniface*. Perivale Wood, *P. J. Edwards*. **7.** Ken Wood, *Mackay*, *T. & D.*, 402.

BRYACEAE

ORTHODONTIUM Schwaegr.

O. lineare Schwaegr.
O. gracile (Wils.) Schwaegr. var. *heterocarpum* W. Watson
Colonist on tree stumps, banks, etc., in woods. Very rare, but increasing, and confined to acid soils.
1. Harrow Weald Common, 1951, *Boniface*; *Rose*, *Trans. Brit. Bryol. Soc.* 3: 778. Bayhurst Wood, Ruislip, 1952, *Boniface*. Mimmshall Wood, 1954. **7.** Ken Wood, 1949, *P. Bell*.

LEPTOBRYUM (Bruch, Schimp. & Guemb.) Wils.

L. pyriforme (Hedw.) Wils.
Colonist in ditches, on banks, cinder-heaps, etc. Very rare, and sporadic.
2. Rubbish-heap, Poyle, abundant, 1923, *Richards Ec.* Gravel pit, Shepperton, 1967, *Townsend*. **4.** Highgate golf course, 1916, *Andrewes*, *L.N.* 40: 61. Hendon, 1926, *Richards Ec.* **5.** East Acton,

1910, *Knightley* and *Sherrin.* **7.** Hammersmith, 1790, *E.B.*, 389. Clapton, *E. Forster*, *B.G.*, 413. Pots in greenhouses, Buckingham Palace grounds, *Rose*, *Fl. B.P.G.*

POHLIA Hedw.

P. nutans (Hedw.) Lindb.
Webera nutans Hedw.
Native in woods, on heaths, banks, etc. Formerly locally common on acid soils, but now rare or overlooked.
1. Pinner, *Hind*, *T. & D.*, 402. Ruislip Woods; Stanmore Common, very abundant; near South Mimms, *Richards Ec.* Harrow Weald Common; near Potters Bar, 1923; Mimmshall Wood, 1925; Northwood, *Richards.* **2.** Fulwell Park, 1969, *Townsend.* **3.** Hounslow Heath, *Richards Ec.* **4.** Hampstead Heath, *R. Braithwaite*, *T. & D.*, 402; North-west Heath, Hampstead; Bishop's Wood; Highgate golf course, *Andrewes MSS.* Wood Lane, Brockley Hill,† *Richards Ec.* **5.** Perivale Wood, *P. J. Edwards* and *A. R. Hall.* **6.** Queen's Wood, Highgate, *Richards Ec.* **7.** East Heath, Hampstead, *Richards Ec.* Buckingham Palace grounds, *Rose*, *Fl. B.P.G.* Kensington Gardens, 1967, *J. C.* and *M. W. Gardiner.*

Var. **longiseta** Bruch, Schimp. & Guemb.
1. Harrow Weald Common, 1925, *Richards.* **7.** East Heath, Hampstead, 1948, *Boniface.*

P. annotina (Hedw.) Loeske
Webera annotina (Hedw.) Bruch
Colonist on damp acid soils. Very rare, or overlooked.
7. Buckingham Palace grounds, 1962, *Rose*, det. *Warburg*, *Fl. B.P.G.*

P. proligera (Lindb.) Limpr.
Webera proligera (Lindb.) Kindb.
Perhaps formerly a colonist on damp sandy ground, but extinct if ever it occurred.
1. Banks by pond, Little Common, Stanmore, *Richards Ec.* Possibly an error; *P. annotina* may have been seen, *Richards.*

P. wahlenbergii (Web. & Mohr) Andr.
P. albicans (Wahlenb.) Lindb., *Webera albicans* (Wahlenb.) Schimp.
Colonist on banks, etc., on acid soils, on stonework by water, etc. Very rare, and perhaps extinct.
1. Canal walls, Uxbridge and Harefield, *Benbow*, *J.B.* (*London*) 32: 106. Stanmore Common, *De Cresp. Fl.*, 94, South Mimms, 1923, *Richards.* **4.** Hampstead Heath, *De Cresp. Fl.*, 94.

P. delicatula (Hedw.) Grout
Webera carnea Schimp.

Colonist on banks, etc., particularly by water. Very rare, or overlooked. **1.** Margin of Swakeleys lake,† *Benbow, J.B. (London)* 32: 107. New Years Green; Breakspeares, *Richards Ec.* **2.** Banks of the Thames near Shepperton, occasional; bank by Penton Hook backwater, *Richards Ec.* **4.** Hendon; Edgwarebury, *Richards Ec.* Kingsbury Green, 1923,† *Richards.* **7.** East Heath, Hampstead, *Richards*; 1948, *Boniface.* Buckingham Palace grounds, *Fl. B.P.G.*

BRYUM Hedw.

B. pendulum (Hornsch.) Schimp.
Formerly native in wet places on sandy soils, but now extinct.
1. Northwood, *Benbow, J.B. (London)* 32: 370.

B. inclinatum (Brid.) Bland.
Formerly native in damp places on sandy soils, but now extinct.
1. Uxbridge; Ickenham; Northwood, *Benbow, J.B. (London)* 32: 106. **7.** Highgate, 1923, *Sutton, Rep. Brit. Bryol. Soc.* 1924, 74.

B. pallens Sw.
Native in wet heathy places, etc. Very rare, and perhaps extinct.
1. Ruislip; Northwood, *Benbow, J.B. (London)* 32: 106. Ruislip Woods; Harefield; Stanmore Common, *Richards Ec.* Harrow Weald Common, 1879, *De Crespigny* (MANCH). **4.** Hampstead, 1927, *Richards.*

B. turbinatum (Hedw.) Turn. **1.** Canal between Denham and Harefield Moor Locks, *Benbow, J.B. (London)* 32: 370. Probably an error, fide *Warburg.*

B. pseudotriquetrum (Hedw.) Schwaegr. var. **bimum** (Brid.) Lilj.
Native in marshes, by water, etc. Very rare.
1. Uxbridge; Ruislip Reservoir, *Benbow, J.B. (London)* 32: 106. Ruislip Woods, *Peterken, L.N.* 40: 62; 1949, *Boniface.*

B. creberrimum Tayl.
B. affine (Bruch) Schultz
Formerly established on canal walls, but now extinct.
1. Canal wall between Denham and Harefield Moor Locks, *Benbow, J.B. (London)* 34: 400.

B. intermedium (Brid.) Bland.
Colonist on bare ground, etc., on sandy and gravelly soils. Very rare, or overlooked.
2. Gravel pit between Chertsey and Shepperton, 1967; gravel pits, Ashford, 1967, *Townsend.*

B. caespiticium Hedw.
Native, or colonist, on banks, bare ground, etc., established on wall-tops, etc. Common.

7. Hampstead!, *Richards*; walls, Hampstead Lane. Hampstead Churchyard. Primrose Hill!, abundant, *Richards*. Regents Park. Camden Town, Ken Wood. Highgate. Hammersmith. Kensington Gardens, *J. C.* and *M. W. Gardiner*. Euston Square. Bombed sites, Cripplegate, E.C., abundant, *Castell*.

B. argenteum Hedw.
Colonist on bare ground, walls, banks, paths, etc. Very common.
7. Abundant in the district.

Var. **lanatum** (P. Beauv.) Hampe
1. Uxbridge, frequent, *Benbow*, *J.B.* (*London*) 32: 307. Near Ruislip, *Richards Ec.* Canal side near Harefield Moor, 1922, *Richards*. **4.** Golders Green; Finchley, 1922, *Richards*.

B. bicolor Dicks.
B. atropurpureum Web. & Mohr
Colonist on bare ground, banks, heaths, wall-tops, etc. Common.
7. East Heath, Hampstead, *Syme*, *T. & D.*, 402. Bombed site, Moorgate, E.C., 1948, *Castell*, *L.N.* 40: 62. Buckingham Palace grounds, *Fl. B.P.G.*

Var. **gracilentum** Tayl. ex Braithw.
2. Gravel pit, Shepperton, *Townsend*.

B. donianum Grev. **1.** Sandy bank in lane by Highway Farm, Harefield (barren), *Benbow*, *J.B.* (*London*) 32: 307. Probably an error, fide *Warburg*.

B. radiculosum Brid.
Colonist on limestone steps. Very rare.
7. On steps (limestone) of a house in Gordon Square, [W.C.1.], 1966, *M. O. Hill*, *Rep. Brit. Bryol. Soc.* 5: 639.

B. ruderale Crundw. & Nyh.
Colonist on paths, etc. Frequency not yet known.
2. Sandy gravel just inside the main gates, Queen Mary Reservoir, Littleton, 1971, *Townsend*. **7.** Garden path, Highgate, 1966, *M. O. Hill*, *Trans. Brit. Bryol. Soc.* 5: 640.

B. violaceum Crundw. & Nyh.
Colonist on cultivated ground, etc. Frequency not yet known.
7. Raspberry bed, garden, Highgate, 1966, *M. O. Hill*, *Trans. Brit. Bryol. Soc.* 5: 640.

B. rubens Mitt.
Colonist on bare sandy or gravelly soils. Frequency not yet known.
2. Among loose gravel by a working gravel pit by the Sheepwalk, between Shepperton and Chertsey, 1967, *Townsend*, *Trans. Brit.*

Bryol. Soc. 5: 640. Gravel pits, Ashford, 1964, *Townsend.* **3.** Garden ground, Twickenham, 1969, *Townsend.*

B. capillare Hedw.
Colonist on tree trunks, banks, walls, etc. Common.
7. East Heath, Hampstead. Buckingham Palace grounds, *Rose, Fl. B.P.G.*

MNIACEAE

MNIUM Hedw.

M. hornum Hedw.
Native on banks, tree stumps, etc., in woods. Common on acid soils.
7. Hampstead!, *Curt. F.L.* East Heath, Hampstead. Ken Wood!, *Boniface.* Holland House grounds, Kensington.

M. longirostrum Brid.
M. rostratum (Sm.) Röhl.
Native in woods, etc. Very rare, and mainly confined to calcareous soils.
1. Near Harefield, 1923, *Richards*; 1955, *Boniface.* Near South Mimms, *Richards Ec.* **4.** Harrow Weald, *T. & D.*, 403. **5.** Perivale Wood, *P. J. Edwards* and *A. R. Hall.*

M. affine Bland.
Native in woods, on heaths, etc. Very rare, or overlooked.
1. Harrow Weald Common; Stanmore Common, *Richards Ec.* Harefield, *Sherrin.* **3.** Hounslow Heath, *Richards Ec.* **5.** Perivale Wood, *P. J. Edwards* and *A. R. Hall.*

M. rugicum Laur.
M. affine Bland. var. *rugicum* (Laur.) Bruch, Schimp. & Guemb.
Formerly native in wet heathy places, but now extinct.
1. Harrow Weald Common, 1907, *Sherrin.*

M. undulatum Hedw.
Native in most woods, on banks, etc. Formerly locally common, but now very rare, or overlooked.
1. Pinner Hill, *Hind, T. & D.*, 403. Near South Mimms, rare; near Batchworth Heath; near Harefield, *Richards Ec.* Northwood; Park Wood, Ruislip, 1927, *Richards.* Stanmore Common, 1930, *C. A. Cooper* and *Binstead, Trans. Herts. N.H.S.* 19: 134; 1949, *Boniface.* **3.** Hatch End, *Richards.* **4.** Harrow, *Hind*; Hendon, *W. Davies, T. & D.*, 403. Kingsbury, *De Cresp. Fl.*, 91. Hampstead Woods, very rare; near Edgwarebury, *Richards Ec.* **5.** Osterley Park, *Boniface.* Perivale Wood, *T. G. Collett!*, teste *Swinscow.* **7.** Ken Wood, 1922, *Richards.*

M. punctatum Hedw.
Native in damp woods, etc. Very rare, and perhaps extinct.
1. Margins of water courses in Mad Bess Wood, Ruislip, *Benbow*, *J.B. (London)* 37: 441. **4.** Hampstead Heath, *De Cresp. Fl.*, 91. **6.** Queen's Wood, Highgate, 1916, *Andrewes MSS.*

AULACOMNIACEAE

AULACOMNIUM Schwaegr.

A. palustre (Hedw.) Schwaegr.
Native in wet heathy places. Very rare, or overlooked.
1. Stanmore Common!; Harrow Weald Common!, *De Cresp. Fl.*, 84. Ruislip Common, *Richards Ec.* Poor's Field, Ruislip, *Boniface.* **4.** Hampstead Heath!, *Huds. Fl. Angl.*, 403; West Heath, Hampstead, *Syme, T. & D.*, 402; 1916, *Andrewes MSS.*; 1949, *Boniface*; 1962. **7.** West Meadow Bog, Ken Wood, *Bangerter* and *H. C. Harris*!

A. androgynum (Hedw.) Schwaegr.
Native, or colonist, on peaty banks, rotten tree stumps, etc., in woods, on heaths, etc. Formerly locally plentiful, but now apparently rare or overlooked.
1. Ruislip Common!; Stanmore Common!; Harrow Weald Common, locally abundant; near Harefield, *Richards Ec.* Near Springwell; Northwood; Mimmshall Wood, *Richards.* Old Park Wood, Harefield! *L.N.* 40: 63. Garett Wood, Springwell, 1955,† *Boniface.* **2.** Bushy Park!, *Boniface.* **4.** Hampstead Heath, *Huds. Fl. Angl.*, 403. Near Hampstead, *R. Braithwaite, T. & D.*, 402. Colindeep Lane, Hendon,† *Richards Ec.* **5.** Osterley Park, *Boniface.* Perivale Wood!, *Lillford* and *Ing, Groves Veg.* **7.** Between Tottenham and Walthamstow,† *De Cresp. Fl.*, 84. Ken Wood, *Boniface.*

BARTRAMIACEAE

BARTRAMIA Hedw.

B. pomiformis Hedw.
Native on banks, etc. Very rare, and mainly confined to acid soils.
1. Between Boreham Wood station and Stanmore Common, *De Cresp. Fl.*, 84; [= Wood Lane, Brockley Hill], 1925,† *Richards.* Near Harefield, *Richards Ec.* Springwell, 1950, *Boniface.* **4.** Hampstead Heath,† *Dillenius, Huds. Fl. Angl.*, 404. Highgate golf course, 1915, *Andrewes, L.N.* 40: 63.

PHILONOTIS Brid.

P. fontana (Hedw.) Brid.
Native in bogs, wet heathy places, etc. Very rare, and perhaps extinct.
1. Ruislip Common, 1927, *Richards Ec.* **4.** Bog on West Heath, Hampstead, *W. Davies, T. & D.*, 403; c. 1877,† *De Cresp. Fl.*, 84.

ORTHOTRICHACEAE

ZYGODON Hook. & Tayl.

Z. viridissimus (Dicks.) R.Br.
Colonist on tree stumps, walls, etc. Very rare, or overlooked.
1. Uxbridge; Harefield!, *Benbow, J.B. (London)* 32: 106. Near Mimms Hall, 1927, *Richards.* Ruislip Woods, *Benbow, J.B. (London)* 32: 369.

ORTHOTRICHUM Hedw.

O. anomalum Hedw.
Colonist on the river-wall of the Thames. Very rare, or overlooked.
2. Stone embankment of river Thames between Kingston and Hampton Court Bridges, 1967, *J. C. Gardiner, Trans. Brit. Bryol. Soc.* 5: 642.

O. affine Brid.
Formerly a colonist on tree trunks, but now extinct.
1. Pinner, *Hind, T. & D.*, 402. **4.** Harrow Weald; Kenton, *De Cresp. Fl.*, 92.

O. lyellii Hook. & Tayl.
Colonist on tree trunks. Very rare, and perhaps extinct.
1. Park Wood, Swakeleys; Uxbridge, *Benbow, J.B. (London)* 32: 369. Chalkpit near South Mimms, very sparingly, 1925, *Richards.*

O. diaphanum Brid.
Colonist on tree trunks, stones, etc., often near water. Very rare.
1. Near Harefield, *Richards Ec.* New Years Green, 1922; Springwell, 1927, *Richards.* **2.** On alders and willows by the Thames between Shepperton and Penton Hook, occasional; walls about Shepperton, locally abundant, *Richards Ec.*; 1952, *Boniface.* **4.** Near Hendon, 1879, *De Crespigny* (MANCH).

ULOTA Brid.

U. crispa (Hedw.) Brid.
Formerly a colonist on tree trunks, but now extinct.
1. Pinner, *Hind, T. & D.*, 402.

FONTINALACEAE

FONTINALIS Hedw.

F. antipyretica Hedw.
Native attached to stones and tree roots in rivers, streams and pools, etc. Very rare, or overlooked.
1. Brooks in woods near Ruislip!, *Richards Ec.* **2.** Between Shepperton and Penton Hook!, abundant, *Richards Ec.* Pool by Walton Bridge, abundant, 1952!, *Boniface.* **3.** Twickenham, *T. & D.*, 404. **4.** Harrow,† *Hind, T. & D.*, 404. Harrow Weald,† *De Cresp. Fl.*, 86. **5.** Lake, Gunnersbury Park, Brentford, 1968.

Var. **cymbifolia** Nicholson
2. Thames between Shepperton and Penton Hook Lock, abundant, *Richards Ec.*

F. squamosa Hedw. Cited for Middlesex, almost certainly in error, in the British Bryological Society's *Census Catalogue*, edition 1. I have been unable to trace the source of the record.

CLIMACIACEAE

CLIMACIUM Web. & Mohr

C. dendroides (Hedw.) Web. & Mohr
Native in bogs, marshes, etc. Very rare, and perhaps extinct.
1. Uxbridge; Ruislip Reservoir; Swakeleys, *Benbow, J.B. (London)* 32: 106. Stanmore Common, *Richards Ec.*

CRYPHAEACEAE

CRYPHAEA Mohr

C. heteromalla (Hedw.) Mohr
Formerly a colonist on tree trunks, but now extinct.
4. Kenton Lane, near Harrow Weald, *De Cresp. Fl.*, 85. **6.** Enfield Chase, *Dillenius, Huds. Fl. Angl.*, 396; c. 1795, *Dicks. Hort. Sicc.*

LEUCODONTACEAE

LEUCODON Schwaegr.

L. sciuroides (Hedw.) Schwaegr.
Formerly a colonist on tree trunks, but now extinct.

1. Harrow Weald Common, *De Cresp. Fl.*, 91. Pinner, *Melv. Fl.* 2: 134. **4.** Harrow, *Hind, T. & D.*, 403. Kenton Lane, near Harrow Weald,† *De Cresp. Fl.*, 91.

PTEROGONIUM Sm.

P. gracile (Hedw.) Sm.
Formerly a colonist on tree trunks, but now long extinct.
4. Hampstead, *Coop. Fl.*, 103. **6.** On beeches in Enfield Chase, abundantly, *Huds. Fl. Angl.*, 430.

ANTITRICHIA Brid.

A. curtipendula (Hedw.) Brid.
Formerly a colonist on tree trunks, but now long extinct.
6. On stumps in Enfield Forest, near Southgate, *With. Bot. Arr.* 2: 134.

NECKERACEAE

NECKERA Hedw.

N. crispa Hedw. **1.** Ruislip Nature Reserve, *Moxey Br.*
4. Hampstead Heath, *Burn. Med. Bot.*, 105 and *Coop. Fl.*, 103. Almost certainly errors, teste *Warburg.*

N. complanata (Hedw.) Hüben.
Native on banks, tree trunks in woods, on calcareous soils, established on walls, etc. Formerly local, but now very rare, or overlooked.
1. Near South Mimms; near Harefield, *Richards Ec.* Harefield; Garett Wood, Springwell, 1950,† *Boniface.* **4.** Hampstead,† *Dillenius, Huds. Fl. Angl.*, 419. Bishop's Wood,† *Coop. Fl.*, 103. Harrow,† *Hind, T. & D.*, 404. Harrow Weald,† *De Cresp. Fl.*, 91. Hendon;† near Edgwarebury, *Richards Ec.*

OMALIA (Brid.) Bruch, Schimp. & Guemb.

O. trichomanoides (Hedw.) Bruch, Schimp. & Guemb.
Colonist on tree trunks, banks, walls, etc. Very rare, or overlooked.
1. Hillingdon, *R. Braithwaite, T. & D.*, 404. Near Batchworth Heath, *Richards Ec.* Harefield, 1964. **4.** Near Kingsbury, 1874, *De Crespigny* (MANCH). **7.** Tottenham, *De Cresp. Fl.*, 86.

THAMNIUM Bruch, Schimp. & Guemb.

T. alopecurum (Hedw.) Bruch, Schimp. & Guemb.
Porotrichum alopecurum (Hedw.) Mitt.

Native on stones and tree trunks in shady woods, etc. Very rare.
1. Garett Wood, abundant, 1955,† *Boniface.* Habitat destroyed by chalk-quarrying operations. **4.** Bishop's Wood, *Coop. Fl.*, 103. Harrow, *Hind, T. & D.*, 403.

HOOKERIACEAE

HOOKERIA Sm.

H. lucens (Hedw.) Sm.
Pterygophyllum lucens (Hedw.) Brid.
Formerly native in moist woods on sandy soils, but now long extinct.
4. Bishop's Wood, *Coop. Fl.*, 103.

LESKEACEAE

LESKEA Hedw.

L. polycarpa Hedw.
Native on bases and roots of trees by water. Very rare, or overlooked.
1. Ickenham, *A. B. Jackson* and *Sherrin* (SLBI). **2.** On alders and willows by the Thames between Shepperton and Penton Hook Lock, occasional, *Richards Ec.*; 1955, *Boniface.* **4.** Harrow Weald, *Hind, T. & D.*, 403. On trees by the Brent, Stonebridge,† *De Cresp. Fl.*, 90. **5.** On trunks of willows, Chiswick, *Teesdale, E.B.*; 1922, *Richards*; 1952, *Boniface.* Willow tree roots, Brentford, 1879, *De Crespigny* (MANCH).

THUIDACEAE

ANOMODON Hook. & Tayl.

A. viticulosus (Hedw.) Hook. & Tayl.
Native on shady banks, bases and roots of trees, etc. Very rare, and mainly confined to calcareous soils.
1. Ickenham; Harefield; Springwell, *Benbow, J.B. (London)* 32: 106; Garett Wood, Springwell, 1955,† *Boniface.* Near South Mimms, *Richards Ec.* **4.** Near Kingsbury,† *De Cresp. Fl.*, 83.

THUIDIUM Bruch, Schimp. & Guemb.

T. tamariscinum (Hedw.) Bruch, Schimp. & Guemb.
Native in woods, on banks, etc. Formerly rather common, but now rare, or overlooked.
1. Pinner, *Hind, T. & D.*, 404. Stanmore Common!, *De Cresp. Fl.*, 94.

Ruislip Woods, locally frequent!; Harrow Weald Common!; Watt's Common, Harefield; near Harefield; Ruislip Common, *Richards Ec.* Old chalkpit, Harefield, *Peterken, L.N.* 40: 66. Mimmshall Wood!, *Richards.* **4.** Bishop's Wood, *T. & D.*, 404. Highgate, 1916, *Andrewes MSS.* **5.** Perivale Wood, *Lillford* and *Ing*, *Groves Veg.*; 1965, but not seen since, *P. J. Edwards* and *A. R. Hall.*

T. philibertii Limpr.
Native on grassy slopes, etc. Very rare, and confined to calcareous soils.
1. Harefield, 1907, *Sherrin.* Springwell!, *Richards*; 1949!, *Boniface*; 1963. Copper Mill Down, between Harefield and Springwell, 1967, *Coker*, fide *Stott, L.N.* 47: 14.

T. recognitum (Hedw.) Lindb. **1.** Chalk downs, Harefield, *Benbow, J.B. (London)* 32: 369. Error: *T. philibertii* intended.

HYPNACEAE

CRATONEURON (Sull.) Spruce

C. filicinum (Hedw.) Spruce
Amblystegium filicinum (Hedw.) De Not.
Native on banks and stones by water. Formerly local, but now very rare, or overlooked.
1. Pield Heath;† canal walls and banks, Uxbridge! and Harefield, *Benbow, J.B. (London)* 32: 107; c. 1927; Ruislip Common, *Richards Ec.* **2.** Banks of the Thames near Shepperton, abundant, *Richards Ec.* **4.** Near Kingsbury, *De Cresp. Fl.*, 88. **7.** Regents Park, 1921, *Richards.*

CAMPYLIUM (Sull.) Mitt.

C. protensum (Brid.) Kindb.
Hypnum stellatum Brid. var. *protensum* (Brid.) Röhl.
Native in marshes, wet heathy places, etc. Very rare, and perhaps extinct.
1. Stanmore Common; Ruislip Common, *Richards Ec.*

C. chrysophyllum (Brid.) J. Lange
Hypnum chrysophyllum Brid.
Native in grassy places, etc. Very rare, and confined to calcareous soils.
1. Harefield, *Benbow, J.B. (London)* 32: 107; 1907, *Sherrin*; c. 1927, *Richards Ec.* Springwell, *Benbow, J.B. (London)* 32: 107; plentiful in an old chalkpit, 1950, *Boniface.* **4.** Hampstead, *Bagnall, T. & D.*, 403. An error, the record refers to Hamstead, Warwickshire, *Benbow, J.B. (London)* 32: 107.

C. calcareum Crundw. & Nyh.
C. sommerfeltii (Myr.) Bryhn., *Hypnum hispidulum* Brid. var. *sommerfeltii* (Myr.) Dix.
Formerly native on tree roots, stones, etc., in woods on calcareous soils, but now extinct.
1. Harefield; Springwell, *Benbow*, *J.B.* (*London*) 32: 107.

LEPTODICTYUM (Schimp.) Warnst.

L. riparium (Hedw.) Warnst.
Native, or colonist, on tree roots, stones, bare ground, etc., usually near water. Formerly locally common, but now rare, or overlooked.
1. Pinner, *Hind, T. & D.*, 403. Gravel pits, Watt's Common, Harefield, *Richards Ec.* **2.** River wall between Shepperton and Penton Hook Lock, frequent!, *Richards Ec.*; 1952, *Boniface*; c. 1960, *Peterken, L.N.* 40: 66; 1965. **4.** Wall round Leg of Mutton Pond, West Heath, Hampstead, *Richards Ec.* Near Monk's Mere, Hendon, 1921;† Edgwarebury, 1923, *Richards.* **5.** Southall, 1907; Acton, 1909,† *Wroughton*, teste *Richards.* **7.** Plentiful on the banks of the Lee [Navigation] Canal, *De Cresp. Fl.*, 87; 1923, *Richards*; Hackney, c. 1960, *Peterken*; Regents Park, 1920, *Pegler*, teste *Sherrin, L.N.* 40: 66. Marshy edge of lake, Buckingham Palace grounds, *Rose, Fl. B.P.G.*

HYGROAMBLYSTEGIUM Loeske

H. tenax (Hedw.) Jenn.
Amblystegium irriguum (Wils.) Bruch, Schimp. & Guemb.
Formerly native on stones, etc., by water, but now extinct.
1. Tumbling Bay, near Harefield; canal walls near Ruislip Reservoir, *Benbow, J.B.* (*London*) 32: 370.

AMBLYSTEGIUM Bruch, Schimp. & Guemb.

A. serpens (Hedw.) Bruch, Schimp. & Guemb.
Native, or colonist, on bare ground, stones, rotten wood, etc. Formerly common, but now very rare, or overlooked.
1. Near Batchworth Heath; near South Mimms, frequent; South Mimms, *Richards Ec.* Ruislip Nature Reserve, *Moxey Br.* **2.** Bank by the Thames near Shepperton, occasional, *Richards Ec.* **4.** Harrow, *Hind, T. & D.*, 403. Highgate golf course, *Andrewes MSS.* Hampstead Woods, occasional; near Edgwarebury, occasional, *Richards Ec.* Bank on Hampstead Heath Extension, frequent, *Richards.* **7.** On stones by the canal about Tottenham; tombstones in Kensal Green Cemetery, *De Cresp. Fl.*, 87.

A. juratzkanum Schimp.
Formerly native, or colonist, on stones, tree roots among grass, etc., by water, but now extinct.
1. Harefield, 1907, *Sherrin* (SLBI). **4.** Stanmore, 1907, *Loydell* (SLBI).

A. varium (Hedw.) Lindb.
Native or colonist, on stones, tree roots, etc., by water. Very rare, and perhaps extinct.
1. By Swakeleys lake, Ickenham,† *Benbow, J.B. (London)* 32: 307. **2.** Bank by Penton Hook backwater, very rare, *Richards Ec.* **4.** Mill Hill, *Blundell*, teste *Sherrin, L.N.* 40: 67.

DREPANOCLADUS (C. Müll.) Roth

D. aduncus (Hedw.) Warnst.
Hypnum aduncum Hedw.
Native in marshes, ditches, on pond and lake verges, etc. Very rare, or overlooked.
1. Ruislip Reservoir; Fray's Meadows, Uxbridge; Harefield Moor; Northwood, *Benbow, J.B. (London)* 37: 441. **5.** Abundant in stream, Perivale Wood, 1968, *P. J. Edwards* and *A. R. Hall.* Floor of disused reservoir, Ealing, 1969, *P. J. Edwards*, det. *Townsend.*

D. fluitans (Hedw.) Warnst.
Hypnum fluitans Hedw.
Native in bogs, shallow pools on acid heaths, ditches, etc. Very rare.
1. Stanmore Common; Harrow Weald Common, *Richards, L.N.* 40: 67. Pinner, *Melv. Fl.* 2, 136. **4.** West Heath, Hampstead, *W. Davies* and *R. Braithwaite, T. & D.*, 404; c. 1877, *De Cresp. Fl.*, 88; c. 1927, *Richards Ec.*; 1950, *Boniface.* Harrow, *Hind, T. & D.*, 404. **5.** Alperton, *Melv. Fl.* 2: 136. **7.** Ditches near Hackney, *With. Bot. Arr.* 2: 120. Ditches by the railway above Lee Bridge station, *De Cresp. Fl.*, 88.

D. exannulatus (Bruch, Schimp. & Guemb.) Warnst.
Hypnum exannulatum Bruch, Schimp. & Guemb.
Native in bogs, shallow pools on acid heaths, etc. Very rare, and perhaps extinct.
1. Northwood; Ruislip Common and Reservoir, *Benbow, J.B. (London)* 32: 107. Harrow Weald Common, *De Cresp. Fl.*, 88; c. 1927; Stanmore Common, *Richards Ec.*

D. uncinatus (Hedw.) Warnst., *Hypnum uncinatum* Hedw. **4.** Harrow Weald, *Hind, T. & D.*, 404. Almost certainly an error, fide *Warburg.*

HYGROHYPNUM Lindb.

H. luridum (Hedw.) Jenn.
Hypnum palustre Huds.
Formerly native on stones, wood, etc., by water, but now extinct.
4. Hampstead Heath, *Burn. Med. Bot.*, 105 and *Coop. Fl.*, 103. **6.** Stones by Lee [Navigation] Canal, 1879, *De Crespigny* (MANCH).

ACROCLADIUM Mitt.

A. stramineum (Brid.) Rich. & Wall.
Hypnum stramineum Brid.
Native in bogs. Very rare, and perhaps extinct.
4. West side of Hampstead, *Dickson*, *B.G.*, 405. Bog, West Heath, Hampstead, sparingly, *Syme*, *T. & D.*, 403; scarce, c. 1877, *De Cresp. Fl.*, 88; sparingly, 1916, *Andrewes*, *L.N.* 40: 21; 1921, *Richards*. **7.** East Heath, Hampstead, 1928, *Richards*.

A. cordifolium (Hedw.) Rich. & Wall.
Hypnum cordifolium Hedw.
Native in bogs, marshes, etc. Very rare, and perhaps extinct.
1. Pinner, *Hind*, *T. & D.*, 404. Stanmore Common, *Richards*, *Rep. Brit. Bryol. Soc.* 1923, 30; 1930, *C. A. Cooper* and *Binstead*, *Trans. Herts. N.H.S.* 19: 134. Harrow Weald Common, *De Cresp. Fl.*, 88. Copse Wood, Northwood, *Richards*. **4.** Harrow, *Hind*, *T. & D.*, 404.

A. giganteum (Schimp.) Rich. & Wall.
Hypnum giganteum Schimp.
Formerly native in marshes, but now extinct.
4. Stanmore Marsh (= Whitchurch Common), 1922, *Richards*, *Rep. Brit. Bryol. Soc.* 1923, 30.

A. cuspidatum (Hedw.) Lindb.
Native in marshes, wet grassy places, damp woodland rides, etc. Common, especially on heavy soils.

ISOTHECIUM Brid.

I. myurum Brid.
Eurhynchium myurum (Brid.) Dix.
Native or colonist on tree trunks, flints, etc., in woods. Very rare.
1. Ruislip Woods, abundant, *Richards Ec.*

I. myosuroides Brid.
Eurhynchium myosuroides (Brid.) Schimp.
Native or colonist on tree trunks, stumps, etc., in woods on acid soils. Very rare.

1. Harrow Weald Common, *De Cresp. Fl.*, 87. Ruislip Woods, 1949, *Boniface*; 1960, *Peterken*; Harefield!, *L.N.* 40: 68. Springwell. **4.** Hampstead, *Dillenius*, *With. Bot. Arr.* 2: 138. Bishop's Wood, *Coop. Fl.*, 103. Harrow, *Hind*, *T. & D.*, 403. **5.** Perivale Wood, *T. G. Collett*!, det. *Swinscow*.

CAMPTOTHECIUM Bruch, Schimp. & Guemb.

C. sericeum (Hedw.) Kindb.

Native, or colonist, on tree trunks, walls, etc. Very rare, or overlooked.

1. Near South Mimms; near Batchworth Heath; near Harefield; near Ruislip, *Richards Ec.* Springwell Lane, Harefield, 1955, *Boniface.* **2.** Hampton Court!; Shepperton, frequent, *Richards Ec.* **4.** Harrow, *Hind*, *T. & D.*, 403. Wood Lane, Brockley Hill; near Edgwarebury; Canons Park, *Richards Ec.* **6.** Enfield Chase, *Coop. Fl.*, 106. **7.** Ken Wood ponds, *Richards Ec.* Hampstead Lane, 1916, *Andrewes MSS.*

C. lutescens (Hedw.) Bruch, Schimp. & Guemb.

Native or colonist in grassy places, etc. Very rare, and confined to calcareous soils.

1. Chalkpits, Harefield, rare!, *Richards Ec.* Springwell, 1951, *Boniface.*

BRACHYTHECIUM Bruch, Schimp. & Guemb.

B. albicans (Hedw.) Bruch, Schimp. & Guemb.

Native on heaths, commons, in short turf, etc. Rare, or overlooked, and mainly confined to sandy soils.

1. Stanmore Common!, *De Cresp. Fl.*, 86. Poor's Field, Ruislip, locally abundant!, *Richards Ec.* Village Green, Harefield, 1950, *Boniface.* Springwell, locally common, *Stott*, *L.N.* 47: 14. **2.** Colnbrook, 1900, *Loydell* (D.). Causeway, Staines Reservoir, 1971, *Townsend.* **4.** Hampstead Heath, *R. Braithwaite*, *T. & D.*, 403; c. 1877, *De Cresp. Fl.*, 86.

B. glareosum (Spruce) Bruch, Schimp. & Guemb.

Native on banks, etc. Very rare, and confined to calcareous soils.

1. Springwell, 1951, *Boniface.* **4.** Harrow, *Hind*, *T. & D.*, 403; c. 1877,† *De Cresp. Fl.*, 86. Probably errors. **7.** Tottenham,† *De Cresp. Fl.*, 86. Probably an error.

B. mildeanum (Schimp.) Milde

Native in marshes, wet grassy places, etc. Very rare, and perhaps extinct.

1. Uxbridge Common;† Park Wood, Swakeleys; between Denham and Harefield Moor Locks, *Benbow*, *J.B.* (*London*) 37: 441.

B. rutabulum (Hedw.) Bruch, Schimp. & Guemb.
Native on bare ground, tree roots, in woods, established on wall-tops, etc. Common in shady places.
7. East Heath, Hampstead.

B. rivulare Bruch, Schimp. & Guemb.
Native in wet shady places. Very rare.
1. By a spring, Old Park Wood, Harefield, 1951, *Boniface*. Stanmore Common, 1930, *C. A. Cooper* and *Binstead, Trans. Herts. N.H.S.* 19: 134.

B. velutinum (Hedw.) Bruch, Schimp. & Guemb.
Native, or colonist, on tree roots, stumps, etc. Common.
7. Ken Wood.

B. populeum (Hedw.) Bruch, Schimp. & Guemb.
Native on flints, bare ground, etc. Very rare, and perhaps extinct.
1. Stones by the canal, Harefield, *Richards Ec.* **4.** Hendon, *Richards Ec.* **6.** Tottenham, *De Cresp. Fl.*, 86.

B. plumosum (Hedw.) Bruch, Schimp. & Guemb. **6.** On trunks of trees in Enfield Chase, *Huds. Fl. Angl.*, 423. The record probably refers to *B. mildeanum*, teste *Warburg*. **7.** By the Lee [Navigation] Canal, and by a drain adjoining, *De Cresp. Fl.*, 86. Almost certainly an error, fide *Warburg*.

SCLEROPODIUM Bruch, Schimp. & Guemb.

S. caespitosum (Wils.) Bruch, Schimp. & Guemb.
Brachythecium caespitosum (Wils.) Dix.
Formerly native on stones and tree roots by water, but now extinct.
1. Side of brook, Swakeleys, *Benbow, J.B. (London)* 34: 400. **4.** Tree roots by the Brent, Neasden, 1879, *De Crespigny* (MANCH).

CIRRIPHYLLUM Grout

C. piliferum (Hedw.) Grout
Eurhynchium piliferum (Hedw.) Bruch, Schimp. & Guemb.
Native in woods, etc. Very rare.
1. Ruislip Woods!, *Richards Ec.* Ickenham, 1907, *A. B. Jackson* and *Sherrin*.

C. crassinervum (Tayl.) Loeske & Fleisch.
Eurhynchium crassinervum (Tayl.) Bruch, Schimp. & Guemb.
Formerly native on tree roots, bare ground, etc., on calcareous soils, but now extinct.
1. Harefield Park, *Benbow, J.B. (London)* 34: 400.

EURHYNCHIUM Bruch, Schimp. & Guemb.

E. striatum (Hedw.) Schimp.
Native in shady woods, etc. Very rare, or overlooked.
1. Pinner, *Hind*, *T. & D.*, 403. Pinner Wood, *De Cresp. Fl.*, 87. Ruislip Woods, locally frequent, *Richards Ec.* Chalkpit, Harefield, 1949, *Boniface.* **5.** Perivale Wood, *Lillford* and *Ing*, *Groves Veg.*: 1968, *P. J. Edwards.*

E. praelongum (Hedw.) Hobk.
Native in woods, grassy places, on hedgebanks, in ditches, etc. Very common.
7. Common in the area.

E. swartzii (Turn.) Curn.
Native in woods, on heaths, cultivated and waste ground, etc. Formerly locally plentiful, but now rare, or overlooked.
1. Uxbridge!; Harefield!, *Benbow*, *J.B.* (*London*) 32: 107; 1949, *Boniface*; 1967; Springwell, *Stott*, *L.N.* 47: 14. Ruislip Woods; Stanmore Common; Harrow Weald Common, *Richards Ec.* South Mimms, *Richards.* **4.** Monk's Mere, Hendon;† Hampstead Heath Extension, *Richards Ec.* Golders Hill Park!, *Richards*; plentiful on streambanks, 1950, *Rose*! West Heath, Hampstead, *Boniface.* Kingsbury Green; Edgwarebury, *Richards.* **5.** Perivale Wood, *P. J. Edwards* and *A. R. Hall.* **6.** Near Enfield, *Richards.* **7.** Ken Wood, *Boniface.* Buckingham Palace grounds, *Rose*, *Fl. B.P.G.*

E. schleicheri (Hedw. f.) Lor.
E. abbreviatum (Turn.) Schimp.
Native on hedgebanks, etc. Very rare, or overlooked.
1. Harefield, *Benbow*, *J.B.* (*London*) 32: 370. Hedgebanks, Springwell Lane, locally abundant, 1951, *Boniface.*

E. speciosum (Brid.) Milde
Native on stones and tree roots by water. Very rare, and perhaps extinct.
1. Lake in Swakeleys Park;† Frays Meadows, Uxbridge, *Benbow*, *J.B.* (*London*) 32: 370. **2.** Banks of the Thames near Shepperton, locally abundant, *Richards Ec.* **4.** Kingsbury, *De Cresp. Fl.*, 88. Wembley, 1906,† *Sherrin.*

E. riparioides (Hedw.) Rich.
E. rusciforme (Neck.) Milde
Native on stones, established on concrete and brickwork, etc., in and near water. Formerly locally common, but now very rare, or overlooked.
1. Canal at Harefield; stream at Breakspeares, 1925, *Richards.* **2.** Staines, *R. Braithwaite*, *T. & D.*, 403. Shingle by the Thames between

Shepperton and Penton Hook!, *Richards Ec.*; 1952, *Boniface*; 1965. **4.** Harrow, *Hind, T. & D.*, 403. **5.** In and around the fast-flowing spring stream water at 'waterfall', The Hermitage, Hanwell, 1971, *P. J. Edwards*, teste *Townsend.* **6.** Lee Navigation Canal, Ponders End, 1923, *Richards.* **7.** Canal banks, Tottenham, *De Cresp. Fl.*, 87. Highgate Ponds, 1916, *Andrewes, L.N.* 40: 69; 1923, *Richards.*

E. murale (Hedw.) Milde

Native, or colonist, on stones, tree roots, bare ground, bases of walls, etc. Very rare, or overlooked.

1. Uxbridge; Ruislip; Garett Wood, on stones,† *Benbow, J.B.* (*London*) 32: 107. **7.** East Heath, Hampstead; Tottenham, *De Cresp. Fl.*, 87. Wall at edge of Hampstead Heath, *Richards Ec.* Hampstead Lane; Highgate Ponds, *Andrewes MSS.*

E. confertum (Dicks.) Milde

Native, or colonist, on stones and tree bases in shady places. Very rare, or overlooked.

1. Near Stanmore Common, *De Cresp. Fl.*, 87. Near Harefield; near South Mimms, *Richards Ec.* Harrow Weald Common, 1922, *Richards.* Harefield, *Peterken, L.N.* 40: 70. Edge of Hampstead Heath, *Richards Ec.* Monk's Mere, Hendon, 1926,† *Richards.* **5.** Perivale Wood, *P. J. Edwards.* **7.** Between Highgate and Hampstead; Tottenham, *De Cresp. Fl.*, 87.

E. megapolitanum (Bland.) Milde

Formerly native on banks, etc., on sandy soils, but now extinct.

1. Sandy bank between Highway Farm and Harefield, *Benbow, J.B.* (*London*) 32: 107.

RHYNCHOSTEGIELLA (Bruch, Schimp. & Guemb.) Limpr.

R. pumila (Wils.) E. F. Warb., *R. pallidirostra* (A.Br.) Loeske, *Eurhynchium pumilum* (Wils.) Schimp. Cited for Middlesex in the British Bryological Society's *Census Catalogue*, edition 1, but I have been unable to trace the source of the record.

R. tenella (Dicks.) Limpr.

Eurhynchium tenellum (Dicks.) Milde

Formerly native, or a colonist, of stones and flints in woods, etc., on calcareous soils, and on limestone walls, but now extinct.

1. Harefield; Springwell, *Benbow, J.B.* (*London*) 32: 107. **7.** Tottenham, *De Cresp. Fl.*, 87.

Var. **litorea** (De Not.) Rich. & Wallace

Eurhynchium tenellum (Dicks.) Milde var. *scabrellum* Dix., *Hypnum litoreum* De Not.

1. Near Uxbridge, *Benbow* (BM), det. *Dixon.*

PSEUDOSCLEROPODIUM (Limpr.) Fleisch.

P. purum (Hedw.) Fleisch.
Brachythecium purum (Hedw.) Dix.
Native in open woods, on heaths, banks, etc. Formerly locally common, but now very rare, or overlooked.
1. Harefield!; Stanmore Common, plentiful; Ruislip Woods; near South Mimms; Poor's Field, Ruislip, locally abundant, *Richards Ec.* Copse Wood, Northwood; near Elstree, *Richards.* Springwell, *Stott, L.N.* 47: 14. **3.** Hounslow Heath, *Richards Ec.* **4.** Golders Hill Park, *Richards Ec.* Harrow Weald, *T. & D.*, 404. Stanmore Marsh (= Whitchurch Common), *Richards.*

PLEUROZIUM Mitt.

P. schreberi (Brid.) Mitt.
Hypnum schreberi Brid.
Native in open woods, on heaths, etc. Very rare, or overlooked, and confined to acid soils.
1. Pinner, *Hind, T. & D.*, 404. Ruislip Woods!, locally abundant; Harrow Weald Common!; Stanmore Common, rare!; Watt's Common, Harefield; Poor's Field, Ruislip, locally abundant, *Richards Ec.* Copse Wood, Northwood, *Richards.*

ISOPTERYGIUM Mitt.

I. elegans (Hook.) Lindb.
Plagiothecium elegans (Hook.) Sull.
Native, or colonist, on banks, tree bases, in woods, on heaths, etc. Formerly locally common on acid soils, but now rare, or overlooked.
1. Stanmore Common!; Harrow Weald Common, *Richards Ec.*; 1949, *Boniface.* Near Batchworth Heath; near Harefield, *Richards Ec.* Old Park Wood, Harefield. Ruislip Nature Reserve, *Moxey Br.* **4.** Hampstead Woods, *Richards Ec.* Bishop's Wood; North-west Heath, Hampstead, *Andrewes MSS.* Brockley Hill, rare; Stanmore Marsh (= Whitchurch Common), *Richards.* **5.** Osterley Park; Chiswick House grounds, 1949, *Boniface.* **6.** Highgate and Queen's Woods, *Andrewes MSS.* **7.** Ken Wood, 1922, *Sutton*, teste *Richards*; sparingly, 1926, *Richards*; on damp banks, 1950!, *Boniface.*

PLAGIOTHECIUM Bruch, Schimp. & Guemb.

P. latebricola Bruch, Schimp. & Guemb.
Native, or colonist, on tree stumps, decaying wood, etc. Very rare.
1. Ruislip Woods, rare, 1921, *Richards.* **4.** Near Edgwarebury, *Richards Ec.* Scratch Wood, Edgwarebury, sparingly, 1950, *Boniface.*

Forma **geniculosis** Ryan & Trogen
1. Park Wood, Ruislip, 1927, *Richards.* **4.** Near Edgwarebury, on hawthorn, 1926, *Richards.*

P. denticulatum (Hedw.) Bruch, Schimp. & Guemb.
Native, or colonist, in woods, on ditch banks, etc. Common on acid soils.
7. Ken Wood!; Ken Wood Fields!, *Boniface.*

P. sylvaticum (Brid.) Bruch, Schimp. & Guemb.
Native, or colonist, on bases of trees in woods, etc. Formerly local, but now very rare or overlooked.
1. Harrow Weald Common!, *De Cresp. Fl.*, 89. Ruislip Woods, occasional; near Batchworth Heath; near Harefield, *Richards Ec.* **4.** Harrow, *Hind, T. & D.*, 404. Hampstead Woods, very rare; near Edgwarebury, *Richards Ec.* Monk's Mere, Hendon,† *Richards.* **5.** Acton, 1899,† *Sherrin.* Perivale Wood, rare, *P. J. Edwards* and *A. R. Hall.* **7.** Ken Wood, 1922, *Sutton*, fide *Richards.*

P. undulatum (Hedw.) Bruch, Schimp. & Guemb.
Native, or colonist, in shady places in woods. Very rare, and confined to acid soils.
1. Pinner, *Hind, T. & D.*, 404. Harrow Weald Common; Stanmore Common, *Richards Ec.*; two small patches, 1949, *Boniface.* **4.** Bishop's Wood, *Buddle, Huds. Fl. Angl.*, 420. **5.** Perivale Wood, *T. G. Collett!*, det. *Swinscow.*

HYPNUM Hedw.

H. cupressiforme Hedw. var. **cupressiforme** Hedw.
Native, or colonist, on trees, stones, bare ground, etc. Common.
7. Ken Wood, *Rose.*

Var. **resupinatum** (Wils.) Schimp.
1. Ruislip Woods; Garett Wood, Springwell,† *Benbow, J.B.* (*London*) 32: 107. Copse Wood, Northwood. Ruislip Nature Reserve, *Moxey Br.* Mimmshall Wood, 1923, *Richards.* **4.** Edgwarebury, 1923, *Richards.* **6.** Whitewebbs Park, Enfield, *Peterken, L.N.* 40: 71.

Var. **filiforme** Brid.
1. Ruislip Nature Reserve, *Moxey Br.*

Var. **ericetorum** Bruch, Schimp. & Guemb.
1. Stanmore Common, *C. A. Cooper* and *Binstead, Trans. Herts. N.H.S.* 19: 134.

CTENIDIUM (Schimp.) Mitt.

C. molluscum (Hedw.) Mitt.
Native on heaths, etc., in woods on calcareous soils, etc. Very rare.
1. Stanmore Common; Ruislip Common; Harefield!, *Richards Ec.* Springwell!, *Richards.*

RHYTIDIADELPHUS (Lindb.) Warnst.

R. triquetrus (Hedw.) Warnst.
Hylocomium triquetrum (Hedw.) Bruch, Schimp. & Guemb.
Formerly native in woods, etc., but now extinct.
4. Harrow, *Hind, T. & D.*, 404.

R. squarrosus (Hedw.) Warnst.
Hylocomium squarrosum (Hedw.) Bruch, Schimp. & Guemb.
Native in moist grassy places, woodland rides, on banks, etc. Formerly local, but now very rare, or overlooked.
1. Ruislip Woods!, Stanmore Common, plentiful!; Watt's Common, Harefield, abundant, *Richards Ec.*; 1955, *Boniface.* Springwell, *Stott, L.N.* 47: 14. Ruislip, locally abundant, *Richards Ec.* Mimmshall Wood, *Richards.* Pinner, *Melv. Fl.* 2: 136. Harrow Weald Common, *Boniface.* **3.** Hounslow Heath, *Richards Ec.* **4.** Harrow, *Hind, T. & D.*, 404. Harrow Weald!, *De Cresp. Fl.*, 88. Golders Hill Park, *Richards Ec.* Stanmore Marsh (= Whitchurch Common), *Richards.*

R. loreus (Hedw.) Warnst.
Hylocomium loreum (Hedw.) Bruch, Schimp. & Guemb.
Formerly native in woods and on heaths, but now extinct.
1. Harrow Weald Common, *De Cresp. Fl.*, 88. Pinner, *Melv. Fl.* 2: 136. **4.** Harrow, *T. & D.*, 404.

HYLOCOMIUM Bruch, Schimp. & Guemb.

H. splendens (Hedw.) Bruch, Schimp. & Guemb.
Native in woods, on heaths, banks, etc. Very rare, or overlooked.
1. Stanmore Common, *De Cresp. Fl.*, 88; 1927; Ruislip Woods, very rare; Watt's Common, Harefield, frequent, *Richards Ec.*

PTERIDOPHYTA

LYCOPSIDA

LYCOPODIACEAE

LYCOPODIELLA Holub

L. inundata (L.) Holub Marsh Clubmoss.
Lepidotis inundata (L.) C. Börn., *Lycopodium inundatum* L.
Formerly native on damp heaths, in bogs, etc., but now extinct.
First record: *Ray*, 1670. Last record: *Trimen* and *Dyer*, 1869.
1. Harefield Common, rather plentiful, *T. & D.*, 344. **3.** Hounslow Heath, *Huds. Fl. Angl.* 2: 463. **4.** Hampstead Heath, *Ray Cat.*, 215; c. 1700 (DB); 1828, *Pamplin, T. & D.*, 344; c. 1838, *Irv. Lond. Fl.*, 83; c. 1842, *E. Newman, Phyt.* 1: 50.

LYCOPODIUM L.

L. clavatum L. Stag's-horn Moss, Common Clubmoss.
Formerly native on heaths, but now extinct.
First record: *Gerarde*, 1597. Last record: *Grugeon*, 1865.
3. Hounslow Heath, abundantly, *Huds. Fl. Angl.*, 394; c. 1780 (LT). **4.** Hampstead Heath, *Ger. Hb.*, 1373, and later authors; 1865, *Grugeon, T. & D.*, 344. Bishop's Wood, *Budd. MSS.*

SPHENOPSIDA

EQUISETACEAE

EQUISETUM L.

E. fluviatile L. Water Horsetail.
E. limosum L., *E. heleocharis* Ehrh.
Native in marshes, swamps, watery places, etc., especially on heaths and commons. Local, and decreasing.
First record: *T. Johnson*, 1638.
1. Breakspeares,† *Blackst. Fasc.*, 26. Harefield!; Northwood;† Ickenham;† Uxbridge,† *Benbow MSS.* Ruislip Reservoir! and Common!, *Melv. Fl.*, 96. Near Yiewsley,† *Newbould*; Stanmore Common!, *T. & D.*, 336. Denham!, *A. H. G. Alston* (BM). **2.** Near Sunbury,

T. & D., 336. Laleham,† *Benbow MSS.* Hampton Court!, *Welch*; Shortwood Common!, *K. & L.*, 326. Staines Moor!, *Briggs.* Near Shepperton Green. **3.** Near Hatton,† *T. & D.*, 336. **4.** Hampstead Heath, *Merr. Pin.*, 35; West Heath, 1868, *Trimen* (BM). Between Hampstead and Finchley,† *Irv. Lond. Fl.*, 84. Kingsbury,† *Benbow MSS.* Whitchurch Common!,† *L. B. Hall* and *C. S. Nicholson*, *K. & L.*, 326. Near Mill Hill. **5.** Near Greenford,† *Benbow MSS.* Syon Park. **6.** Whetstone; Edmonton!, *T. & D.*, 336. Near Ponders End, *De Cresp. Fl.*, 81. Palmers Green,† *Sherrin* (SLBI). Finchley Common!,† *Benbow MSS.* Enfield!, *L. M. P. Small.* **7.** Tottenham, *Johns. Cat.* Hammersmith,† *Merr. Pin.*, 35. Isle of Dogs,† *E. Newman*, *Phyt.* 1, 691. Ken Wood grounds!, *Trimen* (BM). Highgate Ponds!, *Fitter*, *K. & L.*, 326. East Heath, Hampstead.

E. palustre L. Marsh Horsetail.
Native by stream- and riversides, in marshes, watery places, etc. Rather common, especially in the valleys of the Thames, Colne and Lee.
First evidence: *Buddle*, c. 1705.
7. Fields near Ken Wood, near a lane from Kentish Town, 1789,† *Forst. MSS.* Ken Wood!, *Andrewes MSS.* East Heath, Hampstead! *T. & D.*, 336. Hackney Marshes, 1797,† *Salt* (S). Tottenham.

E. sylvaticum L. Wood Horsetail.
Native on damp heaths, in wet heathy woods on acid soils, etc. Very rare.
First record: *l'Obel*, c. 1600.
1. Harefield,† *Blackst. Fasc.*, 26. Harrow Weald Common!, *Hind. Fl.*; still plentiful over a small area, 1973. Stanmore Common, very scarce!,† *K. & L.*, 325. **4.** Meadows by Hampstead Heath,† *Merr. Pin.*, 35, and later authors. West Heath, Hampstead,† *T. & D.*, 335. Bishop's Wood!, 1866, *George* (LNHS); c. 1869, *T. & D.*, 335. Turner's Wood, 1861, *Trimen* (BM); meadows near Bishop's and Turner's Woods, abundant!, *T. & D.*, 335; part of this area is now contained within Highgate golf course, where the plant is still locally plentiful, 1970. Golders Hill, 1904,† *L. B. Hall*, *K. & L.*, 325. **6.** Highgate,† *Lob. Stirp. Ill.*, 143 and *Park. Theat.*, 1201. The Alders, near Whetstone, 1867, *Mill*, *T. & D.*, 335; 1887,† *Benbow* (BM). **7.** Ken Wood!, . . . shown me by Mr J. Field . . . , *J. Andrews* (BM); now confined to a single locality where it is scarce, 1967. Fields near Ken Wood, *Rayer*, *B.G.*, 413. Between Kentish Town and Ken Wood,† *E. Forster* (BM).

Var. **capillare** Hoffm.
1. Harrow Weald Common!, *Loydell* (D). **7.** Ken Wood!, *Dillenius* (DILL).

E. arvense L. Common Horsetail.
Lit. *Proc. Bot. Soc. Brit. Isles* **7**: 202 (1968) (*Abstr.*)
Native on river banks, hedgebanks, cultivated and waste ground, in ditches, fields, by roadsides, etc., established on railway tracks, canal paths, etc. Common.
First record: *T. Johnson*, 1638.
7. Isle of Dogs!, *Newbould*; Highgate!; Fulham!, *T. & D.*, 335. East Heath, Hampstead. Ken Wood. Portland Place, W.1, *Holme*; Eaton Square, S.W.1!; Buckingham Palace grounds, 1956!,† *L.N.* 39: 43. Hyde Park, *Brewis*, *L.N.* 44: 19. Brompton Cemetery. Regents Park. Grosvenor Road, S.W.1. Chelsea. St John's Wood. Kentish Town. Islington. Hoxton. Wapping. Bethnal Green. Hackney. Golden Lane, E.C.
An extremely variable species.
The hybrid *E.* × *litorale* Kühlew. ex Rupr. (*E. arvense* × *fluviatile*) has not yet been noted in the vice-county but is probably present.

E. telmateia Ehrh. Great Horsetail.
E. maximum auct.
Denizen in damp woods, marshes, on heaths, roadbanks, pond verges, etc., established in cemeteries, on railway banks, etc. Locally frequent, chiefly in the northern parts of the vice-county.
First evidence: *Blake*, 1821.
1. Near Harrow Weald Common!; near Pinner Wood,† *Benbow MSS.* Stanmore Common, locally abundant!, *L.N.* 27: 35. Ruislip Common!, *K. & L.*, 325. **2.** Near Teddington!, *T. & D.*, 335. **3.** Roadside near Sudbury Town station, 1967. **4.** Hampstead Heath, 1823, *J. J. Bennett*; 1886,† *Benbow* (BM). West Hampstead Cemetery, 1966. Turner's Wood!, *Irv. MSS.* Marylebone Cemetery, East Finchley, 1966. Bentley Priory!, *Hind*, *Melv. Fl.*, 96; 1972. Hendon Wood Lane, near Barnet, locally abundant, 1950→. Dennis Lane, Stanmore, 1966→. Brockley Hill, 1966, *Kennedy*! Near Edgwarebury!, *T. & D.*, 335. **6.** The Alder's Copse, Whetstone, abundant, *T. & D.*, 335; scarce, 1887,† *Benbow* (BM). By Finchley Common,† *Benbow*, *J.B.* (*London*) 25: 19. Railway banks, Oakleigh Park, abundant, 1960→, *J. G. Dony*! Disused railway, Muswell Hill, abundant, 1961!, *Bangerter* and *Raven*. Near East Barnet, 1965, *Bangerter*. **7.** Highgate Archway, 1821, *Blake* (OXF). Highgate Cemetery, abundant!, *Trim. MSS.*; still there in profusion, 1974. Near West End, Hampstead,† *Trimen* (BM). Near Ken Wood (= Highgate Ponds)!, *T. & D.*, 335; still plentiful, 1974. Brompton Cemetery, 1950→. Kensal Green Cemetery, 1950→.
The status of *E. telmateia* in Middlesex is of some interest. It was not discovered until the early nineteenth century, and most of the pre-1880 records are from the Hampstead–Highgate–Barnet areas. These

districts were subjected to detailed examination by many seventeenth- and eighteenth-century botanists, and had such a conspicuous species been present then it would surely have been gathered or recorded. The apparent absence of *E. telmateia* in the vice-county at this period is confirmed by the absence of specimens in the Sloane herbarium, although there are numerous gatherings of *E. fluviatile*, *E. palustre*, *E. sylvaticum* and *E. arvense* made by Buddle, Petiver, Doody and others from Middlesex and the London area. A survey of first records for adjacent vice-counties reveals that the species was not reported until the early or mid-nineteenth century. In East Kent it was first recorded from between Margate and Sandwich by Thomas Johnson in 1632. It is possible, therefore, that the species was native in marshy coastal areas of south-east England, and has in historical times spread inland either by natural means or through the agency of man. A further feature of the species is its unusual preference for cemetery habitats.

FILICOPSIDA

OPHIOGLOSSACEAE

OPHIOGLOSSUM L.

O. vulgatum L. Adder's Tongue.

Native in damp meadows, moist grassy places, open woodland rides, etc. Formerly rather common, but now rare and decreasing.

First record: *Gerarde*, 1597.

1. Near Harefield!, abundantly, *Blackst. Fasc.*, 66. Harefield Moor!; Ruislip; Pinner,† *Benbow MSS.* Cowley,† *Wilmer*, *Blackst. Spec.*, 62. Ruislip Moor, abundant, 1862,† *Trimen* (BM). Ickenham, 1912,† *Green* (SLBI). Near Northwood, 1902, *Whiting*, *K. & L.*, 331. Northwood!, *Pickess*; near Harrow Weald Common!, *P. Taylor*, *K. & L.*, 332. Potters Bar, 1898, *L. B. Hall* (LNHS); 1965, *J. G.* and *C. M. Dony*. **2.** Colnbrook!,† *Ger. Hb.*, 327. West of Shepperton, *Welch*, *L.N.* 26: 65. **3.** Wyke Farm, Sion Lane, Isleworth,† *Francis*, 55. Between Alperton and Harrow, 1835,† *J. Morris* (BM). Roxeth,† *Benbow MSS.* **4.** Hampstead Heath, *Huds. Fl. Angl.*, 382; 1865,† *Syme*, *T. & D.*, 343. Between Hampstead and Hendon, *J. J. Bennett* (BM). Harrow, *Hind Fl.*; c. 1864, *Melv. Fl.*, 96. Edgware, 1925, *J. E. Cooper*, *K. & L.*, 331. Mill Hill, *Sennitt.* **5.** Osterley Park,† *Beevis*, *Francis*, 55. Brentford!, *Hemsley*, *T. & D.*, 343; Syon Park!, *K. & L.*, 331. Abundant in several meadows by the Brent near Perivale,† *Lees*, *T. & D.*, 343; now extinct, *Brown Chron.*, 137; railway bank by Perivale Wood, 1961!, *Welch*; 1965; increasing, and extending to the borders of Perivale Wood, 1969, *P. J. Edwards*. Greenford,† *Melv.*

Fl., 96. Acton, 1852,† *S. O. Gray* (BM). Acton Green, 1841 and 1858,† *Donald, Gard. Chron.*, 1872, 1392. **6.** Enfield, *H. & F.*, 149; near Botany Bay, 1915, *R. W. Robbins, K. & L.*, 332. **7.** Meadows near the preaching Spittle adjoining to London;† the Mantells [near Islington], by London,† *Ger. Hb.*, 327. Near the Mill by Bow, Hackney Marsh, *J. Newton, Ray Syn.* 3: 128; c. 1750,† *Alch. MSS.* St Marylebone, 1818† (G & R). Lee Bridge, 1845,† *R. Pryor* (LNHS). Wet meadows between Parsons Green and the Thames,† *Pamplin, T. & D.*, 343. Ken Wood grounds!, *Trimen* (BM); 1949!, *H. C. Harris, K. & L.*, 332. Lower part of Branch Hill Lodge, Hampstead,† *Jewitt, T. & D.*, 343.

OSMUNDACEAE

OSMUNDA L.

O. regalis L. Royal Fern.

Formerly native in wet heathy places. It no longer occurs as a native species but may be seen on damp heaths, peaty pond verges, etc., as an established introduction. Regeneration has been noted in two localities. Very rare.

First record: *Gerarde*, 1597.

1. Verge of lake, Grimsdyke grounds, Harrow Weald, planted but regenerating. Highgrove, Eastcote, 1965, *Jeffkins, J. Ruislip & Distr. N.H.S.* 15: 14. **2.** Bushy Park!, about six plants known for many years, 1949, *Boucher, K. & L.*, 331; 1967. **4.** Hampstead Heath, *Ger. Hb.*, 969, *Johns. Enum.* and *Park. Theat.*, 1039; of late it is all destroyed, *Johns. Ger.*, 1131; towards the north side of Hampstead Heath, *Pet. Midd.*; on the low part of the heath sparingly,† *Forst. MSS.* and *Alch. MSS.* Golders Hill, regenerating. **6.** Beyond Highgate,† *Lob. Stirp. Ill.*, 68. **7.** Ken Wood,† *Hunter, Park Hampst.*, 30.

ADIANTACEAE

ADIANTUM L.

A. capillus-veneris L. *sensu lato.* Maidenhair-fern.

Lit. *Brit. Fern Gaz.* **9**: 350 (1967).

An introduction which is established on old walls and other brickwork as sporelings originating from spores from cultivated pot plants, also established as a 'weed' in old greenhouses. Very rare.

First record: *Whittaker*, 1950.

5. Concrete wall, South Ealing, 1965!, *L.N.* 45: 20. Frequent 'weed' of greenhouses, Heston, 1960→. **6.** 'Weed' in old nurseries at Enfield† and Bush Hill Park,† where it unfortunately no longer exists,

W. S. Williams, Brit. Fern Gaz. 9: 350. Whitewebbs Park, Enfield, a single plant, 1962, *Halligey, L.N.* 42: 12. **7.** Bombed site near Holborn, 1950,† *Whittaker*. 'Weed' of greenhouses, Chelsea Physic Garden!, *W. S. Williams, Brit. Fern Gaz.* 9: 350. Frequent 'weed' of old greenhouses, Fulham Palace grounds, 1966, *Brewis*.

Some of the above records may be referable to the closely allied species *A. raddianum* C. Presl (*A. cuneatum* Langsd. & Fisch., non G. Forst.), from S. America, and *A. venustum* Don, from India, both of which are commonly grown as pot plants.

PTERIDACEAE

PTERIS L.

P. cretica L. Ribbon Fern.

Lit. *Proc. Bot. Soc. Brit. Isles* **5**: 121 (1963).

Alien. S. Europe, etc. Long cultivated as a pot plant, and now established on brickwork as an 'escape'. Very rare.

First record: *Kent*, 1968.

5. Established on brick footings of old greenhouses, Heston, 1968. **7.** Established on brickwork of basement area, Dean Street, W.1, 1968!, *McLean*, conf. *Jarrett*; still present, 1973. Basement, Vauxhall Bridge Road, S.W.1, 1973.

HYPOLEPIDACEAE

PTERIDIUM Scop.

P. aquilinum (L.) Kuhn Bracken.

Pteris aquilina L.

Native on heaths, commons, waste ground on acid soils, in woods, etc., established on railway banks, old walls, canal paths, etc. Common.

First certain record: *Merrett*, 1666.

7. Mature plants are common on waste ground, railway banks, etc., and sporelings are common on old walls.

THELYPTERIDACEAE

THELYPTERIS Schmidel

T. limbosperma (All.) H. P. Fuchs Mountain Fern.

T. oreopteris (Ehrh.) Slosson, *Dryopteris oreopteris* (Ehrh.) Maxon, *Lastrea oreopteris* (Ehrh.) Bory, *Aspidium oreopteris* (Ehrh.) Sw.

Formerly native in woods and on heaths, but now extinct.

First record: *Rayer*, pre-1797. Last record: *Westcott*, 1864.
1. Harrow Weald Common, *Westcott, Melv. Fl.*, 94. By the hedge or fence separating the farm buildings and yard of the Priory from Stanmore Common, 1827–30, *Varenne, T. & D.*, 337. **4.** On the edge of Hampstead Heath, *Rayer, B.G.*, 413; 1818, *J. J. Bennett* (BM). Bishop's Wood, plentiful, *Irv. Lond. Fl.*, 80; 1858 (WD).

ASPLENIACEAE

ASPLENIUM L.

A. adiantum-nigrum L. Black Spleenwort.
Formerly native on hedgebanks, but now extinct; established on old walls and other brickwork. Very rare.
First record: *Blackstone*, 1737.
1. Lane from Harefield to Rickmansworth, abundantly,† *Blackst. Fasc.*, 2. **2.** On Teddington Church,† *Francis, Coop. Suppl.*, 12. Hampton Court, 1902, *Green* (SLBI); 1949, *Welch, K. & L.*, 328. Stanwell, 1948!,† *L.N.* 28: 35. Staines, 1968, *Briggs* and *Missen*. **3.** Headstone Lane,† *Stuart, Melv. Fl.*, 95. **4.** Hedgebank, Harrow Weald,† *Wilkinson, Hind Fl.* Hampstead,† *Hunter, Park Hampst.*, 30. Hendon;† between Hendon and Finchley,† *Mart. Pl. Cant.*, 66. Plentiful on stonework of terrace, North London Collegiate School, Edgware, 1967! *Horder*. **5.** Heston, 1901;† Ealing, 1905,† *Green* (SLBI). **6.** On brickwork of ha-ha, Groveland's Hospital Southgate!, 1959, *A. Vaughan*; 1966, *Kennedy*; 1967, *Kennedy*!; 1973.

A. trichomanes L. Maidenhair Spleenwort.
Lit. *Brit. Fern Gaz.* **9**, 147 (1964): *Proc. Bot. Soc. Brit. Isles* **6**: 66 and 121 (1965).
Formerly native on hedgerows, but now extinct: established on old walls and other brickwork, in surface-drains, etc. Rare, and decreasing.
First record: *Blackstone*, 1737.
1. Breakspears;† Harefield!, *Blackst. Fasc.*, 100; still occurs in very small quantity on the wall of Harefield Churchyard, and on an old wall in Church Hill, 1972. Hillingdon!,† *Warren, T. & D.*, 340; survived until 1963 when the wall was demolished. Ruislip!; Ickenham!, *Benbow MSS.* Uxbridge!, *K. & L.*, 328; still plentiful on an old wall by the demolished High Street railway station, 1970. **2.** Hampton Court!, *W. Davies, T. & D.*, 340; 1964. River wall, Staines, one plant, 1966. **3.** Old wall, Isleworth, one plant, 1966, *Boniface*. Near Hounslow, *L. M. P. Small*. **4.** Hampstead,† *Wheeler, Park Hampst.*, 30; near 'The Spaniards', Hampstead,† *Irv. Lond. Fl.*, 81. Sudbury,† *Wilkinson*, fide *Hind, T. & D.*, 340. **5.** Chiswick,† *Benbow MSS.* Ealing, 1903,† *Green* (SLBI). Railway bridge, Perivale

Park, 1969, *L. M. P. Small.* Surface-drain, Syon Park, 1946, *Welch*!, *K. & L.*, 328; 1953. **7.** Walls of Chelsea College . . . ,† *Nichols, Blackst. Spec.*, 99.

All Middlesex specimens seen have been referable to subsp. **quadrivalens** D. E. Meyer emend. Lovis.

A. ruta-muraria L. Wall-Rue.

Denizen established on old walls, especially church and churchyard walls, brick tombs, stone and brick bridges, etc. Rather common.

First evidence: *S. Dale*, 1691.

7. Wall of Chelsea Garden, 1691,† *S. Dale* (BM). On an old stone conduit between Islington and Jack Straw's Castle [Hampstead],† *Pet. Midd.* Fulham Churchyard!, *Blackst. Spec.*, 1; 1820, *J. J. Bennett* (BM); still plentiful on an old tomb, 1973. Highgate!, *Church*; brickwork of kitchen area, Bloomsbury Street, W.C.1, 1866;† kitchen garden wall, Ken Wood!; East Heath (= Heath) Street, Hampstead!, *T. & D.*, 340; still plentiful at Ken Wood, and in Heath Street, near the Whitestone Pond, 1973. Buckingham Palace grounds, 1960,† *Fl. B.P.G.*

CETERACH DC.

C. officinarum DC. Rusty-back Fern.

Asplenium ceterach L.

Denizen established on old walls, and other brick- and stonework. Very rare.

First evidence: *S. Dale*, 1691.

1. Harefield, *Webb & Colem. Fl.*, 495; 1906,† *C. S. Nicholson, K. & L.*, 331. **2.** Hampton Court, *Hooper, T. & D.*, 341. Harmondsworth Churchyard, 1843 (WD); 1850, *J. Lloyd*, fide *Pamplin, T. & D.*, 341; 1920,† *P. H. Cooke* (LNHS). Stanwell Churchyard, 1904,† *Green* (SLBI). Staines, 1968, *Briggs* and *Missen.* **3.** Old wall, Isleworth, a single plant, 1969, *Sandford.* **5.** Perivale Churchyard!, 1881, *Masters* (BM); still present in very small quantity on a brick tomb, 1970. Syon Park, in crevices of a stone obelisk!, *L.N.* 40: 21; increasing, 1967; apparently eradicated during the construction of the Gardening Centre, 1968.† **7.** Wall of Chelsea Garden, 1691,† *S. Dale* (BM). Clapton,† *J. Miller, Blackst. MSS.* Hackney,† *Wilmer, Blackst. Spec.*, 5.

PHYLLITIS Hill

P. scolopendrium (L.) Newman Hart's-tongue Fern.

Scolopendrium vulgare Sm., *Asplenium scolopendrium* L.

Native on banks, in woods, etc., established in chalkpits, on old walls and other brickwork. Very rare in native habitats, but rather common

as sporelings on old walls, etc. Mature plants are rare in the latter habitat.
First record: *Gerarde*, 1597.
7. Ken Wood, *Irv. MSS.*; c. 1910, *Whiting Fl.* Bombed sites, Cripplegate, E.C.,† *A. W. Jones, K. & L.*, 327. Basement wall, Gillingham Street, S.W.1, 1957, *Fitter*; old wall, Marylebone, 1958!, *L.N.* 39: 43; now known on several walls at Marylebone and St John's Wood, 1973. Buckingham Palace grounds, 1961, *Fl. B.P.G.* Old wall, St Katherine's Dock, Stepney, 1963, *Fitter, L.N.* 43: 21. Area wall, Red Lion Square, W.C.2, one plant, c. 4 inches long, 1965, *Burton, L.N.* 46: 27. Old wall, Hornsey, 1966. Old wall by railway between Hammersmith and Baron's Court, 1967. In a drain, Shadwell, 1967; in a grating, Chancery Lane, W.C.2, 1968, *Lousley*. Old wall, Bow, 1968.

ATHYRIACEAE

ATHYRIUM Roth

A. filix-femina (L.) Roth Lady-fern.
Asplenium filix-femina (L.) Bernh.
Native in woods, shady places, on heaths, etc., established on old walls. Rare, but apparently increasing.
First record: *Petiver*, 1695.
1. Harefield!, *Benbow, J.B.* (*London*) 22: 280. Ickenham; Uxbridge!, *Benbow* (BM). Damp wood near Denham Lock, one plant, 1958, *I. G. Johnson, Donovan* and *Pickess.* Near Harrow Weald Common, *Wilkinson, Melv. Fl.*, 95. Harrow Weald Common, 1966. Stanmore Common, *T. G. Collett*! Pinner Hill, 1965. Near South Mimms, 1966. **2.** Bushy Park!, *A. W. Jones, K. & L.*, 328. Teddington Churchyard, 1965. Hampton golf course, 1966. **3.** Harrow Grove, *Harley Fl.*, 1. **4.** Woods about Hampstead! and Highgate, *Pet. Midd.*; c. 1912, *Whiting Fl.* Hampstead Heath!, *Irv. MSS.* Stanmore!, *Varenne, T. & D.*, 339. Mill Hill, *Sennitt.* **5.** Syon Park!, 1927, *Tremayne, K. & L.*, 328; 1946, *Welch*!, *L.N.* 26: 65; 1962. Osterley Park, *L. G. Payne, K. & L.*, 328. Long Wood, Wyke Green. Chiswick House grounds!, *L.N.* 28: 35. **6.** Winchmore Hill Wood,† *Church, T. & D.*, 339. **7.** Ken Wood, 1931, *Tremayne.* Red Lion Square, W.C.2, a number of fine plants in a well-basement, 1965!; eradicated, 1966,† *D. N. Turner, L.N.* 46: 27. Canal side between Marylebone and Primrose Hill, 1968.

CYSTOPTERIS Bernh.

C. fragilis (L.) Bernh. Brittle Bladder-fern.
Formerly established as an introduction on old walls, but now extinct.

First record: *Hind* and *W. A. Tooke*, 1871. Last evidence: *Green*, 1904.

1. Pinner Hill, *Hind* and *W. A. Tooke, J.B. (London)* 9: 272. **5.** Gunnersbury House, Acton, 1904, *Green* (SLBI).

ASPIDIACEAE

DRYOPTERIS Adanson

D. filix-mas (L.) Schott Male Fern.

Lastrea filix-mas (L.) C. Presl, *Aspidium filix-mas* (L.) Sw.

Native in woods, shady places, etc., established on old walls and other brickwork. Very common.

First certain evidence: *Doody*, c. 1705.

7. Frequent as sporelings on old walls. Mature plants have been noted at East Heath, Hampstead; Ken Wood; Holland House grounds, Kensington, etc.

D. carthusiana (Vill.) H. P. Fuchs Narrow Buckler-fern.

D. spinulosa Watt, *D. lanceolatocristata* (Hoffm.) Alston, *Lastrea spinulosa* C. Presl, *Aspidium spinulosum* (O. F. Müll.) Sw.

Native in woods, damp shady ditches, on shady banks, etc. Very rare.

First record: *Pamplin*, c. 1830.

1. Harefield!, *Benbow MS.* Mad Bess Wood, Ruislip!, *Collenette, L.N.* 27: 35. Furzefield Wood, Potters Bar, 1967, *Kennedy*! **4.** Hampstead Heath!; hedgebanks near Hendon!, *Pamplin, T. & D.*, 338. Near Mill Hill, 1957, *Hinson*. Scratch Wood, *Welch*!, *K. & L.*, 329. **6.** Winchmore Hill Wood, 1859,† *Church, T. & D.*, 338. **7.** Ken Wood, 1945, *Fitter, K. & L.*, 329.

D. dilatata (Hoffm.) A. Gray Broad Buckler-fern.

D. austriaca auct., *Lastrea dilatata* (Hoffm.) C. Presl, *Aspidium dilatatum* (Hoffm.) Sm.

Native in woods, shady places, on heaths, etc., occasionally established on old walls. Common, except near the Thames, and in the south-west parts of the vice-county where it is rare.

First record: *Buddle*, v. 1705.

7. Ken Wood! (DILL). By pond in Vale of Health, Hampstead,† *Irv. MSS.*; East Heath, Hampstead. Highgate Ponds!, *Fitter*. Holland House grounds, Kensington!, *Sandwith*. Buckingham Palace grounds, *Fl. B.P.G.* Well-basement, Red Lion Square, W.C.2, 1965!; eradicated, 1966,† *D. N. Turner, L.N.* 46: 27.

POLYSTICHUM Roth

Lit. *Brit. Fern Gaz.* **8**: 159 (1956).

P. falcatum (L.f.) Diels
Cyrtomium falcatum (L.f.) C. Presl
Alien. E. Asia, etc. Established on brickwork, the plants originating from spores from cultivated plants, also as a 'weed' in old greenhouses. Very rare.
First record: *Brewis*, 1966.
5. Sparingly on a disused railway bridge, Brentford, 1967; destroyed when bridge wall demolished, 1969.† **7.** 'Weed' in derelict greenhouses, Fulham Palace grounds, 1966, *Brewis*, det. *Crabbe*.

P. setiferum (Forsk.) Woynar Soft Shield-fern.
P. aculeatum auct., non (L.) Roth, *P. angulare* (Kit. ex Willd.) C. Presl, *Aspidium angulare* Kit. ex Willd.
Native in woods and shady places. Very rare.
First evidence: *S. O. Gray*, 1851.
1. Between Pinner and Harrow Weald, *Hind Fl.* Harrow Weald Common, two clumps, 1955!, *K. & L.*, 329. Ickenham, 1863 and 1896;† Uxbridge, 1890,† *Benbow* (BM). Damp wood near Denham Lock, one clump, 1958, *I. G. Johnson*, *Donovan* and *Pickess*. **3.** Lampton,† *Moore* 1: 148. Harrow Grove, perhaps planted, *Harley Fl.*, 1. **5.** Lane (= Tentelow Lane) between Hanwell and Norwood, 1851, *S. O. Gray* (BM); 1905,† *Green* (SLBI). Osterley Park,† *S. O. Gray*, *Moore* 1: 148. Near Osterley, 1862,† *Benbow* (BM). Brentford,† *Moore* 1: 140.

P. aculeatum (L.) Roth Hard Shield-fern.
P. lobatum (Huds.) Chevall., *Aspidium lobatum* (Huds.) Sw., *A. aculeatum* (L.) Sw.
Native in woods, on shady banks, etc. Very rare, and almost extinct.
First record: *Newton*, c. 1690.
1. Old Park Wood, plentifully,† *Blackst. Fasc.*, 29. Between Harefield and Ruislip,† *T. & D.*, 338. Pinner, *Melv. Fl.*, 94; 1866,† *Trimen*; Cowley, 1870,† *Benbow*; South Mimms!, 1867, *Trimen* (BM); 1965. **4.** Woods about Hampstead† and Highgate,† *Newt. MSS.* and *Pet. Midd.* Hendon, *Irv. Lond. Fl.*, 80; c. 1868, *W. Davies*; between Finchley and Hendon, 1868,† *Newbould*; Stanmore,† *T. & D.*, 338. Bentley Priory, 1866,† *Trimen* (BM). Harrow Park!, *Harley Fl.*, 1; perhaps planted. **5.** Osterley Park;† Lampton Lane;† Syon Lane,† *Beevis*, *Francis*, 28. Norwood,† *S. O. Gray*, *Moore* 1: 132. **6.** Between Colney Hatch and Whetstone, 1845;† near Colney Hatch, 1848† (HE). **7.** Hornsey Lane,† *Hill Fl. Brit.*, 535. Ken Wood,† *Hunter*, *Park Hampst.*, 30.

GYMNOCARPIUM Newman

G. robertianum (Hoffm.) Newman Limestone Fern, Limestone Polypody.

Thelypteris robertiana (Hoffm.) Slosson, *Dryopteris robertiana* (Hoffm.) C. Chr., *Phegopteris robertiana* (Hoffm.) A. Braun, *Carpogymnia robertiana* (Hoffm.) Löve & Löve, *Polypodium calcareum* Sm.
Introduced. Established on a wall. Very rare.
First record: *R. M. Payne*, 1964.
2. Wall by river, Staines, a large patch, 1964!, *R. M. Payne*, *L.N.* 44: 16; still plentiful, 1970.

BLECHNACEAE

BLECHNUM L.

B. spicant (L.) Roth Hard-fern.
B. boreale Sw., *Lomaria spicant* (L.) Desv.
Lit. *Proc. Bot. Soc. Brit. Isles* **7**: 203 (1968). (*Abstr.*).
Native on acid heaths, peaty pond and lake verges, in peaty woods, etc. Very rare, and almost extinct.
First record: *Gerarde*, 1597.
1. Harefield Common, abundantly,† *Blackst. Fasc.*, 54. Harrow Weald Common, *Hind Fl.*; 1866, *Trimen* (BM); 1884, *Benbow MSS.* Grimsdyke grounds, Harrow Weald, one plant, 1957, *I. G. Johnson*, *L.N.* 46: 20. Ickenham, 1907,† *Green* (SLBI). **3.** Hounslow Heath, rare,† *Hemsley*, *T. & D.*, 341. **4.** Hampstead Heath, *Ger. Hb.*, 979 and many later authors; c. 1868; Bishop's Wood, *T. & D.*, 341. Wood near Highgate† (DILL). **7.** Ken Wood,† *J. Hill* (BM) and *Curt. F.L.* Margin of pond in Vale of Health, Hampstead,† *Irv. MSS.*; East Heath, Hampstead, one plant, 1949!; grubbed up a few days after its discovery, *H. C. Harris*, *K. & L.*, 327; it is likely that other plants survive in some of the many enclosures on the East Heath.

POLYPODIACEAE

POLYPODIUM L.

Lit. *J. Linn. Soc. Bot.* **58**: 27 (1961).

P. vulgare L. Polypody.
Native on hedgebanks, tree stumps, etc., established on old walls and other brickwork. Formerly rather common, but now rare and decreasing.
First record: *T. Johnson*, 1638.

1. Harefield!, *Blackst. Fasc.*, 80; Harefield Grove, 1945!,† *K. & L.*, 330. Pinner,† *Hind Fl.* Near Uxbridge!, *Benbow MSS.*; railway bridge between Uxbridge and Denham, 1966, *T. G. Collett*! Hillingdon Churchyard, very fine, 1966→. Northwood!, 1963, *J. A. Moore*; 1965. Near Denham, 1963, *J. A. Moore*. Elstree,† *Trimen* (BM). **2.** Near Staines!; Tangley Park,† *T. & D.*, 337. Staines!, 1968, *Briggs* and *Missen*; in two localities, 1969→. West Drayton, *Benbow MSS.*; 1902,† *Green* (SLBI). Laleham Park!,† *Welch*, *L.N.* 26: 67. **3.** Twickenham,† *T. & D.*, 337. Hayes Churchyard!, *Green* (SLBI); 1961, *L. M. P. Small*; still present, 1973. West End, Northolt,† *Hind Fl.* Near Teddington, *Benbow MSS.* **4.** Hampstead Heath, 1826–27, *La Gasca Hort. Sicc.*; 1842,† *Stocks* (BM). Hampstead,† *Irv. MSS.* Lane from Hendon to Finchley Road, 1845,† *J. Morris*, *T. & D.*, 337. Harrow Churchyard, 1966, *Temple*!; 1973. **5.** Between Norwood and Hanwell, 1851,† *S. O. Gray* (BM). Osterley Park!, *Green* (SLBI); 1955, *Welch*!, *K. & L.*, 330; 1962. Old wall, Heston, 1960!, *L.N.* 40: 21; still plentiful, 1968; destroyed, 1971.† **6.** Between Edmonton and Winchmore Hill,† *Mart. Pl. Cant.*, 65. Highgate,† *Coop. Fl.*, 104. Edmonton;† Enfield; Winchmore Hill Wood,† *T. & D.*, 337. Plentiful on wall of ha-ha, Groveland's Park Hospital, Southgate!, 1958, *A. Vaughan*; 1966, *Kennedy*; 1967, *Kennedy*!; 1972. **7.** Tottenham,† *Johns. Cat.* Chelsea† (DILL). Kentish Town,† *Button*, *Coop. Suppl.*, 11. Ken Wood,† *Whiting Fl.*

All Middlesex specimens seen have been referable to **P. vulgare** L. *sensu stricto*.

MARSILEACEAE

PILULARIA L.

P. globulifera L. Pillwort.

Formerly native in bogs, on wet heaths, peaty pond verges, etc., but long extinct.

First record: *Doody*, 1696. Last record: *J. E. Smith*, 1800.

1. Near Uxbridge, *Alch. MSS.* Hillingdon Heath, *Lightf. MSS.* and *Smith Fl. Brit.*, 1143. The two records probably refer to the same locality. **3.** Hounslow Heath, *Doody*, *Ray Syn.* 2: 344; c. 1745, *W. Watson*, *Blackst. Spec.*, 28; c. 1762, *Huds. Fl. Angl.*, 393; c. 1780, *Lightf. MSS.* **4.** Hampstead Heath, *Huds. Fl. Angl.*, 393. **6.** Enfield Chase, *Alch. MSS.*

AZOLLACEAE

AZOLLA Lam.

A. filiculoides Lam. Water Fern.

A. caroliniana auct., non Willd., *A. pinnata* auct., non R.Br.

Lit. *J.B. (London)* **52**: 209 (1914).

Alien. N. America. Naturalised in the slow parts of streams and brooks, in ponds, etc., established in canals. Very rare, and varying in quantity from year to year, sometimes occurring in such abundance as to completely choke small waterways, at others times being very scarce. Warm summers appear to be favourable to the species.

First record: *O'Dell*, 1884.

1. Pinner,† *O'Dell, Sci. Goss.* 19: 279. '*Azolla pinnata*, a water plant from Carolina, first thrown on the pond at Eastcote House by an American visitor five years ago is now well established',† *Buckinghamshire Advertiser*, 15 Oct. 1887. Frays river near Uxbridge Common, 1945!, *Welch*; abundant in the canal and Colne from Denham Lock to Uxbridge, 1946!, *L.N.* 26: 65; locally abundant in the canal, Colne and Frays river from Denham Lock to West Drayton, 1972. Home Farm, Wrotham Park, 1965, *Fl. Herts*, 54. **2.** Sunbury, 1910,† *C. E. Britton, Rep. Bot. Soc. & E.C.* 2: 609. Abundant in the Colne, Bonehead ditch and adjacent streams on Staines and Stanwell Moors!, *L.N.* 28: 35; 1948–55!, *K. & L.*, 332; Bonehead ditch near Stanwell, scarce, 1959. Poyle, 1959. Canal, West Drayton, scarce, 1967!, *Burton*. Canal backwater between West Drayton and Dawley, plentiful, 1967. **3.** Ditches, Whitton, 1926, and subsequently, *H. Banks, L.N.* 26: 65. **6.** Sparingly in the Lee Navigation Canal from the county boundary to Enfield Lock, *Pierson, J.B. (London)* 52: 209; 1920,† *J. E. Cooper, K. & L.*, 332. Enfield, *G. C. Druce, Rep. Bot. Soc. & E.C.* 2: 406; New river, Enfield, 1973, *Kennedy*.

GYMNOSPERMAE

CONIFEROPSIDA

PINACEAE

PINUS L.

P. sylvestris L. Scots Pine.

Lit. *J. Ecol.* **56**: 269 (1968).

Introduced. Planted on heaths, commons, in parks, etc., where it sometimes regenerates. Rare.

First record (of regeneration): *Melvill*, 1864.

1. Spontaneously on Harrow Weald Common, the seeds from the plantations of Harrow Weald parks!, *Melv. Fl.*, 74; 1968. Seedlings on Harefield Common,† *Newbould, T. & D.*, 265. Chalkpit, Harefield, seedlings, 1946.† Pinner Hill, seedlings, 1966. **2.** East Bedfont, seedlings!, *Tremayne.* **5.** Syon Park, seedlings. **6.** Dyrham Park, seedlings.

The records given above refer only to localities where regeneration has been noted.

CUPRESSACEAE

CHAMAECYPARIS Spach

C. lawsoniana (A. Murr.) Parl. Lawson's Cypress.

Cupressus lawsoniana A. Murr.

Alien. N. America. Planted in parks, gardens, cemeteries, etc., where it sometimes regenerates. Seedlings and saplings have been noted on old walls. Very rare.

First record (of regeneration): *D. H. Kent,* 1958.

2. Near West Drayton Church, seedlings and saplings on an old wall, 1958!, *L.N.* 38: 21; 1968. **3.** Isleworth, a solitary large seedling on an old wall, 1965→. **4.** Harrow, a solitary sapling on an old wall, 1958!, *L.N.* 38: 21. **7.** Fulham Palace grounds, seedlings on old walls, 1966!, *Brewis.*

JUNIPERUS L.

J. communis L. subsp. **communis** Juniper.

Formerly native on heaths and commons, but long extinct.

First record: *Gerarde,* 1597. Last record: *Blackstone,* 1746.

1. Harefield Common, abundantly, *Blackst. Fasc.*, 48 and *Blackst. Spec.*, 41. **4.** Hampstead Heath, *Ger. Hb.*, 85, copied by many later authors. **6.** Finchley Common, *Coles,* 387, copied by many later authors.

TAXACEAE

TAXUS L.

T. baccata L. Yew.

Introduced. Planted in woods, hedgerows, churchyards, etc. It readily regenerates, and seedlings are common on churchyard, and other, walls, brick tombs, etc. Rather common.

First record (of regeneration): *D. H. Kent,* 1944.

7. Buckingham Palace grounds, *Fl. B.P.G.* Old walls, Highgate. Hampstead Churchyard walls. Old walls, Fulham Palace grounds.

The records given above refer only to localities where regeneration has been noted.

ANGIOSPERMAE

DICOTYLEDONES

RANUNCULACEAE

CALTHA L.

C. palustris L. Kingcup, Marsh Marigold, May Blobs.
Lit. *New Phyt.* **70**: 173 (1971); *BSBI Abstr.* **3**: 45 (1973).
Native in marshes, wet fields and woods, by streamsides, etc. Formerly common, but now local and decreasing and chiefly confined to areas near the Thames and Colne.
First record: *T. Johnson*, 1629.
1. Harefield!, *Blackst. Fasc.*, 13. Pinner, 'introduced',† *Hind, Melv. Fl.*, 4. Yiewsley,† *Newbould*; Ruislip Moor,† *T. & D.*, 20. Uxbridge!,† *W. West.* Denham. Springwell. Drayton Ford. **2.** Locally plentiful near the Thames and Colne. **3.** By the Thames, Twickenham; by the Crane, Hounslow Heath!; Isleworth,† *T. & D.*, 20. **4.** Hampstead Heath, *Johns. Eric.*; last seen on the West Heath in 1904.† Harlesden Green;† Hendon, *A. B. Cole, T. & D.*, 20; 1922, *Richards.* Near Finchley, *Newbould*; Turner's Wood!, *T. & D.*, 20. Golders Hill Park,† *Whiting Fl.* Mill Hill, *Foley*, 12. Edgwarebury!, *J.* and *S. G. Alletson.* Harrow Park!, probably introduced, *Horton Rep.* **5.** Greenford,† *Coop. Fl.*, 108. Hammersmith,† *A. B. Cole, T. & D.*, 20. Brentford Ait!, *Henrey*, 17. Syon Park, *H. Banks*!. Chiswick House grounds. Sparingly in a marsh near Boston Manor, probably introduced, 1966; 1972. **6.** Edmonton;† The Alder's, Whetstone,† *T. & D.*, 20. Colney Hatch,† *Trimen* (BM). **7.** Tottenham,† *Johns. Cat.* Between Westminster and Chelsea,† *Pet. Bot. Lond.* Side of London Canal, Hackney,† *Cherry*; Eel-brook Common, Parsons Green;† Ken Wood pastures, abundant!, *T. & D.*, 20. East Heath, Hampstead, plentiful. Highgate Ponds!, *Fitter.* By the lake, Buckingham Palace grounds, probably planted, *Codrington, Lousley* and *McClintock*!, *L.N.* 36: 15. Regents Park,† *Webst. Reg.*
Plants with double flowers are recorded from **1.** Harefield, *Blackst. Fasc.*, 13.

HELLEBORUS L.

H. viridis L. Green Hellebore, Bear's-foot.
Formerly a denizen in woods, copses, etc., but long extinct.

First record: *M. Collinson*, 1765. Last record: *Hunter*, 1814.
1. Near Harefield, 1791, *Baynes* (LINN). **3.** Down Barn Hill, near Harrow, *Woods, B.G.*, 406. **4.** In Caper's Wood . . . near Mill Hill, 1765, *M. Collinson, Coll. MSS.* In a small wood near Finchley, *Rayer, Curt. F.L.* The latter record probably refers to Collinson's locality. **7.** Ken Wood, *Hunter, Park Hampst.*, 30.

ERANTHIS Salisb.

E. hyemalis (L.) Salisb. Winter Aconite.
Alien. S. Europe, etc. Planted in copses, thickets, etc., where it sometimes becomes naturalised. Very rare.
First record: *Welch*, 1946.
2. Laleham Park!, *Welch, L.N.* 26: 57. **5.** Syon Park!, *L.N.* 26: 57.

ANEMONE L.

A. nemorosa L. Wood Anemone.
Lit. *J.B.* (*London*) **70**: 325 (1933).
Native in woods, thickets, on heaths, etc., established on railway banks, etc. Locally plentiful in the oak–hornbeam woods in the north of the vice-county, but gradually decreasing southwards to become almost absent from the immediate vicinity of the Thames.
First record: *T. Johnson*, 1629.
1. Locally common. **3.** Roxeth, 1863† (ME). **4.** Hampstead Heath, *Johns. Eric.*; survived on the West Heath until 1926.† Kingsbury,† *A. B. Cole*; near Finchley, *Newbould*; Bentley Priory!; Bishop's and Turner's Woods!, *T. & D.*, 13. Edgware!, *Benbow MSS.* Scratch Wood. Hendon!, *Richards.* Highgate golf course, *Andrewes MSS.* Harrow Park, *Harley Add.* Mill Hill, *Sennitt.* **5.** Between Brentford and Hanwell,† *Hemsley*; Alperton,† *A. B. Cole*; Wormholt (= Wormwood) Scrubs,† *Britten, T. & D.*, 13. Osterley Park, *Montg. MSS.*; 1966, *Bickerstaff.* Perivale Wood, *Selb. Mag.* 24: 27; 1932,† *Bartlett.* Horsendon Wood,† *Trim. MSS.* Walpole Park, Ealing, probably planted. Thickets, Hanger Lane, Ealing!, *L. M. P. Small*; a small colony, threatened by building operations, 1965, *H. Britton*; still present in very small quantity, 1966!, *McLean*; 1967. Disused Fox Reservoir, Ealing, several colonies, 1968,† *McLean.* **6.** Hadley!, *Newbould*; Bush Hill, Edmonton; Winchmore Hill Wood,† *T. & D.*, 13. Barnet Gate Wood!, *Benbow MSS.* Southgate, 1902, *L. B. Hall.* Enfield!, *H. & F.*, 148. Queen's Wood, Highgate, *Benbow MSS.* **7.** Marylebone Fields, 1817† (G & R). Ken Wood!, *T. & D.*, 13. Hornsey Wood, 1780,† *Pamplin MSS.* Near Crouch End, 1876,† *Trim. MSS.*

A. apennina L. Blue Anemone.
Alien. S. Europe, etc. Formerly naturalised in woods and copses, but now extinct.
First record: *Du Bois*, 1724. Last record: *G. C. Druce*, 1910.
4. Harrow-on-the-Hill, *Du Bois*, *Ray Syn.* 3: 259. Bishop's Wood, 1863, *Jewitt*, *T. & D.*, 421. **7.** Ken Wood, *Montg. MSS.* and *Druce Notes.*

CLEMATIS L.

C. vitalba L. Traveller's Joy, Old Man's Beard, Wild Clematis.
Native, and locally abundant on wood borders, climbing over hedges and bushes, and scrambling over bare ground on calcareous soils about Harefield and South Mimms, and on alluvial and gravel soils near the Thames and Colne. Less common and probably adventive elsewhere, though frequently seen about railway premises even in very heavily built-up areas.
First record: *Blackstone*, 1736.
7. Between Chelsea and Fulham!, *Forst. Midd.*; Bishop's Park, Fulham, *Welch*! Edgware Road, *Varenne*, *T. & D.*, 12. Finchley Road!, 1870, *Newbould*, *Trim. MSS.*; 1968. East Heath, Hampstead!, probably planted, *Richards*. Highgate!, *Fitter*. Bombed site, Portugal Street, W.C.2!, *Whittaker* (BM). Gordon Square, W.C.1!; Wellington Road, N.W.8!; King's Road, S.W.3!,† *L.N.* 39: 43. Cheltenham Terrace, S.W.3; by Knightsbridge Barracks, S.W.1, *Brewis*; Hoxton!, *L.N.* 46: 27. Regents Park; Cripplegate, E.C.!,† *Brewis*. Isle of Dogs. About railway premises at Notting Hill, Hornsey, Stoke Newington, Holloway and Bow. Upper Clapton.

C. viticella L.
Alien. S. Europe, etc. Garden outcast. Naturalised in a wood, and formerly naturalised on the banks of the Thames. Very rare.
First evidence: *W. H. Morris*, 1897.
2. Bank of the Thames opposite Ditton Ferry, a luxurious plant . . . , 1897,† soon afterwards the plant was rooted up, *W. H. Morris* (BM). Bank of Thames near Hampton Court, 1922;† believed to have spread from a known locality at Esher, Surrey, *C. E. Britton*, *Rep. Bot. Soc. & E.C.* 6: 715. **4.** Scratch Wood!, *Johns.*, *L.N.* 27: 30.

RANUNCULUS L.

Subgenus RANUNCULUS

R. acris L. subsp. **acris**
R. acer auct. plur., *R. boraeanus* Jord.
Lit. *J.B.* (*London*) **38**: 379 (1900): *Rep. Bot. Soc. & E.C.* **9**: 472 (1931): *J. Ecol.* **45**: 289 (1957): *Watsonia* **8**: 237 (1971).

Native in meadows, grassy places, on waste ground, etc., established on railway banks, etc. Very common.
First record: *Gerarde*, 1597.
7. Frequent throughout the district.

Subsp. **friesianus** (Jord.) Rouy & Fouc.
R. acris L. var. *vulgatus* Jord. ex Bor., *R. friesianus* Jord., *R. steveni* auct.
1. Uxbridge, *Druce Notes*. **5.** Acton (D), teste *E. F. Drabble, Rep. Bot. Soc. & E.C.* 9: 478. **7.** Primrose Hill, *Warren, Trim. MSS.*
If correctly identified the plants referred to above would probably have been introduced from Europe.

R. repens L. Creeping Buttercup.
Lit. *J. Ecol.* **45**: 314 (1957).
Native on banks of rivers, streams, ditches, etc., on cultivated and waste ground, in damp grassy places, by roadsides, etc., also a frequent and pernicious weed of lawns. Very common.
First record: *T. Johnson*, 1638.
7. Frequent throughout the district.
A large-flowered, glabrous plant which deserves further study occurs commonly by the Thames and its tributaries. A small-flowered form by the Thames near Chelsea (**7**) is recorded by *Dillenius, Dill. MSS.*, a similar form grew on the walls of the Temple Gardens (**7**) prior to the embankment of the Thames, *T. & D.*, 18. A plant with double flowers is recorded from **4.** Harrow, 1861, *Melv. Fl.*, 13.

R. bulbosus L. subsp. **bulbosus** Bulbous Buttercup.
Lit. *Rep. Bot. Soc. & E.C.* **10**: 242 (1933): *J. Ecol.* **43**: 207 (1955): *loc. cit.* **45**: 325 (1957): *Watsonia* **9**: 207 (1973).
Native in dry grassy places, rich meadows, old pastures, etc., chiefly on basic soils. Common, but much less widespread than *R. acris* and *R. repens*.
First record: *T. Johnson*, 1632.
7. Kensington Gardens, *Warren*; Hyde Park!; Highbury!; Isle of Dogs!, *T. & D.*, 19. Highgate!, *Andrewes MSS.* Ken Wood!, *H. C. Harris*. East Heath, Hampstead. Regents Park!, *Holme*; Lord's Cricket Ground!;† Ranelagh Gardens, Chelsea!, *L.N.* 39: 43. Stoke Newington. Tottenham. Hackney Marshes. Hackney Wick. Buckingham Palace grounds, *Fl. B.P.G.* Holloway, *D. E. G. Irvine*. Brompton Cemetery; Hurlingham, *E. Young*.
The var. **macrorhizus** Godr. is not uncommon, and an apetalous form of it was collected at **4.** Finchley, 1933, *E. F. Drabble* (BM). A very distinct form with white flowers is recorded from **7.** Brompton Cemetery, 1859–63, by *Britten, Phyt. N.S.* 6: 592 and

J.B. (*London*) 41: 249, and a fasciated plant was recorded from **1.** Yiewsley, 1911, by *J. E. Cooper*.

R. arvensis L. Corn Crowfoot, Hedgehogs.
Formerly common in cornfields and on cultivated ground, etc., as a colonist, but now very rare and sporadic on waste ground, rubbish-tips, by roadsides, etc.
First record: *T. Johnson*, 1638.
1. Harefield, frequent, *Blackst. Fasc.*, 84. Uxbridge, *A. B. Cole*, *T. & D.*, 19; 1884, *Benbow MSS.* Woodhall, Pinner, abundant; near Elstree, *T. & D.*, 19. Harrow Weald Common, 1866, *Trimen* (BM); 1939. South Mimms, 1929, *J. C. Robbins*. Cowley Peachey, one plant, 1946!, *L.N.* 26: 57. **2.** Near Staines, 1841, *J. F. Young* (K). Stanwell Moor, 1867; Staines, 1866, *T. & D.*, 19; 1885, *Fraser* (K). Near West Drayton, 1862, *Benbow* (BM). Laleham, *Benbow MSS.* Shepperton, *Newbould*, *T. & D.*, 421. **3.** Near Hounslow, *Hemsley*; Isleworth, *A. B. Cole*; near Heston, *T. & D.*, 19. **4.** Hampstead, *Coop. Fl.*, 101. Stanmore, 1858; Pinner Drive, *Melv. Fl.*, 3. Harrow Weald, *T. & D.*, 19. East Finchley, 1883, *J. E. Cooper*, *Sci. Goss.* 1898, 223. **5.** Greenford, abundant, *Melv. Fl.*, 3. Perivale, 1938!; Hanwell, 1949!, *K. & L.*, 6; 1964–65. Brentford, 1945, *Welch* and *Lousley*!, *L.N.* 26: 57. **6.** Near Edmonton, 1838, *E. Ballard* (BM). Edmonton; Enfield, *T. & D.*, 19. Near East Barnet, one plant, 1966. **7.** Tottenham, *Johns. Cat.* St John's Wood, 1815 (G & R). Islington, *Ball. MSS.* Chelsea, *Newbould*, *T. & D.*, 19. Hackney Marshes, 1912, *Coop. Cas.* Kentish Town (HE).

R. sardous Crantz Hairy Buttercup, Pale Buttercup.
R. hirsutus Curt.
Formerly native on damp moors, waste ground, etc., and in rough grassy places on heavy soils, but now extinct.
First record: *Merrett*, 1666. Last evidence: *J. E. Cooper*, 1912.
1. In profusion on the moors about Uxbridge towards Harefield, 1884, *Benbow* (BM); 1910, *Green* (K). Habitat destroyed by gravel-digging operations. Railway bank, Uxbridge, 1912, *J. E. Cooper* (BM). **3.** Hounslow Heath (DILL; SH; SLO). **4.** Hendon, *Champ. List.* Almost certainly an error. **7.** Near Hampstead, *Merr. Pin.*, 102; between Hampstead and Kentish Town, *Button*, *Coop. Suppl.*, 11. Paddington, *Doody MSS.* Chelsea College, 1861, *Britten*, *T. & D.*, 19. Near Brompton Cemetery, 1860, *Britten*, *Trim. MSS.* Isle of Dogs, 1839, *R. Pryor* (BM).
Forma **parvulus** (L.) Hegi
R. parvulus L.
5. Acton Green, 1867, *Newbould*, *T. & D.*, 19; 1871, *Warren* (W); 1876, *Sim* (BM); 1911 (D).
This form is now regarded as a dwarf state of *R. sardous*.

R. parviflorus L. Small-flowered Buttercup.
Lit. *Ann. Bot.* **45**: 539 (1931).
Formerly native in dry sunny places, but long extinct.
First record: *Doody*, 1695. Last record: *Benbow*, 1885.
1. Several places near Harefield, plentifully, *Blackst. Fasc.* 84. Uxbridge, 1839, *Kingsley* (ME); 1884, *Benbow* (BM). Uxbridge Moor, 1885, *Benbow*, *J.B.* (*London*) 23: 338. Cowley, *Benbow* (BM). **3.** Near Isleworth, *Doody*, *Pet. Midd.* **4.** Kingsbury, 1872, *Reeves*; 1884, *F. H. Ward* (BM). **7.** Hackney, *Blackst. Spec.*, 80. Kentish Town, *Mart. Pl. Cant.*, 72.

R. auricomus L. Goldilocks, Wood Crowfoot.
Native in woods, shady places, etc., rarely amongst grass in open situations. Locally common except in heavily built-up areas, though it still persists in a rough field adjoining Willesden Old Churchyard (**4**), it is also very rare in the south-west parts of the vice-county where it is known only from a small wood near Stanwell Moor (**2**); it is also absent from the immediate vicinity of the Thames.
First record: *T. Johnson*, 1638.
7. Marylebone† (G & R). Isle of Dogs† (BM). Near Fulham,† *Spencer*, 318. Perhaps an error. Near Hampstead, *T. & D.*, 18. East Heath, Hampstead, *White Hampst.*, 362. Finsbury Park, 1875,† *French* (BM).
Most Middlesex specimens have flowers with imperfect petals (cf. var. **apetalus** Wallr.), but plants with kidney-shaped lower leaves (cf. var. **reniformis** Kit.) were found at **3.** Southall, *E. F.* and *H. Drabble* (BM), and at **5.** Alperton, *Rep. Bot. Soc. & E.C.* 3: 7.

R. lingua L. Greater Spearwort.
Introduced. Planted on verges of lakes and ponds, banks of streams, etc., where it has now become naturalised. Very rare.
First certain evidence: *Lightfoot*, c. 1780.
1. Uxbridge Moor† (LT). Near Stanmore !, *Helsby*, *Rep. Wats. B.E.C.* 1926–27, 371; this record refers to a pond near The Grove, Stanmore Common, where the plant flourished in abundance until recent years, the pond has now become very overgrown, but a good colony of the species is still present, *Roebuck*. It is known also by an adjacent pond. Verge of lake, Grimsdyke grounds, Harrow Weald, *Welch* !, *L.N.* 28: 32. Potters Bar, 1955, *Phelp, K. & L.*, 335. **4.** Hampstead Heath, *Coop. Fl.*, 101. An error, *T. & D.*, 17. **5.** Plentiful on the verge of a lake, Syon Park !, *L. G.* and *R. M. Payne*, *L.N.* 27: 30; 1968. **6.** Brook near Frith Manor, Finchley, 1928,† probably carried down from the well-known station at Totteridge, Herts, *Moring MSS.*
Large forms of *R. flammula* are sometimes mistaken for *R. lingua*.

R. flammula L. subsp. **flammula** Lesser Spearwort.
Lit. *Watsonia* **4**: 19 (1957).

Native in wet heathy places, on peaty pond verges, etc. Locally plentiful, but decreasing.
First record: *T. Johnson*, 1629.
7. East Heath, Hampstead!, *T. & D.*, 17. Ken Wood!, *Whiting Fl.* Highgate Ponds, *Fitter.* Beyond Hornsey Wood,† *Ball. MSS.* Hackney Marshes, 1913,† *P. H. Cooke.*

Var. **major** Schulte
1. Harefield Moor, 1918, *Redgrove* (BM).

R. sceleratus L. Celery-leaved Crowfoot.
Native in muddy places on the verges of ponds, lakes, rivers, streams, etc., in ditches, etc., also established on river walls. Common.
First record: *T. Johnson*, 1629.
7. Near Brompton,† *J. Banks* (BM). Parsons Green;† near Chelsea Hospital; Kensington Gardens!; Hyde Park!; Highbury;† Victoria Park!; Isle of Dogs!; Kentish Town, *T. & D.*, 16. East Heath, Hampstead. Islington. Clapton. Hackney Marshes. Tottenham.
The common Middlesex plant is glabrous, but pubescent plants (cf. var. **pubescens** Rouy & Fouc.) are sometimes encountered.

Subgenus BATRACHIUM (DC.) A. Gray
Lit. *Proc. Bot. Soc. Brit. Isles* **7**: 35 (1967) (*Abstr.*).

R. hederaceus L. Ivy-leaved Water Crowfoot.
Native in wet muddy places, shallow pools, etc., particularly on heaths. Formerly common, but now very rare, and almost extinct.
First record: *T. Johnson*, 1632.
1. Harefield,† *Blackst. Fasc.*, 83. Harrow Weald Common, 1863† (ME). Stanmore Common!;† Eastcote,† *T. & D.*, 16. Hillingdon, 1839, *Kingsley* (ME); 1884,† *Benbow* (BM). Uxbridge, 1884;† Ruislip Common!, *Benbow MSS.* Northwood, 1928,† *B. T. Ward, K. & L.*, 5. **2.** Bushy Park, *Benbow MSS.* West Drayton, 1900,† *Green, K. & L.*, 5. **3.** Hatch End;† Hatton;† Hounslow Heath!;† Heston;† Whitton,† *T. & D.*, 16. **4.** Hampstead Heath, *Johns. Enum.*; c. 1868,† *T. & D.*, 16. **5.** Chiswick,† *M. A. Lawson*; Acton Green,† *Newbould, T. & D.*, 16. **6.** Hadley!, *Warren, T. & D.*, 16. **7.** Between Westminster and Chelsea,† *Pet. Bot. Lond.* Chelsea Common,† *J. Banks* (BM). Hyde Park, *E.B.*; 1817,† *J. F. Young* (BM). Tothill Fields, Westminster,† *E.B.* Walham Green,† *J. Morris*; New West End;† East Heath, Hampstead,† *T. & D.*, 16.

R. fluitans Lam. Water Crowfoot.
Native in rivers and streams. Formerly locally abundant, but now decreasing.
First certain evidence: *Dillenius*, c. 1730.
1. Near Uxbridge! *Fraser* (K). Frays river, Cowley Peachey!, teste *Butcher*, *L.N.* 28: 32. **2.** Formerly abundant in the Colne from Staines to Colnbrook, but now much reduced owing to the pollution

of the water. Formerly abundant in the Longford river, Stanwell, but virtually eradicated when the river was 'cleaned out'. **3.** Hounslow Heath river (= Crane)! (DILL). Duke's river south of Isleworth!,† *Wilmott* (BM). **6.** Lee Navigation Canal near Edmonton, 1947.†
The old published records of *R. fluitans* are confused with *R. penicillatus* and are omitted unless substantiated by specimens.

R. circinatus Sibth. Water Crowfoot.
Native in ponds, lakes, ditches and slow streams. Very rare, and decreasing.
First certain evidence: *Dillenius*, c. 1730.
1. Uxbridge, *Blackst. Fasc.*, 83; 1938, *Kingsley* (ME); 1871,† *Warren, Trim. MSS.* Swakeleys!,† *Benbow* (BM). Cowley,† *Benbow MSS.* Harefield;† Moorhall, 1907,† *Green*; near Denham!,† *K. & L.*, 3. Yiewsley,† *Lester-Garland* (K). **2.** Bushy Park!, *Newbould*; Staines!; Queen's river, Hampton, *T. & D.*, 15. Hampton Court!, *R. W. Robbins, K. & L.*, 3; still plentiful in the Long Water, and various ponds. Below Kingston-on-Thames, 1845,† *Biden* (K). Bedfont,† *Westrup*; Colnbrook!,† *K. & L.*, 3. West Drayton,† *Warren, J.B. (London)* 11: 208. **3.** Between Pinner and Harrow,† *Melv. Fl.*, 3. Southall,† *Benbow* (BM). **6.** Edmonton,† *T. & D.*, 15; backstreams of river Lee, and ditches below Prickett's Lock, 1888,† *Benbow* (BM). Gravel pit, north of Enfield Lock, 1966, *Kennedy*! **7.** Isle of Dogs† (DILL). Tothill Fields,† *Blackst. MSS.* Serpentine, Kensington Gardens,† *Warren, J.B. (London)* 13: 336. Hackney,† *Cherry, T. & D.*, 15. Between Hackney Wick and Lee Bridge,† *Benbow MSS.*

R. trichophyllus Chaix Water Crowfoot.
R. drouetii auct.
Native in ponds and lakes, on wet mud, etc. Very rare, and decreasing.
First certain evidence: *E. Ballard*, 1837.
1. Harefield!,† *Newbould*; Ruislip, *Trimen*; between Ruislip and Harrow,† *Benbow* (BM). Cowley, 1852,† *Syme* (W). **2.** West Drayton,† *Benbow*; near Staines!, *Trimen* (BM). Staines!, *L.N.* 27: 30. Hampton Court!, *Pugsley*. Poyle, *Montg. MSS.* **3.** Gravel pit west of Hounslow Heath, 1947!,† teste *Butcher L.N.* 27: 30. Southall.† **4.** Harrow,† *Benbow*; Willesden, 1875,† *Blow*; Finchley, 1877,† *H.* and *J. Groves* (BM). East Finchley, 1942,† *Bedford*, teste *Wilmott.* **5.** Greenford, 1861,† *Hind* (W). Near Acton, 1886, *Gade* (BM); 1909,† *Green, K. & L.*, 3. Near Westbourne Park station, 1882,† *Warren, Newb. MSS.* **6.** Whetstone, 1867,† *Trimen* (BM). Near Hadley!,† *Warren, T. & D.*, 14. Near Barnet, *Blow, J.B. (London)* 13: 177. **7.** Islington, 1837,† *E. Ballard*; near Highgate, 1875,† *French*; Newington,† *Newbould* (BM). Tottenham, c. 1925,† *Butch. MSS.*
Most early published records are omitted unless substantiated by specimens.

R. aquatilis L. Water Crowfoot.
R. heterophyllus Weber, *R. radians* Revel.
Formerly native in ponds and lakes, but now extinct.
First certain evidence: *Dillenius*, c. 1730. Last evidence: *Redgrove*, 1918.
2. West Drayton, 1852, *Syme* (W). Staines Moor, *Benbow* (BM). **3.** Twickenham, *Twining* (W). Southall, 1902, *E. F.* and *H. Drabble* (BM). **4.** Hampstead (DILL); c. 1910, *Druce Notes*. Near Willesden, 1875, *Blow*; Harrow Weald, 1918, *Redgrove* (BM). **5.** Near Wormwood Scrubs, 1889; Wormwood Scrubs, 1890, *Benbow* (BM).

R. peltatus Schrank Water Crowfoot
R. aquatilis L. subsp. *peltatus* (Schrank) Syme
Native in ponds, lakes and slow streams. Locally common, but decreasing.
First certain evidence: *J. Banks*, 1800.
7. Chelsea Common,† *J. Banks*; between Crouch End and Hornsey, 1875,† *French* (BM). Near Hornsey Wood,† *T. & D.*, 15. Crouch End,† *C. S. Nicholson* (LNHS). Clapton,† *Lond. Nat. MSS.* Highgate Ponds!, *H. C. Harris*. Hampstead Ponds.

R. penicillatus (Dumort.) Bab. var. **calcareus** (Butcher) C. Cook Water Crowfoot.
R. pseudofluitans (Syme) Newb. ex Bak. & Foggitt *pro parte*, *R. peltatus* Schrank subsp. *pseudofluitans* (Syme) C. Cook *pro parte*, *R. calcareus* Butcher
Native in swift flowing streams, etc. Formerly locally plentiful, but now much reduced in quantity.
First record: *Blackstone*, 1737.
1. Harefield!, *Blackst. Fasc.*, 83. Formerly frequent in the Colne from Springwell to Yiewsley, but now much reduced owing to the pollution of the river. Fray's river, Cowley!, teste *Butcher* (as *R. calcareus*), *K. & L.*, 4. **2.** Formerly plentiful in the Colne from Yiewsley to Colnbrook, but now scarce owing to the pollution of the river. **5.** Alperton, 1863† (ME).

FICARIA Huds.

Ficaria verna Huds. Lesser Celandine, Pilewort.
F. ranunculoides Roth, *Ranunculus ficaria* L.
Lit. *J. Linn. Soc. Bot.* **50**: 39 (1935): *Biol. J. Linn. Soc.* **1**, appendix 1, x (1969) (*Abstr.*): *Ann. Bot.* (*Oxford*) **36**: 31 (1972).
Native in woods and shady places, on banks, by stream- and riversides, etc. Common and widely distributed except in, and near, Central London, where owing to the absence of suitable habitats it is very rare or extinct.

First record: *T. Johnson*, 1629.

7. Marylebone Fields, 1817† (G & R). Hornsey Wood;† Kentish Town,† *T. & D.*, 17. East Heath, Hampstead. Highgate Ponds. Hackney, *Cherry*; Kensington Gardens!, *Warren, T. & D.*, 17. Hyde Park Corner, introduced with turves, 1963,† *Brewis, L.N.* 44: 19. Hurlingham. Garden weed, Lower Clapton, *L. M. P. Small*.

Two chromosome races are known in Southern England, a diploid, lacking axillary bulbils, and producing fertile seeds, the other, a tetraploid, bearing small axillary bulbils which reproduce the plant vegetatively. The tetraploid is the common Middlesex plant, and Gill, Jones, Marchant, McLeish and Ockendon in their study of the distribution of chromosome races of *F. verna* in the British Isles (*Ann. Bot. (Oxford)* 36: 31–47 (1972)) found that it occurred in the vice-county to the exclusion of the diploid. The two cytotypes have been treated both as subspecies and varieties by modern authors, but Walker and Heywood (*Nature* 189: 604 (1962)) have pointed out that the taxonomic value of the morphological characters separating them are as yet uncertain.

MYOSURUS L.

M. minimus L. Mousetail.

Formerly native on damp waste ground, in bare muddy places, wet fields, cornfields, by roadsides, etc., but now extinct.

First record: *Gerarde*, 1597. Last evidence: *Cooper*, 1914.

1. Uxbridge, *Lightf. MSS.*; 1894, *Benbow*; near Potters Bar, 1914, *J. E. Cooper* (BM). **2.** Hampton, *J. Harris* (SY); 1873, *E. C. White* (BM). Between Hampton and Sunbury, 1896, *White*, comm. *Ellis*, *Rep. Bot. Soc. & E.C.* 12: 265. **4.** Pinner Drive, a single plant, *Farrar, Melv. Fl.*, 2. Near Bishop's Wood, 1829, *Heathfield* (K). Finchley, *Benbow MSS.* **5.** Acton, 1841 '*W.D.*' (BM, K). Perivale, *Lees, T. & D.*, 14. **6.** Near Edmonton, *Ger. Hb.*, 346. Edmonton, *Woods, B.G.*, 403. Between Edmonton and Enfield, 1819, *E. T. Bennett* (BM). Enfield, *H. & F.*, 149. **7.** Way from London to Hampstead, on a barren ditch bank, *Ger. Hb.*, 346. Chelsea, *Merr. Pin.*, 24. Near Hornsey, *Pet. Midd.* Hornsey, *Blackst. MSS.*; 1840, *W. H. F. Talbot* (K). Near Marylebone Park; meadows behind Kentish Town Chapel; lane from Copenhagen House to Kentish Town, *Wilmer, Blackst. Spec.*, 57. Kentish Town, *Dicks. Hort. Sicc.* Islington; Pancras; Paddington, *Curt. F.L.*; wet fields near the Harrow Road . . . abundantly, 1825, *Pamplin, T. & D.*, 14; 1833 (G & R). Ken Wood, *Hunter, Park Hampst.*, 29. Perhaps an error.

AQUILEGIA L.

A. vulgaris L. Columbine.

Formerly native in woods, etc., on calcareous soils, but long extinct. Still present in meadows, thickets, on waste ground, etc., as an escape from cultivation. Very rare.

First record: *Blackstone*, 1737.

1. Harefield, rare,† *Blackst. Fasc.*, 7. Ruislip, *Batko*!, *L.N.* 27: 30. **2.** Near Ashford, *Hasler, K. & L.*, 7. **5.** Ealing, 1946,† *Boucher, K. & L.*, 7. Hanwell, 1962. **7.** Bombed site, Cripplegate, E.C.,† *Wrighton, K. & L.*, 7. Buckingham Palace grounds, casual, 1962,† *Fl. B.P.G.*

Some of the records given above may be referable to other cultivated species of *Aquilegia*.

THALICTRUM L.

T. flavum L. Common Meadow Rue.

Native on banks of streams and rivers, in marshes, bogs, etc. Formerly local in scattered localities in various parts of the vice-county, but now rare and decreasing and chiefly confined to the Colne valley and the immediate vicinity of the Thames.

First record: *Blackstone*, 1736.

1. Harefield!,† *Blackstone* in litt. to *Richardson*. Springwell. Drayton Ford. Near Northwood,† *Green, K. & L.*, 1. Uxbridge, 1839, *Kingsley* (ME); 1884;† Swakeleys, 1884,† *Benbow* (BM). Denham!,† *Druce Notes*. Between Yiewsley and West Drayton. **2.** Sparingly by the Thames in various localities between Staines and Hampton. Between Strawberry Hill and Teddington,† *T. & D.*, 12. Marsh near Walton Bridge!,† *L.N.* 27: 30. Harmondsworth, *P. H. Cooke*. Colnbrook.† Poyle. Yeoveney. Staines Moor!, *Whale Fl.*, 29. Near Stanwell. West Drayton!,† *Benbow MSS.* **3.** Twickenham,† *Firminger, T. & D.*, 13. Isleworth† (G & R). **4.** Hampstead,† *Irv. MSS.* **5.** Greenford,† *Coop. Fl.*, 108. Hammersmith, *A. B. Cole, T. & D.*, 13. Strand-on-the-Green, 1887,† *Benbow* (BM). **6.** Near Ponders End,† *C. S. Nicholson, K. & L.*, 1. Edmonton, 1889,† *Benbow* (BM). Trent Park, Southgate, 1966, *T. G. Collett* and *Kennedy*. **7.** Between Chelsea and Fulham,† *Britten*; Kilburn,† *Varenne*; Hackney Marshes, 1865,† *Grugeon, T. & D.*, 13.

Var. **sphaerocarpum** Lej.

2. Teddington, 1867,† *Dyer* (BM).

Var. **riparium** Jord.

1. Uxbridge, 1884,† *Benbow* (BM).

The taxonomic value of the above varieties is uncertain.

BERBERIDACEAE

BERBERIS L.

B. vulgaris L. Barberry.

Alien. Europe, etc. Naturalised in hedges and thickets, etc. Very rare.

First certain record: *T. Martyn*, 1763.

1. Between Cowley and Hillingdon,† *Benbow* (BM). Eastcote,† *J. A. Brewer, Melv. Fl.* 2: 4. Northwood, 1934,† *P. H. Cooke* (LNHS). **2.** Near Ashford, 1866,† *Trimen* (BM). Lower Halliford, 1905, *Green* (SLBI). **3.** Many places about Twickenham!;† near Isleworth,† *T. & D.*, 21. Near Hounslow, 1870,† *A. Wood, Trim. MSS.* **4.** Hampstead Heath,† *Mart. Pl. Cant.*, 66. Harrow! (ME). Wembley Park, 1889,† *Benbow*; Mill Hill!, *E. F. Drabble* (BM). Bentley Priory, *Benbow MSS.* Finchley, 1891,† *C. S. Nicholson* (LNHS). Edgware, 1904,† *Green* (SLBI). Near Totteridge, *Lousley*!, *L.N.* 26: 57. Golders Hill,† *Whiting Fl.* Hendon, *P. H. Cooke, K. & L.*, 8. **5.** Ealing, 1821† (G & R). Perivale,† *Lees, T. & D.*, 21. **6.** Between East Barnet and Whetstone,† *Webb & Colem. Suppl.*, 1. Near North Finchley, 1875,† *French* (BM). Enfield, 1949, *Johns, K. & L.*, 8. **7.** Between Hampstead Heath and Belsize,† *Button, Coop. Suppl.*, 11. Primrose Hill, rare,† *Irv. Lond. Fl.*, 186. Near Stoke Newington, 1839,† *J. Reynolds* (W). Kentish Town, 1845† (HE).

MAHONIA Nutt.

M. aquifolium (Pursh) Nutt. Oregon Grape.

Berberis aquifolium Pursh

Alien. N. America. Planted in woods, etc., where it sometimes regenerates; it occurs also on hedgebanks, in ditches, etc., where it is presumably bird-sown from gardens. Rare.

First evidence: *Green*, 1901.

1. Between Denham and Uxbridge, 1901, *Green* (SLBI). Near Denham!, *K. & L.*, 8. Uxbridge and Hillingdon, regenerating in several places, 1966. Near Potters Bar, 1965!, *J. G.* and *C. M. Dony*. Eastcote, regenerating, 1966. Harefield Grove, regenerating, 1966, *I. G. Johnson* and *Kennedy*!. **2.** Laleham Park!, *Welch, K. & L.*, 8; 1967. Stanwell Moor Village, 1966. **3.** Hounslow Heath!, *H. Banks, K. & L.*, 8; 1967. Hatton Cross, regenerating, 1966, *Kennedy*!. **4.** Harrow, regenerating, *Temple*!. **6.** Enfield, 1966, *Kennedy*!.

Ahrendt (*J. Linn. Soc. Bot.* **57,** 1 (1961)) points out that much British material referred to *M. aquifolium* is in fact the hybrid *M. aquifolium* × *repens*. Some of the records given above may, therefore, be in need of revision.

NYMPHAEACEAE

NYMPHAEA L.

N. alba L. White Water-lily.
Lit. *J. Ecol.* **43**: 719 (1955).
Native in slow parts of streams and in ponds and lakes, but frequently planted for ornamental purposes. The native plant is smaller flowered than the planted form, and is rare; the cultivated form is common.
First record: *T. Johnson*, 1638.
7. Tottenham, *Johns. Cat.* East Heath, Hampstead. Ken Wood!, *Hunter, Park Hampst.*, 30. Fulham,† *Mart. Pl. Cant.*, 66. Buckingham Palace grounds, planted.

NUPHAR Sm.

N. lutea (L.) Sm. Yellow Water-lily, Brandy-bottle.
Lit. *J. Ecol.* **43**: 344 (1955).
Native in rivers, streams, lakes, ponds, etc. Rather common, though sometimes planted for ornamental purposes.
First record: *T. Johnson*, 1638.
7. Moat at Fulham,† *Pet. Midd.* Ken Wood!, *T. & D.*, 23. Buckingham Palace grounds, planted. Tottenham!, *Johns. Cat.* Lee Navigation Canal between Lee Bridge and Tottenham.

CERATOPHYLLACEAE

CERATOPHYLLUM L.

Lit. *Ann. Rep. & Proc. Bristol Nat. Soc.* **6**: 303 (1926).

C. demersum L. Horn-wort.
Native in ponds, lakes, ditches, slow streams, etc. Rather common.
First certain evidence: *Lightfoot*, c. 1780.
7. Isle of Dogs,† *E. Forster* (BM). Clapton Common!, *C. S. Nicholson*; Vale of Health pond, East Heath, Hampstead!, *K. & L.*, 261.

C. submersum L. Unarmed Horn-wort.
Native in similar habitats to *C. demersum*. Very rare.
First certain evidence: *Benbow*, 1896.
1. Swakeleys;† between Harefield and Denham, 1896, *Benbow* (BM). **2.** Between West Drayton and Staines, *Benbow MSS.* Near East Bedfont, *Welch* and *Rose*!, *K. & L.*, 261. **3.** Near Southall!,† *K. & L.*, 261. **7.** British Medical Association pond, Tavistock Square, W.C.1, 1947, *Burges* (L). Vale of Health pond, East Heath, Hampstead, *H. C. Harris*!.

PAPAVERACEAE

PAPAVER L.

P. rhoeas L. Common Poppy, Field Poppy.
Lit. *J. Ecol.* **52**: 767 (1964).
Colonist. Formerly common in cornfields, but now sporadic on cultivated and waste ground, railway banks, by roadsides, etc. Common, but rarely seen in quantity.
First record: *T. Johnson*, 1638.
7. East Heath, Hampstead!, *T. & D.*, 24. Ken Wood fields. Westbourne Park. Near Paddington!, *L.N.* 39: 44. St John's Wood. Eaton Square, S.W.1!, *McClintock*; Hyde Park!, *D. E. Allen*, *L.N.* 39: 44. Near Dorchester Hotel, W.1, *D. E. Allen*; Paddington!; churchyard of All Hallows Berkynge, E.C.3!, *L.N.* 46: 27. Parsons Green. South Kensington. Bayswater. Holloway. Hoxton. Wapping. Isle of Dogs. Hackney. Clapton. Stepney. Fulham Palace grounds!, *Brewis*.

Var. **strigosum** Boenn.
1. Harefield. **7.** Hyde Park, 1962, *Allen Fl.*
This 'variety' may prove to be a hybrid between *P. dubium* and *P. rhoeas*.

Var. **pryori** Druce
1. Harefield, 1908, *A. B. Jackson*, *K. & L.*, 9; 1919, *Tremayne*.

Var. **wilkesii** Druce Shirley Poppy.
3. Near Hayes, 1967. **5.** Hanwell, 1952.

P. dubium L. Pale Red Poppy, Long-headed Smooth Poppy.
Lit. *J. Ecol.* **52**: 780 (1964).
Colonist on cultivated and waste ground, railway banks and tracks, disturbed ground, by roadside, etc. Rather common, especially on calcareous soils, and near the Thames.
First record: *Petiver*, 1695.
7. Chelsea,† *Pett. Midd.* Islington† (HE). The tops of the numerous high brick walls in the W. suburbs of town bear a large crop of this plant annually,† *T. & D.*, 24. Weed in churchyard of All Hallows Berkynge, E.C.3!, *L.N.* 46: 27.

P. lecoqii Lamotte Babington's Poppy.
Lit. *J. Ecol.* **52**: 784 (1964).
Formerly a colonist on cultivated ground, etc., but now extinct.
First record: *Hind*, 1861. Last record: *Benbow*, 1888.
1. Pinner, *Hind*, *T. & D.*, 24. Near Denham Lock, *Benbow MSS.* **3.** Roxeth, *Hind Fl.* **5.** Chiswick, 1888, *Benbow MSS.*

P. hybridum L. Round Prickly-headed Poppy.
Lit. *J. Ecol.* **52**: 789 (1964).
Formerly a colonist in cornfields, etc., also noted in waste places and tops of old walls, but now extinct.
First record: *Lorkin* and *T. Johnson*, 1633. Last record: *Green*, 1907.
1. Frequent among corn, Harefield, *Blackst. Fasc.*, 71; waste ground above the cement works, 1907, *Green*, *K. & L.*, 10. **3.** Isleworth, 1866, *T. & D.*, 23. **7.** Chelsea and Hammersmith fields, Mr *Robert Lorkin* and I, *Johns. Ger.*, 373. Between London and Chelsea, *Moris. Hist.* 2: 279. Chelsea, *Pet. Midd.*

P. argemone L. Long Prickly-headed Poppy.
Lit. *J. Ecol.* **52**: 786 (1964).
Formerly a colonist in cornfields and cultivated ground, now sporadic on cultivated and waste ground, railway banks, etc., chiefly on light soils. Very rare.
First record: *Lorkin* and *T. Johnson*, 1633.
1. Harefield, *Blackst. Fasc.*, 71; c. 1868, *T. & D.*, 23; 1884, *Benbow* (BM). Between Harefield and Uxbridge, *A. B. Cole*, *T. & D.*, 23; 1906, *Green* (SLBI); 1923, *Tremayne*; 1930, *P. H. Cooke*, *K. & L.*, 10. Uxbridge Moor,† *Benbow MSS.* Yiewsley, 1909, *Green* (SLBI). **2.** Between West Drayton and Yiewsley, *Newbould*; Sunbury, *T. & D.*, 23. West Drayton, 1866, *French* (BM). Between West Drayton and Colnbrook!, *Benbow MSS.* Near Staines, 1938!, *K. & L.*, 10. Between Staines and Poyle, 1958, *Briggs.* **4.** Near Brent Reservoir,† *Warren*, *T. & D.*, 431. Northwick Park, *L. M. P. Small.* **5.** Chiswick, *Newbould*, *T. & D.*, 23; 1887,† *Benbow* (BM). Acton,† *Newbould*, *T. & D.*, 23. Bedford Park,† *Cockerell Fl.* Near Brentford, a large patch, 1973, *Mullin.* **7.** Chelsea and Hammersmith fields, Mr *Robert Lorkin* and I, *Johns. Ger.*, 373; *Ibid.* copiously,† *Moris. Hist.* 2: 279. Chelsea,† *Pet. Midd.* Ken Wood fields, one plant, 1949,† *H. C. Harris* and *Welch*!, *K. & L.*, 10.

P. somniferum L. subsp. **hortense** (Hussenot) Corb. Opium Poppy, White Poppy.
Alien of hortal origin. Garden escape, but also introduced with bird seed. Waste ground, roadsides, rubbish-tips, and as a garden weed. Common.
First record: *A. Irvine*, c. 1830.
7. Frequent throughout the district.

Subsp. **setigerum** (DC.) Corb.
1. New Years Green, 1962. **5.** Hanwell, 1955. **7.** Hackney Marshes, 1967. Red-flowered plants are quite common.

P. atlanticum (Ball) Cosson
Lit. *Watsonia* **1**: 117 (1949).

Alien. N. Africa. Garden escape. Formerly established on bombed sites, waste ground, etc., but now extinct.
First record: *Lousley*, 1946. Last record: *Lousley*, 1957.
7. Plentiful on rubble in the vicinity of Gresham Street and Aldermanbury, E.C., 1946!, *Lousley*, *L.N.* 27: 49; during the summer of 1947 it increased considerably, and by the autumn was plentiful on debris over an area of about a quarter of a square mile about Wood Street and London Wall!, *Lousley*, *Watsonia* 1: 117; Cripplegate, 1949, *L.N.* 29: 86; still plentiful in the Wood Street area, 1953, *Scholey*; still present, but rapidly decreasing as its habitat is obliterated by re-building, 1957, *Lousley*.

CHELIDONIUM L.

C. majus L. Greater Celandine.
Alien. Europe, etc. Garden escape. Naturalised on hedgebanks, in ditches, by roadsides, etc., established on old walls, etc. Common, especially near the Thames.
First record: *T. Johnson*, 1638.
7. Tottenham!, *Johns. Cat.* Fulham!, *Britten*; Parsons Green; Ken Wood!, *T. & D.*, 25. Near Downshire Hill, *Burn. Med. Bot.* 2: 86. Hampstead, *H. C. Harris*. Garden weed, St John's Wood, *Holme*, *L.N.* 39: 44. Hyde Park Corner, weed in shrubbery, *Brewis*, *L.N.* 46: 27. Crouch End, Hornsey, 1833, *J. F. Young* (K). Regents Park, 1966, *Brewis*; 1968, *Coward*. Upper Thames Street, E.C., 1969, *L. M. P. Small*. Hurlingham, *E. Young*.
A plant with semi-double flowers is recorded from **5.** Turnham Green, *T. & D.*, 25, and a plant with double flowers was noted at **2.** East Bedfont, 1966.

FUMARIACEAE

CORYDALIS Vent.

C. bulbosa (L.) DC.
C. solida (L.) Sw., *Capnoïdes solida* (L.) Mill.
Alien. Europe, etc. Formerly naturalised in woods, but now long extinct.
First evidence: *W. James*, 1840. Last record: *Syme*, 1863.
1. Near Uxbridge, 1840, *W. James* (W); naturalised . . . near Uxbridge *E.B.* 3: 1, 101.

C. claviculata (L.) DC. White Climbing Fumitory.
Native or denizen in rough grassy places, etc. Very rare.
First certain evidence: *Whittaker*, 1950.
4. Plentiful in rough field, Mill Hill, 1958!, *Hinson*, *L.N.* 38: 21.

7. In the hedges near Bonner's Row, Bethnal Green, *Alch. MSS.* Possibly a rampant *Fumaria* mistaken for it, *T. & D.*, 26. Under bracken on a bombed site, Theobalds Road, W.C.1, 1950,† *Whittaker* (BM). No doubt an introduction.

C. lutea (L.) DC. Yellow Fumitory.
Alien. S. Europe, etc. Garden escape. Established on old walls and other brickwork, railway tracks, waste ground, etc. Rather rare.
First record: *Varenne*, 1827.
1. Uxbridge!, *A. B. Cole, T. & D.*, 25. Harefield, 1839,† *Kingsley* (ME). Ruislip Reservoir, 1949. Near Harrow Weald Common, 1966. **2.** Harmondsworth!; Sunbury!, *T. & D.*, 25. Teddington, 1902, *Green* (SLBI). Staines. Ashford Churchyard. West Drayton Churchyard. **3.** Isleworth. Whitton, *L. M. P. Small.* **4.** Harrow Weald, 1827, *Varenne, T. & D.*, 25. Harrow! (ME). Near Barnet Gate, 1910, *Cooke* (LNHS). Finchley. East Finchley. **5.** Ealing!, *Irv. Lond. Fl.*, 131; on several garden walls in Ealing and West Ealing, 1973. Brentford,† *A. Wood, Trim. MSS.* Railway bridge, Golden Manor, Hanwell, plentiful. Osterley. **6.** Edmonton, *T. & D.*, 25. Enfield, *Kennedy*!. Disused railway, Muswell Hill, 1961, *Bangerter* and *Kennedy*!. Palmers Green, 1967. **7.** Highgate, *De Cresp. Fl.*, 18. St John's Wood!, *L.N.* 39: 44; 1973. Bombed sites, Gresham Street† and Queen Victoria Street areas,† *Jones Fl.* Hyde Park, 1962,† *Allen Fl.*

FUMARIA L.

Lit. Suppl. to *J.B. (London)* **50** (1912): *J. Linn. Soc. Bot.* **44**: 233 (1919): *loc. cit.* **47**: 427 (1927): *loc. cit.* **49**: 93 (1932): *loc. cit.* **49**: 517 (1934): *loc. cit.* **50**: 541 (1937): *Proc. Bot. Soc. Brit. Isles* **5**: 358 (1964). (*Abstr.*)

F. purpurea Pugsl.
Formerly native on cultivated ground, but now extinct.
First, and only evidence: *Benbow*, 1892.
1. Uxbridge, 1892, *Benbow* (BM), teste *Pugsley.*

Var. **brevisepala** Pugsl.
1. Uxbridge Common, 1892, *Benbow* (BM), teste *Pugsley.*

F. muralis Sond. ex Koch subsp. **boraei** (Jord.) Pugsl.
F. boraei Jord.
Native on cultivated ground, etc., chiefly on light soils. Rare, and decreasing.
First certain evidence: *Trimen*, 1866.
1. Uxbridge, 1888,† *Benbow* (BM), teste *Pugsley.* **2.** Fulwell, 1867, *Dyer*; between Teddington and Bushy Park, 1870, *Moggeridge*; between Hampton Wick and Teddington, 1889; Teddington, 1889, *Benbow* (BM), all teste *Pugsley.* Fulwell golf course, 1973, *Townsend.*

Near Stanwell Moor, 1945, *Welch*!; near Heathrow, 1945, *Rose* and *Welch*!, *L.N.* 26: 57. Near Perry Oaks, 1948!, *L.N.* 28: 32. Staines!; Harmondsworth!, *K. & L.*, 11. Poyle. East Bedfont. West Bedfont. Near Hampton. Sunbury, 1967, *Kennedy*! **3.** Whitton, 1886,† *Trimen*; Twickenham!,† *Dyer* (BM), both teste *Pugsley*. Hounslow Heath, 1947, *H. Banks*!, teste *Pugsley*, *L.N.* 27: 30. **5.** Chiswick!,† *Benbow* (BM), teste *Pugsley*. Hanwell, 1946!,† teste *Pugsley*, *L.N.* 26: 57. **6.** Edmonton, 1869,† *Dyer* (BM), teste *Pugsley*. **7.** Fulham, casual, 1947!,† teste *Pugsley*, *L.N.* 27: 30.

Var. **gracilis** Pugsl.
1. Uxbridge;† near Uxbridge, 1892,† *Benbow* (BM), teste *Pugsley*.

F. densiflora DC.
F. micrantha Lag.
Native on cultivated ground, etc., chiefly on light soils. Rare, and decreasing.
First certain evidence: *Newbould*, 1866.
1. Near Harefield, *Benbow* (BM), teste *Pugsley*. Harefield, 1966, *I. G. Johnson* and *Kennedy*!. Near Uxbridge!, *Druce Notes*. **2.** Halliford, 1886, *Benbow* (BM), teste *Pugsley*. Harmondsworth!, *Benbow*, *J.B. (London)* 25: 15; 1967. Heathrow, abundant, *Welch*!, *L.N.* 26: 57; largely eradicated by the extension of London Airport. Near Staines!, *K. & L.*, 12. East Bedfont, plentiful, 1960. Stanwell, 1965. West Bedfont, 1965, *Kennedy*!. Laleham, a single plant, 1965!, *L.N.* 45: 20. Near Hampton, 1966, *Tyrrell*. **3.** Near St Margaret's, 1886;† near Twickenham, 1886;† between Hanworth and Twickenham, 1886,† *Benbow* (BM), all teste *Pugsley*. Whitton,† *Benbow MSS*. **5.** Near Shepherds Bush, 1866;† between Turnham Green and Brentford,† *Newbould* (BM), both teste *Pugsley*. Near Acton, 1867,† *Newbould* (SY), teste *Pugsley*. Chiswick,† *Benbow MSS*. South Ealing, 1965!, *L.N.* 45: 20; 1973. **6.** Tottenham, 1966, *Kennedy*.

F. officinalis L. subsp. **officinalis.** Common Fumitory.
Native on cultivated and waste ground, etc., chiefly on light soils. Rather common.
First certain evidence: *Trimen*, 1866.
7. Near Chelsea, 1866, *Trimen* (BM), teste *Pugsley*. Eaton Square, S.W.1!, *McClintock*, *L.N.* 39: 44. Kensington Gardens, 1961, *D. E. Allen*, *L.N.* 41: 17. Hyde Park, 1962, *Allen Fl.* Buckingham Palace grounds, 1962, *Fl. B.P.G.* Holland House grounds, Kensington, 1965. Near Poplar, 1966. Chelsea, 1969, *E. Young*.

Var. **elegans** Pugsl.
1. Uxbridge, 1884, *Benbow* (BM), teste *Pugsley*.

CRUCIFERAE

BRASSICA L.

B. napus L. Rape, Cole, Swede.
Alien of uncertain origin. Naturalised on waste ground, field borders, etc. Common, though the many cultivated races and varieties are complex and sometimes difficult to identify.
First record: *Blackstone*, 1737.
7. Marylebone Fields, 1815 (G & R); Regents Park, 1966, *Brewis*. Kentish Town (HE). South Kensington. Bayswater. Notting Hill. Hornsey. Hoxton. Wapping. Isle of Dogs. Hackney. Tottenham.

B. rapa L. var. **sylvestris** H. C. Watson ex Briggs Turnip, Navew.
Lit. *J.B. (London)* **7**: 346 (1869): *loc. cit.* **8**: 369 (1870): *loc. cit.* **61**: 104 (1923).
Denizen or native. Frequent by the Thames and its tributaries, also on waste ground.
First certain record: *Winch*, 1829.
7. Side of Lee between Lee Bridge and Tottenham!, *Cherry*, *T. & D.*, 35. River wall of Thames between Hammersmith and Grosvenor Bridge, S.W.1. Hyde Park, *Allen Fl.*

Var. **sativa** H. C. Watson
Common in arable fields as a relic of cultivation.

Var. **briggsii** H. C. Watson ex Briggs
2. West Drayton, *Druce Notes*.

B. nigra (L.) Koch Black Mustard.
Sinapis nigra L.
Denizen on waste ground, hedgebanks, by stream- and riversides and roadsides, established on rubbish-tips, in gravel pits, etc. Rather common.
First record: *T. Johnson*, 1633.
7. Banks at the back of Old Street;† and in the way to Islington, *Johns. Ger.*, 245. On ditch sides in many places about London, *Merr. Pin.*, 113. Highgate; Gospel Oak Fields! (HE); 1967. Camden Town!, 1818, *E. T. Bennett* (BM); 1966. Chelsea, 1815 (G & R); c. 1868, *Newbould*, *T. & D.*, 36. South Kensington,† *Dyer List*. Upper Clapton, *Cherry*; East Heath. Hampstead,† *T. & D.*, 36. Hackney Marshes!, *Coop Cas.* Shadwell!, 1952, *Henson* (BM); locally abundant, 1967. Bombed sites, Finsbury Square† and Gresham Street areas,† and near the Tower of London!, *A. W. Jones*, *L.N.* 37: 195. Regents Park. Old Ford, abundant, 1967. Tottenham.

B. integrifolia (West) O. E. Schulz var. **carinata** (A.Br.) O. E. Schulz
Lit. *J.B.* (*London*) **79**: 86 (1941).
Alien. N.-E. Africa, etc. Naturalised on waste ground, etc. Very rare.
First record: *F. Druce* and *Sandwith*, 1941.
4. Cricklewood, *Farend*, *K. & L.*, 333. **7.** Air-raid shelter, Chelsea Embankment Gardens, 1941, *F. Druce* and *Sandwith*, *J.B.* (*London*) 79: 86. Trinity Gardens, E.C., 1950–51!, *Lousley*, *K. & L.*, 337; still present in small quantity, 1967.

SINAPIS L.

S. arvensis L. Charlock, Wild Mustard.
Brassica arvensis (L.) Rabenh., non L., *B. sinapis* Vis., *B. kaber* (DC.) Wheeler
Lit. *J. Ecol.* **38**: 415 (1950).
Colonist on cultivated ground, etc., also on waste ground, by roadsides, etc. Very common.
First record: *T. Johnson*, 1638.

Var. **orientalis** (L.) Koch & Ziz
Forma *orientalis* (L.) Aschers., *Sinapis orientalis* L.
1. Yiewsley. **5.** Hanger Hill, Ealing, 1965–66. **7.** Hyde Park, 1962, *Allen Fl.* This variety is probably more common than the few records given above suggest.

S. alba L. White Mustard.
Brassica alba (L.) Boiss., *B. hirta* Moench
Colonist on cultivated ground, etc., also on waste ground, by roadsides, etc., rather common, but often merely casual.
First record: *T. Johnson*, 1633.
7. Banks at the back of Old Street;† near Islington, *Johns. Ger.*, 245. Upper Clapton, *Cherry*; near East Heath, Hampstead; Kensington Gore, *T. & D.*, 37. South Kensington, *Dyer List.* Hyde Park!; Kensington Gardens!, *L.N.* 39: 44. Haverstock Hill, 1867 (SY). Ken Wood fields. Highgate. Hackney Marshes, *J. E. Cooper* (BM). Bombed site near Cannon Street, E.C., 1954,† *Welch.* Wapping!, *Henson* (BM). Buckingham Palace grounds, *Fl. B.P.G.* Tottenham, *Kennedy.*

HIRSCHFELDIA Moench

H. incana (L.) Lagr.-Foss. Hoary Mustard.
Brassica adpressa Boiss., *B. incana* (L.) Meigen., non Ten.
Alien. Mediterranean region, etc. Established on waste ground, rubbish-tips, canal banks, etc. Rare, but increasing.
First record: *Welch*, 1954.
2. Near West Drayton, 1965→. Near Harlington, 1968!, *Clement* and *Ryves.* **3.** Hounslow Heath, 1954!, *Welch*, *L.N.* 34: 6; abundant, and

increasing, 1972. **5.** Rubbish-tip, Hanwell, one plant, 1954!,† *G. W.* and *T. G. Collett*. Canal bank between Hanwell and Brentford, 1961!, *L.N.* 41: 13; still present in several places, 1972, but gone, 1973;† disturbed ground near the railway between Boston Manor and Brentford, plentiful, 1966→. Brentford, 1967→. **7.** Car park, East Smithfield, E.C.1, 1966, *Brewis*, *L.N.* 46: 27; 1967→.

DIPLOTAXIS DC.

D. muralis (L.) DC. Wall Rocket, Sand Rocket.
Brassica muralis (L.) Huds.
Colonist on cultivated ground, etc., also by roadsides, in waste places, on railway tracks, old walls, etc. Common and increasing.
First record: *H. C. Watson*, 1837.
7. Chelsea, *Britten*; near Cremorne, *Fox*; Upper Clapton, *Cherry*; Parsons Green!; Brompton Cemetery; Fulham!; Brook Green; entrance to Grosvenor Canal!; East Heath, Hampstead!, *T. & D.*, 38. Finchley Road, *Trim. MSS.* Railway tracks, Victoria Station!, *L.N.* 39: 44. Hackney. Camden Town. Dalston. Highbury. Isle of Dogs. Regents Park, *Holme*. Tottenham. Cripplegate, E.C., *Brewis*.
The forma **caulescens** Kit. (var. *babingtonii* Syme), which is sometimes mistaken for *D. tenuifolia*, is common.

D. tenuifolia (L.) DC. Perennial Wall Rocket.
Brassica tenuifolia (L.) Fr.
Denizen on waste ground, rubbish-tips, about railway premises, by roadsides, etc. Common on walls in many parts of central London until the end of the nineteenth century when it became very rare. During the Second World War it spread into the eastern part of the vice-county via the railway systems from areas in west Kent and south Essex, where it has long been locally abundant, and began colonising bombed sites, waste ground, etc., soon reaching the bombed areas of the City of London. Since the end of the war it has spread along the railway systems and is locally abundant about railways, on waste ground, canal paths, etc., in the eastern part of Middlesex, particularly in the East End and the Lee Valley; it is now also locally plentiful in other parts of the vice-county and is continuing to spread.
First record: *Gerarde*, 1597.
1. Harefield, *Benbow MSS.* Yiewsley, locally plentiful. Near Harrow Weald Common, 1966. **2.** Sunbury, *De Cresp. Fl.*, 20. Gravel pits at East and West Bedfont. **3.** Isleworth, 1942!, *L.N.* 26: 58; 1967. Spring Grove, *Montg. MSS.* Hounslow!, *Lousley*, *L.N.* 26: 58; 1966. Hounslow Heath!, *Philcox* and *Townsend*. Hatton. Feltham. **4.** Hampstead, *Cochrane Fl.* Probably an error. East Finchley, *J. E. Cooper*, *Sci. Goss.* 1887, 223. Kenton!, *L.N.* 26: 58. Scratch Wood

railway sidings, 1957. Disused railway between Stanmore and Belmont, 1968, *Roebuck*. Railway banks, Cricklewood and Harlesden, 1965. Brent Reservoir, 1966. **5.** Near Greenford, 1945!,† *L.N.* 26: 58. Brentford!, *Welch, K. & L.*, 22; frequent about railways, on waste ground, etc., 1973. Between Brentford and Hanwell. Hanwell, 1952, *Brenan*!; 1968. Chiswick!, *Murray List.* Old Oak Common railway sidings, abundant and spreading rapidly, 1961→. Southall, 1966→. **6.** Frequent near the Lee Valley Navigation Canal from Tottenham to near Enfield Lock. Near Trent Park, a single plant, 1965. Bowes Park, 1965. **7.** Most brick and stone walls about London and elsewhere covered with it,† *Ger. Hb.*, 192. On the walls of the City in great plenty,† *Mill. Bot. Off.*, 189. Chelsea, Tower ditch, etc., *Newt. MSS.* On London wall between Cripplegate and Bishopsgate,† *Petiver* (SLO). Walls of the Charterhouse,† *Pet. Midd.* Walls round the Tower;† back of Bedlam;† near Hyde Park!, *Curt. F.L.* London Bridge, *Jones, With. Bot. Arr.* 3: 593; 1820,† *Pocock*; Green Park wall, 1774,† *J. Banks* (BM). Wall near Hyde Park Corner, 1817† (G & R). Walls in Church Lane† and Silver Street, Kensington,† *Mag. Nat. Hist.* 9: 90. Westminster School wall, *Macreight Man.*, 22; 1867,† *T. & D.*, 37. Near Kentish Town, *Irv. MSS.* Brickfield north of New Market [Islington], *Ball. MSS.* Kensington Vicarage wall, Church Street,† *Warren* (BM). Upper Clapton!, *Cherry*; near Parsons Green, *Britten*; Hammersmith; between Chelsea Hospital and the river, *T. & D.*, 137. In quantity over a limited area on both sides of Upper Thames Street from Darkhorse Lane to near Bush Lane!; rare by Dean Street, Chancery Lane,† *Lousley Fl.*

At the present time the species is locally abundant, and often in profusion, on waste ground, etc. from Bow Creek, northwards along the Lee Valley to Tottenham, and westwards to the Euston–Clerkenwell area.

RAPHANUS L.

R. raphanistrum L. Wild Radish.

Colonist on cultivated ground, also on waste ground, rubbish-tips, by roadsides, etc. A variable species, particularly in the colour of the flowers which range from pale yellow to deep violet. Common.

First record: *Blackstone*, 1737.

7. Side of London Canal [Hackney], *Cherry, T. & D.*, 44. South Kensington, *Dyer List.* Behind St Paul's Terrace [Islington], *Ball. MSS.* Westbourne Park. East Heath, Hampstead!, *H. C. Harris.* Near British Museum, *Shenst. Fl. Lond.* Paddington. Buckingham Palace grounds, *Codrington, Lousley* and *McClintock*!; Regents Park!, *Holme, L.N.* 39: 44. Hyde Park!, *D. E. Allen, L.N.* 41: 17. Tower Hill!, *Brewis, L.N.* 44: 20. St John's Wood!, *L.N.* 46: 28. Clerkenwell.

Holloway. Shoreditch. Stepney. Wapping. Clapton. Hackney. Hackney Marshes. Stamford Hill. Tottenham.

Forma **hispida** Lange
6. Southgate, 1858, *Brocas* (K).

Var. **aureus** Wilmott
7. Shadwell, 1952, *Henson* (BM).

Var. **ochroleucus** (Stokes) Peterm.
7. Hyde Park, 1960, *D. E. Allen*; 1961, *D. E. Allen* and *McClintock*, *L.N.* 41: 17 (as var. *flavus* Schub. & Mart.).

Var. **violaceus** Woerlein
5. Ealing Common, 1966. **7.** Near Ravenscourt Park!,† *Sandwith*, *Watsonia* 1: 117.

RAPISTRUM Crantz

R. rugosum (L.) All. subsp. **rugosum**
Alien. Mediterranean region, etc. Naturalised on waste ground, etc., but often merely casual. Usually introduced with bird-seed. Rare.
First record: *J. E. Cooper*, 1898.
1. Hillingdon, *P. H. Cooke*, *Rep. Bot. Soc. & E.C.* 11: 240. Yiewsley!, *L.N.* 28: 32. Harefield, 1966, *I. G. Johnson* and *Kennedy*!; 1967. **2.** West Drayton, *Druce Notes*. Stanwell, 1953, *Westrup*. East Bedfont, 1960, *T. G. Collett*!. Shepperton, 1969, *Clement* and *Ryves*. **5.** Hanwell, well established on waste ground!, *L.N.* 28: 32; still present, 1969. Ealing, 1959 and 1963. Brentford, 1960. Horsendon Hill, 1947!, *L.N.* 27: 30. Greenford, 1953, 1958 and 1960. Chiswick, 1949, *Boniface*, *K. & L.*, 25. **7.** Crouch End, 1898, *J. E. Cooper*, *Sci. Goss.* 1898, 223. Hackney Marshes, 1913, *J. E. Cooper* (BM). Zoological Gardens, Regents Park, *Shove Fl.* Regents Park, 1966, *Potter*.

Subsp. **orientale** (L.) Arcangeli
R. orientale (L.) Crantz
1. Uxbridge, *Loydell*, *Rep. Bot. Soc. & E.C.* 5: 641. Yiewsley, *Coop. Cas.* **2.** West Drayton!, *C. S. Nicholson* (LNHS). **3.** Cranford, 1954, *Bigg-Wither*. **4.** Fortis Green, 1911, *J. E. Cooper* (BM). **6.** Muswell Hill, 1913, *J. E. Cooper* (BM). **7.** Hackney Marshes, 1912 and 1924, *J. E. Cooper* (BM).

Subsp. **linnaeanum** Rouy & Fouc.
R. hispanicum (L.) Crantz, *R. linnaeanum* Boiss. & Reut., *nom. illegit.*
7. Bedford College, Regents Park, 1966, *Brewis*, teste *McClintock* (as *R. hispanicum*), *L.N.* 46: 28.

LEPIDIUM L.

L. campestre (L.) R.Br. Pepperwort, Field Cress.
Native in fields, on hedgebanks, cultivated and waste ground, etc., on calcareous and sandy soils, adventive on railway tracks, waste ground, etc., on other soils. Formerly rather common but now rather rare and sometimes casual.
First record: *W. Turner*, 1548.
1. Harefield!, *Blackst. Fasc.*, 98. Between Harefield and Ruislip!; between Pinner and Harrow Weald, *T. & D.*, 42. New Years Green, plentiful in old sandpits. Near Uxbridge!, *Benbow* (BM). Cowley; Yiewsley,† *Benbow MSS.* Pinner! (ME). Near Potters Bar, *J. G.* and *C. M. Dony*. **2.** Between Staines and Twickenham, *Wats. MSS.* West Drayton,† *De Cresp. Fl.*, 37. Near Shepperton, *Benbow MSS.* **3.** Wood End,† *Hind Fl.* Near Isleworth!; between Fulwell and Whitton, *Benbow MSS.* Hanworth,† *Wright* (BM). Near Northolt. Between Southall and Hayes. **4.** Hampstead,† *Irv. MSS.* Near Brondesbury,† *De Cresp. Fl.*, 37. Pinner Drive,† *Melv.*, 9. Edgwarebury, *Welch*!, *K. & L.*, 23. **5.** Syon,† *Turn. Names.* Ealing,† *Curt. F.L.* Near Alperton,† *Newbould, T. & D.*, 42. Between Hanwell and Southall, one plant, 1946,† *Welch*!, *L.N.* 26: 58. Hanwell, one plant, 1951,† *T. G. Collett*!. Railway side near Brentford, a small colony, 1966. Near Kew Bridge, 1968. Chiswick,† *Benbow MSS.* North Acton.† **6.** Tottenham,† *Johns. Cat.* New Enfield!, *T. & D.*, 42. Hadley Common,† *Benbow MSS.* Near Barnet, *F. J. Hanbury* (BM). New Edmonton, 1966. **7.** 'In London it groweth in Master Riche's garden† and in Master Morganne's† also', *Turn. Hb.* 2: 152; perhaps cultivated there, *T. & D.*, 42. Ball's Pond,† *Ball. MSS.* Hackney Marshes, 1913–15 and 1918,† *J. E. Cooper, K. & L.*, 23.

Forma **longistylum** A. G. More
1. Harefield, 1914, *Tremayne* (BM).

L. heterophyllum (DC.) Benth. var. **canescens** Godr. Smith's Cress.
Native on hedgebanks, by roadsides, in grassy places, etc., chiefly on light soils. Very rare, and decreasing and now confined to a few localities in the south-west parts of the vice-county.
First record: *Trimen* and *Dyer*, 1866.
1. Harefield, 1902, *Green, Williams MSS.*; 1948,† *Welch, K. & L.*, 24. Uxbridge Moor!,† *Benbow* (BM). **2.** Near Charlton!; between Staines and Hampton, *T. & D.*, 42; 1884,† *Benbow MSS.* Littleton, 1889; Ashford Common, *Benbow* (BM). Between Staines and Ashford, 1949!, *K. & L.*, 24. Staines, *Briggs.* West Bedfont,† *Welch* and *Rose*!, *L.N.* 26: 58. **3.** Near Hounslow!;† near Hanworth, *T. & D.*, 42; 1886, *Benbow* (BM); 1905,† *Montg. MSS.* Near Whitton, *Benbow* (BM); c. 1910,† *Green* (SLBI). Between Hounslow and Feltham!,†

Benbow MSS. Hounslow Heath!,† *H. Banks, L.N.* 27: 30; habitat destroyed by gravel-digging operations. **4.** Hampstead, *Cochrane Fl.* An error, *L. campestre* probably intended. Harrow, *Horton Rep.* An error. **5.** Osterley Park,† *Flippance* (K). Near Brentford, 1973, *Mullin.* **6.** Near Trent Park, *Benbow MSS.*

L. ruderale L. Narrow-leaved Pepperwort.

Colonist on bare waste ground, canal paths, rubbish-tips, railway tracks. Formerly rare, but now rather common.

First evidence: *A. Irvine*, c. 1830.

7. Near Highgate Archway, *A. Irvine*; South Kensington (BM). Hackney, *Cherry, T. & D.*, 42. Kensington, 1874, *Warren, Trim. MSS.* Two plants just west of Kensington Palace, *Warren, J.B. (London)* 9: 228. Kilburn, 1857, *Hind, Trim. MSS.* Hornsey, 1875, *French*; bombed site, Shadwell, 1954, *Henson* (BM). Crouch End, *J. E. Cooper, Sci. Goss.* 1898, 223. Fulham, *Johns. Nat.*, 63. East Heath, Hampstead!, *H. C. Harris.* City of London bombed sites.† Regents Park, 1961, *Holme, L.N.* 41: 17. Bethnal Green. Holloway. Stepney.

L. latifolium L. Dittander, Broad-leaved Pepperwort.

Introduced. Naturalised on the banks of rivers and streams, and established by canals, in gravel pits, on waste ground, railway banks, etc. Locally abundant and spreading.

First record: *D. H. Kent*, 1939.

1. Banks of the Colne near Drayton Ford, 1939!, *L.N.* 26: 58; canal bank in several places near Harefield, 1945!, *Welch, L.N.* 26: 58; in profusion by the canal from above Springwell Lock to near Jack's Lock, Harefield, 1967. Colne banks and gravel pits, Springwell, in profusion!, *K. & L.*, 23. Hill End, Harefield, 1954.† Canal bank and adjacent gravel pits, Harefield Moor, in quantity, 1955→. Canal bank between Harefield Moor and Denham Lock, 1956→. Near Uxbridge, 1957!, *L.N.* 41: 13; extensive colonies by the canal south of Western Avenue, 1967. Canal side between Uxbridge Moor and Cowley, in abundance, 1965. Canal bank, Cowley, 1967. **2.** Rough field between Hayes and West Drayton, 1967. **3.** Canal bank, Frogmore Green, near Southall, a small colony, 1961!, *L.N.* 41: 13; plentiful and increasing, 1973. Railway bank, Feltham, 1961, *Clement*; 1968.

I first noted this species, which is native in coastal saltmarshes, at Rickmansworth, Herts, in 1937, where it had possibly been introduced with ballast from the long-boats that operate on the canal. Shortly before World War Two it spread into Middlesex and has now extended its range throughout the upper Colne valley. It is probable that it will eventually reach the Thames at Brentford via the canal, and at Staines via the Colne.

CORONOPUS Zinn

C. squamatus (Forsk.) Aschers. Swine-cress, Wart-cress.
C. procumbens Gilib., *C. ruellii* All., *Senebiera coronopus* (L.) Poir.
Native on waste ground, bare muddy places, paths, by roadsides, etc. Very common.
First record: *Gerarde*, 1597.
7. Frequent throughout the district.

C. didymus (L.) Sm. Lesser Swine-cress.
Senebiera didyma (L.) Pers.
Alien. S. America. Established on disturbed ground, by roadsides, etc. Formerly rare, but now common and increasing.
First record: *J. E. Smith*, 1795.
7. Highgate Archway, *Irv. MSS.* In a manner naturalised in Chelsea Gardens, *E.B.*, 248. Chelsea College, 1861, *Britten*; Parsons Green, *Irvine*; Isle of Dogs, plentiful!, *Cherry*, *T. & D.*, 43. Kensington Gardens!, *Warren Fl.* Brompton. Highgate; near Parliament Hill Fields, *Andrewes MSS.* East Heath, Hampstead!, *H. C. Harris.* Bromley. Buckingham Palace grounds, *Codrington*, *Lousley* and *McClintock*!; Hyde Park!; St James's Park, *D. E. Allen*, *L.N.* 39: 44. Lord's Cricket Ground!, *L.N.* 41: 17. Tower Hill!, *Brewis*, *L.N.* 44: 20. Regents Park. Islington. Stepney. Holloway. Tottenham.

CARDARIA Desv.

C. draba (L.) Desv. Hoary Cress, Hoary Pepperwort, Whitlow Pepperwort.
Lepidium draba L.
Lit. *J. Ecol.* **50**: 489 (1962): *Proc. Bot. Soc. Brit. Isles* **5**: 243 (1964). (*Abstr.*)
Alien. Europe, etc. Established in fields, on waste ground, railway banks, by roadsides, etc. Formerly very rare, but now very common.
First record: *Newbould*, 1866.
7. Railway bank, Edgware Road station, abundant, 1866, *Newbould*, *T. & D.*, 41; 1888, *Montg. MSS.* South Kensington, *Dyer List.* Finchley Road, *T. & D.*, 422. Isle of Dogs!, 1886, *Benbow* (BM). Fulham. East Heath, Hampstead!, *H. C. Harris.* Hackney Marshes!, *Coop. Cas.* Primrose Hill!, *L.N.* 39: 44. Tower Hill; Chelsea Hospital grounds, *Brewis*, *L.N.* 44: 20. Notting Hill. Kilburn. Bayswater. Kentish Town. Highgate. Holloway. Islington. Finsbury Park. Hornsey. Dalston. Tottenham. Wapping. Bethnal Green. Clapton. Hackney Wick. Bromley.

THLASPI L.

T. arvense L. Field Penny-cress.
Colonist on cultivated and waste ground, by roadsides, etc. Common, though rarely seen in quantity.
First record: *Gerarde*, 1597.
7. Tothill Fields;† between Tottenham and Newington, *Woods, B.G.*, 407. Near Highgate, *How MSS.* Newington Common . . . , 1865, *Grugeon*; Chelsea!, *Britten, T. & D.*, 40. Adelaide Road [Hampstead], *Newbould*; Green Park, sown with grass . . . , *T. & D.*, 41. South Kensington!, *Dyer List.* Kensington Gardens, casual, *Warren. Fl.* Notting Hill. Bombed site, Ebury Street, S.W.1,† *McClintock.* Bombed site, Suffolk Lane, E.C., 1946.† Hackney Marshes!, *J. E. Cooper.* Bishop's Park, Fulham, *Welch*!. Near 'The Spaniards', Hampstead Heath, *Curt. F.L.* Regents Park; St Marylebone, 1957, *Holme, L.N.* 39: 44. Near Dorchester Hotel, W.1, 1964, *D. E. Allen* and *Potter*; Knightsbridge, one plant, 1966!;† Marylebone!, 1966, *L.N.* 46: 28.

TEESDALIA R.Br.

T. nudicaulis (L.) R.Br. Shepherd's Cress.
Formerly native in bare sandy and gravelly places but now extinct.
First certain record: *Doody*, 1696. Last record: *Westrup*, 1932.
1. Near Uxbridge, 1893, *F. J. Hanbury* (HY). **2.** Hampton Court Park, *Doody, Ray Syn.* 2: 344. Between Hampton and Hampton Court, *Lamb. MSS.* Teddington, *Blackst. Spec.*, 64. Dawley, 1907, *Green, K. & L.*, 25. **3.** Hounslow Heath, *Dicks. Hort. Sicc.*; 1932, *Westrup, K. & L.*, 25. Whitton, *Francis, Coop. Suppl.*, 12; Whitton Park Inclosure, 1887, *Benbow* (BM). Near Twickenham, 1842, *Twining, T. & D.*, 41.

CAPSELLA Medic.

C. bursa-pastoris (L.) Medic. Shepherd's Purse.
Native on cultivated and waste ground, in fields, by roadsides, etc., established on railway tracks, etc. Very common.
First record: *T. Johnson*, 1638.
7. Frequent throughout the district.
The following segregates are on record:—

C. abscissa (E.At.) Wilmott
2. West Drayton, *Druce, Rep. Bot. Soc. & E.C.* 8: 725.

C. batavorum (E.At.) Wilmott
2. West Drayton, *Druce, Rep. Bot. Soc. & E.C.* 6: 197.

C. druceana (E.At.) Wilmott
2. West Drayton, *Druce, Rep. Bot. Soc. & E.C.* 8: 725.

C. patagonica (E.At.) Wilmott
2. West Drayton, *Rep. Bot. Soc. & E.C.* 8: 725.

C. turoniensis (E.At.) Wilmott
5. Brentford, *Druce, Rep. Bot. Soc. & E.C.* 6: 197.

COCHLEARIA L.

C. danica L. Danish Scurvy-grass.
Introduced. Established on railway tracks. Very rare.
First record: *J. G. Dony*, 1950.
4. Plentiful between the metals on railway main line by Scratch Wood sidings!, *J. G. Dony, K. & L.*, 17; 1963.

C. anglica L. Long-leaved Scurvy-grass.
Formerly native on the muddy shores of the tidal Thames but long extinct.
First record: *J. Martyn*, 1763. Last record: *Warren*, 1869.
7. Isle of Dogs, *Mart. Pl. Cant.*, 65; 1869, *Warren, T. & D.*, 422.

BUNIAS L.

B. orientalis L.
Alien. S.-E. Europe, etc. Established on waste ground, railway and canal banks, etc. Rare, but increasing.
First evidence: *Dyer*, 1866.
1. Uxbridge,† *Benbow* (BM). Yiewsley, 1917, *Lester-Garland* (K); 1919–20,† *J. E. Cooper, K. & L.*, 25. Between Ruislip and Harefield, 1928, *Turrill* (K). Harefield, 1956, *T. G. Collett* and *Pickess*; 1960–62. **3.** Sudbury, 1881,† *De Crespigny* (BM). Railway bank, Southall, 1950!, *Welch, K. & L.*, 25; plentiful and spreading, 1967. Between Hayes and Southall, 1966. Between Southall and Greenford, 1968, *L. M. P. Small.* **4.** Willesden, 1883,† *W. R. Linton, Rep. Bot. Soc. & E.C.*, 1: 201. Hendon, 1903, *E. J. Salisbury* (K); 1912,† *P. H. Cooke* (LNHS). Neasden, 1957. Cricklewood, 1966. **5.** Chiswick!, *Benbow* (BM); still on waste ground by the Thames, 1967. Bedford Park,† *Cockerell Fl.* Acton,† *Druce Notes.* **6.** Ponders End!, *Benbow* (BM); 1911, *Essex Nat.* 16: 319; still by the Lee Navigation Canal, 1967. Enfield, 1927,† *Davy* (K). **7.** Near Shepherds Bush, 1866,† *Dyer*; Isle of Dogs!, 1887, *Benbow* (BM); still very abundant on waste ground, railway banks, etc., throughout the 'island', 1971. Thames bank, Wapping, 1967. Bombed sites, Shadwell, plentiful, 1967. Bombed site, Bunhill Row, E.C., 1950!,† *Lawfield, K. & L.*, 25.

Bombed site, Upper Thames Street, E.C., 1952,† *Lousley, L.N.* 32: 82. Bombed site, Cripplegate, E.C., 1952,† *Scholey*. Canal bank, Camden Town, a large colony, 1966. Bombed sites and waste ground, Bow, Poplar and Limehouse, 1967.

EROPHILA DC.

E. verna (L.) Chevall. subsp. **verna** Whitlow Grass.
Draba verna L.
Native in dry fields, stony places, on banks, etc., adventive on railway tracks, etc., also formerly common on old walls. Rare and decreasing and chiefly confined to the vicinity of the Thames.
First record: *T. Johnson*, 1632.
1. Elstree Reservoir, 1902,† *L. B. Hall, Lond. Nat. MSS.* Pinner Hill,† *W. A. Tooke, Trim. MSS.* Harefield!; Potters Bar!; near South Mimms!, *K. & L.*, 17. **2.** Shepperton!, *Newbould, T. & D.*, 422. Between Staines and Hampton, *T. & D.*, 39. West Drayton, 1926,† *Tremayne*; Harmondsworth, 1909,† *P. H. Cooke, K. & L.*, 17. Dawley,† *Warren, J.B. (London)* 8: 209. Teddington,† *Phyt. N.S.* 2: 279. **3.** Isleworth,† *T. & D.*, 39. **4.** Stanmore,† *T. & D.*, 39. Whitchurch Common,† *C. S. Nicholson, K. & L.*, 17. Neasden;† Kingsbury,† *T. & D.*, 39. **5.** Osterley Park,† *Masters*; Turnham Green,† *T. & D.*, 39. Hanwell,† *Bacot, K. & L.*, 17. Horsenden,† *Hind Fl.* Acton,† *Loydell* (D). **6.** Muswell Hill,† *A. B. Cole*; between Whetstone and Totteridge;† Edmonton;† Enfield,† *T. & D.*, 39. **7.** Hampstead Heath,† *Johns. Enum.* Near the Town,† *Rand* (DB). Chelsea, 1820,† *Pamplin MSS.* Marylebone, 1827,† *Varenne*; Haverstock Hill,† *T. & D.*, 39. Kensal Green Cemetery, 1869,† *Warren* (BM).

Subsp. **spathulata** (Láng) S. M. Walters
Erophila spathulata Láng, *E. boerhaavii* (Van Hall) Dumort., *E. inflata* (Bab.) F. J. Hanb., *E. duplex* Winge
1. Near Denham Lock!; near Uxbridge Common!, *L.N.* 26: 58. Springwell!, *Welch.* Cowley, 1903,† *Green* (SLBI). Pinner, 1866,† *Trimen* (BM). Disused railway between Uxbridge and Denham. **2.** Hampton Court! (W). Stanwell!, *Benbow* (BM). Near Staines. Thames towing path in various places between Staines and Shepperton Lock. Staines Moor. **4.** Stanmore, 1904,† *Green* (SLBI). Neasden, 1867,† *Fox* (D). **5.** Greenford,† *J. F. Young* (BM).
An extremely variable species. The subsp. *spathulata* shows a marked preference for calcareous and alluvial soils, although the two subspecies are not always readily separable. It is likely that some of the records given above are in need of revision.

ARMORACIA Gilib.

A. rusticana Gaertn., Mey. & Scherb. Horse-radish.
Cochlearia armoracia L., *Armoracia lapathifolia* auct.
Alien. S.-E. Europe. An escape or outcast from cultivation now extensively naturalised by stream- and riversides, roadsides, in fields, grassy places, on waste ground, etc., established on canal banks, railway banks, etc. Very common.
First record: *Bredwell* and *W. Martin*, 1597.
7. Frequent throughout the district.

CARDAMINE L.

C. pratensis L. Cuckoo Flower, Lady's Smock.
Lit. *Watsonia* **3**: 170 (1955): *Proc. Bot. Soc. Brit. Isles* **7**: 210 (1968). (*Abstr.*)
Native in damp grassy places, marshes, open woods, etc., established on lawns, railway banks, etc. Common, but apparently decreasing, and rare or extinct in heavily built-up areas.
First record: *T. Johnson*, 1629.
7. Primrose Hill;† Kentish Town;† Hackney Marshes!, *T. & D.*, 31. Islington,† *Ball. MSS.* Newington,† *Spencer*, 318. River wall, Hurlingham, 1970, *D. S. Warren.*

Subsp. **pratensis**
7. Regents Park!, *Holme, L.N.* 39: 45.

Subsp. **palustris** (Wimm. & Grab.) Janchen
7. Highgate Ponds!, *Fitter*. East Heath, Hampstead.

A very variable collective species; subsp. *pratensis* is usually a slender plant with pale lilac flowers, while subsp. *palustris* is a more robust plant, with paler, often white flowers, favouring wetter habitats than the former. The collective species has been extensively studied in Europe by various workers with differing results and the segregates have been described as species, subspecies, and varieties by different specialists. Although careful studies have been made on British populations, particularly by D. E. Allen, little has so far been published on the complex in these islands. Specimens collected by the author have been determined by Allen as follows:—

C. dentata Schultes
1. Harefield. Uxbridge. **2.** Staines. **3.** Hounslow Heath. **4.** Harrow. **5.** Hanwell. Heston.

C. dentata × fragilis
3. Hounslow Heath. Northolt.

C. fragilis (Lloyd) Bor.
2. Staines. **3.** Hounslow Heath. **5.** Syon Park. **6.** Finchley Common. Enfield.

C. amara L. Large-flowered Bitter Cress.
Native by sides of streams, rivers, canals, lakes, etc., in marshes, wet woods, etc. Very rare, and decreasing and now mainly confined to the Colne valley and the immediate vicinity of the Thames.
First record: *W. Sherard*, 1690.
1. Sparingly by the Colne and canal in various places between Springwell and Uxbridge. Cowley, 1862,† *Benbow* (BM). Yiewsley!,† *Benbow* (MSS). **2.** Thames-side between Hampton and Hampton Court,† *Newbould, T. & D.*, 31. Near Colnbrook, *C. S. Nicholson, K. & L.*, 15. Between Colnbrook and Harmondsworth!,† *Lowne* (LNHS). Longford, *J. E. Cooper, K. & L.*, 15. Near Staines!.† West Drayton,† *Benbow MSS.* and *Druce Notes.* Between Hampton and Sunbury, 1872,† *J. W. White* (BM). **3.** Hanworth, *De Cresp. Fl.*, 10; 1884,† *C. A. Wright* (BM). **4.** Wood near 'The Spaniard's' [= Turner's Wood], Hampstead, *Irv. MSS.*; 1862,† *Trimen* (BM). Kingsbury, 1827–30,† *Varenne, T. & D.*, 32. Bishop's Wood,† '*Sigma*', *Phyt. N.S.* 6: 47. North End, Hampstead, 1882,† *De Ves. MSS.* **5.** Greenford, *Hind. Fl.*; c. 1920,† *M. Taylor.* Osterley Park, *L. G. Payne, L.N.* 28: 32. **7.** Osier holts by the Thames-side . . . Fulham, *W. Sherard, Ray Syn.* 1, 238; c. 1820,† *Sweet*, teste *Pamplin, T. & D.*, 32. Near Chelsea,† *Pet. Midd.* and *Curt. F.L.* Isle of Dogs,† *Woods, B.G.*, 408.

Var. **erubescens** Peterm.
1. Near Denham, 1942!,† *L.N.* 26: 58. **2.** Shepperton, 1967, *Kennedy*!.

C. impatiens L. Narrow-leaved Bitter-cress.
Formerly as a denizen in shady places, woods, etc., but now long extinct.
First record: *Merrett*, c. 1666. Last record: *Whichelmore*, 1901.
4. Near Harrow, *Whichelmore, J.B. (London)* 39: 245. **7.** Primrose Hill, *Merr. MSS.* Near London, by the first stone going from St Ægidius' Church [= St Giles-in-the-Fields] towards Hampstead, *Moris. Hist.* 2: 222. On the moat-sides . . . at Fulham, *Pet. Midd.* By the Thames, near the Physic Garden, Chelsea, plentifully, *W. Watson, Hill Fl. Brit.*, 334. By the ditch sides in Hell [= Eel] Brook, at Parsons Green, plentifully, *Alch. MSS.*
Trimen and Dyer (*Fl. Middx.*, 30) were of the opinion that some other plant may have been mistaken for *C. impatiens*, e.g. *C. flexuosa* or *C. amara*. Johnson, however, refers to it (*Johns. Ger.*, 260) as being cultivated in his time and it may have been an escape from physic

gardens, furthermore there are correctly named but unlocalised specimens in Herb. Sloane.

C. flexuosa With. Wood Bitter-cress.
Native in damp woods, moist shady places, etc., and as a weed of shrubberies and garden paths. Common.
First record: *J. Newton*, c. 1680.
7. Near Cane (= Ken) Wood!; between Pancras and Kentish Town, *Newt. MSS.* New River banks between Canberry (= Canonbury) House and Newington, *Pet. Midd.* Lane opposite Mother Redcap's, *W. Watson, Hill Fl. Brit.*, 334. Chelsea; Highgate!; Hampstead!, *Curt. F.L.* Walham Green, *J. Morris, T. & D.*, 31. Kensington Gardens, *Warren Fl.*; 1949!, *L.N.* 39: 45. Buckingham Palace grounds, 1961, *Fl. B.P.G.*

C. hirsuta L. Hairy Bitter-cress.
Native on dry hedgebanks, cultivated ground, by waysides, etc. Local.
First certain record: *Hind*, 1860.
1. Ruislip!, *Melv. Fl.*, 7. Pinner!, *T. & D.*, 31. Uxbridge!, *Benbow* (BM). Harefield. Northwood. Denham. Eastcote. Potters Bar!, *Tremayne.* **2.** Near Kingston Bridge!, *Pugsley* (BM). Hampton Green!, *Williams MSS.* Bushy Park. Bedfont, *Tremayne.* Staines. West Drayton. **3.** Hanworth, *Montg. MSS.* Southall, *C. A. Wright* (BM). Isleworth. Twickenham. Northolt. Near Wood End. **4.** Harrow!, *Hind Fl.* Stanmore, *T. & D.*, 31. Highgate, *Andrewes MSS.* **5.** Near Willesden Junction,† *Warren*; Ealing Common;† Twyford,† *Newbould, T. & D.*, 31. Syon Park. Osterley. Perivale!, *Tremayne.* **6.** Hadley!, *Warren, T. & D.*, 31. **7.** St John's Wood, 1957, *Holme*; Regents Park!, *L.N.* 39: 45. Hyde Park, 1958!, *D. E. Allen, L.N.* 41: 17. Hampstead Heath Extension!, *H. C. Harris.* Fitzroy Park, Highgate, *Andrewes MSS.* Buckingham Palace grounds, *Fl. B.P.G.* Brompton Cemetery, *Brewis.*

C. bulbifera (L.) Crantz Coral-wort.
Dentaria bulbifera L.
Native, and locally plentiful in woods and copses at the junction of chalk and clay soils over a very small area in the extreme north-west part of the vice-county.
First record: *Blackstone*, 1734.
1. Harefield!, *Blackstone*, in litt. to Richardson, 1734; Old Park Wood, Harefield, abundantly!, *Blackst. Fasc.*, 23; still plentiful, 1973. Grove by Harefield Church! (HE); still present in small quantity, 1972. Garett Wood, Springwell!,† *Phyt. N.S.* 1: 62; abundant there until the wood was destroyed by chalk quarrying about 1955. Copse near Jack's Lock, Harefield!, *Benbow* (BM); still plentiful, 1967.

BARBAREA R.Br.

B. vulgaris R.Br. Winter Cress, Yellow Rocket.
Lit. *J.B.* (*London*) **54**: 208 (1916).
Native on the banks of rivers, streams, ditches, etc., in damp grassy places, on waste ground, etc. Very common.
First record: *T. Johnson*, 1629.
7. Common in the district but rare in Central London owing to the absence of suitable habitats. It has, however, been recorded from Regents Park!, *T. & D.*, 29. Hyde Park, *Allen Fl.* Hyde Park Corner, 1963, *Brewis*, *L.N.* 44: 20.
The common Middlesex plant appears to be the var. **campestris** Fr., a robust plant with siliquae up to 25 mm. long.

Var. **sylvestris** Fr.
3. Between Isleworth and Twickenham, 1916, *Wilmott* and *Jackson* (BM).

Var. **arcuata** (Opiz ex Presl) Fr. sub-var. **brachycarpa** A. B. Jackson
2. Thames-side between Kingston Bridge and Hampton Court, 1912, *C. E. Britton*, *J.B.* (*London*) 54: 210.

B. stricta Andrz. Small-flowered Yellow Rocket.
Lit. *J.B.* (*London*) **46**: 109 (1908).
Alien. Europe. Naturalised by river- and streamsides, in ditches, marshes, on damp waste ground, etc. Very rare, and mainly confined to the immediate vicinity of the Thames.
First certain record: *Dyer*, 1867.
2. Between Hampton Court and Kingston Bridge, in seed, 1867, *Bloxam*, *T. & D.*, 29. This may have been a form of *B. vulgaris*. West Drayton, 1923,† *Worsdell* (K). **3.** Chase Bridge, between Twickenham and Hounslow, 1867, *Dyer*, *J.B.* (*London*) 9: 304; 1871,† *Dyer*; Thames at Twickenham, 1892, *Wolley-Dod* (BM). Banks of the Crane near Twickenham, 1933, *Watts*, *K. & L.*, 114. Near Isleworth Church, *Baker*, *J.B.* (*London*) 9: 213; 1916, *Wilmott* (BM). Between Richmond and Isleworth, on both sides of the river, *Britten*, *Dickens*, 31. Duke's river, Isleworth, 1878, *Britten* (BM). **5.** Syon Park!, *A. B. Jackson*. Chiswick House grounds!, *L.N.* 32: 81; 1965.

B. intermedia Bor. Intermediate Yellow Rocket.
Alien. Europe, etc. Naturalised on waste ground, etc. Very rare, though perhaps sometimes confused with forms of *B. vulgaris*.
First evidence: *A. Irvine*, 1835.
1. Near Uxbridge, 1885–91; Uxbridge, 1888, *Benbow* (BM). Near Harefield, 1955,† *T. G. Collett*. Ruislip, *Moxey List*. Perhaps *B.*

vulgaris intended. Near Denham, 1966. **3.** Near Hounslow, 1866–67, *Trimen* (BM). Twickenham, *Butch. MSS.* **5.** Chiswick!,† *A. B. Jackson, K. & L.*, 14. **6.** Southgate, 1966, *D. E. G. Irvine.* **7.** Little Chelsea, 1835,† *A. Irvine* (STR). Crouch End, 1875,† *French* (BM).

B. verna (Mill.) Aschers. Early-flowering Yellow Cress, American Cress.
B. praecox (Sm.) R.Br.
Alien. Mediterranean region, etc. Naturalised on cultivated and waste ground, etc. Rare, and sometimes merely casual.
First record: *A. Irvine*, c. 1830.
1. Ruislip, 1866, *Trimen*; between Uxbridge and Cowley, 1884; Yiewsley!; Hillingdon, *Benbow* (BM). Uxbridge Moor, 1946!,† *L.N.* 26: 57. Harefield, 1953. Ickenham, 1955, *T. G. Collett.* **2.** Near Laleham, *Benbow* (BM). West Drayton, 1923, *Worsdell* (K). **3.** Twickenham, *T. & D.*, 29. Near Whitton, *Benbow MSS.* Near Hounslow, 1880, *H. Groves* (BM); 1905, *Montg. MSS.* Hounslow Heath, 1961, *Philcox* and *Townsend.* Feltham, *P. H. Cooke*; Sutton, 1950!, *K. & L.*, 14. Roxeth, *Melv. Fl.*, 7. **4.** Hampstead, *Irv. MSS.* Harrow Weald, *T. & D.*, 29. Near Stanmore, 1878, *De Crespigny* (BM). Near Mill Hill, 1966, *Kennedy*!. **5.** Northolt, *Melv. Fl.*, 7. Brentford, *Irv. Ill. Handb.*, 694. Alperton; Turnham Green, *T. & D.*, 29. Acton, 1902, *Green* (SLBI). Near Ealing, plentiful, 1967, *McLean.* **6.** Southgate, *A. B. Cole, T. & D.*, 29. **7.** Chelsea,† *Newbould, T. & D.*, 29. South Kensington, 1865, *Trimen* (BM); 1867,† *Dyer* (K).

ARABIS L.

A. glabra (L.) Bernh. Tower Mustard.
Turritis glabra L.
Formerly native on dry hedgebanks, waste ground, by roadsides, etc., chiefly on light soils, but long extinct.
First record: *Gerarde*, 1597. Last record: *Newbould*, c. 1860.
2. Between Hampton Wick and Sunbury, 1835, *A. Irvine, Trim. MSS.* Between Staines and Twickenham, *Wats. MSS.* **3.** Near Isleworth, *Doody, Pet. Midd.* **5.** Between Turnham Green and Brentford, c. 1860, *Newbould, T. & D.*, 30. **6.** At Pyms, by a village called Edmonton, *Ger. Hb.*, 213. **7.** Kensington Park (= Gardens), *Forst. Midd.*

NASTURTIUM R.Br.

N. officinale R.Br. Watercress.
Rorippa nasturtium-aquaticum (L.) Hayek
Lit. *Watsonia* **1**: 228 (1950): *J. Ecol.* **40**: 228 (1952).
Native in watery places. Common.
First certain evidence: *Children*, 1837.

7. East Heath, Hampstead. Highgate Ponds. Hackney Marshes. Hackney Wick.

N. microphyllum (Boenn.) Reichb. One-rowed Watercress.
N. uniseriatum Howard & Manton, *Rorippa microphylla* (Boenn.) Hyland.
Lit. *Watsonia* **1**: 228 (1950): *J. Ecol.* **40**: 228 (1952).
Native in watery places. Rather rare, though sometimes confused with *N. officinale* with which it sometimes grows.
First certain evidence: *E. Metcalfe*, 1835.
1. Uxbridge, *Benbow* (BM). Stanmore Common!, *K. & L.*, 13. South Mimms, *J. G. Dony*!. **2.** Shortwood Common, Staines. Staines Moor. Knowle Green. Bushy Park. Hampton Court Park. **3.** Near Twickenham. Thames-side, Twickenham. **4.** Hampstead Heath!, 1835, *E. Metcalfe* (K); still locally plentiful near the bog behind Jack Straw's Castle on the West Heath. North End Fields, Hampstead† (HPD). Mill Hill!, *Hinson*. **5.** Syon Park!, *Turrill* (K). **6.** Finchley Common!, *K. & L.*, 13. Near Enfield. Southgate. Trent Park, *T. G. Collett* and *Kennedy*. **7.** East Heath, Hampstead.

N. microphyllum × **officinale** = **N.** × **sterile** (Airy Shaw) Oefelein
Rorippa × *sterilis* Airy Shaw
Lit. *Watsonia* **1**: 228 (1950): *J. Ecol.* **40**: 228 (1952).
2. Staines, '*J.F.G.*' (BM). **4.** West Heath, Hampstead!, *T. Moore* (BM). **6.** Finchley Common!, *K. & L.*, 13.

RORIPPA Scop.

Lit. *Proc. Bot. Soc. Brit. Isles* **7**: 410 (1968). (*Abstr.*)
R. sylvestris (L.) Besser Creeping Yellow-cress.
Nasturtium sylvestre (L.) R.Br.
Native on the banks of rivers and streams, verges of ponds, damp waste ground, by roadsides, etc. Common, and increasing.
First record: *Petiver*, 1695.
7. Between Whitechapel and Mile End, *Pet. Midd.* Tothill Fields, Westminster, *Blackst. Spec.*, 20; very abundant there, *Curt. F.L.*; c. 1780 (LT); c. 1800, *E. Forster* (BM); 1811,† *Garry Notes*. Hammersmith!, *E.B.*, 2324. Fulham!, *Britten*; near Chelsea, *Pamplin*; Kensington Gardens!; East Heath, Hampstead!, *T. & D.*, 28. Regents Park!, *Holme*; Hyde Park!; Primrose Hill!, *L.N.* 39: 45. Buckingham Palace grounds, *Fl. B.P.G.* Tower of London gardens, 1966!, *L.N.* 46: 28. Clapton. Hackney. Tottenham. Hurlingham, *L. M. P. Small*.
R. sylvestris is an extremely variable species, and Jonsell (*Symbol. Bot. Upsal.* (**19**) **2** (1968)) has pointed out that tetraploid, hexaploid and pentaploid races occur, though these are not morphologically distinct

from each other. In Scandinavia the aggressive 'weedy' race is usually hexaploid, and it seems likely that the Middlesex plant which readily invades grassy road verges and waste ground often far from damp habitats comes under this category.

R. palustris (L.) Besser subsp. **palustris** Marsh Yellow-cress.
R. islandica auct. eur., non (Oeder ex Murray) Borbás, *Nasturtium palustre* (L.) DC., non Crantz
Native on the banks of rivers and streams, verges of ponds, damp waste ground, by roadsides, etc., also established on damp old walls. Common.
First record: *Curtis*, 1783.
7. Common throughout the district.

R. amphibia (L.) Besser Great Yellow-cress, Water Rocket.
Nasturtium amphibium (L.) R.Br., *Armoracia amphibia* (L.) G. F. W. Mey.
Native on the banks of rivers and streams, verges of ponds, etc. Locally common, especially by the Thames and its tributaries.
First record: *Gerarde*, 1597.
7. In the chinks of a stone wall upon the river Thames by the Savoy in London,† *Ger. Hb.*, 186. Tothill Fields, Westminster† (LT). Near Mile End,† *Spencer*, 318. Isle of Dogs, 1817 (G & R). Westbourne Green, 1827–30,† *Varenne*; Hackney Marshes!; brick wall of a house fronting the river (now in the embankment) at the bottom of Surrey Street, Strand, 1867,† *T. & D.*, 40. River wall near Chelsea Bridge, 1967. Bishop's Walk, Fulham, *Britten*; river bank at Cremorne, *T. & D.*, 40. Kensal Green!,† *Warren, Trim. MSS.* By Lee Navigation Canal, Upper Clapton.
Both diploid and tetraploid races occur in the vice-county, though these cannot be distinguished morphologically. Jonsell (*Symbol. Bot. Upsal.* **19** (2) (1968)) has pointed out, however, that the Thames-side plant is exclusively diploid.

R. amphibia × palustris
5. Strand-on-the-Green!, *F. Ballard* (K). Near Barnes Bridge!, *Bangerter, K. & L.*, 14. Brentford.
This interesting sterile hybrid, which spreads vegetatively, was first recorded by Britten (*J.B.* (*London*) 47: 430 (1909)), from the river wall near Hammersmith Bridge, Surrey. It is a tall robust plant, with deeply pinnatifid leaves, and pale yellow flowers, smaller than those of *R. amphibia*, but larger than those of *R. palustris*, and is possibly more widespread than the few records suggest.

R. amphibia × sylvestris
2. Thames bank below Hampton Court, 1909, *Pugsley* (BM).
A fertile hybrid, usually with pinnatifid-pinnatisect cauline leaves, the terminal segment $\frac{1}{4}$–$\frac{2}{3}$ of the total leaf-length; styles nearly always

longer than 1·2 mm; ripe pods at least c. 1·5 mm broad, and most pedicels deflexed from the base in fruit. It is probably more frequent than the solitary record suggests.

R. austriaca (Crantz) Besser Austrian Yellow-cress.
Alien. Europe, etc. Established on waste ground. Very rare.
First record: *Simpson* and *Sandwith*, 1942.
5. Several large patches for a distance of half a mile on waste ground by the Thames, Chiswick!, *Simpson* and *Sandwith*, *Rep. Bot. Soc. & E.C.* 12: 480; still plentiful, 1963; considerably reduced by mowing, and the parking of cars on the habitats, 1964; not seen, 1965; a solitary small patch still present, 1967; increasing considerably, 1968.

HESPERIS L.

H. matronalis L. Dame's Violet.
Alien. Europe, etc. Garden escape. Naturalised in grassy places and established on waste ground, etc. Rather rare.
First record: — 1856.
1. Near Uxbridge, *Benbow*; Yiewsley, 1908, *J. E. Cooper* (BM). Abundant in an old gravel pit between Harefield and Ickenham, 1944, *Payton*!, *K. & L.*, 18. Chalkpit, Springwell, 1965, *T. G. Collett*, *Kennedy* and *Pickess*!. Pinner, 1867, *Hind*, *Melv. Fl.* 2: 29. Northwood, *Benbow MSS.* South Mimms, 1965, *C. M. Dony*!. **2.** Near Hampton Court, *Benbow MSS.* **3.** Hayes, 1966. Twickenham, 1966. **4.** Bentley Priory, *Melv. Fl.*, 18. West Heath, Hampstead, one plant, 1949, *E. C. Wallace*!, *K. & L.*, 18. Near East Finchley, 1909, *Green* (SLBI). **5.** Chiswick, *Sandwith*, *L.N.* 26: 58. Near Kew Bridge, in quantity, 1967–69.† **6.** Near Barnet, 1856, *Phyt. N.S.* 1: 407. Near Muswell Hill, 1886, *Benbow* (BM). **7.** Hyde Park, 1940, *Gibson*, *K. & L.*, 18; 1961, *Brewis*, *L.N.* 41: 17. Regents Park, 1956, *Holme*, *L.N.* 39: 45. Bombed site between Silver Street and Cheapside, E.C.!,† *Sladen* and *McClintock*, *Lousley Fl.* Bombed site, Cripplegate, E.C., 1953,† *Lawfield.* Shadwell!, 1954, *Henson* (BM); 1967. St John's Wood, 1956, *Holme.* Upper Clapton, *L. M. P. Small.*
Both the subsp. **matronalis,** with red-mauve to violet flowers, and the subsp. **nivea** (Baumg.) Kulcz., with white flowers occur. It is likely also that the hybrid between the two subspecies is present.

ERYSIMUM L.

E. cheiranthoides L. Worm Seed, Treacle Mustard.
Colonist on cultivated and waste ground, by roadsides, etc. Common, especially on light soils and near the Thames.
First record: *Merrett*, c. 1670.
7. Frequent in the district.

CHEIRANTHUS L.

C. cheiri L. Wallflower, Gilliflower.

Alien. S. Europe, etc. Garden escape. Established on old walls, in chalkpits, etc. Very rare.

First record: *T. Johnson*, 1638.

1. Harefield!, *Blackst. Fasc.*, 51. Cowley, 1862,† *Benbow* (BM). Uxbridge. Ruislip, *I. G. Johnson.* **2.** Colnbrook!, *K. & L.*, 13. West Drayton churchyard. **3.** Isleworth,† *T. & D.*, 27. Cranford, 1945!,† *K. & L.*, 13. **4.** Harrow,† *Hind Fl.* Hampstead,† *Irv. MSS.* Stanmore, 1966. **5.** Chiswick,† *A. B. Cole, T. & D.*, 27. **6.** Wood Green,† *A. B. Cole*; Edmonton,† *T. & D.*, 27. Hornsey† (HE). Enfield!, *H. & F.*, 273; 1967. **7.** Tottenham,† *Johns. Cat.* Newington, *Ball. MSS.*; Stoke Newington, 1967. Old Bailey, E.C., deliberately introduced on to bombed sites,† *Lousley Fl.* Bombed sites, Cripplegate, E.C., 1950,† deliberately introduced. Waste ground near Euston Station, 1963.†

ALLIARIA Heist. ex Fabr.

A. petiolata (Bieb.) Cavara & Grande Garlic Mustard, Jack-by-the-Hedge.

A. officinalis Bieb., *Sisymbrium alliaria* (L.) Scop.

Native on hedgebanks, in ditches, shady places, by wood borders, roadsides, etc. Very common except in heavily built-up areas where owing to the lack of suitable habitats it is rare or has been eradicated.

First record: *T. Johnson*, 1629.

7. Blackstock Lane;† Green Lanes;† Highgate!, *T. & D.*, 34. Regents Park!, *L.N.* 39: 45. Fulham Palace grounds. Parsons Green. St John's Wood. East Heath, Hampstead. Ken Wood. Near Holborn. Finsbury Park. Hornsey. Stoke Newington. Tottenham. Isle of Dogs. Tufnell Park, *L. M. P. Small.*

SISYMBRIUM L.

S. officinale (L.) Scop. Hedge Mustard.

Native on hedgebanks, waste ground, by roadsides, etc. Very common.

First record: *T. Johnson*, 1638.

7. Frequent throughout the district.

The var. **leiocarpum** DC., with light-green coloured foliage and glabrous silicules, is common though probably adventive.

S. irio L. London Rocket.

Alien. Mediterranean region, etc. Naturalised on waste ground, by roadsides, etc., though sometimes merely casual. Locally plentiful in

London from the mid-seventeenth to the early nineteenth century when it apparently became extinct. It reappeared in the vice-county at the end of the Second World War and is now well established in a few localities near central London.

First record: *Goodyer*, c. 1650.

5. Rubbish-tip, Greenford, abundant, 1953!,† *Welch, L.N.* 33: 54. Carville Hall Park South, Brentford, 8 plants, 1973. **7.** Near Whitechapel,† *Goodyer, How MSS.* Almost everywhere in the suburbs of London,† *Merr. Pinn.*, 66; especially on earth mounds between the City and Kensington;† in 1667 and 1668, after the City was burnt, it grew very abundantly on the ruins round St Paul's,† *Ray Cat.*, 104. Copiously about Chelsea, *Moris. Hist.* 2: 219; c. 1695, *Pet. Midd.*; in Chelsea Garden and all that neighbourhood a troublesome weed, *E.B.*, 1631; 1807,† *D. Turner* (K). New Chelsea,† *E. Forster* (BM). Near Islington,† *Pet. Bot. Lond.* Goswell Street,† *Hill Fl. Brit.*, 338. Frequent enough about London,† *Curt. F.L.* Brompton,† *Borrer*; Shoreditch,† *Dillwyn, B.G.*, 408. Kensington,† *Lawson MSS.* Between Earls Court and Walham Green, 1832;† between Little Chelsea and Hyde Park Corner,† *Haworth*, teste *Pamplin, T. & D.*, 33. Haggerston,† *E. Forster* (BM). Strand, *Johns. Nat.* Error, the plant seen was *S. altissimum.* Near the British Museum, *Shenst. Fl. Lond.* Error, the plant seen was probably *S. loeselii.* Walham Green† (BM). Hornsey,† *Ball. MSS.* Trinity Gardens, E.C., 1945!, *Evetts* (BM); spreading, 1952!, *Lousley, L.N.* 33: 53; plentiful, 1966. Tower of London gardens!, *Lousley*; abundant, and increasing, 1967. Bombed site, Great Tower Street, E.C.!,† *A. W. Jones, L.N.* 33: 53. Regents Park!, *Holme, L.N.* 36: 15; still present at the side of the Outer Circle behind the Zoological Gardens, 1968. Near the Snake Pit, Zoological Gardens, Regents Park, 1966, *Kennedy, L.N.* 46: 28; 1967.

The abundant spread of *S. irio* over the ruins of the City of London following the Great Fire of 1666 suggests a 'population explosion' which may in some degree be compared with the spread of *Epilobium angustifolium* and *Senecio squalidus* over the London bombed sites from 1941 onwards. The following account of the spread of the species has been translated from *Moris. Prael. Bot.*, 498 (1669):—'The spring after the conflagration at London all the ruins were overgrown with an herb or two, but especially one with a yellow flower: and on the south side of St Paul's Church (= Cathedral) it grew as thick as could be; nay on the very top of the tower. The herbalists call it *Ericoleoris neopolitana* (*sic*) Small Bank Cresses of Naples; which plant Thos. Willisel told me he knew before but in one place about the town and that was at Battle Bridge (= King's Cross) by the "Pindar of Wakefield" and that in no great quantity.'

Other species of *Sisymbrium* have frequently been mistaken for *S. irio*, in particular *S. altissimum*, *S. orientale* and *S. loeselii*. The diagnostic characters of these, and other species known to have occurred in the London area, are, however, well defined by Bangerter and Welch (*L.N.* **31,** 13 (1952)).

S. loeselii L.

Alien. Central Asia, etc. Naturalised on waste ground, by roadsides, established on rubbish-tips, etc. Locally common near the North Circular Road, rather rare, but increasing elsewhere.

First evidence: *Benbow*, 1883.

1. Uxbridge Common, 1883,† *Benbow*; Yiewsley, *J. E. Cooper* (BM). Uxbridge Moor!,† *L.N.* 27: 30. Stanmore Common, 1957.† New Years Green, 1961, *J. G. Dony*!; 1964, *T. G. Collett*!. **2.** West Drayton!, *P. H. Cooke* (LNHS). Harmondsworth, locally plentiful, 1967. Shepperton, 1969, *Clement* and *Ryves*. **3.** Hounslow Heath, locally plentiful. Northolt. **4.** Hampstead!, 1910, *J. E. Cooper* (BM); 1966. Stonebridge, frequent!, *L.N.* 27: 30; 1967. Near Park Royal, 1967, *McLean*. Locally abundant on waste ground near the North Circular Road from Staples Corner, Cricklewood to the Brent Cross Flyover. Willesden Green, 1963.† The Hyde, Hendon, 1965. Brent Reservoir, frequent, 1966. Edgware Road, Cricklewood, 1966. Between Kenton and Wealdstone, 1966. **5.** Osterley Park!,† *L.N.* 27: 30. Hanwell!, *K. & L.*, 19; 1967. Acton!; Park Royal, *Druce Notes*. Near Brentford, 1966!, *Boniface*. **6.** Near Finchley!, *L.N.* 27: 30. Palmers Green!, *K. & L.*, 19. **7.** Russell Square, W.C.!, *H. T. White* (K); frequent there for nearly sixty years but now almost eradicated; old wall near Russell Square, 1964, *Wurzell*. Bombed sites, Bedford Row, W.C., abundant, 1950!,† *Whittaker* (BM). By Southwark Bridge, 1945–51,† *Lousley*. Hackney Marshes!, 1957, *B. T. Ward* and *Lousley*; 1967. Bombed sites, Barbican, E.C. and Stepney, 1966. Slum clearance area near Paddington Basin, 1966!, *L.N.* 46: 28. Isle of Dogs, 1966.

Var. **ciliatum** Beck

7. Russell Square, 1946, *David* (K).

S. orientale L. Eastern Rocket.

S. columnae Jacq.

Alien. Mediterranean region, etc. Naturalised on waste ground, by roadsides, etc., established on rubbish-tips, railway tracks, etc. Common, and increasing, though sometimes merely casual.

First evidence: *Benbow*, 1883.

7. Westminster!, 1912, *Drummond* (K); 1946. Westbourne Park!, *L.N.* 28: 32. Near Mornington Crescent, N.W.1!; Regents Park!, *Holme*, *L.N.* 39: 45. Hyde Park, 1961, *Brewis*; Kensington Gardens, 1961, *D. E. Allen*, *L.N.* 41: 17. Near Euston Station, 1965!; roadside, Stanhope Terrace, W.2, 1966!, *L.N.* 46: 28. Abundant on several bombed sites in the City, 1944→!, *Lousley*. Island at bottom of Park

Lane, W.1; Tower Hill, 1965!, *Brewis*; Euston Road, N.W.1, 1966, *Holme*, *L.N.* 46: 28. Bombed site, Mile End Road, E.1,† *Lovis*. Hackney Marshes!, *J. E. Cooper* (BM). Hackney, *Henson*. Holloway!, 1965, *D. E. G. Irvine*; 1966. Isle of Dogs, formerly frequent but now scarce. Weed in flower-bed, Buckingham Palace grounds, 1967, *McClintock*, *L.N.* 47: 9. Earls Court, 1965, *Brewis*. Chelsea, 1969, *E. Young*.

S. altissimum L. Tall Rocket, Tumbling Mustard.
S. pannonicum Jacq., *S. sinapistrum* Crantz
Alien. E. Europe, etc. Naturalised on waste ground, by roadsides, etc., established on rubbish-tips, etc. Common.
First evidence: *Benbow*, 1862.
7. Waste ground at the back of the Tate Gallery, S.W.1,† *H.* and *J. Groves*; Brompton Road, 1922,† *Kirkpatrick* (BM). Abundant on vacant building sites between the Strand and Aldwych, W.C., *C. E. Britton*, *J.B.* (*London*) 47: 431; 1920,† *F. Druce* (BM). City of London bombed sites, 1945→!, *Lousley*. Bombed sites, Tottenham Court Road!† and Great Russell Street, W.C.1!,† *L.N.* 28: 32. Bombed site, Red Lion Square, W.C., 1950,† *Whittaker* and *Bangerter*; bombed sites, Shadwell!, *Henson* (BM); 1967. Mornington Crescent, N.W.1, 1956, *Holme*, *L.N.* 39: 45. Piccadilly, W.1, 1961!,† *D. E. Allen* and *McClintock*, *L.N.* 41: 17. Near Hyde Park Corner, 1961!,† *D. E. Allen* and *McClintock*, *L.N.* 41: 13. Isle of Dogs, 1887, *Benbow* (BM). St Katherine's Dock, Stepney, 1963, *Fitter*; 1966!, *Brewis*. Hackney!, *Renson*. Hackney Marshes, abundant. Hornsey. Slum clearance areas between Royal Oak and Paddington, abundant, 1966!; site of Harrow Road–Edgware Road Flyover, 1966!;† Hoxton, 1966!, *L.N.* 46: 28. Bishop's Walk, Fulham, 1966. Regents Park, *Brewis*.

ARABIDOPSIS (DC.) Heynh.

A. thaliana (L.) Heynh. Thale Cress, Common Wall Cress.
Sisymbrium thalianum (L.) Gay
Native on dry hedgebanks, cultivated and waste ground, in fields, by roadsides, etc., established on old walls, railway tracks, etc., also as a weed of flower beds in parks, cemeteries, etc. Common on light soils in the southern part of the vice-county, especially near the Thames. Less frequent elsewhere except on railway tracks and as a weed of flower beds, etc.
First record: *T. Johnson*, 1638.
7. Tottenham!, *Johns. Cat.* Near Marylebone; near Islington, *Merr. Pin.*, 93. Highgate, *Andrewes MSS*. Hackney Marshes, *J. E. Cooper*; bombed site, Bedford Way, W.C.,† *Whittaker* (BM). Hyde Park!; Kensington Gardens!, *Allen Fl.* Buckingham Palace grounds, *Fl. B.P.G.* Holloway, *D. E. G. Irvine*. Garden weed, Chelsea, 1969; Hurlingham, *E. Young*.

RESEDACEAE

RESEDA L.

R. luteola L. Dyer's Rocket, Weld.

Native in fields, on waste ground, etc., and established on railway banks, old walls, etc. Common, especially on the chalk and near the Thames.

First record: *Blackstone*, 1737.

7. Parsons Green, *A. Irvine, Phyt. N.S.* 2: 168. Hyde Park!, 1815 (G & R); 1962!, *Allen Fl.* St Pancras!, 1816, *E. T. Bennett, T. & D.*, 45; St Pancras Goods Yard, 1966. South Kensington, *T. & D.*, 45. Kentish Town!, *Irv. MSS.* Kensington Gardens, 1958!, *D. E. Allen*; near Paddington!, *L.N.* 39: 45. Tufnell Park, 1870, *French, Trim. MSS.* Highbury, *Ball. MSS.* Bombed site, Fish Street Hill, E.C., 1945!,† *Lousley*. Wapping. Near Shadwell!, *Henson*; Isle of Dogs!, *Benbow*; bombed sites, Leather Lane, E.C.† and Red Lion Square, W.C.,† *Whittaker* (BM). Bombed sites, Finsbury Square† and Cripplegate areas, E.C.,† *Jones Fl.* Fulham. Building site, Hampstead Road, N.1, 1966!, *L.N.* 46: 28. Hammersmith. Shoreditch. Tottenham, *Kennedy*.

R. lutea L. Wild Mignonette.

Native on banks, cultivated ground, etc., on calcareous and alluvial soils in the north-west and south-west parts of the vice-county respectively. In other areas it occurs about railway premises, on waste ground, etc., as an established introduction. Local as a native species but rather common as an introduction.

First record: *Newbould*, 1868.

1. Harefield!, *Newbould, T. & D.*, 44. Springwell!, *Benbow, J.B. (London)* 23: 36. Uxbridge!, *Benbow*; Yiewsley, *J. E. Cooper* (BM). Denham!, *Coop. Cas.* Disused railway between Uxbridge and Denham, frequent. **2.** Shepperton, sparingly, *Newbould, T. & D.*, 422. Yeoveney!, *Benbow* (BM). Near Chertsey Lock. Staines!, *Welch, L.N.* 27: 30. Colnbrook. Near Ashford, *Hasler*. West Drayton. **3.** Twickenham, *Trim. MSS.* Hounslow. Hounslow Heath. Hanworth, *Westrup*. Southall. Northolt. South Ruislip!, *Wrighton, L.N.* 26: 58. Near Hayes. **4.** Finchley, *J. E. Cooper* (BM). Highgate, *Andrewes MSS.* Railway banks, Harlesden. Near Scratch Wood; disused railway, Mill Hill, *Kennedy*!. Disused railway between Stanmore and Belmont, *Roebuck*. **5.** Hanwell!, *L.N.* 27: 30. Greenford!, *L.N.* 26: 58. Between Hanwell and Brentford. Ealing!, *Benbow* (BM). West Ealing Goods Yard. Acton Goods Yard. Old Oak railway sidings. Gunnersbury, 1872,† *Britten, Trim. MSS.* Chiswick, *Murray List.* Near Osterley. Canal bank, Brentford!, *Boniface*. **6.** Grange

Park, *Tremayne*. Enfield!, *H. & F.*, 150. Wood Green!, *Scholey*. Edmonton. Railway side near Hadley Wood!, *J. G.* and *C. M. Dony*. Tottenham!, *Kennedy*. **7.** Finchley Road!, *T. & D.*, 422. South Kensington, 1870,† *Dyer*; Hackney Marshes, 1924,† *J.E. Cooper*; bombed sites, Shadwell!, *Henson* (BM). Railway side, West Brompton!, *L.N.* 26: 58. Highgate!, *L.N.* 27: 30. Railway side near Paddington!, *L.N.* 39: 45. Hyde Park, one plant, 1960, *D. E. Allen*, *L.N.* 41: 17. Isle of Dogs!, *Benbow*, *J.B.* (*London*) 25: 365; 1967. Railway side between Holloway and Finsbury Park!, and at Finsbury Park!, *Le Tall*, *Whitwell MSS.* Clapton, *Lond. Nat. MSS.* Bombed site, Bunhill Row area, E.C.,† *Jones Fl.* Highgate. Camden Town!; St Pancras Goods Yard!; building site, Hampstead Road, N.1!; Paddington!,† *L.N.* 46: 28. Railway side, Bromley.

Var. **pulchella** J. Muell.
7. Bombed site, Hammersmith, 1952!,† *Sandwith*, *L.N.* 32: 81.

VIOLACEAE

VIOLA L.

Subgen. VIOLA
Lit. E. S. Gregory, *British Violets*. Cambridge. 1912.

V. odorata L. Sweet Violet.
Lit. *Rep. Bot. Soc. & E.C.* **12**: 834 (1946).
Denizen, though obviously often an escape or outcast from gardens. Hedgebanks, thickets, wood borders, etc. Rather rare, and decreasing.
First record: *T. Johnson*, 1638.
1. Harefield!, *Blackst. Fasc.*, 109. Between Woodhall and Pinner,† *T. & D.*, 45. Ickenham!; Uxbridge Common!, *Benbow* (BM). Ruislip!, *Benbow MSS.* Near Bentley Priory!; Denham, *Tremayne*; New Years Green!; Pinner Hill!; Northwood!, *K. & L.*, 27. Cowley† (SY). Near Potters Bar. South Mimms. Bentley Heath, *J. G.* and *C. M. Dony*. **2.** Between Kingston Bridge and Hampton Court!, *T. & D.*, 45. Between Shepperton and Laleham, *Welch*. Laleham. Stanwell!, *Briggs*. **3.** Roxeth,† *Melv. Fl.*, 11. Worton;† Twickenham,† *T. & D.*, 45. Isleworth. Wood near Hounslow Heath!, 1965, *P. J. Edwards*; 1967. **4.** Near Harrow, 1836, *Fowle* (BM). Harrow Park, *Melv. Fl.*, 11. Hampstead Heath, *Benbow* (BM); c. 1910,† *Whiting Fl.* New Hampstead, 1819† (G & R). Near Highwood, *Evans Hist.*, 16. Mill Hill. Stanmore!, *K. & L.*, 27. Willesden Old Churchyard, 1965. **5.** Alperton,† *A. B. Cole*; Syon Park, *T. & D.*, 45. Near Brentford, 1883,† *Fraser* (K). Greenford,† *Green*, *Williams MSS.* Ealing, 1946!,† *Boucher*. Horsendon Hill, 1951!, *L. M. P. Small*. Chiswick,† *Green*, *Williams MSS.* **6.** Tottenham,† *Johns.*

Cat. Edmonton,† *T. & D.*, 45. Enfield!, *H. & F.*, 150. Near Winchmore Hill, *Hansen*.

Var. **dumetorum** (Jord.) Rouy & Fouc.
1. Near Uxbridge Common!, *K. & L.*, 27. **2.** Wood near Sunbury, 1967, *Kennedy*!. **3.** Wood near Hounslow Heath!, 1965, *P. J. Edwards*; 1967. **4.** Harrow Park, *Melv. Fl.*, 11. **5.** Hanwell Churchyard, 1945!,† *K. & L.*, 27.

Var. **imberbis** (Leight.) Henslow
1. Garett Wood, Springwell!;† Old Park Wood, Harefield!, *K. & L.*, 27.

V. hirta L. Hairy Violet.
Native on hedgebanks, wood borders, shady places, etc. Locally common on the chalk, very rare on other soils.
First record: *Blackstone*, 1737.
1. Locally common on the chalk at Harefield and Springwell. **2.** By towing path, Hampton Court, 1963, *Clement*, *L.N.* 43: 22. **4.** Whitchurch Common, 1906,† *L. B. Hall*, *K. & L.*, 27. **5.** Perivale Churchyard, *Green*; 1934!,† *K. & L.*, 27.

V. riviniana Reichb. subsp. **riviniana** Common Violet, Dog Violet.
Lit. *New Phyt.* **40**: 189 (1941).
Native in woods, on heaths, hedgebanks, etc., and established on railway banks, etc. Locally common, especially in the north of the vice-county.
First certain evidence: *S. P. Woodward*, 1844.
1. Frequent. **2.** Between Staines and Hampton. Between Ashford Common and Staines. Hampton Court. Spout Wood, Stanwell!, *Briggs*. **3.** Hounslow Heath, 1866, *Trimen*; 1887,† *Benbow* (BM). Near Hounslow;† Whitton, *T. & D.*, 46; 1887,† *Benbow* (BM). Railway banks near Wood End, abundant, *Payton*!; 1966. Hayes Park, *Royle*!. **4.** Harrow!, *Melv. Fl.*, 11. Deacon's Hill!; Bentley Priory!; Bishop's! and Turner's Woods, *T. & D.*, 46. Hampstead Heath, 1865 (SY); 1882† (HPD). Highgate, *Andrewes MSS*. Kingsbury Churchyard. Scratch Wood. Mill Hill, *Harrison*. Brockley Hill, *Kennedy*!. East Finchley, 1966. Edgware, 1967, *Horder*!. Disused railway between Stanmore and Belmont, *Roebuck*. **5.** Osterley Park!, *Welch*. Hanger Lane, Ealing, 1965→. Railway bank near Park Royal, 1966, *McLean*. **6.** Hadley!, *Warren*; Edmonton; Winchmore Hill Wood,† *T. & D.*, 46. Queen's Wood, Highgate!, *Bangerter*. Crew's Hill golf course, Enfield; near Forty Hill, Enfield, *Kennedy*!. **7.** Ken Wood!, *H. C. Harris*. Buckingham Palace grounds, 1961,† *Fl. B.P.G.* Tottenham, *Kennedy*.

Forma **villosa** Neum., Wahlst. & Murb.
1. Ruislip Woods!; near Northwood, *Riddelsdell* (BM).

Subsp. **minor** (Gregory) Valentine
Var. *minor* Gregory
1. Uxbridge Common;† Ruislip Common!; Pield Heath, near Hillingdon;† Ickenham Green!; near Northwood, *Benbow MSS.* **4.** Harrow† (ME). Hampstead, 1844,† *S. P. Woodward* (CYN). Hampstead Heath;† Bishop's Wood, *T. & D.*, 46. Edgwarebury!, *K. & L.*, 27.

V. reichenbachiana Jord. ex Bor. Pale Wood Violet.
V. sylvestris auct.
Native in woods, shady places, on hedgebanks, etc. Locally common, especially on the chalk, though much less widespread than *V. riviniana* with which it sometimes grows.
First certain evidence: *Twining*, 1841.
1. Springwell. Harefield. Northwood. Elstree.† North of Bentley Heath, Potters Bar, *Kennedy*. **2.** Bedfont, overrunning the churchyard, 1919,† *Tremayne, K. & L.*, 28. **3.** Near Twickenham, 1841,† *Twining* (K). Southall, 1883,† *Wright* (BM). **4.** Harrow!, *Hind, Melv. Fl.*, 11. Bishop's Wood!; Scratch Wood!, *K. & L.*, 28. Stanmore, *Tyrrell*!. **5.** Hanwell, 1887,† *Benbow* (BM). Horsendon Wood,† *A. Wood, Trim. MSS.* Perivale Wood,† *Shenst. Fl.* **6.** Near Finchley, *Lousley*; Hadley!, *K. & L.*, 28. Highgate Wood, 1947.

Forma **pallida** Neum.
1. Harefield, 1950!, *K. & L.*, 28.

Var. **punctata** (Rouy & Fouc.) Wilmott
1. Old Park Wood, Harefield!; Ruislip Woods!; Park Wood, Ickenham!, *K. & L.*, 28. Stanmore Common. **3.** Gutteridge Wood, Northolt!,† *K. & L.*, 28. **4.** Scratch Wood!, *K. & L.*, 28.

V. reichenbachiana × riviniana = V. × intermedia Reichb., non Krock.
1. Mad Bess Wood, Ruislip, 1950!, *K. & L.*, 28. Copse Wood, Northwood, 1958.

V. canina L. Heath Dog Violet.
Native on dry sandy heaths, grassy places on light soils, etc. Very rare and sometimes confused with forms of *V. riviniana* and *V. reichenbachiana*.
First certain evidence: *Trimen*, 1866.
1. Near Harefield, 1866,† *Trimen* (BM). Ruislip Common!; Ickenham Green!; Harrow Weald Common,† *Benbow MSS.* Stanmore Common!, *Richards*. Grimsdyke golf course, 1965, *Kennedy*!. **2.** Near Staines, 1866,† *Trimen* (BM). **3.** Hounslow Heath!,† *Newbould* (BM). **4.** Hampstead Heath,† *Whiting Fl.* Perhaps *V. riviniana* subsp. *minor* intended. Harrow Park, *Horton Rep.* Error, *V. riviniana* intended. **5.** Near Hatton;† Heston,† *T. & D.*, 46.

Var. **ericetorum** (Schrad.) Reichb.
Viola ericetorum Schrad., *V. flavicornis* Sm., non Forst.
1. Uxbridge Common,† *Benbow MSS.* Ruislip Common!, *Green* (SLBI). Northwood,† *Green, Williams MSS.* Ickenham Green!; Stanmore Common!, *K. & L.*, 29. **3.** Near Hounslow Heath,† *Benbow MSS.*

Var. **sabulosa** Reichb.
var. *pusilla* Bab.
1. Harefield† (D).

Forma **alba** Greg.
1. Ruislip Common, 1902, *Green* (SLBI).

V. canina × **lactea** = **V.** × **militaris** Savouré
3. Hounslow Heath!, 1868, *Newbould* (BM); 1947!, *Welch, L.N.* 27: 30; 1948–60.† Habitat destroyed by gravel digging operations.

V. lactea Sm. Pale Heath Violet.
Lit. *J. Ecol.* **46**: 527 (1958).
Formerly native on a heath but now extinct.
First, and only evidence: *Newbould*, 1868.
3. Hounslow Drilling Ground (= Heath), 1868, *Newbould* (BM).

V. palustris L. subsp. **palustris** Marsh Violet.
Formerly native in bogs, wet heathy places, etc., but now extinct.
First record: *Hudson*, 1762. Last record: *Whiting*, 1912.
1. On a bank at Pinner Hill, 1871–72, *W. A. Tooke, Williams MSS.* An unusual habitat but no reason to doubt the authenticity of the record. **4.** Hampstead Heath, *Huds. Fl. Angl.*, 330; 1790, *Forst. MSS.*; 1815 and 1817 (G & R); 1821, *E. T. Bennett* (BM); 1826, *Heathfield* (K); still in good quantity in the bog behind 'Jack Straw's Castle', but flowers now very sparingly, *T. & D.*, 45; in flower, 1891, *Benbow* (BM); dwindling, 1898, *Garlick W.F.*; a small patch in flower, 1901, *Whiting Notes*; still present but did not flower, 1912, *Whiting Fl.*; not there in 1920, *Richards.*

Subgen. MELANIUM (DC.) Hegi

V. tricolor L. subsp. **tricolor** Wild Pansy.
Native or denizen on cultivated and waste ground, etc. Very rare, and decreasing.
First record: *T. Johnson*, 1638.
1. Harefield!, *A. B. Cole, T. & D.*, 46; 1960. Springwell!; Uxbridge!, *Benbow MSS.* Near Potters Bar, 1965. **2.** Near Hampton Court, *Blackst. Spec.*, 110. Teddington, *Francis, Coop. Suppl.*, 12. West Drayton; near Colnbrook; Staines, *Benbow MSS.* Near Ashford,

Hasler. **3.** Isleworth, 1815† (G & R). Twickenham, *T. & D.*, 47. Feltham, *Pettet.* **4.** Hampstead,† *Irv. MSS.* Near Brent Reservoir,† *Warren, T. & D.*, 422. **5.** Hanwell.† **6.** Edmonton,† *T. & D.*, 47. **7.** Tottenham,† *Johns. Cat.* Hornsey,† *Ball. MSS.* Victoria Street, S.W.1,† *T. & D.*, 47. East Heath, Hampstead, 1948,† *H. C. Harris.*
Some of the above records are probably referable to the degenerate garden pansy—**V.** × **wittrockiana** Gams.
The following segregates are on record:—

V. lejeunei Jord.
1. Harefield, *E. F. Drabble, Rep. Bot. Soc. & E.C.* 8: 195. **5.** Greenford Green, *Rep. Bot. Soc. & E.C.* 8: 195.

V. lloydii Jord. var. **insignis** Drabble
4. Mill Hill, *E. F. Drabble, Rep. Bot. Soc. & E.C.* 8: 195.

V. arvensis Murr. Field Pansy.
Native on cultivated and waste ground, etc. Locally common, but decreasing.
First record: *Blackstone*, 1737.
1. Harefield!, *Blackst. Fasc.*, 110. Ruislip!; Eastcote; Pinner!, *Melv. Fl.*, 11. Harrow Weald Common, *T. & D.*, 47. Northwood, *C. S. Nicholson, Williams MSS.* Uxbridge!; Hillingdon; Ickenham, *Benbow MSS.* Potters Bar, *J. G. Dony*!. South Mimms. Near Elstree. **2.** Teddington; Fulwell; near Strawberry Hill; Tangly Park,† *T. & D.*, 47. West Drayton; Yeoveney; Staines!; Ashford; Halliford!, *Benbow MSS.* Harlington, *Cooke.* Stanwell. East Bedfont. Hampton. **3.** Twickenham, *T. & D.*, 47. Northolt. *Hind Fl.* **4.** Hampstead, *Irv. MSS.* Harrow, *Melv. Fl.*, 11. Mill Hill, *Harrison.* Disused railway between Stanmore and Belmont, *Roebuck.* **5.** Chiswick, *Newbould*; Acton, *Tucker, T. & D.*, 47. **6.** Colney Hatch; Edmonton, *T. & D.*, 47. Cockfosters, *F. Clarke.* Enfield, *Kennedy.* **7.** Paddington, *Warren*; Parsons Green, *T. & D.*, 47. Eaton Square, S.W.1, 1954, *L.N.* 34: 5; 1958, *McClintock.* Kensington Gardens, 1961, *D. E. Allen, L.N.* 41: 17. Weed in flower-bed, Buckingham Palace grounds, 1967, *McClintock.*
The following segregates are on record.

V. agrestis Jord.
2. Fulwell, *W. H. Brown*, det. *E. F. Drabble*, Suppl. *J.B.* (*London*) 47: 18.

V. obtusifolia Jord.
1. Pinner, *Walkin*, det. *E. F. Drabble, Rep. Bot. Soc. & E.C.* 9: 458.

V. deseglisei Jord.
3. Hounslow Heath, *E. F. Drabble, Rep. Bot. Soc. & E.C.* 8: 195. **4.** Golders Green; Finchley, *E. F. Drabble, Rep. Bot. Soc. & E.C.* 8: 195.

V. arvatica Jord.
2. West Drayton, *Sherrin*, det. *E. F. Drabble*, *Rep. Bot. Soc. & E.C.* 8: 195.

V. arvensis × tricolor
Lit. *Watsonia* **6**: 51 (1964).
1. Harefield!, *Bishop* (BM). Ruislip, *Pettet*, *Watsonia* 6: 60.
The following segregates are on record.

V. contempta Jord.
1. Harefield, *E. F. Drabble*, *Rep. Bot. Soc. & E.C.* 8: 195.

V. variata Jord. var. **sulphurea** Drabble
1. Harefield, *E. F. Drabble*, *Rep. Bot. Soc. & E.C.* 11: 320.

POLYGALACEAE

POLYGALA L.

P. vulgaris L. Common Milkwort.
Native in meadows, dry grassy places, etc., chiefly on calcareous soils. Very rare and decreasing.
First record: *Blackstone*, 1737.
1. Harefield!, *Blackst. Fasc.*, 79; 1968. Springwell!, *Benbow MSS.*; 1968. Uxbridge,† *Druce Notes*. **6.** Edmonton,† *T. & D.*, 48.

P. serpyllifolia Hose Heath Milkwort.
P. serpyllacea Weihe, *P. depressa* Wender.
Native on heaths and damp grassy places on acid soils. Rare and decreasing.
First certain evidence: *Dillenius*, c. 1730.
1. Ruislip Common!, *Hind Fl.* Below Springwell!, *Benbow* (BM). Drayton Ford!;† Northwood;† Harefield Moor!, *Benbow MSS.*; still locally plentiful in the last-mentioned locality. West of Ruislip,† *Wrighton*. Harrow Weald Common!, *Hind Fl.* Stanmore Common!, *T. & D.*, 148. Grimsdyke grounds, Harrow Weald, *Welch*!, *K. & L.*, 31. Pinner Hill, *Trim. MSS.*; c. 1884,† *Benbow MSS.* **2.** Staines,† *Whale Fl.*, 22. Between Staines and Colnbrook,† *Benbow MSS.* **3.** Hounslow Heath, *T. & D.*, 48; c. 1884;† between Hanworth and Hampton,† *Benbow MSS.* **4.** Near Highwood Hill,† *Blackst. Spec.*, 77. Deacon's Hill,† *T. & D.*, 48. Edgwarebury!, *Benbow MSS.* Hampstead Heath (DILL); 1848, *T. Moore* (BM); 1864 (SY); c. 1869, *T. & D.*, 48; almost extinct, 1901, *Whiting Notes*; once common, but now (1912)† almost extinct, *Whiting Fl.* **6.** Southgate (DILL). **7.** East Heath, Hampstead,† *T. & D.*, 48.

HYPERICACEAE

HYPERICUM L.

H. androsaemum L. Tutsan.

Native in woods, copses, thickets and shady places in the northern parts of the vice-county, occurs also in thickets, ditches, etc., as a bird-sown introduction from gardens. Very rare, and almost extinct.

First record: *Gerarde*, 1597.

1. Harefield!, *Blackst. Fasc.*, 44; last seen in Old Park Wood about 1947, and possibly now extinct. Pinner Hill, 1874,† *W. A. Tooke, Melv. Fl.* 2: 20. Mimmshall Wood, 1908,† *Green* (SLBI). **3.** Near Hounslow Heath, 1946!,† *K. & L.*, 41; presumably bird-sown from a garden. **4.** Stanmore, 1827, *Varenne, T. & D.*, 63; 1876,† *Benbow MSS.* Harrow, garden escape, *Harley Fl.*, 5. Hampstead Wood, *Ger. Hb.*, 435; woods about Hampstead† and Highgate,† *Pet. Midd.*; common there till these were grubbed up and the land subjected to cultivation, *Loud. Arb.* 1: 403. Near Child's Hill, 1829,† Heathfield (K). **5.** Spontaneously in a garden, Ealing, presumably bird-sown, 1965!, *T. G. Collett*; 1968. **6.** Hadley Common, 1873,† *P. Gray, Trim. MSS.* Near Enfield, *A. Irvine, T. & D.*, 63. Enfield Chase!, *Benbow* (BM). Not seen there for many years and perhaps extinct. Between Highgate and Muswell Hill, 1745,† *Blackst. Spec.*, 38.

H. inodorum Mill.

H. elatum Ait.

Alien. Madeira, Canaries, etc. Garden escape. Naturalised in grassy places. Very rare.

First record: *McLean*, 1967.

5. Floor of disused reservoir, Ealing, 1967→, *McLean*, conf. *Robson*.

H. calycinum L. Great St John's Wort, Rose of Sharon.

Alien. Europe, etc. Garden escape or outcast. Naturalised in woods, shady places, ditches, on hedgebanks, etc. Very rare.

First record: *Lowe*, 1836.

2. Near West Drayton,† *Benbow MSS.* **3.** Osier holt near Hounslow Heath, 1929,† *A. R. Horwood* (K); roadside at edge of Hounslow Heath, 1947!, *H. Banks, L.N.* 27: 31; still present, 1969. Hayes Park!, *Royle.* **4.** Harrow!, *Hind Fl.* **5.** Chiswick House grounds,† *T. & D.*, 62. Hedgebank, Southall, plentiful, 1965→. **7.** Highgate,† *Lowe, Coop. Fl.*, 110.

The species apparently rarely sets seed in Britain (cf. *Watsonia* **5**, 368 (1963)).

H. perforatum L. Common St John's Wort.

Native on heaths, in dry grassy places, by roadsides, etc., established on railway banks, etc. Common.

Lit. *Proc. Bot. Soc. Brit. Isles* **7**, 212 (1968) (*Abstr.*)
First record: *T. Johnson*, 1638.
7. Tottenham!, *Johns. Cat.* Edgware Road, 1815† (G & R). South Kensington,† *Dyer List.* Regents Park!, *Holme, L.N.* 39: 45. Kensington Gardens, one plant, 1960, *D. E. Allen, L.N.* 41: 17. Buckingham Palace grounds, *Fl. B.P.G.* Highgate. Stepney. Shoreditch. Holloway. Hackney. Hackney Wick.

Var. **angustifolium** DC.
1. Harefield!, *Druce Notes.* Springwell!, *K. & L.*, 62.
This 'variety' may be a mere state induced by growing on calcareous soils.

H. maculatum Crantz subsp. **obtusiusculum** (Tourlet) Hayek Imperforate St John's Wort.
H. quadrangulum auct., *H. dubium* Leers
Lit. *Proc. Bot. Soc. Brit. Isles* **2**: 237 (1957): *loc. cit.* **3**: 99 (1958): *loc. cit.* **7**: 212 (1968) (*Abstr.*)
Introduced. Established on a railway bank. Very rare, and perhaps extinct.
First certain record: *McLean*, 1967.
2. One plant under the hedge by the towing-path of the river between Kingston Bridge and Hampton Court, 1867, *Bloxam*; requires confirmation, *T. & D.*, 64. Near Kingston Bridge, 1944, *Welch, L.N.* 26: 59. Editorial error, the plant intended was *H. tetrapterum*, *L.N.* 28: 32. **4.** Harrow, *Hind Fl.* 'Mr Hind thinks he gathered it somewhere near Harrow before 1860, but he has kept no specimen; nor are there any in Herb. Harrow. Requires confirmation,' *T. & D.*, 64. **5.** Railway bank, West Acton, 1967!, *McLean*, conf. *Robson.*

H. tetrapterum Fr. Square-stemmed St John's Wort.
H. acutum Moench, *H. quadrangulum* auct., *H. quadratum* Stokes.
Native in damp, heathy places, marshes, on pond verges, waste ground, by stream- and riversides, etc. Formerly common, but now local, and usually in small quantity except near the Thames where it is more frequent.
First record: *Turner*, 1548.
1. Harefield!, *Blackst. Fasc.*, 44. Pinner; Stanmore Common!; Elstree,† *T. & D.*, 63. Swakeleys!; Uxbridge, *Benbow* (BM). Cowley!; South Mimms!, *K. & L.*, 40. Wrotham Park, Potters Bar, *J. G.* and *C. M. Dony*. Eastcote. **2.** Locally common near the Thames. Hanworth Park,† *T. & D.*, 63. Stanwell!, *P. H. Cooke*; near Colnbrook!, *J. E. Cooper*; Yeoveney!, *K. & L.*, 42. **3.** Roxeth (ME). Hounslow Heath;† Whitton;† Twickenham, *T. & D.*, 63. **4.** Harrow,† *Hind Fl.* Edgware, *E. M. Dale* (SLBI). Disused railway between Stanmore and Belmont, *Roebuck*. Kingsbury, 1910,† *P. H. Cooke* (LNHS). Mill Hill!, *Atkins* (BM). Between Hampstead and Barnet, *Mann*, *Coop. Fl.*,

116. Highgate, *Andrewes MSS.* **5.** Syon Park!, *Turn. Names.* Near Lampton,† *T. & D.*, 63. Bedford Park,† *Cockerell Fl.* Near Hanwell,† *Hemsley*; Perivale!,† *Lees, T. & D.*, 63. Garden weed, Ealing, 1962! *T. G. Collett*; still present, 1969. Brentford!, *A. Wood, Trim. MSS.* **6.** Highgate Wood, *Johns. Eric.*; Coldfall Wood, Highgate,† *E. M. Dale, K. & L.*, 42. Wood Green,† *Munby, Nat.* 1867, 181. Edmonton,† *T. & D.*, 63. Finchley Common, *Payton*!, *K. & L.*, 42, Enfield!, *Johns, L.N.* 27: 31; 1967. Muswell Hill, 1961, *Kennedy* and *Bangerter*!; 1966. **7.** Hammersmith Marshes,† *Coop. Fl. MSS.* East Heath, Hampstead,† *T. & D.*, 63. Bombed site, Ebury Street, S.W.1,† *McClintock.* Regents Park, 1956, *Holme, L.N.* 39: 45. Near Temple Mills,† *Cherry, T. & D.*, 63. Kilburn,† *Warren, T. & D.*, 422. Buckingham Palace grounds, *Fl. B.P.G.*

H. humifusum L. Trailing St John's Wort.
Native on heaths, in dry woods, etc. Local and decreasing.
First record: *Blackstone*, 1737.

1. Harefield!, *Blackst. Fasc.*, 44. Harrow Weald Common!; Stanmore Common!, *Melv. Fl.*, 17. Ruislip Common!, *T. & D.*, 64. Near Northwood!, *Benbow MSS.* Northwood!, *J. E. Cooper* (BM). Uxbridge Common;† near Swakeleys, *Benbow MSS.* Pinner Hill, *W. A. Tooke, Trim. MSS.* Mimmshall Wood!, *Benbow MSS.* **2.** Fulwell;† near Staines;† Tangley Park,† *T. & D.*, 64. **3.** Near Hounslow!;† near Hatton,† *T. & D.*, 64. Hounslow Heath!; Whitton Park,† *Benbow MSS.* **4.** Hampstead Heath, 1840, *E. Metcalfe* (K); c. 1912,† *Whiting Fl.* Highgate, *Andrewes MSS.* Temple Fortune, 1925,† *J. E. Cooper* (BM). Deacon's Hill, *T. & D.*, 64. Mill Hill. Scratch Wood!, *Benbow MSS.* Barnet Gate Wood, *Benbow* (BM). **5.** Hanwell, 1846,† *J. Morris, T. & D.*, 64. Horsendon Hill,† *Williams MSS.* Syon Park, 1966, *Boniface.* **6.** Hadley!, *Warren, T. & D.*, 64. Enfield Chase!, *Benbow MSS.* Winchmore Hill Wood,† *C. S. Nicholson, Lond. Nat. MSS.* **7.** Near Kilburn, c. 1820,† *Sneyd, Trim. MSS.* Hyde Park, 1815† (G & R). East Heath, Hampstead,† *T. & D.*, 64. Buckingham Palace grounds, *Fl. B.P.G.*

H. pulchrum L. Slender St John's Wort.
Native in dry woods, on heaths, etc. Local and decreasing.
First record: *Johnson*, 1633.

1. Harefield!, *Blackst. Fasc.*, 44. Ruislip. Pinner, *Melv. Fl.*, 17. Hillingdon,† *Warren, T. & D.*, 65. Uxbridge, *Benbow MSS.* Harrow Weald Common!; Stanmore Common!, *T. & D.*, 65. Northwood!, *Welch*; Mimmshall Wood!, *K. & L.*, 42. **2.** Between Staines and Hampton, *T. & D.*, 64. Near Hampton, *Tyrrell*! **3.** Hounslow Heath!, *Curt. F.L.* **4.** Hampstead,† *Blackst. Spec.*, 38. Hampstead Heath, *T. & D.*, 65; c. 1912,† *Whiting Fl.* Highwood Hill, *Blackst. Spec.*, 38. Scratch Wood!; Deacon's Hill!, *K. & L.*, 42. **5.** Bedford Park,† *Cockerell Fl.*

Probably an error. **6.** Highgate,† *Coop. Fl.*, 104. Southgate, *A. B. Cole*; Hadley, *Warren*; Colney Hatch;† Edmonton,† *T. & D.*, 65. **7.** St John's Wood, [Highbury],† *Johns. Ger.*, 540. Hornsey Wood, 1815† (G & R). Ken Wood,† *Burn. Med. Bot.*, 105.

H. hirsutum L. Hairy St John's Wort.
Native in grassy places, open woods, etc., established on railway banks, etc., chiefly on basic soils. Local.
First record: *Petiver*, 1715.
1. Harefield!, *Blackst. Fasc.*, 43. Pinner!, *T. & D.*, 64. Uxbridge,† *Benbow* (BM). Ickenham!; New Years Green!, *Tremayne*. Ruislip!, *Welch*. Northwood. Mimmshall Wood, *J. G. Dony*!. **2.** Staines!, *T. & D.*, 64. Banks of reservoirs, Stanwell. **4.** Hampstead,† *Pet. H.B. Cat.* Turner's Wood, 1829,† *Heathfield* (K). Between Golders Green and Finchley, 1861,† *Phyt. N.S.* 5: 319. Harrow,† *Hind Fl.* Bentley Priory; Scratch Wood!; Edgware!, *T. & D.*, 64. Whitchurch;† Kingsbury, *P. H. Cooke*. Mill Hill!, *L.N.* 28: 32. Stonebridge, *L. M. P. Small*. **5.** Horsendon Hill,† *T. & D.*, 65. Perivale!, *Welch*. Hanwell,† *Warren*, *T. & D.*, 65. Shepherds Bush, 1870,† *Warren*, *Trim. MSS.* **6.** North Finchley, 1883,† *J. E. Cooper*. Enfield!, *H. & F.*, 149. **7.** Between Islington and Hornsey, 1816,† *E. T. Bennett*; Regents Park, 1827–30,† *Varenne*, *T. & D.*, 65. South End, Hampstead† (HE). South Kensington,† *Dyer List.*

H. elodes L. Marsh St John's Wort.
Formerly native in boggy places on heaths, but long extinct.
First record: *Doody*, c. 1696. Last record: *Pamplin*, c. 1826.
3. Hounslow Heath, *Doody MSS.* and *Mart. Tourn.* 2: 32. **4.** Hampstead Heath, *Blackst. Spec.*, 39; c. 1825–26, *Pamplin MSS.*

CISTACEAE

HELIANTHEMUM Mill.

H. nummularium (L.) Mill. Common Rockrose.
H. vulgare Gaertn., *H. chamaecistus* Mill.
Lit. *J. Ecol.* **44**: 701 (1956).
Formerly native on dry banks and grassy slopes, mainly on calcareous soils, but now extinct. Apparently deliberately reintroduced but did not persist.
First record: *W. A. Tooke*, 1871. Last record (as a native): *Green* 1907.
1. Pinner Hill, barely six feet on the right side of the Middlesex boundary, *W. A. Tooke*, *J.B.* (*London*) 9: 272. Habitat apparently destroyed by road-widening operations. Chalkpit under Garett Wood, Springwell, 1901, *Benbow* (BM); 1907, *Green*, *Williams MSS.* Ruislip, 1960, *Pickess*; ? extinct, *Moxey List.* Deliberately introduced, *Pickess.*

CARYOPHYLLACEAE

SILENE L.

S. dioica (L.) Clairv. subsp. **dioica** Red Campion.
Lychnis dioica L., *pro parte*, *L. diurna* Sibth., *Melandrium rubrum* (Weigel) Garcke, *M. dioicum* (L.) Coss. & Germ.
Lit. *J. Ecol.* **36**: 96 (1948).
Native in woods, thickets, shady places, etc. Common in the northern parts of the vice-county, less frequent, though sometimes locally plentiful in the southern districts.
First record: *T. Johnson*, 1638.
7. Marylebone fields, 1817 (G & R); Regents Park!, *L.N.* 39: 46. Kentish Town,† *T. & D.*, 51. Islington,† *Ball. MSS.* Highgate Ponds!; Ken Wood!, *Fitter.* Hyde Park!, *Watsonia* 1: 297; still plentiful in the Bird Sanctuary, 1973. Holland House grounds, Kensington, plentiful!, *Sandwith*; still frequent, 1973. Bombed site, Cripplegate,† *Wrighton, L.N.* 28: 383. Clapton. Stoke Newington. Finsbury Park. Tottenham!, *Johns. Cat.* Hackney; Hackney Wick, *L. M. P. Small.*
The colour of the flowers in this species is extremely variable, ranging from deep rose through pale pink to white. White-flowered plants are rare, but have been recorded from **1.** Harefield!, *T. & D.*, 52. **4.** Bentley Priory, *Melv. Fl.*, 13.

S. alba (Mill.) E. H. L. Krause White, or Evening Campion.
Lychnis alba Mill., *L. vespertina* Sibth., *Melandrium album* (Mill.) Garcke
Lit. *J. Ecol.* **35**: 274 (1947): *loc. cit.* **36**: 96 (1948).
Native on cultivated and waste ground, hedgebanks, by roadsides, etc., established on railway banks, canal paths, etc. Common.
First record: *T. Johnson*, 1638.
7. Frequent in the district.
Plants completely glabrous in all parts, which were shining as if varnished, and coloured with deep purple anthocyanin, especially on stems, bracts and calyces, which were also sticky owing to the presence of minute white papillae, but with the petals of a normal white colour (*Melandrium album* (Mill.) Garcke var. *viscosum* (F. Aresch.) Ahlfr., *M. album* (Mill.) Garcke var. *glabrum* (De Vries) Sandwith, *Lychnis vespertina* Sibth. var. *glabra* De Vries) occurred in **5.** On waste ground by the Thames, Chiswick, 1942 and 1945!,† *Sandwith* (BM, K) (cf. *Nat.* **1948**; 45 (1948); *loc. cit.* **1949**; 47 (1949); *Watsonia* **2**: 113 (1951) (*Abstr.*)). A monstrous form, apetalous, with numerous long dirty white styles was also found in **5.** Waste ground near the Thames, Chiswick, 1942,† *Sandwith* (BM, K).

S. alba × dioica This putative hybrid has been reported from **1.** Harefield, *Druce Notes*; 1912, *Barton* (BM). Near Northwood, 1957.

Near Uxbridge, 1966. **3.** Near Northolt, *Payton*!, *K. & L.*, 34. **4.** Harrow, *Harley*. **5.** Brentford!, *Welch*, *K. & L.*, 34. **6.** Enfield, 1960. **7.** Near Chelsea,† *Rand* (DB). Holloway, 1965, *D. E. G. Irvine*.

S. noctiflora L. Night-flowering Campion, Night-flowering Catchfly.

Melandrium noctiflorum (L.) Fr.

Colonist or native on cultivated and waste ground, usually on light soils, and as a garden weed. Very rare, and often merely casual.

First record: *J. Morris*, 1848.

1. Between Denham and Harefield, 1888, *Benbow* (BM). Uxbridge! *Benbow MSS*. Yiewsley, 1909 and 1913, *Coop. Cas.* **2.** West Drayton!, *Benbow*, *J.B.* (*London*) 23: 36. Between Poyle and Yeoveney, 1951!, *K. & L.*, 34. **3.** Hounslow Heath, a single plant, 1866, *T. & D.*, 50; 1954, *Welch*. **4.** Near Stonebridge, 1866, *Benbow MSS*. **5.** Hanwell, 1950!, *K. & L.*, 34. Acton, 1902, *Worsdell* (K). Garden weed, Ealing, 1948!, *Cattley*, *L.N.* 28: 30; garden weed, Haven Green, Ealing, c. 20 plants, 1967, *McLean*. Brentford, 1957. **6.** Muswell Hill, 1902, *Cooper* (BM). **7.** Between Walham Green and Hammersmith, 1848, *J. Morris*; Chelsea, 1858–63; Parsons Green, 1864, *Britten*, *T. & D.*, 51. Finchley Road, *T. & D.*, 422; 1891, *F. P. Thompson*; Crouch End, 1897, *J. E. Cooper* (BM). Hackney Marshes, 1910, *Coop. Cas.*; 1918 and 1924, *J. E. Cooper*, *K. & L.*, 34. Bombed site by Mark Lane, E.C.,† *Graham*, *L.N.* 28: 30.

S. vulgaris (Moench) Garcke Bladder Campion.

S. cucubalus Wibel, *S. inflata* (Salisb.) Sm.

Native on cultivated and waste ground, hedgebanks, by roadsides, etc., and established on canal paths, railway banks and tracks, etc. Formerly rather rare, but now common.

First record: *Doody*, c. 1696.

7. South Kensington!, *Dyer List*. Newington, *Ball. MSS*. Kensington Gardens!, *Watsonia* 1: 297. Paddington. Highgate. Bishop's Park, Fulham, *Welch*!. Bombed site, St Alban's Church, W.C., 1950,† *Whittaker* (BM). Bombed site, Cripplegate, E.C. Holland House grounds, Kensington!, *Sandwith*. Hyde Park!, *D. E. Allen*; near Paddington Basin!, *L.N.* 39: 46. Near Tower Hill, *Brewis*, *L.N.* 44: 21. Parsons Green. East Heath, Hampstead!, *H. C. Harris*. Islington. Highbury. Finsbury Park. Hornsey. Dalston. Stepney. Bromley by Bow. Isle of Dogs. Tottenham. Hackney. Hackney Wick. Hackney Marshes. Bow. Holloway, *D. E. G. Irvine*. Old Ford.

Hairy forms, which have received a number of varietal names (cf. Jones, E. M. & Turrill, W. B. *Bladder Campions*, 287 (1957)), have been recorded from **1.** Harefield!, *Druce Notes*. **3.** Hounslow!, *K. & L.*, 33. **5.** Hanwell.

Var. **commutata** (Guss.) Coode & Cullen
Silene commutata Guss., *S. vulgaris* subsp. *macrocarpa* Turrill, *S. linearis* auct., *S. angustifolia* auct.
1. Yiewsley, 1909, *J. E. Cooper* (BM). **6.** Muswell Hill, 1913, *J. E. Cooper* (BM).

LYCHNIS L.

L. flos-cuculi L. Ragged Robin.
Native in marshes, wet meadows, damp grassy places, by streamsides, etc. Formerly common, but now rather rare, and rapidly decreasing.
First record: *T. Johnson*, 1638.
1. Harefield!, *Blackst. Fasc.*, 55. Harefield Moor, still locally abundant. Springwell. Uxbridge. Yiewsley. Ruislip. Pinner, *Kennedy*!. Stanmore Common. Mimmshall Wood. Wrotham Park, Potters Bar, *J. G.* and *C. M. Dony*!. **2.** Staines. Laleham. Poyle, Yeoveney. Littleton. Hampton Court. **3.** Roxeth,† *Hind Fl.* Hounslow Heath.† Near Wood End!, *Payton*. Near Northolt. **4.** Harrow, *Horton Rep.* Whitchurch Common,† *C. S. Nicholson* (LNHS). Hampstead,† *Ray Cat.* 2: 76. Highgate,† *Andrewes MSS.* Near Mill Hill!, *Pigott.* Hendon, *Champ. List.* **5.** Chiswick Eyot, *Corn. Nat.* Syon Park. Perivale Wood!, *Shenst. Fl.* **6.** Southgate!; Enfield!, *L. B. Hall, Lond. Nat. MSS.* **7.** Tottenham,† *Johns. Cat.* South Kensington,† *F. Naylor*; East Heath, Hampstead;† Ken Wood,† *T. & D.*, 51. Kensington Gardens,† *Warren, J.B.* (*London*) 13: 336. Clapton Marsh, 1896† (LNHS).
White-flowered plants are sometimes encountered.

AGROSTEMMA L.

A. githago L. Corn Cockle.
Lychnis githago (L.) Scop.
Alien. Mediterranean region, etc. Formerly locally common as a colonist in cornfields, etc., now very rare and sporadic on waste ground, rubbish-tips, etc., usually as an introduction with bird-seed.
First record: *T. Johnson*, 1638.
1. Harefield!, *Blackst. Fasc.*, 55. Ruislip, *Melv. Fl.*, 13. Uxbridge, *Benbow* (BM). Swakeleys; Hillingdon, *Benbow MSS.* Pinner, *Druett.* Yiewsley, 1909 and 1913, *Coop. Cas.*; 1920, *Tremayne*; c. 1927, *M. & S.* Denham!, *Benbow MSS.* **2.** Between Kingston Bridge and Hampton Court, *T. & D.*, 52. Sunbury; between Laleham and Staines, *Benbow MSS.* Near Ashford, 1940–42, *Hasler.* Laleham. Hampton Wick, 1894, *Bradley* (BM). **3.** Wood End, *Melv. Fl.*, 13. Twickenham, a single plant, *T. & D.*, 52. Near Feltham; Hayes, *Lond.*

Nat. MSS. Hounslow, *A. Wood, Trim. MSS.* Alperton, one plant, *T. & D.*, 52; 1912, *Wilmott* (BM). **4.** Stanmore, 1815, *Varenne*; near Edgware, a single plant, *T. & D.*, 52. Harrow!, *Benbow MSS.* Kenton, *Benbow* (BM). Between Elstree and Stanmore, *Benbow MSS.* Edgware, 1899, *E. M. Dale* (LNHS). East Finchley, 1908–9, *Coop. Cas.* Finchley, 1925, *J. E. Cooper.* Golders Green, 1909, *P. H. Cooke.* Mill Hill, *Harrison.* Golders Hill, *Whiting.* Bishop's Wood, *Benbow MSS.* **5.** Near Ealing; near Acton, *Newbould, T. & D.*, 52. Greenford, 1940!; Hanwell, two plants, 1946!, *L.N.* 26: 53; six plants, 1953. Horsendon Hill, one plant; Chiswick, *Benbow MSS.* Bedford Park, *Cockerell Fl.* **6.** Tottenham, *Johns. Cat.* Edmonton, *T. & D.*, 52. Hadley; between Ponders End and Edmonton, *Benbow MSS.* Southgate, 1900, *L. B. Hall*; Palmers Green, *C. S. Nicholson, Lond. Nat. MSS.* **7.** South Kensington, *Dyer List.* Kensington Gardens, *Warren Fl.* Near Finchley Road, *Champ. List.* Highgate Cemetery, *Trim. MSS.* Ken Wood fields, 1949, *H. C. Harris*!, *K. & L.*, 34. Islington, *Ball. MSS.* Isle of Dogs, *Benbow MSS.* Hornsey, *C. S. Nicholson, Lond. Nat. MSS.*

CUCUBALUS L.

C. baccifer L. Berry Catchfly, Berry-bearing Chickweed.
Formerly native or denizen in thickets, on banks, etc., but long extinct.
First evidence: *Luxford*, 1837. Last evidence: *Syme*, 1852.
7. Isle of Dogs, 1837, *Luxford* (LINN); 1838, *E. Forster* (BM); 1839, *Borrer* (K); 1840, *Coleman* (LINN); 1841, *Freeman* (BM); 1842, *Syme* (SY); 1844, *R. Brown*; 1846, *J. A. Brewer* and *Dennes* (BM); 1850, *J. A. Brewer* (K); in considerable abundance, 1852, *Westcombe, Phyt.* 4: 605; 1852, *Syme* (BM); the part including that where it grew is now covered by new docks, a railway, etc., *T. & D.*, 50.

DIANTHUS L.

D. armeria L. Deptford Pink.
Native in dry sandy and gravelly places, grassy places on light soils, etc. Very rare, and perhaps extinct.
First record: *J. Newton*, c. 1680.
1. On rough gravelly banks of the railway between Uxbridge and West Drayton, *Benbow* (BM); 1886–90,† *Benbow MSS.* **3.** Near Hayes, 1862,† *Benbow* (BM). **4.** East Finchley, 1883, *J. E. Cooper* (BM). **7.** Beyond Highgate, *Newt. MSS.*; foot of Parliament Hill, 1948, *H. C. Harris, L.N.* 28: 29. I have searched the latter locality on numerous occasions since 1950 but the grass is frequently mown, and if the plant survives there it has no opportunity to flower.

D. deltoides L. Maiden Pink.

Native in dry grassy places on light soils, also on other soils as a garden escape. Very rare, and perhaps extinct as a native species, but still occurring from time to time as an escape from gardens.

First record: *Parkinson*, 1640.

2. Hampton Court Park, 1691, *Brewer MSS.*; c. 1700, *Buddle* (SLO); abundantly, and in the fields thereabouts, *Doody MSS.*; on a mound of earth in Hampton Court Park . . . not one plant, however, could I find away from the mound . . . nor are there any of the same variety growing wild in the neighbourhood, *Jesse Glean.*, 143; precincts of Hampton Court Park, 1906, *Green* (SLBI); 1939 (LNHS). I have searched in vain for this attractive species in the Hampton Court area, and it may persist as there are still many habitats suitable for the plant. Teddington, *Merr. Pin.*, 22; 1838, *D. Cooper*; 1841,† *Seeley* (BM). **3.** Wood End, garden escape, 1940–48!,† *Payton, K. & L.*, 31. **4.** Hampstead Heath,† *Huds. Fl. Angl.*, 161. **5.** Near Kew Bridge, in quantity, a garden escape, 1967. **7.** Growing in the thick grass in our meadows about London, namely towards Tottenham Court,† *Park. Theat.*, 1338. Bombed site, Cheapside, E.C.,† deliberately introduced, *Farenden, K. & L.*, 32. Bombed sites, Cripplegate,† *Jones Fl.* Holland Walk, W.8, a small patch, garden escape, 1965.†

SAPONARIA L.

S. officinalis L. Soapwort, Bouncing Bet.

Alien. Europe, etc. Garden escape. Naturalised in rough grassy places, by roadsides, etc., and established on waste ground, railway banks, etc. Rather rare.

First record: *T. Johnson*, 1638.

1. Uxbridge, *Blackst. Fasc.*, 90. Northwood. Ruislip Reservoir. **2.** East Bedfont, *Welch, K. & L.*, 32. Teddington. **3.** Near Hayes, 1904, *Green* (SLBI). Near St Margaret's!; near Hounslow Heath, abundant!, *K. & L.*, 32. Near Sudbury Hill, *Temple*!. **4.** Child's Hill, 1948!;† railway bank near Willesden Junction, abundant!, *Boniface, K. & L.*, 32. Wembley Exhibition grounds, abundant. Stonebridge Park power station grounds, abundant. Wealdstone. Edgware. Scratch Wood railway sidings.† **5.** Ealing, 1945!;† near Greenford, abundant;† near the Thames, Chiswick, abundant, 1945–48!,† *K. & L.*, 32. Near Norwood Green. Park Royal.† Near Brentford. **6.** Tottenham, *Johns. Cat.* Whetstone Stray, *J. G. Dony*!. **7.** Near Cremorne,† *Britten, Bot. Chron.*, 21. Near Kingsland Turnpike,† *Alch. MSS.* South Kensington!, *Dyer* (BM); 1967. Near West India Docks!, 1875, *Britten, Trim. MSS.*; 1966. Millwall Docks, abundant. Hackney. Clapton, *P. H. Cooke* (LNHS). Well established on bombed sites between Well Street and Monkwell Street;† brick rubble, corner of Wood Street and Addle Street!,† *Lousley Fl. City.* Bombed sites, Cripplegate, E.C.!, *Wrighton, L.N.* 29: 86. Persistent by Route 11 in the City, 1967, *Lousley, L.N.* 47: 10. Bombed site, Mile End Road, E.1, 1952,† *Henson* (BM). White City. Primrose Hill. St Katherine's Dock, Stepney, 1966, *Brewis.*

Double-flowered plants are frequent.

PETRORHAGIA (Ser.) Link

Lit. *Bull. Brit. Mus.* (*Bot.*) **3**: 122 (1964).

P. nanteulii (Burnat) P. W. Ball and Heywood Proliferous Pink.

Kohlrauschia prolifera auct., *Dianthus prolifera* auct., *Tunica prolifera* auct.

Perhaps formerly native in grassy places on sandy and gravelly soils, but long extinct if ever it occurred.

First record: *How*, 1650. Last record: *Macreight*, 1837.

2. In the grounds 'twixt Hampton Court and Teddington, *How Phyt.*, 10 and *Merr. Pin.*, 10; Hampton Court, *Macreight Man.*, 28. The latter record may have been copied from earlier authors, and it is possible that the records refer to *Dianthus deltoides.*

GYPSOPHILA L.

G. paniculata Bieb.

Alien. Asia Minor, etc. Garden escape naturalised by a river, casual on waste ground, etc. Very rare.

First record: *J. E. Cooper*, 1913.

5. Near Brentford, casual, 1966.† Bank of Brent near Greenford!, 1967, *L. M. P. Small*, det. *Lousley*; 1970. **7.** Hackney Marshes, 1913,† *Coop. Cas.*

CERASTIUM L.

C. arvense L. Field Mouse-ear Chickweed.

Native in fields, dry grassy places, on banks, etc. Very rare, and now mainly confined to sandy and gravelly soils near the Thames. Strangely absent from calcareous soils.

First record: *Merrett*, 1666.

1. Ruislip, *Moxey List.* Almost certainly introduced. **2.** Between Hampton Wick and Teddington,† *Merr. Pin.*, 63. Banks of the Thames half a mile beyond Kingston Bridge, going to Staines,† *L. Brown* (DILL). By the Thames below Hampton Court Bridge!,† *H. C. Watson, Wats. New Bot.*, 98. Teddington, 1936!;† between Ashford Common and Staines!, *K. & L.*, 35; 1963. **3.** Field adjoining Hounslow Heath, abundant, 1945–48!,† *Westrup, K. & L.*, 35.

Habitat destroyed by gravel-digging operations. **4.** Near Hampstead, 1845,† *J. Morris*, *T. & D.*, 58. Side of railway between Harrow and Kenton, 1863,† probably introduced with ballast (ME). **5.** Sandy field by the Thames, Chiswick, abundant, 1866, *T. & D.*, 58; 1888,† *Benbow* (BM). Habitat destroyed by gravel-digging operations.

C. tomentosum L. Dusty Miller, Snow-in-Summer.
Lit. *Proc. Bot. Soc. Brit. Isles* **7**: 37 (1967). (*Abstr.*)
Alien. S.-E. Europe, etc. Garden escape. Naturalised in fields, on waste ground, etc., established on railway banks and tracks, etc. Rare.
First record: *D. H. Kent*, 1946.
1. Railway banks, Uxbridge, plentiful. Grassy banks on calcareous soil, Harefield, abundant, 1966, *I. G. Johnson*!. **2.** Thames bank near Chertsey Lock, 1965→. Near West Drayton, 1965. Sunbury, 1967, *Kennedy*!. **4.** Harrow!, *Harley Fl.*, 5. Hampstead Heath Extension, 1949. Field near Stonebridge Park station, 1962→. **5.** Waste ground by the Thames, Chiswick, abundant, 1946!, *L.N.* 26: 59; 1947–48.† **6.** Enfield, 1966!, *Kennedy*. **7.** Bombed sites, Cripplegate!,† *Wrighton*, *L.N.* 29: 86.
It is possible that some of the above records may be referable to the closely allied species *C. biebersteinii* DC.

C. fontanum Baumg. subsp. **triviale** (Link) Jalas Common Mouse-ear Chickweed.
C. holosteoides Fr., *C. vulgatum* auct., *C. triviale* Link
Native in grassy places, on cultivated and waste ground, etc., established on lawns, etc. Very common.
First record: *Merrett*, 1666.
7. Very frequent throughout the district.

C. glomeratum Thuill. Sticky Mouse-ear Chickweed.
C. vicosum auct.
Native in dry grassy places, on cultivated and waste ground, etc. Common.
First record: *T. Johnson*, 1638.
7. Tottenham, *Johns. Cat.* Hampstead; Hyde Park, *Merr. Pin.*, 6. Islington, *Ball. MSS.* East Heath, Hampstead!; Parliament Hill Fields!, *Fitter*. Ken Wood. Highgate. Regents Park!, *Holme*, *L.N.* 39: 46.
Apetalous forms have been noted at **2.** Ashford Common!, teste *Morton*, *K. & L.*, 35. **7.** Brompton, *J. Banks* (BM).

C. diffusum Pers. subsp. **diffusum** Dark-green Mouse-ear Chickweed.
C. atrovirens Bab., *C. tetrandrum* Curt.
Introduced. Established on railway tracks. Very rare, and perhaps extinct.

First record: *J. G. Dony*, 1951.
1. Railway tracks near Potters Bar, 1965, *J. G.* and *C. M. Dony*. **4.** Between the metals of the main line by Scratch Wood sidings, Edgwarebury!, *J. G. Dony*, *L.N.* 31: 12; 1960. **5.** Railway tracks, Brentford, 1951.

C. pumilum Curt. Curtis's Mouse-ear Chickweed.
Introduced. Established on railway tracks. Very rare, and perhaps extinct.
First and only record: *Milne-Redhead* and *J. G. Dony*, 1951.
4. Railway tracks, Mill Hill, *Milne-Redhead* and *J. G. Dony*, *K. & L.*, 35.

C. semidecandrum L. Little Mouse-ear Chickweed.
Native on dry banks, heaths, etc., established on railway banks and tracks, tops of old walls, etc. Very rare, and chiefly confined to light soils near the Thames.
First certain record: *Curtis*, 1778.
1. Uxbridge, *Benbow* (BM). Tracks of disused railway between Denham and Uxbridge, 1966→. **2.** Hampton Court!, *Newbould*, *T. & D.*, 58. Thames bank and river wall in several places between Kingston Bridge and Hampton Court, 1967. Between Ashford Common and Staines!, *K. & L.*, 35. Staines Moor, 1963. Shepperton, 1967. **3.** Hounslow Heath!, 1867, *Trimen* (BM); 1950,† Twickenham, *T. & D.*, 58. Isleworth, 1879;† between Isleworth and Twickenham, 1892,† *Wolley-Dod* (BM). Hanworth,†. Sudbury,† *Melv. Fl.*, 15. Hatch End, 1967, *Kennedy*. **4.** Some parts of Hampstead Heath, abundant, 1868,† *Warren* (BM). **5.** Chiswick, *J. Sowerby* (BM); abundant in a sandy field, c. 1868,† *T. & D.*, 58. **7.** About Hackney, particularly abundant,† *Curt. F.L.* Kensington, 1863,† *Newbould*, *T. & D.*, 58. Ken Wood,† *Fitter*, *K. & L.*, 35.

MYOSOTON Moench

M. aquaticum (L.) Moench Water Chickweed.
Stellaria aquatica (L.) Scop., *Malachium aquaticum* (L.) Fr.
Native by river- and streamsides, in marshes, wet woods, on damp waste ground, etc. Common, especially by the Thames and its tributaries.
First certain evidence: *Dillenius*, c. 1730.
7. In a ditch beyond Stepney† (DILL). Between Chelsea Hospital and the river, 1861; Hackney Wick, in several places, *T. & D.*, 57. Pimlico, 1832† (CGE). Finchley Road, 1870,† *Newbould*, *Trim. MSS.* Bombed site, City of London, 1946,† *Lousley*. Bombed site, north of St Paul's, 1950,† *Whittaker* (BM). By the New river, Clissold Park, Stoke Newington, 1967.

STELLARIA L.

Lit. *New Phyt.* **66**: 769 (1967).

S. media (L.) Vill. Common Chickweed.
Native on cultivated and waste ground, by roadsides, etc. Very common.
First record: *T. Johnson*, 1638.
7. Abundant throughout the district.

S. pallida (Dumort.) Piré
S. boraeana Jord., *S. apetala* auct., non Ucria
Native in grassy places, on banks, waste ground, etc., usually on light soils. Rare, or overlooked.
First evidence: *Trimen*, 1866.
1. Harefield!; Cowley, *Druce Notes*. Hillingdon. **2.** Thames bank above Kingston Bridge!, *Pugsley* (BM). Hampton Court, abundant!, *Rose, L.N.* 26: 59; 1968. Staines Moor. East Bedfont. West Drayton. **3.** Hounslow Heath!; Isleworth!, *K. & L.*, 36. **5.** Chiswick, *Trimen* (BM). Hanwell!; Syon Park!; Osterley Park!, *K. & L.*, 36. West Ealing. **6.** Enfield!, *Kennedy*. **7.** Isle of Dogs, *Newbould, T. & D.*, 55. Regents Park, *Holme, L.N.* 37: 185.

S. neglecta Weihe Greater Chickweed.
S. umbrosa Opiz & Rufr.
Native on banks, in shady places, etc. Rare, or overlooked.
First record: *Trimen* and *Dyer*, 1869.
1. Harefield!; Uxbridge!; Cowley!, *K. & L.*, 36. **2.** Shepperton!, *K. & L.*, 36. Walton Bridge. Laleham. **3.** Sudbury Hill. **4.** Near Stanmore, *Loydell* (BM). West Heath, Hampstead, *E. C. Wallace*!, *K. & L.*, 36. Golders Hill Park. Edgware, *Druce Notes*. **5.** Chiswick, *T. & D.*, 422. Hanwell!, *K. & L.*, 36. **7.** Isle of Dogs, *T. & D.*, 422.

S. holostea L. Greater Stitchwort.
Lit. *Rep. Bot. Soc. & E.C.* **12**: 840 (1946).
Native on hedgebanks, wood borders, heaths, commons, etc., established on railway banks, etc. Common, except in the vicinity of central London where owing to the absence of suitable habitats it is rare.
First record: *T. Johnson*, 1638.
7. Tottenham, *Johns. Cat.* Islington, *Ball. MSS.* Hornsey Wood,† *T. & D.*, 55. Marylebone Fields, 1817† (G & R). Near Lee Bridge,† *Cherry, T. & D.*, 55. East Heath, Hampstead. Railway bank near Paddington.

S. palustris Retz. Marsh Stitchwort.
S. dilleniana Moench, non Leers, *S. glauca* With.
Native in marshes, bogs, wet grassy places, etc. Very rare, and decreasing.
First evidence: *Kingsley*, 1839.

1. Cowley;† Harefield Moor!, *Benbow* (BM); 1967. Uxbridge Moor, 1839,† *Kingsley* (ME). Opposite Denham,† *Druce Notes*. Springwell,† *Pugsley*. **2.** Ditch by the roadside leading from Buckinghamshire to Staines, 1860,† *Phyt. N.S.* 4: 263. Near Walton Bridge!,† *F. N. Williams* (BM). Survived in the latter locality until c. 1963, when the marsh was drained. Shortwood Common!,† *Benbow*, *J.B.* (*London*) 25: 15. Survived in the latter locality until 1955 when the damp hollows in which it grew were filled with rubbish and levelled. Staines Moor, 1967, *Boniface*. Yeoveney,† *Benbow* (BM). Stanwell, 1878,† *Montg. MSS*. Plentiful in wet meadows between Chertsey Lock and Shepperton, 1966→. **3.** Near Hatton, 1867,† *Trimen*; Duke's river between Twickenham and Isleworth, 1916,† *Wilmott* (BM). **4.** Near Stonebridge,† *Hind*, *T. & D.*, 56. Hampstead Heath, *White Hampst.*, 364. An error. Near Golders Green, *Champ. List*. An error.

S. graminea L. Lesser Stitchwort.
Native on heaths, commons, in pastures, woods, by roadsides, etc., established on railway banks, etc. Common, except in heavily built-up areas where it is rare or extinct.
First record: *T. Johnson*, 1629.
7. Between Kentish Town and Hampstead, *Johns. Eric.* Marylebone Fields, 1815† (G & R). Kensington Gardens, 1845,† *J. Morris*; Hornsey,† *Cherry*; East Heath, Hampstead!; Hackney Wick, *T. & D.*, 56. South Kensington,† *Dyer List*. Bombed site, Ebury Street, S.W.1,† *McClintock*. Ken Wood. Hyde Park, two plants on a soil dump, 1961,† *D. E. Allen* and *McClintock*, *L.N.* 41: 17. Stag Place, S.W.1, 1963, *McClintock*, *L.N.* 43: 20. Buckingham Palace grounds, *Fl. B.P.G.*

Var. **grandiflora** Peterm.
1. Copse Wood, Ruislip, 1947!, *L.N.* 27: 31.

S. alsine Grimm Bog Stitchwort.
S. uliginosa Murr.
Native in bogs, marshes, wet woods, on damp heaths, pond verges, etc. Rare, and mainly confined to the northern parts of the vice-county, and absent from the immediate vicinity of the Thames.
First record: *Petiver*, 1695.
1. Uxbridge Moor, 1839,† *Kingsley* (ME). Ruislip!, *Melv. Fl.*, 4. Harefield,† *Blackst. Fasc.*, 4. Eastcote;† Harrow Weald Common!; Elstree,† *T. & D.*, 56. Northwood!; Stanmore Common, frequent!, *K. & L.*, 27. South Mimms, *C. S. Nicholson* (LNHS). **3.** New Hounslow,† *T. & D.*, 56. Harlington,† *Burkill*, *K. & L.*, 37. Cranford,† *Montg. MSS*. **4.** Near Harrow,† *Melv. Fl.*, 14. Wetter parts of Hampstead Woods, *Pet. Midd.* Near Finchley, *Newbould*; Deacon's Hill,† *T. & D.*, 56. Bishop's Wood, 1860; Turner's Wood, 1861, *Trimen* (BM). West Heath, Hampstead!. *Syme* (SY). Brockley Hill,

1966, *Kennedy*!. **6.** Hadley!, *Warren*; Edmonton;† near Whetstone, *T. & D.*, 56. Highgate Wood, *C. S. Nicholson*; Coldfall Wood, Highgate, *E. M. Dale, K. & L.*, 37. **7.** Kensington Gardens, 1817† (G & R). East Heath, Hampstead!, *T. & D.*, 56. Ken Wood!, *Andrewes MSS.* Parliament Hill Fields!, *K. & L.*, 37.

MOENCHIA Ehrh.

M. erecta (L.) Gaertn., Mey. & Scherb. Upright Chickweed.
Native on heaths, in short dry turf, etc. Very rare, and decreasing, and confined to sandy soils.
First record: *Merrett*, 1666.
1. Harefield Common;† Uxbridge Common!,† *Blackst. Fasc.*, 42. Hillingdon Heath, 1890;† Ruislip Common!, *Benbow* (BM). Northwood. **2.** Bushy Park!, *Barton* (BM); 1973. Golf course, Hampton Court Park!, *Welch, L.N.* 26: 58; 1973. Near Staines!, *K. & L.*, 36. **3.** Hounslow Heath!,† *Trimen*; west of Hounslow,† *Benbow* (BM). Hatton,† *T. & D.*, 58. Near Hanworth, 1950.† **4.** Highwood Hill,† *Blackst. Spec.*, 38. Harrow,† *Harley Add.* Hampstead Heath, *Pet. Midd.*, and many later authors; 1868,† *Warren* (BM). Whitchurch Common!,† *C. S. Nicholson* (LNHS). **5.** Acton, 1871,† *Warren, Trim. MSS.* **6.** Enfield,† *P. H. Cooke, Lond. Nat. MSS.* **7.** Old wall, King's Road, Chelsea,† *Pamplin, Wats. New Bot.*, 98. Hyde Park, *Merr. Pin.*, 63; c. 1790, *Dicks. Hort. Sicc.*; 1820,† *E. T. Bennett* (BM).

SAGINA L.

Lit. *Rep. Bot. Soc. & E.C.* **5**: 190 (1918).

S. apetala L. subsp. **apetala** Ciliate Pearlwort.
S. ciliata Fr.
Native on heaths, dry grassy and stony places, usually on light soils, also established on garden paths, in crevices at the bases of walls, etc. Rare.
First certain record: *A. Irvine*, 1858.
1. Harrow Weald Common, 1866, *Trimen*; between Uxbridge and West Drayton, *Benbow* (BM). Pinner, *W. A. Tooke* and *Hind, J.B. (London)* 9: 272. Ruislip!, *Melv. Fl.* 2: 10. Uxbridge Common; between Uxbridge Common and Swakeleys!; Hillingdon!, *Benbow* (BM). **2.** Between Staines and Hampton, in several places, 1866, *Trimen*; towpath near Hampton Court!, *Benbow* (BM); river wall, Hampton Court, 1967. Railway tracks, Staines to Colnbrook, 1951!, *K. & L.*, 38. Hounslow Heath, abundant, *T. & D.*, 53; 1948, *Westrup, K. & L.*, 38. **4.** Holders Hill, Hendon, 1924, *Moring* (HPD). **7.** Chelsea Hospital, *Irv. Ill. Handb.*, 769. Hyde Park, *Warren Fl.* Abundant in the gravel paths of Chelsea College, *Britten* (BM). Bombed

site near Tower of London,† *Rose, K. & L.*, 38. Bombed sites, Cripplegate, *Jones Fl.*

Forma **reuteri** (Boiss.) Elliston Wright
Sagina reuteri Boiss.
2. [West] Drayton, *Druce Notes.*

A variant, referable to **S. filicaulis** auct., non Jord., allied to *S. apetala* L. subsp. *apetala*, was found at **3.** Hounslow Heath, *Trimen* (BM).

Subsp. **erecta** (Hornemann) F. Hermann Common Pearlwort.
Sagina apetala auct.
1. Eastcote, *Melv. Fl.*, 13. Harefield!, *Lond. Nat. MSS.* Uxbridge; Uxbridge Common; Hillingdon!, *Benbow MSS.* Ruislip Common. Cowley. South Mimms. Potters Bar. **2.** Sunbury; Hampton, *T. & D.*, 53. [West] Drayton, *Druce Notes.* Near Dawley. Staines. Teddington. **3.** Twickenham!, *T. & D.*, 53. Near Harlington!, *Lond. Nat. MSS.* Isleworth. St Margaret's Wood, Hounslow Heath, *W. R. Linton* (BM). Hayes Park. **4.** Harrow, *Hind Fl.* Harrow Weald Churchyard;† Edgware† and Whitchurch Churchyards!, *T. & D.*, 53. Hampstead; Hampstead Heath!, *De Cresp. Fl.*, 61. Mill Hill, *Britten*, *Trim. MSS.* Hendon, *Lond. Nat. MSS.* Near Willesden Green;† Child's Hill† (HPD). Near Neasden;† Wembley Park,† *Benbow MSS.* Kingsbury. Oakleigh Park. **5.** Near Ealing!;† Hanwell Churchyard!,† *Newbould*; Chiswick!, *T. & D.*, 53. Heston and Norwood Churchyards. Osterley Park. Syon Park. Near Kew Bridge. Ealing Common,† *Benbow MSS.* **6.** Edmonton, *T. & D.*, 53. Hadley, *French* (BM). Highgate!, *Trim. MSS.* Woodside Park. **7.** Hornsey, *French* (BM). Kensington Gardens, one plant, 1961, *Allen Fl.* Bombed sites, Cripplegate,† *Wrighton, L.N.* 29: 87. Buckingham Palace grounds, *Fl. B.P.G.*

Var. **prostrata** S. Gibs.
1. Ruislip, *Hind*, *Melv. Fl.*, 13.

S. procumbens L. Procumbent Pearlwort.
Native on bare stony ground, etc., and established at bases of walls, in crevices between paving stones, on gravel paths, lawns, etc. Very common.
First record: *Merrett*, 1666.
7. Frequent throughout the district.

S. glabra (Willd.) Fenzl
Alien. Europe, etc. A rock garden plant which has escaped and become naturalised in turf, etc. Very rare.
First record: *Studley*, 1959.
1. Plentiful on a dry roadside bank near houses, Stanmore Hill, 1966.

Pinner Cemetery, originally planted on graves but now plentifully established in adjacent turf, 1967, *Kennedy*!. **4.** Pinner Road Cemetery, Harrow, originally planted on graves but now established in adjacent turf, 1959, *Studley*, *L.N.* 39: 40.

S. subulata (Sw.) C. Presl Awl-leaved Pearlwort.
Formerly native on sandy heaths, etc., but long extinct.
First record: *Lightfoot*, c. 1783. Last record: *Winch*, 1825.
1. Uxbridge Moor, *Lightfoot*, *Curt. F.L.* Ruislip Common, 1908, *Druce Notes*. An error, the specimen (D) on which the record was based is *Moenchia erecta*. **7.** A weed in Chelsea Physic Garden, 1825, *Winch MSS*. No doubt an introduction.

S. nodosa (L.) Fenzl Knotted Pearlwort.
Formerly native on damp heaths, in moist sandy and gravelly places, etc., but now extinct.
First record: *J. Hill*, 1736. Last record: *Lovell*, 1906.
1. Uxbridge Moor, abundantly, 1736, *J. Hill*, *Blackst. MSS.*; c. 1760, *Alch. MSS.* Harefield Moor, *J. Hill* (BM), and *Blackst. Spec.*, 3. By the canal between Harefield and Uxbridge, *De Cresp. Fl.*, 61. Moors north of Uxbridge, *Benbow*, *J.B.* (*London*) 22: 56. Meadow west of Denham Lock; towpath of canal near Harefield Moor Lock, 1892, *Benbow* (BM). **2.** Hampton Court, 1837 (W). **3.** Hounslow Heath, *Curt. F.L.* Isleworth, 1904, *Green*, *K. & L.*, 340; 1906, *Lovell*, *Benn. MSS.* **4.** Hampstead Heath, *Huds. Fl. Angl.*, 178.

MINUARTIA

M. hybrida (Vill.) Schischk. Fine-leaved Sandwort.
Arenaria tenuifolia L., *Alsine tenuifolia* (L.) Crantz, *Minuartia tenuifolia* (L.) Hiern, non Nees ex Mart.
Native or denizen in bare stony places on light soils, also established on railway tracks. Very rare.
First evidence: *Francis*, 1869.
1. Railway tracks between Denham and Uxbridge!, *K. & L.*, 38; the railway has now become disused, the tracks have been partially removed and much of the area has been levelled; the plant, however, still persists in small quantity, 1971. **2.** Hampton Court, 1869, *Francis* (BM). Frequent visits to the area have failed to result in the rediscovery of the species though many apparent suitable habitats for it still exist. **6.** Tottenham,† *De Cresp. Fl.*, 3. Probably an error.

MOEHRINGIA L.

M. trinervia (L.) Clairv. Three-nerved Sandwort.
Arenaria trinervia L.
Native in woods, shady places, etc. Common, except in heavily built-up areas where it is very rare or extinct.
First record: *Petiver*, 1695.
7. Marylebone fields, 1817† (G & R). East Heath, Hampstead. Ken Wood. Hyde Park, 1947!, *Watsonia* 1: 297. Kensington Gardens, 1961, *D. E. Allen*, *L.N.* 41: 18.

ARENARIA L.

A. serpyllifolia L. subsp. **serpyllifolia** Thyme-leaved Sandwort.
Native in dry fields, stony places, etc., and established on walls, railway tracks, etc. Common.
First certain evidence: *Kingsley*, 1839.
7. Clapton, *J. Morris*; Hampstead, *T. & D.*, 55. South Kensington,† *Dyer List.* Whitehall,† *Cottam.* Highgate!, *C. S. Nicholson* (LNHS). Bombed site, near Cannon Street Station,† *Sladen*, *Lousley Fl.* Bombed site near Tower of London, 1945.† Bombed site, Cripplegate,† *Wrighton*, *L.N.* 29: 86. Hackney, 1840, *H. Newman* (BM). Eaton Square, S.W.1!; Regents Park!, *L.N.* 39: 46.

Subsp. **tenuoir** (Mert. & Koch) Arcangeli Lesser Thyme-leaved Sandwort.
Arenaria leptoclados (Reichb.) Guss., *A. serpyllifolia* subsp. *leptoclados* (Reichb.) Nyman
7. Kensington Gardens, 1871, *Warren* (BM). Regents Park!, *K. & L.*, 38. Near Ken Wood!, *Druce Notes.*

SPERGULA L.

S. arvensis L. Corn Spurrey.
S. sativa Boenn.
Lit. *J. Ecol.* **49**: 205 (1961).
Native on heaths, cultivated and waste ground, etc., chiefly on light soils, but by no means confined to them. Common, especially in the south-western parts of the vice-county.
First record: *T. Johnson*, 1638.
7. Tottenham, *Johns. Cat.* Near Ken Wood!, *J. Hill* (BM). Cremorne;† Kensington!; Shepherds Bush; Green Park;† Hampstead; Isle of Dogs, *T. & D.*, 59. Piccadilly, W.1., 1868,† *T. & D.*, 60. Hyde Park! and Kensington Gardens!, casual . . . sown with grass seed, *Warren Fl.* Haverstock Hill, 1867 (SY). Tollington Park, 1875, *French* (BM). Hackney Marshes, 1918, *J. E. Cooper.* Buckingham Palace grounds, *Fl. B.P.G.* Chelsea, *L. M. P. Small.*

SPERGULARIA (Pers.) J. & C. Presl

S. rubra (L.) J. & C. Presl Sand-spurrey.
Buda rubra (L.) Dumort., *Lepigonum rubrum* Fr.
Native on heaths, commons, bare waste ground, dry grassy places, etc., and established on gravel paths, railway tracks, etc. Locally plentiful, especially on sandy and gravelly soils.
First record: *Johnson*, 1633.
1. Harefield!, *Newbould*; Harrow Weald Common; Stanmore Common, *T. & D.*, 59. Northwood!, *Green* (SLBI). Uxbridge Common!, *Benbow* (BM). Hillingdon, *Trim. MSS.* Near Denham. Ruislip Common. Yiewsley, *J. E. Cooper*. Pinner, *Hind*, *Trim. MSS.* Wrotham Park, *J. G.* and *C. M. Dony*. **2.** Near Teddington, *T. & D.*, 59. Hampton Court, abundant!, *Welch*. Bushy Park, locally abundant. Kempton Park. Between Colnbrook and Stanwell, *Benbow* (BM). Ashford Common, *Monckt. Fl.*, 19. Staines, *Briggs*. **3.** Whitton, 1843, *Twining* (BM). Hounslow Heath!; Hatton, *T. & D.*, 59. **4.** Harrow!, *Hind Fl.* Hampstead Heath, *Blackst. Spec.*, 94; 1861, *Trimen* (BM); c. 1915, *Andrewes MSS.* North-west Heath, Hampstead, 1921, *Richards*. Stanmore, *T. & D.*, 59. Mill Hill golf course, abundant. Mill Hill School grounds, *Sennitt*. **5.** Near Hanwell, *T. & D.*, 59. Near Ealing,† *A. Wood*; Brentford!, *Britten*, *Trim. MSS.* Syon Park, locally abundant. Greenford Park Cemetery, abundant on gravel paths. **6.** Highgate, *Coop. Fl.*, 104. Hadley, *Warren*; Edmonton, *T. & D.*, 59. Enfield, *J. E. Cooper*; Winchmore Hill, *P. H. Cooke*. **7.** Tothill Fields, Westminster,† *Johns. Ger.*, 1125. Tottenham, *Johns. Cat.* Hyde Park, *Blackst. MSS.*; c. 1835, *Forsyth*, *Trim. MSS.*; 1961–62, *D. E. Allen* and *McClintock*, *L.N.* 41: 13. East Heath, Hampstead; Kensington Gardens, 1866,† *T. & D.*, 59. Fulham Palace grounds, 1966, *Brewis*.

HERNIARIA L.

H. glabra L. Glabrous Rupture-wort.
H. hirsuta auct., non L.
Denizen on bare waste ground, etc. Very rare, and almost extinct.
First record: *Hudson*, 1762.
1. Harefield, 1875,† *Browner*, teste *A. Bennett*, *Trans. Norf. & Norw. Nat. Soc.* 8: 528. Almost certainly an introduction. **4.** Kenton!, 1943, *Potts* (BM); 1949!, *L.N.* 29: 6; occurred annually until 1962, often in abundance; reappeared in very small quantity in 1966, *Kennedy*!; 1967. Undoubtedly an introduction. **5.** Garden ground, Ealing Common, 1817, *Burchell*, teste *Britten*, *J.B.* (*London*) 9: 271. An error, the specimen (K), on which the record was based, is the European

adventive *H. hirsuta* L. **6.** Colney Hatch, *Huds. Fl. Angl.*, 94; c. 1792,† *B.G.*, 455. Finchley,† *B.G.* 455. Finchley Common, 1795, *E.B. Suppl.* 3: 2857.

SCLERANTHUS L.

S. annuus L. subsp. **annuus** Annual Knawel.

Native on heaths, commons, cultivated and waste ground, etc., chiefly on light soils. Rare, and decreasing.

First record: *Blackstone*, 1737.

1. Harefield!, *Blackst. Fasc.*, 79. Pinner, *Hind, Melv. Fl.*, 65. Hillingdon, *Warren*; Harrow Weald Common!, *T. & D.*, 115. Uxbridge, *Benbow, MSS.* North of Ruislip Reservoir, *Welch*; South Mimms!, *K. & L.*, 231. Near South Mimms, *J. G.* and *C. M. Dony*! **2.** Tangley Park,† *Dyer* (BM). Teddington; between Teddington and Bushy Park, *T. & D.*, 115. Staines!; Sunbury!; Halliford; Ashford Common, *Benbow MSS.* **3.** Hounslow!; near Hanworth,† *T. & D.*, 115. Hayes!, *Benbow MSS.* Feltham!, *K. & L.*, 231. **4.** Hampstead Heath,† *Irv. MSS.* Kingsbury,† *Varenne, T. & D.*, 115. Abundant near Brent Reservoir,† *Warren, T. & D.*, 424. The latter record may refer to Varenne's locality. **6.** Colney Hatch,† *Benbow MSS.* **7.** Crouch End, *J. E. Cooper, Sci. Goss.* 1898: 223. Kensington Gardens, 1962, *Allen Fl.* Weed in flower-bed, Buckingham Palace grounds, 1967, *McClintock, L.N.* 47: 9.

The subsp. **polycarpos** (L.) Thell., which is smaller than subsp. *annuus* and calcifuge has not been recorded from Middlesex. It is not, however, well understood, and it is possible that some of the above records may be referable to it.

PORTULACACEAE

MONTIA L.

Lit. *Watsonia* **3**: 1 (1953).

M. fontana L. subsp. **chondrosperma** (Fenzl) Walters Blinks.

M. verna auct., *M. minor* auct.

Native on wet heaths, commons, in damp woods on acid soils, etc. Rare, and decreasing.

First certain evidence: *J. Banks*, c. 1800.

1. Harefield Common.† Ruislip Common!, *Benbow* (BM). Potters Bar. **2.** Near Walton Bridge!;† Staines Moor!, *K. & L.*, 340. Bushy Park. Hampton Court Park!, *Welch, K. & L.*, 340. **3.** Hounslow Heath!,† *Fisher* (BM). **4.** Hampstead Heath, 1842, *Tuck* (BM); 1866, *Parsons* (CYN). **6.** Hadley Common. Enfield. **7.** Chelsea Common,† *J. Banks* (BM).

Subsp. **amporitana** Sennen

Subsp. *intermedia* (Beeby) Walters, *Montia lusitanica* Sampaio, *M. rivularis* auct.

1. Uxbridge Common, 1884,† *Benbow* (BM). Stanmore Common. Lawn, Wrotham Park, 1965, *J. G.* and *C. M. Dony, Fl. Herts*: 64.

All published records have been omitted unless substantiated by specimens.

CLAYTONIA L.

C. perfoliata Donn ex Willd.

Montia perfoliata (Willd.) Howell

Alien. N. America. Naturalised in grassy places, thickets, etc., and established on cultivated and waste ground. Occurs also as a persistent weed in parks and gardens, particularly on light soils. Rare, but apparently increasing.

First record: *E. Newman*, 1853.

1. The Grove, Stanmore Common!, *Melv. Fl.* 2: 40. Near Uxbridge!, *W. West.* Several places about Hillingdon in quantity, 1966→. Northwood!, 1950, *Graham, K. & L.*, 40; 1962. **2.** Hampton, 1950, *Westrup*. **4.** Hampstead Heath Extension, 1949!, *H. C. Harris*; West Heath, Hampstead!, *K. & L.*, 40. Edgware, *Horder*. Mill Hill, *L. M. P. Small.* **5.** Weed of flower beds, Hanger Hill Park, and as a garden weed in various other parts of Ealing. Weed of flower beds, Gunnersbury, 1967→. **6.** Cockfosters, *Absolon, K. & L.*, 40. Whetstone, 1967. **7.** Introduced . . . into the Botanic Garden at Chelsea, where it soon became a most troublesome weed, and remains so at the present day (1853), coming up spontaneously by thousands in various parts of the garden, *E. Newman, Phyt.* 4: 983; still abundant, 1869, *T. Moore, T. & D.*, 114. Chelsea Embankment Gardens, 1953, *McClintock, L.N.* 39: 46. Garden weed, Old Church Street, Chelsea, 1964, *Clement, L.N.* 44: 21. Chelsea Hospital grounds, 1973. Buckingham Palace grounds, 1957!, *McClintock.* Hyde Park, 1959, *D. E. Allen, L.N.* 39: 46. Kensington Gardens, 1962, *Allen Fl.* Regents Park, 1965, *Wurzell, L.N.* 45: 20.

C. sibirica L. Spring Beauty.

C. alsinoides Sims, *Montia sibirica* (L.) Howell

Alien. N. America. Naturalised in woods, copses, shady places by streams, etc., particularly on light sandy and gravelly soils, but by no means confined to them. Occurs also as a garden weed. Rare.

First evidence: *Fitter*, 1945.

1. Harefield!,† *L.N.* 28: 32. Garden weed near Ruislip, 1955. **4.** Abundant by a small stream in Turner's Wood!, 1945, *Fitter* (LNHS);

1967. Hampstead Heath!, *Welch, L.N.* 28: 32; still present in several places on and near the West Heath, 1968. Hampstead Heath Extension!, *H. C. Harris, L.N.* 28: 29; 1966. Golders Hill Park, 1947, *Whidborne* (BM). Harrow!, *Horton Rep.*; still present in several places on the hill, 1969. **5.** Hanger Hill, Ealing, 1945,† *Boucher, K. & L.*, 40. **6.** Queen's Wood, Highgate, *Bangerter* and *A. W. Jones*. A well-naturalised colony in a wood at Enfield, 1967, *Kennedy*!, *L.N.* 47: 9. **7.** Ken Wood, *L. M. P. Small*. Garden weed, St John's Wood, *Holme, L.N.* 39: 46. Weed in churchyard near Porchester Road, Paddington, 1969, *L. M. P. Small*.

AMARANTHACEAE

AMARANTHUS L.

A. retroflexus L.

Alien. N. America. Naturalised on waste ground, by roadsides, etc., established on canal paths, etc. Occurs also as a garden weed. Rather rare, and usually merely casual.

First record: *A. Irvine*, 1858.

1. Uxbridge!; near Ruislip; between Denham and Harefield, *Benbow MSS*. Yiewsley!, 1908 and 1910–11, *Coop. Cas.*; 1917, *Lester-Garland*; 1923, *J. E. Cooper*; 1926, *G. C. Brown* (K); 1962. Springwell, 1945–46, *L.N.* 26: 68. Rubbish-tip, New Years Green, 1969, *Clement* and *Ryves*. **2.** Hampton Court, *T. & D.*, 230. Near Walton Bridge; Stanwell Moor, *Benbow MSS*. Sunbury, 1881, *De Crespigny*; West Drayton!, *Benbow*; Staines, 1901, *H.* and *J. Groves* (BM). **3.** Twickenham, 1867, *Dyer* (BM). Hounslow Heath, 1948!, *L.N.* 28: 33; 1964. Feltham, 1952, *Russell* and *Welch*!. **4.** Neasden, *Benbow MSS*. Garden weed, Willesden Green, 1953. East Finchley, 1897; Finchley, 1910–11, *J. E. Cooper* (K). Mill Hill, introduced with bird-seed, *Harrison*. **5.** Hanwell!, 1891, *C. B. Clarke* (K); still occurs annually on the canal towpath under the wall of St Bernard's Hospital. Between Greenford and Perivale, *Benbow MSS*. Greenford!, *K. & L.*, 232. Acton, *Druce Notes*. Railway embankment, Kew Bridge station, 1947!, *Lousley, L.N.* 27: 32. Chiswick, 1948, *Boniface, K. & L.*, 232. Near Hanwell, 1960. Garden weed, West Ealing, 1960. Near Shepherds Bush, 1963. **6.** Tottenham, 1893, *Riddelsdell* (BM); 1964, *Wurzell*, teste *Lousley*. By Lee Navigation Canal between Edmonton and Ponders End, 1955. **7.** Chelsea, *Irv. Ill. Handb.*, 390. Camden Town, *T. & D.*, 230; Walham Green, c. 1864, *Britten, Trim. MSS.* Isle of Dogs, 1866, *Newbould*; Hackney Wick, 1867; Kilburn, 1870, *Dyer*; Primrose Hill, 1875, *Blow*; Highgate, 1877; Tollington Park, 1875, *French*; bombed site, Shadwell, *Henson* (BM). Crouch End,

J. E. Cooper, Sci. Goss. 1887, 223. Hackney Marshes, 1912–13, *Coop. Cas.*; 1918 and 1924, *J. E. Cooper, K. & L.*, 232. Bombed sites, Great Tower Street† and Lower Thames Street, E.C. !,† *Lousley Fl.* Bombed site, Beer Lane, E.C., 1946,† *Lousley* (K). Tower of London gardens, *Lousley, L.N.* 39: 46.

Var. **delilei** (Richter-Loret) Thell.
1. Springwell, 1945 !, *Lousley* (L); 1948 !, *Brenan* and *Sandwith, L.N.* 28: 33. **3.** Northolt, 1947, *Welch* and *Lousley*! (L). **5.** Canal bank, Hanwell, 1891, *C. B. Clarke* (K).

CHENOPODIACEAE

CHENOPODIUM L.

C. bonus-henricus L. All-Good, Good King Henry.
Alien. Europe, etc. Formerly cultivated as a pot-herb, and long naturalised by roadsides, on cultivated and waste ground, etc. Very rare.
First record: *T. Johnson*, 1629.
1. Harefield !,† *Blackst. Fasc.*, 11. Hillingdon churchyard,† *M. & G.*, 446. Hillingdon !, *K. & L.*, 233. Cowley;† Moorhall;† Ruislip, *Benbow MSS.* South Mimms, 1965, *J. G.* and *C. M. Dony*!. **2.** Staines !,† *M. & G.*, 446. Hampton Court ! (G & R). Bushy Park !, *T. & D.*, 326. Stanwell. West Drayton,† *Benbow* (BM). **3.** Harrow,† *Melv. Fl.*, 65. Near Richmond Bridge;† between Twickenham and Isleworth,† *T. & D.*, 236. Northolt Churchyard !,† *K. & L.*, 233. Hayes, 1965. Near Northolt, one plant, 1966.† **4.** Stanmore,† *Varenne, T. & D.*, 236. Mill Hill,† *Benbow MSS.* Finchley, 1912–14, *Coop. Cas.*; 1916–19 and 1925, *J. E. Cooper*,† *K. & L.*, 233. **5.** Lampton,† *T. & D.*, 236. Hanwell !,† *K. & L.*, 233. Ealing, 1876,† *Britten, Trim. MSS.* Acton,† *Druce Notes.* **6.** Edmonton;† Enfield Green,† *T. & D.*, 236. Enfield, 1962, *Kennedy*. **7.** Between Kentish Town and Hampstead,† *Johns. Eric.* Tottenham,† *Johns. Cat.*

C. polyspermum L. All-seed.
Native on cultivated and waste ground, by roadsides, etc. Common.
First record: *Petiver*, 1710.
7. Frequent in the district.

C. vulvaria L. Stinking Goosefoot.
Denizen on waste ground, under walls, established on rubbish-tips, etc. Formerly common in London and the inner suburbs, but now very rare in the vice-county.
First record: *T. Johnson*, 1632.

1. Harefield,† *Blackst. Fasc.*, 8. Pinner, plentiful, *Hind, J.B.* (*London*) 9: 272; a troublesome weed in the garden of Pinner Parsonage,† *Hind, Melv. Fl.* 2: 84. Near Uxbridge Common, 1885,† *Benbow MSS.* Yiewsley, 1963–65. **2.** Under the wall of Hampton Court, 1886, *Benbow, J.B.* (*London*) 25: 18. Hampton Wick, 1919, *E. K.* and *W. K. Robinson.* Near West Drayton, 1965!, *Wurzell.* **3.** Isleworth,† *Francis* and *G. Lloyd* (K). North Hyde, 1920, *J. E. Cooper*; Hounslow Heath, *Westrup, K. & L.*, 236; 1964, *Wurzell, Clement* and *Mason.* Yeading, 1951, *Boniface, K. & L.*, 236. **4.** Hampstead Heath,† *Johns. Enum.* Hampstead,† *Morley MSS.* Willesden,† *De Cresp. Fl.*, 16. **5.** Sandy field near Chiswick, *Fox, T. & D.*, 231; 1887, *Benbow MSS.*; near Chiswick House!, *Bangerter, K. & L.*, 236; 1965. Grove Park, 1894, *Fraser*; Brentford, 1927, *Hubbard*; Hammersmith, 1847,† *Stevens* (K). Canal path between Hanwell and Southall!, *Welch, K. & L.*, 236; occurred annually under the wall of St Bernard's Hospital until the last few years when it has not been seen. **6.** Tottenham, *Johns. Cat.* **7.** About the mud walls in the fields about London, plentifully,† *Turn. Bot.*, 14, and many later authors. Hackney; between St James's Park and Hyde Park,† *Newt. MSS.* Hampstead, *Morley MSS.*; c. 1800, *Woods, B.G.*, 40; 1845,† *Barham* (K). Between London and Hampstead;† Hyde Park;† Pancras Churchyard,† *M. & G.*, 448. Belsize Park,† *Irv. Lond. Fl.*, 121. St John's Wood, 1822,† *E. T. Bennett*; Park Lane, W.1,† *H. Davies* (BM). Islington,† *Woods, B.G.*, 401. Kensington Gore,† and under the wall of Hyde Park,† *T. & D.*, 231. Paddington, 1855,† *Hind, Trim. MSS.* St Paul's Churchyard,† *Crowe MSS.* Camden Town, 1840† (HE). Hackney Marshes, 1914–16, *J. E. Cooper* (BM). A single plant in a flower tub, Marsham Street, S.W.1, 1965, *McClintock, L.N.* 45: 21.

C. album L. White Goosefoot, Fat Hen.
Lit. *Watsonia* **5,** 47 (1961): *loc. cit.* **5,** 117 (1962): *J. Ecol.* **51,** 711 (1964).
Native on cultivated and waste ground, by roadsides, etc., established on rubbish-tips, etc. Very common, and extremely variable in growth habit, size, leaf shape and texture, etc. Numerous varieties and forms have been recorded from Middlesex but the records of most of these are omitted on account of the confusion of the taxonomy and nomenclature of this very plastic species.
First certain evidence: *Buddle*, c. 1710.
7. Frequent throughout the district.

Var. **viridescens** St. Amans
var. *paganum* (Reichb.) Syme
1. Yiewsley!, *M. & S.* Uxbridge. **2.** West Drayton!, *Rep. Bot. Soc. & E.C.* 6: 143. **7.** Hackney, *Melville* (K). Canal side between Paddington and Kensal Green.

Forma **cymigerum** (Koch) Schinz & Thell.
3. Northolt, *Brenan*!, *K. & L.*, 235. **7.** Bombed site, Neville Court, E.C.,† *Lousley* (L), teste *Brenan*.

Subsp. **album** var. **borbasii** (J. Murr) Aell.
5. Hanwell, 1952, *Brenan*!, *K. & L.*, 235.

C. suecicum J. Murr
C. viride auct., non L. ?
Native or denizen on cultivated and waste ground, etc. Very rare, or confused with *C. album*.
First, and only certain record, *M. Cole*, 1962.
6. Enfield, *M. Cole*, *Watsonia* 5: 120.

C. opulifolium Schrad. ex Koch & Ziz
Alien. Europe, etc. Established on waste ground, rubbish-tips, by roadsides, etc. Very rare, or confused with *C. album*.
First record: *A. Irvine*, 1862.
1. Uxbridge!; Yiewsley!, *Benbow* (BM). Near Denham, *Druce Notes*. Harefield!, 1955, *Graham* and *Harley*; 1966. New Years Green, 1969, *Clement* and *Ryves*, teste *Brenan*. **2.** West Drayton!, *Druce Notes*. Near Staines Moor, 1966, *Boniface*!. **5.** Alperton, 1867, *Trimen*; Chiswick!,† *Benbow* (BM). Greenford, 1952, *Brenan*!; 1953, *K. & L.*, 234; 1954–57.† **7.** Parsons Green, 1862,† *A. Irvine*; Finchley Road, Hampstead,† *T. & D.*, 233. Near West End, Hampstead, 1870,† *Dyer* (K). Isle of Dogs, *Benbow* (BM).

Var. **mucronulatum** G. Beck
7. Paddington, 1869 (SY).

C. ficifolium Sm. Fig-leaved Goosefoot.
Native on cultivated and waste ground, by roadsides, etc. Common.
First record: *Buddle*, 1710.
7. Common in the district.

Var. **dolichophyllum** J. Murr
1. Between West Drayton and Iver, 1928 (D).

C. murale L. Nettle-leaved Goosefoot.
Native or denizen on cultivated and waste ground, chiefly on light soils, by roadsides, etc., established about dung heaps in farmyards, etc. Rare, and often merely casual.
First certain record: *Curtis*, c. 1780.
1. Springwell, 1946!, *K. & L.*, 234. **2.** Near Sunbury, a single plant, 1945, *Welch*, *L.N.* 26: 63. Near Walton Bridge, 1945, *Welch*. Hampton Court, *A. R. Pryor* (BM). Hampton Wick, *E. K.* and *W. K. Robinson*. West Drayton, *Rep. Bot. Soc. & E.C.* 4: 500. East Bedfont, 1964, *Wurzell*; 1965, *Clement*, conf. *Lousley*. **3.** Feltham!; between

Hounslow and Lampton; Whitton, *T. & D.*, 234. Hounslow Heath, 1959, *D. Bennett.* **4.** East Finchley, 1906, *J. E. Cooper*; near Finchley, 1923, *Redgrove* (BM). Golders Green, 1910, *P. H. Cooke* (LNHS). Willesden Green, 1961. Near The Hyde, Hendon, 1965. **5.** Near Acton, *Newbould*, *T. & D.*, 234. Hanwell, 1950!, *L.N.* 30: 7. Turnham Green, *Knowlton* (BM). Greenford, 1954, *Welch*, *K. & L.*, 234. Brentford, 1965, *Boniface.* **6.** Muswell Hill, 1896–97, *J. E. Cooper*, *Sci. Goss.* 1898, 223. North of Cockfosters, 1945, *Welch*, *L.N.* 26: 63. Enfield, 1965, *Kennedy*!; 1966. Pymmes Park, Edmonton, 1966, *Kennedy.* **7.** Plentifully on most of the great roads leading from the metropolis; Edgware Road, abundant, *Curt. F.L.* Kentish Town, 1846; back of Camden Villas, 1847 (HE). Kensington Gardens, 1868; Kilburn, *Warren*; Chelsea!, *T. & D.*, 234; Ranelagh Gardens, Chelsea, in quantity, 1959!, *L.N.* 39: 46; by Chelsea Old Church, 1968→, *E. Young.* Marylebone Fields, 1817 (G & R). Islington, *Ball. MSS.* Bombed site, Causton Street, S.W.1, 1943, *McClintock*, *Rep. Bot. Soc. & E.C.* 13: 749. Bombed site, The Temple, 1944, *Sandwith* (K). Derelict garden, Pimlico, 1956, *Codrington*, *L.N.* 36: 15. Hyde Park, 1962, *Allen Fl.* Shadwell, 1966.

C. hybridum L. Sowbane, Maple-leaved Goosefoot.
Alien. Europe, etc. Naturalised on cultivated and waste ground, etc. Very rare, and sometimes merely casual.
First certain record: *A. Irvine*, 1858.
2. West Drayton, 1969, *Clement* and *Ryves.* **3.** Twickenham, a single plant, 1867, *Trimen* (BM). Yeading, 1949!, *Boniface*, *L.N.* 29: 13. **5.** Hanwell, 1950!, *K. & L.*, 234; 1952!, *Grigg*, *L.N.* 32: 82. Brentford, 1965, *Boniface*; 1966. **6.** Grange Park, *L. B. Hall*, *Lond. Nat. MSS.* **7.** Chelsea, *Irv. Ill. Handb.*, 386; Chelsea College, 1864; Parsons Green, *Britten*, *T. & D.*, 234. Bombed site, Shadwell, 1952!, *Henson* (BM); 1965→.

C. rubrum L. Red Goosefoot.
Lit.: *J. Ecol.* **57**: 831 (1969).
Native on cultivated and waste ground, pond verges, in muddy places, etc., established on railway tracks, etc. Common, especially about farm yards, etc.
First record: *T. Johnson*, 1629.
7. Between Kentish Town and Hampstead, *Johns. Eric.* About London, *Merr. Pin.*, 12. Among the new buildings near Bedford Square, 1789, confirmed by Curtis to be *rubrum* (LINN). Walham Green, *Francis*, *Coop. Suppl.*, 12. Between Kensal Green and Notting Hill, *Britten*; Eel Brook Meadow, Parsons Green; East Heath, Hampstead!; Primrose Hill; Camden Town; behind Adelaide Road [, Hampstead]; Thames embankment opposite Somerset House, 1866; Hackney Wick!, abundant; Isle of Dogs!, *T. & D.*, 235. Pimlico,

Pamplin; Brompton!; Euston Square Gardens, 1870, *Trim. MSS.* Near Hornsey!; near Highgate, *C. S. Nicholson, Lond. Nat. MSS.* Bombed sites, Monument Street!,† Bush Lane,† Upper Thames Street!,† Mark Lane!,† Hart Street† and near St Giles, Cripplegate, E.C.!, *Lousley Fl.* Near Chelsea, *Newbould*; Haverstock Hill; Gospel Oak Fields, *Syme* (SY). Bombed site, Mile End, E.1, 1951, *Lovis.* Bombed site, Kirby Street, W.C.,† *Whittaker* and *Morton*; Shadwell!, *Henson* (BM). Bombed sites, Gresham Street,† Queen Victoria Street† and Finsbury Square areas;† near the Tower of London, *Jones Fl.* Clerkenwell. Kensington Gardens!; St Pancras!, *L.N.* 39: 46. South Kensington. Shoreditch. Hoxton. Dalston. Stoke Newington. Holloway. Bethnal Green. Buckingham Palace grounds, *Fl. B.P.G.* Car park, Lord's Cricket Ground, 1967.

Var. **pseudo-botryodes** H. C. Wats. ex Syme
2. Staines, 1897, *C. E. Britton* (BM). **4.** Finchley, 1909, *J. E. Cooper, K. & L.*, 233. **7.** Crouch End, 1899, *J. E. Cooper, K. & L.*, 233.

Var. **blitoides** Wallr.
1. Railway sidings, Uxbridge, abundant, 1947.† **4.** Formed over 70% of the dense plant covering on the mud floor of the Brent Reservoir during the drought of 1919, *E. J. Salisbury, Nat.* 1921, 329.

Var. **kochiiforme** J. Murr
5. Greenford, 1952, *Brenan*!, *K. & L.*, 223.

C. glaucum L. Glaucous Goosefoot, Oak-leaved Goosefoot.
Native on muddy shores of lakes and ponds, rich cultivated ground, waste ground, etc. Very rare, and sometimes merely casual.
First certain evidence: *Petiver*, c. 1710.
1. Yiewsley, 1909, *J. E. Cooper* (BM). Ruislip Reservoir, scarce, 1949!,† *K. & L.*, 235. **2.** Staines Moor, abundant, 1867, *Dyer* (BM, K). **3.** Roxeth,† *Melv. Fl.* Sudbury, 1881,† *De Crespigny* (MANCH). **4.** Kenton,† *Hind Fl.* Beyond Kilburn towards Edgware, 1825,† *Pamplin, Wats. New Bot.*, 101. Cricklewood, 1870,† *Syme*; Willesden, 1884,† *W. R. Linton* (BM). Brent Reservoir!, *J. E. Cooper, K. & L.*, 235; still present, but varies in quantity from year to year. Hendon, *Champ. List.* **6.** Near Edmonton, 1885,† *F. J. Hanbury* and *W. R. Linton*; near Palmers Green, 1885,† *W. R. Linton*; Muswell Hill, 1906,† *J. E. Cooper* (BM). **7.** Tothill Fields, *Petiver* (SLO); c. 1775† (LT). Between Newington and Clapton;† near Kentish Town,† *E. Forster* (BM). Whitechapel Fields,† *Forst. Midd.* Paddington, 1801,† *D. Turner* (LINN). Walham Green,† *Francis, Coop. Suppl.*, 12. Abundant on waste ground behind Euston Square,† *Irv. MSS.* Downshire Hill, Hampstead, 1825† (HE). Roadsides . . . between Little Chelsea and Hyde Park Corner, 1770–95,† *Haworth*, teste

Pamplin, T. & D., 235. In the greatest abundance at the back of houses in Adelaide Road, Hampstead, 1867,† *Syme* (SY). Kentish Town, 1847,† *T. Moore* (BM). Between Hampstead and Hornsey, 1850,† *Syme* (SY). Hoxton,† *Woods, B.G.*, 401. West End, Hampstead,† *De Cresp. Fl.*, 16. Eel Brook Common, 1828† (BO). Edgware Road, 1842,† *Stocks*; between Hampstead and Hornsey, 1850,† *Syme* (BM).

BETA L.

B. vulgaris L. subsp. **maritima** (L.) Arcangeli Sea Beet.
B. maritima L.
Formerly native on the muddy shores of the tidal Thames, but now extinct.
First and only evidence: *Benbow*, 1887.
7. Isle of Dogs, 1887, *Benbow* (BM).

ATRIPLEX L.

A. triangularis Willd. Hastate Orache.
A. hastata auct., non L.
Native on cultivated and waste ground, by roadsides, etc. Very common.
First record: *T. Johnson*, 1629.
7. Common throughout the district.
A very plastic and variable species.

A. patula L. Iron-root, Common Orache.
A. erecta Huds.
Native on cultivated and waste ground, by roadsides, etc. Common, but less frequent than *A. triangularis*.
First record: *T. Johnson*, 1629.
7. Common throughout the district.
An extremely plastic and variable species. The var. **linearis** (Gaud.) Moss & Wilmott and the var. **erecta** (Huds.) Lange both appear to be common.

A. glabriuscula Edmondst. Babington's Orache.
A. babingtonii Woods
Introduced. Naturalised in a marshy field where it was probably introduced with shingle ballast from the coast, elsewhere as a casual on waste ground, rubbish-tips, etc. Very rare.
First certain record: *Hind*, 1857.
3. Roxeth, 1860,† *Hind Fl.* 'This may have been *A. hastata*'. *Trim. MSS.* Naturalised in a marshy field near Yeading!, *Boniface, K. & L.*, 237; the field has been disturbed, and was enclosed by a high fence in 1966 but the plant probably survives. **7.** Plentifully between

Kilburn and Kensal Green, 1857,† *Hind, Phyt. N.S.* 5: 204. Mr Hind may have mistaken forms of *A. hastata* for it, *T. & D.*, 238. 'The plant at Kilburn was among sea gravel and I have no doubt of the species', *Hind, Trim. MSS.* Hackney Marshes, 1960.†

TILIACEAE

TILIA L.

T. platyphyllos Scop. Large-leaved Lime.
T. grandifolia. Ehrh. ex Hoffm.
Introduced. Planted in woods, parks, etc., where it sometimes regenerates. Very rare.
First record (of regeneration): *D. H. Kent*, 1964.
2. Hampton Court, seedlings, 1964.

T. cordata Mill. Small-leaved Lime.
T. parvifolia Ehrh. ex Hoffm.
Introduced. Planted in woods, parks, etc., where it sometimes regenerates. Very rare.
First record (of regeneration): *Boucher*, 1946.
2. Bushy Park, seedlings, 1966. **5.** Hanger Hill, Ealing, a single small sapling, 1946!, *Boucher.* Osterley Park, seedlings, 1967.

T. cordata × platyphyllos = T. × vulgaris Hayne Common Lime.
T. europaea L., *T. intermedia* DC.
Introduced. Planted in woods, parks, etc., and frequently as a street tree. Regeneration appeared rarely to take place in Middlesex until a decade ago, but self-sown seedlings are now fairly common.
First record (of regeneration): *D. H. Kent*, 1960.
1. Uxbridge, 1970. Hillingdon, 1971. **2.** Staines, 1969. **3.** Cranford, seedlings, 1965. **4.** Harrow, seedlings, 1960→. **5.** Osterley Park, seedlings, 1960. Syon Park, seedling, 1960. Ealing; W. Ealing; Hanwell, 1970→. **6.** Enfield Chase, seedlings, 1961→. **7.** East Heath, Hampstead; Ken Wood, seedlings, 1962.

MALVACEAE

MALVA L.

M. moschata L. Musk Mallow.
Native in meadows, grassy places, etc., established on railway banks, etc. Locally plentiful, especially on heavy soils.
First record: *Gerarde*, 1597.

1. Harefield!, *Blackst. Fasc.*, 3. Pinner Hill!, *Hind Fl.* Woodready;† Stanmore Common;† near Elstree,† *T. & D.*, 60. Ruislip, *Melv. Fl.* 2: 16. Uxbridge;† Hillingdon;† near Denham!, *Benbow MSS.* Yiewsley, *J. E. Cooper.* Between Cowley and West Drayton;† Springwell!; Northwood!, *Benbow MSS.* Borders of Bayhurst Wood, *Jeffkins* and *J. A. Moore, J. Ruisl. & Distr. N.H.S.* 15: 14. Potters Bar!, *Tremayne.* South Mimms. Warren Gate. **2.** Between Stanwell Moor and Staines. **3.** Isleworth, 1845,† *Parker* (K). Twickenham Common, one plant,† *T. & D.*, 60. Southall!, *Wright* (BM). Near Hounslow Heath. **4.** Hampstead Heath, *Johns. Enum.*; c. 1912,† *Whiting Fl.* Between Turner's Wood and North End,† *Bliss, Park Hampst.*, 30. Harrow!; Kingsbury!, *Varenne*; Stonebridge, *Farrar*; Stanmore, *T. & D.*, 60. Railway banks between Preston Road and Harrow. Rayners Lane!,† *Schmolle.* Wembley. Hendon, *Bedford.* **5.** Osterley Park!, *A. Irvine*; Perivale,† *Lees*; Horsendon Hill!;† railway bank, Hanwell!,† *T. & D.*, 60. Near Boston Manor. Between Hanwell and Osterley. Railway banks between Brentford and Isleworth,† *Britten, Trim. MSS.* Ealing Common, casual, 1950.† **6.** Wood Green, *Munby, Nat.* 1867, 181. Edmonton, *T. & D.*, 60. Enfield!; Winchmore Hill, *C. S. Nicholson, Williams MSS*, Friern Barnet,† 1883, *J. E. Cooper.* **7.** Near Tyburn [= Marble Arch];† between London and Old Ford;† near Hackney,† *Ger. Hb.*, 786. Between Hampstead and London, *Park. Theat.*, 306. Near Highgate,† *Blackst. MSS.* New churchyard, Hampstead,† *Button, Coop. Suppl.*, 11. Ken Wood,† *Hunter, Park Hampst.*, 30.

Var. **heterophylla** Lej. & Court.

3. Near Hounslow Heath, 1947!, *L.N.* 27: 31. **4.** Stanmore, 1859, *Trimen* (BM). **5.** Southall, 1882, *Wright* (BM). Hanwell, 1947, *Wrighton*!, *L.N.* 27: 31. **6.** Enfield, 1966, *Kennedy*!.

Plants with white flowers have been noted at **1.** Uxbridge, 1903; Northwood, 1907, *Green, Williams MSS.*; 1914, *J. E. Cooper, K. & L.*, 144. **4.** Finchley, 1911, *J. E. Cooper* (BM). **5.** Hanwell, 1937!,† *K. & L.*, 44.

M. sylvestris L. Common Mallow.

Native in waste places, by roadsides, etc. Very common.

First record: *T. Johnson*, 1638.

7. Common in the district.

Var. **matrinii** Rouy

5. Chiswick, 1948, *A. H. G. Alston* and *Simpson.*

Var. **angustiloba** Čelak.

5. Chiswick, 1948, *A. H. G. Alston* and *Simpson.*

M. neglecta Wallr. Dwarf Mallow.
M. rotundifolia auct.
Native in waste places, by roadsides, on banks, etc., established on canal paths, etc. Common.
First record: *T. Johnson*, 1638.
7. Tottenham!, *Johns. Cat.* Between Westminster and Chelsea!, *Pet. MSS.* Islington, *Ball. MSS.* Kensal Green, *A. B. Cole*; Chelsea!, *Britten*; East Heath, Hampstead!, *T. & D.*, 61. Bombed site, Ebury Street, S.W.1,† *McClintock.* Highgate!, *Andrewes MSS.* Parsons Green. Kensington Gardens, *D. E. Allen*; Green Park!; Regents Park!, *L.N.* 39: 47. Fulham. Hammersmith. Brompton. Notting Hill. Isle of Dogs. Shoreditch. Hoxton. Stepney. Old Ford. Hackney Wick. Hackney Marshes. Bromley by Bow. Upper Clapton, *L. M. P. Small.*

ALTHAEA L.

A. officinalis L. Marsh Mallow.
Formerly native by the tidal Thames and as an established introduction elsewhere, but long extinct.
First record: *Blackstone*, 1737. Last record: *A. Irvine*, 1838.
7. By the Thames side at the Isle of Dogs . . . , *Alch. MSS.* I once (1737) gathered it by the side of Chelsea Waterworks, *Blackst. Spec.*, 4. Naturalised in a lane leading up to the Wood House, Hornsey, from Newington, *Irv. Lond. Fl.*, 181.

LINACEAE

LINUM L.

L. bienne Mill. Narrow-leaved Flax.
L. angustifolium Huds.
Introduced. Naturalised in dry grassy places, established on waste ground, etc. Very rare, and perhaps extinct.
First certain record: *Welch*, 1946.
1. Harefield!, *Welch, L.N.* 27: 31. **3.** Near Hounslow Heath, 1953,† *Welch, K. & L.*, 341. Habitat destroyed by the tipping of rubbish. **4.** Hendon, *Champ. List.* Almost certainly an error. Finchley, 1908, *J. E. Cooper.* Probably an error.

L. usitatissimum L. Cultivated Flax.
Alien. N. Africa, etc. Introduced with cage-bird seed, etc. Sporadic on rubbish-tips, waste ground, by roadsides, in gardens, etc. Common.
First record: *T. Johnson*, 1638.
7. Common in the district.

Plants with pure white flowers are recorded from **3.** Hounslow Heath, 1961. **5.** Greenford, 1956.

L. catharticum L. Fairy Flax, Purging Flax.
Native on heaths, in dry grassy places, etc. Rather common on calcareous and alluvial soils, rare on other soils.
First record: *Merrett*, 1666.
1. Harefield!, *Blackst. Fasc.*, 53. Harrow Weald Common!; Stanmore Common!, *T. & D.*, 70. Pinner, 1866,† *Trimen* (BM). Ruislip!; Northwood!, *Benbow, J.B. (London)* 25: 15. Ickenham!, *K. & L.*, 46, Springwell!, *Benbow MSS.* Railway banks, Denham to Uxbridge. Harefield Moor, locally abundant. South Mimms!, *Benbow* (BM). Warren Gate, *Benbow MSS.* **2.** Near Littleton, *Welch, K. & L.*, 46. Near Chertsey Lock. Railway banks, West Drayton, *Benbow MSS.* Bushy Park. Hampton Court Park. **4.** Hampstead,† *Irv. MSS.* Near Harrow,† *Melv. Fl.*, 15. **6.** Between Highgate and Barnet,† *Merr. Pin.*, 25. Finchley Common, 1901,† *C. S. Nicholson, K. & L.*, 45. Whetstone,† *Benbow, J.B. (London)* 25: 15. **7.** Bombed site, Great Turnstile Street, W.C.1, 1950,† *Whittaker* (BM). Buckingham Palace grounds, *Fl. B.P.G.*

RADIOLA Hill

R. linoides Roth All-seed.
R. millegrana Sm.
Formerly native in damp sandy and gravelly places on heaths, but now extinct.
First record: *Petiver*, 1695. Last evidence: *Benbow*, 1890.
1. Harefield Common, *Blackst. Fasc.*, 61; abundant, 1866, *Trimen*; abundant, 1890, *Benbow* (BM). Hillingdon Heath, *Blackst. Fasc.*, 61. **3.** Hounslow Heath, *Pet. Midd.*; c. 1800, *J. Banks* (BM). **4.** Hampstead Heath (BO).

GERANIACEAE

GERANIUM L.

G. pratense L. Meadow Cranesbill.
Native in damp meadows, on river banks, by roadsides, etc., established on railway banks, waste ground, in gravel pits, etc. Locally plentiful on alluvial soils near the Thames and Colne, rare, and probably adventive elsewhere.
First record: *J. E. Smith*, 1797.
1. Pinner Hill,† *Hind, Melv. Fl.*, 18. Harefield Moor!; Uxbridge Moor!, *L.N.* 26: 59. Between Uxbridge and Cowley!, *K. & L.*, 46. Near Drayton Ford, abundant, 1965; near Northwood, a single plant,

1963, *Pickess*. Ruislip, *L. M. P. Small*. **2.** Frequent near the Thames and Colne. Railway banks, Poyle to Colnbrook, abundant!, *K. & L.*, 46. Harmondsworth!, *Green* (SLBI); by river Colne, 1967. Near Littleton. Ashford. **3.** Twickenham!, *T. & D.*, 66. Hounslow Heath, *Hubbard*!, *L.N.* 28: 32. Hayes, *Tremayne*, *K. & L.*, 46. Thames bank between Richmond Bridge and Twickenham. **4.** Harrow!, *J. E. Smith*, *E.B.*, 404. Stanmore,† *O'Dell*, *Williams MSS*. Kingsbury Churchyard, one plant,† *Farrar*, *T. & D.*, 126. Near Wild Hatch, Hendon, one plant,† *Champ. List*. West Heath, Hampstead, 1967, *L. M. P. Small*. **5.** Syon Park!, *K. & L.*, 46. Hanwell, 1933.† Ealing and Old Brentford Burial Ground, South Ealing, locally plentiful. Waste ground, Brentford, 1967. Chiswick, 1965, *Murray*. Hammersmith,† *Borrer MSS*. **6.** Whetstone Stray, 1960, *J. G. Dony*!; several colonies, 1966; well naturalised in meadows, Myddleton House grounds, Enfield, 1966, *Kennedy*!. **7.** Regents Park, *Holme*, *L.N.* 39: 57. Tottenham, 1960, *Kennedy*.

G. endressii Gay

Alien. Europe, etc. Garden escape. Naturalised in grassy places. Very rare.

First record: *Harley*, 1958.

4. Harrow, *Harley Add.* **6.** Well naturalised and abundant in meadows in Myddleton House grounds, Enfield, 1966, *Kennedy*!. **7.** Regents Park, 1958, *Holme*.

G. versicolor L. Striated Cranesbill.

G. striatum L.

Alien. Europe, etc. Garden escape. Formerly naturalised on hedgebanks, etc., but now extinct.

First evidence: *Hind*, 1863. Last record: *D. H. Kent*, 1946.

1. Pinner Hill, 1863, *Hind* (BM). Near Elstree, 1946!, *L.N.* 26: 59.

G. phaeum L. subsp. **phaeum** Dusky Cranesbill.

Alien. Europe, etc. Garden escape. Naturalised in grassy places, thickets, etc. Very rare.

First record: *Hind*, 1860.

1. Uxbridge Common, 1906,† *Green* (SLBI). **2.** Bushy Park, 1879, *Grubbe* (K). Hampton Wick, *E. K.* and *W. K. Robinson*, *Country-side* 1920, 78 and 1921, 66. **4.** Harrow!, *Hind Fl.* Roadside near Hampstead,† *Garlick*. **7.** Enclosure near Ken Wood, 1948, *Soper*; 1949!, *H. C. Harris*, *K. & L.*, 47: 1960.

Subsp. **lividum** (L'Hérit.) Pers.

5. Syon Park, 1946!,† *L.N.* 26: 59.

G. sanguineum L. Bloody Cranesbill.

Introduced. Garden escape. Naturalised in grassy places, on waste ground, etc. Very rare.

First certain record: *L. M. P. Small*, 1950.
1. Uxbridge, 1950,† *L. M. P. Small*, *K. & L.*, 46. **4.** Bentley Priory, 1961, *Warmington*, *L.N.* 41: 13. Railway bank between Willesden and Neasden, 1957–64.† Wembley, 1965. Stonebridge Park, 1968, *McLean*. **5.** Kew Bridge railway yard, 1966→. **7.** Bombed sites, Cripplegate,† *Jones Fl.* Regents Park, *Holme*, *L.N.* 39: 47. Holloway, 1965, *D. E. G. Irvine*.

G. macrorrhizum L.
Alien. Europe, etc. Garden escape. Naturalised in grassy places, etc. Very rare.
First record: *G. C. Druce*, 1910.
2. Hampton, *Druce Notes*. **5.** Hanwell, 1932–35,† *K. & L.*, 149. **6.** Well naturalised and plentiful in meadows, Myddleton House grounds, Enfield, 1966, *Kennedy*!. **7.** Regents Park, *Holme*, *L.N.* 39: 47.

G. ibericum Cav. × **platypetalum** Fisch. & Mey. = **G.** × **magnificum** Hyl.
G. ibericum auct. eur., *pro parte*, non L., *G. platypetalum* auct. eur. *pro parte*, non Fisch. & Mey.
Lit. *Proc. Bot. Soc. Brit. Isles* **7**: 389 (1968).
Alien of hortal origin. Garden escape. Naturalised in grassy places. Very rare.
First record: *D. H. Kent*, 1932.
4. Edgware, 1967, *Horder*!. **5.** Hanwell, 1932.† Railway yard, Kew Bridge, 1965→. **6.** Well naturalised and abundant in meadows, Myddleton House grounds, Enfield, 1966, *Kennedy*!. **7.** Regents Park, 1960.

G. pyrenaicum Burm. f. Mountain Cranesbill.
G. perenne Huds.
Denizen on hedgebanks, in grassy places, etc., established on railway banks, in cemeteries, etc. Locally plentiful, and increasing.
First record: *Hudson*, 1762.
1. Harefield!, *Cole*, *T. & D.*, 66; locally plentiful, 1968. Eastcote!, *Hind*, *T. & D.*, 66. Ruislip!, *Hind* (BM). Uxbridge Moor!, *Benbow MSS.* Denham!, *L.N.* 26: 59; locally common, 1968. Northwood!, *P. H. Cooke*, *K. & L.*, 47. Ickenham, Tremayne. Pinner Hill, *Hind*, *Melv. Fl.*, 18. An error. **2.** Hampton Wick!, *De Cresp. Fl.*, 28. East Bedfont!; near Hampton Court!, *Welch*, *L.N.* 26: 59; locally frequent, 1968. Near Walton Bridge!; Yeoveney!, *L.N.* 26: 59. Near Sunbury!; Ashford!; Staines Moor!, *K. & L.*, 47. Teddington. **3.** Twickenham, 1838, *J. F. Young* (K); 1887, *Benbow MSS.* Near Hounslow Heath, *H. Banks*!, *L.N.* 26: 59. Hanworth, *L. M. P. Small*. Southall. Near Cranford. Hayes Churchyard, abundant. **5.** Acton,† *Tucker*;

Turnham Green!; Chiswick!, *T. & D.*, 66. Brentford, frequent!, *K. & L.*, 47. Railway bank, Ealing. Garden weed, Ealing!, *T. G. Collett.* Ealing and Old Brentford Burial Ground!, 1943, *Welch*; frequent, 1968. **6.** Near Enfield!, *Huds. Fl. Angl.*, 245. Enfield!, *E. Forster* (BM); 1967. Near Finchley, *L. B. Hall, K. & L.*, 47. Railway banks, Oakleigh Park; Whetstone Stray, 1960, *J. G. Dony*!; 1967. **7.** Between Hyde Park and Little Chelsea,† *Huds. Fl. Angl.*, 265. Chelsea, *Curt. F.L.*, c. 1858, *Irv. Ill. Handb.*, 750; c. 1868,† *Newbould, T. & D.*, 67. Brompton!, *Huds. Fl. Angl.* 2: 303; Brompton Cemetery!, 1865, *Britten, Nat.* 1865, 113; 1967. Eccleston Street, S.W.1, 1957, *Codrington, L.N.* 39: 47. Fulham Cemetery, abundant. Near Fulham Palace.

Plants with white flowers are recorded from **7.** Chelsea,† *Curt. F.L.*

G. columbinum L. Long-stalked Cranesbill.

Native in dry sunny places, established on railway tracks, etc. Very rare.

First record: *Goodyer*, 1654.

1. Harefield!, *Blackst. Fasc.*, 32. Near Uxbridge!, *Gawler, Coop. Fl.*, 117. Hillingdon;† Cowley;† railway banks, Uxbridge to West Drayton, *Benbow MSS.* Springwell!, 1948, *Boniface, K. & L.*, 48; 1966. Railway tracks, Denham to Uxbridge!, *K. & L.*, 48; railway now disused but the plant was abundant near Denham in 1966. **2.** Railway tracks, Colnbrook, 1951!, *K. & L.*, 48; 1967. **7.** . . . in the streets near Whitechapel, east of Aldgate, 1654,† *Goodyer, How MSS.* Marylebone Park, 1753,† *Hill Fl. Brit.*, 853.

G. dissectum L. Cut-leaved Cranesbill, Jagged Cranesbill.

Native in grassy places, on cultivated and waste ground, etc. Very common.

First record: *Blackstone*, 1737.

7. Frequent in the district.

G. rotundifolium. Round-leaved Cranesbill.

Native on dry hedgebanks, in fields, by roadsides, etc., and established on railway banks, canal paths, etc. Very rare, and decreasing.

First certain record: *T. Martyn*, 1763.

1. Harefield, 1945!,† *L.N.* 26: 59. Probably introduced. **3.** Roxeth, 'introduced',† *Hind Fl.* **5.** Chiswick!; between Turnham Green and Brentford,† *Newbould, T. & D.*, 68. Formerly abundant on the railway bank by Barnes railway bridge, but not recently seen. **7.** Between Islington and Canonbury,† *Mart. Pl. Cant.*, 72. Islington,† *Woods, B.G.*, 109. Near Holloway, 1831,† *Varenne*; Upper Clapton,† *Wollaston*; Chelsea,† *A. Irvine, T. & D.*, 68. Between Hackney and Homerton, 1839† (HE). Hackney, *E. Forster* (BM); 1865,† *Reeves, T. & D.*, 68.

Plants with white flowers are recorded from **7.** Hackney,† *J. Hill, Druce Notes.*

G. molle L. Doves-foot Cranesbill.
Native in meadows, on cultivated ground, hedgebanks, etc., and established on railway banks, etc. Very common.
First record: *T. Johnson*, 1629.
7. Common in the district.

Var. **aequale** Bab.
4. Hendon, 1912, *E. F.* and *H. Drabble, Rep. Bot. Soc. & E.C.* 8: 259.

Var. **grandiflorum** Vis.
2. Hampton Court, *C. E. Britton, J.B. (London)* 48: 331. Stanwell!, *L.N.* 28: 32. **7.** 'In the lawn before Chelsea Hospital I have noticed this plant almost as large as the *pyrenaicum* of Linnaeus',† *Curt. F.L.* Probably refers to this variety.
Plants with white flowers have been noted at **2.** Near Staines Moor.

G. pusillum L. Small-flowered Cranesbill, Soft Cranesbill.
Native on dry hedgebanks; in dry pastures, waste places, etc. Common, especially near the Thames.
First record: *Dillenius*, 1724.
7. West side of London, in neglected gardens and fallow fields near Little Chelsea . . . ,† *Curt. F.L.* Paddington† (G & R). East Heath, Hampstead, *Lawson*; Parsons Green;† Brompton Cemetery!; Isle of Dogs!,† *T. & D.*, 67. Kensington Gardens, 1871,† *Warren* (BM). Hyde Park, 1946,† *Welch, L.N.* 39: 47. Bombed site, Brewer's Lane, E.C., 1946,† *Lousley*.
Plants with white flowers have been recorded from **2.** Near Penton Hook Lock, *Benbow MSS.* **7.** Hackney,† *Dillenius, Ray Syn.* 3: 359.

G. lucidum L. Shining Cranesbill.
Native on dry hedgebanks, etc., established on old walls, etc., also as an introduced garden weed. Very rare and almost extinct.
First record: *Merrett*, 1666.
1. Harefield, *Blackst. Fasc.*, 32; c. 1869,† *Newbould, T. & D.*, 68. Near Harefield Moor, 1917,† *J. E. Cooper*; Cowley, 1884, *Benbow* (BM); 1917,† *J. E. Cooper*. Between Moorhall and Ickenham, 1915, *Tremayne, K. & L.*, 49; *J. E. Cooper* (BM). **3.** Near Whitton,† *Hemsley*; between Worton and Whitton;† Twickenham,† *T. & D.*, 68. Near Twickenham, 1872† (W). **4.** Mill Hill, *P. H. Cooke* (LNHS); 1916, *J. E. Cooper* (BM); near Mill Hill, in very small quantity, 1965, *Hinson, L.N.* 45: 20. **5.** Between Brentford and Syon House, *Merr. Pin.*, 46; near Brentford, 1871† (HY). Between Hanwell and Brentford, 1873,† *Britten*; Osterley,† *Shepheard, Trim. MSS.* Greenford, *Melv. Fl.*, 18; 1902, *Green* (SLBI); 1912,† *P. H. Cooke, K. & L.*, 49. Norwood,†

Masters; between Turnham Green and Ealing,† *Newbould, T. & D.*, 68. Garden weed, Ealing, introduced with soil from Wiltshire!, *T. G. Collett*; 1967. **7.** Newington Green,† *Buddle, Newt. MSS.*

G. robertianum L. subsp. **robertianum** Herb Robert.
Lit. *Watsonia* **3**: 270 (1956).
Native on hedgebanks, in woods and shady places, etc. Common, except in heavily built-up areas where, owing to the absence of suitable habitats, it is very rare or extinct.
First record: *T. Johnson*, 1629.
7. Between Kentish Town and Hampstead,† *Johns. Eric.* Fields between Hagbush Lane and Highgate,† *Ball. MSS.* Waste ground, Wapping, 1966. Garden weed, Chelsea, 1969, *E. Young*. Plants with white flowers have been recorded from **1.** Near Ruislip, 1735, *Blackst. Fasc.*, 33; 1745, *Blackst. Spec.*, 27. **3.** The Grove, Harrow (ME). **5.** Ealing, 1947!, *L.N.* 27: 31; 1967. Persistent garden weed, West Ealing, though originally introduced. Perivale Churchyard!, *Welch, K. & L.*, 49; abundant, 1968. **6.** Winchmore Hill, 1944–47, *Scholey, K. & L.*, 49. Enfield, in several places, 1966, *Kennedy*!.
The seashore plant (cf. subsp. **maritimum** (Bab.) H. G. Baker) is mentioned by Smith (1827, *Engl. Fl.* **3**, 326), as being 'a weed in (7) Chelsea Botanic Garden'.† It is not so now.

ERODIUM L'Hérit.

Lit. *Watsonia* **1**: 170 (1949). (*Abstr.*)

E. cicutarium (L.) L'Hérit. subsp. **cicutarium** Common Storksbill.
E. cicutarium (L.) L'Hérit. subsp. *arvale* Andreas, *E. triviale* Jord., *E. ballii* Jord.
Native in fields, on waste ground, by roadsides, etc., established on railway tracks, canal paths, etc. Rare, and decreasing, and now mainly confined to the light soils in the vicinity of the Thames.
First record: *T. Johnson*, 1629.
1. Uxbridge!, *A. B. Cole*; Pinner, *T. & D.*, 69. Ickenham Green,† *Benbow, J.B. (London)* 25, 15. Yiewsley, *Benbow MSS.*; 1919, *J. E. Cooper* (BM). Railway banks between Denham and Uxbridge, 1953, *T. G.* and *M. G. Collett*!; 1962. Harefield, 1957, *I. G. Johnson*. **2.** Between Staines and Hampton, *T. & D.*, 69. Staines, *Briggs*. Poyle, *L. M. P. Small*; West Drayton, *P. H. Cooke* (LNHS). Between Kingston Bridge and Hampton Court!, *Warren, Trim. MSS.*; also small colonies on the river wall, 1968. Hampton Court Park. Bushy Park. **3.** Beyond Twickenham, *De Cresp. Fl.*, 23. Near Teddington, *Tremayne*. Feltham, *L. M. P. Small*. Between Whitton and Hounslow, *Benbow* (BM). Hounslow Heath, a single plant, 1965, *Clement, Mason* and *Wurzell*. **4.** Willesden,† *De Cresp. Fl.*, 108. Finchley, 1913,†

Champ. List. Hampstead,† *Cochrane Fl.* Stonebridge,† *Benbow* (BM). **5.** Chiswick!, *Hemsley*; Kew Bridge station, *Newbould, T. & D.*, 69. Bedford Park, *Cockerell Fl.* Hanwell, 1933!, *K. & L.*, 49; a small patch on a lawn, 1967. Railway bank, Brentford!, *K. & L.*, 49; factory lawn, Great West Road, Brentford, six patches, 1967→. **6.** Edmonton, *T. & D.*, 69. Enfield, 1973, *Kennedy*. **7.** Between Kentish Town and Hampstead,† *Johns. Eric.* Near Chelsea,† *Buddle* and *Doody* (SLO). Chelsea Common, c. 1800, *J. Banks* (BM); 1806,† *Garry Notes.* Hyde Park, 1817† (G & R). Near Fulham,† *J. F. Young* (BM). Near Hackney,† *Dillenius, Ray Syn.* 3, 358. Hackney Marshes, 1915,† *J. E. Cooper, K. & L.*, 49. Homerton, a single plant,† *Cherry*; Kensington Gardens, a single plant, 1868,† *Warren, T. & D.*, 69. Highgate, 1919,† *Andrewes MSS.*

OXALIDACEAE

OXALIS L.

Lit. *Watsonia* **4**: 51 (1958).

O. acetosella L. Wood Sorrel.

Native in woods, shady places, etc. Local, and confined to the northern parts of the vice-county.

First record: *T. Johnson*, 1629.

1. Locally frequent. **3.** Weed in flower beds, West Middlesex Hospital, Isleworth, 1946,† introduced. **4.** Highgate, *Johns. Eric.* Deacon's Hill, *Newbould*; Bishop's Wood; Turner's Wood; West Heath, Hampstead, *T. & D.*, 72; 1901,† *Whiting Notes*; extinct before 1920, *Richards.* Scratch Wood. Near Mill Hill. **6.** Hadley, *Newbould*; Winchmore Hill Wood,† *T. & D.*, 72. Queen's Wood, Highgate!, *Bangerter.* Enfield!, *H. & F.*, 273. **7.** Hampstead Heath, 1844,† *Parsons* (CYN). Ken Wood,† *T. & D.*, 72.

Var. **caerulea** Pers.

var. *subpurpurascens* DC.

1. Pinner, *W. A. Tooke, J.B. (London)* 9: 271. Old Park Wood, Harefield, *Rose*!, *L.N.* 26: 59; 1966.

O. corniculata L. Procumbent Yellow Sorrel.

Alien. Cosmopolitan. Established as a persistent garden weed, also by roadsides, on waste ground, etc. Rare.

First certain evidence: *T. Moore*, 1840.

1. Harefield, *McIvor* (SY). Uxbridge, 1872; Hillingdon, 1876, *Benbow* (BM). Cowley, *Benbow MSS.* Eastcote, 1966. **2.** Staines!, *K. & L.*, 50; 1968. Stanwell. Hampton Wick. Bushy Park!, *Clement.* **4.** Harrow!, *Harley Fl.*, 9. Harrow Weald. Finchley, *Wurzell.* Mill Hill, 1969, *Harrison.* **5.** West Ealing!, *K. & L.*, 50; 1968. Osterley. Heston, 1967. Acton, 1967. Chiswick, 1967. **6.** Muswell Hill, 1897, *J. E. Cooper* (BM). Enfield, 1965, *Bangerter* and *Kennedy*!. **7.** Weed in Chelsea Botanic Garden, 1840, *T. Moore* (D). Brompton, 1861, *Britten* (BM). Chelsea College, 1862, *Britten, Phyt. N.S.* 6: 349. Fulham Palace grounds, *Welch*!. St John's Wood, *Holme, L.N.* 39: 47. Buckingham Palace grounds, *Fl. B.P.G.* Earls Court, 1965, *Brewis.*

Var. **repens** (Thunb.) Zucc.
1. Garden weed, Wrotham Park, *Fl. Herts*, 66.

Var. **atropurpurea** Van Houtte
3. Twickenham, 1966. **4.** Harlesden, 1965, *Casey.* **5.** South Ealing, 1967. Brentford, 1967. **6.** Myddleton House grounds, Enfield, 1966, *Kennedy*!. **7.** Tottenham, 1965, *Ing.* Stoke Newington, *Kennedy.*

O. exilis A. Cunn.
O. corniculata L. var. *microphylla* Hook. f.
Alien. New Zealand. Established as a persistent garden weed, by roadsides, etc. Very rare, or overlooked.
First evidence: *L. M. P. Small*, 1949.
2. Hampton Court, 1965. **5.** Ealing, 1949, *L. M. P. Small* (BM), teste *D. P. Young* (as *O. corniculata* var. *microphylla*). Brentford, 1967→. **6.** Tottenham, *Wurzell.*

O. europaea Jord. Upright Yellow Sorrel.
O. stricta auct., non L.
Alien. N. America. Established as a persistent garden weed, also by roadsides, on waste ground, etc. Locally common on light soils in the south-western parts of the vice-county, and near the Thames, rare elsewhere.
First record: *D. Cooper*, 1836.
1. Harefield, 1867, *Shuttleworth* (BM). Cowley, *Benbow MSS.* Ruislip, *Wrighton, Spooner* and *H. Banks*!, *K. & L.*, 50; 1966. **2.** Locally common. **3.** Locally common about Hounslow, Twickenham, etc. **4.** Stanmore, 1850 (BM). Willesden Green!, *K. & L.*, 50. Hendon. Near Hampstead, *Boniface, K. & L.*, 50. Mill Hill, 1969, *Harrison.* **5.** Turnham Green, *Newbould, T. & D.*, 72. Brentford!, *Britten, Trim. MSS.*; 1968. Syon Park!; Ealing!, *K. & L.*, 50; 1968. West Ealing, 1968. Hanwell, 1968. **6.** Edmonton, *T. & D.*, 72. Myddleton House grounds, Enfield, 1966, *Kennedy*!. **7.** Hollywood School playground, abundant, 1863,† *Britten*; Tottenham, 1839, '*L.S.*' (BM). Parsons Green; Chelsea, *Britten, T. & D.*, 72. South Kensington, 1875, *Trimen, J.B. (London)* 14: 275. Hampstead, 1846–66 (SY). Fulham Palace grounds, *Welch*!; 1966, *Brewis.*

Forma **pilosella** Wiegand
6. Church End, Finchley, 1925 and 1927, *J. E. Cooper* (BM), teste *D. P. Young.*

O. corymbosa DC.
O. floribunda Lehm.
Alien. S. America. Formerly cultivated as a border plant, but now a persistent and obnoxious weed in flower-beds in parks and gardens. Very common.
First evidence: *Loydell*, 1900.
7. Frequent throughout the district.

O. incarnata L.
Alien. S. Africa. Formerly cultivated as a border plant, and now established as a garden weed. Very rare.
First evidence: *Murison*, 1912.
5. Hammersmith, 1912, *Murison* (K). West Ealing, 1967!, *L.N.* 47: 9; 1972.

BALSAMINACEAE

IMPATIENS L.

I. capensis Meerburgh Orange Balsam, Jewel Flower.
I. biflora Walt., *I. fulva* Nutt.
Alien. N. America. Garden escape. Naturalised by the sides of rivers and streams, in marshes, by ponds and lakes, etc., and established by canals, in gravel pits, etc. Locally abundant, and increasing, especially by the Thames and its tributaries.
First record: *Ralph*, 1838.
1. Abundant by water throughout the Colne valley. **2.** Frequent by all the waterways, especially the Thames and Colne, in gravel pits, etc. **3.** Locally common by the Thames about Twickenham and Isleworth, and by the Crane at Twickenham and Hounslow Heath. Cranford Park!, *Westrup*, *L.N.* 33: 22; abundant, 1968. Feltham!, *Welch*, *K. & L.*, 31. Canal side in various places between Southall and Hayes. **4.** Near Harrow† (HY). **5.** Near Kew Bridge!, *Jewitt*; Chiswick!, *F. Naylor*, *T. & D.*, 71. Syon Park!, *Higgins* (K). Canal side between Greenford and Perivale, between Hanwell and Brentford, and at Frogmore Green, near Southall. **6.** Lee Navigation Canal and adjacent gravel pits from near Waltham Abbey to Edmonton. **7.** Sparingly by a ditch near the Vale of Health Pond, Hampstead, before 1866,† *T. & D.*, 71. Near Hammersmith Bridge, a small colony on rotting barges, 1965!, *L.N.* 45: 20. Canal bank, Regents Park, 1966, *Brewis*, *L.N.* 46: 29.
This attractive species originated from the gardens of Albury Park, Surrey, where it was first noted as an escape by J. S. Mill in 1822 (*Phyt.* **1**: 40 (1847)). A small stream, the Tillingbourne, flows through these gardens and enters the Wey above Guildford, this river in turn flows into the Thames nearly opposite Shepperton. In this way the

seeds have been carried by water currents and by human agency, particularly by anglers, throughout the waterways of Middlesex.

I. parviflora DC. Small Balsam.
Lit. *J. Ecol.* **44**: 701 (1956).
Alien. Northern Asia. Naturalised by rivers and streams, in woods, on cultivated and waste ground, by roadsides, etc., also as a persistent weed in parks and gardens. Common, and increasing.
First evidence: *Syme*, 1857.
7. Weed in the garden of 70 Adelaide Road, N.W., 1857; 1866–67, *Syme* (SY). Fields between Hampstead and Swiss Cottage, abundant on rubbish, 1869, *T. & D.*, 422. Chelsea!, *Britten, Bot. Chron.*, 57. Kensington; Bayswater, 1860 (ME). Notting Hill, 1867, *F. Naylor, T. & D.*, 72. Regents Park, common!, *Trim. MSS.*; 1968. Hyde Park, 1871,† *Newb. MSS.* Stoke Newington, 1903; Highbury, 1905, *E. M. Dale* (LNHS). Crouch End, 1897, *J. E. Cooper* (BM). Bombed site, Finsbury Pavement, E.C. 1945.† Bombed site, Red Lion Square, W.C.1, 1950,† *Whittaker*; bombed sites, Mile End Road, E.1† and Shadwell!, 1952, *Henson* (BM); still locally abundant, also a frequent weed in the churchyard and grounds of St Georges-in-the-East, 1968. Between Camden Town and Hampstead. Primrose Hill!, *Fitter*. Highgate Cemetery, abundant, also in other localities about Highgate and Hampstead. St John's Wood!, *Knipe, L.N.* 39: 47. East Heath, Hampstead!, *H. C. Harris, L.N.* 28: 32. By Metropolitan railway line just west of Baker Street, 1964!, *Johnson, L.N.* 44: 17. Garden weed, Basil Street, S.W.1, *McClintock*. Christ Church, Greyfriars, E.C., 1966, *Brewis, L.N.* 46: 29. Hackney, *Renson*. Tottenham, *Scholey*. Hurlingham, 1970, *D. S. Warren*.

I. glandulifera Royle Himalayan Balsam, Policeman's Helmet.
I. roylei Walp.
Lit. *J.B. (London)* **38**: 50, 87, 278 and 445 (1900): *loc. cit.* **39**: 184 (1901).
Alien. Himalaya. Garden escape. Naturalised on the banks of rivers, streams and ponds, damp waste ground, etc., established by canals and gravel pits, on railway banks, etc. Locally abundant, and increasing.
First record: *A. Irvine*, 1855.
1. Colne between Harefield and Denham!, *A. Irvine, Phyt. N.S.* 1: 166; 1967. Yiewsley!, *M. & S.*; abundant by a canal backwater, 1968→. New Years Green!, *Wrighton*. Banks of the Pinn, Eastcote, and as a weed in Eastcote House grounds, abundant, 1967→. **2.** Canal bank, Dawley!, *K. & L.*, 52; 1967. West Drayton. Thames bank between Hampton Court and Kingston Bridge, 1967→. Near the Thames, Staines, 1967→. **3.** Thames banks and aits from Twickenham to Syon Park, in abundance. St Margaret's. Banks of the Crane near

Cranford. Yeading Brook near Southall. Feltham, *L. M. P. Small.* Banks of the Crane, Hounslow Heath. Waste ground, Sudbury, casual, 1965.† **4.** Frequent by the Silkstream from Edgware to Brent Reservoir. Brent banks, Stonebridge!, *K. & L.*, 52; 1973. West! and North-west Heaths, Hampstead!; Kenton!, *K. & L.*, 52. Mill Hill, casual, 1960,† *Sennitt.* Hendon. **5.** Abundant by the Brent from Stonebridge to Hanwell. Canal side between Hanwell and Brentford, in profusion. Thames side and aits, Syon Park to Chiswick, abundant. Osterley Park. Ealing Common, casual, 1950.† Railway bank near Ealing Common station, 1966. Perivale Wood, 1958, *Bartlett.* **6.** Pymmes Brook, Edmonton!, *R. M. Payne*; abundant, 1967. Lee Navigation Canal between Tottenham and Edmonton!, *L.N.* 27: 31; 1966. Enfield!, *Kennedy.* **7.** Highgate Ponds, abundant!, *Fitter*; East Heath, Hampstead!, *K. & L.*, 52. Primrose Hill, 1954–56,† *Holme, L.N.* 39: 47. Railway side near Westbourne Park!, *K. & L.*, 52: 1968. Plentiful by Regents Canal, Islington!, *L.N.* 39; 47; 1972. River Lee, Hackney Marshes, abundant!, *B. T. Ward*; 1967. Bombed site near Tower of London!, *Jones Fl.* Churchyard of St Dunstan's At Hill, Idol Lane, E.C.3, abundant, 1966. Disturbed ground, Westbourne Grove, Paddington, 1967.† Waste ground, Tottenham, 1968, *Kennedy.*

SIMAROUBACEAE

AILANTHUS Desf.

A. altissima (Mill.) Swingle Tree-of-Heaven.
A. glandulosa Desf.
Alien. China, etc. Commonly planted in London parks and squares and elsewhere, where it readily regenerates. Seedlings and saplings are common on waste ground, etc., particularly in inner London.
First record (of regeneration): *Lousley*, 1944.
7. Frequent in the district.

ACERACEAE

ACER L.

A. pseudoplatanus L. Sycamore, Great Maple.
Lit. *J. Ecol.* **32**: 215 (1945).
Alien. Europe, etc. Frequently planted in woods, parks, recreation grounds, gardens, cemeteries, etc., and as a street tree. It regenerates freely, and seedlings and saplings are very common in woods, on waste ground, by roadsides, etc.

First record (of regeneration): *D. H. Kent*, 1932.
7. Frequent in the district.

A. platanoides L. Norway Maple.
Lit. *J. Ecol.* **32**: 238 (1945).
Alien. Northern Europe. Planted in parks, woods, by roadsides, etc., where it frequently regenerates. Common.
First record (of regeneration): *D. H. Kent*, 1955.
1. Pinner Hill, 1967. Eastcote, 1967. Hillingdon; Hillingdon Heath; **2.** River-wall, Hampton Court, 1958!, *Knipe*, *L.N.* 39: 40; 1969. Bushy Park, 1967. Near Hanworth, 1966. Staines, 1969. **3.** Hounslow, 1966. Isleworth, *L. M. P. Small.* **4.** West Heath, Hampstead, 1966. Near Mill Hill, 1966, *Kennedy*!. Colindale; Wembley Park; near Totteridge, *L. M. P. Small.* **5.** Brentford, 1955. Syon Park, 1957. Gunnersbury Park, Acton, 1968. Ealing, 1968. Chiswick, 1968. Osterley; Greenford; Perivale; Hanger Hill, Ealing, *L. M. P. Small.* **6.** Enfield. Southgate. **7.** Holland House grounds, Kensington!, *Fitter*, *L.N.* 40: 21; 1969. East Heath, Hampstead, 1966. Ken Wood, 1966. Upper Clapton, *L. M. P. Small.*

A. campestre L. Field Maple.
Lit. *J. Ecol.* **32**: 215 (1945).
Native in hedges, on wood borders, etc. Common, except in heavily built-up areas where owing to the eradication of its habitats it is very rare or extinct.
First record: *T. Johnson*, 1638.
7. Tottenham!, *Johns. Cat.* About the northern outskirts of London; Kentish Town;† Hornsey Wood;† Green Lanes,† etc., *T. & D.*, 65.

HIPPOCASTANACEAE

AESCULUS L.

A. hippocastanum L. Horse Chestnut.
Alien. S.-E. Europe, Asia Minor, etc. Planted in parks, on heaths, commons by roadsides, in woods, cemeteries, etc., where it sometimes regenerates. Common.
First record (of regeneration): *D. H. Kent*, 1933.
7. Hyde Park!; Kensington Gardens!; Buckingham Palace grounds!, *L.N.* 39: 48. Regents Park. Paddington. Stoke Newington. Hackney Wick. Tottenham. Poplar.

AQUIFOLIACEAE

ILEX L.

I. aquifolium L. Holly.
Lit. *J. Ecol.* **55**: 841 (1967).
Native in woods, hedges, etc., chiefly in the northern parts of the vice-county, also widely planted for ornamental purposes in parks, hedges, cemeteries, etc., where it readily regenerates. Local as a native, but common as an introduction.
First record: *T. Johnson*, 1638.
7. Ken Wood. East Heath, Hampstead. Kensington Gardens, seedlings!, *D. E. Allen*, *L.N.* 39: 48. Bombed sites, Cripplegate area, E.C., seedlings!,† *Jones Fl.* Buckingham Palace grounds, seedlings, *Fl. B.P.G.* Fulham Palace grounds, seedlings!, *Brewis*.
Plants with variegated leaves, probably always planted, have been recorded from **4.** Hampstead, *How MSS.* **6.** Near Highgate, *Merr. Pin.*, 3. Tottenham, *M. & G.*, 301.

CELASTRACEAE

EUONYMUS L.

E. europaeus L. Spindle Tree.
Native on wood borders, in hedges, copses, etc. Local, and decreasing, and mainly confined to calcareous soils at Harefield and South Mimms, and alluvial soils near the Thames and Colne.
First record: *T. Johnson*, 1638.
1. Harefield!, *Blackst. Fasc.*, 27. Near Yiewsley!, *Newbould*; Pinner, *T. & D.*, 173. North-west of Uxbridge!; Ickenham!; Eastcote!; Ruislip!, *Tremayne*; Springwell!; South Mimms!, *K. & L.*, 53. Hillingdon; Swakeleys,† *Benbow MSS.* Near Denham Lock!, *Williams MSS.* Bentley Heath, *J. G.* and *C. M. Dony*. Border of Mimmshall Wood, 1966, *J. G.* and *C. M. Dony*!. **2.** Staines!, *Newbould*; between Hampton Court and Kingston Bridge!, *T. & D.*, 73. Laleham. Penton Hook. Near Colnbrook!, *Druce Notes*. Yeoveney!,† *Welch*, *L.N.* 27: 31. Poyle. **3.** Harrow Grove,† *Melv. Fl.*, 20. **4.** Bishop's Wood, *M. A. Lawson*, *T. & D.*, 73. Near Fortune Green, one tree;† between Willesden and Acton,† *De Cresp. Fl.*, 23. West Heath, Hampstead, 1905,† *L. B. Hall*, *K. & L.*, 53. Golders Hill,† *Whiting Fl.* Kinsgbury;† Stonebridge,† *Benbow MSS.* Hendon. **5.** Alperton;† Greenford,† *Melv. Fl.*, 20. Horsendon Hill,† *T. & D.*, 73. By the Brent between Hanwell and Brentford,† *Benbow MSS.* Syon Park, *Jacks. Cat.* Perivale Wood!, *Benbow MSS.* **6.** Totten-

ham,† *Johns. Cat.* Queen's Wood, Highgate, one tree, 1916, *J. E. Cooper.* Hadley, *P. H. Cooke, K. & L.*, 53. **7.** Kensington Gardens,† *Loud. Arb.*, 497.

BUXACEAE

BUXUS L.

B. sempervirens L. Box.
Introduced. Naturalised in woods and copses where it was originally planted. Very rare.
First record: *T. Johnson*, 1638.
1. Harefield!, *Blackst. Fasc.*, 12. **2.** Laleham Park. **3.** Hayes Park. **4.** Appears to grow wild in some places about Hampstead, *Whiting Notes.* Not so now. **7.** Tottenham, *Johns. Cat.*

RHAMNACEAE

RHAMNUS L.

R. catharticus L. Common Buckthorn.
Lit. *J. Ecol.* **31**: 66 (1943).
Native in woods, hedges and thickets, etc. Rare, and now mainly confined to calcareous soils at Harefield, and alluvial soils near the Thames and Colne.
First record: *Blackstone*, 1737.
1. Harefield!, *Blackst. Fasc.*, 86. Near Uxbridge,† *R. & P.*, 219. By Elstree Reservoir,† *T. & D.*, 73. Between Cowley and Hillingdon, *Benbow, J.B. (London)* 23: 36. Hillingdon, *L. M. P. Small.* Yiewsley!, *J. E. Cooper* (BM). North of Uxbridge!,† *Tremayne*; Ickenham, 1949, *Wrighton*; between Yiewsley and Iver!, *K. & L.*, 53. **2.** Between Hampton Court and Sunbury,† *H. C. Watson, Wats. New Bot.*, 98. Between Hampton and Hampton Court!, *Newbould*; between Sunbury and Walton Bridge!;† near Staines!, *T. & D.*, 73. Shepperton, *Benbow MSS.* Harmondsworth!,† *P. H. Cooke* (LNHS). West Drayton!,† *Druce Notes.* Poyle!; Yeoveney!; near Stanwell Moor!, *K. & L.*, 53. Littleton!, *Welch.* **3.** Twickenham, 1908,† *Sprague* (K). **4.** By the Brent between Hendon and Hampstead,† *Irv. MSS.* By the Brent in many places between Hendon and Stonebridge,† *Benbow MSS.* Hampstead!, 1864 (SY); North-west Heath, Hampstead, a single bush, 1949. East Finchley, a single bush, 1888,† probably planted, *J. E. Cooper* (BM). Near Woodcock Farm [Kenton],† *Gunn, Melv. Fl.* 2, 24. Whitchurch Common, *Welch*!, *L.N.* 28: 32. **5.** Near Heston,† *Masters*; by the Brent about Perivale,† *Lees, T. & D.*, 73. Alperton;† Greenford,† *Benbow MSS.* **6.** Enfield Chase,

Benbow MSS. **7.** Between Paddington and Notting Hill,† 1820, *J. J. Bennett* (BM). Between Belsize and West End Lane, 1870,† *Trim. MSS.* Buckingham Palace grounds, one bush, perhaps planted, *Codrington, Lousley* and *McClintock*!, *L.N.* 39: 48.

FRANGULA Mill.

F. alnus Mill. Alder Buckthorn.
Rhamnus frangula L.
Lit. *J. Ecol.* **31**: 77 (1943).
Native in damp woods, thickets and heaths. Rare, and decreasing, and mainly confined to acid soils.
First record: *Gerarde*, 1597.
1. Harefield,† *Blackst. Fasc.*, 4. Pinner,† *Hind Fl.* Near Harrow Weald Common, *T. & D.*, 73. Harrow Weald Common!, *Welch, K. & L.*, 53. Ruislip!, *Druce Notes.* Near Denham Lock, 1956, *Minnion.* Copse Wood! and Duck's Hill, Northwood, *Benbow MSS.* **2.** Shepperton,† *Druce Notes.* Perhaps *Rhamnus catharticus* intended. **4.** Hampstead Heath!, *Johns. Enum.*; one bush left, 1901, *Whiting Notes*; 1927, *Richards*; North-west Heath, Hampstead, one bush, 1948!, *K. & L.*, 53; 1960. Woods at Hampstead, *Ger. Hb.*, 1286; Bishop's Wood, plentiful, *T. & D.*, 73. East Finchley, 1913, *J. E. Cooper* (BM). By the Brent near Kingsbury,† *Farrar, T. & D.*, 73. Perhaps *Rhamnus catharticus* intended. Stanmore, 1815,† *Finch* (D). Harrow Park,† *Melv. Fl.*, 20. **5.** Horsendon Wood,† *Lees, T. & D.*, 74. **6.** Highgate Wood!, 1827–30, *Varenne, T. & D.*, 74; c. 1900, *C. S. Nicholson* (LNHS); 1948!, *K. & L.*, 53. Colney Hatch, *Newbould, T. & D.*, 74; Coppett's Wood, Colney Hatch, *Benbow, J.B. (London)* 25: 15. Winchmore Hill Wood, *T. & D.*, 74; c. 1900,† *L. B. Hall, K. & L.*, 53. Muswell Hill, *Benbow, MSS.* Queen's Wood, Highgate, 1963, *L. Martin.* **7.** Between Islington and Hornsey,† *Ger. Hb.*, 1286. St John's Wood, Hornsey,† *Park. Theat.*, 240. Hornsey Wood,† *Mill. Bot. Off.*, 25, and later authors.

VITACEAE

VITIS L.

V. vinifera L. Vine.
Alien. S. Europe, etc. Naturalised on waste ground, climbing over hedges, etc., originating from rejected grape stones. Rare, but increasing.
First record: *Shenstone*, 1912.
1. Yiewsley!, *M. & S.*; 1967. New Years Green. **2.** Staines, *Clement.* Harlington. Near West Drayton. **3.** Hounslow Heath, 1944–51!, *K. & L.*, 54; 1967. Northolt, 1947!, *Lousley* (L); 1951!, *K. & L.*, 54;

1967. Canal side, Hayes, 1966. **4.** Hendon, 1962, *Brewis*. **5.** Horsendon Lane, Perivale, climbing trees, 1939–51!, *K. & L.*, 54. South Acton, frequent on waste ground, 1965. Brentford, 1966. Spontaneously in a garden, West Ealing, 1968→. **6.** Enfield, 1970, *Kennedy*!. **7.** Building site behind the British Museum, Bloomsbury,† *Shenst. Fl. Lond.* Bombed site, Bedford Way, 1950;† bombed site, Shadwell, 1952, *Henson* (BM). Regents Park, 1954, *Holme, L.N.* 39: 48. Hyde Park, 1962, *Allen Fl.* Bombed churchyard of St Dunstan's At Hill, Idol Lane, E.C.3, 1966. Fulham Palace grounds, 1966, *Brewis*. Near Limehouse, 1966.

PARTHENOCISSUS Planch.

Lit. *Proc. Bot. Soc. Brit. Isles*, **7**: 218 (1968). (*Abstr.*)

P. quinquefolia (L.) Planch. Virginia Creeper.
Hedera quinquefolia L., *Parthenocissus pubescens* (Schlecht.) Graebn., *Vitis hederacea* Ehrh.
Alien. N. America. Garden outcast. Established on waste ground, climbing over hedges, etc. Rare.
First certain record: *D. H. Kent*, 1933.
1. Yiewsley. **2.** Staines. Harmondsworth. **3.** Hounslow Heath. Yeading. **4.** Brent Reservoir. **5.** West Ealing. Chiswick. **7.** Bombed site, Theobalds Road, W.C., 1950,† *Whittaker*; bombed site, Shadwell!, *Henson* (BM); 1967.

P. inserta (A. Kern.) Fritsch. Virginia Creeper.
Vitis inserta A. Kern., *Cissus quinquefolia* Soland., ex Sims, *Parthenocissus vitacea* A. S. Hitchcock, *P. inserens* Hayek, *P. dumetorum* (Focke) Rehd., *P. quinquefolia* auct., non (L.) Planch.
Alien. N. America. Garden outcast. Established on waste ground, climbing over hedges, etc. Rare.
First certain record: *D. H. Kent*, 1967.
1. Yiewsley. **2.** Near Hampton Wick. **3.** Hounslow Heath. **5.** Northolt. Chiswick. **6.** Edmonton, *Kennedy*!. **7.** Near Limehouse.

PAPILIONACEAE

LUPINUS L.

L. arboreus Sims. Tree Lupin.
Alien. N. America. Planted on railway banks, etc., where it sometimes regenerates. Very rare.
First record (of regeneration): *D. H. Kent*, 1945.
5. Railway banks between Northfields and Boston Manor, seedlings.

L. polyphyllus Lindl. Garden Lupin.
Alien. N. America. Garden escape or outcast. Established on railway banks, etc. Common.
First record: *D. H. Kent*, 1945.
7. Common on railway banks in the district.

LABURNUM Fabr.

L. anagyroides Medic. 'Laburnum'.
Alien. S. Europe, etc. Planted in gardens, parks, on railway banks, etc., where it readily regenerates. Very common.
First record (of regeneration): *D. H. Kent*, 1946.
7. Seedlings and saplings are frequent in the squares of Central and Inner London.

GENISTA L.

G. tinctoria L. subsp. **tinctoria** Dyer's Greenweed.
Native in rough grassy places on heavy soils. Locally plentiful.
First record: *T. Johnson*, 1632.
1. Harefield!, *Blackst. Fasc.*, 31. Elstree, 1866†, *Trimen* (BM). Harrow Weald Common,† *Hind*, *Melv. Fl.*, 21. Mad Bess Wood, Ruislip; Northwood!, *Benbow* (BM). By Bayhurst Wood, near Ruislip!, *Benbow MSS.* Pole Hill, Hillingdon, 1905,† *J. O. Braithwaite* (LNHS). Near Pinner Hill, *W. C. R. Watson*; South Mimms!, *K. & L.*, 56. Woodhall,† *Hind*, *Trim. MSS.* Ickenham Green. **4.** Kingsbury,† *Irv. MSS.* Hampstead Heath,† *Johns. Enum.* Between Finchley and Bishop's Wood,† *Coop. Suppl.*, 12. Near Scratch Wood, abundant!, *Trimen* (BM). Near Brent Reservoir,† *Farrar*, *T. & D.*, 75. **5.** Northolt,† *Melv. Fl.*, 21. Horsendon Hill, abundant!, *Benbow*, *J.B. (London)* 25: 15. Ealing, a single plant, 1945,† *Boucher*, *K. & L.*, 56. **6.** Between Southgate and East Barnet, 1843,† *E. Warner*, *Trim. MSS.*

G. anglica L. Needle Furze, Petty Whin.
Native on heaths and commons. Very rare, decreasing.
First record: *Gerarde*, 1597.
1. Harefield!, *Blackst. Fasc.*, 31. Ruislip Common!, *Melv. Fl.*, 21. Harrow Weald Common;† Stanmore Common!, *T. & D.*, 75. Uxbridge Common,† *Benbow* (BM). Hillingdon Heath,† *Benbow MSS.* Warren Gate,† *P. H. Cooke*, *K. & L.*, 55. Northwood!, *C. S. Nicholson* (LNHS). **3.** Hounslow Heath!, *J. Banks*; near Twickenham, 1844,† *Twining* (BM). Whitton Park,† *Benbow MSS.* **4.** Hampstead Heath!, *Ger. Hb.*, 1139; now confined to the West Heath, where it is very scarce. Whitchurch Common!,† *C. S. Nicholson* (LNHS). **5.** Old Oak Common,† *Britten*; near Willesden Junction,† *Newbould*, *T. & D.*, 75. The two records probably refer to the same

locality. Ealing Common,† *Benbow MSS.* **6.** Hadley!,† *Warren*; Enfield, *T. & D.*, 75; c. 1890,† *Benbow MSS.* **7.** East Heath, Hampstead,† *T. & D.*, 75.

ULEX L.

U. europaeus L. Furze, Gorse, Whin.

Denizen or native on heaths and commons, in rough grassy places, etc., usually on light soils, established on railway banks, etc. Common and widely distributed except in heavily built-up areas, where owing to the absence of suitable habitats, it is rare. Sometimes planted for ornamental purposes.

First record: *T. Johnson*, 1632.

7. East Heath, Hampstead!, *T. & D.*, 74. South Kensington,† *Dyer List.* Parliament Hill Fields!, *Andrewes MSS.* Hyde Park, casual seedlings, *Warren Fl.* Railway bank, Gloucester Road, S.W., planted. Thames bank, Pimlico, planted. Hackney Wick, *L. M. P. Small.*

U. minor Roth Dwarf Furze.

U. nanus T. F. Forst.

Native on heaths and commons, formerly local, but now very rare.

First record: *Gerarde*, 1597.

1. Near Pinner Hill,† *Mrs Tooke*, teste *Hind*; Hillingdon,† *Warren*; Harefield Common!, *T. & D.*, 74. Uxbridge Common,† *Benbow* (BM). Ruislip Common!; Pield Heath;† Ickenham Green;† Northwood!; Duck's Hill, *Benbow MSS.* **2.** Hampton Court, 1808,† *Winch MSS.* Fulwell,† *T. & D.*, 74. Ashford Common,† *Benbow MSS.* **3.** Hounslow!, *Dill. MSS.*; *Planchon, Bot. Gaz.* 1, 289; Hounslow Heath!, 1820, *E. T. Bennett* (BM); still present, 1968; roadside from drilling ground to cemetery,† *T. & D.*, 74. Feltham,† *Benbow MSS.* **4.** Hampstead Heath, *Ger. Hb.*, 1139; c. 1710, *Budd. MSS.*; 1809, *Winch MSS.*; 1901,† *Whiting Notes.* **5.** Wyke Green,† *Masters*; Old Oak Common,† *Britten, T. & D.*, 74.

CYTISUS L.

C. scoparius (L.) Link subsp. **scoparius** Broom.

Sarothamnus scoparius (L.) Wimmer ex Koch subsp. *scoparius*

Native on heaths, waste ground, in rough pastures, open woods, etc. chiefly on light soils, established on railway banks, etc. Rather common, though often planted for ornamental purposes.

First record: *T. Johnson*, 1638.

7. Tottenham, *Johns. Cat.* South Kensington,† *Dyer List.* Poplar, *Newbould*; railway banks near Dalston; East Heath, Hampstead!; Kilburn,† *T. & D.*, 75. Bombed site, Ebury Street, S.W.1,† *McClintock.* Bombed site, St Dunstan's Hill, E.C., 1945,† *Lousley.* Bombed site, Notting Hill Gate, 1966→. Fulham Palace grounds!, *Brewis*. Railway banks near Gloucester Road, S.W., planted. Railway banks, West Brompton, planted.

Var. **andreanus** (Puiss.) Dippel
3. Hounslow Heath, 1951!,† *K. & L.*, 56. **4.** West Heath, Hampstead!, 1948, *Welch*, *K. & L.*, 56; 1964. **5.** Railway bank, Hanwell!,† *Druce Notes*.

SPARTIUM L.

S. junceum L.
Alien. Europe, etc. Planted on railway banks, where it sometimes regenerates. Very rare.
First record (of regeneration): *D. H. Kent*, 1955.
5. Railway side near Hammersmith, regenerating, 1955→. Railway banks, Perivale, regenerating, 1960→. **6.** Railway banks near Southgate, regenerating, 1965→.

ONONIS L.

O. repens L. subsp. **repens** Smooth Restharrow.
O. arvensis auct.
Native in dry grassy places, on banks, etc., chiefly on calcareous and alluvial soils, adventive by railways, in fields, on waste ground, etc., on other soils. Rare.
First record: *Blackstone*, 1737.
1. Harefield!, *Blackst. Fasc.*, 7. Ruislip and Harrow Weald Commons, *Melv. Fl.*, 21. Errors; *O. spinosa* intended. Uxbridge Common;† Springwell!; South Mimms!; near Warren Gate, *Benbow MSS*. Railway bank near Uxbridge, 1965→. Railway banks, Potters Bar, *J. G. Dony*! **2.** Staines!; between Hampton and Sunbury!, *Newbould*; between Hampton Court and Kingston Bridge; Teddington!, *T. & D.*, 76; 1967. Penton Hook. Railway banks between Colnbrook and West Drayton!, *K. & L.*, 57; 1967. **4.** Hampstead,† *Irv. MSS*. Scratch Wood railway sidings.† **5.** Osterley, 1933!;† Horsendon Hill!, *K. & L.*, 57. Railway yard, Brook Green, 1956, *Murray*. **6.** Bush Hill Park, 1966, *Kennedy*.

O. spinosa L. subsp. **spinosa** Spiny Restharrow.
O. campestris Koch & Ziz
Native on commons and rough pastures, etc., on heavy soils. Locally abundant.
First record: *J. Newton*, 1680.
1. Harefield,† *Blackst. Fasc.*, 6. Pinner Hill,† *Hind*; near Yiewsley,† *Newbould*; near Harrow Weald Common, *T. & D.*, 76. Ruislip! and

Harrow Weald Commons, *Melv. Fl.*, 21 (as *O. arvensis*). Haste Hill, Northwood, *Wrighton*!, *L.N.* 27: 31; 1967. Uxbridge Moor!, *K. & L.*, 57; 1968. **2.** Staines!, *T. & D.*, 76; abundant on Shortwood Common, Knowle Green and Staines Moor. Between Kingston Bridge and Hampton Court!, *T. & D.*, 76. Colnbrook;† Stanwell Moor!, *Benbow MSS.* **4.** Hampstead!, *Irv. MSS.* Wembley;† near Brent Reservoir,† *Farrar*; Golders Green,† *T. & D.*, 76. Harrow Weald, 1866, *Trimen* (BM). Willesden,† *De Cresp. Fl.*, 109. Hendon, *Champ. List.* Near Barnet Gate!, *K. & L.*, 57. Mill Hill. **5.** Between Perivale and Ealing,† *Lees, T. & D.*, 76. Hanwell!,† *K. & L.*, 57. **6.** Finchley Common!; Hadley Common!; Enfield Chase!, *Benbow MSS.* Railway banks between Hadley Wood and Barnet!, *J. G. Dony* and *Hinson.* **7.** Between Battle Bridge (= King's Cross) and Highgate,† *Newt. MSS.* East Heath, Hampstead!, *Trimen* (BM); 1968.

MEDICAGO L.

M. sativa L. subsp. **sativa** Lucerne, Alfalfa.

Alien. Mediterranean region, etc. Cultivated as a fodder crop, and now naturalised in fields, by roadsides, on waste ground, etc., and established on railway banks, rubbish-tips, etc. Common.

First record: *Hind*, 1864.

7. South Kensington, *T. & D.*, 76; 1870,† *Dyer* (BM). Reservoir banks, Fulham, in abundance, *Corn. Surv.* Bishop's Park, Fulham, *Welch*!. Hackney Marshes!, *J. E. Cooper.* Kensington Gardens!; Primrose Hill!, *L.N.* 39: 48. Regents Park!, *Holme, L.N.* 41: 18. Hyde Park, one plant, 1962,† *Allen Fl.* City of London bombed sites, *Lousley.* Paddington, 1947. Westbourne Park, 1948. Highgate, *Andrewes MSS.* Hammersmith. Earls Court. Holloway. Islington. Limehouse. Wapping. Isle of Dogs. Poplar.

Plants with white flowers were noted at **5.** Hanwell, 1951.

Subsp. **falcata** (L.) Arcangeli

M. borealis Grossh., *M. falcata* L.

Introduced. Naturalised in grassy places, on waste ground, etc. Very rare, and often merely casual.

1. Uxbridge Common, 1888, *Benbow* (BM); 1903, *Loydell* (D); 1912,† *P. H. Cooke* (SLBI). North of Uxbridge, 1908,† *Sprague* (K). The latter record may refer to the Uxbridge Common station. **2.** Near Hampton Court, 1922,† *C. E. Britton* (K). Near West Drayton!, 1950, *H. Banks* and *Boniface*, *K. & L.*, 58; 1965!, *Wurzell.* **3.** Southall,† *Druce Notes.* Feltham, 1952,† *Welch.* Hounslow Heath, 1961, *Philcox* and *Townsend.* **5.** Alperton, 1911,† *Wilmott* (BM). Ealing, 1938!,† *K. & L.*, 58. **6.** Muswell Hill, 1904; North Finchley, 1925, *J. E. Cooper, K. & L.*, 58. **7.** South Kensington,† *Dyer* (BM). Isle of Dogs, *Benbow*, *J.B.* (*London*) 25: 365. Hackney Marshes, 1921,†

J. E. Cooper (BM). Bombed site between Milton Street and Whitecross Street, E.C., 1949,† *Wrighton, L.N.* 29: 14.

M. lupulina L. Black Medick, Nonsuch.
Native in fields, by roadsides, on cultivated and waste ground, etc., and established on railway banks, etc. Very common.
First record: *Blackstone*, 1737.
7. Frequent in the district.

Var. **scabra** Gray
1. Harefield, 1923, *Bishop, K. & L.*, 59. Uxbridge. **2.** West Bedfont. **3.** Hounslow Heath. **5.** Railway tracks between Brentford and Southall. Chiswick. **6.** Railway tracks near Muswell Hill. **7.** East Heath, Hampstead. Paddington.

Var. **willdenowiana** Koch
1. Harefield, 1923, *Bishop, K. & L.*, 59. Springwell, *Sandwith*!, *L.N.* 26: 59. Yiewsley, 1909, *J. E. Cooper* (BM). Uxbridge, *Druce Notes*. Ruislip. **5.** Chiswick, 1937–39, *Polunin*; 1947. Hanwell, *Sandwith*!, *K. & L.*, 59. Ealing.
This variety is probably more widespread than the records given above suggest. Plants with foliaceous calyces have been noted at **1.** Springwell, *Welch, K. & L.*, 59. **2.** West Drayton, *L. B. Hall, K. & L.*, 59. **5.** Chiswick, 1944!, *K. & L.*, 59.

M. polymorpha L. Hairy Medick.
M. denticulata Willd., *M. hispida* Gaertn., *M. lappacea* Gaertn., *M. nigra* Willd.
Introduced. Naturalised in grassy waste places, etc., usually on light soils. Very rare, and often merely casual.
First record: *Doody*, c. 1700.
1. Springwell, 1946!,† *K. & L.*, 58. Yiewsley, 1908 and 1913, *J. E. Cooper* (BM). Pinner, one plant,† *Hind, Melv. Fl.*, 21. **2.** Near Hampton Court, *Rand* (DB); 1894 (BM). Hampton Court Park, plentiful, 1908, *Montg. MSS.* Gravel pit, East Bedfont, abundant, 1946, *Brenan, J. G. Dony, Lousley* and *Woodhead*!, *L.N.* 26: 59; 1960. **3.** Near Hounslow; between Whitton and Isleworth, 1886, *Benbow* (BM). **4.** Hendon, 1912, *P. H. Cooke* (LNHS). **5.** Field between Turnham Green and Acton, plentiful,† *Lawrence, Phyt.* 2: 811. Near Acton, 1846,† *McIver* (SY). Alperton, 1918† (K). **7.** Parsons Green, 1856,† '*A. J.*', *Phyt. N.S.* 2: 168. Hackney Marshes, 1912–13, *Coop. Cas.*; 1915 and 1920,† *J. E. Cooper, K. & L.*, 58. Crouch End,† *Druce Notes*.

Var. **polymorpha**
7. Hackney Marshes, 1912, *J. E. Cooper* (BM).

Var. **vulgaris** Shinners forma **apiculata** (Willd.) Shinners
Medicago apiculata Willd., *M. polymorpha* L. var. *apiculata* (Willd.) Ooststroom & Reichgelt
1. Yiewsley, 1909, *Coop. Cas.*; 1912, *J. E. Cooper* (BM). **2.** West Drayton, *Druce Notes*. **3.** Isleworth, 1876,† *Montgomery* (MY). Near Worton, 1886,† *Benbow* (BM). **6.** Behind Ponders End, near Enfield,† *Doody MSS*. **7.** Cornfields near Paddington,† *Doody MSS*. Between Paddington and Kensington,† *Budd. MSS*. Hackney Marshes, 1910 and 1912,† *J. E. Cooper* (BM).

Forma **tuberculata** (Gren. & Godr.) Shinners
Medicago apiculata Willd. var. *confinis* W. Koch, *M. hispida* Gaertn. var. *confinis* (W. Koch) Burnat
4. Finchley, 1909, *E. F. Drabble* (BM). **7.** Hackney Marshes *M. & S.*

Var. **polygra** (Urban) Shinners
Medicago hispida Gaertn. var. *polygra* Urban, *M. reticulata* Benth.
7. Hackney,† *Druce, Melville* and *R. L. Smith, Rep. Bot. Soc. & E.C.* 8: 303.

M. arabica (L.) Huds. Spotted Medick.
M. maculata Sibth.
Native in fields, waste places, on roadside verges, banks, etc., on light soils near the Thames, probably adventive and often merely casual elsewhere. Local.
First record: *T. Johnson*, 1632.
1. Uxbridge Common,† *Blackst. Fasc.*, 101. Cowley, 1884,† *Benbow*; Yiewsley, *J. E. Cooper* (BM). Denham Lock!, 1962, *Davidge*; 1968. **2.** In scattered localities by the Thames from Staines to Teddington. East Bedfont!, *L.N.* 26: 59; 1965. West Bedfont, 1965, *Goom* and *Kennedy*!. **3.** Twickenham; Isleworth;† near Richmond Bridge,† *T. & D.*, 77. Near Hounslow Heath!,† *K. & L.*, 58. **4.** Hampstead Heath, *Johns. Enum.*; c. 1912,† *Whiting Fl.* Mill Hill, 1965, *Hinson*. **5.** Near Hanwell,† *Warren*; near Kew Bridge,† *Jewitt*; Chiswick,† *T. & D.*, 77. Brentford, casual, 1949,† *H. C. Harris*!, *K. & L.*, 58. East Acton,† *Druce Notes*. **6.** Palmers Green, 1909, *P. H. Cooke*, *K. & L.*, 58. **7.** Green Park, sown with grass seed,† *T. & D.*, 77. Hackney Marshes,† *M. & S.* Near Marble Arch, casual, 1962,† *Allen Fl.*

Var. **longispina** Rouy
5. Acton, 1902, *Loydell* (D).
A variable species worthy of further study. Some of the adventive plants were possibly of foreign origin.

MELILOTUS L.

M. altissima Thuill. Tall Melilot.
M. officinalis auct.
Denizen in fields, waste places, on wood borders, by roadsides, etc., established on railway banks, etc. Common.
First record: *T. Johnson*, 1638.
7. Tottenham, *Johns. Cat.* Brook Green Marshes, 1852† (SY). Homerton; Upper Clapton!, *Cherry*; near Chelsea Hospital, 1861; Westminster, *T. & D.*, 78. Finchley Road, *T. & D.*, 422. Near Newington, *Ball. MSS.* South Kensington, *Dyer List.* Bombed site, Basinghall Street, E.C.,† *McClintock*, *Lousley Fl.* Bombed site, St Bartholomew's Close, W.C., 1950,† *Whittaker* (BM). Hornsey. Hackney Wick. Stepney. Old Ford.

M. officinalis (L.) Pall. Common Melilot.
M. arvensis Wallr.
Alien. Europe. Naturalised in fields, on waste ground, by roadsides, etc., established on railway banks, rubbish-tips, etc. Very common.
First evidence: *Benbow*, 1862.
7. Frequent throughout the district.

M. alba Medic. White Melilot.
Alien. Europe, etc. Naturalised in fields, on waste ground, by roadsides, etc., established on railway banks, rubbish-tips, etc. Common.
First record: *Britten*, 1858.
7. Common in the district.

M. indica (L.) All. Small-flowered Melilot.
M. parviflora Desf.
Alien. Mediterranean region, etc. Naturalised on waste ground, in fields, by roadsides, etc. Rare, and often merely casual.
First record: *Britten*, 1861.
1. Harefield!; near Denham Lock, 1884; near Uxbridge, 1886; Uxbridge, 1890, *Benbow*; Yiewsley, *J. E. Cooper* (BM). Northwood, 1911, *J. E. Cooper*; near Denham, 1919, *L. B. Hall*, *K. & L.*, 60. New Years Green, 1953, *T. G. Collett*!. Warren Gate, *P. H. Cooke.* **2.** West Drayton, 1909, *P. H. Cooke* (LNHS). Between Staines and Poyle, 1958, *Briggs.* **3.** Twickenham, 1886, *Benbow MSS.* Northolt, 1947!, *K. & L.*, 60. Hounslow Heath, 1961, *Philcox.* **4.** Finchley, *Coop. Cas.* Near Finchley, 1920, *J. E. Cooper* (BM). Hendon, 1915, *J. E. Cooper*, *K. & L.*, 60. Kenton, 1966. Mill Hill, 1966. **5.** Hanwell!; between Hanwell and Brentford, 1944!; South Greenford, 1947!, *K. & L.*, 60. **6.** Muswell Hill, 1902, *Coop. Cas.* **7.** Chelsea!; Parsons Green, *Britten*, *T. & D.*, 78. Fulham, 1955, *Welch*!. Isle of Dogs, 1887, *Benbow* (BM). Hackney Marshes, 1909–14, *Coop. Cas.* Holloway, 1965, *D. E. G. Irvine.* Limehouse, 1966.

TRIFOLIUM L.

T. ornithopodioides L. Birdsfoot Fenugreek.
Trigonella ornithopodioides (L.) DC.
Formerly native in dry sandy and gravelly places but now extinct.
First record: *Doody*, 1690. Last evidence: *Benbow*, 1885.
1. Hillingdon, 1871, *Warren*; Uxbridge Common, 1872 and 1884, *Benbow* (BM). **2.** Hampton Court, *Borrer MSS.* **3.** Waste land west of Hounslow Heath, abundant, 1885, *Benbow* (BM). **4.** Hampstead Heath, *Woods*, *B.G.*, 109. **5.** Hanwell Heath, *Goodenough*, *Coop. Fl.*, 108. **7.** Tothill Fields, Westminster, *Doody*, *Ray Syn.*, 246; c. 1700, *Buddle* (SLO); there abundantly, *Curt. F.L.*; c. 1780 (LT); c. 1800, *J. Banks* (BM).

T. micranthum Viv. Slender Trefoil.
T. filiforme L., *nom. ambig.*
Native on commons, heaths, in meadows, by roadsides, etc. Locally common, especially on light soils near the Thames.
First evidence: *Buddle*, c. 1710.
1. Harefield Common, *Blackst. Fasc.*, 102; 1887,† *Benbow* (BM). Ruislip Reservoir, *Hind*, *Melv. Fl.*, 22. Between Harefield and Ruislip!; Harrow Weald Common!; between South Mimms and Potters Bar!, *T. & D.*, 82. Uxbridge!; Hillingdon Heath,† *Benbow MSS.* South Mimms!, *J. E. Cooper*, *K. & L.*, 64. **2.** Common near the Thames. **3.** Hounslow Heath!, *T. & D.*, 82. **4.** Grimsdyke grounds, Harrow Weald!, *L.N.* 26: 60. Near Stonebridge Park, *Summerhayes* and *Milne-Redhead* (K). Hampstead Heath, 1780,† *Pamplin MSS.* Hampstead, 1831, *Tuck* (BM); 1865† (SY). Hendon, *P. H. Cooke* (LNHS). Finchley. Mill Hill. Edgware, *Horder*!. **5.** Opposite Mortlake!, *T. & D.*, 82. Acton,† *Druce Notes.* Ealing Common, abundant,† *Benbow MSS.* Clayponds Hospital grounds, South Ealing. Syon Park, *Welch*!, *L.N.* 26: 60. Hanwell. **6.** Enfield Green, *Benbow MSS.* Grounds of Forty Hall and Myddleton House, Enfield, *Kennedy*!. **7.** Tothill Fields, Westminster,† *Buddle* (SLO). Hyde Park, 1815 (G & R) near Marble Arch, casual, 1962,† *Allen Fl.* Kensington Gardens, 1871,† *Trim. MSS.* Hornsey,† *Ball. MSS.* Railway bank, Hackney Marsh, *Grugeon*, *T. & D.*, 82.

T. dubium Sibth. Suckling Clover, Lesser Yellow Trefoil.
T. minus Sm., *T. filiforme* L. subsp. *dubium* (Sibth.) Gams
Native in fields, waste places, by roadsides, etc., established on railway banks and tracks, lawns, etc. Very common.
First certain record: *A. Irvine*, c. 1830.
7. Frequent in the district.

T. campestre Schreb. Hop Trefoil.
T. procumbens auct.
Native in dry grassy places, by roadsides, on waste ground, etc., established on railway tracks, etc. Formerly rather common, but now local and decreasing.
First record: *T. Johnson*, 1638.
1. Harefield!, *Blackst. Fasc.*, 101. Pinner; South Mimms!, *T. & D.*, 82. Uxbridge Moor. Disused railway between Uxbridge and Denham. **2.** Near Staines!; Hampton; Tangley Park,† *T. & D.*, 82. Harlington, *Newb. MSS.* Ashford, *Hasler.* Staines Moor. Between Kingston Bridge and Hampton Court. East Bedfont. **3.** Wood End,† *Melv. Fl.*, 22. Twickenham Park, *T. & D.*, 82. Spring Grove,† *A. Wood*, *Trim. MSS.* Southall. **4.** Harrow Weald Churchyard,† *T. & D.*, 82. Cricklewood, *P. H. Cooke* (LNHS). West Heath, Hampstead, *H. C. Harris.* Scratch Wood railway sidings. **5.** Brentford!, *Cherry*; Kew Bridge railway station,† *T. & D.*, 82. Railway between Southall and Brentford. Syon Park. Hanger Hill, Ealing.† **6.** Tottenham, *Johns. Cat.* Edmonton; Enfield!, *T. & D.*, 82. **7.** Green Lanes, Newington,† *Newbould*, *T. & D.*, 82. Chelsea, casual, 1944.† Railway bank, Finchley Road,† *Trim. MSS.*

T. hybridum L. subsp. **hybridum** Alsike Clover.
Alien. Europe, etc. Cultivated as a fodder crop, and now naturalised in grassy places, on waste ground, by roadsides, etc., established on railway banks, canal paths, etc. Common.
First record: *Trimen* and *Dyer*, 1866.
7. Common in the district.

Subsp. **elegans** (Savi) Aschers. & Graebn.
T. elegans Savi
Lit. *Watsonia* **1**: 119 (1949).
7. Clapton, 1909, *P. H. Cooke* (LNHS). Isle of Dogs. Hackney Marshes, 1957.
Proliferous forms have been recorded from **1.** Harefield, 1910; Yiewsley, *J. E. Cooper*, *K. & L.*, 63. **4.** Finchley, *J. E. Cooper*, *K. & L.*, 63.

T. repens L. White Clover, Dutch Clover.
Native in grassy places, etc., established on lawns, railway banks, etc. Very common.
First record: *T. Johnson*, 1638.
7. Abundant throughout the district.
Proliferous forms have been recorded from **1.** Uxbridge, *J. E. Cooper*, *K. & L.*, 63. **2.** West Drayton, *J. E. Cooper*; East Bedfont, *Westrup*, *K. & L.*, 63. **4.** Near Mill Hill, 1892, *W. H. Hudson*, *Sci. Goss.* 28, 22. East Finchley, *J. E. Cooper*, *K. & L.*, 63. **7.** Hornsey, *J. E. Cooper*, *K. & L.*, 63. Crouch End, *J. E. Cooper* (BM). Chester Square, S.W.1.

T. glomeratum L. Clustered Clover.
Formerly native in grassy places on sandy and gravelly soils, but now extinct.
First certain record: *Goodenough*, c. 1782. Last record: *Newbould*, 1871.
2. Hampton Court (BO). **4.** Hampstead Common, 1809, *Winch MSS.* **5.** Hanwell Heath, *Goodenough, Curt. F.L.* **7.** Hyde Park, a single plant, 1871, *Newbould, Warren Fl.*

T. fragiferum L. Strawberry Clover.
Native in damp pastures, by river- and streamsides, by roadsides, etc. Formerly rather common, but now very rare, and decreasing, and mostly confined to clay and alluvial soils near the Thames.
First record: *T. Johnson*, 1638.
1. Harefield,† *Blackst. Fasc.*, 101. Breakspeares;† near Cowley,† *Benbow* (BM). Yiewsley!,† *K. & L.*, 63. **2.** Between Hampton and Hampton Court!, *Newbould*; by the Thames between Hampton Court and Kingston Bridge; Staines!, *T. & D.*, 81. Littleton!; Shepperton!, *Benbow MSS.* Chertsey Bridge!, *C. S. Nicholson, K. & L.*, 62; locally plentiful in meadows between Chertsey Bridge and Shepperton, 1967. **3.** Thames bank between Twickenham and Richmond Bridge,† *T. & D.*, 81. Roxeth,† *Hind Fl.* **4.** Between Willesden and Neasden,† *Farrar*; Harrow;† Harrow Weald,† *T. & D.*, 81. Cricklewood, 1870,† *Trim. MSS.* Child's Hill, 1870,† *Dyer* (BM). Kingsbury,† *P. H. Cooke* (LNHS). **5.** Hanwell.† **6.** Tottenham,† *Johns. Cat.* **7.** Hornsey;† near Pancras,† *Curt. F.L.* Hyde Park,† *Dicks. Hort. Sicc.* Eel Brook Meadow, Parsons Green, '*J. A.*', *Phyt. N.S.* 1: 464; c. 1869;† East Heath, Hampstead, *T. & D.*, 81. Between Marylebone and Hampstead, 1870,† *Dyer* (BM). Regents Park, 1957, *Holme*; Coach Mound, Lord's Cricket Ground!,† *L.N.* 39: 48; eradicated by the construction of the Warner Stand in 1957. Lee Marshes, near Tottenham, 1899,† *R. W. Robbins* (LNHS).

T. medium L. Zigzag Clover.
Native in grassy places, established on railway banks, etc. Locally plentiful, especially on heavy soils.
First record: *Buddle*, c. 1710.
1. Pinner Hill!, *Hind Fl.* Stanmore Common!; Elstree!, *Trimen* (BM). Near South Mimms!, *T. & D.*, 79. Near Ruislip, *F. J. Hanbury* (HY). Harefield!; Northwood!, *Benbow* (BM). South Mimms!, *P. H. Cooke, K. & L.*, 60. Potters Bar. Bentley Heath. Ruislip Common, *Westrup.* **2.** River bank between Richmond Bridge and Hampton Court, *Newbould, T. & D.*, 79. Hampton Court. Near Ashford, *Hasler.* **3.** Near Isleworth, *Boniface.* Northolt!, *K. & L.*, 60. **4.** Common on the London Clay even in heavily built-up areas, e.g. railway banks at Neasden and Willesden. **5.** Between Acton and Turnham Green,† *Newbould, T. & D.*, 79. Horsendon Hill, plentiful!, *L.N.* 27: 31.

Between Perivale and Ealing. Near Greenford, 1973. **6.** Frequent on the London Clay. **7.** About London, *Budd. MSS.* East Heath, Hampstead, 1866, *Trimen* (BM); c. 1912, *Whiting Fl.* Finchley Road,† *T. & D.*, 79. Highgate Ponds!, *H. C. Harris*, *K. & L.*, 60. Bombed site, Red Lion Square, W.C., 1950,† *Whittaker* (BM). Regents Park!, *L.N.* 39: 48. Railway bank, Kilburn, 1967.

T. arvense L. Hare's-foot Trefoil.

Native in fields, on waste ground, by roadsides, etc., established on railway banks, etc. Rare, and decreasing, and mainly confined to light soils, mostly near the Thames.

First record: *Blackstone*, 1737.

1. Harefield!, *Blackst. Fasc.*, 102. Springwell, *Pickess.* Uxbridge Common,† *Benbow MSS.* Yiewsley, *J. E. Cooper* (BM). Warren Gate, *J. E. Cooper*, *K. & L.*, 61. **2.** Teddington, *Dyer* (BM). Teddington Park; between Staines and Hampton, one plant, *T. & D.*, 79. Between Sunbury and Hampton, *Benbow*, *J.B. (London)* 25: 16. Near Sunbury!, *Cund. Guide*, 99. Between Shepperton and Walton Bridge; Yeoveney!; Staines Moor!, *Benbow* (BM). East Bedfont!, *P. H. Cooke*; Ashford, *Hasler*, *K. & L.*, 61. **3.** Spring Grove [Isleworth],† *A. Wood*, *Trim. MSS.* Near Twickenham, 1867 (SY). Near Hounslow Heath, 1965, *Boniface.* Hayes Park, 1966. **4.** Hampstead Heath, *Coop. Fl.*, 99; c. 1890,† *Cochrane Fl.* Near Brent Reservoir,† *Warren*, *T. & D.*, 422. Mill Hill, *Moring MSS.* **5.** Gravel pits near Hanwell!,† *Warren*, *T. & D.*, 79. Meadow near Boston Manor.† Elthorne Park, Hanwell!,† *L.N.* 26: 60. Between Hanwell and Greenford, 1954.† Between Castle Bar and Greenford, 1960.† Chiswick!, *K. & L.*, 61. **6.** North Finchley, 1908,† *J. E. Cooper* (BM). **7.** Chelsea, 1860,† *Britten*, *Bot. Chron.*, 58. South Kensington,† *Dyer List.* Hackney Marshes, 1916,† *J. E. Cooper* (BM). Brompton Cemetery, 1965, *R. Alston.*

T. scabrum L. Rough Trefoil.

Formerly native in dry pastures, on waste ground, etc., on sandy and gravelly soils, but now extinct.

First record: *A. Irvine*, 1834. Last evidence: *Wright*, 1892.

1. Hillingdon Heath, 1872, *Benbow* (BM). **2.** Thames side below Hampton Court Gardens, 1839, *H. C. Watson* (W); 1843, *H. C. Watson*; 1892, *Wright* (BM). Teddington, 1847, *Stevens* (K). **7.** Between Bayswater and Paddington, 1834, *Irv. MSS.* and *Irv. Lond. Fl.*, 178.

T. striatum L. Soft Trefoil.

Native in fields, short turf, bare places, and on waste ground on sandy and gravelly soils. Locally common near the Thames, very rare in other parts of the vice-county.

First evidence: *Stonestreet*, c. 1700.

1. On Oliver's Mount on Uxbridge Moor, plentifully,† *Blackst. Fasc.*, 103. Uxbridge Common!; Frays Meadows, *Benbow* (BM). Cowley,† *Benbow MSS.* South Mimms!, *Green* (SLBI). **2.** Locally common near the Thames. Roadsides about Ashford and Charlton, abundant, *T. & D.*, 80. Between Staines and Poyle, *Briggs.* **3.** Twickenham, 1867 (SY); 1903,† *E. F.* and *H. Drabble*; Hounslow Heath!, *Trimen* (BM). Near Hatton,† *T. & D.*, 80. Near East Bedfont!, *K. & L.*, 62. **4.** Hampstead Heath, 1840,† *T. Moore* (BM). Stonebridge Park, 1898,† *Druce Notes.* Willesden, on ballast, 1864,† *Parsons* (CYN). **5.** Sandy field by the river, Chiswick, 1867, *Trimen*; 1888,† *Benbow* (BM). Habitat destroyed by gravel-digging operations. Osterley Park,† *Britten, Trim. MSS.* Hanwell!,† *Benbow MSS.* Syon Park, 1965. **6.** North Finchley, 1926,† *J.E. Cooper* (BM). **7.** Chelsea,† *Stonestreet* (DB). Between Southampton Row and Hampstead,† *Blackst. Spec.*, 100. On an unused bridge over the North-Western railway near Edgware Road station, abundant, 1869,† *Warren, T. & D.*, 423.

Var. **erectum** Leight.
1. Mimmshall, *Green* (SLBI).

T. pratense L. subsp. **pratense** Red Clover.
Native in pastures, on waste ground, by roadsides, etc., established on railway banks, etc., also as an introduction as a fodder plant. Very common, but the native plant (var. **sylvestre** Syme) is much less frequent than the larger fodder plant (var. **sativum** Sturm).
First record: *T. Johnson*, 1638.
7. Very common in the district.

Var. **parviflorum** Bab.
1. Yiewsley, 1920, *J. E. Cooper* (BM). Harefield, *J. E. Cooper, K. & L.*, 61. **2.** Staines, *Green, Druce Notes.* **4.** Edgware, *Green* (SLBI). **5.** Park Royal, *Druce Notes.*

Var. **americanum** Harz
1. Near Harefield, *Druce Notes.* **4.** Finchley, 1913, *E. F.* and *H. Drabble* (BM).

Var. **leucochraceum** Aschers. & Prantl
4. East Finchley, 1908, *J. E. Cooper* (BM).

Plants with white flowers have been collected at **1.** Springwell, 1890, *Benbow* (BM). **2.** Harmondsworth, 1906, *J. E. Cooper* (BM). **4.** East Finchley, 1908, *J. E. Cooper* (BM). A plant with pedunculate flower heads was gathered at **4.** East Finchley, 1907, *J. E. Cooper* (BM). Plants with twin flower heads merging into each other were collected at **4.** Finchley, 1907; East Finchley, 1908, *J. E.* Cooper (BM), and specimens with foliaceous calyces were gathered at **1.** Harefield, 1910, *J. E. Cooper* (BM). **4.** East Finchley, 1908, *J. E. Cooper* (BM).

T. squamosum L. Sea Clover.
T. maritimum Huds.
Formerly native on the bank of the tidal Thames, but now long extinct.
First, and only evidence: *J. Andrews* and *Field*, pre-1721.
7. 'I gathered it by the Thames side near the Earl of Peterborough's Palace, in company with Mr John Field . . . ' *J. Andrews* (BM).

T. subterraneum L. Subterranean Trefoil.
Native on heaths and commons, in meadows and grassy places, etc. Very rare, and mainly restricted to sandy and gravelly soils in the immediate vicinity of the Thames.
First record: *Plukenet*, c. 1670.
1. Harefield Common;† Uxbridge Common!, *Blackst. Fasc.*, 102. Hillingdon Heath, 1889,† *Benbow* (BM). **2.** Between Hampton and Hampton Court!, *Wats. New Bot.*, 98. Hampton Court Park!, *Tremayne, K. & L.*, 62; and on the adjacent towpath, 1968. Common by Walton Bridge!;† between Staines and Ashford!,† *T. & D.*, 80. Near Sunbury, *Cund. Guide*, 99. Staines, *W. R. Linton* (K). **3.** Hounslow Heath, 1867,† *Trimen* (BM). Near Hatton,† *T. & D.*, 80. **5.** Field by the river, Chiswick, *T. & D.*, 80; 1888,† *Benbow* (BM). Habitat destroyed by gravel-digging operations. Syon Park, 1955. **7.** Tothill Fields, Westminster, *Pluk. MSS.*; c. 1695,† *Budd. MSS.* and *Pet. Midd.* Hyde Park, 1780,† *J. E. Smith* (LINN). Kensington,† *Forsyth* (K).

ANTHYLLIS L.

A. vulneraria L. Kidney-vetch, Ladies Fingers.
Lit. *J.B.* (*London*) **71**: 207 (1933).
Introduced. Naturalised in dry grassy places, on waste ground, by roadsides, etc., established on railway banks, etc. Very rare.
First certain record: *H. C. Watson*, 1873.
1. Railway banks, Uxbridge!, *P. H. Cooke* (LNHS); plentiful until 1962 when the railway became disused, the tracks removed, and much of the area levelled; the plant, however, still survives in small quantity, 1968. Disused railway near Denham, 1967→. Abundant in an old chalkpit near Old Park Wood, Harefield, 1955→!, *Pickess* and *I. G. Johnson*. **2.** Thames bank near Kingston Bridge, 1873,† *H. C. Watson, Trim. MSS.* Near Staines Moor, 1910–13,† *Coop. Cas.* **3.** Roadside near Southall!,† *L.N.* 26: 60. Habitat destroyed by road-widening operations. **5.** Railway bank near Greenford!, *K. & L.*, 64; 1967. **6.** Enfield, *Hanson* (LNHS). **7.** Bombed site, Cripplegate, E.C., 1952,† *Scholey, L.N.* 32: 82.
All Middlesex specimens seen have been referable to subsp. **vulneraria**.

LOTUS L.

L. corniculatus L. Birdsfoot-trefoil, Bacon and Eggs.
Native in meadows, grassy places, etc., established on railway banks, etc. Common.
First record: *Blackstone*, 1737.

Var. **hirsutus** Koch
Var. *incanus* Gray
1. Harefield Common, *Benbow* (BM). **4.** Scratch Wood. **5.** Near Boston Manor.†

L. tenuis Waldst. & Kit. ex Willd. Slender, or Narrow-leaved Birdsfoot-trefoil.
L. tenuifolius (L.) Reichb.
Denizen in meadows, grassy waste places, by roadsides, etc., established on railway banks, etc., Very rare, and sometimes merely casual.
First evidence: *Newbould*, 1866.
1. Railway banks between Uxbridge and West Drayton, 1884,† *Benbow* (BM). Between Denham Lock and Harefield, a single plant, 1890,† *Benbow MSS.* Yiewsley, 1917, *Lester-Garland* (K); 1919, *J. E. Cooper* (BM). **4.** Golders Green, 1921, *Richards*, *K. & L.*, 65. Hampstead Heath, 1947 (BM). **5.** Between Hanwell and Southall, 1947,† *Wrighton*!, *L.N.* 27: 31. Habitat destroyed by building operations, 1952. Chiswick, 1866,† *Newbould* (BM). **7.** South Kensington,† *T. & D.*, 83. Hackney Marshes, 1912 and 1915,† *J. E. Cooper* (BM).

L. uliginosus Schkuhr Large, or Marsh Birdsfoot-trefoil.
L. major auct., *L. pedunculatus* auct., non Cav.
Native in marshes, ditches, on wet heaths, etc. Formerly common, but now local, though usually plentiful where it occurs.
First record: *Buddle*, c. 1710.
1. Common. **2.** Locally frequent. **3.** Twickenham, *T. & D.*, 83. Sudbury, 1859,† *Trimen* (BM). Yeading. Northolt. Hounslow Heath. **4.** Hampstead! (DILL); West Heath, Hampstead!, *T. & D.*, 83. Harrow!, *Melv. Fl.*, 22. Mill Hill. Hendon. **5.** Near Twyford;† Ealing Common;† between Acton and Turnham Green,† *Newbould*, *T. & D.*, 83. Hanwell.† Boston Manor.† Perivale.† Syon Park. **6.** Edmonton; Enfield!, *T. & D.*, 83. Coldfall Wood, Highgate, *Cherry*, *Trim. MSS.* Hadley Common. Near East Barnet. **7.** Belsize,† *Irv. MSS.* Kilburn,† *Varenne*; Finchley Road,† *Newbould*; East Heath, Hampstead!, *T. & D.*, 83. Highgate Ponds!, *Fitter*. Ken Wood Fields.

Var. **glabriusculus** Bab.
Var. *glaber* Bréb.
1. Uxbridge, *Benbow* (BM). Harefield. Springwell. Pinner. Stanmore Common. **2.** Colnbrook. West Drayton. **4.** Hampstead (DILL). **5.** Hanwell.† Perivale.† **7.** Near Chelsea,† *Budd. MSS*. In the fields behind Mother Huff's,† *Doody*, *Ray Syn.*, 334. Ken Wood fields.

GALEGA L.

G. officinalis L. French Lilac, Goat's Rue.
Alien. Europe, etc. Garden escape. Naturalised in grassy places, on waste ground, etc., established on railway banks, etc. Common, and increasing.
First record: *G. C. Druce*, 1903.
7. Thames bank, Hammersmith!; Isle of Dogs!; Primrose Hill, plentiful!, *K. & L.*, 65. Regents Park!, *Holme*, *L.N.* 39: 48. Highgate Ponds!, *Fitter*. Bombed sites, Cripplegate, E.C., *A. W. Jones*. Parsons Green. Kilburn. Hampstead. Highbury. Holloway. Hornsey. Stoke Newington. Clapton. Hackney. Hackney Wick. Canal side between Marylebone and Primrose Hill, frequent, 1968.
The white-flowered plant (var. **albiflora** (Boiss.) Brenan) is equally as common as the lilac-flowered plant, and the two often grow together.

ROBINIA L.

R. pseudoacacia L. False Acacia, Black Locust.
Alien. N. America. Commonly planted in parks, squares and gardens, where it readily regenerates. Seedlings and saplings are common on waste ground, railway banks, etc.
First record (of regeneration): *D. H. Kent*, 1950.
7. Seedlings and saplings are common in the squares of Central London.

COLUTEA L.

C. arborescens L. Bladder Senna.
Alien. Mediterranean region, etc. Planted on railway banks, etc., where it sometimes regenerates, also naturalised on waste ground, by roadsides, etc. Common, especially in the East End, and the immediate northern suburbs of London.
First record: *C. S. Nicholson*, 1900.
7. Frequent on railway banks, especially in the eastern parts of the district.

ORNITHOPUS L.

O. perpusillus L. Birdsfoot.

Native on heaths, commons, cultivated and waste ground, etc., chiefly on sandy and gravelly soils. Formerly common, but now local and decreasing.

First record: *Gerarde*, 1597.

1. Harefield!, *Blackst. Fasc.*, 70. Ruislip Common!, *Melv. Fl.*, 23. Harrow Weald Common!, *T. & D.*, 88. Uxbridge Common!, *Benbow* (BM). Hillingdon Heath,† *Benbow MSS.* Northwood golf course, *J. Ruisl. & Distr. N.H.S.* 15: 15. Stanmore Common. South Mimms. **2.** Towing-path between Kingston Bridge and Hampton Court!, *Bloxam*; Teddington; common by Walton Bridge!;† roadside between Staines and Hampton, abundant, *T. & D.*, 88. Near Sunbury, *Cund. Guide*, 99. Hampton Court Park!, *Welch.* Bushy Park. **3.** Hounslow Heath!; near Hatton, *T. & D.*, 88. **4.** Hampstead Heath, *Ger. Hb.*, 1061, and many later authors; c. 1869,† *T. & D.*, 88. Mill Hill. Near Brent Reservoir,† *Warren*; Stanmore, *T. & D.*, 88. **5.** Field near Wyke House Lane, Brentford,† *Cherry*; Hanwell, *Warren, T. & D.*, 88; 1884,† *Benbow* (BM). **6.** Hadley, *Warren, T. & D.*, 88. Railway tracks, Muswell Hill, 1961, *Bangerter* and *Raven.* Crew's Hill, Enfield, 1966, *Kennedy*!. **7.** Hyde Park, plentifully, 1773, *Hill Veg. Syst.* xxii; 1790, *J. E. Smith* (LINN); 1815, *J. F. Young* (BM); 1816 (G & R); near Marble Arch, casual, 1962,† *Allen Fl.* Weed of flower-bed, Buckingham Palace grounds, 1956, *McClintock, L.N.* 39: 49. St James's Park, casual, 1962,† *Brewis.*

CORONILLA L.

C. varia L. Crown Vetch.

Alien. Europe, etc. Garden escape. Naturalised in grassy places, on waste ground, etc. Very rare.

First record: *A. Irvine*, 1858.

1. Uxbridge, 1903,† *Green* (SLBI). Yiewsley, 1918, *A. D. Webster* (BM). **2.** Near Staines Moor!,† *Westrup.* **3.** Near river Crane, 1905, *Sprague* (K). Hounslow Heath, 1952, *Westrup.* **4.** The Hyde, Hendon!, *L.N.* 27: 31; 1948→. **7.** Chelsea,† *Irv. Ill. Handb.*, 684. Isle of Dogs, 1955.

ONOBRYCHIS Mill.

O. viciifolia Scop. Sainfoin.

O. sativa Lam.

Introduced. Formerly cultivated as a fodder crop, and now naturalised in grassy places, on waste ground, etc., established on railway banks, etc. Rare, and sometimes merely casual.

First record: *Blackstone*, 1737.

1. Harefield!, *Blackst. Fasc.*, 66; abundant in the fields above the canal, *T. & D.*, 88; canal path, a few plants, 1945–48. Railway bank, Pinner,† *Melv. Fl.*, 97. By disused railway between Uxbridge and Denham, 1966. South Mimms!, *Welch, L.N.* 26: 60. Warren Gate!, *P. H. Cooke, K. & L.*, 68. **2.** Near Sunbury, *Cund. Guide*, 99; 1903, *Green* (SLBI). Near Chertsey Bridge, 1947!, *Welch, K. & L.*, 68. Staines Moor, 1955, *Briggs*. **4.** Railway bank, Golders Green, 1909,† *P. H. Cooke* (LNHS). **5.** Railway bank between Ealing and Acton, 1946–49!;† Hanwell, 1945–50!,† *K. & L.*, 68. Railway bank, Acton, 1960→. **6.** Bank of reservoir, Enfield, 1966. **7.** Chelsea, 1860,† *Britten, Bot. Chron.*, 58. Finsbury Park, 1882,† *Le Tall, Whitwell MSS.* Bombed sites, Cripplegate,† *Jones Fl.*

VICIA L.

V. hirsuta (L.) Gray Hairy Tare.

Native in meadows, on wood borders, waste ground, etc., and established on railway banks, etc. Common.

First record: *T. Johnson*, 1629.

7. Between Kentish Town and Hampstead,† *Johns. Eric.* Thames Embankment;† South Kensington,† *T. & D.*, 84. Kensington Gardens!; Hyde Park, *Warren Fl.* Eaton Square, S.W.1, 1954, *McClintock*; Lord's Cricket Ground, 1959!, *L.N.* 39: 49. St James's Park, 1962, *Brewis, L.N.* 44: 22. Crouch End, 1899, *J. E. Cooper* (BM). Highgate, *Andrewes MSS.* East Heath, Hampstead!, *H. C. Harris*. Bombed site, Ebury Street, S.W.1,† *McClintock*.

V. tetrasperma (L). Schreb. Smooth Tare.

Native in meadows, on wood borders, waste ground, etc., and established on railway banks, etc. Rather common, but less widespread than *V. hirsuta*.

First record: *T. Johnson*, 1638.

7. Tottenham, *Johns. Cat.* Shoreditch;† Bethnal Green,† *Dill. MSS.* East Heath, Hampstead, a single plant,† *T. & D.*, 88. Near Edgware Road station,† *Warren, T. & D.*, 423. Hackney Marshes, 1924,† *J. E. Cooper*. Buckingham Palace grounds, *Codrington, Lousley* and *McClintock*!; garden weed, St John's Wood, 1956, *Holme, L.N.* 39: 49. Near Hyde Park, 1965, *D. E. Allen, McClintock* and *Rönaasen*; site of Luxborough Lodge, Marylebone Road, N.W.1, 1966, *Brewis, L.N.* 46: 30. Highgate.

V. cracca L. Tufted Vetch.

Native in rough grassy places, on waste ground, etc., particularly on heavy soils, established on railway banks, etc. Very common.

First record: *Blackstone*, 1737.
7. Frequent in the district.

V. sepium L. Bush Vetch.
Native on hedgebanks, in woods, shady places, etc. Common and widely distributed except in heavily built-up areas, where owing to the absence of suitable habitats it is very rare or extinct.
First record: *Blackstone*, 1737.
7. Hornsey Wood† (HE). Millfield Lane [Highgate],† *T. & D.*, 85. Vale of Health [Hampstead], 1886,† *W. G. Smith* (BM). Hyde Park!, *Watsonia* 2: 119.
Plants with white flowers (cf. var. **alba** Rouy) are recorded from **4.** Hendon, *Irv. Lond. Fl.*, 175.

V. sativa L. subsp. **nigra** (L.) Ehrh. Narrow-leaved Vetch.
V. angustifolia L.
Native on heaths, commons, in fields, by roadsides, etc., and established on railway banks, etc. Common.
First record: *T. Johnson*, 1629.
7. Common in the area.
Plants with pure white flowers have been noted in **2.** Field near Dawley, abundant!, *L.N.* 28: 32. **4.** Waste ground, The Hyde, Hendon, 1965, *Kennedy*!.

Subsp. **sativa** Common Vetch.
Alien. Europe. Cultivated as a fodder crop, and now naturalised in fields and waste places, and established on railway banks, etc. Rather rare, and sometimes confused with subsp. *nigra*.
1. Uxbridge!, *Druce Notes*. Yiewsley, *M. & S.* **2.** Shepperton. Colnbrook; West Drayton, *Druce Notes*. Between Harmondsworth and Stanwell. Near Heathrow. Hampton Court Park. **3.** Hayes. **4.** Harrow, *Harley*. Edgware. Mill Hill. Hendon. **5.** Bedford Park, casual,† *Cockerell Fl.* **6.** Enfield, *Kennedy*!. **7.** Hackney Marshes, *M. & S.* Bombed site, Ebury Street, S.W.1,† *McClintock*. Site of Luxborough Lodge, Marylebone Road, N.W.1, 1966, *Brewis*; Finsbury Square, E.C.2, 1966!, *L.N.* 46: 30. Buckingham Palace grounds, 1965, *McClintock*.

V. lathyroides L. Spring Vetch.
Formerly native in grassy places on sandy and gravelly soils but now extinct.
First record: *J. E. Smith*, 1829. Last evidence: *Trimen*, 1866.
1. Ruislip, *Hind Fl.* An error. **2.** Hampton Court, *Borr. MSS.* **3.** Twickenham, *De Cresp. Fl.*, 117. Probably an error. **5.** Chiswick, 1866, *Trimen* (BM). **7.** Hyde Park, *Smith Engl. Fl.* 3: 283.
Small forms of *V. sativa* subsp. *nigra* are sometimes mistaken for this species. It is possible that it may be rediscovered in the Hampton Court area.

LATHYRUS L.

L. aphaca L. Yellow Vetchling.

Native in rough grassy places and on cultivated ground in the north-eastern parts of the vice-county, elsewhere as a sporadic adventive on waste ground, rubbish-tips, by roadsides, etc. Very rare.

First record: *Doody*, c. 1700.

1. Uxbridge, 1907, *Green* (SLBI). Yiewsley, 1912 and 1914, *J. E. Cooper* (BM); 1917, *Lester-Garland* (K). **2.** West Drayton, *Druce Notes*. **3.** Twickenham, *Doody MSS*. Hounslow, 1945, *H. Banks*. **4.** Finchley, 1909, *J. E. Cooper*; 1915, *E. F. Drabble* (BM). Roadside, Neasden, a solitary fine plant, 1965. **5.** Hanwell, 1963–64. **6.** Enfield, *Doody MSS.*; c. 1789, *Forst. Midd.*; a small colony in a rough field near Enfield Lock, 1966, *Kennedy*!. Frequently about Tottenham and Enfield, *Curt. F.L.* **7.** South Kensington, *Dyer List*. Garden weed, Maida Vale, 1954!, *L.N.* 39: 19. Between Adelaide Road and Belsize Park, *Trim. MSS*. Hackney Marshes, 1909 and 1912, *J. E. Cooper* (BM).

L. nissolia L. Crimson Grass Vetchling.

Lit. *Watsonia* **6**: 28 (1964).

Native in meadows, rough grassy places, etc., especially on heavy soils, also established on railway banks, etc. Rare, but easily overlooked when out of flower.

First record: *T. Johnson*, 1632.

1. Harefield!, *Blackst. Fasc.*, 16. South of Harefield!, *Benbow*, *J.B. (London)* 22: 279. Near Denham. Ickenham, *Davidge*. Breakspeares, 1957, *I. G. Johnson*. **2.** Stanwell Moor, *Welch*!, *L.N.* 26: 60. Staines Moor!, *K. & L.*, 71. Dawley, 1919, *J. E. Cooper*. **3.** Near Sudbury, 1895,† *Bradley* (BM). Whitton, *H. Banks*. Near Hounslow Heath, *Cannon*. South Ruislip, *Wrighton*, *L.N.* 26: 60. **4.** Edgware!, *E. M. Dale* (LNHS). Scratch Wood!, *H. C. Harris*, *L.N.* 28: 32. Mill Hill, *J. E. Cooper*, *K. & L.*, 71. Hampstead Heath,† *Johns. Enum.* Near Kenton, *Longman*, *Melv. Fl.*, 24. Near Finchley, *Button*, *Coop. Suppl.*, 12. Wembley, 1860,† *Hind* (SY). **5.** Alperton,† *Farrar*, *T. & D.*, 86. Near Wyke Green!, *Welch*; Hanwell!; Southall!;† Osterley Park!, *K. & L.*, 71. Chiswick,† *Polunin*. **6.** Muswell Hill; near Tottenham, *Blackst. Spec.*, 59. Tottenham, *Woods*, *B.G.*, 109. Edmonton, *Mart. Pl. Cant.*, 72. Colney Hatch, *Munby*, *Nat.* 1867, 180. Hadley!, *Creasey* (LNHS); railway banks near Hadley Wood station in profusion, 1965, *Kennedy*!. Enfield Chase!, *Benbow* (BM). Near Crew's Hill, Enfield, 1965!, *Kennedy*. **7.** Pancras,† *Johns. Ger.*, 1250, and many later authors. Near Highgate,† *Park. Theat.*, 1079. Tyburn (= Marble Arch);† Marylebone, *J. Dale*, *Merr. Pin.*, 125; site of Luxborough Lodge, Marylebone Road, N.W.1, 1966, *Brewis*, *L.N.*

46: 30. Meadows near Finchley Road, *Garlick W.F.*; 1903, *E. J. Salisbury* (K). Paddington, 1816† (G & R). Ken Wood,† *Hunter, Park Hampst.*, 30.

L. pratensis L. Yellow Meadow Vetchling.
Native in grassy places, on waste ground, etc., and established on railway banks, etc. Very common.
First record: *T. Johnson*, 1638.
7. Frequent in the district.

Var. **gracilis** Druce
5. Acton, *Loydell* (D).

L. sylvestris L. Narrow-leaved Everlasting Pea.
Introduced. Established on waste ground, railway banks, etc. Very rare, or confused with narrow leaved forms of *L. latifolius.*
First certain evidence: *Benbow*, 1900.
1. Near Uxbridge, 1900, *Benbow* (BM). Uxbridge!, *I. G. Johnson*; 1967. **4.** Near Edgwarebury, 1946!, *Simmonds, K. & L.*, 70. **5.** Chiswick, 1955, *Murray List.* An error, *L. latifolius* intended. South Ealing, 1965→.

L. latifolius L. Broad-leaved Everlasting Pea.
Alien. S. Europe, etc. Garden escape. Widely established on railway banks, waste ground, etc. Common.
First record: *D. H. Kent*, 1933.
7. Bombed sites, Cripplegate. Railway bank near Paddington!, *L.N.* 41: 18. Buckingham Palace grounds, 1960–61, *Fl. B.P.G.* St John's Wood Churchyard, 1966!, *Holme, L.N.* 46: 30. Railway banks, Hammersmith, West Brompton, near Earls Court, Kilburn, Highgate, Isle of Dogs, near Euston, Holloway, Finsbury Park, Hornsey, Dalston, Stoke Newington, Tottenham, Hackney, Stepney, Bow and Hackney Wick.

L. linifolius (Reichard) Bässler Bitter Vetch.
var. **montanus** (Bernh.) Bässler
L. montanus Bernh., *L. macrorrhizus* Wimm., *Orobus tuberosus* L.
Native in woods and thickets, on heaths, etc., and established on railway banks. Rare, and decreasing.
First record: *Gerarde*, 1597.
1. Harefield, *Blackst. Fasc.*, 8; c. 1869,† *T. & D.*, 87. Uxbridge,† *R. & P.*, 219. Near Ruislip Reservoir!, *Melv. Fl.*, 24. Near Elstree Reservoir,† *T. & D.*, 87. Bayhurst! and the Ruislip Woods!, *Benbow MSS.* Pinner Wood, *Benbow* (BM). Northwood, *Benbow MSS.* Stanmore Common!, *Webb. & Colem. Suppl.*, 3; very scarce, 1946, *L.N.* 27: 31, 1964. **4.** Hampstead Wood, *Ger. Hb.*, 1057, and many later authors. Hampstead Heath, *Johns. Eric.*; 1839, *Mann* (K); c. 1869,† *T. & D.*, 87. Bishop's Wood (SY). Scratch Wood,† *T. & D.*,

87. Railway bank near Harrow,† *Melv. Fl.*, 24. East Finchley, 1876,† *French* (BM). **5.** Wormholt (= Wormwood) Scrubs, 1815† (G & R). **6.** Barnet† (HE). Hadley!, *Newbould*; Winchmore Hill Wood;† The Alders, near Whetstone,† *T. & D.*, 87. Railway banks near Hadley Wood!, *J. G. Dony* and *Hinson*. Deadman's Wood, Highgate, 1864, *Bywater* (BM). Highgate Woods! *J. E. Cooper*. Coldfall Wood, Highgate, *E. M. Dale, K. & L.*, 71. **7.** Near Ken Wood,† *J. Andrews* (BM). Hornsey Wood,† *Ball. MSS.*

Var. **tenuifolius** (Roth) Garcke
1. Ruislip Woods, *Green* (SLBI). Pinner Wood, *Melv. Fl.*, 24. **4.** Bishop's Wood, *T. & D.*, 87.

ROSACEAE

SORBARIA A.Br.

S. sorbifolia (L.) A.Br.
Spiraea sorbifolia L.
Alien. Himalaya, etc. Garden escape. Established in a disused reservoir. Very rare.
First record: *McLean*, 1968.
1. Near Yiewsley, 1971. **5.** Floor of disused reservoir, Ealing, 1968, *McLean.*

SPIRAEA L.

S. douglasii Hook.
Alien. N. America. Planted in a wood where it has now become naturalised. Very rare.
First record: *G. C. Druce*, 1910.
4. Scratch Wood!, *Druce Notes* (as *S. salicifolia*); 1967.

FILIPENDULA Mill.

F. vulgaris Moench Dropwort.
F. hexapetala Gilib., *Spiraea filipendula* L.
Native in dry grassy places on alluvial soils near the Thames in the south-west part of the vice-county, elsewhere in grassy places, on waste ground, etc., as an escape from gardens. Very rare.
First record: *W. Turner*, 1548.
1. Pinner, 1967. **2.** Between Hampton and Sunbury!, *Newbould*; Bushy Park, *T. & D.*, 90. Hampton Court Park!, *Tremayne, K. & L.*, 74; 1968. Near Charlton, 1866, *Trimen* (BM). **3.** Meadow on south side of Richmond Bridge, abundant, 1867,† *Dyer*; near Twickenham, 1885,† *Benbow* (BM). **5.** In the meadows . . . at Syon . . . in great plenty,† *Turn. Names.* Waste ground, Hanwell, 1949–53!,† *K. & L.*, 74.

F. ulmaria (L.) Maxim. Meadow-sweet.
Spiraea ulmaria L.
Native in damp meadows, marshes, ditches, by rivers and streams, etc. Common and widely distributed except in heavily built-up areas where, owing to the eradication of its habitats, it is very rare or extinct.
First record: *T. Johnson*, 1638.
7. Tottenham!, *Johns. Cat.* Isle of Dogs, *Coop. Fl.*, 115. Thames side near Fulham,† *T. & D.*, 90. Buckingham Palace grounds, two plants, *Fl. B.P.G.* Side of Lee Navigation Canal, Upper Clapton, 1967.

RUBUS L.

I am indebted to the late W. C. R. Watson for this account of the brambles of Middlesex which is based on the data he provided for *A Hand List of the Plants of the London Area*, 74–100 (1952), with the addition of further records made by himself, the late C. Avery, and the late J. E. Woodhead. Watson also kindly examined and corrected the naming of many Middlesex specimens in the National Herbaria, and also named my own gatherings. On his advice most early printed records and notes have been omitted unless substantiated by specimens. I am also indebted to R. J. Pankhurst for details of Middlesex Rubi in the herbarium of the Botany School, Cambridge. The species are arranged in the order given in Watson's *Handbook of the Rubi of Great Britain and Ireland* (1958).

Subgen. ANOPLOBATUS Focke

R. parviflorus Nutt.
R. nutkanus Moç. ex Ser.
Alien. N. America. Planted by a lake where it has now become naturalised. Very rare.
First record: *Wurzell*, 1965.
7. A well-established, suckering thicket by the lake opposite Ken Wood House, 1965!, *Wurzell*, teste *Lousley*; 1968.

Subgen. IDAEOBATUS Focke

R. idaeus L. Raspberry.
Lit. *Watsonia* **4**: 238 (1960).
Denizen in woods and thickets, etc., naturalised on waste ground, etc. Common.
First record: *Blackstone*, 1737.
7. Kensington Gardens!, presumably bird-sown, *D. E. Allen*, *L.N.* 39: 49. Holland House grounds, Kensington. East Heath, Hampstead. Isle of Dogs. Hornsey. City of London bombed sites. Stoke Newington. Tottenham. Old Ford. Hackney Wick.

Forma **obtusifolius** (Willd.) W. Wats.
var. *obtusifolius* Willd., var. *anomalus* Arrhen.
1. Near Uxbridge Common, *Benbow* (BM), det. *Watson*.

Subgen. GLAUCOBATUS Dumort. *pro parte*

R. caesius L. Dewberry.
Native in woods, on hedgebanks, damp waste ground, by streamsides, etc. Common.
First certain evidence: *Hind*, 1860.
7. East Heath, Hampstead. Ken Wood. Tottenham.

Var. **pinnensis** W.Wats.
1. Near the river Pinn, near Swakeleys!, *Watson, K. & L.*, 345.

R. caesius × idaeus
2. Near Shepperton, *C. E. Britton, J.B. (London)* 55: 326.

R. caesius × ulmifolius
1. Uxbridge. Harefield. Near Denham. **4.** Harrow. **5.** Hanwell. Southall. Osterley. **6.** Enfield Chase. All teste *Watson*.
This hybrid is probably common.

Subgen. GLAUCOBATUS × RUBUS

R. fruticosus L. agg. Bramble, Blackberry.
Native in woods, thickets, on hedgebanks, waste ground, by roadsides, rivers, etc. Very common.
First record: *T. Johnson*: 1632.
7. Frequent in the district.

Sect. Triviales P. J. Muell.
R. conjungens (Bab.) W. Wats.
Native in woods, on hedgebanks, etc. Rather common.
1. Pinner Hill!; Pinner Wood!; near Ruislip Reservoir, *L.N.* 26: 68. Harefield!, *Benbow* (BM), det. *Watson*. Near Northwood, *Watson*. **2.** Staines, det. *Watson*. **5.** Horsendon Hill, *K. & L.*, 93; 1964, *Miles* (CGE). **6.** Finchley Common, *K. & L.*, 93.

R. sublustris Ed. Lees
Native in woods and thickets, etc. Rather common.
1. Near Swakeleys; Uxbridge, *L.N.* 26: 68. Copse Wood, Northwood, *K. & L.*, 93. Ickenham, *Benbow* (BM), det. *Watson*. **3.** Hounslow Heath, *K. & L.*, 93. **4.** Hampstead Heath, *L.N.* 26: 68.

R. balfourianus Bloxam ex Bab.
Native on hedgebanks, in open woods, etc. Local.
1. Uxbridge, *L.N.* 26: 68. **2.** Between Hatton and Harlington, *Benbow* (BM), det. *Watson*. **3.** Hounslow Heath, *K. & L.*, 93. **5.** Perivale Wood, *Benbow* (BM), det. *Watson*. **5.** Hanger Hill, Ealing!, *K. & L.*, 93.

R. warrenii Sudre
R. dumetorum var. *concinnus* Bak. ex Warren
Native in woods and thickets. Local.

1. Pinner; Pinner Wood!, *L.N.* 26: 68. Copse Wood; Bayhurst Wood, *K. & L.*, 93; 1958, *Woodhead* (CGE). **6.** Finchley Common, *K. & L.*, 93.

R. adenoleucus Chaboiss.
Native in woods and thickets. Local.
1. Park Wood, Ruislip, *K. & L.*, 94. **4.** Scratch Wood, *K. & L.*, 94; 1953, *Woodhead* (CGE). Apex Corner, Edgwarebury, *K. & L.*, 94. **6.** Finchley Common, *K. & L.*, 94.

R. halsteadensis W. Wats.
R. dumetorum var. *raduliformis* A. Ley, *R. raduliformis* (A. Ley) W. Wats., non Sudre
Native in woods and thickets. Local.
1. Near Ruislip Reservoir; Northwood, *L.N.* 26: 68. **4.** Hampstead Heath, *L.N.* 26: 68.

R. purpureicaulis W. Wats.
R. corylifolius var. *purpureus* Bab.
Native in hedgerows and thickets. Very rare.
1. Long Lane, Ickenham,† *Benbow* (BM), det. *Watson*. **6.** Finchley Common!, *K. & L.*, 93; 1951, *Woodhead* (CGE).

R. tuberculatus Bab.
R. dumetorum var. *tuberculatus* (Bab.) Rogers
Native in woods and hedgerows. Common.
7. East Heath, Hampstead. Ken Wood. Both det. *Watson*.

R. babingtonianus W. Wats.
R. althaeifolius Bab., non Host, *R. dumetorum* var. *fasciculatus* Rogers *pro parte*
Native in hedgerows and thickets. Local.
1. Harefield, frequent, *Avery* and *Watson*. **4.** North End, Hampstead, *K. & L.*, 94.

R. scabrosus P. J. Muell.
R. dumetorum var. *ferox* auct.
Native in woods and thickets. Local.
1. Near Ruislip Reservoir; Harrow Weald Common, *L.N.* 26: 68. Copse Wood, Northwood, *K. & L.*, 94. **4.** Hampstead Heath, *K. & L.*, 94.

R. myriacanthus Focke
R. dumetorum var. *diversifolius* Rogers *pro parte*, *R. dumetorum* var. *pilosus* auct.
Native in woods and hedgerows. Local.
1. Stanmore Common; Mimmshall Wood, *L.N.* 26: 68. Duck's Hill, Northwood, *K. & L.*, 94. Between Northwood and Harefield, *Avery* and *Watson*. **4.** Edgwarebury, abundant, *K. & L.*, 94. **6.** Arnos Grove, 1932,† *L.N.* 26: 68.

R. britannicus Rogers
R. dumetorum var. *britannicus* (Rogers) Rogers
Native in hedgerows, on banks, etc. Common.
7. Highgate. Near Ken Wood. Both det. *Watson*.

Subgen. RUBUS (MORIFERI)

Sect. Sylvatici P. J. Muell.
R. gratus Focke
Native in thickets. Very rare.
4. Hampstead Heath, one bush, 1943, *L.N.* 26: 67.

R. laciniatus L.
Denizen in woods, thickets, on hedgebanks, waste ground, etc. Common.
7. Hyde Park, *D. E. Allen*, *L.N.* 41: 18. Buckingham Palace grounds, rare, *Fl. B.P.G.* Hampstead. Tottenham.

R. vulgaris Weihe & Nees
Native in woods and thickets. Very rare.
6. Hadley Wood, *K. & L.*, 345.

R. nitidoides W. Wats.
Native in open scrub, etc. Very rare.
5. Horsendon Hill, *L.N.* 26: 67. **7.** East Heath, Hampstead, *Benbow* (BM), det. *Watson*; 1945, *L.N.* 26: 67; 1962, *Miles* (CGE).

R. lentiginosus Lees
R. carpinifolius Weihe & Nees, non J. & C. Presl
Native in woods and thickets, on hedgebanks, etc. Local.
1. Between Swakeleys and Uxbridge Common!, *Benbow* (BM), det. *Watson*. Stanmore Common; Park Wood, Ruislip; Northwood, *L.N.* 26: 67. Copse Wood, Northwood, *K. & L.*, 77. Bayhurst Wood, near Harefield, *Watson* and *Avery*. **3.** Hounslow Heath, *L.N.* 26: 67. **4.** Hampstead Heath, *L.N.* 26: 67. Scratch Wood, det. *Watson*. **6.** Enfield Chase, det. *Watson*.

R. selmeri Lindeb.
R. nemoralis sensu W. Wats., ? P. J. Muell.
Native in hedges and thickets. Very rare.
1. Near Swakeleys, *L.N.* 26: 67. Uxbridge Common!, *K. & L.*, 75. **5.** Hanger Hill, Ealing, *K. & L.*, 78.

R. oxyanchus Sudre
R. nemoralis auct.
Native in hedgerows. Very rare.
1. Uxbridge Circus, *L.N.* 26: 67.

R. lindleianus Ed. Lees
R. platyacanthus Muell. & Lefèv.
Native on heaths, commons, wood borders, hedgerows, etc. Very common.
7. East Heath, Hampstead. Ken Wood. Highgate. All det. *Watson*.

R. egregius Focke
R. mercicus var. *bracteatus* Bagnall
Native in woods and thickets. Rare.
1. Bayhurst Wood, Ruislip, *Benbow* (BM), det. *Watson*. Park Wood, Ruislip; Mad Bess Wood, Ruislip, *K. & L.*, 79. **6.** Between Botany Bay and Enfield, *Benbow* (BM), det. *Watson*.

R. egregius × ulmifolius
1. Park Wood, Ruislip, one bush, *K. & L.*, 79. Erroneously recorded as *R. danicus* Focke ex Frid. & Gel. in *L.N.* 26: 67.

R. macrophyllus Weihe & Nees
Native in woods and hedgerows. Local.
1. Mimmshall Wood; Park Wood, Ruislip, *L.N.* 26: 67. Mad Bess and Bayhurst Woods, Ruislip, *K. & L.*, 78. Garett Wood, Springwell,† *Benbow*; Stanmore Common, *A. B. Jackson* (BM), both det. *Watson*. Near Harefield, *Watson* and *Avery*. **4.** Bishop's Wood, 1882, *Benbow* (BM), det. *Watson*; 1948, *K. & L.*, 78. East Finchley, det. *Watson*. **6.** Arnos Grove, *L.N.* 26: 67.

R. subinermoides Druce ex W. Wats.
Native in woods and thickets, on heaths, etc. Common.
1. Northwood; Stanmore Common!; near Ruislip Reservoir; near Swakeleys, *L.N.* 26: 67. Harefield, *Watson* and *Avery*. Springwell, det. *Watson*. **4.** Hampstead Heath, *L.N.* 26: 67; 1964, *Miles* (CGE). Harrow, det. *Watson*. **6.** Hadley Common, *L.N.* 26: 67.

R. subinermoides × ulmifolius
1. Duck's Hill, Northwood, 1900, *Benbow* (BM), det. *Watson*; 1948, *K. & L.*, 78.

R. amplificatus Ed. Lees
Native in woods and hedgerows, on heaths, etc. Rather common.
1. Mimmshall Wood; Bayhurst Wood, near Harefield; Uxbridge Circus; Northwood, *L.N.* 26: 67. Harefield, *Avery* and *Watson*. Springwell, det. *Watson*. Harrow Weald Common, *K. & L.*, 78. Near Potters Bar, det. *Watson*. **2.** Between Harmondsworth and Stanwell, *Benbow* (BM), det. *Watson*. **6.** Arnos Grove, *L.N.* 26: 67. Finchley Common, *K. & L.*, 78. **7.** Brondesbury, 1879,† *De Crespigny* (BM), det. *Watson*.

R. pyramidalis Kalt.
Native in scrub. Very rare.
5. Horsendon Hill, one bush, 1946, *L.N.* 26: 67.

Var. **parvifolius** Frid. & Gel.
4. Hampstead Heath, *L.N.* 26: 67.

R. danicus (Focke) Focke
1. Park Wood, Ruislip, one bush, *L.N.* 26: 67. An error, the plant seen was *R. egregius* × *ulmifolius*, *K. & L.*, 79.

R. londinensis (Rogers) W. Wats.
R. imbricatus var. *londinensis* Rogers, *R. daveyi* Rilstone
Native on heaths. Very rare.
3. Hounslow Heath, *K. & L.*, 80. **4.** Hampstead Heath, *L.N.* 26: 67.

R. ramosus Bloxam ex Briggs
Native on heaths. Very rare.
1. Harrow Weald Common, plentiful!, *L.N.* 26: 67.

R. rhodanthus W. Wats.
R. carpinifolius var. *roseus* Weihe & Nees, *R. rhombifolius* auct.
Native on heaths, in thickets, etc. Very rare.
1. Harrow Weald Common; Stanmore Common, *K. & L.*, 76. **3.** Hounslow Heath, one bush, 1933,† *L.N.* 26: 67. **4.** Hampstead Heath, 1942, *L.N.* 26: 67.

R. atrocaulis P. J. Muell.
R. stereacanthos P. J. Muell. ex Genev.
Native on heaths. Very rare.
4. Hampstead Heath, *Rogers* (BM), det. *Watson*; 1945, *Watson* (CGE); 1952, *K. & L.*, 345.

R. polyanthemus Lindeb.
R. pulcherrimus Neum., non Hook.
Native on heaths, in woods, hedgerows, etc. Very common.
7. East Heath, Hampstead. Ken Wood. Regents Park. All det. *Watson*.

R. rubritinctus W. Wats.
C. cryptadenes Sudre, non Dumort.
Denizen in thickets, etc. Very rare.
7. Buckingham Palace grounds, *Fl. B.P.G.*

R. rhombifolius Weihe ex Boenn.
R. argenteus Weihe & Nees, non C. C. Gmel.
Native in woods and hedgerows, on heaths, etc. Very rare.
1. Near Swakeleys; Harrow Weald Common, *L.N.* 26: 67. Copse Wood, Ruislip, *K. & L.*, 79. **3.** Hounslow Heath, *L.N.* 26: 67.

R. alterniflorus Muell. & Lefèv.
Native in thickets, etc. Very rare.
5. Horsendon Hill, 1948, *Avery* (CGE), det. *Watson*; *K. & L.*, 80.

R. separinus Genev.
R. cissburiensis Barton & Riddelsd., *R. gelertii* auct.
Native on heaths, in woods and hedgerows, etc. Local.
1. Near Swakeleys; Park Wood, *L.N.* 26: 67. **3.** Hounslow Heath, *L.N.* 26: 67. **4.** Hampstead Heath, *K. & L.*, 80. **7.** Kensington Gardens!, *Allen*, *L.N.*, 39: 48. Hyde Park!, *Allen Fl.*

R. cardiophyllus Muell. & Lefèv.
Native on heaths, waste ground, in woods, hedgerows, etc. Common.
7. Bombed site near Holborn, 1950,† *Whittaker* (BM), teste *Watson.* East Heath, Hampstead. Ken Wood. Highgate Cemetery. Regents Park. All det. *Watson.*

Sect. Discolores P. J. Muell.
R. ulmifolius Schott f.
Native on heaths, waste ground, in hedgerows, woods, thickets, etc. Very common.
7. Frequent in the district.

R. ulmifolius × vestitus
4. West Heath, Hampstead, det. *Watson.*

R. pseudobifrons Sudre ex Bouv.
Native in woods, thickets, hedgerows, on heaths, waste ground, etc. Very common.
7. East Heath, Hampstead. Ken Wood. Highgate Cemetery. Victoria Park. Isle of Dogs. All det. *Watson.*

R. winteri P. J. Muell. ex Focke
Native in woods, hedgerows, etc. Very rare, or overlooked.
1. Park Wood, Ruislip, 1946, *L.N.* 26: 67. **2.** East Bedfont, *J. G. Dony*!, det. *Watson.* **5.** Near Hanwell, det. *Watson.*

R. discolor Weihe & Nees Himalayan Blackberry.
R. procerus P. J. Muell., *R. armeniacus* Focke
Alien. Asia Minor, etc. Bird-sown from gardens, and naturalised in woods, on hedgebanks, etc., and established on waste ground, railway banks, etc. Common.
7. Common in the area.

R. falcatus Kalt.
R. thyrsoideus auct.
Native in woods, etc. Very rare.
1. Ruislip, *Watson.*

R. hylophilus Rip. ex Genev.
R. brittonii Barton & Riddelsd.
Native in woods and thickets, on heaths, etc. Rather rare.
1. Park Wood, Ruislip, 1936; Poor's Field, Ruislip, *L.N.* 26: 67. Ruislip Common, *K. & L.*, 81. **2.** East Bedfont, *Woodhead*!; Staines, det. *Watson*. **3.** Hounslow Heath, *K. & L.*, 81. **5.** Hanger Hill, Ealing, *K. & L.*, 81.

Sect. Sprengeliani Focke
R. sprengelii Weihe
Native on heaths, etc. Rare.
1. Harrow Weald Common!; Stanmore Common, *L.N.* 26: 67. **4.** Hampstead Heath, *L.N.* 26: 67. Scratch Wood, *K. & L.*, 76; 1964, *Miles* (CGE).

[**R. splendidus** Muell. & Lefèv.
1. Park Wood, Ruislip, *L.N.* 26: 67. Error, fide *Watson*.]

Sect. Appendiculati Genev.
R. lasiostachys Muell. & Lefèv.
R. surrejanus Barton & Riddelsd., *R. hirtior* W. Wats., *R. leucanthemus* auct.
Native in woods and thickets, on heaths, hedgebanks, etc. Rare.
1. Park Wood, Ickenham!, *Benbow* (BM), det. *Watson*. Between Uxbridge Common and Swakeleys, *L.N.* 26: 67. **4.** Hampstead Heath, one bush, 1945, *L.N.* 26: 67.

R. condensatus P. J. Muell.
R. densiflorus Gremli
Native on heaths. Very rare.
1. North end of Harrow Weald Common!, 1941, *Watson* (CGE); *L.N.* 26: 67. The only known British locality for the species.

R. vestitus Weihe
R. diversifolius Lindl., *R. leucostachys* auct.
Native on heaths, in hedgerows, etc. Rather common.

Var. **roseiflorus** Boul.
4. Hampstead Heath, one bush, 1945, *L.N.* 26: 67.

Var. **albiflorus** Boul.
R. leucanthemus P. J. Muell.
1. Common. **3.** Cranford, det. *Watson*. **4.** Hampstead Heath!, *L.N.* 26: 67. Scratch Wood. Harrow. Both det. *Watson*. **5.** Northolt. Horsendon Hill. Perivale. All det. *Watson*. **6.** Hadley Common. Enfield. Both det. *Watson*.

R. leucostachys Schleich. ex Sm.
Native on heaths. Very rare.

1. Harrow Weald Common, *L.N.* 26: 67. **4.** East of Scratch Wood, 1964, *Miles* (CGE).

R. macrothyrsus Lange
Native in woods and thickets, on heaths, etc. Rather rare.
1. Harrow Weald Common; Mimmshall Wood, *L.N.* 26: 67. **4.** Hampstead Heath, *L.N.* 26: 67. Scratch Wood, *K. & L.*, 82. **5.** Hanger Hill, Ealing, *K. & L.*, 82. **6.** Enfield Chase, det. *Watson.* **7.** Ken Wood, det. *Watson.*

R. criniger (E. F. Linton) Rogers
R. gelertii var. *criniger* E. F. Linton
Native in woods. Rare.
1. Copse Wood, Ruislip; Uxbridge Common, *L.N.* 26: 67. Mad Bess Wood, Ruislip, *K. & L.*, 82. **6.** Botany Bay, Enfield Chase, *Benbow* (BM), det. *Watson.*

R. cinerosus Rogers
Native in hedgerows. Very rare.
4. Barn Hill, Wembley Park, 1933, *L.N.* 26: 68; 1956, det. *Woodhead.*

R. atrichantherus E. H. L. Krause
Native in woods. Very rare.
1. West side of Bayhurst Wood!, *K. & L.*, 83.

R. hypomalacus Focke
R. macrophyllus var. *velutinus* Weihe & Nees, *R. hanseni* E. H. L. Krause, *R. mucronatoides* A. Ley
Native in woods. Very rare.
1. Bayhurst Wood, 1933, *L.N.* 26: 67. Error, teste *Watson.* Mimmshall Wood, on the Middlesex–Herts boundary, a rather glandular form with pink flowers, *L.N.* 26: 67.

R. chaerophyllus Sag. & Schultze
Native in hedgerows. Very rare, and perhaps extinct.
6. Arnos Grove, one bush, 1932, *L.N.* 26: 68.

R. gremlii Focke
Native in woods. Very rare.
1. Copse Wood, Ruislip, *K. & L.*, 81. The only known British station; 1950, *Woodhead* (CGE).

R. radula Weihe ex Boenn.
R. decipiens P. J. Muell.
Native in woods and hedgerows, on heaths, etc. Common.
1. Ickenham, *Benbow* (BM), det. *Watson.* Near Ruislip Reservoir; between Uxbridge Common and Swakeleys; Harrow Weald Common, *L.N.* 26: 68. **3.** Hounslow Heath, *L.N.* 26: 68. **4.**

Scratch Wood, *Benbow* (BM), *Watson*; 1948, *K. & L.*, 83. **5.** Horsendon Hill, *L.N.* 26: 67.

R. sectiramus W. Wats.
Native in woods and thickets, on heaths, etc. Local.
1. Pinner Wood, *L.N.* 26: 68. **3.** Hounslow Heath, *L.N.* 26: 68. **4.** Hampstead Heath, *Rogers* (BM), det. *Watson*; 1945, *L.N.* 26: 68. Hanwell, frequent, det. *Watson*.

R. echinatus Lindl.
R. discerptus P. J. Muell.
Native in woods and hedgerows, on heaths, etc. Common.
1. Park Wood, Ruislip; near Ruislip Reservoir; near Swakeleys; Pinner Wood, *L.N.* 26: 68. Old Park Wood!, and elsewhere about Harefield, *Watson* and *Avery*. Springwell, det. *Watson*. **3.** Hounslow Heath, *L.N.* 26: 68. **4.** Harrow, det. *Watson*. **5.** Horsendon Hill, *L.N.* 26: 68. Ealing. Hanwell. Both det. *Watson*.

R. echinatoides (Rogers) Sudre
R. radula var. *echinatoides* Rogers
Native in woods, thickets and hedgerows. Local.
1. Near Swakeleys; Harrow Weald Common; Mimmshall Wood, *L.N.* 26: 68. Uxbridge Common; Mad Bess Wood, Ruislip, *K. & L.*, 84. Northwood; Harefield; Springwell, *Watson* and *Avery*. **5.** Horsendon Hill, one bush, 1946, *L.N.* 26: 68. **6.** Enfield Chase, det. *Watson*.

R. aspericaulis Muell. & Lefèv.
Native in thickets. Very rare.
5. Horsendon Hill, *K. & L.*, 83. **6.** Between Barnet and Elstree, 1952, *Mills* (CGE), conf. *Watson*.

R. rudis Weihe & Nees
Native in woods, on heaths, etc. Rare.
1. Bayhurst Wood, Ruislip, *L.N.* 26: 68. **4.** Hampstead Heath, *L.N.* 26: 68. Scratch Wood, *K. & L.*, 84.

R. radulicaulis Sudre
Native in woods. Very rare.
1. West side of Bayhurst Wood, *K. & L.*, 84.

R. prionodontus Muell. & Lefèv.
Native in woods. Very rare.
1. East side of Bayhurst Wood, 1964, *Miles* (CGE).

R. granulatus Muell. & Lefèv.
R. radula var. *bloxamianus* Coleman ex Purchas
Native in woods and hedgerows. Rare.

1. Near Uxbridge Common, *Benbow* (BM), det. *Watson*. Near Swakeleys; Potter Street, Pinner; Pinner Wood, *L.N.* 26: 68. **6.** Arnos Grove, 1932,† *Watson*.

R. foliosus Weihe & Nees
R. flexuosus Muell. & Lefèv., *R. saltuum* Focke, *R. hyposericeus* Sudre
Native in woods and hedgerows, on heaths, etc. Local.
1. Park Wood, Ruislip, abundant!; Uxbridge Circus, *L.N.* 26: 68. Park Wood, Ickenham, *Benbow* (BM), det. *Watson*. Mad Bess Wood, Ruislip; Copse Wood, Northwood, *K. & L.*, 85. Old Park Wood, Harefield, *Watson* and *Avery*.

R. subtercanens W. Wats.
Native in woods. Very rare.
1. Copse Wood, Ruislip, 1964, *Miles* (CGE).

R. teretiusculus Kalt.
Native in woods. Very rare.
1. Outside the north-east end of Copse Wood, Ruislip, *K. & L.*, 86; 1959, *Woodhead* (CGE).

R. bloxamii Ed. Lees
R. pallidus subsp. *bloxamii* (Ed. Lees) Sudre, *R. multifidus* Boul. & Malbr.
Native in woods. Very rare.
1. Old Park, Harefield, 1889, *Benbow* (BM), det. *Watson*; 1948, *K. & L.*, 85.

R. acutipetalus Muell. & Lefèv.
Native in woods and on heaths. Very rare.
1. Mad Bess Wood; Ruislip Common, with white flowers, *K. & L.*, 85.

R. trichodes W. Wats.
R. foliosus auct., *R. hirtus* subsp. *rubiginosus* Rogers *pro parte*
Native in woods and hedgerows. Local.
1. Mad Bess Wood, Ruislip, *Benbow* (BM), det. *Watson*; 1948, *K. & L.*, 85. Park Wood, Ruislip, abundant; Uxbridge Circus, *L.N.* 26: 68. Old Park, Harefield, *Watson* and *Avery*. Northwood, *K. & L.*, 85. Poor's Field, Ruislip, 1964, *Miles* (CGE). **5.** Perivale Wood!, *Benbow* (BM), det. Watson.

R. adamsii Sudre
R. babingtonii var. *phyllothyrsus* Rogers *pro parte*.
Native in woods. Local.
1. Copse below Uxbridge Common, *Benbow* (BM), det. *Watson*. Park Wood; Mad Bess Wood, Ruislip; east of Harefield Park, *K. & L.*, 85.

R. pallidus Weihe & Nees
R. cernuus P. J. Muell.
Native in moist woods. Local.
1. Near Uxbridge Common, 1902, *Benbow* (BM), teste *Watson*; 1948, *Watson* and *Avery*, *K. & L.*, 86. Bayhurst Wood, 1889, *Benbow* (BM), det. *Watson*. Park Wood, Ruislip; Swakeleys, abundant, *L.N.* 26: 68.

R. drymophilus Muell. & Lefèv.
Native in woods and thickets, on heaths, etc. Very rare.
1. Park Wood, Ruislip, *K. & L.*, 85. **5.** Hanger Hill, Ealing; Horsendon Hill, *K. & L.*, 85. **7.** East Heath, Hampstead, *K. & L.*, 85.

R. argutifolius Muell. & Lefèv.
R. glareosus E. S. Marshall ex Rogers, *R. monachus* auct.
Native on heaths. Very rare.
4. Hampstead Heath, *K. & L.*, 86.

R. euryanthemus W. Wats.
R. pallidus var. *leptopetalus* Frid. ex Rogers
Native in woods and thickets, on heaths, etc. Local.
1. Park Wood, Ruislip; Pinner Hill, *L.N.* 26: 68. **4.** Hampstead Heath, *Benbow* (BM); 1945, *Avery* and *Watson*, *L.N.* 26: 68; 1962, *Miles* (CGE). Scratch Wood, *K. & L.*, 86. **6.** Hadley Wood, *L.N.* 26: 68. Cherry Tree and Queen's Woods, Highgate. Highgate Wood. All det. *Watson*. **7.** Ken Wood, teste *Watson*.

R. insectifolius Muell. & Lefèv.
R. fuscus var. *nutans* Rogers, *R. nuticeps* Barton & Riddelsd.
Native in hedgerows. Local.
1. Abundant between Uxbridge Circus and Swakeleys, *K. & L.*, 86.

R. microdontus Muell. & Lefèv.
Native on heaths, in woods, etc. Very rare.
1. Top Wood, north of Harefield, *Benbow* (BM), det. *Watson*. Lower part of Stanmore Common, *L.N.* 26: 68.

R. thyrsiflorus Weihe & Nees
R. hirtus subsp. *flaccidifolius* Rogers *pro parte*
Native in woods and hedgerows. Local.
4. Edgwarebury, abundant!, *K. & L.*, 87. **6.** Arnos Grove, 1932, *Watson*.

R. scaber Weihe & Nees
R. dentatus (Bab.) Bloxam
Native in woods. Local.
1. West of Ruislip Reservoir; Park Wood, Ruislip, *L.N.* 26: 68. Copse! and Mad Bess Woods, Ruislip, *K. & L.*, 87. Old Park Wood, Harefield, 1947, *Avery* and *Watson*.

R. rufescens Muell. & Lefèv.
R. velatus Lefèv., *R. rosaceus* subsp. *infecundus* (Rogers) Rogers
Native in woods, on heaths, etc. Common.
7. Ken Wood. Highgate Cemetery. Both det. *Watson.*

R. grypoacanthus Muell. & Lefèv.
Native in woods. Very rare.
1. Copse Wood, Ruislip, *K. & L.*, 346; 1953, *Watson* (CGE).

R. apiculatus Weihe & Nees
R. anglosaxonicanus Gelert, *R. curvidens* A. Ley
Native in woods. Very rare.
1. North-east of Ruislip Reservoir, one bush, 1932, *L.N.* 26: 68.

R. melanoxylon Muell. & Wirtg.
Native in woods. Very rare.
1. Copse Wood, Ruislip, *K. & L.*, 89.

R. phaeocarpus W. Wats.
R. babingtonii auct.
Native in hedgerows. Very rare.
1. Near Swakeleys, *L.N.* 26: 68.

[**R. euanthinus** W. Wats.
R. apiculatus var. *vestitiformis* (Rogers) Riddelsd.
1. Park Wood, Ruislip, *L.N.* 26: 68. Error, fide *Watson.*]

R. leightonii Ed. Lees ex Leighton
R. ericetorum Lefèv. ex Genev.
Native in woods. Very rare.
1. Old Park Wood, Harefield, *Watson* and *Avery.*

R. disjunctus Muell. & Lefèv.
R. moylei Barton & Riddelsd., *R. cuneatus* (Rogers & Ley) Druce, *R. ericetorum* auct.
Native on heaths. Very rare.
4. Hampstead Heath, *K. & L.*, 346.

R. diversus W. Wats.
R. hirtus subsp. *kaltenbachii* Rogers *pro parte*
Native in woods. Very rare.
6. Hadley Wood, *L.N.* 26: 68.

R. hostilis Muell. & Wirtg.
Native in woods and on heaths. Local.
1. Ruislip; Pinner; Northwood; near Swakeleys; Harrow Weald Common, *L.N.* 26: 68. Harefield, abundant, *Watson* and *Avery*, *L.N.* 27: 31. Poor's Field, Ruislip, 1964, *Miles* (CGE).

R. hostilis × ulmifolius
1. Roadside by Mad Bess Wood, Ruislip, in two places, *K. & L.*, 89.

R. rosaceus Weihe & Nees
Native in woods and thickets. Rather common.
1. Park Wood, Ruislip, *L.N.* 26: 68. 4. Headstone Farm;† between Wood End and Harrow, *Benbow* (BM), det. *Watson*, Hampstead Heath, *L.N.* 26: 68. Mill Hill, det. *Watson*. 5 Perivale Wood!, *Benbow* (BM), det. *Watson*. Horsendon Hill, abundant!, *L.N.* 26: 68; 1964, *Miles* (CGE). Hanwell, det. *Watson*.

Sect. Glandulosi P. J. Muell.
R. murrayi Sudre
R. rosaceus subsp. *adornatus* Rogers, *R. histrix* var. *adornatus* auct.
Native in woods and on heaths. Local.
3. Hounslow Heath, *K. & L.*, 92. 4. Hampstead Heath, *L.N.* 26: 68. Scratch Wood, *K. & L.*, 92.

R. ochrodermis Ley
Native in thickets, etc. Very rare.
5. Horsendon Hill, 1964, *Miles* (CGE).

[**R. coronatus** Boulay var. **cinarescens** W.Wats.
1. Duck's Hill, Northwood, *Benbow* (BM), det. *Watson*, *K. & L.*, 92. Error, fide *Watson*, *K. & L.*, 346.]

R. pygmaeopsis Focke
Native in woods. Very rare.
1. Park Wood, Ruislip, *K. & L.*, 91.

R. adornatus P. J. Muell. ex Wirtg.
Native in woods. Very rare.
1. Park Wood, Ruislip, *L.N.* 26: 68. Copse Wood, Ruislip, *K. & L.*, 91; 1950, *Woodhead*; 1951, *Mills* (CGE).

R. spinulifer Muell. & Lefèv.
R. tumulorum Rilstone, *R. koehleri* auct.
Native in thickets. Very rare.
5. Brabsden Green, one bush, 1932,† *L.N.* 26: 68. Horsendon Hill; Hanger Hill, Ealing, *K. & L.*, 91; *Watson* (CGE).

R. spinulifer × ulmifolius
5. Hanger Hill, Ealing, *K. & L.*, 91.

R. apricus Wimm. var. **sparsipulus** W.Wats.
Native in woods. Local.
1. Mimmshall Wood, *L.N.* 26: 68. Park Wood, Ruislip, *K. & L.*, 91. 4. Scratch Wood, Edgwarebury, *K. & L.*, 91; 1964, *Miles* (CGE).

R. hylocharis W. Wats.
R. rosaceus var. *silvestris* R. P. Murray ex Rogers
Native on heaths and in woods. Local.
4. Hampstead Heath, *L.N.* 26: 68; 1962, *Miles* (CGE). **7.** Ken Wood, *L.N.* 26: 68.

R. histrix Weihe & Nees
Native in woods. Very rare.
1. Park Wood, Ruislip, *K. & L.*, 90.

R. infestus Weihe ex Boenn.
R. setulosus Muell. & Lefèv.
Native on heaths. Very rare.
3. Hounslow Heath, *L.N.* 26: 68.

R. adenolobus W. Wats.
R. cognatus N. E. Brown
Native in woods and hedgerows, on heaths, etc. Rather common.
1. Park Wood, Ickenham; Duck's Hill, Northwood, *Benbow* (BM), det. *Watson*. Ruislip, frequent; Harrow Weald Common, *L.N.* 26: 68. Mimmshall Wood, *Benbow* (BM), det. *Watson*; 1945, *L.N.* 26: 68. Harefield, *Watson* and *Avery*. **3.** Whitton Park, *Benbow* (BM), det. *Watson*. **4.** Hampstead Heath; Harrow, *Benbow* (BM), det. *Watson*. **7.** Ken Wood, det. *Watson*.

R. koehleri Weihe & Nees subsp. **dasyphyllus** Rogers
Native in hedgerows and woods, on heaths, etc. Common.
1. Park Wood, Ruislip, *Benbow* (BM), det. *Watson*; 1943; near Swakeleys, *L.N.* 26: 68. Harefield; Northwood, *Watson* and *Avery*. **3.** Whitton Park, *Benbow* (BM), det. *Watson*. Hounslow Heath, *L.N.* 26: 68. **4.** Hampstead Heath, *L.N.* 26: 68. **6.** Hadley Wood, *L.N.* 26: 68.

R. marshallii Focke & Rogers
Native in woods. Very rare.
1. Park Wood, Ickenham, *Benbow* (BM), det. *Watson*. Near Swakeleys, one bush, 1933, *L.N.* 26: 68.

R. fuscoater Weihe & Nees
Native on heaths. Very rare.
4. Hampstead Heath, one bush, *L.N.* 26: 68.

R. oigocladus Muell. & Lefèv.
Native in woods. Very rare.
1. Park Wood, Ruislip, *K. & L.*, 90.

R. hirtus Waldst. & Kit.
Native in woods and thickets. Local.
5. Horsendon Hill, 1848, *Lees* (SLBI); 1899, *Benbow* (BM), both det.

Watson; abundant, 1946!; Brabsden Green, 1932, *L.N.* 26: 68. Perivale Wood!, *Benbow* (BM), det. *Watson.*

R. guentheri Weihe & Nees
R. hirtus subsp. *guentheri* (Weihe) Sudre
Native in woods. Rare.
1. Copse and Mad Bess Woods, Ruislip, 1950, *Woodhead*; 1951, *Watson* (CGE); *K. & L.*, 93.

R. leptadenes Sudre
R. serpens subsp. *leptadenes* Sudre, *R. echinatus* P. J. Muell., non Lindl.
Native in woods. Very rare.
1. Park Wood, Ruislip, *Avery*, det. *Watson*, *L.N.* 27: 31.

R. analogus Muell. & Lefèv.
Native in woods. Very rare.
1. Bayhurst Wood, Ruislip, 1948, *Watson* (CGE); *K. & L.*, 346; 1958, *Woodhead*; 1964, *Miles* (CGE). The only known British locality for the species.

POTENTILLA L.

P. palustris (L.) Scop. Marsh Cinquefoil.
Comarum palustre L.
Introduced. Formerly naturalised in a bog, but now long extinct.
First record: *Blackstone*, c. 1734. Last record: *T. Martyn*, 1807.
4. Hampstead Heath, planted there by Mr Rand in the bogs, *Blackst. MSS.*; c. 1760, *J. Hill* (BM); a few plants there, *Mart. Mill. Dict.*

P. sterilis (L.) Garcke Barren Strawberry.
P. fragariastrum Ehrh.
Native in woods, on hedgebanks, heaths, etc., and established on railway banks, etc. Locally common, but apparently decreasing, and absent from the immediate vicinity of the Thames and inner London.
First record: *T. Johnson*, 1638.
1. Common. **3.** Whitton,† *T. & D.*, 94. Southall. **4.** Harrow!, *Melv. Fl.*, 27. Bishop's Wood. West Heath, Hampstead,† *T. & D.*, 94. Hampstead,† *Pet. H.B.C.* Mill Hill! (HE). Scratch Wood. Edgware. **5.** Near Lampton,† *T. & D.*, 94. Horsendon,† *A. Wood*, *Trim. MSS.* Railway banks, Southall to Brentford. Osterley Park. **6.** Hadley!, *Warren*; Colney Hatch!; Edmonton, *T. & D.*, 94. Oakleigh Park, *J. G. Dony*! Enfield!, *Bacot* (BM). Highgate Woods. **7.** Tottenham,† *Johns. Cat.* Islington,† *Ball. MSS.* West End Lane,† *T. & D.*, 94.

P. anserina L. Silverweed.
Native by roadsides, on damp waste ground, etc., and established on railway banks, etc. Common.

First record: *T. Johnson*, 1638.

7. Tottenham!, *Johns. Cat.* Isle of Dogs!; Thames embankment, *T. & D.*, 92. Hyde Park!; Kensington Gardens, *D. E. Allen*; Buckingham Palace grounds. Hammersmith. Brompton. St John's Wood. Holloway. Finsbury Park. Stoke Newington. Hackney Wick.

P. argentea L. Hoary Cinquefoil.

Formerly native in dry grassy places on sandy and gravelly soils, but now extinct.

First record: *Blackstone*, 1737. Last record: *J. E. Cooper*, 1916.

1. Harefield, *Blackst. Fasc.*, 74. Uxbridge Common, 1888, *Benbow*; 1892, *Slatter* (BM). Yiewsley, 1901, *Coop. Cas.*; 1916, *J. E. Cooper*, *K. & L.*, 101. Between Potters Bar and North Mimms, *Phyt. N.S.* 1: 407. **2.** Between Staines and Hampton, 1866; near Hampton Court, *Trimen* (BM). Ashford, one plant, *Phyt. N.S.* 4: 264. Near Strawberry Hill, *T. & D.*, 93. Teddington, 1914, *Flippance* (K). Near Sunbury, 1842, *R. S. Hill* (BM). **3.** Isleworth, 1845, *Cubitt* (K). **4.** Hampstead Heath, *Coop. Fl.*, 99. **5.** Chiswick, *Fox*; near Hanwell, *Warren*, *T. & D.*, 93; 1884, *Benbow* (BM). **7.** Highgate, 1887–88, *Coop. Cas.* Regents Park, *Webster Reg.* Probably an error. Isle of Dogs, *Benbow MSS.*

P. recta L.

Alien. Europe, etc. Naturalised in grassy places, etc. Very rare, and often only casual.

First record: *A. Irvine*, 1858.

1. Hillingdon, 1889, *Benbow* (BM). Northwood, 1954, *Graham.* **2.** Teddington. 1905, *B.B. Thompson* (K). **4.** Temple Fortune, 1915, *J.E. Cooper* (K). Golders Hill, 1958, *Hinson.* **5.** Ealing, 1905, *Druce Notes.* Error, the specimen (D) on which the record was based is *P. intermedia.* Ealing, 1935, *Bull*, *K. & L.*, 102. Bedford Park, *Cockerell Fl.* Well established on railway bank near Acton Town, 1965!, *L.N.* 45: 20; 1968. **7.** Parsons Green, *Irv. Ill. Handb.*, 624; 1862, *Britten*, *J.B.* (*London*) 1: 375. West India Docks, *Cherry*, *T. & D.*, 92. Near British Museum, Bloomsbury, *Shenstone Fl. Lond.* Hyde Park, 1945, *Welch*, *L.N.* 26: 60. Regents Park, *Holme*, *L.N.* 39: 49.

P. erecta (L.) Räusch. subsp. **erecta** Common Tormentil.

P. tormentilla Neck.

Lit. *Biol. J. Linn. Soc.* **1**, appendix 1: x (1969): *New Phyt.* **69**: 171 (1970).

Native on heaths, commons, in grassy places, dry woods, etc., especially on acid soils. Locally common.

First record: *T. Johnson*, 1629.

7. East Heath, Hampstead!, *T. & D.*, 93. Ken Wood. Bombed site, New Oxford Street, W.C.1, 1950,† *Whittaker* (BM). Kensington Gardens, 1958, *D. E. Allen*, *L.N.* 39: 49. Error, *D. E. Allen*, *L.N.* 41: 18.

P. erecta × **reptans** = ? **P.** × **mixta** Nolte ex Reichb.
P. × *italica* Lehm.
Lit. *New Phyt.* **69**: 171 (1970).
7. Between Hampstead and Hornsey, 1865 (SY). Bombed site, Bedford Way, W.C., 1950,† *Whittaker*; in grass in front of Natural History Museum, South Kensington, 1943,† *Wilmott* (BM), both det. *D. E. Allen*. Buckingham Palace grounds, *Fl. B.P.G.* Hyde Park!; Kensington Gardens!, *Allen Fl.*; 1966. Zoological Gardens, Regents Park, 1966!, *Holme, L.N.* 46: 30. Hackney, 1968.
This extremely variable putative hybrid is common throughout the vice-county and frequently occurs in the absence of both parents. *Matfield et al.* (*New Phyt.* 69: 171 (1970)) point out that as experimental studies showed that fertile back cross progenies were raised from both *P.* × *suberecta* and *P.* × *mixta* the possibility of widespread introgression in nature cannot be ruled out.

P. anglica Laicharding Trailing, or Creeping Tormentil.
P. procumbens Sibth.
Native on heaths, in dry grassy places, open woods, etc. Very rare, and apparently frequently confused with *P. erecta* × *reptans.*
First record: *Petiver*, 1695.
1. Uxbridge; Harefield; Ruislip; Northwood; Mimmshall Wood; Warren Gate, *Benbow MSS.* Harrow Weald Common, *C. S. Nicholson, K. & L.*, 102. **3.** Hounslow Heath!, *K. & L.*, 102. Error, the plant seen was *P. erecta* × *reptans.* **4.** Hampstead, *T. & D.*, 93; 1926, *R. W. Robbins, K. & L.*, 102. Edgwarebury, *Benbow MSS.* **5.** Horsendon Hill, *T. & D.*, 93. **6.** Winchmore Hill Wood,† *T. & D.*, 93. **7.** Near Islington, *Pet. Midd.* South Kensington, 1943, *Wilmott, Rep. Bot. Soc. & E.C.* **12**: 718. Error, the specimen (BM) on which the record was based is *P. erecta* × *P. reptans*, det. *D. E. Allen.*
It is possible that most, if not all, of the above records are referable to *P. erecta* × *reptans.* This suggestion is strengthened by the fact that *P. erecta* grows, or formerly grew, in almost all of the localities given.

[**P. anglica** × **reptans** = **P.** × **suberecta** Zimmeter
3. Hounslow Heath!, *K. & L.*, 102. Error, the plant seen was *P. erecta* × *reptans.* **4.** Stonebridge Park, *Druce Notes.* Error.]

P. reptans L. Creeping Cinquefoil.
Native on waste ground, in fields, by roadsides, etc., and established on railway tracks, etc. Very common.
First record: *Gerarde*, 1597.
7. Common throughout the district.
Var. **microphylla** Tratt.
1. Harefield!, *L.N.* 26: 60. **3.** Wood End. **7.** Hyde Park (DILL).
Plants with semi-double flowers were collected at **3.** Isleworth, 1893, *Wolley-Dod* (BM).

FRAGARIA L.

F. vesca L. Wild Strawberry.

Native in woods and shady places, on hedgebanks, etc., and established on railway banks, etc. Locally plentiful, except near the Thames, where it is very rare, and in heavily built-up areas where owing to the eradication of its habitats it is extinct.

First record: *Merrett*, 1666.

1. Common. **2.** Hampton Court; Strawberry Hill,† *T. & D.*, 94. **3.** Harrow Grove,† *Melv. Fl.*, 26. **4.** Harrow Park!, *Melv. Fl.*, 26. Stanmore!, *Warren*; Wembley Park,† *Farrar*; Deacon's Hill;† Bishop's Wood, *T. & D.*, 94. Edgware!, *Benbow MSS.* Scratch Wood. Hendon. **5.** Wormholt (= Wormwood) Scrubs, 1815† (G & R). Osterley Park. Perivale;† Horsendon Hill!,† *Benbow MSS.* **6.** Hadley!, *Warren*; near Enfield!, *Tucker*; Winchmore Hill Wood,† *T. & D.*, 94. Enfield Chase!, *Benbow MSS.* **7.** Hampstead Heath, 1780, *Pamplin MSS.*; East Heath, Hampstead, 1949, *H. C. Harris.* Hyde Park,† *Merr. Pin.*, 39. Regents Park, 1966. Holloway, 1965, *D. E. G. Irvine.*

Var. **sylvatica** Rupr.

5. Near Hanwell,† *Druce Notes.*

F. chiloensis (L.) Duchesne × **virginiana** Duchesne = **F.** × **ananassa** Duchesne Garden Strawberry.

F. grandiflora Ehrh.

Alien of hortal origin. Escape, or outcast, from cultivation. Established on railway banks, etc. Rare.

First record: *D. H. Kent*, 1951.

1. Uxbridge. **2.** Kempton Park. Laleham Park, *Kennedy*!. **4.** Harrow, *Harley Add.* Mill Hill, *Kennedy*!. **5.** Brentford. Syon Park. **6.** Edmonton, *Kennedy.* **7.** Bombed sites, Cripplegate,† *Jones Fl.* Hyde Park, a single plant, 1961, *Allen Fl.*

GEUM L.

G. urbanum L. Herb Bennett, Wood Avens.

Native on hedgebanks, in woods, shady places, etc. Very common.

First record: *T. Johnson*, 1638.

7. East Heath, Hampstead!, *H. C. Harris.* Highgate Ponds. Hyde Park. Kensington Gardens!, *D. E. Allen, L.N.* 39: 49. Holloway.

G. rivale L. Water Avens.

Native in wet meadows, ditches, etc. Very rare, and almost extinct.

First record: *Lightfoot*, c. 1780.

2. 'In a boggy meadow on the left hand of the road from Colnbrook to

London on the skirts of the county of Middlesex, about a mile from the Great Turnpike (= Bath) Road, and about a mile from Colnbrook', *Lightf. MSS.*; 1910, *Wallis, Druce Notes*; near Perry Oaks, several small colonies in a ditch, 1953!, *Briggs, L.N.* 38: 21; 1960.

AGRIMONIA L

A. eupatoria L. Common Agrimony.

Native on hedgebanks, field and wood borders, by roadsides, etc., established on railway banks, etc. Locally abundant, especially on calcareous soils at Harefield, Springwell and South Mimms, and on alluvial soils near the Thames.

First record: *Morley*, 1677.

1. Common. **2.** Common. **3.** Twickenham, *T. & D.*, 91. Hounslow Heath. Southall. **4.** Hampstead!, *Morley MSS.* Harrow!, *Melv. Fl.*, 27. Wembley!, *A. B. Cole*; between Whitchurch and Stanmore!, *T. & D.*, 91. Harrow Weald. Edgware. Scratch Wood. Mill Hill. Hendon. Hatch End, Kenton. **5.** Greenford!, *Coop. Fl.*, 108. Hanwell,† *Newbould*; Alperton!; Lampton, *T. & D.*, 91. Wyke Green. Osterley Park. Railway banks, Southall to Brentford. Syon Park. Perivale. Horsendon Hill. Ealing. Acton. North Acton. Park Royal. Bedford Park!, *Cockerell Fl.* Near Chiswick. Strand-on-the-Green. **6.** Edmonton!; Enfield!, *T. & D.*, 91. Southgate. Near Hadley Wood station, abundant, *Kennedy*!. **7.** Hornsey,† *Ball. MSS.* Near Fortune Green,† *Trim. MSS.* Highgate Ponds!, *H. C. Harris*. Hyde Park!, *Watsonia* 1: 297. Kensington Gardens!, *D. E. Allen, L.N.* 39: 49. Weed of flower-bed, Lincoln's Inn Fields, W.C. 2, 1965, *Burton, L.N.* 46: 30. Holloway, 1965, *D. E. G. Irvine*.

A. procera Wallr. Fragrant Agrimony.

A. odorata auct., non Mill.

Formerly native on hedgebanks but now extinct.

First, and only evidence: *Montgomrey*, 1874.

2. Littleton, 1874, *Montgomrey* (MY).

ALCHEMILLA L.

Lit. *Watsonia* **1**: 6 (1949).

A. filicaulis Buser subsp. **vestita** (Buser) M. E. Bradshaw Lady's-Mantle.

A. vestita (Buser) Raunk., *A. pseudominor* Wilmott, *A. minor* auct., non Huds.

Native in damp grassy places. Very rare, and confined to the north-west part of the vice-county.

First certain evidence: *Kingsley*, 1839.

1. Harefield!, *Kingsley* (ME). Meadows and moors from Denham to Harefield Moor Lock!; Harefield Moor!, *Benbow* (BM); still locally plentiful in both localities, 1970. Between Ruislip Reservoir and Harefield, 1866,† *Griffiths*; near Uxbridge, 1885, *Benbow* (BM). Harefield Grove, 1966, *I. G. Johnson.*

APHANES L.

Lit. *Watsonia* **1**: 163 (1949).

A. arvensis L. Parsley Piert.
Alchemilla arvensis (L.) Scop.
Native on cultivated and waste ground, heaths, etc. Chiefly on light soils. Rare.
First certain evidence: *E. Ballard*, 1839.
1. Harefield!, *K. & L.*, 104. Ruislip. Northwood. South Mimms. Potters Bar, 1970, *Widgery*. **2.** Harmondsworth!; Staines!, *K. & L.*, 104. Stanwell. Laleham. Near Kempton Park, *Tyrrell*!. **3.** Hanworth, 1884, *Wright* (BM). Hounslow Heath. **4.** Near Edgware, 1871, *F. J. Hanbury* (HY). **6.** North Finchley, 1875, *French* (BM). Trent Park, 1966, *T. G. Collett* and *Kennedy*. **7.** Newington, 1839,† *E. Ballard* (BM). Buckingham Palace grounds, *Codrington*, *Lousley* and *McClintock*!, *L.N.* 39: 49. Error, the plant seen was *A. microcarpa.* Hurlingham, *E. Young*.

A. microcarpa (Boiss. & Reut.) Rothm. Parsley Piert.
Native on heaths, dry grassy places, etc., chiefly on light soils. Rare.
First certain evidence: *Twining*, 1842.
1. Ruislip Common!, *K. & L.*, 104; 1967. Uxbridge Common. Disused railway between Uxbridge and Denham, 1966. Harefield!, *I. G. Johnson.* **2.** Hampton Court Park!, *Sandwith*; 1968. **3.** Near Whitton, 1842, *Twining* (BM). Hounslow Heath!, *K. & L.*, 104. **5.** Syon Park. **6.** Enfield Chase (BO). **7.** Buckingham Palace grounds, *Codrington*, *Lousley* and *McClintock*!, *L.N.* 39: 49 (as *A. arvensis*). Hyde Park; Kensington Gardens, 1962, *Allen Fl.* Tower Hill, 1963, *Brewis*, *L.N.* 44: 22. Weed of flower-beds, Zoological Gardens, Regents Park, 1966, *Brewis*, *L.N.* 46: 30. Fulham Palace grounds, 1966, *Brewis*. No doubt an introduction in some of the metropolitan localities.

SANGUISORBA L.

Lit. *Proc. Bot. Soc. Brit Isles* **7**: 421 (1968). (*Abstr.*)

S. officinalis L. Great Burnet.
Native in damp grassy places, waste ground, etc., on heavy soils, and established on railway banks, etc. Rare, and decreasing.
First record: *W. Turner*, 1548.

1. Uxbridge, 1913,† *P. H. Cooke* (LNHS). Probably an introduction. **4.** In field five or six miles from London on the road to Harrow, 1817 (G & R). Whitchurch, 1827–30, *Varenne*, *T. & D.*, 90; 1906, *C. S. Nicholson* (LNHS). Stanmore, 1827–30, *Varenne*, *T. & D.*, 91; one plant, 1964→, *Roebuck*. Edgware, 1894, *Pugsley* (BM); 1910; *P. H. Cooke* (LNHS); 1912, *J. E. Cooper*; Kenton,† *Benbow* (BM). Wembley Park, 1882, *Find. MSS.*; abundant, 1890, *Benbow* (BM); sparingly in fields between Wembley Park and Kingsbury, 1966. Kingsbury!, *T. & D.*, 91; Brent Reservoir, locally abundant, especially near the Edgware Road, 1973. Near Finchley, 1845, *J. Morris*; near Stonebridge!, *Wharton*, teste *Farrar*, *T. & D.*, 91; waste ground by North Circular Road in very small quantity, 1966. Plentiful by the Brent, Tokyngton Recreation Ground, near Stonebridge Park, 1968, *L. M. P. Small*. Harrow, 1961, *Knipe*. North-east of Mill Hill, 1965, *Hinson*. **5.** Syon,† *Turn. Names*. Perivale!, *Benbow* (BM); still present on railway banks at Perivale Park, 1970. Ealing, 1905, *Green* (SLBI); 1960;† site now converted into a recreation ground. Between Greenford and Hanwell, 1960;† site now built on. **7.** Between Paddington and Lysons Green (cf. Lisson Grove),† *Ger. Hb.*, 889. Tottenham,† *Johns. Cat.* Probably an error. Pancras,† *Park. Theat.*, 583 and *Baylis*, 300. Ken Wood, *Hunter*, *Park Hampst.* Error. Near Fortune Green, 1870,† *Trim. MSS.* Railway side near West Hampstead, one plant, 1948,† *Graham*, *L.N.* 28: 32.

S. minor Scop. subsp. **minor** Salad Burnet.
Poterium sanguisorba L.
Lit. *Proc. Bot. Soc. Brit. Isles* **7**: 421 (1968). (*Abstr.*)
Native in dry fields, by roadsides, etc., on calcareous soils at Harefield and Springwell, and on alluvial soils near the Thames, elsewhere in fields, on waste ground, railway banks and tracks, etc., as an introduction. Rare.
First record: *Blackstone*, 1737.
1. Harefield!, *Blackst. Fasc.*, 77. Springwell!, *Benbow* (BM). Railway banks, Uxbridge, abundant;† railway disused and site now destroyed by levelling. Railway banks near Denham. Ickenham, *Tremayne*. **2.** Between Sunbury and Hampton!, *Newbould*; Hampton; Teddington; between Teddington and Strawberry Hill,† *T. & D.*, 91. Penton Hook!; railway banks, West Drayton to Colnbrook!; railway banks between Colnbrook and Staines!, *Benbow MSS.* Longford,† *J. E. Cooper*. Stanwell Moor. Between Hampton Court and Kingston Bridge!, *Trimen* (BM). Near Littleton, *Welch*. Ashford, *Hasler*. **3.** Riverside near Richmond Bridge,† *T. & D.*, 91. Twickenham,† *Benbow MSS.* Railway bank, Northolt, 1962. **4.** Bishop's Wood,† *Lawson*, *T. & D.*, 91. Hampstead,† *Whiting Fl.* **5.** Field near Greenford, one patch, 1965. **6.** Railway side, Crew's Hill, Enfield, one

clump, 1965, *Kennedy*!. **7.** Brompton Cemetery, 1861;† Kensal Green Cemetery, 1862,† *Britten, Bot. Chron.*, 58.

Subsp. **muricata** (Gremli) Briq. Fodder Burnet.
Poterium muricatum Spach, *P. polygamum* Waldst. & Kit.
1. Between Harefield and Springwell; near Uxbridge!, *Benbow MSS.* **2.** West Drayton, *Benbow MSS.* Near Teddington, 1935. Near Harmondsworth, 1967, *Lousley*!. **3.** Near Hounslow Heath, 1960, *Buckle.*

ROSA L.

Lit. *Supplement J.B.* (*London*) **67** and **68** (1930–31).
I am greatly indebted to the late N. Y. Sandwith for considerable assistance in compiling an account of this critical genus. He kindly named all my gatherings, and provided a number of additional records based on material in the national herbaria. The records are arranged in accordance with A. H. Wolley-Dod's *Revision of the British Roses* (*Supplement J.B.* (*London*) **67** and **68** (1930–31)).

R. arvensis Huds. Field Rose.
Native on wood borders, heaths, in thickets, scrub, etc. Common, especially on heavy soils, except in heavily-built up areas, where owing to the destruction of its habitats it is extinct.
First record: *Merrett*, 1666.
7. Between Hackney and London,† *Merr. Pin.*, 105. Marylebone Fields, 1815† (G & R). East Heath, Hampstead, *T. & D.*, 104.

Var. **vulgaris** Ser.
The most common variety in the vice-county.

Forma **hispida** Lej. & Court.
1. Copse Wood, Northwood!, *K. & L.*, 105.

Forma **major** Coste
1. Harefield!, *Druce Notes.* Stanmore Common!, *K. & L.*, 105.

Var. **ovata** (Lej.) Desv.
1. Ruislip Reservoir, 1942, *Sandwith* (K). **6.** Finchley Common!, *K. & L.*, 105.

Var. **biserrata** Crép.
1. Harefield, *Green* (SLBI), det. *Sandwith.* **4.** Brent Reservoir, det. *Sandwith.*

R. multiflora Thunb.
Alien. Japan, etc. Garden outcast. Naturalised in hedges and scrub, etc. Very rare.
First evidence: *Kiddle*, 1950.

1. Watts Common, Harefield, 1960→, det. *Sandwith.* **3.** Hayes, 1950, *Kiddle* (K), det. *Sandwith.*

R. pimpinellifolia L. Burnet Rose.
R. spinosissima auct.
Formerly native on heaths and commons on sandy soils but now extinct.
First record: *Gerarde*, 1597. Last evidence: *Benbow*, 1887.
2. Teddington, *Blackst. Spec.*, 84. **3.** Hounslow Heath (DILL); c. 1830, *Wats. New Bot. Suppl.*, 588; 1887, *Benbow* (BM). Twickenham Common, plentifully, *Waring.* **4.** Hampstead Heath, *Johns. Enum.* **6.** Tottenham, *Johns. Cat.* **7.** Between Knightsbridge and Fulham, *Ger. Hb.*, 1088.

R. pimpinellifolia × **villosa** = **R.** × **sabinii** Woods
3. Twickenham, *Castles, Wats. New Bot. Suppl.*, 588; 1843,† *Castles* (BM).

R. rugosa Thunb.
Alien. China, Japan, etc. Garden outcast. Naturalised in thickets, etc. Very rare.
First record: *Buckle*, 1960.
2. Between West Drayton and Hayes, 1967. **3.** Near Hounslow Heath, 1960, *Buckle.* **4.** Brent Reservoir, several large bushes near the Edgware Road, 1967→. **6.** North of Enfield Lock, 1967, *Kennedy*!.

R. stylosa Desv.
Formerly native in hedges, etc., on heavy soils but now extinct.*
First evidence: *E. Forster*, c. 1800. Last evidence: *Benbow*, 1887.
1. Pinner, *T. & D.*, 103. Ruislip, *Moxey List.* Almost certainly an error. **4.** Between Child's Hill and Hendon, *De Cresp. Fl.*, 58. Harrow, 1887, *Benbow* (BM). **7.** Clapton, *E. Forster* (BM).

Var. **systyla** (Bast.) Bak.
R. systyla Bast.
4. Hampstead, 1865, *Syme* (BM, K). **7.** Near Hornsey; Walham Green, 1815; Stoke Newington, 1820, *Woods* (LINN).

R. canina L. Dog Rose.
Native in hedges, thickets, on wood borders, commons, waste ground, etc. Common.
First certain evidence: *Woods*, 1810.
7. East Heath, Hampstead. Canal side between Lisson Grove and Regents Park, 1968.

Group 1. Lutetianae
Var. **lutetiana** (Lem.) Bak.
The most common variety in the vice-county.

*Recently rediscovered in the Potters Bar area (1).

Forma **lasiostylis** Borbás.
6. Finchley Common!, *K. & L.*, 106.

Var. **sphaerica** (Gren.) Dumort.
4. Neasden, 1881,† *De Crespigny* (MANCH).

Var. **flexibilis** (Déségl.) Rouy
3. Northolt!, *K. & L.*, 107. **6.** Finchley Common, plentiful!, *K. & L.*, 107.

Var. **senticosa** (Aschers.) Bak.
Frequent in the vice-county.

Forma **mucronulata** (Déségl.) W.-Dod
3. Northolt, *K. & L.*, 107.

Group 2. Transitoriae
Var. **spuria** (Pug.) W.-Dod
1. Harefield Moor, *Sandwith*!; Ruislip!; Stanmore Common!; Yiewsley!, *K. & L.*, 107. **2.** Dawley!, *K. & L.*, 117. **3.** Hayes!; Northolt!, *K. & L.*, 107. **4.** Hendon!, *K. & L.*, 107. Edgware; Mill Hill, det. *Sandwith.* **5.** Hanwell!; Hanger Hill, Ealing!, *K. & L.*, 107. **6.** Finchley Common!, *K. & L.*, 107. Enfield Chase, det. *Sandwith.*

Forma **syntrichostyla** (Rip.) Rouy
3. Near Yeading!, *K. & L.*, 107.

Var. **ramosissima** Rau
1. Stanmore Common!; Eastcote!; between Iver and Yiewsley!, *K. & L.*, 107. South Mimms; Potters Bar, det. *Sandwith.* **2.** Dawley!; Colham Green!; Staines Moor!, *K. & L.*, 107. **3.** Cranford!; Northolt,! *K. & L.*, 107. Hounslow, det. *Sandwith.* **5.** Twyford,† *Druce Notes.*

Var. **dumalis** (Bechst.) Dumort.
Frequent in most parts of the vice-county.

Forma **viridicata** (Pug.) Rouy
1. Stanmore Common!; Northwood!, *K. & L.*, 107. **3.** Cranford!; Roxeth!,† *K. & L.*, 107. **4.** Harrow!; Kenton!, *K. & L.*, 107. **6.** Finchley Common!, *K. & L.*, 107. Enfield Chase, det. *Sandwith.*

Forma **cladoleia** (Rip.) W.-Dod
1. Pield Heath, near Hillingdon!, *K. & L.*, 107. **2.** Laleham!, *K. & L.*, 107. **5.** Brentford!; Horsendon Hill!, *K. & L.*, 107. **6.** Finchley Common!, *K. & L.*, 107. **7.** Green Lanes, 1815,† *Woods* (LINN), det. *Sandwith.*

Var. **carioti** (Chab.) Rouy
2. Between Yeoveney and Poyle, det. *Sandwith.*

Var. **sylvularum** (Rip.) Rouy
1. Uxbridge Moor!, *K. & L.*, 108.

Group 4. Andegavenses
Var. **andegavenses** (Bast.) Desp.
1. Potters Bar, *Woods* (BM). **6.** Finchley Common!, *K. & L.*, 108. **7.** Clapton, *Woods* (BM), conf. *Sandwith.*
Treated as a species by Klasterský in *Flora Europaea* vol. 2.

Forma **agraria** (Rip.) W.-Dod
1. Potters Bar, 1815 (BO), det. *Sandwith.*

Forma **surculosa** (Woods) Hook.
7. Stoke Newington, 1820,† *Woods* (BM, K, LINN).

Var. **verticillicantha** (Mérat) Bak.
1. Harefield Moor, *Sandwith*!, *K. & L.*, 108. **4.** Willesden,† *Warren* (BM), det. *Sandwith.* **7.** Tottenham, 1823,† *Woods* (BM), det. *Sandwith.*

Forma **clivicola** (Rouy) W.-Dod
5. Greenford!, *K. & L.*, 108.

Var. **schottiana** Ser.
1. Uxbridge Moor!, *K. & L.*, 108.

Var. **pouzinii** (Tratt.) W.-Dod forma **wolley-dodii** Sudre
1. Ruislip Lido Common, 1942, *Sandwith* (K).

R. corymbifera Borkh. Dog Rose.
R. dumetorum Thuill., *R. canina* L. subsp. *dumetorum* (Thuill.) Hartm.
Native in hedges, thickets, on wood borders, commons, etc. Common.
First certain evidence: *Woods*, 1815.
7. Buckingham Palace grounds, *Codrington*, *Lousley* and *McClintock*, *L.N.* 39: 49.
The following are treated as varieties or forms of *R. dumetorum* by Wolley-Dod in *A Revision of the British Roses* (1930–31), but in view of the confusion in the taxonomy and nomenclature of European *Rosa*, it has not been considered advisable to transfer them to new combinations under *R. corymbifera.*

Group 1. Pubescentes
R. dumetorum Thuill. forma **urbica** (Lém.) W.-Dod
The most frequent form encountered in Middlesex.

Forma **semiglabra** (Rip.) W.-Dod
1. Harefield!, *K. & L.*, 108. **3.** Hatch End!, *K. & L.*, 108.

Var. **ramealis** (Pug.) W.-Dod
1. Old Park Wood, Harefield, *I. G. Johnson*, det. *Melville.*

Var. **gabrealis** (F. Gér.) R. Kell.
1. Cowley Peachey!, *K. & L.*, 108. **4.** Bentley Priory!, *Druce Notes.*

Var. **calophylla** (Rouy) W.-Dod
1. Harefield Moor, *Sandwith*! (K).

Var. **platyphylla** (Rau) W.-Dod
4. Near Stanmore, *Druce Notes.*

Var. **hemitricha** (Rip.) W.-Dod
Frequent in many parts of the vice-county.

Var. **erecta** W.-Dod
2. Near Shepperton, 1877–78, *H.* and *J. Groves* (BM), det. *Wolley-Dod.*

Group 2. Déséglisei
Var. **incerta** (Déségl.) W.-Dod
1. Ruislip Common, 1946; near Ruislip Lido, 1946, *Sandwith* (K).

Forma **laevistyla** W.-Dod
1. Elstree!, *K. & L.*, 109. **4.** Deacon's Hill, *K. & L.*, 109.

R. obtusifolia Desv.
R. borreri Woods, *R. tomentella* Lém.
Native in hedges and thickets, etc. Rare.
First certain evidence: *Woods*, c. 1800.
1. Springwell, *Lousley*!, *K. & L.*, 109. **2.** Stanwell Moor!, *K. & L.*, 109. **3.** Hounslow Heath!, *K. & L.*, 109. **5.** Horsendon Hill!, *K. & L.*, 109.

Var. **typica** W.-Dod
1. Harefield Moor, *Sandwith*!, *K. & L.*, 109. **2.** West Drayton!, *K. & L.*, 109. **3.** Hounslow Heath!, *K. & L.*, 109.

Forma **concinna** (Bak.) W.-Dod
1. Harefield, *Sandwith* (K).

Var. **tomentella** (Lem.) Bak.
1. Potters Bar, Woods (LINN), det. *Sandwith.* Drayton Ford, *Lousley*!, *K. & L.*, 109.

Forma **glandulosa** (Crép.) W.-Dod
4. Hendon, 1879, *De Crespigny* (BM), det. *Sandwith.*

Var. **borreri** (Woods) W.-Dod
6. Southgate, *Woods* (LINN), det. *Sandwith.* **7.** Upper Clapton,† *E. Forster* (BM), det. *Sandwith.* Stoke Newington,† *Woods* (LINN), det. *Sandwith.*

R. tomentosa Sm. Downy Rose.
Native in woods, hedges and thickets. Very rare, and mostly confined to heavy soils.
First certain evidence: *J. Harris*, c. 1726.
4. Hampstead, *Irv. MSS.* **5.** Near Ealing, 1820, *Sneyd, Trim. MSS.*; 1887,† *Benbow MSS.* Near Alperton,† *Newbould, T. & D.*, 102. **6.** Highgate, 1827–30,† *Varenne, T. & D.*, 102. Between Highgate and Hampstead,† *Groult, Garry Notes.* **7.** Stoke Newington,† *Woods, B.G.*, 406. Ken Wood,† *Hunter, Park Hampst.*, 30.

Var. **typica** W.-Dod
1. Near Potters Bar, *Woods* (LINN), det. *Sandwith*. Mimmshall Wood!, *Benbow* (BM), det. *Sandwith*, 1967. **3.** Roxeth (SY); 1887,† *Benbow* (BM), det. *Sandwith*. **4.** Bishop's Wood!, *J. Harris* (DILL), det. *Sandwith*; 1948, *Andrewes*!; 1966. Harrow, 1859,† *Hind* (SY), det. *Sandwith*. Near Totteridge, *Pugsley* (BM). **5.** Ealing,† *Benbow* (BM), det. *Sandwith*. Horsendon Hill, very rare!, *W. C. R. Watson, L.N.* 28: 32. Perivale!,† *Benbow* (BM), det. *Sandwith*.

Var. **scabriuscula** Sm.
7. Kilburn, *Warren* (BM), det. *Sandwith*.

R. mollis Sm.
R. villosa auct., non L., *R. mollissima* Willd.
Introduced. Formerly naturalised in hedges but now extinct.
First record: *Hind*, 1860. Last evidence: *Benbow*, 1887.
3. Roxeth, *Hind Fl.*; 1887, *Benbow* (BM), det. *Sandwith* (as *R. villosa*). **4.** Harrow, *Hind, Melv. Fl.*, 28. **5.** Near Twyford; between Twyford and Alperton, *Lees, T. & D.*, 101. The latter two records are probably errors, and it is likely that *R. tomentosa* was intended.

R. rubiginosa. Sweetbriar.
R. eglanteria auct.
Native on heaths, in scrub, hedges, etc. Very rare, and perhaps extinct.
First certain evidence: *Dillenius*, c. 1730.
1. Harefield, *Blackst. Fasc.*, 87; c. 1869,† *T. & D.*, 102. Between Rickmansworth and Harefield, on the common moor;† Stanmore Common, *Shute, Webb & Colem. Suppl.*, 10; c. 1869, *T. & D.*, 102. Northwood, *Benbow* (BM). Pinner,† *Hind, Trim. MSS.* **3.** Hounslow!† (DILL). Roxeth,† *Melv. Fl.*, 27. **4.** Hampstead,† *Morley MSS.* Perhaps an error; c. 1830. Harrow,† *Melv. Fl.*, 27. Stanmore, *Varenne, T. & D.*, 103. **5.** Hanger Hill, Ealing (hortal form),† det. *Sandwith*. **6.** Highgate! (hortal form), det. *Sandwith, K. & L.*, 110. **7.** Stoke Newington,† *Sowerby* (BM).

R. micrantha Borrer ex Sm. Sweetbriar.
Formerly native in hedges and thickets, chiefly on calcareous and alluvial soils, and established on the river wall of the Thames, but now extinct.

First certain evidence: *Woods*, 1815. Last evidence: *Green*, 1910.
1. South Mimms, 1910, *Green* (SLBI), det. *Sandwith.* **2.** Hampton Court, *Castles*, *Wats. New Bot. Suppl.*, 588; river wall, 1885, *Benbow* (BM), conf. *Sandwith.*

Var. **operta** (Pug.) W.-Dod
1. Potters Bar, 1815, *Woods* (K, LINN), det. *Sandwith.*

R. agrestis Savi
R. sepium Thuill., non Lam.
Formerly native in hedges, but now long extinct.
First, and only, evidence: *Woods*, 1818.
7. Green Lanes [Newington], 1818, *Woods* (LINN), conf. *Sandwith.*

PRUNUS L.

P. spinosa L. Blackthorn, Sloe.
Native in hedges, on wood borders, etc. Common, except in heavily built-up areas where it is very rare or extinct.
First record: *T. Johnson*, 1632.
7. West End Lane;† Finchley Road, etc.,† *T. & D.*, 88. Near Kilburn. Kensington Gardens, *D. E. Allen*, *L.N.* 39: 49: 'error, the locality should read Hyde Park!', *Allen Fl.*

Var. **macrocarpa** Wallr.
2. West Drayton!, *L.N.* 26: 60.

P. domestica L. subsp. **insititia** (L.) C. K. Schneid. Bullace.
P. insititia L.
Alien. Probably of hortal origin. Bird-sown from gardens, etc., and naturalised in hedges, thickets, on wood borders, etc. Rare.
First certain record: *A. Irvine*, c. 1830.
1. Harefield!, *Newbould*; near Pinner Wood, *T. & D.*, 89. Between Cowley and Hillingdon, 1884,† *Benbow* (BM). Between Yiewsley and Cowley!, *L.N.* 26: 60. Uxbridge!, *K. & L.*, 73. Near Eastcote, *Green* (SLBI). **2.** Between Kingston Bridge and Hampton Court!; Tangley Park,† *T. & D.*, 89. West Drayton!, *P. H. Cooke*; Dawley!, *K. & L.*, 73. **3.** Roxeth,† *Melv. Fl.*, 24. Twickenham,† *T. & D.*, 89. **4.** Dollis Hill,† *Fox*; Sudbury,† *T. & D.*, 89. Harrow,† *Melv. Fl.*, 24. Hampstead Heath, two trees, 1906, *C. S. Nicholson*, *K. & L.*, 73. **7.** Primrose Hill,† *Irv. MSS.* Kentish Town,† *T. & D.*, 89.

Subsp. **domestica** Plum.
P. domestica L.
1. Between Harefield and Ruislip,† *Fox*, *T. & D.*, 89. Between Cowley and Hillingdon,† *Benbow* (BM). Ruislip Woods, *P. H. Cooke*, *K. & L.*, 72. South Mimms, 1966. **3.** Roxeth,† *Hind Fl.* Isleworth!,

L.N. 26: 60. **4.** Neasden;† near Willesden, 1863,† *J. Morris, T. & D.*, 89. **7.** Bombed sites, Cripplegate area, E.C., *Jones Fl.*

P. cerasifera Ehrh. Cherry-Plum.
P. divaricata Ledeb.
Alien. Central Asia, etc. Bird-sown (?), or planted, in hedges, etc., where it has now become naturalised. Possibly common, but confused with *P. spinosa* and *P. domestica*.
First record: *Wrighton*, 1946.
7. East Heath, Hampstead. Ken Wood.

P. avium (L.) L. Gean, Wild Cherry.
Native in woods, hedgerows, on heaths, etc., particularly in the northern parts of the vice-county, also naturalised in hedges, on waste ground, etc., from discarded 'stones' of cultivated cherries. Local as a native species, but rather common as an introduction.
First certain record: *Blackstone*, 1737.
7. Ken Wood. East Heath, Hampstead. Kensington Gardens, seedlings!, *Allen Fl.* Notting Hill Gate, seedlings. Lord's Cricket Ground, seedlings. Shadwell, seedlings and saplings.
Many of the plants resulting from the discarded stones of cultivated cherries may be referable to the hybrid *P. avium* × *cerasus*.

P. cerasus L. Sour, or Dwarf Cherry.
Alien of unknown origin. Naturalised in hedges, on waste ground, etc. Either originally planted, or resulting from discarded 'stones' of the cultivated Morello Cherry. Rare, and much confused with *P. avium* and *P. avium* × *cerasus*.
First certain evidence: *J. Banks*, c. 1800.
1. Harefield!, *Newbould*; near Harrow Weald Common,† *T. & D.*, 89. Northwood, 1938, *P. H. Cooke* (LNHS). Yiewsley!; Cowley!, *K. & L.*, 72. Elstree,† *Trimen* (BM). Uxbridge Moor. **2.** Tangley Park,† *T. & D.*, 89. West Drayton, Staines. **3.** Near Isleworth,† *T. & D.*, 89. Southall!, *K. & L.*, 72. Hounslow Heath, Northolt. **4.** Mill Hill, *J. Banks* (BM). Edgware, *Varenne*; near Finchley, *Newbould, T. & D.*, 90. Willesden. **5.** Near Ealing,† probably planted, *Hemsley, T. & D.*, 90. Wyke Green!, *K. & L.*, 72. Hanwell. **6.** Coldfall Wood, *E. M. Dale*; Highgate Woods, *C. S. Nicholson, K. & L.*, 72. Near Muswell Hill, *Benbow* (BM). **7.** East Heath, Hampstead; Highgate, *Irv. MSS.* Buckingham Palace grounds, *Fl. B.P.G.*

COTONEASTER Medic.

C. horizontalis Decne.
Alien. China, etc. Bird-sown from gardens and naturalised in scrub, and established on old walls, etc. Very rare.

First record: *D. H. Kent*, 1946.
1. Harefield!, *K. & L.*, 114; increased considerably on the face of an old chalkpit, and among scrub at the top of the pit, between 1950 and 1964, but has now possibly been eradicated as the floor of the pit and part of the slopes were bulldozed in 1965. The area in which the plant grew is now inaccessible and has been severely 'scarred'. Bank of the river Pinn, Ickenham, one plant amongst scrub, 1963, *Davidge*. **5.** Old wall, West Ealing, a solitary large seedling, 1966. **7.** Wall adjoining Horse Guard's Parade, S.W.1, one plant, 1965, *Brewis*, *L.N.* 46: 30.

PYRACANTHA M. J. Roem.

P. coccinea M. J. Roem. Fiery Thorn.
Alien. S. Europe, etc. Bird-sown from gardens. Naturalised on a heath, and formerly established in an old chalkpit. Very rare.
First record: *T. G. Collett*, 1953.
1. Old chalkpit, Harefield, 1953!, *T. G. Collett*, teste *Melderis*, *K. & L.*, 347; 1954–64;† eradicated in 1965 when the floor of the chalkpit was bulldozed. **4.** West Heath, Hampstead, a single plant, 1962→.

CRATAEGUS L.

Lit. *Rep. Bot. Soc. & E.C.* **12**: 847 (1946).

C. laevigata (Poir.) DC. subsp. **laevigata** Midland Hawthorn.
C. oxyacantha L., *nom. ambig.*, *C. oxyacanthoides* Thuill.
Native in woods, thickets, on commons, etc., chiefly on heavy soils. Locally common, especially in the northern parts of the vice-county.
First record: *Macreight*, 1837.
1. Common. **2.** Near Staines, *T. & D.*, 104. Teddington!, *Benbow MSS.* **3.** Yeading!, *T. Moore*, *Trim. MSS.* Roxeth!, *Benbow MSS.* Northolt. Hatch End. Near Hounslow Heath!, *P. J. Edwards*. Hatton Cross, *Kennedy*!. **4.** Common. **5.** Common, except in the immediate vicinity of the Thames where it is rare. **6.** Common. **7.** East Heath, Hampstead. Ken Wood. Highgate.

Var. **eriocalyx** (Freyn)
1. Near Ruislip Common!, *Batko*, *K. & L.*, 113. **4.** Mill Hill, *E. F.* and *H. Drabble*, *Rep. Bot. Soc. & E.C.* 8: 259.
Bushes bearing very small fruits were noted in **1.** Near Ruislip Common!, *Batko*, *K. & L.*, 113.

Subsp. **palmstruchii** (Lindm.) Franco
C. palmstruchii Lindm.
1. Near Ruislip Common!, 1944, *Batko* (BM); 1946, *Batko*!; enclosed within the garden of a new house, 1951; 1962.

C. laevigata subsp. **laevigata** × **monogyna** = **C.** × **media** Bechst.
Frequent wherever the two parents grow together.

C. monogyna Jacq. subsp. **nordica** Franco Hawthorn.
C. oxyacantha L., *nom. ambig.*
Native in hedges, thickets, on wood borders, etc. Common, though frequently planted, particularly for hedging.
First certain record: *Hind*, 1860.
7. East Heath, Hampstead. Ken Wood. Hyde Park, seedlings, *D. E. Allen, L.N.* 39: 50. Kensington Gardens. Buckingham Palace grounds. Bishop's Park, Fulham. Brompton Cemetery, seedlings. Hammersmith. Frequent in the squares of Pimlico, Belgravia, Chelsea, etc.

Var. **laciniata** Stev.
1. Ruislip Common!, *K. & L.*, 113. **4.** Finchley, *E. F.* and *H. Drabble, Rep. Bot. Soc. & E.C.* 8: 529. **6.** Finchley Common!, *K. & L.*, 113. Enfield.

Var. **cuneata** Druce
1. Uxbridge!, *Druce Notes.* **4.** Hendon.

Var. **splendens** (Druce) Druce
1. Ickenham, *Groves* (BM). Near Ruislip Common!, *Batko, Rep. Bot. Soc. & E.C.* 12: 859.

Var. **urceolata** Hobk.
1. Near Ruislip Common!, *Batko, Rep. Bot. Soc. & E.C.* 12: 859.

MESPILUS L.

M. germanica L. Medlar.
Pyrus germanica (L.) Hook. f.
Alien. S.-E. Europe, etc. Planted for its fruits, and sometimes naturalised in hedges. Very rare.
First record: *Merrett*, 1666.
2. Near Staines Moor, two trees!,† *Welch, L.N.* 27: 31. Both trees were felled c. 1950. **3.** Harrow Grove,† *Melv. Fl.*, 28. **4.** Hedges between Hampstead and Highgate,† *Merr. Pin.*, 77. **5.** Near Heston, five bushes, 1965!, *L.N.* 45: 20; 1968.

SORBUS L.

S. aucuparia L. Rowan, Mountain Ash.
Pyrus aucuparia (L.) Ehrh.
Native in woods, hedges, on heaths, etc., in the northern parts of the vice-county, and extensively planted in hedges, gardens, parks, cemeteries, etc., elsewhere, where it readily regenerates. Locally common as a native, and common as a naturalised introduction.

First record: *T. Johnson*, 1629.

7. Many small trees in a little wood a mile beyond Islington,† *Johns. Ger.*, 1470. Ken Wood!, *E. Ballard* (BM). East Heath, Hampstead. Bombed sites, Finsbury Square,† Cripplegate,† Gresham Street† and Queen Victoria Street areas,† *Jones Fl.* Fulham, seedlings.

S. intermedia (Ehrh.) Pers.
Pyrus intermedia Ehrh.

Alien. N. Europe. Planted in woods, hedges, on heaths, etc., where it sometimes regenerates. Rare.

First record (of regeneration): *Higgins*, 1917.

4. Hampstead Heath!, *Higgins, Rep. Bot. Soc. & E.C.* 4: 485; regenerating freely and well naturalised on the West and North-west Heaths, 1969. **5.** Between Ealing and Kew Bridge, regenerating, 1967→. **6.** Near Enfield, regenerating, 1966, *Kennedy*. **7.** East Heath, Hampstead and Ken Wood, regenerating. Bombed site, Cripplegate, a seedling,† *Jones Fl.* Wall of car park by Westminster Abbey, a large seedling, 1962, *Brewis, L.N.* 43: 22; 1966. St Botolph's-without-Aldgate, E.C., a seedling, 1966, *Brewis, L.N.* 46: 30. Shoreditch, several large seedlings on waste ground, 1966.

S. aria (L.) Crantz White Beam.
Pyrus aria (L.) Ehrh.

Native in woods, hedges, etc., on calcareous soils at Harefield and Springwell, elsewhere planted in woods, hedges, on heaths, commons, etc., where it sometimes regenerates. Very rare.

First record: *T. Johnson*, 1629.

1. Harefield!, *Blackst. Fasc.*, 96; still present but now very rare, 1969. Uxbridge Common, *Blackst. Fasc.*, 96; c. 1877,† *De Cresp. Fl.*, 55. Garett Wood, Springwell!,† *Benbow MSS.* Habitat destroyed by chalk-quarrying operations. Wrotham Park, South Mimms, regenerating, 1965, *J. G.* and *C. M. Dony*. **4.** Hampstead Heath, *Johns. Eric.*; still present, and regenerating on the West Heath, 1969. Bishop's Wood, *Mart. Pl. Cant.*, 68; rare, 1912, *Hampst.* Between Hendon and Finchley, *Irv. Lond. Fl.*, 187; c. 1869, *T. & D.*, 106; 1887, *Benbow MSS.*; one tree, 1918,† *Tremayne*. Harrow,† *Benbow MSS.* **5.** Near Alperton,† *Newbould, T. & D.*, 107. **6.** Tottenham,† *Johns. Cat.* Colney Hatch, one tree!,† *K. & L.*, 112. **7.** Hornsey,† *Sowerby, Winch MSS.* Ken Wood!, *R. W. Robbins*!, *K. & L.*, 112.

S. torminalis (L.) Crantz Wild Service Tree.
Pyrus torminalis (L.) Ehrh.

Native in woods, hedges, etc., particularly on heavy soils. Rare and decreasing, and mainly confined to the northern parts of the vice-county.

First record: *Gerarde*, 1597.

1. Harefield!, *Blackst. Fasc.*, 96; a number of fine old trees near Harefield Grove Farm, 1966!, *I. G. Johnson*. Pinner, *Webb & Colem. Suppl.*, 10; Pinner Hill, *W. A. Tooke, Trim. MSS.*; Pinner Wood, *Melv. Fl.* 2: 37. Ruislip Woods!, *Benbow MSS*. Northwood, 1930, *R. C. B. Gardner* (BM). **4.** Hampstead Heath, *Johns. Eric.*; 1916, *J. E. Cooper*; 1926,† *R. W. Robbins, K. & L.*, 112. Woods adjacent to Hampstead Heath, *Budd. MSS*. Bishop's Wood, *Mart. Pl. Cant.*, 68; 1866, *Trimen* (BM); 1914, *J. E. Cooper, K. & L.*, 112. Between Highgate and Hampstead (DILL). Turner's Wood; near Fortune Green,† *Trim. MSS*. By Brent Reservoir;† Cricklewood,† *Warren, T. & D.*, 423. Mill Hill; East Finchley!, *J. E. Cooper, K. & L.*, 112. Big Wood, Finchley; Scratch Wood!, *Trimen* (BM). Bentley Priory, *Tremayne*. Harrow,† *Melv. Fl.*, 29. Near Finchley, 1885, *W. R. Linton* (BM). **5.** Between Perivale and Greenford, one tree, *Lees, T. & D.*, 107; 1909,† *Green* (SLBI). Perivale Wood!, *Webb Br. Vall.* Horsendon Hill, very rare!, *W. C. R. Watson, L.N.* 28: 32; one small tree left, 1955. **6.** Queen's and Coldfall Woods, Highgate, *L. B. Hall, K. & L.*, 112. Whitewebbs Park, Enfield!, *Welch*. Hillyfields Park, near Enfield, one tree, 1952, *L. M. P. Small*; 1961. Grovelands Park, Southgate, two trees, 1958, *A. Vaughan*. **7.** Near Islington,† *Ger. Hb.*, 1290. Near Hackney,† *Spencer*, 318. Hornsey Wood,† *T. & D.*, 107. Near Hampstead, *Merr. Pin.*, 115. Between Regents Park and Hampstead, 1818,† *J. J. Bennett* (BM). Primrose Hill† (HE). This is probably Bennett's locality. Ken Wood!, *Mart. Pl. Cant.*, 68. Millfield Lane, Highgate,† *T. & D.*, 107. Near Kilburn, 1866,† *Parsons* (CYN). Between East Heath Road and Ken Wood, 1947, *August*.

PYRUS L.

P. pyraster Burgsd. Pear.
P. communis auct., non L., *P. communis* var. *achras* Wallr.
Alien. Europe, etc., Naturalised in woods, hedges, etc. Rare.
First record: *Blackstone*, 1737.

1. Harefield!, frequent, *Blackst. Fasc.*, 82; now rare. Near Ruislip Reservoir, *Benbow MSS*. Pinner Hill!, *W. A. Tooke, Trim. MSS.* Yiewsley. **2.** Staines Moor!,† *Benbow MSS*. Near Ditton Ferry, 1906,† *Green* (BM). Near Stanwell!,† *Briggs*. Between Hampton and Kingston,† *Druce Notes*. **3.** Near Twickenham,† *T. & D.*, 105. **4.** Between Child's Hill and Hendon,† *Irv. Lond. Fl.*, 187. Between Neasden and Blackfoot (=Blackbird) Hill (Kingsbury),† *Farrar, T. & D.*, 105. Between Mill Hill and Totteridge, frequent, *Lousley*!, *L.N.* 26: 60. Near Hendon! (SY). Scratch Wood, a single tree!, *Boniface, K. & L.*, 111. Hampstead, *Whiting Notes*. Northwick Park golf course, a single tree!,† *Harley Add.* Between Stonebridge and Kingsbury,† *Benbow MSS*. **5.** Ealing,† *Jacks. Ann.*, 341. Alperton,†

Benbow MSS. Thames bank, Chiswick, a single tree, 1948 !,† *Boniface, K. & L.*, 111. Perivale Wood, 1902,† *Green* (SLBI). **6.** North of Finchley !, *Loud. Arb.*, 882. Highgate Wood, 1850 (HE). Queen's Wood, Highgate, 1958. Hadley, *Warren, T. & D.*, 105. **7.** Between Primrose Hill and Adelaide Road (Hampstead), a solitary tree, 1862,† perhaps remains of a garden, *T. & D.*, 105. Bombed site, Cripplegate area, E.C.,† *Jones Fl.*

MALUS Mill.

M. sylvestris (L.) Mill. Crab.
Pyrus malus L. var. *sylvestris* L.
Native in woods, hedges, on heaths, etc. Formerly rather common, but now local.
First certain evidence: *Goodger* and *Rozea*, 1815.
1. Locally frequent. **2.** Staines, *T. & D.*, 105. Dawley. **3.** Wood End, *Payton* !. Northolt. Hounslow Heath. **4.** Locally common, except in heavily built-up areas, where it is very rare or extinct. **5.** Wormholt (=Wormwood) Scrubs,† *Britten*; near Kew Bridge railway station,† *Newbould, T. & D.*, 106. Ealing !, *Jacks. Ann.*, 341. Gunnersbury Park, Acton !, *Henrey*, 34. Syon Park, probably planted. **6.** Colney Hatch; Edmonton, *T. & D.*, 106. Hadley !, *Tremayne*. Between Enfield and Potters Bar !, *Benbow MSS*. Whitewebbs Park !, *Johns*. Highgate Woods !, *Benbow MSS*. **7.** Marylebone Fields, 1815† (G & R). Hornsey Wood,† *Ball. MSS*. Kilburn, 1869,† *Warren, Trim. MSS*. Ken Wood.

M. domestica Borkh. Cultivated Apple.
M. sylvestris (L.) Mill. subsp. *mitis* (Wallr.) Mansf., *Pyrus malus* L. var. *mitis* Wallr.
Alien. S.-E. Europe, etc. Naturalised in hedges, fields, on waste ground, etc., originating from discarded apple cores. Common.
First evidence: *Melvill*, c. 1863.
7. Common in the district.

PLATANACEAE

PLATANUS L.

P. occidentalis L. × **orientalis** L. = **P.** × **hispanica** Muenchh. London Plane.
P. acerifolia (Ait.) Willd., *P.* × *hybrida* Brot.
Lit. *Proc. Bot. Soc. Brit. Isles* **7**: 507 (1968).
Alien of cultivated origin. Extensively planted about London as a street

tree. It regenerates freely, and seedlings and saplings are common in the squares and parks of Central London and elsewhere.
First record (of regeneration): *Lousley*, 1944.
7. Common in the district.

CRASSULACEAE

SEDUM L.

Lit. *J. Roy. Hort. Soc.* **46**: 1 (1921).

S. telephium L. subsp. **telephium** Orpine, Livelong.
Native on wood borders, hedgebanks, etc., established on railway banks, etc. Very rare, and sometimes possibly originating from gardens.
First record: *Petiver*, c. 1710.
1. Harefield!,† *Blackst. Fasc.*, 97. Springwell!,† *Benbow, J.B.* (*London*) 22: 280; persisted in small quantity on a bank in Springwell Lane until c. 1953. Between Potters Bar and South Mimms!, *T. & D.*, 115. Railway banks, Potters Bar, plentiful!, *K. & L.*, 118; still frequent between Potters Bar station and the Herts boundary, 1973. **2.** Between Hampton and Sunbury, 1824† (W). **3.** Between Headstone and Pinner, 1872,† *W. A. Tooke, Trim. MSS.* **4.** Hampstead Wood,† *Irv. MSS.* and *Coop. Fl.*, 102. Hampstead Heath,† *Pamplin MSS.* Between Hampstead and Hendon,† *Button, Coop. Suppl.*, 11. Stanmore, 1827–30,† *Varenne, T. & D.*, 115. Scratch Wood, 1888,† *Benbow* (BM). **5.** Near Heston, 1937!,† *K. & L.*, 118. **7.** Between Westminster and Chelsea,† *Pet. Bot. Lond.* Ken Wood,† *Hunter, Park Hampst.*

S. spurium Bieb.
S. stoloniferum auct.
Alien. Caucasus, etc. Garden escape. Naturalised in grassy places, on waste ground, etc. Very rare.
First record: *G. C. Druce*, 1910.
1. Near Potters Bar, 1966, *Kennedy*. **2.** Colnbrook, *Druce Notes.* **4.** Waste ground, Stonebridge Park power station, in profusion, 1966. **5.** Hanger Hill, Ealing, an extensive patch, 1958!, *L.N.* 38: 21. Railway embankment, Perivale, 1967. **6.** Enfield, 1966, *Kennedy*!. **7.** Bombed site, Queen Victoria Street, E.C., 1959,† *McClintock*.

S. dasyphyllum L. Thick-leaved Stonecrop.
Alien. S. Europe, etc. Garden escape. Established on the tops of old walls. Very rare.
First evidence: *Dillenius*, c. 1730.
1. Hillingdon, 1872, *Trim. MSS.*; on walls for about a mile around

Hillingdon, *J. Clarke*, teste *Trimen*, *J.B.* (*London*) 12: 247; 1884,† *Benbow* (BM). Near Uxbridge, 1907,† *Green* (SLBI). This probably refers to the Hillingdon locality. **3.** Twickenham,† *Irv. Lond. Fl.*, 171. Syon Lane, Isleworth, plentifully for about twenty yards . . ., 1870, *A. Wood*, *Trim. MSS.*; 1900,† *Montg. MSS.* Nazareth House, Isleworth, 1958; very scarce, 1964. **5.** Chiswick, 1843† (HE). **7.** Chelsea (DILL); c. 1763, *Mart. Pl. Cant.*, 65; near Chelsea Hospital . . .;† between Kensington gravel-pits and Acton,† *Curt. F.L.* Wall in the lane leading from Kensington Church to the gravel-pits,† *E. Forster* (BM). This may refer to Curtis's locality.

S. album L. subsp. **album** White Stonecrop.
Alien. Europe, etc. Garden escape. Naturalised on waste ground, etc., established on old walls, and on the river-wall of the Thames. Rare, but increasing.
First record: *Blackstone*, 1746.
1. Cowley, 1855,† *Phyt. N.S.* 1: 65. **2.** River-wall between Kingston Bridge and Hampton Court, in two places, 1961!, *Boniface*; locally abundant, 1973. River-wall, Laleham, 1965→. **3.** Twickenham!, *Borrer*, *B.G.*, 405; sparingly on the river-wall, 1966→. Abundant on a heap of gravel ballast near Sudbury Hill, 1944–46!,† *K. & L.*, 118. Rubbish-tip, Hounslow Heath, 1960, *Philcox* and *Townsend*. Old wall, Hayes Park, plentiful, 1966→. **4.** Stanmore, 1827–30,† *Varenne*, *T. & D.*, 116. Harrow, *Harley Fl.*, 13. Stone terrace of North London Collegiate School, Edgware, plentiful, 1967!, *Horder*. **5.** Brentford, 1836, *Trimmer* (BM); abundant on the high wall of Syon House grounds . . . beyond Brentford, *T. & D.*, 116; near Brentford, 1871,† *F. J. Hanbury* (HY). Ealing,† *Lees*, *T. & D.*, 116. River-wall, Duke's Meadows, Chiswick, several patches, 1961!, *Boniface*; abundant, 1971. **7.** Kentish Town, *Blackst. Spec.*, 91; c. 1775;† Bromley,† *Curt. F.L.* Fulham,† *Pamplin*, *Wats. New Bot.* 99 and *Coop. Fl.*, 113. Hackney, 1850,† *Freeman* (BM).

S. acre L. Wall-pepper, Yellow Stonecrop.
Denizen on waste ground, etc., on sandy and gravelly soils, etc., established on old walls, railway tracks, etc. Common.
First record: *T. Johnson*, 1638.
7. Tottenham!, *Johns. Cat.* Old walls about Islington and Newington, *E. Ballard* (BM). Kilburn!, *A. B. Cole*; Haverstock Hill;† King's Road, Chelsea;† Brompton!; Parsons Green!, *T. & D.*, 116. Old walls, Fulham Palace grounds, plentiful, *Welch*!. Bombed site, Portland Place, W.1, 1956,† *Holme*. Bombed site, Upper Thames Street, E.C. Bombed sites, Finsbury Square† and Cripplegate areas!, and near the Tower of London, *Jones Fl.* Top of brick wall enclosing a bombed site used as a car park, south side of Ludgate Hill, E.C., 1965, *Burton*, *L.N.* 46: 30. Hammersmith. Earls Court. Hampstead. Hornsey.

S. reflexum L. Large Yellow Stonecrop.
Alien. Europe, etc. Garden escape. Naturalised on waste ground, by roadsides, etc., established on old walls, river-wall of the Thames, etc. Rare, but apparently increasing.
First record: *Blackstone*, 1737.
1. Harefield!, *Blackst. Fasc.* 92. Near Elstree,† *Rudge, T. & D.*, 117. Pinner,† *De Cresp. Fl.*, 66. Old walls of 'The Cedars', Hillingdon, 1884,† *Benbow* (BM). South Mimms, 1957, *Warmington.* **2.** Laleham!, *Druce Notes*; established on the river-wall, 1965→. River-wall, Hampton Court, 1964→. Harlington, *Benbow MSS.* **3.** Roxeth,† *Melv. Fl.*, 32. Hounslow Heath, 1960, *Philcox* and *Townsend.* **4.** Mill Hill, 1841 (HE). Stanmore, 1830,† *Varenne, T. & D.*, 117. Finchley, 1964, *Wurzell.* **5.** Hanwell, 1946.† Hanger Hill, Ealing, 1946.† **6.** Highgate, 1819† (G & R). Near Barnet, 1967, *Kennedy*!.

SAXIFRAGACEAE

SAXIFRAGA L.

S. spathularis Brot. × **umbrosa** L. = **S.** × **urbium** D. A. Webb London Pride.
S. umbrosa L. var. *crenatoserrata* Bab.
Alien of hortal origin. Garden escape. Naturalised in grassy places, on waste ground, etc., established on railway banks, etc. Very rare.
First record: *Harley*, 1953.
1. Watts Common, Harefield, plentiful, 1963!, *L.N.* 42: 12. Old sandpit, Springwell, 1966. **2.** Railway bank, Staines, 1966, *J. E. Smith*!. **4.** Harrow, *Harley Fl.*, 13.

S. tridactylites L. Rue-leaved Saxifrage.
Native in dry stony places, established on old walls and the river-wall of the Thames, also on roofs and in gutters of old houses. Formerly common, but now very rare, and decreasing, and mainly confined to the vicinity of the Thames.
First record: *Gerarde*, 1597.
1. Harefield,† *Blackst. Fasc.*, 72. Pinner, 1860,† *Hind* (BM). Eastcote,† *Melv. Fl.*, 33. Ruislip, *T. & D.*, 118; 1964→, *I. G. Johnson.* Uxbridge, 1862;† Cowley;† Hillingdon,† *Benbow MSS.* **2.** Staines, 1867, *Trimen* (BM). Sunbury; Halliford!, *T. & D.*, 118; 1962. Shepperton!, *C. S. Nicholson, K. & L.*, 115; 1972. River-wall near Hampton Court, *Welch*!, *L.N.* 27: 31; 1968. West Drayton, 1926,† *Tremayne, K. & L.*, 115. **3.** Headstone, 1860,† *Hind* (BM). Abundant about Isleworth,† *T. & D.*, 118. **4.** Hampstead Heath, *Johns. Enum.*; by Jack Straw's Castle;† Stanmore, *T. & D.*, 118; c. 1910,† *P. H. Cooke* (LNHS). Harrow,† *Melv. Fl.*, 33. Mill Hill, 1908,† *C. S. Nicholson* (LNHS).

5. Osterley Park,† *Masters*; Drayton Green,† *A. B. Cole*; Heston;† walls at Brentford,† Turnham Green,† etc., abundant, *T. & D.*, 118. Chiswick, 1872,† *Britten, Trim. MSS.* Walls of Syon House,† *De Cresp. Fl.*, 63; brickwork of ha-ha, by the Thames, Syon House grounds, 1955, *Sandwith*!; 1966. Walls of Manor House, East Acton, 1903,† *Middleton* (BM). **6.** Enfield; Edmonton;† Whetstone,† *T. & D.*, 118. Walls of Highgate Cemetery,† *Hampst.* **7.** Plentifully upon the brick wall in Chancery Lane, belonging to the Earl of Southampton,† *Ger. Hb.*, 500. Canonbury, 1837,† *E. Ballard*; Chelsea, 1845,† *T. Moore* (BM). Hammersmith;† Fulham, *Britten, T. & D.*, 118; 1874,† *Trim. MSS.* Between Hammersmith and Fulham,† *De Cresp. Fl.*, 63. Shepherds Bush;† Haverstock Hill, 1866,† *T. & D.*, 118.

Atmospheric pollution may be responsible for the decline of this attractive species.

S. granulata L. Meadow Saxifrage.

Native in meadows, on banks, etc., on light soils. Very rare, and now confined to the immediate vicinity of the Thames.

First record: *Gerarde*, 1597.

1. Near Moor Hall, Harefield, *Blackst. Fasc.*, 96; 1839,† *Kingsley* (ME). **2.** Between Twickenham and Staines!, *Wats. MSS.* Hampton Court Park!, *Green* (SLBI); 1973. **4.** Near Child's Hill, *Irv. MSS.*; 1882,† *Whiting Fl.* **5.** Chiswick, 1869, *Warren*; 1873, *Dyer*; very abundant, 1888, *Benbow* (BM); formerly grew in meadows by the Thames but the habitat was destroyed by gravel-digging operations; near Chiswick House grounds, 1965, *Murray*. Near Osterley, 1933.† Syon Park, *A. B. Jackson, Spooner* and *Welch*!, *L.N.* 26: 60; still in meadows by the Thames, 1971. **7.** Near Islington,† *Ger. Hb.*, 693. Meadows and grassy sandy places on the back side of Grays Inn . . .,† *Park. Theat.*, 23. Meadows near Marylebone,† *J. Banks* (BM).

CHRYSOSPLENIUM L.

C. oppositifolium L. Opposite-leaved Golden Saxifrage.

Native in wet shady places. Very rare.

First record: *Blackstone*, 1746.

1. Side of a ditch in a meadow just below Coney's Farm at Harefield, plentifully,† *Blackst. Spec.*, 89. Harefield (=Old Park) Wood!, *Kingsley* (ME); still grows there plentifully in a boggy area near a spring and along the banks of a brook, 1972. Abundant by the sides of brooks in woods in Harefield Grove, 1966!, *I. G. Johnson.* **4.** Hampstead Heath, *Huds. Fl. Angl.*, 156; bank of a ditch by the roadside, 1824, *Pamplin, T. & D.*, 119; extinct by 1912,† *Whiting Fl.* Turner's Wood, *Irv. MSS.* Bishop's Wood, *Macreight Man.*, 93; by the stream in the south part of Bishop's Wood, abundant, 1869, *T. & D.*, 119; 1887, *Benbow* (BM); abundant, 1902, *C. S. Nicholson, K. & L.*, 116;

1914, *Pugsley*; the wood was partially felled and enclosed in the gardens of large houses in The Bishop's Avenue shortly after the First World War, but it is possible that the plant survives. Scratch Wood, *Welch*!, *L.N.* 28: 32; 1963. **5.** Osterley Park!, *Welch, L.N.* 28: 32. Long Wood, Wyke Green!, *K. & L.*, 116.

C. alternifolium L. Alternate-leaved Golden Saxifrage.
Formerly native in wet shady places, but now extinct.
First, and only evidence: *Kingsley*, 1838.
1. Harefield (=Old Park) Wood, 1838, *Kingsley* (ME).

TELLIMA R.Br.

T. grandiflora (Pursh) Dougl. ex Lindl.
Alien. N. America. Garden escape. Naturalised by roadsides, etc. Very rare.
First record: *Studley*, 1955.
3. In small quantity by a footpath between Isleworth and Osterley, 1968. **4.** Harrow, in two localities!, *Studley, L.N.* 39: 40; 1960; a thorough search of the two localities in 1966 and 1967 failed to reveal a single plant.

PARNASSIACEAE

PARNASSIA L.

P. palustris L. Grass of Parnassus.
Formerly native in marshes, wet meadows and moors in a few localities in the upper part of the Colne valley, but now extinct owing to the eradication of its habitats by gravel-digging operations.
First record: *Ligo*, 1735. Last record: *Pugsley*, 1900.
1. In the moist meadows near Harefield Mill . . . observed by Mr *Ligo* . . . 1735, *Blackst. MSS.* Meadows between Harefield and Rickmansworth, 1868, *Newbould*; wet meadows in the extreme north-west angle of the county beyond Springwell Lock, close to the Herts boundary and Drayton Ford, 1884, *Benbow* (BM); 1900, *Pugsley, K. & L.*, 116. Uxbridge Moor, *J. Hill* (BM).

GROSSULARIACEAE

RIBES L.

R. rubrum L. Red Currant.
R. sylvestre auct.
Denizen in woods and thickets, by streamsides, on waste ground, etc. Rare, and possibly always bird-sown from cultivated plants.

First record: *Blackstone*, 1746.

1. Sparingly in some coppices near Harefield, *Blackst. Spec.*, 82. Chalk-pit in the Old Park, Harefield, *T. & D.*, 118; Old Park Wood, Harefield. Near Uxbridge Common, 1891,† *Benbow* (BM). Denham, *Druce Notes*, Cowley Peachey, 1945,† *Welch*!, *L.N.* 26: 60. Ruislip, *Moxey List*. **2.** Staines, 1958, *Briggs*. **3.** Cranford Park!, *Westrup*. Wood near Hounslow Heath!, 1965, *P. J. Edwards*; 1966. Canal bank near Southall, from a garden, 1965. **4.** Harrow Park,† doubtfully wild, *Melv. Fl.*, 33. Bishop's Wood, 1862, *Trimen* (BM). Turner's Wood, 1925, *Moring MSS*. Scratch Wood; Mote Mount Park, Edgwarebury, *Welch*!, *L.N.* 28: 32. Waste ground, Willesden Green, from a garden, 1965. **5.** Hanwell, from a garden, 1953.† Norwood Green, from a garden, 1965. **6.** Copse near Warren Lodge, Edmonton,† *T.& D.*, 118. **7.** Bombed site, Cripplegate, one plant,† *Jones Fl.*

R. nigrum L. Black Currant.

Denizen by streamsides, in woods, thickets, on waste ground, etc. Very rare, and possibly always bird-sown from garden plants.

First evidence: *Dillenius*, c. 1730.

1. In a meadow near the Warren Pond at Breakspeares, plentifully,† *Blackst. Fasc.*, 87. Harefield, 1907, *Tremayne*, *K. & L.*, 117. Near Uxbridge Common,† *Benbow MSS*. **2.** Between Teddington and Hampton Court, in several places, 1885, *Benbow* (BM); 1886–87,† *Benbow MSS*. **3.** Isleworth (DILL); 1969, *Sandford*. By the Crane, Hounslow Heath, one bush,† *T. & D.*, 118. **4.** Bishop's Wood, *Benbow MSS*. Harrow, one bush, 1966, *Temple*!. **5.** Horsendon Hill, 1962→. Hanwell, from a garden, 1965. **6.** Bank of old railway, Muswell Hill, 1966, *P. Moxey*, *L.N.* 46: 12. Wood, Enfield, 1967, *Kennedy*!.

R. uva-crispa L. Gooseberry.

R. grossularia L.

Denizen in woods, thickets, on waste ground, etc. Rare, and possibly always bird-sown from cultivated plants.

First record: *A. Irvine*, c. 1834.

1. Uxbridge!, *K. & L.*, 117. Harefield, 1958, *Pickess*. Ruislip, *Moxey List*. Harrow Weald Common, *Welch*, *K. & L.*, 117. Wrotham Park, 1965, *J. G.* and *C. M. Dony*. **2.** West Drayton, on a pollarded willow, *Druce Notes*. Harmondsworth, 1911, *P. H. Cooke* (LNHS). Hampton Hill, 1965, *Clement*. **3.** Grove wall, Harrow,† *Melv. Fl.*, 33. Cranford, 1953, *Westrup*. Hayes Park, 1966. **4.** Hampstead, *Irv. MSS*. **5.** Osterley Park, 1958. Near Drayton Green, Ealing, 1968. **6.** Near Barnet, *Fox*; Edmonton, *T. & D.*, 117. **7.** Between Belsize Park and Hampstead,† *T. & D.*, 117. Hyde Park, one plant, 1919,† *Tremayne*. Kensington Gardens, one plant, 1958!, *D. E. Allen*, *L.N.* 39: 50. Fulham Palace grounds, 1966, *Brewis*.

DROSERACEAE

DROSERA L.

D. rotundifolia L. Round-leaved Sundew.

Formerly native in bogs on heaths, etc., but extinct by 1936. Deliberately reintroduced in 1960. Very rare.

First record: *T. Johnson*, 1629.

1. Battles-well near Harefield,† *Blackst. Fasc.*, 87. Harefield Common, abundant, *T. & D.*, 47; extinct before 1884,† *Benbow MSS*. Ruislip Common,† *M. & G.*, 473. Low parts of Harrow Weald Common in plenty, *T. & D.*, 47; very scarce, c. 1935!,† *K. & L.*, 119. Bushey Heath,† *Webb & Colem. Suppl.*, 6. Stanmore Common, 1878,† *O'Dell, Williams MSS*. **4.** Hampstead Heath, *Johns. Eric.*, and many subsequent authors; in some plenty, amongst *Sphagnum*, in the bog behind 'Jack Straw's Castle', 1866, *Trimen* (BM); 1881, *De Ves. MSS.*; 1883, *C. Andrews* (BM); extinct by 1901,† *Whiting Notes*. **7.** Deliberately planted in West Meadow Bog, Ken Wood, 1960, teste *Hillaby*.

D. intermedia Hayne Long-leaved Sundew.

D. longifolia auct.

Formerly native in a bog on Hampstead Heath, the only authentic record being made in 1746. Deliberately reintroduced into another bog in 1960.

First certain record: *Blackstone*, 1746.

1. Bogs on Ruislip and Harrow Heaths in great abundance: . . . between Pinner and Stanmore (?=Harrow Weald Common), *M. & G.*, 474. Errors; *D. rotundifolia* intended. **4.** Hampstead Heath,† *Blackst. Spec.*, 84. **7.** Deliberately planted in West Meadow Bog, Ken Wood, 1960, teste *Hillaby*.

LYTHRACEAE

LYTHRUM L.

L. salicaria L. Purple Loosestrife.

Native on the verges of rivers, streams, ponds, etc., in marshes and wet places. Formerly very common, but now local and decreasing except by the Thames where it is still locally plentiful.

First record: *T. Johnson*, 1638.

1. Harefield!, *Blackst. Fasc.*, 56. Springwell. Yiewsley. Uxbridge. Uxbridge Moor. Cowley. **2.** Locally common by the Thames and Colne. **3.** Locally common by the Thames. By the Crane in various places between Twickenham and Hounslow Heath. Hayes. **4.**

Harrow Park!, *Harley Fl.*, 13. Hendon, *C. S. Nicholson*. **5.** Locally common by the Thames. Canal side near Brentford. Gunnersbury Park, Brentford. Chiswick House grounds. Marsh near Boston Manor. **6.** Tottenham,† *Johns. Cat.* Enfield. **7.** Chelsea,† *Blackst. Spec.*, 62. Fulham;† Hackney Marshes,† *T. & D.*, 107. By the lake, Buckingham Palace grounds!, probably planted, *Eastwood*. Lea Navigation Canal, Tottenham.

L. hyssopifolia L. Grass Poly.
Formerly native on damp heaths, in wet fields, etc., on gravelly soils, but long extinct.
First record: *Merrett*, 1666. Last certain record: *W. Hudson*, 1778.
1. Yiewsley, casual, *M. & S.* The adventive species *L. junceum* Banks & Sol. may have been seen. **2.** Marshy field between Staines and Laleham, *Nicholls, Blackst. Spec.*, 23. **3.** Hounslow Heath, 1766 (LT); c. 1778, *Huds. Fl. Angl.* 2: 59. **5.** Between Acton and Uxbridge, amongst the corn, *Merr. Pin.*, 59.

L. portula (L.) D. A. Webb subsp. **portula** Water Purslane.
Peplis portula L.
Native on pond margins, in wet places on heaths and commons, wet woods, etc. Rare.
First record: *T. Johnson*, 1633.
1. Harefield!, *Blackst. Fasc.*, 5. Uxbridge Common, 1839, *Kingsley* (ME); 1884,† *Benbow* (BM). Harrow Weald Common!; Ruislip Common!, *Melv. Fl.*, 30. Stanmore Common!; South Mimms!, *T. & D.*, 108. Eastcote,† *Trim. MSS.* Wrotham Park, 1965, *J. G.* and *C. M. Dony*!. **2.** Between Staines and Twickenham, *Wats. MSS.* Feltham Green,† *T. & D.*, 108. **3.** Hounslow Heath!;† near Hatton,† *T. & D.*, 108. **4.** Harrow, *Hind Fl.*, West Heath, Hampstead (SY); c. 1869;† Turner's Wood, *T. & D.*, 108. Mill Hill, *P. H. Cooke* (LNHS). Grimsdyke grounds, Harrow Weald!, *K. & L.*, 121. **6.** Hadley!, *Warren*; Edmonton, *T. & D.*, 108. Weed of rock garden, Enfield, 1965!, *Kennedy*. Gravel pit north of Enfield Lock, 1966, *Kennedy*!. **7.** Between Kentish Town and Hampstead,† *Johns. Ger.*, 615. Hyde Park, 1803,† *I'Anson* (BM).

THYMELAEACEAE

DAPHNE L.

D. laureola L. Spurge Laurel.
Introduced in woods, copses, thickets, etc., where it sometimes becomes naturalised. Very rare.
First record: *W. Turner*, 1551.

1. In a little grove near Breakspears, plentifully,† *Blackst. Fasc.*, 51. Woods between Uxbridge and Harefield,† *De Cresp. Fl.*, 120. Probably an error. **2.** Laleham Park!, *Green* (SLBI); 1968. **4.** Hampstead,† *Irv. MSS.* Harrow Park, one plant,† *Melv. Fl.*, 67. **5.** Syon,† *Turn. Hb.* 1, 198. **6.** Hadley Common,† *De Cresp. Fl.*, 120. Enfield!, *H. & F.*, 149; Whitewebbs Park!, *Eagles, K. & L.*, 246. **7.** Kentish Town,† *Mart. Pl. Cant.*, 72.

ONAGRACEAE

EPILOBIUM L.

E. angustifolium L. Rosebay, French Willow, Fireweed.
Chamaenerion angustifolium (L.) Scop., *Chamerion angustifolium* (L.) Holub
Denizen in woods and on heaths, and established on waste ground, railway banks, old walls, etc. Formerly rare, and known from only a few scattered localities in the vice-county until the end of the nineteenth century when it began its spectacular spread which accelerated in the first thirty years of the present century. It is now widespread and very common in all parts of the vice-county.
First record: *J. Hill*, 1760.
The records which follow are those known prior to 1910.
1. Harrow Weald!, introduced, *Hind, Melv. Fl.*, 29. Stanmore Common!, 1909, *C. S. Nicholson* (LNHS). **2.** Bank by the pier of the aqueduct which carries the water of the Queen's river across the Thames Valley Ry. not far from Hampton Station, in some abundance; solitary plants in two places in the unused roads, Fulwell, *T. & D.*, 108. **4.** East Finchley, 1909, *J. E. Cooper*. **6.** Coldfall Wood, Highgate!, 1901, *L. B. Hall, Lond. Nat. MSS.*; 1906; Muswell Hill!, 1907, *J. E. Cooper*. **7.** Ken Wood!, *Hill Fl. Brit.*, 201; c. 1813, *Bliss, Park Hampst.* On soil brought from elsewhere near Paddington Cemetery!, where it has held ground for four years (1868), *Warren, T. & D.*, 108. Site of the International Exhibition, South Kensington!, 1872, *Dyer List.* Whitehall!, 1891, *Cottam.*

E. hirsutum L. Great Hairy Willow-herb, Codlins and Cream.
Native by stream- and riversides, on pond verges, waste ground, in ditches, etc., established by canals, on railway banks, etc. Very common.
First record: *T. Johnson*, 1638.
A plant with very pale pink flowers was noted in **4.** Disused railway between Stanmore and Belmont, 1968, *Roebuck*. Plants with pure white flowers are uncommon but have been recorded from **1.** Potters Bar, 1960, *J. G. Dony*!. Near Yiewsley. **2.** Staines. **3.** Hounslow

Heath, *H. Banks*!, *K. & L.*, 122. **4.** Finchley, 1928, *J. E. Cooper*, *K. & L.*, 122. Hampstead Heath Extension, 1948, *H. C. Harris*. West Hampstead Cemetery, 1966. Harrow!, 1959, *Knipe*; 1966. Near Rayners Lane, 1966. **5.** Railway bank, Ealing, 1934–52!,† *K. & L.*, 122. South Greenford, 1967. **6.** Highgate Woods, 1901; North Finchley, 1917, *J. E. Cooper*; Enfield!, 1926, *Tremayne*, *K. & L.*, 122; 1966, *Kennedy*!. Muswell Hill, *J. E. Cooper*. **7.** Highgate Ponds!, *Fitter*. Highgate Hill, 1966. **7.** Kensington Gardens, 1961, *Allen Fl.* Swan Walk, Chelsea, 1967.

E. hirsutum × **parviflorum** = **E.** × **subhirsutum** Genn.
E. intermedium Ruhmer, non Mérat
1. Between Cowley and Hillingdon, *Benbow* (BM). **2.** West Drayton, *Druce Notes*. **5.** Ealing, 1968, *McLean*. **6.** Finchley Common. Trent Park, 1966!, *T. G. Collett* and *Kennedy*.

E. hirsutum × **roseum** = **E.** × **goerzii** Rubner
1. Between Cowley and Hillingdon, 1889, *Benbow* (BM).

E. parviflorum Schreb. Small-flowered Willow-herb.
Native by river- and streamsides, on pond verges, damp waste ground, in ditches, marshes, etc. Common.
First certain record: *Varenne*, 1830.
7. Regents Park!, 1830, *Varenne*; Eel-Brook Meadow, Parsons Green;† Hackney Marsh!, *T. & D.*, 109. South Kensington,† *Dyer List*. Whitehall,† *Cottam*. Hyde Park. Ken Wood. East Heath, Hampstead. Bombed sites, Cripplegate, Finsbury Square,† Gresham Street† and Queen Victoria Street areas,† *Jones Fl.* Kensington Gardens, one plant, 1961, *D. E. Allen*, *L.N.* 41: 18. Grounds of Chapel of St Barnabas, Manette Street, W.1, 1967, *McLean*.

E. parviflorum × **roseum** = **E.** × **persicinum** Reichb.
1. Near Cowley, 1889. *Benbow* (BM).

E. montanum L. Broad-leaved Willow-herb.
Native in woods, shady places, hedgebanks, on cultivated and waste ground, etc., and as a persistent garden weed. Very common.
First record: *T. Johnson*, 1633.
7. Common throughout the district.

E. montanum × **parviflorum** = **E.** × **limosum** Schur
1. Yiewsley, 1891, *Benbow* (BM).

E. montanum × **roseum** = **E.** × **mutabile** Boiss. & Reut.
7. Railway bank, Clapton, 1890, *F. J. Hanbury* (BM). Kensington Gardens, 1958!, *D. E. Allen*, det. *Ash*, *L.N.* 39: 50.

E. montanum × obscurum = E. × aggregatum Čelak.
1. Near Ruislip Common, 1890, *Benbow* (BM). **7.** Railway bank, Clapton, 1890, *F. J. Hanbury*, det. *E. S. Marshall* (BM).

E. lanceolatum Seb. & Mauri Spear-leaved Willow-herb.
Introduced. Naturalised on waste ground, by roadsides, etc., and as a weed of flower-beds, etc. Very rare, or overlooked.
First record: *Lousley*, 1945.
5. Ealing, 1969, *McLean*. **6.** Tottenham, 1967, *Kennedy*. **7.** Near Shadwell Basin, 1945, *Lousley*, *L.N.* 25: 14. Buckingham Palace grounds, 1961, *Fl. B.P.G.* Holloway, 1965, *D. E. G. Irvine*. Abundant weed of flower-beds, St Paul's Churchyard, Covent Garden, W.C.!, 1967, *McLean*; still present, but in reduced quantity, 1968→.
The presence of *E. lanceolatum* in the eastern half of the vice-county suggests that it may have originated from wind-borne seed from colonies in West Kent where the species is native and locally abundant in woods and shady places.

E. lanceolatum × montanum = E. × neogradiense Borbás
6. Near Edmonton, 1947!, det. *Ash*, *L.N.* 27: 31.

E. roseum Schreb. Small-flowered Willow-herb.
Native in woods, shady places, on hedgebanks, waste ground, etc., and established on railway banks, wall-tops, etc. Locally plentiful, especially in and near Central London, but absent from large areas of the vice-county.
First record: *Varenne*, 1830.
1. Common. **2.** East Bedfont!, *Welch*, *K. & L.*, 123. Hampton Hill. **3.** Hounslow. Hounslow Heath. **4.** Hampstead! (SY). Hampstead Heath!, *Bradley* (BM). Hendon, *P. H. Cooke* (LNHS). Mill Hill. Willesden. Cricklewood. Harrow!, *Harley Add.* Harrow Weald. **5.** Near Acton, 1846, *McIvor* (SY). Brentford!; Hanwell!; Ealing!, *K. & L.*, 123. Park Royal. Heston. **6.** Edmonton, *T. & D.*, 109. Southgate. Enfield. **7.** Locally common, especially in Central London and the East End.

E. roseum × tetragonum = E. × borbásianum Hausskn.
1. Near Ickenham, 1884; Breakspeares, 1890, *Benbow* (BM).

E. adenocaulon Hausskn.
Lit. *J.B.* (*London*) **73**: 177 (1935): *loc. cit.* **73**: 327 (1935).
Alien. N. America. Naturalised on heaths, cultivated and waste ground, in woods, marshes, by rivers, streams and ponds, etc., and established on railway banks, railway tracks, wall-tops, etc., also a persistent garden weed. Very common, and by far the most frequent species of *Epilobium* now encountered in the vice-county. Plants with white flowers are quite frequent.

First record: *Lousley*, 1945.
7. Abundant throughout the district.
E. adenocaulon appears to have entered Middlesex early in the Second World War, probably originating from colonies in Surrey, where it had long been naturalised, and to have spread with amazing rapidity into numerous natural and artificial habitats. The var. **occidentale** Trel. also appears to be present.

E. adenocaulon × hirsutum = E. × novae-civitatis Smejkal
3. Hounslow Heath, *Clement*, teste *Pennington*. **7.** Near Regents Park Underground Station, 1966!, *D. E. Allen*, *L.N.* 46: 30.

E. adenocaulon × montanum
1. Uxbridge, 1947!, det. *Ash*, *L.N.* 27: 31. Harefield, 1965. Pinner, 1967. **2.** Near Colnbrook, 1964. Bushy Park, 1967. *Clement*. **5.** Ealing, 1956. West Ealing, 1966. **6.** Muswell Hill, 1961!, *Bangerter* and *Raven*, *L.N.* 41: 14. Near Enfield, 1966; Southgate, 1967, *Kennedy*. **7.** Regents Park, 1964. Tottenham, 1967!, *Kennedy*.
This hybrid is probably common.

E. adenocaulon × parviflorum
6. Finchley Common!, 1948, *Welch*, teste *Ash*, *K. & L.*, 123; 1966. Trent Park, 1966!, *T. G. Collett* and *Kennedy*.

E. adenocaulon × roseum
7. Tottenham, 1967, *Kennedy*.

E. adenocaulon × tetragonum
4. Mill Hill, 1966, *Kennedy*!. **7.** Tottenham, 1967, *Kennedy*.

E. tetragonum L. subsp. **tetragonum** Square-stemmed Willow-herb.
E. adnatum Griseb.
Native in woods, by rivers and streams, on waste ground, etc., and established on railway banks, etc. Common.
First record: *Curtis*, 1778.
7. In the lane leading from Newington to Hornsey Wood,† *Curt. F.L.* Lanes beyond Hornsey Wood,† *Ball. MSS.* Regents Park, 1830, *Varenne*, *T. & D.*, 110; 1873† (BM). Sandy End, Fulham;† South Kensington,† *T. & D.*, 110. Kilburn,† *Warren*, *T. & D.*, 423. Finchley Road,† *Trim. MSS.* Kensington Gardens,† *Warren Fl.* Hampstead. Holloway. Dalston.

Subsp. **lamyi** (F. W. Schultz) Nyman
E. lamyi F. W. Schultz.
2. Near Hampton Court, *Haussknecht*, *J.B.* (*London*) 27: 145. Gravel pit, East Bedfont, abundant, *Brenan*!, *L.N.* 26: 60; 1965. **4.** Burgess Hill, Hampstead, 1866, *Trimen*; East Finchley, 1909, *J. E. Cooper*

(BM). Finchley, 1912, *E. F.* and *H. Drabble, Rep. Bot. Soc. & E.C.* 8: 530. **5.** Hanwell!, det. *Ash, K. & L.*, 123. **6.** Near Woodside Park, 1965, *Hinson*, det. *Pennington, L.N.* 45: 20.

E. obscurum Schreb.
Native by stream- and riversides, in woods, shady places, on damp waste ground, etc., and as a garden weed. Common.
First evidence: *Dillenius*, c. 1730.
7. East Heath, Hampstead!; Finchley Road, *T. & D.*, 110. Kensington Gardens!, *Warren Fl.* Hyde Park!; Buckingham Palace grounds!; Ranelagh Gardens, Chelsea!, *L.N.* 39: 50. Bombed site, Cripplegate, *Wrighton.* Hammersmith, Earls Court. Kilburn. Highgate. Stepney. Old Ford.

E. obscurum × **parviflorum** = **E.** × **dacicum** Borbás
1. Near Ruislip, 1890, *Benbow* (BM).

E. obscurum × **palustre** = **E.** × **schmidtianum** Rostk.
1. Cowley, 1889, *Benbow* (BM).

E. palustre L. Marsh Willow-herb.
Native in marshes, bogs, by streamsides, etc., especially on acid soils. Formerly local, but now very rare and decreasing.
First certain evidence: *La Gasca*, 1826.
1. Cowley, 1887,† *Benbow* (BM). Moorhall,† *Benbow, J.B. (London)* 23: 37. Bayhurst Wood, Ruislip, *Wrighton, K. & L.*, 124. **2.** Between Hampton and Hampton Court,† *Newbould, T. & D.*, 110. Hampton Court! (W). Near Poyle!,† *L.N.* 26: 60. Walton Bridge!,† *K. & L.*, 124. Yeoveney,† *Benbow MSS.* **3.** Hounslow Heath, 1866,† *Trimen* (BM). Gutteridge Wood, Northolt!,† *K. & L.*, 124. **4.** Hampstead Heath, 1826, *La Gasca Hort. Sicc.*; c. 1869, *T. & D.*, 110; 1939,† *Fitter.* Brent Reservoir, 1884,† *Wright MSS.* Brent Park, Hendon!,† *K. & L.*, 124. Harrow, *Harley Fl.*, 14. **5.** Near Greenford,† *Benbow MSS.* **6.** Edmonton,† *T. & D.*, 111. Queen's Wood, Highgate,† *C. S. Nicholson, K. & L.*, 124. Whitewebbs Park, Enfield, 1965, *Kennedy*!. **7.** Railway bank, Baker Street, N.W.1,† *Trim. MSS.* Perhaps an error.

E. brunnescens (Cockayne) Raven & Engelhorn subsp. **brunnescens**
E. pedunculare A. Cunn. var *brunnescens* Cockayne, *E. nerterioides* A. Cunn.
Lit. *J. Ecol.* **49**: 753 (1961).
Alien. New Zealand. Garden escape. Naturalised in damp grassy places. Very rare, and perhaps extinct.
First, and only evidence: *Libbey*, 1945.
1. Potters Bar golf course, 1945, *Libbey* (L).

OENOTHERA L.

O. biennis L. Evening Primrose.

Alien. N. America. Garden escape. Established on waste ground, railway banks, etc. Rather rare, and sometimes confused with *O. erythrosepala*.

First record: *A. Irvine*, 1838.

1. Uxbridge!, *P. H. Cooke* (LNHS). Yiewsley!, *M. & S.* Eastcote. **2.** West Drayton!, *Druce Notes*. East Bedfont, 1965. **3.** Twickenham, one plant, *T. & D.*, 111. **4.** Hendon!, *Irv. Lond. Fl.*, 149. Harrow!, *Horton Rep.* Cricklewood, 1966. Mill Hill, 1966, *Kennedy*!. Stonebridge Park Power Station. **5.** Chiswick, 1887. *Benbow* (BM); 1965!, *Murray*. Brentford!, *Henrey*, 85. **6.** North Finchley, 1909, *J. E. Cooper*; 1945, *Fitter*. Edmonton, 1966. Finchley Common, 1966. **7.** Chelsea, *Britten*; Cremorne, *Fox*; Paddington, *Warren*; Shepherds Bush; Victoria Street, S.W.1,† *T. & D.*, 111. Regents Park!, *L.N.* 39: 50. South Kensington, *T. & D.*, 111; 1907,† *Boulger*, det. *de Vries*, *J.B.* (*London*) 45: 355. Bloomsbury, *Shenstone Fl. Lond.* Canonbury; Hoxton; railway banks, West Brompton and Walham Green, *Fitter*. East Heath, Hampstead!, *H. C. Harris*. Bombed sites, Cripplegate,† *Wrighton*, *L.N.* 29: 87. Hyde Park Corner, 1961–62,† *Allen Fl.* St Katherine's Dock, Stepney!, 1963, *Fitter*; 1966. Tower Hill, 1965, *Brewis*, *L.N.* 46: 30. Isle of Dogs, 1966. Old Ford, 1967. It is probable that many of the older records given above are referable to *O. erythrosepala*.

O. erythrosepala Borbás Large-flowered Evening Primrose.

O. lamarckiana auct.

Alien. N. America. Garden escape. Established on railway banks, waste ground, in gravel pits, etc. Common, and increasing.

First record: *G. C. Druce*, 1910.

7. Chelsea!, *L.N.* 39: 50. A large colony by the railway near Paddington, 1964!, *L.N.* 44: 23. Tower Hill, 1965, *Brewis*; waste ground, Paddington, a large colony, 1965!;† Camden Town, 1966!; St Pancras Goods Yard, 1966!, *L.N.* 46: 30. Hammersmith. West Brompton. Earls Court. Isle of Dogs. Islington. Highbury. Finsbury Park. Hornsey. Dalston. Stoke Newington. Hoxton. Clapton. Stepney. Tottenham. Old Ford.

O. parviflora L. Small-flowered Evening Primrose.

O. muricata L.

Alien. North America. Garden escape. Established on waste ground, in gravel pits, etc. Very rare.

First record: *D. H. Kent*, 1945.

1. Eastcote, 1966. **4.** Hendon, 1945. Harlesden, 1973. **6.** Near Prickett's Lock, Edmonton, 1966, *Kennedy*!. **7.** Hyde Park Corner, 1962!,† *Brewis*, *L.N.* 42: 12.

CIRCAEA L.

C. lutetiana L. Common Enchanter's Nightshade.
Lit. *Watsonia* **5**: 262 (1963).
Native in woods, shady places, on hedgebanks, etc., also a persistent garden weed. Very common.
First record: *Blackstone*, 1737.
7. Frequent throughout the district.
In view of the present-day abundance of *C. lutetiana* as a weed in Central London and the East End it is curious that the species was not recorded by the seventeenth-century botanists. It seems probable that had the plant been present it would have been noted by Gerarde, Johnson, Petiver, Buddle or Doody, whose joint activities covered the entire area of what is now Central London. This, together with the very few records for the Metropolitan district cited by Trimen and Dyer, suggests that the spread of the species into built-up areas may have taken place during the last hundred years.

HALORAGACEAE

MYRIOPHYLLUM L.

M. verticillatum L. Whorled Water-milfoil.
Native in slow streams, ponds, lakes, etc. Very rare.
First record: *Lightfoot*, c. 1780.
1. Near Uxbridge,† *Lightf. MSS.* Uxbridge,† *Benbow MSS.* Harefield Moor!, 1884, *Benbow* (BM); 1907, *Green* (SLBI); 1966. Uxbridge Moor,† *Benbow*, *J.B.* (*London*) 23: 37. Gravel pit, Springwell, 1965, *Pickess*. **2.** West Drayton,† *Newbould*; pool on common by Walton Bridge,† *T. & D.*, 111. Shortwood Common;† Staines Moor!, *Benbow*, *J.B.* (*London*) 25: 16. Near Staines, 1882, *H.* and *J. Groves*. Stanwell Moor, 1913, *J. E. Cooper* (BM). Longford river, Stanwell!,† *Grigg*. Longford river near East Bedfont, in small quantity, 1966. **3.** Canal backwater north of Southall, 1947!,† *L.N.* 27: 31. **4.** Harrow Park lake, *Horton Rep.* Error. **5.** Greenford,† *Coop. Fl.*, 108. Near Greenford,† *Farrar*, *T. & D.*, 111. **7.** Stoke Newington,† *Woods*, *B.G.*, 12. Near Newington, 1839,† *E. Ballard* (BM). Floating in the Hackney Canal,† *T. & D.*, 112.

M. spicatum L. Spiked Water-milfoil.
Native in ponds, lakes, slow streams, etc. Locally plentiful.
First record: *Merrett*, c. 1670.
1. Near Harefield, *Blackstone*; Ruislip Reservoir!, *Benbow* (BM). Ruislip Common, *Wrighton*!, *L.N.* 27: 31. Wrotham Park, *J. G.* and *C. M. Dony*. **2.** Hampton Court!, *Pet. Midd.* Between Teddington

and Bushy Park,† *Newbould*; Bushy Park!; river between Hampton Court and Kingston Bridge,† *T. & D.*, 112. Shortwood Common, abundant; near Yeoveney, abundant, *Welch*!; Thames at Walton Bridge!, *L.N.* 27: 31. Colnbrook, *Benbow MSS.* Shepperton. East Bedfont!, *Welch, K. & L.*, 120. **3.** Hounslow Heath. Longford river, Feltham to Hanworth. **4.** Bentley Priory, *T. & D.*, 112. Hampstead Heath!, *Hampst.* Between Finchley and Mill Hill, *Newbould, T. & D.*, 423. **5.** In the river (=Brent) betwixt Harrow and London, running to Brentford,† *Merr. MSS.* Greenford,† *Hind Fl.* Syon Park!, *T. & D.*, 112. **6.** Edmonton, *T. & D.*, 112. New river, Enfield, abundant, *Bangerter* and *Kennedy*!. Town Park, Enfield!, *Kennedy.* **7.** Near Poplar,† *Pet. Midd.* Isle of Dogs, 1792,† *E.B.*, 83. Round Pond, Kensington Gardens!, 1868, *Warren* (BM); 1958!, *D. E. Allen, L.N.* 39: 50. Fulham,† *Britten*; Walham Green, 1845,† *J. Morris*; East Heath, Hampstead!, *T. & D.*, 112. Lake in Buckingham Palace grounds!,† *Eastwood, L.N.* 39: 50.

M. alterniflorum DC. Alternate-leaved Water-milfoil.
Native in ponds, lakes, slow streams, etc. Very rare, and almost extinct.
First record: *Doody*, c. 1696.
1. Harefield,† *Blackst. Fasc.*, 81. Between Ruislip Reservoir and Northwood;† pond on Duck's Hill Farm, Northwood,† *Benbow, J.B. (London)* 22: 279. Swakeleys Park, 1884,† *Benbow* (BM). Uxbridge.† **2.** Staines Moor!, *K. & L.*, 120. Shortwood Common. **3.** In the river (=Crane) on Hounslow Heath, *Doody MSS.*; c. 1705,† *Buddle* (SLO). **4.** Near Edgware,† *Benbow, J.B. (London)* 25: 364.

HIPPURIDACEAE

HIPPURIS L.

H. vulgaris L. Mare's-tail.
Native in ponds, lakes, ditches and slow streams. Very rare.
First record: *Blackstone*, 1737.
1. Uxbridge Moor;† Harefield!, *Blackst. Fasc.*, 26; still occurs in small quantity in at least one gravel pit on Harefield Moor, 1970. Uxbridge, 1793,† *Dicks Hort. Sicc.* Between Yiewsley and Iver Bridge, *Newbould, T. & D.*, 113; 1884,† *Benbow MSS.* Still occurs just over the vice-county boundary in the Slough arm of the Grand Junction Canal, Buckinghamshire. Many places between Denham Lock and Harefield, *Benbow MSS.*; not so now, though it may survive in small quantity in a few isolated areas. Cowley,† *Benbow MSS.* **2.** Thames above Shepperton,† *Newbould, T. & D.*, 424. Thames side just below Walton Bridge, 1838† (W). Above Chertsey Bridge, 1907,† *Green* (SLBI). Bushy Park!, *Benbow* (BM); 1965. Staines,† *P. H. Cooke.*

West Drayton,† *P. H. Cooke* (LNHS). **3.** Old cut, Southall, abundant,† *Benbow MSS.* **4.** Hampstead,† *Coop. Fl.*, 100. **6.** Enfield!, *L. B. Hall, K. & L.*, 119. **7.** Near Hornsey,† *Huds. Fl. Angl.*, 2. Near Stoke Newington;† Highgate,† *M. & G.*, 8. Between Stamford Hill and Hornsey,† *E. Forster* (BM). Viaduct Pond, East Heath, Hampstead, *Hampst.* Error; *Myriophyllum spicatum* possibly intended.

CALLITRICHACEAE

CALLITRICHE L.

Lit. *Watsonia* **3**: 186 (1955): *Proc. Bot. Soc. Brit. Isles* **3**: 28 (1958): *loc. cit.* **6**: 380 (1967): *loc. cit.* **7**, 424 (1968). (*Abstr.*)

C. stagnalis Scop.
Native in ponds, streams, ditches, on wet mud, etc. Common.
First certain evidence: *J. F. Young*, 1831.
7. Paddington, 1831,† *J. F. Young* (K). Ken Wood. Viaduct Pond, East Heath, Hampstead. Hackney Wick.

C. obtusangula LeGall
Native in ditches, streams, ponds, etc. Very rare, or confused with *C. stagnalis* from which it is scarcely indistinguishable when sterile, though very different in fruit.
First record: *H.* and *J. Groves*, 1877.
1. Uxbridge; Harefield, *Druce Notes.* **2.** Stream near Sunbury, *H.* and *J. Groves, Rep. Bot. Loc. Rec. Club* 1877 *Rep.*, 215. Near Staines; Poyle; West Drayton; ditch at Colnbrook, 1900, *Druce Notes.*

C. platycarpa Kütz.
C. polymorpha auct., *C. palustris* auct., non L., *C. verna* auct., *C. vernalis* auct.
Native in ponds, streams, ditches, etc. Common, but confused with *C. stagnalis*.
First certain evidence: *Hind*, 1861.
7. East Heath, Hampstead.

C. hamulata Kütz. ex Koch.
C. intermedia auct.
Native in ponds, lakes, slow streams, ditches, etc. Very rare, or overlooked.
First certain evidence: *Hind*, 1860.
1. Near Ruislip Reservoir, *Hind* (BM). **2.** Shortwood Common!; near Staines Moor!, *K. & L.*, 121. Bushy Park. Longford river, Stanwell!,† *K. & L.*, 121. **5.** Chiswick House grounds. **6.** Pond, Enfield Chase, 1966.

LORANTHACEAE

VISCUM L.

V. album L. Mistletoe.
Native or denizen. Parasitic on trees and shrubs. Rare.
First record: *S. Brewer*, 1691.
1. Harefield, on various trees!, *Blackst. Fasc.*, 111; Knightscote Farm, rare on *Crataegus* and *Malus*, 1958!; Breakspeares!, *Pickess.* Northwood, on *Crataegus* and *Malus*, *W. A. Tooke*, *J.B. (London)* 9: 271; 1950!, *Graham*, *K. & L.*, 247; 1965. Waxwell Farm, Pinner, *Minnion.* Bayhurst Wood, near Harefield!, *Milner.* **2.** Hampton Court!, in great plenty, *Brewer MSS.* Bushy Park!, on the following trees, *Tilia* sp.!, *Acer opulus*, *A. rubrum*, *Robinia pseudoacacia*!, *Laburnum* sp., *Sorbus aucuparia*, *Aesculus hippocastanum*, *Ulmus* sp., *Jesse*, *Gentleman's Mag. n.s.* 1: 72; on *Crataegus*, 1867, *Trimen* (BM). Dawley, on *Malus*, 1865,† *Benbow MSS.* **3.** Hayes, on *Malus*,† 1865, *Benbow MSS.* Cranford Park!, *Smith*; on *Tilia* sp.!, and other trees!, *Boniface*, *K. & L.*, 247; 1967. **4.** Canons, Edgware, on *Tilia* sp.; Mill Hill, on *Juglans* sp., *Knowlton*, teste *P. Collinson*, *Dillwyn Hort. Coll.*, 57. c. 1926, *Moring MSS.* **5.** Ealing, on *Malus*, 1965!→, *T. G. Collett.* **6.** Bone (=Bohunt or Barnet) Gate, on *Tilia* sp., *Knowlton*, teste *P. Collinson*, *Dillwyn Hort. Coll.*, 57. Forty Hill, Enfield, 1940, *Johns*, *K. & L.*, 247; 1967, *Kennedy.* Myddleton House grounds, Enfield, on *Crataegus*, 1966, *Kennedy*!. **7.** Clarendon House, St James's,† *Pet. Midd.* Waterlow Park, Highgate, on *Tilia* sp. and *Populus* sp., *Andrewes MSS.* Holland Park.

CORNACEAE

CORNUS L.

C. sanguinea L. Dogwood, Wild Cornel.
Thelycrania sanguinea (L.) Fourr., *Swida sanguinea* (L.) Opiz
Native on wood borders, hedges, thickets, etc. Common, except in heavily built-up areas where, owing to the eradication of its habitats, it is very rare or extinct.
First record: *T. Johnson*, 1638.
7. Marylebone fields, 1815† (G & R); Hornsey Wood;† East Heath, Hampstead!, *T. & D.*, 137. Near Kensal Green, *Warren*, *T. & D.*, 424.

ARALIACEAE

HEDERA L.

H. helix L. subsp. **helix** Ivy.
Native in woods, hedges, thickets on waste ground, etc. Very common, though frequently planted for ornamental purposes.
First record: *T. Johnson*, 1638.
7. Common in the area.

BEGONIACEAE

BEGONIA L.

B. evansiana Andr.
Alien. Malaysia, China, Japan. Garden outcast or escape. Naturalised by the Thames. Very rare.
First record: *Street*, 1973.
2. Banks of the Thames, Staines, *Street*, *Gard. News*, 11 May 1973; 25.

HYDROCOTYLACEAE

HYDROCOTYLE L.

H. vulgaris L. Marsh Pennywort, White-rot, Sheep-rot.
Native in marshes, bogs, wet places, etc., particularly on heaths and acid soils. Locally common.
First record: *T. Johnson*, 1629.
1. Harefield Common!,† *Blackst. Fasc.*, 21. Harefield Moor!; Uxbridge Common;† Hillingdon Heath;† Swakeleys,† *Benbow MSS.* Pinner,† *Tooke*, *Trim. MSS.* Harrow Weald Common!, *Hind Fl.* Ruislip Common!; Ruislip Reservoir!. *Melv. Fl.*, 33. Stanmore Common!, *T. & D.*, 120. Potters Bar!, *Benbow MSS.* **2.** Bushy Park!, *Newbould*, *T. & D.*, 120. Verges of Hampton Wick pond, Hampton Court Park. Staines Moor!, *L.N.* 28: 33. East Bedfont,† *P. H. Cooke*, *K. & L.*, 126. Weed in flower tub, London Airport, 1965, *Perring*. **3.** Hounslow Heath!; Whitton Park Inclosure, *T. & D.*, 120; c. 1884,† *Benbow MSS.* **4.** Hampstead Heath, *Johns. Eric.*; West Heath, c. 1869, *T. & D.*, 120; 1922,† *Richards.* Grimsdyke grounds, Harrow Weald!, *K. & L.*, 126. Brockley Hill, 1966, *Kennedy*!. **6.** Finchley Common,† *Benbow MSS.* Whitewebbs Park, Enfield!, *L. M. P. Small.* Trent Park, *Kennedy*. **7.** East Heath, Hampstead!,

T. & D., 120; 1968. Ken Wood!, *Andrewes MSS.* Tottenham,† *Johns. Cat.* Isle of Dogs,† *Cowper*, 107. Edge of lake, Buckingham Palace grounds!, *Eastwood, K. & L.*, 126.

UMBELLIFERAE

SANICULA L.

S. europaea L. Wood Sanicle.

Native in woods and shady places. Rare, and confined to the northern parts of the vice-county.

First record: *Petiver*, 1713.

1. Harefield!, *Blackst. Fasc.*, 90. North of Harefield. Garett Wood, Springwell!,† *Benbow MSS.* Northwood, *J. A. Moore.* Pinner Wood, *Hind Fl.* Stanmore Common!, *K. & L.*, 126; locally plentiful, 1967. Harrow Weald Common, a single plant, 1948.† Potters Bar!; Mimmshall Wood!, *Benbow MSS.* **4.** Hampstead, *Pet. H.B.C.*; c. 1912,† *Whiting Fl.* Bishop's Wood, 1860, *Trimen* (BM). Bentley Priory, *Melv. Fl.*, 33. Stanmore, *Varenne, T. & D.*, 120. Scratch Wood!, *Benbow MSS.*; 1967. Brockley Hill, 1966, *Kennedy*!. **6.** Highgate Woods, *Coop. Fl.*, 104; c. 1900,† *C. S. Nicholson, K. & L.*, 120. Hadley Wood!, *Newbould, T. & D.*, 120. Enfield!, *Benbow MSS.* Hadley Common. **7.** Ken Wood,† *Mart. Pl. Cant.*, 68.

ASTRANTIA L.

A. major L. Master Wort.

Alien. Europe. Naturalised in grassy places. Very rare.

First record: *Welch*, 1946.

1. Plentiful on the sides and bed of a dried-up pond, Eastcote House grounds, and sparingly in grassy places in Haydon Hall grounds, Eastcote, 1966!, *L.N.* 46: 13. **2.** Near the Canal, Hampton Court Park, 1946!, *Welch, K. & L.*, 126. **4.** Waste ground by railway, East Finchley, 1968, *L. M. P. Small.* **7.** Regents Park, 1953, *Holme, K. & L.*, 328.

CHAEROPHYLLUM L.

C. temulentum L. Rough Chervil.

Native on hedgebanks, waste ground, in ditches, fields, etc. Common, except in heavily built-up areas where it is very rare or extinct.

First record: *Blackstone*, 1737.

7. Upper Clapton;† between Lee Bridge and Tottenham, *Cherry, T. & D.*, 134. Islington, common,† *Ball. MSS.* Lee Bridge, 1839,† *R. Pryor* (BM). Kensington Gardens, 1942!,† *L.N.* 41: 19. Finsbury Park, 1966.

C. aureum L.
Alien. Europe, etc. Established in grassy places. Very rare.
First record: *McClintock*, 1950.
7. Well established and regenerating vigorously in the grounds of Buckingham Palace!, though no doubt originally planted, *McClintock*, *L.N.* 39: 51: *loc. cit.* 41: 19.

ANTHRISCUS Pers.

A. caucalis Bieb. Bur Chervil.
A. scandacina (Weber) Mansf., *A. vulgaris* Pers., non Bernh., *A. neglecta* Boiss. & Reut., *Chaerophyllum anthriscus* (L.) Crantz
Denizen on heaths, railway banks, in waste places, etc. Rare, and often merely casual.
First record: *T. Johnson*, 1632.
1. Harefield,† *Blackst. Fasc.*, 17. Between Harefield and Ruislip!;† Elstree;† South Mimms, *T. & D.*, 134. Near Uxbridge Common,† *Benbow, J.B. (London)* 25: 16. Yiewsley, *J. E. Cooper* (BM). Near Pield Heath, 1946!,† *K. & L.*, 131. Ruislip!, *Wrighton, L.N.* 27: 32. **2.** Shepperton, *Monckt. Fl.*, 35. West Drayton!, *Druce Notes*; 1967. **3.** Railway embankment, Hayes, 1902, *C. S. Nicholson, K. & L.*, 131; 1909, *P. H. Cooke* (LNHS). Northolt!,† *K. & L.*, 131. Hounslow Heath, two plants, 1947!,† *Welch, L.N.* 27: 32. **4.** Hampstead Heath,† *Johns. Enum.* **5.** Chiswick, *T. & D.*, 134; abundant, 1888,† *Benbow* (BM). Railway bank, Kew Bridge, 1872,† *Britten, Trim. MSS.* **6.** Between Edmonton and Ponders End,† *Cherry, T. & D.*, 134. **7.** Brompton,† *J. Banks*; Tollington Park, 1875,† *French* (BM). Stoke Newington, 1800† (BO). Plentiful about Fulham, towards Hammersmith,† *De Cresp. Fl.*, 15. Hackney Marshes, 1864,† *T. & D.*, 134. Palace gardens, Kensington, 1872,† *Warren, Trim. MSS.* Hyde Park, one plant, 1961,† *D. E. Allen*; Green Park, 1961,† *D. E. Allen* and *McClintock, L.N.* 41: 19.

A. sylvestris (L.) Hoffm. Cow Parsley, Hedge Parsley, Keck.
Chaerophyllum sylvestre L.
Native on hedgebanks, field borders, by roadsides, in ditches, woods, etc., and established on railway banks, waste grounds, etc. Very common.
First record: *T. Johnson*, 1638.
7. Frequent in the district.

Var. **latisectum** Druce
1. Uxbridge, *Druce Notes.*

SCANDIX L.

S. pecten-veneris L. Shepherd's Needle.
Formerly as a colonist in cornfields and cultivated ground, but now sporadic and little more than a casual on waste ground, canal paths, by roadsides, etc. Very rare.
First record: *T. Johnson*, 1638.
1. Harefield!, *Blackst. Fasc.*, 73. Uxbridge!; near Springwell, *Benbow MSS.* Pinner, *Druett.* Northwood, *Bain, L.N.* 28: 30. South Mimms. **2.** West Drayton!; Colnbrook!; between Colnbrook and Staines!; between Staines and Laleham, *Benbow MSS.* **3.** Wood End, *Hind Fl.* Near Hounslow, *Newbould, T. & D.*, 133. Hounslow Heath. Southall!, *Wright MSS.* **4.** Hampstead,† *Irv. MSS.* Kenton,† *Hind Fl.* Harrow,† *Harley Fl.*, 15. Whitchurch Common,† *L. B. Hall, Lond. Nat. MSS.* Near Brent Reservoir, *Warren*,† *T. & D.*, 424. **5.** Near Ealing!; Acton,† *Newbould, T. & D.*, 133. Hanwell, 1935.† Canal path between Hanwell and Southall, almost annually, sometimes in quantity. **6.** Tottenham,† *Johns. Cat.* Finchley Common† (HE). Wood Green, 1937, *P. H. Cooke.* **7.** Marylebone† (G & R). Brompton,† *J. Banks* (BM). Kentish Town† (HE). Ball's Pond† and Newington,† *Ball. MSS.* Hackney Marshes, 1924,† *J. E. Cooper.* Bombed site, Bush Lane, E.C., 1945!;† Old Jewry, E.C., 1948,† *Lousley, L.N.* 28: 30. Bombed sites, Finsbury Square† and Cripplegate areas,† *Jones Fl.*

MYRRHIS Mill.

M. odorata (L.) Scop. Sweet Cicely.
Introduced. Cultivated as a pot herb and now established on railway banks, etc. Very rare.
First record: *Scholey*, 1950.
1. Deliberately sown on the banks of the canal, Harefield, teste *Donovan, L.N.* 37: 18. **2.** Well-established colonies on both sides of the railway embankment near Hampton Wick station, 1963, *Ryves*; 1966. **5.** Osterley, 1969, *L. M. P. Small.* **6.** Southgate, 1950, *Scholey, L.N.* 30: 7; 1951–53,† *Scholey.*

TORILIS Adans.

T. japonica (Houtt.) DC. Upright Hedge-parsley.
T. anthriscus (L.) C. C. Gmel., *Caucalis anthriscus* (L.) Huds.
Native on hedgebanks, wood borders in ditches, by roadsides, etc. Common, except in heavily built-up areas where it is rare or extinct.
First record: *T. Johnson*, 1638.
7. Tottenham, *Johns. Cat.*; 1839, *Pryor* (BM). Newington; Hornsey,

Ball. MSS. Side of Hackney Canal, *Cherry*; Hackney Wick, *T. & D.*, 132. Kensington Gardens, 1951 !,† *L.N.* 39: 51.

T. arvensis (Huds.) Link subsp. **arvensis** Spreading Hedge-parsley.
T. infesta (L.) Spreng., *Caucalis arvensis* Huds.
Colonist in cornfields, on cultivated ground, etc., chiefly on light soils. Very rare, and decreasing.
First record: *Blackstone*, 1737.
1. Harefield, *Blackst. Fasc.*, 16; c. 1869, *T. & D.*, 133; 1913–19,† *J. E. Cooper, K. & L.*, 136. Uxbridge, 1889,† *Benbow MSS.* Uxbridge Moor,† *Benbow, Druce Fl.*, 164. Springwell,† *Benbow MSS.* Near Potters Bar, 1899, *L. B. Hall* (BM). South Mimms !, *L.N.* 27: 32; 1966. **2.** Near Sunbury, *H. C. Watson, Wats. New Bot.*, 99. Staines, 1885, *Broadway*; near Colnbrook, 1913, *J. E. Cooper* (BM); Harmondsworth; between Harmondsworth and Stanwell, 1945, *Welch* !, *L.N.* 26: 61. **3.** Near Hounslow, 1866,† *Trimen* (BM). Hounslow,† *Benbow MSS.* **4.** Hampstead,† *Irv. MSS.* **7.** Paddington, 1830,† *Varenne, T. & D.*, 133.

Subsp. **leptophylla** (Guss.) Thell.
T. leptophylla (L.) Reichb.
1. Uxbridge, 1925,† *A. Cobbe, Rep. Bot. Soc. & E.C.* 7: 574.

T. nodosa (L.) Gaertn. Knotted Hedge-parsley.
Caucalis nodosa (L.) Scop.
Native in dry sunny places. Locally common by the Thames, very rare elsewhere.
First record: *T. Johnson*, 1633.
1. Harefield Park, 1888;† Uxbridge Moor, 1883,† *Benbow* (BM). Uxbridge, 1907,† *Green* (SLBI). Yiewsley, 1922,† *J. E. Cooper, K. & L.*, 136. **2.** Locally common by the Thames between Staines and Hampton. Staines Moor, 1966, *J. E. Smith* !. West Drayton, 1914,† *Coop. Cas.* **4.** Hampstead,† *Irv. MSS.* Fortis Green, 1911,† *Coop. Cas.* Mill Hill, *Harrison*, teste *Sandwith.* **5.** Canal path between Hanwell and Southall, 1946, *Welch* !, *L.N.* 26: 61; 1947–60.† **6.** Finchley Common, 1826† (HE). Tottenham, 1962. **7.** Banks about St James's† and Piccadilly,† *Johns. Ger.*, 1023. Isle of Dogs, 1866, *Trimen* (BM); 1896,† *Warren, T. & D.*, 133. Finchley Road, 1869,† *T. & D.*, 423. Holloway, 1875,† *French*; Hackney Marshes, 1915,† *J. E. Cooper* (BM). Hornsey,† *J. E. Cooper, K. & L.*, 136. By the lake, Buckingham Palace grounds, *Codrington, Lousley* and *McClintock* !, *L.N.* 39: 51.

Var. **pedunculata** (Rouy & Fouc.) Druce
2. Penton Hook !, *C. E. Britton, Rep. Bot. Soc. & E.C.* 5: 381. Laleham !, *L.N.* 26: 61. **4.** Finchley, 1912,† *E. F.* and *H. Drabble* (BM).

SMYRNIUM L.

S. olusatrum L. Alexanders.

Alien. S. Europe, etc. Formerly cultivated as a pot-herb, and now naturalised on hedgebanks, in ditches, by roadsides, etc. Very rare.

First record: *Blackstone*, 1736.

1. Harefield, *Blackstone in litt. to Richardson*, 8 Dec. 1736; 1907,† *Loydell* (SLBI). Cowley, *Blackst. Fasc.*, 42; abundant in several places about Cowley and Cowley Peachey, known here for nearly seventy years, 1902, *Benbow MSS.*; between Cowley and Uxbridge, two plants, 1946!,† *K. & L.*, 127. Uxbridge,† *Sibth. MSS.* **2.** Hampton Court!, *Francis, Coop. Suppl.*, 12; still grows in abundance under the wall of Hampton Court Park and gardens, and in various other localities by the river between Hampton Court and Kingston Bridge, 1973. Near Charlton, a small colony in a ditch, 1956. West Drayton, 1910,† *P. H. Cooke* (LNHS). **3.** Hounslow, 1947,† *Westrup, K. & L.*, 127. **4.** Among the ruins of old Stanmore Church,† *Longman, Melv. Fl.*, 36. **7.** Millbank, by the side of the river;† one plant on the wall at Whitehall Stairs,† *M. & G.*, 431.

S. perfoliatum L.

Alien. S.-E. Europe, etc. Naturalised in grassy places, on waste ground, etc. Very rare.

First record: *Talbot*, 1932.

1. Eastcote, 1954, *Milner, L.N.* 34: 15. **4.** Harrow Park, 1954, *Harley Add.* **7.** Burton's Court, Chelsea!, *Talbot, Rep. Bot. Soc. & E.C.* 11: 30; still there in fair quantity, 1969; a single plant seen, 1973. Abundant weed in Chelsea Physic Garden, 1968, *L. M. P. Small.*

CONIUM L.

C. maculatum L. Hemlock.

Native by stream- and riversides, roadsides, in fields, on waste ground, etc. Common, especially on heavy soils, though less frequent near the Thames.

First record: *T. Johnson*, 1638.

7. Tottenham!, *Johns. Cat.* Marylebone, 1815 (G & R). Between Haverstock Hill and Kentish Town!, *Irv. MSS.* Gospel Oak! and Five-acre Fields between Hampstead and Kentish Town (HE). South Kensington,† *T. & D.*, 135. Finchley Road, 1870;† West End [Hampstead], 1870,† *Trim. MSS.* Islington!, *Ball. MSS.* Isle of Dogs!, *Cowper*, 107. Hackney. Hyde Park, in two places . . ., *D. E. Allen, L.N.* 39: 51. Green Park, Piccadilly, on disturbed ground, 1961,† *D. E. Allen* and *McClintock, L.N.* 41: 19. Canal side near Paddington Basin, frequent, 1966!; canal side, Camden Town, 1966!, *L.N.* 46: 31. Notting Hill Gate. Clapton.

BUPLEURUM L.

B. rotundifolium L. Hare's-ear, Thorow-wax.
Formerly a colonist in cornfields, on cultivated ground, etc., but now extinct.
First record: *Blackstone*, 1735. Last record: *Garlick*, 1918.
1. In 1735, in a field of corn near Harefield Mill, but not plentifully, *Blackst. Fasc.*, 75. Pinner, a single plant, *Hind, J.B. (London)* 9: 271. **4.** Hampstead, 1918, *Garlick, Rep. Hampst. Sci. Soc.* 1920, 45. **5.** Between Hanwell and Ealing, *E. C. White, J.B. (London)* 9: 271. **6.** Whetstone, 1906, *Coop. Cas.*
The European adventive *B. lancifolium* Hornem., commonly introduced with bird-seed, is sometimes mistaken for *B. rotundifolium*, and the two twentieth-century records may be in need of revision.

B. tenuissimum L. Slender Hare's-ear.
Formerly a denizen in rough grassy places, on waste ground, etc., but now extinct.
First record: *Merrett*, 1666. Last evidence: *Lees*, c. 1860.
5. Ealing Common, abundant, *Lees* (BM). **6.** Fields to the north of Highgate, 1842, *Mitten* (K). **7.** Paddington, *Merr. Pin.*, 17.
The former occurrence of this species in the vice-county is curious, and it may perhaps have been introduced with sea gravel.

APIUM L.

A. graveolens L. subsp. **graveolens** Wild Celery.
Native on banks and mud, and established on river walls, of the tidal Thames. Very rare.
First record: *Petiver*, c. 1710.
2. River wall of the Thames near Kingston Bridge, 1885, *Benbow* (BM); 1886,† *Benbow MSS.* **5.** Brentford!, *A. Wood, Trim. MSS.*; river wall, 1957. Thames bank between Barnes Bridge and Chiswick in several places, 1887,† *Benbow MSS.* **7.** Between Westminster and Chelsea,† *Pet. Bot. Lond.* In Tower ditch,† and on the walls;† on a wharf at Millbank;† in the crevices of the tombs in Pancras Churchyard,† *M. & G.*, 429. The latter record was almost certainly an error. Isle of Dogs, 1866, *Trimen* (BM); 1869, *Warren, T. & D.*, 121. River wall, London Docks, *Lousley, Watsonia* 1: 250.

Subsp. **dulce** (Mill.) Lemke & Rothm. Cultivated Celery.
A. dulce Mill.
An escape from cultivation naturalised on the banks of streams, rivers, ponds, etc. Rare.
2. Shortwood Common, 1949. Near Harmondsworth, 1967. **3.** Canal bank near Southall!, *L.N.* 27: 32; 1965. Hounslow Heath,

1965; *Clement*, *Mason* and *Wurzell*. **5.** Greenford; between Perivale and Harrow, *Lees*, *T. & D.*, 121. Canal bank near Alperton, 1965. **6.** Canal backwater, Ponders End, a large colony, 1967. **7.** Site of Luxborough Lodge, Marylebone Road, N.W.1, 1966, *Brewis*, *L.N.* 46: 31.

A. nodiflorum (L.) Lag. Fool's Watercress, Mudwort.
Helosciadum nodiflorum Koch
Native in ditches, marshes, by stream- and riversides, on pond verges, etc. Common, except in heavily built-up areas, where it is very rare or extinct.
First record: *T. Johnson*, 1633.
7. Almost in every watery place about London, *Johns. Ger.*, 251. Paddington Canal, 1815 (G & R). Clapton, 1830, *Varenne*; East Heath, Hampstead!; Millfield Lane;† Blackstock Lane, Highbury;† ditch between Hyde Park and Kensington Gardens;† Eel Brook Meadow, Parsons Green;† Hackney Wick; Isle of Dogs!, *T. & D.*, 122. Near Hornsey Wood,† *Ball. MSS.* Hyde Park,† *Warren Fl.* Haverstock Hill, 1867,† *Syme* (SY). Clapton, 1872 (HY). Ken Wood. Finsbury Park, *Bangerter*!. Established in a lawn at the back of the National Gallery (Herrick Street), S.W.1, 1965, *McClintock*, *L.N.* 46: 31. The locality should read at the back of the Tate Gallery; a most unusual habitat for the species. Old Ford. Hackney Marshes. Upper Clapton, *L. M. P. Small*.

Var. **ochreatum** DC.
Helosciadum repens auct., non Koch
2. Between Kingston Bridge and Hampton Court, *Bloxam*, *T. & D.*, 122. **6.** Finchley Common,† *Woods*, *Winch MSS.* **7.** Eel Brook Meadow,† '*J.A.*', *Phyt. N.S.* 1: 464. East Heath, Hampstead!, *Irv. MSS.* Highgate Ponds!; Parliament Hill Fields!, *Fitter*. Near Hornsey Wood,† *Ball. MSS.*

Var. **pseudo-repens** (H. C. Wats.) Druce
Helosciadum repens auct., non Koch
2. Tothill Fields, Westminster,† *E. Forster* (BM).

A. inundatum (L.) Reichb. f. Least Marshwort.
Native on verges of ponds, lakes, etc., often submerged. Rare.
First record: *Petiver*, 1695.
1. In the bogs on Harefield Common, plentifully,† *Blackst. Fasc.*, 94. Pinner Hill, 1863,† *Hind* (BM). Harrow Weald Common,† *Melv. Fl.*, 34. Stanmore Common!, *T. & D.*, 122. Uxbridge Common, 1872,† *Benbow* (BM). Pield Heath;† Cowley;† Ruislip!; Ickenham Green;† Northwood,† *Benbow MSS.* Elstree Reservoir. **2.** Bushy Park!, *T. & D.*, 122. Shortwood Common!, *J. E. Cooper*; Staines Moor!, *M. G.*

Collett, *K. & L.*, 128; 1966, *J. E. Smith*!. **3.** Hounslow Heath!, *Pet. Midd.* Near Hatton,† *T. & D.*, 122. **4.** Between Kingsbury and Harrow,† *M. & G.*, 411. Hampstead Heath,† *Forst. Midd.* Golders Green,† *Irv. MSS.* Between Whitchurch and Edgware, *Benbow MSS.*; 1909,† *P. H. Cooke* (LNHS). **5.** Greenford,† *Coop. Fl.*, 108. Alperton,† *Melv. Fl.*, 34. **6.** Finchley Common!; Hadley Common!; Enfield Chase, *Benbow MSS.* Hadley Green!, *Fl. Herts*, 76. **7.** Between Westminster and Chelsea,† *Pet. Bot. Lond.* Between Newington and Hornsey,† *Huds. Fl. Angl.*, 104. In several parts of the New River,† *M. & G.*, 411. Eel Brook Meadow, Parsons Green, 1867,† *T. & D.*, 122.

PETROSELINUM Hill

P. crispum (Mill.) Nyman Parsley.
P. sativum Hoffm., *Carum petroselinum* (L.) Benth.
Alien. S. Europe, etc. Long cultivated for culinary purposes and now naturalised in grassy places, on waste ground, by roadsides, etc., as an escape. Very rare.
First evidence: *Benbow*, 1862.
1. Cowley, 1862;† near Uxbridge, 1890,† *Benbow* (BM). Yiewsley!, *M. & S.* Canal path, Harefield, 1934–52!, *K. & L.*, 129; 1953→. **2.** Between West Drayton and Harmondsworth, *Druce Notes.* **3.** Roxeth!,† *K. & L.*, 129. Hounslow Heath, 1947!, *L.N.* 27: 32; 1948–52!, *K. & L.*, 129; 1953→. **4.** Harrow, *Hind Fl.*; c. 1864,† *Melv. Fl.*, 33. **5.** Hanwell, 1946.† **7.** Building site, Bloomsbury,† *Shenstone Fl. Lond.* Bombed site, Ebury Street, S.W.1,† *McClintock*, *K. & L.*, 129. Bombed site between Queen Victoria Street and Upper Thames Street, E.C.,† *Sladen*, *Lousley Fl.* Portland Place, W.1, 1956, *Holme.* Holland Walk, Kensington, 1965.

P. segetum (L.) Koch Corn Caraway.
Carum segetum (L.) Benth. ex Hook. f.
Formerly native on hedgebanks, in waste places, etc., but now extinct.
First record: *Blackstone*, 1744. Last record: *M. Taylor*, c. 1930.
1. Harefield, 1908, *A. B. Jackson*, *K. & L.*, 129. **3.** Between Harlington and Hayes, 1883, *Newb. MSS.* **5.** Hanwell, 1912, *P. H. Cooke* (LNHS). Near Perivale Church, c. 1930, *M. Taylor*, teste *Welch*, *K. & L.*, 129. **7.** Brickfield adjoining to Tyburn turnpike, 1744, *Blackst. Spec.*, 93. Hampstead Heath; Hyde Park; Tothill Fields, *Forst. Midd.* Isle of Dogs, 1836, *J. F. Young* (K); 1866, *Trimen* (BM); 1869, *Warren*, *T. & D.*, 121.

SISON L.

S. amomum L. Stone Parsley, Stonewort.

Native on hedgebanks, field borders, in ditches, by roadsides, etc. Common, especially on heavy soils.

First certain record: *Morison*, 1672.

7. Ditch banks round London, *Moris. Pl. Umb.*, 15. Marylebone fields, 1815† (G & R). Roadsides about Newington† and Hornsey,† *Ball. MSS.* Islington (HE). Kentish Town, 1840, *Slatter* (BM). Abundant near Hornsey Wood;† Kilburn, *T. & D.*, 123. Shepherds Bush, 1870,† *Warren*; Paddington Canal towards Kensal Green Cemetery!, *A. R. Pryor*; Finchley Road,† *Trim. MSS.*

CICUTA L.

C. virosa L. Cowbane, Water Hemlock.

Formerly native on margins of ponds and streams, but long extinct.

First record: *How*, c. 1650. Last record: *Lightfoot*, c. 1780.

1. Uxbridge, *How MSS.* and *Mart. Pl. Cant.*, 74. In the river by Mercer's Mill, near Cowley . . . found by Dr *Wilmer*, 1746, *Blackst. MSS.* Between Cowley and Hillingdon, *Pluk. MSS.*; at the tail of Briggs' Mill, near Cowley, *Lightf. MSS.* It is likely that all the records refer to the same locality. **2.** Near Staines Bridge, and in the road to Uxbridge, *Doody MSS.* About [West] Drayton, *Blackst. MSS.*; near [West] Drayton Mill, between that and Colnbrook, *Lightf. MSS.* **3.** In a shallow pool of water by the highway side on Hounslow Heath, near the town's end and in pools of water about Isleworth, *Ray Cat.*, 285. In one of the ponds near the road at Hayes, three miles from Uxbridge, gathered once by Dr *Wilmer*, *Blackst. Spec.*, 1792.

AMMI L.

A. visnaga L.

Alien. Europe, etc. Naturalised on waste ground. Very rare, and sometimes merely casual.

First evidence: *Benbow*, 1886.

3. Hounslow Heath!, 1953, *D. Bennett*, *L.N.* 33: 54; plentiful, 1954!, *Welch*, *L.N.* 34: 6; very rare, 1964, *Wurzell*, teste *Lousley*; 1965!, *Wurzell*, *Clement* and *Mason*; 1966. **4.** Near Stonebridge, 1886,† *Benbow* (BM). Mill Hill, *Harrison*, teste *Sandwith*. **5.** West Acton, 1968, *McLean*. **7.** Hackney Marshes, 1924, *J. E. Cooper*, *K. & L.*, 128. Shoreditch, 1965, *D. Bennett*. St Luke's Estate, Old Street, E., 1969, *L. M. P. Small*.

FALCARIA Bernh.

F. vulgaris Bernh.
Prionitis falcaria (L.) Dumort.
Alien. Europe, etc. Naturalised in grassy places, and established on a railway bank. Very rare.
First record: *D. Bennett*, 1950.
2. Railway bank near Yeoveney, several large colonies, 1951!, *L.N.* 31: 12; still present, 1971. **3.** Near Hounslow Heath, 1950!, *D. Bennett*, *L.N.* 30: 7. The locality was threatened by building in 1963 and a number of specimens were transplanted on adjacent waste ground, *D. Bennett*.

CARUM L.

C. carvi L. Caraway.
Introduced. Cultivated for culinary purposes and now naturalised in grassy places, on waste ground, etc., as an escape. Very rare, and often only casual.
First certain evidence: *Stevens*, 1847.
1. Railway bank, Pinner,† *Farrar*, *T. & D.*, 124. Yiewsley, *J. E. Cooper* (BM); 1915–16 and 1920,† *J. E. Cooper*, *K. & L.*, 128. Near Cowley, 1944.† **2.** Waste ground by the canal, West Drayton!, *Newbould*, *T. & D.*, 124; 1901, *P. H. Cooke*; 1932!, *K. & L.*, 128; 1959. **3.** Southall, 1902, *E. F.* and *H. Drabble* (BM). Roxeth, 1942.† **4.** Highgate, 1887;† East Finchley, 1909,† *Coop. Cas.* **5.** Hanwell, 1847,† *Stevens* (K). Near Osterley, 1885;† Chiswick, 1888,† *Benbow* (BM). Bombed site, West Ealing, 1944–46!, *L.N.* 26: 60; 1947–48.† **6.** Near Colney Hatch, plentiful on a railway bank, 1867,† *Trimen*; North Finchley, 1909, *E. F.* and *H. Drabble* (BM); 1917 and 1927,† *J. E. Cooper*, *K. & L.*, 128. Muswell Hill, *Druce Notes.* **7.** Hyde Park, *De Cresp. Fl.*, 14; 1944,† *Sladen*, *L.N.* 24: 12. By the Paddington Canal near Bloomfield Road [Maida Hill], 1869,† *Warren*, *T. & D.*, 424. Hackney Marshes, 1914, *Coop. Cas.*; 1917–18, *J. E. Cooper*, *K. & L.*, 128; c. 1927,† *M. & S.* Crouch End, 1887,† *J. E. Cooper*. Bombed site near the Tower of London,† *Jones Fl.*

CONOPODIUM Koch

C. majus (Gouan) Loret Pignut, Earthnut.
C. denudatum Koch, *Bunium flexuosum* Stokes
Native in meadows, rough grassy places, on heaths, banks, in open woods, etc. Common, except near the Thames, where it is local.
First record: *Gerarde*, 1597.
7. . . . field adjoining to Highgate; field . . . by Marylebone! near the

way that leadeth to Paddington, by London, *Ger. Hb.*, 906; near Marylebone Park!, *Merr. Pin.*, 17; Marylebone!, *J. Banks* (BM); Regents Park!, 1957, *Welch, L.N.* 41: 19. Kensington; *Turn. Bot.*; Kensington Gardens, abundant, but stunted in growth, *T. & D.*, 123; in late mown grass, 1961, *D. E. Allen, L.N.* 41: 19. Lawn weed, Burlington Gardens, W.1, 1952!; garden weed, Sloane Square, 1947!; Chelsea Hospital grounds!, *L.N.* 39: 51. Tower Hill, 1963, *Brewis*; Vincent Square, S.W.1, 1964, *McClintock, L.N.* 44: 23. Buckingham Palace grounds, *Fl. B.P.G.* Ken Wood grounds!; East Heath, Hampstead!; Highgate Ponds!, *Fitter*. Parliament Hill Fields. Kilburn. Holloway. Bethnal Green. Stoke Newington.

PIMPINELLA L.

P. saxifraga L. Burnet Saxifrage.

Native in dry grassy places, and established on railway banks, in cemeteries, churchyards, etc. Common.

First record: *Morley*, 1677.

7. East Heath, Hampstead!; Hackney Wick, *T. & D.*, 125. Clapton (HY). Formerly at Chelsea,† *Dickens*. In turf, Finsbury Square, E.C.2, 1966!, *L.N.* 46: 31. Bombed site, Shoreditch, 1957.† Brompton Cemetery. Waste ground, Horseferry Road, S.W.1, 1967. Tottenham.

Var. **dissecta** (Retz.) Spreng.

1. Harefield!, *Druce Notes*. **2.** Sunbury. Hampton Court. **5.** Syon Park. Near Southall. **6.** Enfield.

P. major (L.) Huds. Greater Burnet Saxifrage.

P. magna L.

Native on hedgebanks, in grassy places, etc., on heavy soils. Confined to the north, chiefly the north-east, parts of the vice-county where it is very local.

First record: *Morley*, 1677.

4. Hampstead,† *Morley MSS.* Hampstead Heath,† *M. & G.*, 413. Near Kingsbury,† *Vardy, Doody MSS.* Between Totteridge Green and Finchley!, *Irv. MSS.* Near Scratch Wood!, *H. C. Harris, L.N.* 28: 33; road banks and ditches, Hendon Wood Lane, near Barnet Gate, abundant!, *L.N.* 37: 185; 1967. **5.** Plentiful between Osterley and Trumper's Crossing!, 1971→ *Gilbert*. **6.** Colney Hatch, 1867; Hadley, 1868, *Trimen* (BM). Abundant in meadows between Barnet and Hadley Green and Cockfosters!, *Benbow J.B.* (*London*) 25: 16. Hedgebanks from Barnet to Beech Hill Park, abundant, *Benbow* (BM). Near Mimms Wash, *J. G. Dony*!. Vicarage Farm, Enfield, 1965, *Kennedy*!. Near Hadley Wood station, 1966.

Var. **laciniata** Gray
var. *dissecta* Druce
1. Springwell, 1954,† *T. G. Collett, K. & L.*, 349.

AEGOPODIUM L.

A. podagraria L. Herb Gerard, Gout Weed, Bishop's Weed, Ground Elder.
Alien. Europe, etc. Cultivated for medicinal purposes in early times and now naturalised in fields, ditches, on waste ground, by roadsides, etc., also established as a pernicious garden weed. Very common.
First record: *Merrett*, 1666.
7. Frequent in the district.

SIUM L.

S. latifolium L. Water Parsnip.
Native by rivers and streams, on lake and pond verges, in osier holts, etc. Very rare.
First record: *Blackstone*, 1746.
1. In several parts of Harefield river (=Colne),† *Blackst. Spec.*, 92. Northwood gravel pits, 1911,† *C. S. Nicholson* (LNHS). Fray's river north of Uxbridge, 1927,† *Tremayne*. Perhaps an error. Sparingly by a lake west of Potters Bar, 1965, *J. G.* and *C. M. Dony*!, *L.N.* 45: 20; 1967. Perhaps planted. **2.** Thames side between Hampton and Hampton Court, *H. C. Watson* (W). In some plenty by the Thames about a quarter of a mile above Kingston Bridge;† a large patch a little below Teddington Lock,† *T. & D.*, 125. Wet copse near Chertsey Bridge, 1910,† *Green* (SLBI). Near Shepperton,† *J. E. Cooper, K. & L.*, 129. Staines, 1893,† *Wolley-Dod* (BM). **3.** By the Thames, Twickenham,† *M. & G.*, 399. Sparingly about midway between Twickenham and Richmond Bridge,† *T. & D.*, 125. Hounslow Heath,† *Spencer*, 318. **7.** Brickfield near Tyburn turnpike,† *Hill Fl. Brit.*, 147. Hyde Park,† *Cockfield Cat.* Perhaps an error. By the Thames near the Botanic Garden, Chelsea, 1825† (G & R).

BERULA Koch

B. erecta (Huds.) Coville Narrow-leaved or Erect Water-parsnip.
Sium erectum Huds., *S. angustifolium* L.
Native by ponds and streams, in ditches, marshes, etc. Very rare, and decreasing.
First record: *Blackstone*, 1737.
1. Harefield!, *Blackst. Fasc.*, 94. Near Yiewsley!,† *Newbould*; Eastcote,† *Farrar, T. & D.*, 125. Fray's river south of Uxbridge!,† *L.N.* 27: 32.

Uxbridge!,† *L. B. Hall*; Springwell!, *K. & L.*, 130. **2.** West Drayton,† *Newbould*; Staines Moor, *T. & D.*, 125. Shepperton!, *Green* (SLBI). Chertsey Bridge,† *C. S. Nicholson*; near Colnbrook, 1910,† *J. E. Cooper, K. & L.*, 130. Yeoveney!,† *Welch, L.N.* 27: 32. **3.** Cranford,† *De Cresp. Fl.*, 68. **4.** Edgware Road, 1830,† *Varenne*; Finchley Road,† *Newbould*; all along the Brent,† *Farrar*; Harrow Weald,† *T. & D.*, 125. Between Hampstead and Golders Green,† *Irv. MSS.* **5.** Canal at Greenford, *Melv. Fl.*, 35; c. 1877,† *De Cresp. Fl.*, 68. Alperton,† *Lees, T. & D.*, 125. Syon Park!, *L. G.* and *R. M. Payne, L.N.* 27: 32. **6.** Near Enfield Lock, 1967, *Kennedy*!. **7.** Near Hornsey Wood,† *Ball. MSS.* Isle of Dogs,† *J. F. Young*; East Heath, Hampstead,† *T. & D.*, 125. Tottenham, 1967, *Kennedy, L.N.* 47: 9.

OENANTHE L.

O. fistulosa L. Water Dropwort.

Native in marshes, wet grassy places, ditches, on pond verges, etc. Formerly local, but now very rare and decreasing.

First record: *Gerarde*, 1597.

1. Harefield, *Blackst. Fasc.*, 66; Harefield Moor; near Moor Hall;† Ickenham Green,† *Benbow MSS.* Ruislip,† *Hind, T. & D.*, 126. Near Denham,† *Druce Notes.* Yiewsley,† *J. E. Cooper, K. & L.*, 133. **2.** Shortwood Common!; common by Walton Bridge!,† *T. & D.*, 126. Staines Moor, abundant!, *Benbow MSS.*; plentiful until c. 1960, but now much reduced in quantity. Yeoveney!; Stanwell Moor!, *Benbow MSS.* Colnbrook, 1774,† *J. Banks* (BM). Knowle Green!, *K. & L.*, 133. West Drayton,† *Druce Notes.* **3.** Hounslow Heath,† *T. & D.*, 126. **4.** Hampstead,† *Irv. MSS.* **5.** Greenford, *Lees, T. & D.*, 126; 1885,† *Benbow, J.B. (London)* 25: 16. Marsh by river Brent between Hanwell and Greenford, 1919,† *Tremayne, K. & L.*, 133. Perivale, 1869,† *A. Wood, Trim. MSS.* **7.** Fulham,† *Ger. Hb.*, 962. Blackwall,† *M. & G.*, 384. Marylebone Fields, 1815† (G & R). Near Homerton;† Hackney,† *Cherry, T. & D.*, 127. Eel Brook Common, Parsons Green,† *Trimen* (BM). Clapton,† *C. S Nicholson, K. & L.*, 133. Tottenham,† *De Cresp. Fl.*, 45.

O. pimpinelloides L.

Formerly native in grassy places, but now extinct.

First, and only evidence: *H. T. White*, 1915.

4. Dry field, Hendon, 1915, *H. T. White* (L).

O. silaifolia Bieb. Sulphur Wort.

O. peucedanifolia auct.

Formerly native in damp grassy places, but now extinct.

First, and only certain evidence: *Goodger* and *Rozea*, 1815.

4. Kingsbury, *Farrar*, *T. & D.*, 127. Error, the plant from this locality was *O. aquatica*. **5.** Banks of the Brent near Greenford, abundant in 1863, but we fear, now extirpated from the bank being cut away, *Melv. Fl.*, 97. Almost certainly an error. **7.** Marylebone Fields, 1815 (G & R).

O. lachenalii C. C. Gmel. Parsley Water Dropwort.
Formerly native in brackish ditches but now extinct, though it has occurred as a casual introduction in comparatively recent times.
First, and only evidence as a native species: *H. C. Watson*, 1847.
2. Ditch by the road, four miles from Staines towards Hampton Court, also in the cross roads thence to Sunbury, 1847, *H. C. Watson* (W): the ditches partly cleared out and no sign of the plant left, 1869, *H. C. Watson*, *Trim. MSS.* **5.** Rubbish-tip, Hanwell, a single plant, 1953, *Bangerter*!, *L.N.* 33: 54.

O. crocata L. Hemlock Water Dropwort.
Native by rivers and streams, in marshes, wet woods, etc. Rather common, especially by the Thames.
First record: *T. Johnson*, 1633.
1. Scattered along the Colne and Grand Junction Canal from Springwell to Cowley. **2.** Common by the Thames. Colnbrook!, *Druce Notes*. West Drayton, *Benbow MSS.* **3.** Frequent by the Thames. **5.** Frequent by the Thames. Brent between Greenford and Hanwell, formerly abundant, but now much reduced in quantity. Canal near Brentford. Long Wood, Wyke Green. Osterley Park. Chiswick House grounds. **6.** Lee Navigation Canal Edmonton to Waltham Abbey!, *F. Clarke*. **7.** In scattered localities by the Thames between Fulham and the Isle of Dogs. Hackney. By the lake, Buckingham Palace grounds!, *Eastwood*; Regents Canal, Regents Park!, *L.N.* 39: 51. Canal near Paddington Basin.

O. aquatica (L.) Poir. Fine-leaved Water Dropwort.
O. phellandrium Lam.
Native on the verges of ponds, lakes, slow streams, etc., with a marked preference for still waters. Very rare.
First record: *Milne* and *Gordon*, 1793.
1. By the Colne, Harefield, *De Cresp. Fl.*, 45. Error, *O. fluviatilis* intended. **2.** Staines!, *Whale Fl.*, 20; still present by the lake on Shortwood Common, 1969. Hampton Court Park!, *J. E. Cooper* (BM); still present by Hampton Wick Pond, 1969. **4.** Kingsbury, *Farrar* (as *O. silaifolia*), *T. & D.*, 127; 1886, *Benbow*, *J.B.* (*London*) 25: 16; 1900,† *Pugsley*. Eradicated when the pond near Woodfield House was filled in. **5.** Banks of the canal between Hanwell and Brentford,† *Hemsley*; Brent at Alperton, abundant,† *Lees*, *T. & D.*, 128.

Near Greenford, abundant;† between Perivale and Ealing,† *Benbow MSS.* Greenford Green,† *Druce Notes.* **7.** Millbank,† *M. & G.*, 435. Near Hornsey, 1815† (G & R). Copenhagen Fields, 1841† (HE). Near Hammersmith, 1850,† *Wing* (BM). Holloway, 1864,† *Munby, Nat.* 1867, 179. Kentish Town,† *T. & D.*, 128.

O. fluviatilis (Bab.) Coleman River Water Dropwort.
Native in swift-flowing shallow rivers and streams; often submerged and sterile. Locally plentiful.
First certain record: *Dillenius*, 1724.
1. Colne from Springwell to Yiewsley: formerly abundant, but now much reduced in quantity owing to the pollution of the waters. Fray's river, Uxbridge to Cowley. **2.** Colne from Yiewsley to Staines, formerly abundant, but now considerably reduced in quantity as a result of the pollution of the waters. Thames from Sunbury Lock to Walton Bridge,† *H. C. Watson, T. & D.*, 128. **3.** Duke's river near Isleworth,† *T. & D.*, 128. Crane from Cranford to Twickenham. **6.** River Lee and its backwaters!, *Benbow MSS.* **7.** In Hackney river (=Lee), abundantly,† *Dillenius, Ray Syn.* 3: 216. Isle of Dogs,† *Hurlock, Blackst. Spec.*, 72. Perhaps O. *crocata* intended. New river head, Clerkenwell,† *Irv. Ill. Handb.*, 592. Ditch near Victoria Park, 1867,† *Cherry*; Lee at Temple Mills, and below that point,† *T. & D.*, 128.

AETHUSA L.

A. cynapium L. Fool's Parsley.
Native on cultivated and waste ground, by roadsides, etc. Very common.
First record: *Blackstone*, 1737.
7. Frequent in the district.

Var. **agrestis** Wallr.
1. Harefield!, *K. & L.*, 123. **2.** Harmondsworth!; Staines!, *K. & L.*, 123. Stanwell. Laleham. **3.** Feltham. Cranford. **5.** Northolt.

FOENICULUM Mill.

F. vulgare Mill. subsp. **vulgare** Fennel.
F. officinalis All.
Alien. Europe, etc. Formerly cultivated as a pot-herb, and now naturalised in grassy places, on waste ground, etc., established on railway banks, etc. Rather rare.
First record: *Waring*, 1790.
1. Cowley!, *Benbow MSS.*; 1973. Uxbridge, 1968!, *L. M. P. Small*, Harefield, 1930,† *P. H. Cooke*; Yiewsley!, *J. E. Cooper, K. & L.*, 132.

New Years Green!, *I. G. Johnson*. **2.** Near Strawberry Hill,† *T. & D.*, 129. West Drayton!, *Druce Notes*. Staines, *Briggs*. Kempton Park. Near Stanwell. Hampton Court!, *Tremayne*. East Bedfont. **3.** Twickenham,† *T. & D.*, 129. Near Hounslow!, *Benbow MSS*. Near Hounslow Heath. Between Feltham and Bedfont!, *L.N.* 26: 61. Feltham. Southall. **4.** The Hyde, Hendon!, *Moring MSS*. Mill Hill, *Harrison*. Near Finchley!, *K. & L.*, 132. Disused railway between Stanmore and Belmont, 1968, *Roebuck*. Railway banks, waste ground and canal side, Harlesden. Near Neasden, a single plant, 1965. Railway bank, West Hampstead. **5.** Railway bank near Hanwell!,† *T. & D.*, 129; survived until c. 1956; waste ground, Hanwell, a large colony, 1968. Railway bank, Brentford!, *A. Wood, Trim. MSS*. Syon Park. Greenford, 1920, *J. E. Cooper*; 1935!, *Bull*; Alperton!,† *K. & L.*, 132. Chiswick!, *Benbow MSS*. South Acton, 1965. Park Royal, in small quantity, 1965. **6.** Near Botany Bay, Enfield Chase, 1886,† *Benbow*; Ponders End!, *P. H. Cooke* (BM). Near Enfield Wash, 1966, *Kennedy*. Near Edmonton, 1966, Kennedy!. **7.** About the gravel pit at Hyde Park Corner,† *Waring*. New Finchley Road, 1840† (HE). Highgate, *Dennes, Wats. New Bot. Suppl.* Shepherds Bush;† Victoria Street, S.W.1,† *T. & D.*, 129. Hyde Park, a single plant!,† *Watsonia* 1: 208. South Kensington,† *Dyer List*. Isle of Dogs!, *L.N.* 26: 61. Clapton, 1909,† *P. H. Cooke*; Hackney Marshes, 1912–21,† *J. E. Cooper*; Tower of London gardens, planted!, *K. & L.*, 132. Bombed sites, Cripplegate† and Queen Victoria Street areas, *Jones Fl.* Bombed site, Bread Street, E.C.,† *Lousley, K. & L.*, 132. Near Bow. Poplar, 1966. Wapping, 1966. Golden Square, W.1, a single seedling, 1969, *L. M. P. Small*.

SILAUM Mill.

S. silaus (L.) Schinz & Thell. Pepper Saxifrage.
Silaus flavescens Bernh., *S. pratensis* (Crantz) Bess.
Native in damp grassy places, on commons, etc., established on railway banks, etc. Locally plentiful.
First record: *Johnson*, 1629.
1. Harefield!, *Blackst. Fasc.*, 99, Stanmore Common!; near Elstree!; Pinner!, *T. & D.*, 129. Uxbridge!; Ickenham!; Ruislip!, *Benbow MSS*. Pinner Hill. **2.** Stanwell Moor!; near Charlton!; common by Walton Bridge,† *T. & D.*, 129. Ashford, *Benbow MSS*. Colnbrook. Laleham. Yeoveney. **3.** Hanworth, *Benbow MSS*. **4.** Harrow!, *Hind Fl.* Mill Hill; between Child's Hill and Hendon,† *Irv. MSS*. Hendon!, *De Cresp. Fl.*, 67. Bentley Priory!, *T. & D.*, 129. Stanmore!; Edgware!, *Benbow MSS*. Near Scratch Wood. Near Hatch End. Northwick Park. **5.** Ealing Common;† near Twyford,† *Newbould*; Horsendon Hill!, *T. & D.*, 129. **6.** Colney Hatch, *Benbow*

MSS. Wood Green, *De Cresp. Fl.*, 67. Near Finchley. Edmonton. *Cooke* (LNHS). Enfield. **7.** Between Kentish Town and Hampstead, *Johns. Eric.*; 1815 (G & R); 1845† (HE).

ANGELICA L.

A. sylvestris L. Wild Angelica.
Native by streamsides, in marshes, wet woods, etc. Common.
First record: *T. Johnson*, 1629.
7. Near Kentish Town, *Park. Theat.*, 940. Duckett's Canal! [Hackney], *Cherry*; Sandy End, Fulham, *T. & D.*, 130. Highgate Ponds!, *Fitter*. Weed in flower-bed, Wellington Road, St John's Wood, 1965!, *L.N.* 46: 31. River-wall, Hammersmith and Fulham. New river near Hornsey. Canal banks, Bethnal Green, Shoreditch, Stepney, Old Ford, Hackney Wick, etc. Thames banks and river-walls, Wapping and Isle of Dogs. By river Lee, Hackney Marshes and Upper Clapton.

Var. **decurrens** Fisch., Mey. & Ave Lall
2. Staines!; Laleham!, *K. & L.*, 134. Shepperton. **3.** Isleworth!, *K. & L.*, 134. Twickenham. **4.** Stanmore (D).

A. archangelica L. subsp. **archangelica** Angelica.
Archangelica officinalis (Moench) Hoffm.
Alien. Europe, etc. Formerly cultivated for culinary purposes and now naturalised on river banks, in marshes, etc., and established by canals. Locally plentiful by the Thames, and by canals in the eastern parts of the vice-county, rare elsewhere.
First record: *Doody*, c. 1700.
2. Hampton Court, planted!, *K. & L.*, 134. **3.** Thames at Isleworth!, *K. & L.*, 134. Twickenham, *L. M. P. Small*. **5.** By the Thames from Syon Park to Chiswick. **6.** Pymmes Brook, Edmonton!, *R. M. Payne*, *K. & L.*, 134. **7.** Frequent about the Tower of London and on the banks of the moats,† *Doody MSS*. Sandy End, Fulham, 1867, *Trimen* (BM); destroyed through the marshy hollows being filled up with earth, 1887,† *Benbow MSS*. Thames at Fulham and Hammersmith. Canal side, Warwick Square, *G. C. Druce*, *Rep. Bot. Soc. & E.C.* 9: 653. Planted in many of the squares and gardens in London for ornament, and in Lincolns Inn Fields now comes up from self-sown seed in abundance,† *T. & D.*, 130. Hackney Marshes!, *J. E. Cooper* (BM). Regents Canal, Islington and Limehouse. By the Thames, Wapping and Isle of Dogs. Hornsey. Canal side, Hackney. Kentish Town,† *Baylis*, 83. By London Bridge, one plant, 1967, *Lousley*, *L.N.* 47: 10.

PASTINACA L.

P. sativa L. subsp. **sylvestris** (Mill.) Rouy & Camus Wild Parsnip.
Peucedanum sativum (L.) Benth. ex Hook. f.
Native by rivers, streams, etc., roadsides, in fields, established on railway banks, etc., on alluvial and calcareous soils. Common, especially near the Thames, Colne and Lee.
First record: *Blackstone*, 1737.
7. Banks of Lee, Hackney Marshes. Side of Lee Navigation Canal, Tottenham.

Subsp. **sativa** Cultivated Parsnip.
P. sativa L. var. *edule* DC.
Alien of cultivated origin. Naturalised on waste ground, by roadsides, established on railway banks, etc. Common.
7. Copenhagen Fields† (HE). Isle of Dogs!, *Newbould*; railway banks, Hampstead, *T. & D.*, 131. South Kensington, *Dyer List*. City of London bombed sites. Kensington Gardens, one plant, 1961, *D. E. Allen*, *L.N.* 41: 19. Holloway, *D. E. G. Irvine*. Stepney. Stoke Newington. Hackney Wick. Buckingham Palace grounds, 1967, *McClintock*, *L.N.* 47: 9.

HERACLEUM L.

H. sphondylium L. subsp. **sphondylium** Cow Parsnip, Hogweed.
Native on cultivated and waste ground, by rivers, streams, roadsides, etc. Very common.
First record: *T. Johnson*, 1638.
7. Frequent in the district.

Var. **angustifolium** Huds.
1. Harefield, *Blackst. Fasc.*, 97. **4.** Cemetery near Kilburn, 1966.

H. mantegazzianum Somm. & Levier *sensu lato* Giant Hogweed.
H. giganteum auct.
Alien. Caucasus. Garden escape. Naturalised on the banks of rivers and streams, by lakes, ponds, on waste ground, etc., and established by canals, on railway banks, etc. Rather rare, but increasing.
First record: *A. Irvine*, 1858.
1. New Years Green, 1958, *I. G. Johnson*. Sandpit north of Harefield, 1965. Harefield Grove, planted, but regenerating freely, 1966, *I. G. Johnson* and *Kennedy*!. **2.** Bushy Park and Hampton Court Park, planted but regenerating freely. Near the railway, West Drayton, abundant. **3.** Gravel pits and roadside verges between Cranford and Feltham, in profusion, 1965→. Hayes Park, 1966, *Royle*!. **4.** Brent banks, Stonebridge!, *L.N.* 27: 32; abundant and spreading on to

adjacent industrial waste ground, 1968. Harrow Park, planted, but regenerating!, *Harley Fl.*, 16. Colindale Hospital grounds, planted, but regenerating, 1955→. Brent Reservoir, 1966→. **5.** Brent banks, Ealing to Hanwell, in profusion!; Thames bank, Chiswick!, *L.N.* 27: 32. Chiswick Ait!, *Fitter*. Chiswick House grounds, *Murray*. Canal side near Brentford, abundant. Waste ground, The Ham, Brentford, a single large plant, 1968. Weed in flower-bed, West Acton, 1965. Weed of allotments, Hanwell, 1965. Near the railway between North Ealing and Park Royal, 1966!, *McLean*; 1973. Hanger Hill, Ealing, a single plant, 1966→. **6.** By Lee Navigation Canal, Tottenham. Railway bank near Enfield, abundant!, *A. Vaughan*. Roadside, Botany Bay, Enfield Chase, *Kennedy*!. Pymmes Park, Edmonton, 1966, *Kennedy*. Trent Park, 1966, *T. G. Collett* and *Kennedy*. **7.** Chelsea!, *Irv. Ill. Handb.*, 598; planted, but regenerating in Ranelagh Gardens, Chelsea!; Regents Park!; Hyde Park!; Kensington Gardens! and Buckingham Palace grounds!, *L.N.* 39: 51. Finchley Road, Swiss Cottage, planted, but regenerating, 1964→. Holland House grounds, Kensington, planted, but regenerating, 1965→. Hackney Marshes!, *M. & S.*; 1968. Grounds of the Natural History Museum, South Kensington, 1949,† *Bangerter* and *J. B. Evans*, *K. & L.*, 134. Bombed sites, Queen Victoria Street† and Gresham Street areas,† *Jones Fl.* Bombed site, Faringdon Street, E.C., a large colony, 1963!, *P. C. Hall*; 1968. Isle of Dogs, 1966→. Railway side near Paddington, a single plant, 1967→. Canal side between Marylebone and Primrose Hill, frequent, 1968. Hackney Wick, *L. M. P. Small*.

Other giant species of *Heracleum* are in cultivation and may occur as naturalised or established garden escapes. D. McClintock (*Gard. Chron.* **164** (7): 8 (1968)) has pointed out that at least two distinct taxa are present. These are designated Taxon A and Taxon B, and differ from each other as follows:—

	Taxon A	Taxon B
duration :	monocarpic	perennial
height :	to 16 ft.	to 8 ft.
no. of stems :	always one	one to many
leaflets :	two pairs	three pairs
leaf outline :	deeply or sharply toothed	more sinuate
leaf colour :	dark green	pale green

Taxon A is frequent, being locally abundant and naturalised in many parts of the British Isles. It is provisionally equated to *H. mantegazzianum*. Taxon B is rare but occurs as an abundant weed of rough places in the garden of Buckingham Palace. No morphological differences have been detected between the flowers and fruits of the two taxa, and taxon B may not necessarily be a distinct species.

H. mantegazzianum × sphondylium

2. West Drayton, 1969. **3.** Gravel pits between Cranford and Feltham!, 1967, *Brummitt*; 1969. **4.** Brent banks, Stonebridge, 1969. **5.** Near the Brent, Ealing!, 1962, *M. G. Collett*; 1969. Near Brent Bridge, Hanwell, 1967, *Brummitt*. Canal backwater near Boston Manor, 1968. Canal side near Brentford, 1967. **7.** Hackney Marshes, 1967.

This hybrid is probably common wherever the two species grow together.

TORDYLIUM L.

T. maximum L. Great Hartwort.

T. officinale auct., non L.

Alien. Europe, etc. Formerly naturalised on hedgebanks and waste ground, but now long extinct.

First record: *Merrett*, 1670. Last evidence: *Wilson*, 1837.

3. About Isleworth, *Doody, Ray Syn.*, 63: 'I have found it in various adjacent localities there for many years', *Doody MSS.*: 'Mr. Doody showed me this growing in two or three places about Isleworth', *Buddle* (SLO); not now to be found in Mr Doody's locality, *Clements, Hill Fl. Brit.*, 137. Near Twickenham, 1827, *Francis*, teste *Pamplin, T. & D.*, 131; 1837, *Wilson*; between Twickenham and Isleworth, *Borrer* (BM); *Castles* (W). **7.** Betwixt St James and Chelsea, *Merr. MSS*. Frequent about London, *Moris. Pl. Umb.*, 40.

DAUCUS L.

D. carota L. Wild Carrot.

Native in grassy places, by roadsides, etc. (as subsp. **carota**); naturalised on waste ground, in fields, by roadsides, etc., as an escape, or outcast from cultivation (subsp. **sativus** (Hoffm.) Arcangeli). Common.

First record: *T. Johnson*, 1638.

7. Tottenham!, *Johns. Cat.* Between Belsize House and New Finchley Road, *Irv. Lond. Fl.*, 196. Paddington!, 1827–30, *Varenne*; railway bank, Hampstead; stray plants by the roadside at Kensington Gore and Constitution Hill, *T. & D.*, 132. South Kensington, *Dyer List*. Turf opposite Kensington Palace, 1872,† *Warren, Trim. MSS*. Bombed site, Ebury Street, S.W.1,† *McClintock*. East Heath, Hampstead, *H. C. Harris*!. City of London bombed sites. Canal side, Paddington!, *L.N.* 39: 51. Stag Place, S.W.1, 1966, *McClintock, L.N.* 46: 12. Railway bank near Westbourne Park. Chelsea, 1967. Hornsey. Islington. Wapping. Hackney. South Bromley.

CUCURBITACEAE

BRYONIA L.

B. dioica Jacq. White Bryony, 'Mandrake'.
B. cretica L. subsp. *dioica* (Jacq.) Tutin
Native. Climbing over hedges, scrambling over bare ground in woods, etc. Common and widely distributed though often strangely absent from apparently suitable habitats, and extinct or very rare in heavily built-up areas.
First record: *T. Johnson*, 1638.
7. Marylebone, 1818† (G & R). Islington;† Hornsey Wood, 1839† (HE). Primrose Hill, 1845,† *J. Morris*; Kentish Town,† *T. & D.*, 113. East Heath, Hampstead. Upper Clapton. Poplar, 1974.

ARISTOLOCHIACEAE

ASARUM L.

A. europaeum L. Asarabacca.
Alien. Europe. Formerly naturalised in a wood but now extinct.
First, and only evidence: *Green*, 1906.
1. The Grove, Stanmore Common, 1906, *Green* (D).

ARISTOLOCHIA L.

A. clematitis L. Birthwort.
Alien. Europe. Naturalised in grassy places. Very rare.
First evidence: *H. C. Watson*, 1838.
2. By the river Hampton Court!, *H. C. Watson* (W); still grows in small quantity in a grassy enclosure of Hampton Court gardens by the river, but is mown down annually. **6.** Chase End, Enfield, 1841, *Warner* (W); c. 1873, *H. & F.*, 149. **7.** Riverside above Hammersmith Bridge,† *J. E. Cooper* (BM).

EUPHORBIACEAE

MERCURIALIS L.

M. perennis L. Dog's Mercury.
Native in woods, on shady banks, etc. Locally abundant in the north of the vice-county, very rare and mostly introduced elsewhere.
First record: *Gerarde*, 1597.
1. Frequent. **3.** Roxeth,† *Hind Fl.* Wood near Hounslow Heath!, *Westrup*. Probably native. **4.** In Hampstead Wood and all the

hedges thereabout, *Ger. Hb.*, 263; c. 1869, *T. & D.*, 250. Big Wood; Little Wood, *Hampst.* Marylebone Cemetery, East Finchley, probably a relict from when the area was covered by woodland. Between Finchley and Hendon!, *Newbould*; Finchley!, *Tucker*; Mill Hill!, *J. Morris, T. & D.*, 250. Scratch Wood. Stanmore, *Tyrrell*!. Edgware. Harrow!, *Harley*. Kingsbury. Near Brent Reservoir. **5.** Horsendon Lane! and Hill!, *Hind Fl.* Syon Park, *Welch*!. Osterley Lane, Norwood Green. Probably native in all four localities. Gunnersbury Park, Acton, introduced with shrubs. Near Castle Bar, Ealing, probably introduced with shrubs. **6.** Hadley!, *Warren*; Edmonton; Whetstone, *T. & D.*, 250. Enfield. Southgate, *Tremayne*. Highgate Woods, *C. S. Nicholson*. **7.** Primrose Hill, 1820† (G & R). Hornsey Wood,† *Ball. MSS.* Between Kentish Town and Highgate,† *T. & D.*, 250. Regents Park, *Webst. Reg.* Buckingham Palace grounds, *Fl. B.P.G.* Probably introduced with shrubs.

M. annua L. Annual Mercury.
Alien. Europe, etc. Naturalised on cultivated and waste ground, and as a persistent garden weed. Common, and increasing, particularly near the Thames, in inner London and in the immediate eastern and northern suburbs.
First record: *W. Turner*, 1562.
7. Very common in the district.

Var. **ambigua** (L.f.) Duby
3. Twickenham, *T. & D.*, 250. **4.** Cemetery near Kilburn, 1966. **5.** Strand-on-the-Green!, *K. & L.*, 250. **6.** Between Wood Green and Hornsey, *Munby, Nat.* 1867, 181. **7.** Parsons Green; South Kensington, *T. & D.*, 250.
Both male and female plants occur commonly, though the former is probably the more frequent of the two. The species is also of some interest as being one of the earliest adventives recorded from Britain.

EUPHORBIA L.

E. lathyrus L. Caper Spurge.
Introduced. Naturalised on cultivated and waste ground, occurs also as a sporadic garden weed. Rather rare, and sometimes merely casual.
First record: *A. Irvine*, c. 1830.
1. Uxbridge, 1884, *Benbow* (BM). Northwood, *L. M. P. Small.* New Years Green, 1956, *Lousley* and *T. G. Collett*!. **2.** Between Kingston Bridge and Hampton Court, *T. & D.*, 249. Hampton Court!, *Welch, L.N.* 26: 63. Ashford, *Hasler*; Shepperton!, *K. & L.*, 249; 1967, *Kennedy*!. East Bedfont, 1946, *Lousley*!; between Sunbury and Walton Bridge, a single plant, 1945, *Welch, L.N.* 26: 63. Stanwell; Perry Oaks, *Westrup, K. & L.*, 249. Near Dawley, 1968. **3.** Roxeth, *Melv. Fl.*, 68.

Twickenham, 1934, *S. King* (K). Hounslow Heath, 1953, *Welch, K. & L.*, 249. **4.** Hampstead!, *Irv. MSS.* Harrow!, *Hind Fl.* Sudbury, *Farrar, T. & D.*, 249. Kenton, 1967. **5.** Hanwell!, *Boniface*; Ealing!, *K. & L.*, 249. Chiswick, one plant, 1965, *Wurzell.* **6.** North Finchley, 1963. Pymmes Park, Edmonton; Enfield, 1966, *Kennedy*!. **7.** Numerous plants on the site of the Exhibition of 1862, South Kensington,† *T. & D.*, 249.

E. dulcis L.
Alien. Europe, etc. Naturalised by a pathside. Very rare.
First record: *Harley*, 1954.
4. Harrow!, *Harley Add.*; increasing, 1966, *Temple*!; 1972.

E. platyphyllos L. Broad, or Warted Spurge.
Formerly a colonist in cornfields, on cultivated ground, etc., but now extinct.
First record: *Doody*, c. 1700. Last evidence: *Benbow*, 1884.
1. In the cornfields leading from Harefield Common to Battleswell, plentifully, *Blackst. Fasc.*, 98; near Harefield (LT); 1793, *Smith Eng. Fl.*, 4: 64; Harefield, 1868, *Newbould*; 1884, *Benbow* (BM). **3.** Amongst the corn near Twickenham Park, over against Richmond, *Doody MSS.* Waste ground, Twickenham, 1861, *Syme* (SY); 1867, *Dyer* (BM). Roadside between Harrow and Eastcote, a single plant, *Hind Fl.* **4.** Stanmore, 1827–30, *Varenne, T. & D.*, 248. **7.** Site of the Exhibition of 1862, South Kensington, 1867, *Dyer* (BM).

E. helioscopia L. Sun Spurge.
Native on cultivated and waste ground, and as a garden weed. Common.
First record: *T. Johnson*, 1638.
7. Common in the district.

E. peplus L. Petty Spurge.
Native on cultivated and waste ground, and as a frequent garden weed. Very common.
First record: *Blackstone*, 1737.
7. Very common in the district.

E. exigua L. Dwarf Spurge.
Native on cultivated and waste ground, chiefly on light soils. Local.
First record: *T. Johnson*, 1638.
1. Harefield!, *Blackst. Fasc.*, 98. Between Yiewsley and Iver, *Newbould*; Pinner, *T. & D.*, 249. Uxbridge; near South Mimms!, *Benbow MSS.* Potters Bar!, *P. H. Cooke, K. & L.*, 249. By Mimmshall Wood, *J. G.* and *C. M. Dony*!. New Years Green, *T. G. Collett*!. Ickenham, *Davidge.* **2.** Between Sunbury and Hampton, *Newbould*; Stanwell Moor!, *T. & D.*, 249. Littleton!; Sunbury!; near Penton Hook;

Shepperton!; Staines!; near Colnbrook!; West Drayton!, *Benbow MSS*. Hampton!, *Benbow* (BM). Harmondsworth!, *P. H. Cooke*; Laleham!, *K. & L.*, 249. Near Poyle, *J. E. Smith*!. **3.** Isleworth, *Doody* (SLO). Twickenham,† *T. & D.*, 249. Wood End,† *Hind Fl.* **4.** Hampstead,† *Irv. MSS*. Stanmore, *Varenne*; by Brent Reservoir,† *Warren*; Harrow Weald!, *T. & D.*, 249. Hendon, *P. H. Cooke, K. & L.*, 249. Harrow, casual, *Harley Fl.*, 16. Willesden, casual, 1959. Near Mill Hill. Railway goods yard between Neasden and Wembley, 1962. East Finchley. Pinner Road Cemetery, near Harrow!, *Wurzell*. **5.** Alperton,† *T. & D.*, 249. Brentford, 1873,† *Britten, Trim. MSS.* Ealing!, *K. & L.*, 249. Park Royal;† Acton,† *Druce Notes*. Heston. **6.** Tottenham, *Johns. Cat.* Between Hadley and Potters Bar, *Benbow MSS*. Enfield!, *P. H. Cooke, K. & L.*, 249. **7.** Railway yard, Finchley Road, 1870,† *Trim. MSS*. Eaton Square, S.W.1 !, *McClintock, K. & L.*, 249. Kensington Gardens, 1962, *Allen Fl.* Shoreditch, casual, 1966.

E. esula L. *sensu lato*
E. uralensis auct., *E. virgata* auct.
Alien. Europe, etc. Naturalised in fields, on hedgebanks, waste ground, by roadsides, etc. Rather rare, but increasing.
First record: *Pamplin*, 1862.
1. Pinner!,† *L.N.* 27: 33. Yiewsley. **2.** Towing path near Hampton Court, 1926,† *Lousley* (L). Field between West Drayton and Dawley, abundant!, *L.N.* 28: 34; 1962. West Drayton, *I. G. Johnson*. Staines, 1956, *Briggs*. Between Stanwell Moor and Staines, 1965. **3.** Yeading!; Southall, in several places!, *L.N.* 27:33; 1973. **4.** Brent Reservoir, 1916,† *J. Hutchinson* (CYN). East Finchley!; railway side, Dollis Hill, Willesden!;† between Neasden and Wembley Park!, *L.N.* 27: 33; plentiful, 1967. Hampstead Heath Extension, 1948!, *H. C. Harris, K. & L.*, 348. Colindale, *L. M. P. Small*. **5.** Norwood Green!, *L.N.* 26: 63; 1973. Wyke Green!,† *L.N.* 26: 63. Hanwell, 1946!,† *K. & L.*, 248. Railway banks, Brentford, abundant, 1960→. Railway banks near Castle Bar, Ealing to Perivale Park, abundant. Chiswick, 1918,† *Burkill, K. & L.*, 248. Railway bank, East Acton, 1957→. **6.** Finchley Common!; Whetstone!, *L.N.* 28: 34. Ponders End, *A. Vaughan*; 1966. Brimsdown, 1966, *Kennedy*!. Near Waltham Abbey!, *F. Clarke, L.N.* 28: 34; 1967. Gordon Hill, Enfield, *L. M. P. Small*. **7.** Railway bank, Kilburn, 1862,† *Pamplin, Phyt. N.S.* 6: 349. Isle of Dogs!, *Benbow MSS.*; 1967. Railway bank near White City station!, *L.N.* 28: 34; 1949→. Railway bank near Hackney, 1966. Northern Outfall Sewer, Old Ford, 1969, *L. M. P. Small*.
An extremely variable taxon, especially in leaf shape and width. The most frequent plant is probably referable to the hybrid between subsp. **esula** and subsp. **tommasiniana** (Bertol.) Nyman, and it is this which has been called *E. uralensis*.

E. cyparissias L. Cypress Spurge.
Alien. Europe, etc. Garden escape or outcast. Naturalised in grassy places, on waste ground, etc., and established on railway banks, etc. Very rare.
First evidence: *Benbow*, 1900.
1. Uxbridge, 1900,† *Benbow* (BM). Road verge east of Potters Bar, 1966. **2.** A large well-established colony in a field near Half Moon Covert, near Kempton Park, 1966, *Tyrrell*!. **4.** Harrow Weald, *Knipe, L.N.* 41: 14. **7.** Railway bank near Kilburn, a small colony, 1946–54!, *K. & L.*, 249; increasing, 1955→. Holloway, 1965, *D. E. G. Irvine*.

E. amygdaloides L. Wood Spurge.
Native in woods, thickets, on shady hedgebanks, etc. Locally common in the north of the vice-county, very rare elsewhere.
First record: *W. Turner*, 1562.
1. Harefield!, abundantly, *Blackst. Fasc.*, 98. Pinner Wood!, *Melv. Fl.*, 69. Ruislip Woods!, *T. & D.*, 248. Near Denham Lock!; Northwood!; Garett Wood, Springwell!,† *Benbow MSS.* East of Potters Bar, *Kennedy*!. **3.** Roxeth,† *Hind Fl.* **4.** Stanmore,† *Varenne*; Mill Hill, 1846, *J. Morris*; Scratch Wood; Bishop's Wood, *T. & D.*, 248. Hampstead Heath,† *Benbow MSS.* East Finchley, *J. E. Cooper.* **5.** Syon Park,† *Turn. Hb.*, 2. Horsendon Lane,† *Hind Fl.* Perivale Wood,† *Benbow MSS.* **6.** Highgate Wood,† *Johns. Eric.* Enfield!; Colney Hatch;† Winchmore Hill Wood,† *T. & D.*, 248. Near Muswell Hill,† *Benbow MSS.* Hadley. **7.** Hornsey Wood, 1780,† *Pamplin MSS.* Lane between Edgware Road and Hampstead, 1809,† *Winch MSS.*

POLYGONACEAE

POLYGONUM L.

Section POLYGONUM
Lit. *Watsonia* **5**: 177 (1962).

P. aviculare L.
P. heterophyllum Lindm.
Native on cultivated and waste ground, by roadsides, etc., established on paths, etc. Very common, especially on light soils.
First certain evidence: *Doody*, c. 1700.
7. Common in the district.

P. rurivagum Jord. ex Bor.
Native on cultivated ground, chiefly on calcareous soils. Very rare or overlooked.

First certain record: *D. H. Kent*, 1962.
1. Harefield. Springwell. Near Mimmshall Wood.

P. arenastrum Bor. Small-leaved Knotgrass.
P. aequale Lindm.
Native on cultivated and waste ground, by roadsides, etc., established on paths, etc. Very common.
First certain record: *Trimen* and *Dyer*, 1869.
7. Frequent in the district.

Section BISTORTA (Mill.) DC.

P. bistorta L. Snake-root, Bistort.
Denizen in grassy places, marshes, woods, on banks, etc. Very rare.
First record: *Blackstone*, 1737.
1. In the meadows near Uxbridge, *Blackst. Fasc.*, 10; c. 1780,† *Lightf. MSS*. Pinner,† *Hind, Melv. Fl.*, 66. Field near Ruislip Reservoir!, *T. & D.*, 242. Near Harefield!,† *Benbow, J.B.* (*London*) 22: 279. Springwell!,† *Benbow MSS*. Eastcote, 1905,† *Green* (SLBI). Harrow Weald Common, 1955→. Between Potters Bar and South Mimms, 1910,† *Green* (SLBI). **2.** Near West Drayton,† *Benbow MSS*. **3.** Headstone Farm,† *Melv. Fl.*, 66. **4.** Plentifully in a meadow by the side of Bishop's Wood, near Hampstead, *Curt. F.L.*; it is now almost eradicated from the meadow but grows on the bank at the border of the wood and extends a little way into the latter, *T. & D.*, 242; 1945, *Fitter, K. & L.*, 239. Hampstead Common (=Heath), 1809, *Winch MSS.*; 1825, *Trim. MSS.*; c. 1912,† *Whiting Fl.* Between Mill Hill and Elstree, *Irv. Lond. Fl.*, 123. Whitchurch Common, *Welch*!, *L.N.* 28: 34. Old orchard, Stanmore, 1961, *D. Morris*. Mill Hill, 1898, *E. M. Dale* (LNHS). Bank by Harrow-on-the-Hill railway station!, *Welch* and *Wrighton, K. & L.*, 239; an enormous patch, 1966, *Temple*!;† eradicated in 1966 by the construction of a car park, *Temple*. **6.** Top of Muswell Hill,† *Woods, B.G.*, 404. Field at Muswell Hill, 1867,† *Cherry*; Edmonton,† *T. & D.*, 242. Abundant in meadows in Myddleton House grounds, Enfield, 1966, *Kennedy*!. **7.** Ken Wood, *Park Hampst.*, 30; c. 1869,† *T. & D.*, 242.

P. amplexicaule D. Don
Alien. Himalaya. Garden escape or outcast. Naturalised in thickets, on waste ground, etc. Very rare.
First record: *Gerrans*, 1947.
2. Teddington, 1947, *Gerrans, K. & L.*, 241. **6.** Enfield, 1955, *Johns*, teste *Bangerter* and *Welch, K. & L.*, 356. **7.** Bombed site north-west of Victoria Park, 1963, *Wurzell*.

Section PERSICARIA (Mill.) DC.

P. amphibium L. Floating Persicaria, Amphibious Bistort.
Native floating on ponds, lakes, streams, ditches, etc. Terrestrial forms occur also on the banks of rivers and streams, in damp grassy places, etc. Common.
First record: *T. Johnson*, 1638.
7. Tottenham!, *Johns. Cat.* Lee Navigation Canal, Upper Clapton. St James's Park,† *Merr. Pin.*, 93. Between Westminster and Chelsea!, *Pet. MSS.*; a small patch among grass by the Thames near Chelsea Bridge, 1967. Bayswater Canal in Kensington Gardens, 1817† (G & R). Abundant in the Serpentine, 1868, *T. & D.*, 243; Kensington Gardens, one patch by the Serpentine, 1961!, *D. E. Allen, L.N.* 41: 19; Round Pond, Kensington Gardens,† *Pamplin*; at the back of Adelaide Road,† *Syme*; Zoological Gardens, Regents Park!; Eel Brook Meadow, Parsons Green;† Isle of Dogs, *T. & D.*, 243. Hyde Park, a patch on Crystal Palace site, 1959, *D. E. Allen*; lake in Buckingham Palace grounds!, perhaps introduced, *L.N.* 39: 52. Hackney Marshes!, *Mart. Pl. Cant.*, 66. Regents Canal, Stepney, Victoria Park and Hackney Wick. East Heath, Hampstead.

P. persicaria L. Persicaria, Red-leg, Willow weed.
Lit. *J. Ecol.* **33**: 121 (1945).
Native on cultivated and waste ground, by roadsides, etc., and established on railway banks and tracks, etc. Very common.
First record: *T. Johnson*, 1638.
7. Very common in the district.

Var. **elatum** Gren. & Godr.
1. Harefield. **3.** Hounslow Heath. **7.** Kilburn, 1870, *Dyer* (BM).

P. lapathifolium L. Pale Persicaria.
P. scabrum Moench
Lit. *J. Ecol.* **33**: 132 (1945).
Native on cultivated and waste ground, by pond verges, roadsides, etc. Very common.
First record: *Petiver*, c. 1713.
7. Common in the district.

Var. **punctatum** Gremli
3. Hayes, *G. C. Druce, Rep. Bot. Soc. & E.C.* 5: 307.

P. nodosum Pers. Large Persicaria.
P. maculatum (Gray) Dyer ex Bab., *P. petecticale* (Stokes) Druce
Lit. *J. Ecol.* **33**: 140 (1945).
Native by river- and streamsides, on pond verges, damp cultivated and waste ground, etc. Rather rare, and mainly confined to river valleys.
First certain record: *Babington*, 1837.

1. Near Uxbridge, *Druce Notes*. Harefield. Springwell. Ruislip Reservoir. South Mimms. **2.** West Drayton, *Druce Notes*. Laleham. Harmondsworth. Near Heathrow. **4.** West Heath, Hampstead. Brent Reservoir!, *L.N.* 29:13. Finchley. **5.** Near Brentford!, *Welch*. Hanwell. Greenford. Northolt. Chiswick. **6.** Tottenham. **7.** Parsons Green, *Irv. Ill. Handb.*, 356; Eel Brook Common;† Sandy End, Fulham;† Victoria Street, S.W.1;† Hampstead!; Hackney Wick!, *T. & D.*, 244. Clapton. Chelsea,† *Newbould*, *T. & D.*, 244.

Forma **densum** Trimen & Dyer
7. Chalk Farm, 1837,† *Babington*, *E.B. Suppl.* Near Swiss Cottage, 1869,† *Dyer* (BM).

Var. **erectum** Rouy
1. Springwell!, *K. & L.*, 240. **4.** Brent Reservoir, *Lousley* (L).

Var. **inundatum** C. E. Britton
2. Hampton Court Park, 1944, *Lousley* (L). **4.** Brent Reservoir, *Lousley* (L), teste *C. E. Britton*.

Forma **salicifolium** Gray
4. Brent Reservoir, *E. J. Salisbury*.
Timson (1963, *Watsonia* **5**: 386) considers that *P. nodosum* is a form of *P. lapathifolium*, but I am of the opinion that it is a valid species; the taxonomic value of the varieties is, however, open to some doubt.

P. hydropiper L. Water-pepper.
Lit. *J. Ecol.* **54**: 815 (1966).
Native by stream- and riversides, pond verges, in marshes, etc. Common, especially near the Thames, though rare or extinct in heavily built-up areas owing to the eradication of its habitats.
First certain record: *W. Hudson*, 1778.
7. Tothill Fields,† *Huds. Fl. Angl.* 2: 171. Hyde Park (G & R); one large plant on a dump of soil, 1960, *D. E. Allen*; 1961, *Brewis*; Kensington Gardens, 1961, *D. E. Allen*, *L.N.* 41: 19. Osier holt, Sandy End;† East Heath, Hampstead!; Kentish Town; Hackney Wick, *T. & D.*, 245. Lane between Highgate and Hampstead, *Ball. MSS.* Parliament Hill Fields. Ken Wood. Highgate. Tottenham. Hackney Marshes. Upper Clapton, *L. M. P. Small.*

P. mite Schrank
P. laxiflorum (Weihe) Opiz
Native by river- and streamsides, pond verges, in ditches, marshes, etc. Very rare, and chiefly confined to areas near the Thames.
First evidence: *Dillenius*, c. 1730.
1. Wet meadow north of Harefield Moor, 1947,† *Sandwith*!, *L.N.* 27: 33. **2.** Bushy Park!; between Hampton Court and Kingston Bridge!, *T. & D.*, 269. Sunbury, 1869, *H. C. Watson*, *Trim. MSS.*

Near West Drayton,† *Benbow, J.B.* (*London*) 23: 340. Yeoveney!, *Benbow MSS.* Near Staines!, *Sherrin* (SLBI). Knowle Green!, *K. & L.*, 240. **3.** Twickenham, on soil brought from elsewhere,† *T. & D.*, 244. **5.** Acton,† *Newbould, T. & D.*, 245. Chiswick House grounds!, *Murray List.* Syon Park, 1966, *Boniface.* **7.** Chelsea (DILL); c. 1826, *La Gasca Hort. Sicc.*; 1843,† *Babington, E.B. Suppl.* Eel Brook Meadow, 1848, *T. Moore* (BM). Near Edgware Road station,† *Warren*; Sandy End, Fulham, 1867,† *T. & D.*, 245.

Plants with white flowers were seen in **1.** Wet meadow north of Harefield Moor, 1947,† *Sandwith*!, *L.N.* 27: 33.

P. minus Huds. Small Persicaria.

Native by streamsides, pond verges, in wet marshy places, etc. Very rare, and now confined to the close vicinity of the Thames.

First record: *Petiver*, 1695.

1. Harefield Moor, *Benbow, J.B.* (*London*) 25: 20. Error, *Benbow, J.B.* (*London*) 25: 340. **2.** Bushy Park!, *T. & D.*, 245. Hampton Court Park!, *Lousley* (L). Staines Moor, *Briggs.* **4.** Golders Green,† *Irv. MSS.* **6.** Finchley Common,† *Dillwyn, B.G.*, 404. **7.** Ditchbanks in the meadows beyond Lord Peterborough's House at Westminster,† *Pet. Mus.* 1: 13. Tothill Fields, *Pet. Midd.*; c. 1745, *Blackst. Spec.*, 70; in the greatest abundance,† *Curt. F.L.* Eel Brook Meadow, 1856, '*J.A.*', *Phyt. N.S.* 1: 464; 1864,† *T. & D.*, 245. Near Chelsea,† *Spencer*, 318. Walham Green, 1842† (BO). Probably refers to the Eel Brook Meadow station. Fulham,† *Kippist* (BM).

Section TINIARIA Meisn.

P. convolvulus L. Black Bindweed.

Bilderdykia convolvulus (L.) Dumort., *Fallopia convolvulus* (L.) A. Löve

Native on cultivated and waste ground, etc. Very common.

First record: *T. Johnson*, 1638.

7. Common in the district.

Var. **subalatum** Lej. & Court.

1. Harefield, *T. & D.*, 246. Yiewsley, *J. E. Cooper*; Uxbridge!, *K. & L.*, 239. **2.** West Drayton, *Druce Notes.* **3.** Twickenham, *T. & D.*, 246. **5.** Acton, *Druce Notes.* **6.** Muswell Hill, *J. E. Cooper*, *K. & L.*, 239. North Finchley, *J. E. Cooper* (BM). **7.** Hackney Wick, *T. & D.*, 246. Crouch End, 1897; Hackney Marshes, 1907, *J. E. Cooper*, *K. & L.*, 239.

P. aubertii Louis Henry Russian Vine.

P. baldschuanicum auct., non Regel, *Bilderdykia aubertii* (Louis Henry) Moldenke, *Fallopia aubertii* (Louis Henry) Holub

Alien. Central Asia. Garden outcast. Naturalised in hedges, on waste ground, etc. Rather rare, but increasing.

First record: *D. H. Kent*, 1944.

1. Warren Gate!, *K. & L.*, 241. Near Potters Bar, 1965, *J. G.* and *C. M. Dony*!. Harefield, 1966, *I. G. Johnson*. **2.** Hampton Wick, 1963, *Ryves*. Near West Drayton, 1965. **3.** Northolt!, *L.N.* 27: 33. Hounslow Heath!, *Welch*, *K. & L.*, 242; 1965. Frogmore Green, near Southall, 1961→. Isleworth, 1966. **4.** Highgate, in several places!, *Fitter*, *K. & L.*, 241. Brent Cross, Hendon, 1959→. Brockley Hill, 1966, *Kennedy*!. **5.** West Ealing, 1944!→; Hanwell, 1945!→; between Hanwell and Southall, 1945!→; Greenford, *K. & L.*, 241. North Hyde, 1965. **7.** Park Road, Marylebone!, *L.N.* 39: 52. Bombed sites, Gresham Street† and Queen Victoria Street areas,† *Jones Fl.* Regents Park, *Holme*. Bombed sites, Shadwell, 1966. Canal side between Marylebone and Primrose Hill, 1968.

Section PLEUROPTERUS (Turcz.) Benth.

P. cuspidatum Sieb. & Zucc.

P. sieboldii hort., *Reynoutria japonica* Houtt.

Alien. Japan. Garden escape or outcast. Naturalised on waste ground, by roadsides, etc. Very common.

First evidence: *A. Irvine*, 1861.

7. Common in the district.

P. sachalinense F. Schmidt

Reynoutria sachalinensis (F. Schmidt) Nakai

Alien. Sakhalin. Garden escape or outcast. Naturalised on waste ground, by roadsides, etc. Rare.

First record: *D. H. Kent*, 1944.

4. Harrow Weald, 1950, *Lousley* (L). Harrow, *Harley Add.* Finchley!, *Mason*. Brent Reservoir. **5.** Osterley Park, *Welch*!. **7.** Bombed sites, Fetter Lane† and Neville Court, E.C., 1950,† *Whittaker* (BM). Hyde Park, two plants, 1961, *D. E. Allen* and *McClintock*; Chelsea Hospital grounds!, *L.N.* 41: 19. Bombed site, Cripplegate area,† *Jones Fl.*

P. polystachyum Wall. ex Meisn.

Alien. Assam, Sikkim, etc. Garden outcast or escape. Naturalised on waste ground, etc. Very rare.

First record: *D. H. Kent*, 1948.

2. Grounds of the National Physical Laboratory, Teddington, *Clement*. **4.** The Hyde, Hendon!, *L.N.* 28: 34; 1968. Edgware, 1967, *Horder*!. **7.** Highgate Ponds, *H. C. Harris*, *L.N.* 28: 34. Kensington Gardens, *Allen Fl.*

P. campanulatum Hook. f.

Alien. Himalaya. Originally planted by water, but now regenerating and naturalised.

First record: *H. C. Harris*, 1948.

7. The Cascade, Ken Wood!, *H. C. Harris*; 1967.

RUMEX L.

Lit. *Rep. Bot. Soc. & E.C.* **12**: 118 (1939): *loc. cit.* **12**: 547 (1944).

Subgenus ACETOSELLA Raf.

R. tenuifolius (Wallr.) A. Löve
R. acetosella L. var. *tenuifolius* Wallr.
Native in grassland on poor soils. Rare, or overlooked.
First record: *D. E. Allen*, 1960.
1. Harefield Moor. Ruislip Common. **2.** Bushy Park. **3.** Hounslow Heath. **4.** West Heath, Hampstead. **7.** Kensington Gardens!, *D. E. Allen*, conf. *Lousley*, *L.N.* 40: 21; 1967. Hyde Park, *D. E. Allen*, *L.N.* 41: 19. Ken Wood!, *D. E. Allen*. East Heath, Hampstead.

R. acetosella L. Sheep's Sorrel.
Native on heaths, commons, dry acid grassland, etc., established on railway banks, etc. Common.
First certain evidence: *Petiver*, c. 1710.
7. Hyde Park!, 1868, *Warren*, *T. & D.*, 242; 1947; East Heath, Hampstead!; Parsons Green!; garden of Lincolns Inn, 1869; Grosvenor Square, 1869,† *T. & D.*, 242. Ken Wood!; Highgate Ponds!, *Fitter*. Fulham Palace grounds, *Welch*!. Kensington Gardens!; Buckingham Palace grounds!; Primrose Hill!, *L.N.* 39: 52. Paddington Green!; railway tracks, Paddington, 1961!, *L.N.* 41: 19. Brompton Cemetery. Kilburn. Kentish Town. Holloway. Finsbury Park. Hornsey. Dalston. Stoke Newington. Tottenham. Stepney. Victoria Park. Hackney Wick. Isle of Dogs.

Subgenus ACETOSA (Mill.) Reching. f.

R. acetosa L. Sorrel.
Native in grassy places, etc. Common.
First record: *T. Johnson*, 1638.
7. Common in the district.

Subgenus RUMEX

R. hydrolapathum Huds. Great Water Dock.
Native by river- and streamsides, on verges of ponds and lakes, etc., established by canals and gravel pits, etc. Common, though sometimes planted for ornamental purposes.
First record: *T. Johnson*, 1638.
7. Between Westminster and Chelsea,† *Pet. MSS*. Hackney Wick; by the reservoirs in Ken Wood grounds!, *T. & D.*, 241. Highgate Ponds!, *Fitter*. By the lake, Buckingham Palace grounds!, *Eastwood*; probably planted. Canal wall near Paddington Basin, a fine clump, 1966!, *L.N.* 46: 31. By Lee Navigation Canal between Upper Clapton and Tottenham.

R. patientia L. subsp. **orientalis** (Bernh.) Danser
Alien. E. Europe, etc. Naturalised in grassy places, on waste ground, etc., established on railway banks, etc. Very rare.
First evidence: *Warren*, 1871.
2. Near Kempton Park, a small colony, 1966, *Tyrrell*!. **5.** Thames side, Chiswick, very abundant!, *Lousley* (L); still present, but reduced in quantity, 1969. Chiswick Ait!, *Fitter*. **6.** A number of isolated colonies on waste ground by the Lee Navigation Canal between Edmonton and Ponders End. **7.** Site of the Exhibition of 1862, South Kensington, 1871,† *Warren* (BM). Isle of Dogs!, *Whellan* (L); still plentiful about Cubitts Town, 1971.

R. cristatus DC.
R. graecus Boiss. & Heldr.
Alien. Mediterranean region. Naturalised on waste ground, etc. Very rare.
First record: *Lousley*, 1950.
5. Thames side, Chiswick, 1950!, *Lousley*; 1968.

R. crispus L. Curled Dock.
Lit. *J. Ecol.* **52**: 754 (1964).
Native on cultivated and waste ground, by roadsides, etc., established on railway banks, etc. Very common.
First record: *Blackstone*, 1737.
7. Common in the district.

Var. **arvensis** Harvey
The common Middlesex plant.

Forma **unicallosus** (Peterm.) Lousley
2. Walton on Thames, *Lousley* (L).

Var. **robustus** Reching.
5. Brentford, 1938, *Lousley* (L).

R. crispus × **cristatus**
5. Thames bank, Chiswick, 1950!, *Lousley*; still present, 1969.

R. crispus × **obtusifolius** = **R.** × **acutus** L.
R. pratensis Mert. & Koch
Common where the parents grow together in quantity and often found growing in the absence of either parent.

R. crispus × **patientia** subsp. **orientalis**
5. Thames bank, Chiswick!, *Lousley* (L); still present, 1969. **6.** By Lee Navigation Canal, Ponders End, 1966, *Kennedy*!. **7.** Cubitts Town, 1967.

R. obtusifolius L. subsp. **obtusifolius** Broad-leaved Dock.
R. obtusifolius L. subsp. *agrestis* (Fr.) Čelak.
Lit. *J. Ecol.* **52**: 737 (1964).
Native on cultivated and waste ground, by roadsides, etc., established on railway banks, etc. Very common.
First certain record: *A. Irvine*, c. 1830.
7. Very common in the district.

Subsp. **transiens** (Simonk.) Reching. f.
R. obtusifolius L. var. *sylvestris* auct. brit., non Fr.
Alien. Europe, etc. Naturalised by riversides, on pond verges, waste ground, etc. Very rare or overlooked.
1. Uxbridge, 1969, *L. M. P. Small*, det. *Lousley*. **2.** River bank between Hampton Court and Kingston Bridge, *C. E. Britton*, *J.B. (London)* 48: 331 (as *R. obtusifolius* var. *sylvestris*). **4.** Brent Reservoir!, *Lousley* (L). **5.** Chiswick!, *Lousley* (L). Hanwell, 1947, *Lousley*!, *L.N.* 27: 33. **6.** Near Enfield Lock, 1966, *Kennedy*!. **7.** Thames bank between Fulham and Hammersmith, 1871, *Warren* (BM).

R. obtusifolius subsp. **obtusifolius** × **palustris** = **R.** × **steinii** Beck.
2. Near Staines Moor, 1966. **4.** Brent Reservoir, 1931 and 1938, *Lousley* (L). **5.** Hanwell, 1948, *Lousley*! (L).

R. obtusifolius subsp. **obtusifolius** × **patientia** subsp. **orientalis** = **R.** × **erubescens** Simonk.
5. Chiswick!, *Lousley* (L); 1969.

R. pulcher L. subsp. **pulcher** Fiddle Dock.
Native in dry sunny places. Rare, and mainly confined to light soils near the Thames.
First certain record: *Ray*, 1670.
1. Yiewsley, 1922,† *J. E. Cooper*, *K. & L.*, 244. Perhaps an error. **2.** Between Hampton and Hampton Court!, *H. C. Watson*, *Wats. New Bot.*, 110. Near Teddington, *Newbould*; Hampton; Sunbury!, *T. & D.*, 240. Near Walton Bridge; Halliford!; Charlton!; Littleton!; Hampton Court!, *Benbow MSS.* Locally common on the river-wall of the Thames between Hampton Court and Kingston Bridge. Near Staines!, *Whale Fl.*, 28. Knowle Green!,† *K. & L.*, 244. Harmondsworth, *Newb. MSS.* Village green, West Drayton, abundant, 1966→. **3.** Twickenham, *T. & D.*, 240. Near Twickenham, *Benbow MSS.* **4.** Hampstead,† *Irv. MSS.* Harrow, *Horton Rep.* Error. Hendon, 1921,† *P. H. Cooke* (LNHS). **5.** Hanwell!;† Syon Park!, *L.N.* 38: 64; much reduced by gravel-digging operations, 1966. Greenford, 1917,† *J. E. Cooper*, *K. & L.*, 244. Perhaps an error. **7.** St James's Fields, Westminster,† *Ray Cat.*, 188. Common about London,†

Merr. MSS. and *Budd. MSS.* Upper Moor-fields† and in the Teynter† and Bunhill-fields,† *Pet. Bot. Lond.* and *Pet. Midd.* Thickly spread over the Palace green, Kensington,† *Warren, J.B. (London)* 13: 276.

R. sanguineus L. var. **sanguineus** Red-veined Dock.
Alien of unknown origin. Formerly cultivated for medicinal purposes, and naturalised as an escape or outcast in woods, thickets, grassy places, etc., but now extinct.
First record: *Merrett*, 1666. Last record: *Hind*, 1869.
1. In an orchard between Harefield and Ruislip, *Blackst. Fasc.*, 50. Pinner Hill, *Hind, T. & D.*, 240. **5.** Near Ealing, plentifully, *Trim. MSS.* **7.** Hampstead, *Hill Fl. Brit.*, 191. Fulham, 1829, *Trim. MSS.*

Var. **viridis** Sibth. Green-veined Dock, Wood Dock.
Rumex condylodes Bieb., *R. nemorosus* Schrad. ex Willd., *R. viridis* (Sibth.) Druce
Native in woods and shady places. Common.
7. Grosvenor Square, 1869,† *Warren*; Finchley Road;† near West End;† East Heath, Hampstead!, *T. & D.*, 239. Highgate Ponds!, *Fitter*. Bishop's Park, Fulham, *Welch*!. Kensington Gardens!, *D. E. Allen*, Buckingham Palace grounds, *Codrington, Lousley* and *McClintock*!; Regents Park!, *L.N.* 39: 52. Hyde Park, plentiful in the Bird Sanctuary!, *D. E. Allen, L.N.* 41: 19. Holland House grounds, Kensington, frequent. Bombed site, Friday Street, E.C.,† *Lousley Fl.* Islington. Holloway. Finsbury Park. Stoke Newington. Hackney Wick.

R. conglomeratus Murr. Sharp Dock.
Native in damp grassy places, woods, on cultivated and waste ground, by roadsides, etc. Common.
First record: *T. Johnson*, 1638.
7. Common in the district.

R. conglomeratus × **crispus** = **R.** × **schulzei** Hausskn.
2. Walton on Thames, *Lousley* (L).

R. conglomeratus × **maritimus** = **R.** × **knafii** Čelak.
4. Child's Hill, 1870, *Dyer* (BM), det. *Rechinger*.

R. conglomeratus × **obtusifolius** L. subsp. **obtusifolius** = **R.** × **abortivus** Ruhmer
2. Staines Moor, 1951, *Lousley*! (L).

R. conglomeratus × **palustris**
4. Brent Reservoir, 1949!, *Lousley* (L). **5.** Hanwell, 1945, *Lousley*! (L). **6.** Whitewebbs Park, Enfield, 1965, *Kennedy*!.

R. palustris Sm. Marsh Dock.
R. limosus auct. angl., non Thuill.

Native in marshes, on pond and lake verges, damp waste ground, etc. Locally abundant, especially in the south-western parts of the vice-county where it is probably more plentiful than in any other comparable area of Britain.

First record: *Rand*, 1700.

1. Yiewsley, 1924–28, *J. E. Cooper, K. & L.*, 244. Rubbish-tip, New Years Green, 1969, *Clement* and *Ryves*. **2.** Near West Drayton, abundant, 1883–85, *Benbow* (BM); scarce, 1887, *Benbow MSS.*; abundant, 1912, *C. S. Nicholson* (BM). Yeoveney!, *Benbow, J.B. (London)* 25: 18. Staines Moor!, *Briggs*. Bulldozed gravel pit areas, Moor Lane, west of Staines Moor from near Staines to the Horton Road, in the greatest profusion, 1966, *Boniface*!; 1967. Shortwood Common, 1964!, *Boniface*; plentiful, 1965→. Gravel pit, West Bedfont, locally plentiful, 1965, *Kennedy*!; 1967. Near Stanwell, 1966, *Boniface*. Stanwell Moor, one plant, 1966, *J. E. Smith*!. Riverside, Hampton Court, one plant, 1966, *Clement*. **3.** Hounslow, one plant,† *Newbould, T. & D.*, 238. Rubbish-tip, Hounslow Heath, 1961, *Philcox* and *Townsend* (K). Gravel pit, Lower Feltham, 1952,† *Russell* and *Welch*!, *K. & L.*, 245. Marshes south of the Western Avenue, Northolt, 1967!, *L.N.* 47: 9. **4.** Brent Reservoir!, 1863, *Parsons* (CYN); 1917, *J. E. Cooper, K. & L.*, 244; 1929 and 1938, *Lousley* (L); still present in quantity, 1973. Hampstead, 1850 and 1852, *Syme, Phyt.* 4: 861; c. 1869;† West Heath, Hampstead;† Child's Hill,† *T. & D.*, 238. The latter record may be an error. Canal side, Harlesden, a single plant, 1965. **5.** Near Chiswick,† *Lindley, Curt. Brit. Ent.* Chiswick,† *Benbow MSS.* Strand-on-the-Green, 1938,† *Lousley* (L). Acton Green,† *Newbould, T. & D.*, 238. Hanwell!, *Welch, L.N.* 26: 63; 1947–52; marsh filled in c. 1953, but a single plant reappeared on adjacent waste ground during the construction of the London–South Wales Motorway in 1964. Between Hanwell and Brentford,† *Lousley*!; near Norwood Green!, *L.N.* 28: 34. Canal wall, Brentford, 1966!, *Boniface*; 1967. **6.** Abundant around a lake, Whitewebbs Park, Enfield, 1965; rough wet field near Edmonton, 1966; rough marshy ground north of Enfield Lock, 1967, *Kennedy*!. **7.** Tothill Fields, Westminster,† *Rand, Pluk. Mant.*, 112. By Lamb's Conduit;† betwixt the back of Montague House and St Giles's Road, and many other places about London,† *Doody MSS.* Hoxton,† *Sherard, Ray Syn.* 3: 142. Camden Town fields, 1827–30,† *Varenne*; Newington, 1867,† *Newbould*; Eel Brook Common, 1862,† *T. & D.*, 239. Shepherds Bush, 1849,† *Wing* (BM). Chelsea, 1842 (BO); bombed site, Chelsea, 1943,† *Sladen, L.N.* 24: 9. Notting Hill† (SY). Between Primrose Hill and Hampstead,† *Dyer* (K). Primrose Hill, 1871, *Dyer* (SY); 1875,† *Blow* (BM). Isle of Dogs, 1852 (SY); 1867, *Cherry*; 1869, *Newbould, T. & D.*, 239; 1887, *Benbow, J.B. (London)* 25: 18. Marylebone,† *J. Banks* (BM). Between Dalston and Hackney,† *Forst. MSS.* East

Heath, Hampstead, rare, 1865† (SY). Hackney Marshes, 1913;† Hornsey, 1897–98;† Crouch End, 1897,† *Coop. Cas.* Canal near Kensal Green, 1948!, *L.N.* 28: 24.

Forma **nanus** (Boenn.) Lousley
4. Brent Reservoir, *Lousley*.

R. maritimus L. Golden Dock.
Formerly native in marshes, on pond verges, damp waste ground, etc., but now extinct.
First record: *Rand*, c. 1700. Last record: *J. E. Cooper*, 1924.
2. New Sunbury, 1877, *H.* and *J. Groves*; near West Drayton, abundant, 1883–85, *Benbow* (BM); habitat filled with rubbish and site partly built over, but a few plants survive in one corner, 1886, *Benbow MSS.*; 1914, *J. E. Cooper* (BM). **4.** Hampstead, a single plant, c. 1870, *Newb. MSS.* **5.** Acton, 1902, *Worsdell* (K). **7.** Moist place near Burlington House; behind Montague House, *Rand, Doody MSS.* St Giles, *Rand, Pet. Bot. Lond.* Conduit Street by Hanover Street [W.1], 1745, *Nicholls, Blackst. MSS.* Hoxton, *Rand, Ray Syn.* 3: 142. Crouch End, in two places, 1887, *J. E. Cooper*; Hackney Marshes, 1918 and 1924, *J. E. Cooper, K. & L.*, 245.

Section AXILLARES Reching. f.

R. triangulivalvis (Danser) Reching. f.
R. salicifolius auct.
Alien. N. America. Naturalised on waste ground, etc., but sometimes merely casual.
First evidence: *J. E. Cooper*, 1909.
1. Uxbridge, 1912,† *Lady Davy, Rep. Bot. Soc. & E.C.* 6: 396. **3.** Hounslow Heath, 1954!, *Welch, L.N.* 34: 6; 1958. **4.** Temple Fortune, 1910,† *J. E. Cooper* (CGE). **6.** Rubbish-heap, Myddleton House grounds, Enfield, a single plant, 1966, *Kennedy*!. **7.** Hackney Marshes, 1909,† *J. E. Cooper* (BM). Shadwell Basin, 1945, *Lousley, L.N.* 25: 14. Waste ground near Shadwell, a few plants, 1951, *L. M. P. Small, L.N.* 31: 12; 1954; increasing, 1957; several large colonies established on waste ground and bombed sites over a distance of at least 100 yards, 1966; decreasing owing to rebuilding operations, 1967.

URTICACEAE

PARIETARIA L.

P. judaica L. Pellitory-of-the-Wall.
P. diffusa Mert. & Koch, *P. officinalis* auct., non L., *P. ramiflora* Moench
Lit. *Proc. Bot. Soc. Brit. Isles* **7**: 228 (1968). (*Abstr.*): *Watsonia* **6**: 365 (1968).

Native on, and at the foot of, old walls, and other brick- and stonework, rarely on waste ground. Locally common near the Thames, and especially on the river wall, elsewhere chiefly in old churchyards, where it is possibly an introduction.

First record: *Petiver*, 1709.

1. Harefield!, *Blackst. Fasc.*, 72; abundant on the walls of Harefield Churchyard. Pinner Churchyard!, *Hind Fl.* Hillingdon! and Ickenham Churchyards!, *Benbow MSS.* Stone embankment of the Fray's river, and elsewhere about Uxbridge. **2.** Frequent near the Thames. West Drayton Churchyard!, *Benbow MSS.* Sunbury, Stanwell and Ashford Churchyards. **3.** Frequent near the Thames. Hanworth Churchyard, *T. & D.*, 253. Northolt. Cranford. **4.** Hampstead!, *Irv. Lond. Fl.*, 119. On the tower of Edgware church, *T. & D.*, 253; 1887, *Benbow MSS.*; Edgware Churchyard, 1967. **5.** Frequent near the Thames. Hedgebank near Turnham Green;† Chiswick Churchyard!, *T. & D.*, 253. Little Ealing!, *A. Wood, Trim. MSS.*; walls of car park of 'The Plough' Inn, and at base of old wall on opposite side of Little Ealing Lane. Hanwell,† *P. H. Cooke.* Railway bank by Brentford station, 1967. **6.** Highgate, *S. Palmer, Mag. Nat. Hist.* 2: 266 and *Irv. MSS.*; Highgate Hill, 1837,† *E. Ballard* (BM). Myddleton House grounds, Enfield, 1966, *Kennedy*!. Probably introduced. **7.** On the Charterhouse Cloisters, etc., *Pet. Bot. Lond.* Walls adjoining the Thames above and below Westminster Bridge,† *Curt. F.L.* Chelsea!; Putney Bridge!, 1817 (G & R); 1968. Wall by Fulham Church!, *De Cresp. Fl.*, 49; frequent on the river wall, Fulham, 1969. Wall of fishpond, Canonbury House, 1818,† *E. T. Bennett, T. & D.*, 253. Wall by railway near Paddington!, *L.N.* 39: 52; 1960→. On the short stretch of exposed 'Roman Wall' by St Alphage Church, E.C.1!, and has now spread in considerable quantity to brick-rubble near Aldermanbury Postern!,† *Lousley Fl.* Bombed sites, Cripplegate,† Gresham Street† and Queen Victoria Street areas,† *Jones Fl.* Bombed site, Ebury Street, S.W.1,† *McClintock.* Brompton Cemetery, plentiful. Marylebone, *Holme, L.N.* 41: 19. Waste ground near the Thames, High Street, Wapping, 1966→.

SOLEIROLIA Gaud-Beaup.

S. soleirolii (Req.) Dandy Mind-your-own-business.
Helxine soleirolii Req.

Alien. Europe. Garden escape. Naturalised in turf, and established at the base of old walls, etc. Rare.

First record: *D. H. Kent*, 1946.

1. Pinner Churchyard, well established in turf, 1965→. **2.** Laleham Churchyard, abundantly established in turf!, *K. & L.*, 253; in profusion, and competing with *Veronica filiformis*, 1966, *Kennedy*!. Hampton Court, 1972. *Mountford*. **3.** Base of an old wall by the Thames near Twickenham Church, where it is daily submerged by the river, 1960→!; base of old wall, St Margaret's, 1965!, *L.N.* 45: 20. **5.** Base of old wall, Acton, 1961. Abundant on brick walls of old greenhouses, Heston, 1962→. Plentiful on brick walls of old greenhouses, Chiswick, 1968. Walls, Osterley, plentiful, 1969. **6.** Myddleton House grounds, Enfield, well naturalised in turf and established on gravel paths, 1966, *Kennedy*!. Old wall, Whetstone, in quantity, 1967. **7.** Buckingham Palace grounds, *Fl. B.P.G.* Fulham Palace grounds, 1966, *Eastwood*. Established at the base of a wall in an old burial ground, Marylebone Road, N.W.1, 1966, *Brewis*, *L.N.* 46: 31.

URTICA L.

U. urens L. Small Nettle.
Lit. *J. Ecol.* **36**: 351 (1948).
Native in cultivated ground, especially as a weed of flower-beds and shrubberies, waste ground, etc. Very common.
First record: *Blackstone*, 1737.
7. Common in the district.

U. dioica L. Stinging Nettle, Great Nettle.
Lit. *J. Ecol.* **36**: 343 (1948).
Native in woods, ditches, on hedgebanks, waste ground, etc. Very common.
First record: *T. Johnson*, 1638.
7. Common in the district.

Var. **angustifolia** Wimm. & Graeb.
1. Uxbridge. Harefield. **2.** West Drayton, *Druce Notes*.

Var. **subinermis** Uechtr.
1. Harefield, 1961, *Harrison*. **5.** Hanwell, 1952!, *K. & L.*, 252. **6.** Near Botany Bay, Enfield Chase, 1965, *Kennedy*!.

CANNABIACEAE

HUMULUS L.

H. lupulus L. Hop.
Native in hedges, thickets, on waste ground, etc., near the Thames, elsewhere in similar habitats but possibly an introduction. Common.
First record: *T. Johnson*, 1638.
7. Tottenham!, *Johns. Cat.* Between Westminster and Chelsea,† *Pet. MSS.* Clapton!, 1837, *Ballard* (BM); 1968. Hedge by Vale of Health, Hampstead, *Benbow MSS.* Highgate. Bishop's Park, Fulham, *Welch*!.

Hammersmith. Regents Park, 1941!, *L.N.* 39: 52. Wellington Road, N.W.8, 1965–66!, *L.N.* 46: 31; 1967→. Notting Hill. Holland House grounds, Kensington. Isle of Dogs. Finsbury Park. Hornsey. Stoke Newington. Wapping. Stepney. Old Ford. Hackney Wick. Bromley-by-Bow. Sale Place, W.2, 1967. Near Kensal Green, *Warren, T. & D.*, 254. Chelsea, 1968, *E. Young*.

ULMACEAE

ULMUS L.

U. glabra Huds. *sensu lato* Wych Elm.
U. scabra Mill., *U. montana* Stokes
Native in hedges, woods, scrub, etc., in the northern parts of the vice-county, especially in the north-east where it is locally plentiful, elsewhere in hedges, woods, etc., possibly as an introduction, where it is rather rare. It is doubtful if pure *U. glabra* occurs in Middlesex today, if it does exist it is probably in the Enfield Chase area. Dr R. Melville considers that the entity regarded as *U. glabra* in Middlesex consists of a fertile hybrid swarm, e.g. *U. carpinifolia* × *glabra*, *U. carpinifolia* × *glabra* × *plotii*, *U. glabra* × *plotii*, etc. This hypothesis would account for the very variable characters noted in the size and shape of trees, the size, shape and texture of leaves, etc., in different populations.
First record: *Petiver*, 1695.
1. Common. **2.** Fulwell!; between Kingston Bridge and Hampton Court!, *T. & D.*, 255. Laleham Park. Ashford. West Bedfont, *Kennedy*!. East Bedfont. Bushy Park. Harlington, *Bangerter*. Between Hampton and Sunbury. **3.** Harrow Grove!, *Melv. Fl.*, 71. Hatton. Hanworth, *L. M. P. Small*. Between Richmond Bridge and Twickenham. **4.** Locally common, except in heavily built-up areas. **5.** Canal bank near Park Royal. Syon Park!, *Jacks. Cat.* Near Brentford. **6.** Common, except in heavily built-up areas. **7.** Hoxton, *Pet. Midd.* Hampstead!; Regents Park!, *T. & D.*, 255. Tottenham, *Kennedy*. Canal side, Islington!, *L.N.* 46: 31. Hammersmith. East Heath, Hampstead. Stoke Newington. Stepney. Hackney Marshes. Fine trees and seedlings in the grounds of Clarence Lodge, Albany Road, N.W.1, to north-east side of Regents Park, 1967, *McClintock*.

U. procera Salisb. English Elm.
U. campestris auct., *U. sativa* auct., *U. suberosa* auct.
Native in hedges, fields, by roadsides, etc. Common; though sometimes planted.
First record: *T. Johnson*, 1638.
7. Buckingham Palace grounds!, *L.N.* 39: 52. Whitehall, S.W.1. Fulham. West Brompton. Holland House grounds, Kensington.

Hyde Park. Chelsea. Brompton. Kensal Green. Notting Hill. Kilburn. Kentish Town. Highgate. Hampstead. Islington. Finsbury Park. Hornsey. Holloway. Dalston. Stoke Newington. Tottenham!, *Johns. Cat.* Wapping. Clapton. Old Ford. Hackney Wick. Bromley-by-Bow. Hurlingham, *L. M. P. Small.*

MORACEAE

FICUS L.

F. carica L. Fig.
Lit. *Rep. Bot. Soc. & E.C.* **13**: 330 (1948).
Alien. S. Europe, etc. Naturalised in hedges, on waste ground, etc., established on railway banks, etc., always originating from seeds from discarded fruit. Rather rare.
First record: *Shenstone*, 1912.
1. Yiewsley, 1956. **2.** Teddington, 1965. Harlington, 1965. **3.** Northolt!, *Lousley, Rep. Bot. Soc. & E.C.* 13: 333. Isleworth, 1962→. Hatton, 1965. **4.** Willesden Junction, 1965→. Cricklewood, 1965 →. Near Kenton, 1966. Queensbury, 1966. Stonebridge, 1968, *McLean.* **5.** West Ealing, 1957→. Brentford, 1960→. Hanwell, 1967. Acton and South Acton, frequent on waste ground, 1965→. Bombed site, Chiswick, 1966→. Hanwell, 1967. **6.** Ponders End, 1958. Enfield, 1972!, *Kennedy.* Winchmore Hill, *L. M. P. Small.* **7.** Building sites, Bloomsbury† and Faringdon Street,† *Shenstone Fl. Lond.* Railway between Bishopsgate and Bethnal Green, 1930, *Tremayne.* Bombed site, St Olaves, Hart Street, E.C.!;† growing out of a wall, Water Lane, E.C.;† bombed site, Holborn Circus, 1945,† *Lousley, Rep. Bot. Soc. & E.C.* 13: 330. Common on bombed sites in Central London—north of Holborn; between Holborn and Fleet Street and in the West End,† *Lousley, Rep. Bot. Soc. & E.C.* 13: 332; still survives on a few of the remaining bombed sites in Central London, but is very rare, 1969. Bombed dried fruit warehouse, London Dock, 1948, *Lousley* (L). Bombed site, Ebury Street, S.W.1!;† bombed site, Chester Square, S.W.1, 1952!,† *K. & L.*, 252. By Wandsworth Bridge, 1965.

MYRICACEAE

MYRICA L.

M. gale L. Bog Myrtle, Sweet Gale.
Formerly native on wet heaths, but now extinct.
First record: *Gerarde*, 1597. Last record: *Whiting*, 1901.

2. Colnbrook, *Ger. Hb.*, 1228. **3.** Hounslow Heath, *Merr. Pin.*, 82. **4.** Hampstead Heath, a single bush, very tall, old, withered, and almost dead, 1887, *Benbow* (BM); 1901, *Whiting Notes.*

BETULACEAE

BETULA L.

B. pendula Roth Silver Birch.
B. verrucosa Ehrh., *B. alba* auct.
Native in woods, on heaths, etc., also commonly planted in parks, gardens, cemeteries, and as a street tree. It readily regenerates and seedlings and saplings may frequently be seen on waste ground, wall-tops, etc. Common.
First certain evidence: *E. Ballard*, 1838.
7. Common in the district.

B. pendula × pubescens = B. × aurata Borkh.
1. Stanmore Common. Harrow Weald Common. Ruislip Woods. Harefield. **3.** Hounslow Heath. **4.** West and North-west Heaths, Hampstead. **7.** East Heath, Hampstead. Hyde Park!; Kensington Gardens!, seedlings, *Allen Fl.*
Many of the planted trees which readily regenerate may be referable to this hybrid which is possibly more frequent than the few records given suggest.

B. pubescens Ehrh. susbp. **pubescens** Birch.
B. glutinosa Wallr., *B. alba* auct.
Native on heaths, in woods, etc., in damper situations than *B. pendula.* Local.
First record: *T. Johnson*, 1632.
1. Harefield!, *Blackst. Fasc.*, 10. Stanmore Common. Northwood. Ruislip Woods. South of Harrow Weald Common, a single tree, *Tremayne.* Near Potters Bar!, *J. G.* and *C. M. Dony.* **3.** Hounslow Heath. **4.** Hampstead Heath!, *Johns. Enum.* Bishop's Wood!, *Warren, T. & D.*, 262. East Finchley. Mill Hill, *Sennitt.*† **5.** Perivale Wood,† *Webb Br.* **6.** Highgate Woods!, *Benbow MSS.* **7.** Holloway, 1965, *D. E. G. Irvine.*

ALNUS Mill.

A. glutinosa (L.) Gaertn. Alder.
Lit. *J. Ecol.* **41**: 447 (1953).
Native by river- and streamsides, on pond verges, in wet woods, etc. Common, though often planted.
First record: *T. Johnson*, 1638.

7. Banks of the Lee!; East Heath, Hampstead!, *T. & D.*, 262. Ken Wood. Highgate Ponds!, *Fitter.* Buckingham Palace grounds, seedlings, rare, *Fl. B.P.G.* River wall, Fulham, frequent. River wall, Chelsea and Pimlico. Thames side, Wapping. By Regent's Canal, Stepney. Lee Navigation Canal, Tottenham. Hackney Marshes.

Var. **macrocarpa** Loud.
1. Harefield. **5.** Hanwell.

CORYLACEAE

CARPINUS L.

C. betulus L. Hornbeam.
Native in woods, hedges, etc. Common, especially on sandy and clay soils, especially in the northern parts of the vice-county, though sometimes planted.
First record: *T. Johnson*, 1629.
1. Frequent. **2.** By Kingston Bridge!; Hampton!, probably planted, *T. & D.*, 264. Ashford, *Hasler.* Hampton Court. Bushy Park. Between Hampton and Sunbury, probably planted. Halliford, probably planted. West Drayton, probably planted. **3.** Near Twickenham; Whitton, *T. & D.*, 264. Cranford!, *Wood Ramb.* Wood End. Roxeth.† Near Hounslow Heath!, *Welch.* Hatton. Hayes Park, *Royle*!. **4.** Frequent, except in heavily built-up areas. **5.** Twyford, *Newbould*; Horsendon Hill!, *T. & D.*, 264. Hanwell. Syon Park. **6.** Common, except in heavily built-up areas. **7.** Near Primrose Hill,† *Merr. Pin.*, 15. Between Regents Park and Hampstead, 1819,† *E. T. Bennett*; Hornsey Wood,† *T. & D.*, 265. Ken Wood!; Parliament Hill Fields!, *Fitter.* Kensington Gardens, seedlings!, *Allen Fl.* Highgate. Hackney and Old Ford, planted.

CORYLUS L.

C. avellana L. Hazel.
Native in woods, hedges, etc. Rather common, but decreasing, and sometimes planted.
First record: *T. Johnson*, 1629.
1. Frequent. **2.** Staines!, *T. & D.*, 264. Colnbrook. Near Ashford!, *Hasler.* Stanwell Moor. Shepperton. Littleton. Near Hanworth. **3.** Twickenham; Worton; Hounslow Heath!, *T. & D.*, 264. Roxeth.† Northolt. Hatton. Cranford. Hayes. Hayes Park, abundant. **4.** Frequent, except in heavily built-up areas. **5.** Horsendon Hill!, *T. & D.*, 264. Perivale Wood!, *Webb Br.* Hanwell. Ealing, planted. **6.** Tottenham, *Johns. Cat.* Edmonton!, *T. & D.*, 264. Enfield. Enfield

Chase. Highgate Woods. Muswell Hill. **7.** Hoxton,† *Spencer*, 318. Hornsey Wood,† *Ball. MSS.* Ken Wood; Parliament Hill Fields, *Fitter*. Holland House grounds, Kensington.

FAGACEAE

FAGUS L.

F. sylvatica L. Beech.

Native in woods and plantations, etc., chiefly on calcareous and sandy soils. Probably always planted on other soils. Local as a native, rather common as a planted tree.

First record: *T. Johnson*, 1638.

1. Harefield!, *Blackst. Fasc.*, 38. Garett Wood, Springwell!,† Park Wood, Ruislip!, *Wrighton*. Harrow Weald Common. The Grove, Stanmore Common. Pinner Wood. Near Uxbridge. South Mimms. Potters Bar. **2.** West Drayton. Near Penton Hook Lock, a single tree. Bushy Park. Hampton Court. Gravel pit, West Bedfont, a solitary large seedling, 1965, *Goom* and *Kennedy*!. Halliford House grounds. **3.** Harrow Grove, *Melv. Fl.*, 73. Whitton Park Inclosure,† *Benbow MSS.* Hayes Park!, *Royle*. **4.** Harrow Park!, *Melv. Fl.*, 73. Barn Hill, Wembley Park. North-west Heath, Hampstead!; Hampstead Heath Extension!, *Fitter*. Scratch Wood. Mill Hill. Brockley Hill, *Kennedy*!. Edgware!, *Horder*. **5.** Syon Park!, *A. Wood, Trim. MSS.* Chiswick. Hanger Hill, Ealing. **6.** Colney Hatch (HE). Hadley!, *Warren*; Winchmore Hill Wood,† *T. & D.*, 263. Highgate Wood. Botany Bay, Enfield Chase!, *Kennedy*. **7.** Ken Wood!, *T. & D.*, 263. East Heath, Hampstead!; Parliament Hill Fields!, *Fitter*. Holland House grounds, Kensington.

CASTANEA Mill.

C. sativa Mill. Sweet Chestnut, Spanish Chestnut.

C. vesca Gaertn.

Alien. S. Europe, etc. Planted in woods, thickets and hedges, etc., where it sometimes regenerates. Rather rare.

First record: *T. Johnson*, 1638.

1. Ruislip Woods!, *Wrighton*. Harefield. Harrow Weald Common. Mimmshall Wood. Near Potters Bar. **2.** Tangley Park,† *T. & D.*, 263. Yeoveney!, *K. & L.*, 255. Stanwell Moor. Halliford House grounds, regenerating, 1966!, *L.N.* 46: 20. **3.** Harrow Grove, *Melv. Fl.*, 73. Side of Duke's river, Isleworth, regenerating, 1967. **4.** Harrow Park, *Melv. Fl.*, 73. Grimsdyke grounds, Harrow Weald, regenerating!, *K. & L.*, 255. North-west Heath, Hampstead. Between Kenton and Wealdstone, a single sapling, 1966. **6.** Tottenham,

Johns. Cat. Enfield Chase!, *Mart. Mill. Dict.* Winchmore Hill Wood,† *Benbow MSS.* Hadley!, *K. & L.*, 255. Near East Barnet, *Bangerter*!. **7.** Ken Wood, *Fitter.* Buckingham Palace grounds, seedlings, *Fl. B.P.G.* Kensington Gardens, seedlings, *Allen Fl.*

QUERCUS L.

Q. cerris L. Turkey Oak.
Lit. *J. Ecol.* **47**: 216 (1959).
Alien. Europe, etc. Planted in woods, parks, hedges, etc., where it readily regenerates. Common, especially in the parks of Central London.
First record (of regeneration): *Welch*, 1954.
7. Rochester Row, S.W.1, seedlings, 1961, *McClintock, L.N.* 41: 19. Hyde Park, waste ground west of Marble Arch, seedlings, 1962, *Allen Fl.* Buckingham Palace grounds, seedlings, *Fl. B.P.G.* Regenerating freely in Hyde Park! and Kensington Gardens!, Bayswater Road, W.2!, etc., *L.N.* 46: 32. Brompton. Notting Hill. East Heath, Hampstead.

Q. ilex L. Evergreen Oak, Holm Oak.
Lit. *J. Ecol.* **47**: 218 (1959).
Alien. Europe, etc. Planted in woods, hedges, parks, etc., where it sometimes regenerates, but apparently much less readily than Q. *cerris.* Very rare.
First record (of regeneration): *D. H. Kent*, 1962.
1. Harefield, numerous seedlings and saplings on the face of an old chalkpit, 1962→; the pit was bulldozed in 1965 and is now being filled with rubbish, but some plants survive on the sheer face of the pit. **2.** Harmondsworth Churchyard, regenerating, 1967. **3.** Hayes Park, regenerating, 1966!, *Royle.* **4.** Harrow, regenerating, 1966, *Temple*!. Edgware, regenerating, 1967, *Horder*!. **5.** Brentford, a solitary large seedling on a disused railway, 1966. Syon Park, regenerating, 1966. **6.** Hadley Common, regenerating, 1963.

Q. robur L. Common Oak, Pedunculate Oak, English Oak.
Q. *pedunculata* Ehrh. ex Hoffm.
Lit. *J. Ecol.* **47**: 169 (1959).
Native in woods, hedgerows, fields, etc. Very common, especially on heavy soils.
First record: *T. Johnson*, 1638.
7. Common in the district.

Var. **heterophylla** Loud.
4. Mill Hill, a single tree, *Loud. Arb.*, 1733. West Heath, Hampstead!, *K. & L.*, 254; 1967. **6.** Highgate Ponds, *Fitter, K. & L.*, 254.

Q. petraea (Mattuschka) Liebl. Sessile Oak, Durmast Oak.
Q. sessiliflora Salisb.
Lit. *J. Ecol.* **47**: 169 (1959).
Native in woods, copses, hedges, fields, etc., chiefly on sandy soils. Local.
First evidence: *Rand*, 1735.
1. Pinner, *Druce Notes*. Ruislip!; Harefield!, *Burkill, K. & L.*, 254. **3.** Near Hayes, a single tree, 1963. **4.** Near Hampstead Heath, 1735, *Rand* (BM). Mill Hill!, *Loud. Arb.*, 1736. Bishop's Wood!; Turner's Wood!; Big Wood!; West Heath, Hampstead!, *Hampst.* Platt's Lane, Hampstead,† *De Crespigny* (MANCH). By Harrow Park lake!, *Harley*. **6.** Highgate Churchyard. Bottom Wood!, *C. S. Nicholson* (LNHS). Highgate Woods. Alexandra Park!, *Trim. MSS.* Grovelands Park!; near Southgate Station; near East Barnet!; Arnos Park!; near Arnos Grove station; Oakwood Park!, *A. Vaughan*. **7.** Ken Wood!, *Loud. Arb.*, 1736. East Heath, Hampstead!, *K. & L.*, 254.
Trees intermediate between **Q. petraea** and **Q. robur**, cf. **Q.** × **rosacea** Bechst. have been recorded from **1.** Ruislip, 1924, *Burkill, K. & L.*, 255. **6.** Grovelands Park; south of Southgate underground station; Bush Hill Park!, *A. Vaughan*.

Q. borealis Michx. f. var. **maxima** (Marsh) Ashe Red Oak.
Q. rubra L. sec. Duroi, *Q. maxima* (Marsh) Ashe
Lit. *J. Ecol.* **47**: 216 (1959).
Alien. N. America. Planted in woods, hedgerows, etc., where it sometimes regenerates. Very rare.
First record (of regeneration): *Castell* and *Kennedy*, 1966.
1. Near Ickenham, regenerating, 1967. **6.** Trent Park, Southgate, regenerating, 1966, *Castell* and *Kennedy*, *L.N.* 46: 14.

SALICACEAE

POPULUS L.

P. alba L. White Poplar, Abele.
Alien. Europe, etc. Planted by streamsides, in wet meadows, etc., Rare, and scarcely naturalised.
First record: *Gerarde*, 1597.
1. Pinner Wood, *De Cresp. Fl.*, 118. Near Pinner Hill!, *Benbow MSS.* Northwood, *Cooke*, *K. & L.*, 260. Harefield,† *Pugsley*!. **2.** Staines Moor Farm!;† Hanworth Park!; Sunbury, *T. & D.*, 261. Between Staines and Laleham!; between Hampton and Sunbury!, *Benbow MSS.* **3.** Harrow Grove, *Melv. Fl.*, 72. Near Fulwell. **4.** Hampstead!, *Irv. MSS.* Between Harrow and Sudbury!, *Benbow MSS.* West Heath, Hampstead!, *Fitter*. Finchley. Harrow!, *Harley Fl.*, 19.

5. Twyford,† *Green, Druce Notes.* **6.** Finchley Common;† Colney Hatch (HE). Enfield!, *Benbow MSS.* **7.** Near Blackwall,† *Ger. Hb.*, 1301. Kilburn, 1815† (G & R).

P. canescens (Ait.) Sm. Grey Poplar.
Denizen in woods, hedges, etc. Common, though often planted.
First record: *Doody*, c. 1700.
7. East Heath, Hampstead. Ken Wood. Kilburn, planted. Regents Park, planted. Upper Clapton.

P. tremula L. Aspen.
Native on heaths, in moist woods, wet fields, by streamsides, etc. Local.
First record: *T. Johnson*, 1629.
1. Harefield!, *Blackst. Fasc.*, 80. Harrow Weald Common!; Potters Bar!, *T. & D.*, 261. East of Potters Bar, *Kennedy*!. Ruislip Reservoir!; between Ruislip and Harefield; Uxbridge Common,† *Benbow MSS.* Ruislip Woods. Stanmore Common. South Mimms. **2.** Colnbrook, *Benbow MSS.* Lower Halliford, *L. M. P. Small.* **3.** Hounslow Heath!, *Newbould*; by the Crane near Hounslow, *T. & D.*, 261. **4.** Hampstead Wood!; Bishop's! and Turner's Woods!, abundant; Edgware, *T. & D.*, 261. Kingsbury,† by the Brent from Kingsbury to Stonebridge;† Golders Green;† Hendon, *Benbow MSS.* Temple Fortune, 1927,† *J. E. Cooper.* Bentley Priory. Harrow Weald. **5.** Perivale Wood!, *Benbow MSS.* **6.** Highgate Wood!; Colney Hatch;† wood between Muswell Hill and Enfield Chase, *Benbow MSS.* Near East Barnet!, *Bangerter.* Enfield Chase, *Kennedy*!. Trent Park, *Kennedy.* **7.** Waste ground alongside Puddle Dock, E.C.4, a sapling, c. 3 feet tall, 1965, *Burton, L.N.* 46: 32. Hornsey Wood,† *T. & D.*, 261.

Var. **villosa** (Lange) Syme
4. East Finchley, *W. R. Linton, Rep. Bot. Soc. & E.C.* 1: 134. Bishop's and Turner's Woods, *Hampst.*

P. nigra L. Black Poplar.
Lit. *Proc. Bot. Soc. Brit. Isles* **7**: 429 (1968). (*Abstr.*)
Denizen in hedges, by streamsides, etc. Rare, and sometimes obviously planted.
First record: *T. Johnson*, 1638.
1. Harefield!, *Blackst. Fasc.*, 80. Pinner Hill, probably planted, *Melv. Fl.*, 73. Woodhall, Pinner, *T. & D.*, 261. Swakeleys; Uxbridge Common; between Northwood and Pinner!, *Benbow MSS.* Potters Bar, 1970, *Widgery.* **2.** Shepperton!, *Benbow MSS.* Between Ashford and Staines. Hampton Court. **3.** A large tree by the Thames near Richmond Bridge,† *T. & D.*, 262. **4.** Bishop's Wood!, *T. & D.*, 262. Hampstead Heath!, *Tremayne, K. & L.*, 260. **5.** Syon Park, planted!,

Jacks. Cat. **6.** Tottenham, *Johns. Cat.* Edmonton, *T. & D.*, 262. Finchley Common!; Winchmore Hill Wood,† *Benbow MSS.* **7.** Highgate Ponds!; East Heath, Hampstead!, *Fitter.*

The subsp. **betulifolia** (Pursh) W. Wettst. occurs near the Thames and may possibly be native; it differs from the subsp. **nigra**, which is always planted, in having a delicate and caducous pubescence on the petioles, inflorescence and young shoots. The var. **italica** Duroi (Lombardy Poplar), believed to be a mutation of subsp. **nigra** which originated in Italy in the late seventeenth or early eighteenth century, is commonly planted. **P. nigra** has been much confused with **P.** × **canadensis,** and some of the records given above may be in need of revision.

P. deltoides Marsh × **nigra** = **P.** × **canadensis** Moench Black Italian Poplar.

P. deltoides auct., non Marsh

Alien of hortal origin. Planted in woods, fields, hedgerows, by river- and streamsides, roadsides, etc. Common, though scarcely naturalised.

First record: *G. C. Druce*, 1910.

7. Common in the district.

The common Middlesex tree is the var. **serotina** (Hartig) Rehder (*P. serotina* Hartig)

SALIX L.

Lit. *Supplement J.B.* (*London*) **51** (1913).

S. pentandra L. Bay Willow.

Introduced. Planted by ponds, streamsides, on heaths, etc. Very rare.

First record: *Blackstone*, 1737.

1. Harefield,† *Blackst. Fasc.*, 89. Swakeleys, 1887, *Benbow, J.B.* (*London*) 25: 364; extinct by 1901,† *Benbow MSS.* Ickenham, 1911,† *P. H. Cooke* (LNHS). **4.** North-west Heath, Hampstead!, *K. & L.*, 256; 1967. Brent Reservoir!, *Fitter.* **5.** Syon Park!, *Jacks. Cat.* Bedford Park,† *Cockerell Fl.* **7.** Chelsea, in the way to Fulham,† *Mart. Pl. Cant.*, 65. East Heath, Hampstead!; Ken Wood, *Fitter.* Highgate Ponds!, 1965, *Wurzell*; 1967.

S. alba L. subsp. **alba** White Willow.

Native by river- and streamsides, ponds, in marshes, wet woods, etc. Common, but often planted.

First record: *Morley*, 1677.

7. Belsize Lane, Hampstead, *T. & D.*, 256. Ken Wood!; Highgate Ponds!; Parliament Hill Fields!, *Fitter.* Bishop's Park, Fulham, *Welch*!. Stepney, planted. Tottenham. Old Ford, planted. Hackney Wick, planted.

Subsp. **vitellina** (L.) Arcangeli Golden Willow.
S. alba L. var. *vitellina* (L.) Stokes
1. Pinner, *Hind, Melv. Fl.*, 71. **2.** Hampton to Hampton Court, *J. Harris, Dill. MSS.* **3.** Hanworth Park,† *T. & D.*, 256. Hatch End;† Twickenham; between Twickenham and Richmond Bridge!, *T. & D.*, 256. **5.** Syon Park, *Tremayne.* **7.** Stoke Newington,† *Woods, B.G.*, 412.

S. alba subsp. **alba** × **babylonica** L. = **S.** × **sepulcralis** Simonk.
3. Thames bank opposite Richmond Bridge, 1926, *Fraser, Rep. Bot. Soc. & E.C.* 9: 731.

S. alba subsp. **alba** × **fragilis** = **S.** × **rubens** Schrank
S. viridis auct.
1. Harefield!; Harefield Moor!; near Pinner!; between Ickenham and Ickenham Green,† *Benbow MSS.* Springwell, *Pugsley*!, *L.N.* 27: 34; 1967. **2.** Between Sunbury and Hampton, *Newbould, T. & D.*, 256. Near Staines!, *Sherrin* (SLBI). **3.** By the Crane at Mother Ive's Bridge, *T. & D.*, 256. Ickenham Marsh!; Roxeth,† *Benbow MSS.* **4.** Harrow, *Hind, Melv. Fl.*, 71. Brockley Hill, *Kennedy*!. **5.** Perivale;† between Greenford and Hanwell,† *Benbow MSS.* **6.** Whitewebbs Park, Enfield, *Kennedy*!.

S. alba subsp. **vitellina** × **babylonica** = **S.** × **chrysocoma** Dode
3. Twickenham, 1928, *Fraser, Rep. Bot. Soc. & E.C.* 9: 721.

S. babylonica L. Weeping Willow.
Alien. Orient. Planted by river- and streamsides, ponds, etc. Rather rare.
First record: *D. H. Kent*, 1939.
1. Ruislip!; Uxbridge!, *K. & L.*, 259. **2.** Teddington. **3.** Twickenham!; near Richmond Bridge!, *K. & L.*, 259. **4.** Harrow. **5.** Chiswick!, *K. & L.*, 259. **7.** East Heath, Hampstead!, *K. & L.*, 259. Highgate Ponds!, *Fitter.*
It is probable that *S. babylonica* is not hardy in Britain, and most likely that the records given above are referable to either *S.* × *sepulcralis* or *S.* × *chrysocoma*, both of which are quite hardy in the vice-county.

S. fragilis L. agg. Crack Willow.
Native by river- and streamsides, pond verges, in wet woods, marshes, fields, on waste ground, etc. Common, though often planted.
First certain evidence: *J. E. Smith*, c. 1790.
7. Common in the district.
Swann (1957, *Proc. Bot. Soc. Brit. Isles* **2**: 339) has pointed out that there are four distinct taxa all named *S. fragilis* at various times by British botanists. These are **S. fragilis** L. sec. Smith (**S. alba** × **fragilis** f. **monstrosa** sec. Floderus), **S. russelliana** Sm., **S. decipiens** Hoffm.

and **S. basfordiana** Scaling ex Salter. Although no attempt has been made to study this critical group in Middlesex all four are known to occur.

S. triandra L. subsp. **triandra** Almond Willow.
Native by river- and streamsides, pond verges, in marshes, etc. Rare, and sometimes planted.
First record: *Petiver*, 1695.
1. Elstree Reservoir, *Walker Ramb.*, 87; 1885, *W. R. Linton* (BM). Uxbridge Common!;† Swakeleys;† near Harefield Moor!; Denham!; Harefield!; Springwell!, *Benbow MSS.* Near Uxbridge, *Druce Notes.* Ruislip, *Moxey List.* **2.** By the Queen's river, Hampton; riverside between Hampton Court and Kingston Bridge, *T. & D.*, 257. Staines!; Laleham!; Shepperton!; near Chertsey Bridge, *Benbow MSS.* **3.** By the Thames between Twickenham and Richmond Bridge!; near Isleworth!; Worton;† between Twickenham and Worton,† *T. & D.*, 257. **4.** Hampstead,† *Irv. MSS.* Wembley Park;† Kenton;† Stonebridge,† *Benbow MSS.* **5.** By the Brent near Alperton,† *Melv. Fl.*, 71. By the Thames, Chiswick,† *Benbow MSS.* **7.** Between Westminster and Chelsea,† *Pet. Midd.* Banks of river Lee, Hackney,† *E. Forster*; near Homerton,† *Fraser* (BM). Between Chelsea and Fulham,† *A. B. Lambert* (K).

Subsp. **hoffmanniana** (Sm.) Bab.
S. hoffmanniana Sm.
1. By Elstree Reservoir, *T. & D.*, 257; c. 1871,† *Walker Ramb.*, 87. **2.** Riverside between Hampton Court and Kingston Bridge, *T. & D.*, 257. **5.** By the Brent near Hanwell,† *Warren, T. & D.*, 257.

S. triandra subsp. **triandra** × **viminalis** = **S.** × **mollissima** Ehrh.
S. undulata Ehrh., *S. hippophaëfolia* Thuill.
2. Between Hampton Court and Kingston Bridge!, *T. & D.*, 257; 1967. Shepperton; near Chertsey Lock, *Benbow MSS.* **3.** Near Richmond Bridge, towards Twickenham, *T. & D.*, 257; c. 1887, *Benbow MSS.* Near Worton,† *T. & D.*, 257. **7.** Osier holts, Sandy End, Fulham,† *T. & D.*, 257. Hammersmith, 1865,† *Parsons* (CYN).

S. purpurea L. Purple Willow.
Native by river- and streamsides, pond verges, etc. Rather common, though often planted.
First evidence: *E. Forster*, c. 1800.
1. Between South Mimms and Potters Bar, *T. & D.*, 258. Swakeleys!;† Denham!; Harefield Moor!; Springwell!; Uxbridge Common,† *Benbow MSS.* Cowley!, *K. & L.*, 257. Ickenham, *R. W. Robbins* (LNHS). **2.** In scattered localities by the Thames between Staines and Teddington. Between West Drayton and Colnbrook!, *Benbow MSS.* West Drayton; Poyle!, *Druce Notes.* Near Littleton, *Welch.* Harmondsworth. Staines Moor, *Boniface*!. East Bedfont. **3.** By

Mother Ive's Bridge; Duke's river near Whitton, *T. & D.*, 258. By the Crane, Isleworth, *De Cresp. Fl.*, 61. Yeading Brook in several places, *Benbow MSS.* Thames bank near Twickenham. **4.** Hampstead, *Irv. MSS.* Wembley Park,† *Benbow MSS.* Brent Reservoir. **5.** Canal between Hanwell and Southall!, *K. & L.*, 257. **6.** By Lee Navigation Canal between Ponders End and Edmonton!, *Benbow MSS.* Near Waltham Abbey. **7.** Near Temple Mills,† *E. Forster* (BM).

S. purpurea × **viminalis** = **S.** × **rubra** Huds.
1. Springwell!; Harefield Moor!; Swakeleys; Park Wood, Ickenham!, *Benbow MSS.* **2.** By the Queen's river, Hampton; Thames side opposite Thames Ditton!, *T. & D.*, 258. Between Hampton Court and Kingston Bridge!, *Benbow* (BM). Poyle. Near Chertsey Bridge!; Shepperton!; near Hanworth, *Benbow MSS.* West Drayton. **3.** Riverside between Twickenham and Richmond Bridge, *T. & D.*, 258. **5.** Southall;† near Hanwell, 1891,† *Benbow MSS.* **6.** Between Finchley and Whetstone; between Waltham Cross and Ponders End, *Benbow MSS.* **7.** Lee Bridge,† *E. Forster*; Newington,† *Woods* (BM).

S. daphnoides Vill. subsp. **acutifolia** (Willd.) Dahl
S. acutifolia Willd.
Alien. E. Europe, etc. Planted by the Thames. Very rare.
First record: *Sandwith*, 1942.
5. Riverside upstream from Barnes Bridge, five or six bushes, 1942!, *Sandwith*; 1943–61; three trees, 1965→.

S. viminalis L. Common Osier.
Native by river- and streamsides, ponds, on damp waste ground, etc. Common, especially by the Thames and its tributaries, though often planted.
First certain evidence: *Lightfoot*, c. 1780.
7. Osier-holt, Sandy End [Fulham], abundant,† *T. & D.*, 259. Bombed sites, New Basinghall Street† and The Temple,† *A. B. Jackson, Lousley Fl.* East Heath, Hampstead!; Ken Wood!, *Fitter.* Shoreditch, 1966.

S. viminalis × ? = **S.** × **stipularis** Sm.
3. Thames bank, Twickenham, 1938, *Melville, Rep. Bot. Soc. & E.C.* 12: 201. **4.** Hampstead, *Irv. MSS.* **7.** Stoke Newington,† *Woods, B.G.*, 413.

S. caprea L. Great Sallow, Goat Willow.
Native in woods, scrub, hedges, on waste ground, etc., established on railway banks, etc. Common.
First certain evidence: *Lightfoot*, c. 1780.
7. Common in the district.

S. caprea × **viminalis** = **S.** × **laurina** Sm.
S. sericans Tausch ex A. Kerner
7. Bombed site, Gresham Street area,† *Jones Fl.*

S. caprea × **cinerea** subsp. **oleifolia** = **S.** × **reichardtii** A. Kerner
1. Ickenham, 1892, *Benbow* (BM). **2.** Thames side, Hampton Court, 1966.

? **S. caprea** × **dasyclados** Wimm. = **S.** × **calodendron** Wimm.
Lit. *Watsonia* **3**: 243 (1952).
4. Brent Reservoir!, 1948, *Meikle* (K); two shrubs on north edge of the reservoir, 1966, *Wurzell.*

S. cinerea L. subsp. **oleifolia** Macreight Common, or Grey Sallow.
S. atrocinerea Brot., *S. cinerea* L. subsp. *atrocinerea* (Brot.) Silva & Sobrinho
Native by river- and streamsides, ponds, in damp woods, hedges, on waste ground, etc., established on railway banks, etc. Common.
First certain evidence: *E. Forster*, c. 1800.
7. Common in the district.

S. cinerea subsp. **oleifolia** × **viminalis** = **S.** × **smithiana** Willd.
S. geminata Forbes, *S. chouardii* Chass. & Görz
1. Pinner, *Melv. Fl.*, 72; c. 1887; Ruislip; Harefield!; Uxbridge; Hillingdon; Swakeleys!; Ickenham; Springwell!, *Benbow, J.B. (London)* 25: 364. **2.** Teddington, *Benbow MSS.* **3.** Roxeth,† *Benbow, J.B. (London)* 25: 364. Near Down Barns [Northolt], *Benbow MSS.* **4.** West Heath, Hampstead!, *T. & D.*, 259. Harrow, *Benbow, J.B. (London)* 25: 364. Near Finchley, *Benbow MSS.* **5.** Near Perivale Church, a single tree,† *Benbow, J.B. (London)* 25: 364. **6.** Whetstone, sparingly, *Benbow, J.B. (London)* 25: 364. **7.** Hackney Wick, 1867, *T. & D.*, 259. South Kensington,† *Dyer List.*

S. aurita L. Eared Sallow.
Native on heaths, in marshes, damp woods, by river- and streamsides, etc., rarely on waste ground. Local.
First evidence: *T. Moore*, 1848.
1. Abundant about heaths and woods between Ruislip Reservoir and Northwood!; north of Harefield, *Benbow, J.B. (London)* 25: 18. Harefield Common!, *Benbow MSS.* **3.** Hounslow Heath!, *T. & D.*, 259. Ickenham Marsh!, *Benbow MSS.* Between Northolt and Ickenham. **4.** Hampstead Heath!, 1848, *T. Moore* (BM); still on the West Heath, 1969. Bishop's Wood; Kingsbury;† between Finchley and Hendon, *T. & D.*, 260. **7.** Waste ground, Holloway, 1965, *D. E. G. Irvine.*

S. aurita × **viminalis** = **S.** × **fruticosa** Doell
5. By the Brent between Hanwell and Greenford, 1891,† *Benbow* (BM).

S. aurita × **caprea** = **S.** × **capreola** J. Kerner ex Anderss.
1. Duck's Hill Heath, near Northwood; Garett Wood, Springwell;† Ickenham, 1892; Uxbridge Common, 1891,† *Benbow* (BM).

S. aurita × **cinerea** subsp. **oleifolia** = **S.** × **multinervis** Doell
S. lutescens A. Kerner
1. Uxbridge Common;† Swakeleys; Park Wood, Ickenham!; Harefield Moor!; Harefield Common!; Duck's Hill, near Northwood!, *Benbow MSS.* Copse Wood, Northwood!, *K. & L.*, 258. **3.** Ickenham Marsh!, *Benbow MSS.* **4.** Hampstead Heath, *Benbow MSS.* **5.** Between Hanwell and Greenford,† *Benbow MSS.* **7.** Hyde Park Corner, two bushes among rubble, 1962!,† *D. E. Allen*, teste *Meikle*, *Allen Fl.*

S. aurita × **repens** subsp. **repens** = **S.** × **ambigua** Ehrh.
1. Harefield Common!,† *Benbow* (BM).

S. repens L. subsp. **repens** Dwarf Willow.
Native on heaths and commons. Rare, and decreasing.
First record: *Gerarde*, 1597.
1. Harefield Common!, *Blackst. Fasc.*, 89. Harrow Weald Common!, *Melv. Fl.*, 72. Stanmore Common!, *T. & D.*, 260. Near Ruislip Reservoir, 1928, *B. T. Ward*; Copse Wood, Northwood!,† *Wrighton*, *K. & L.*, 259. **3.** Whitton,† *Francis, Coop. Suppl.*, 12. Hounslow Heath! (DILL). Near Hatton,† *T. & D.*, 260. Hanworth,† *Benbow MSS.* **4.** Hampstead Heath!, *Ger. Hb.*, 1205; West and Northwest Heaths, 1973. **6.** Southgate,† *Dill. MSS.* **7.** East Heath, Hampstead!, *T. & D.*, 260; 1973.

ERICACEAE

RHODODENDRON L.

R. ponticum L.
Alien. S. Europe, etc. Planted in woods, on heaths and commons, etc., where it sometimes regenerates. Rare. The following records refer to localities where regeneration has been noted.
First record (of regeneration): *D. H. Kent*, 1948.
1. Stanmore Common. Potters Bar, *Kennedy*!. South Mimms. Harefield Grove!, *I. G. Johnson.* **2.** Laleham Park. **3.** Hayes Park!, *Royle.* **4.** Grimsdyke grounds, Harrow Weald, regenerating abundantly!, *L.N.* 28: 32. Scratch Wood, *K. & L.*, 182. Harrow-on-the-Hill Churchyard, *Temple*!. **5.** Chiswick House grounds. **6.** Near Enfield. Whitewebbs Park, Enfield, *Kennedy*!.

CALLUNA Salisb.

C. vulgaris (L.) Hull Ling, Heather.
Lit. *J. Ecol.* **48**: 455 (1960).
Native on heaths, in open woods on acid soils, etc. Rare, and decreasing.
First record: *Gerarde*, 1597.
1. Harefield Common!,† *Blackst. Fasc.*, 27. Ruislip Common!, *Melv. Fl.*, 50. Harrow Weald Common!; Stanmore Common!, *T. & D.*, 181. Hillingdon Heath;† Pinner Wood; roadside by Duck's Hill Farm!; Mimmshall Wood!, *Benbow MSS.* **2.** Roadsides about Staines† and Charlton,† but not commonly, *T. & D.*, 181. Between Hampton and Hanworth,† *Benbow MSS.* **3.** Hounslow Heath!; near Hatton,† *T. & D.*, 181. Hanworth, *Benbow MSS.* **4.** Hampstead Heath, *Ger. Hb.*, 1199; c. 1869, *T. & D.*, 181; scarce, 1871, *Walker Ramb.*, 12; still lingers near the birches on the West Heath, c. 1912, *Whiting Fl.*; one plant, 1927,† *Richards.* Harrow,† *Melv. Fl.*, 50. Grimsdyke grounds, Harrow Weald. Grimsdyke golf course, Harrow Weald, *Kennedy*!. Mill Hill, *Sennitt.* **6.** Winchmore Hill Wood, *T. & D.*, 181; c. 1900,† *C. S. Nicholson, Lond. Nat. MSS.* Railway bank, Colney Hatch,† *T. & D.*, 181. Hadley Common!, *Benbow MSS.* Trent Park, *Castell* and *Kennedy.* **7.** East Heath, Hampstead,† *T. & D.*, 181. Buckingham Palace grounds, introduced with peat blocks!, *Fl. B.P.G.*

Var. **incana** Reichb.
3. Hounslow Heath!, *T. & D.*, 181. **7.** Buckingham Palace grounds, introduced with peat blocks, *Fl. B.P.G.*

ERICA L.

E. tetralix L. Cross-leaved Heath, Bog Heather.
Lit. *J. Ecol.* **54**: 795 (1966).
Native in bogs and wet heathy places. Very rare, and almost extinct.
First record: *Gerarde*, 1597.
1. Harefield,† *Blackst. Fasc.*, 26. Harefield Common,† *Benbow MSS.* Harrow Weald Common!, *Melv. Fl.*, 50. Stanmore Common, *T. & D.*, 182; scarce, 1923; destroyed by fire, 1931,† *Warmington.* **3.** Hounslow Heath!, *T. & D.*, 182. **4.** Hampstead Heath, *Ger. Hb.*, 1199; still west of the Spaniard's Road, 1912, *Whiting Fl.*; very rare, 1914,† *Moring MSS.* Bishop's Wood, sparingly,† *T. & D.*, 182. Grimsdyke grounds, Harrow Weald. Mill Hill, *Sennitt.* **7.** Planted in West Meadow Bog, Ken Wood, 1960, *L.N.* 40: 21. Buckingham Palace grounds, introduced with peat blocks!, *Fl. B.P.G.*

E. cinerea L. Fine-leaved Heath, Bell Heather.
Lit. *J. Ecol.* **53**: 527 (1965).
Native in dry heathy places. Very rare, and almost extinct.
First record: *Gerarde*, 1597.
1. Harefield, *Blackst. Fasc.*, 26; 1866, *Trimen* (BM); c. 1900,† *C. S. Nicholson, K. & L.*, 182. Harrow Weald Common, scarce!, *T. & D.*, 182. Stanmore Common, 1921,† *Richards*. **2.** Near New Hampton, abundant,† *Benbow MSS.* **3.** Hounslow Heath, sparingly!; roadside near the Powder Mills,† *T. & D.*, 182. Between Hanworth and Hounslow;† railway banks between Hounslow Heath and Feltham!; Whitton Park Inclosure,† *Benbow MSS.* **4.** Hampstead Heath, *Ger. Hb.*, 1199; c. 1869, *T. & D.*, 182; 1883, *J. E. Cooper* (BM); one plant, 1909, *Moring MSS.*; still lingers west of the Spaniard's Road, c. 1912,† *Whiting Fl.* Deacon's Hill,† *T. & D.*, 182. Grimsdyke grounds, Harrow Weald. **7.** East Heath, Hampstead, very sparingly,† *T. & D.*, 182.

VACCINIUM L.

V. myrtillus L. Bilberry, Blaeberry, Whortleberry, Huckleberry.
Lit. *J. Ecol.* **44**: 291 (1956).
Native on heaths and commons and in woods on acid soils. Very rare, and almost extinct.
First record: *Gerarde*, 1597.
3. Isleworth,† *A. B. Cole, T. & D.*, 183. **4.** Hampstead Heath!, *Ger. Hb.*, 1230; c. 1869, *T. & D.*, 182; c. 1877, *De Cresp. Fl.*, 104; 1906, *C. S. Nicholson* (LNHS); still west of the Spaniard's Road, c. 1912, *Whiting Fl.*; two small patches on the West Heath, 1921, *Richards* (BM); one small patch left, 1945!, *Richards, K. & L.*, 181; still present, 1967. Bishop's Wood, 1839, *R. S. Hill*; 1860, *Trimen* (BM); 1887, *Benbow MSS.* Turner's Wood (HE). Boggy end of Scratch Wood, *Benbow MSS.* Wood by Highgate called Finchley Wood, *Ger. Hb.*, 1230. The latter record probably refers to Bishop's Wood. Near Hendon, 1823† (G & R). **6.** Southgate, *Cole, T. & D.*, 182. **7.** St John's Wood [Highbury],† *Park. Theat.*, 1458. Ken Wood!, *Blackst. Spec.*, 103; still abundant there, *T. & D.*, 182; still present in small quantity near the lake, 1966. East Heath, Hampstead, near the Vale of Health,† *T. & D.*, 182.

PYROLACEAE

PYROLA L.

P. minor L. Common, or Lesser Wintergreen.
Formerly a denizen in plantation but now extinct.

First evidence: *Hind*, 1874. Last record: *Green*, 1908.

1. In a grove of old trees on the east side of Stanmore Common, 1874, *Hind* (BM); 1908, *Green*, *Williams MSS.*

PRIMULACEAE

PRIMULA L.

P. veris L. Cowslip.

Native in meadows and pastures on basic soils, established on railway banks, etc. Formerly rather common, but now rather rare and usually in small quantity; very rare near the Thames.

First record: *T. Johnson*, 1629.

1. Harefield!, *Blackst. Fasc.*, 81. Railway banks, Elstree!; Potters Bar!; Ruislip, *T. & D.*, 225. Pinner!, *De Cresp. Fl.*, 54. Uxbridge!, *Benbow* (BM). Near Northwood!, *Battley*; between Harefield and Uxbridge!; between Harefield and Ruislip, *Tremayne*, *K. & L.*, 184. Railway banks between Pinner and Northwood. Railway banks between Ruislip and Ickenham, *Pickess*. Springwell!, *K. & L.*, 184. Near Swakeleys. South Mimms, *Brittain*, 5. North of Bentley Heath, *Kennedy*. **2.** Staines,† *T. & D.*, 225. **3.** Twickenham,† *Little*. Near Hounslow Heath, 1935,† *H. Banks*. Gutteridge Wood, Northolt!, *K. & L.*, 184. Sparingly in fields between Hayes and Northolt, 1966. Railway banks near Wood End, 1967. **4.** Hampstead Heath, *Johns. Enum.* Mill Hill!, *J. Morris*; between Bishop's Wood and Finchley;† near Golders Green,† *W. Davies*; Kingsbury,† *Farrar*; Bentley Priory!, *T. & D.*, 225. Harrow, *Coop. Fl.*, 115; c. 1864, *Melv. Fl.*, 62; c. 1877, *De Cresp. Fl.*, 54. Highwood!, *Evans Hist.*, 14. Golders Hill,† *Whiting Notes*. Between Edgware and Elstree!, *E. M. Dale*; Scratch Wood, and district!, *K. & L.*, 184. Field by Willesden Old Churchyard, 1966. **5.** Northolt!, *A. B. Cole*, *T. & D.*, 225. Near Greenford!, *A. Wood*; Perivale, 1872,† *Trim. MSS.* Brent Meadows, Hanwell, 1899,† *Battley*; Syon Park!, *Tremayne*, *K. & L.*, 184. **6.** Tottenham,† *Johns. Cat.* Finchley Common, 1842† (HE). Edmonton,† *T. & D.*, 225. Enfield, *Battley* (LNHS). Near Hadley Wood railway station. **7.** Between Kentish Town and Hampstead,† *Johns. Eric.* Clapton, 1838,† *E. Ballard* (BM). Regents Park,† *Webst. Reg.*, 102. Bombed site, Cripplegate area, very rare,† *Jones Fl.* East Heath, Hampstead, 1964, *Springett*.

P. veris × vulgaris Common, or False Oxlip.

P. variabilis Goupil, non Bast., *P. elatior* auct., non (L.) Hill

1. Harefield, *Blackst. Fasc.*, 81. Pinner Hill,† *Hind*, *Melv. Fl.*, 62. Pinner Wood, 1869, *Trimen* (BM). Park Wood, Ickenham, *Green* (SLBI). **4.** Harrow,† *Hind*; Bentley Priory, one plant,† *Melv. Fl.*, 62. Stanmore,† *Varenne*, *T. & D.*, 225.

P. vulgaris Huds. Primrose.
P. acaulis (L.) Hill
Native in woods, open grassy places, on hedgebanks, etc., established on railway banks, etc. Formerly rather common, but now rare and decreasing rapidly owing to the grubbing up of roots for transfer to gardens.
First record: *T. Johnson*, 1629.
1. Harefield!, *Blackst. Fasc.*, 81. Ruislip Woods!, *T. & D.*, 224. Uxbridge,† *Kingsley* (ME). Swakeleys, *Benbow* (BM). Copse Wood, Northwood, 1945!; Cowley!,† *K. & L.*, 184. Railway banks near Potters Bar. **2.** Spout Wood, Stanwell!, *Briggs*. Probably introduced. **3.** Headstone Farm;† Harrow Grove!; Roxeth!,† *Melv. Fl.*, 62. Railway banks, Wood End, Sudbury and Northolt. Hayes Park!, introduced, *Royle*. **4.** Hampstead Heath,† *Johns. Eric.* Bishop's Wood, nearly eradicated;† Bentley Priory,† *T. & D.*, 224. Highwood,† *Evans Hist.*, 16. Golders Hill,† *Whiting Notes*. Harrow Park!, *Horton Rep.* Railway bank near Wembley Park, 1963, probably a garden outcast. **5.** Near Brentford!, *Hemsley*, *T. & D.*, 224. Syon Park, *Welch*!, *K. & L.*, 184. Probably originally introduced. Horsendon Wood;† between Brentford and Hanwell,† *A. Wood*, *Trim. MSS.* Gunnersbury Park, Acton!, *Henrey*, 34. Probably originally introduced but well naturalised by a lake. Thickets, Hanger Lane, Ealing, 1966!, *McLean*. Probably originally introduced. **6.** Tottenham,† *Johns. Cat.* Hadley!, *Warren*; Winchmore Hill Wood;† Edmonton,† *T. & D.*, 224. Old Park, Southgate, 1902,† *C. S. Nicholson*, *Lond. Nat. MSS.* **7.** Kilburn, 1818† (G & R). Primrose Hill is said to have derived its name from the former abundance of *P. vulgaris* growing there,† *Park Hampst.*, 258. Buckingham Palace grounds, introduced, but regenerating freely.
Plants with red flowers are recorded from **6.** Colney Hatch,† *W. G. Smith*; Edmonton,† *T. & D.*, 225. These may have been referable to forma **hortensis** Pax.

HOTTONIA L.

H. palustris L. Water Violet.
Native in ditches, pools, shallow ponds, etc. Formerly local, but now very rare.
First record: *Blackstone*, 1737.
1. Harefield,† *Blackst. Fasc.*, 60. **2.** Abundant in a ditch by the roadside leading from Bucks. to Staines,† '*J.W.T.* and *A.J.*', *Phyt. N.S.* 4: 263. Many places on Staines Moor!, *T. & D.*, 224; in recent years confined to one ditch in very small quantity, where it has not been seen since c. 1962. Shortwood Common!,† *Benbow MSS.* Formerly in small quantity in a number of small pools, these have now been

filled with rubbish and sown with grass seed. Near Colnbrook,† *Green, Williams MSS.* **3.** In the Crane near Hounslow, *De Cresp. Fl.*, 32. Possibly an error. **4.** Kingsbury, *Irv. Lond. Fl.*, 141; c. 1869, *Farrar, T. & D.*, 224; 1882, *Findon MSS.*; c. 1910,† *Pugsley*. **6.** Ponders End, *Cherry, T. & D.*, 224; c. 1887,† *Benbow MSS.* Ditches, Ponders End to Enfield,† *J. O. Braithwaite, K. & L.*, 183. Pond near Crew's Hill, Enfield, 1965, *Castell* and *Kennedy*. **7.** Chelsea,† *M. & G.*, 275. Fulham,† *Faulkner Hist.*, 22. Walham Green,† *T. Lambert* (K). Tottenham Marshes,† *Cockfield Cat.*, 16. Clapton Marshes, 1896,† *Battley* (LNHS). Ditch near the railway, Clapton,† *C. S. Nicholson, K. & L.*, 183.

LYSIMACHIA L.

L. nemorum L. Yellow Pimpernel.

Native in woods, shady ditches, on hedgebanks, etc. Local, and chiefly confined to the north of the vice-county.

First record: *Gerarde*, 1597.

1. Locally frequent. **4.** Hampstead Wood, *Ger. Hb.*, 497, and many later authors. Hampstead Heath,† *J. Hill* (BM). Bishop's Wood, *T. & D.*, 226. Stanmore, 1815† (G & R). Woods between Edgware and Stanmore,† *Benbow MSS.* Wembley Park,† *A. B. Cole*; Kingsbury;† Stonebridge,† *Farrar, T. & D.*, 226. Wood between Highgate and Hampstead,† *How MSS.* Golder's Hill,† *Whiting Notes.* **5.** Syon Park, rare, 1966, *Boniface*. **6.** Between Muswell Hill and Highgate,† *M. & G.*, 270. Hadley!, *Warren*; Winchmore Hill Wood;† Whetstone,† *T. & D.*, 226. Highgate Wood, *Benbow MSS.*; c. 1900, *C. S. Nicholson*,† *K. & L.*, 185; Coldfall Wood, Highgate,† *Cherry, Trim. MSS.* Enfield Chase!, *Benbow MSS.* Near East Barnet!, *Bangerter*. **7.** Ken Wood,† *Newt. MSS.* and *Mart. Pl. Cant.*, 68.

L. nummularia L. Creeping Jenny, Money Wort.

Native in moist, grassy places, damp woods and waste ground, ditches, etc., on heaths, commons, etc., established on railway banks, etc. Occurs also as a garden escape. Common, except in heavily built-up areas where, owing to the eradication of its habitats, it is rare or extinct.

First record: *Gerarde*, 1597.

7. Banks of Thames, right against the Queen's palace of Whitehall,† *Ger. Hb.*, 505. Between Islington and Hornsey, 1816,† *E. T. Bennett* (BM). Between Primrose Hill and Kensington, 1831;† Copenhagen Fields, 1841† (HE). Near Kensal Green, *Warren*; East Heath, Hampstead!; Highgate!; Ken Wood grounds!, *T. & D.*, 226. Kilburn. Holloway. Finsbury Park. Hackney, garden escape.

L. nummularia rarely sets seed in Britain, and in Middlesex its spread appears to be entirely by vegetative means.

L. vulgaris L. Yellow Loosestrife.
Native by river- and streamsides, on pond- and lake-verges, in marshes, etc. Locally common by the Thames and Colne, though rarely in quantity, rare elsewhere, and sometimes planted for ornamental purposes.
First record: *Blackstone*, 1737.
1. In the meadows near Uxbridge Moor, very frequent,† *Blackst. Fasc.*, 56. Near Uxbridge, 1839,† *D. Cooper* (BM). Near Harefield!, *Blackst. Spec.*, 50. Springwell!; between Cowley and Yiewsley,† *Benbow MSS.* Cowley,† *J. E. Cooper*, *K. & L.*, 184. Ornamental pond, Highgrove House, Eastcote, *I. G. Johnson.* **2.** Small isolated colonies by the Thames between Staines and Strawberry Hill, and by the Colne from near Yiewsley to Staines. West Drayton,† *Benbow MSS.* Wyrardisbury river near Staines Moor. Longford river near Stanwell. **3.** By the riverside near Twickenham!,† *M. & G.*, 264. Between Twickenham and Worton;† by the Crane in several places,† *T. & D.*, 225–226. Hounslow,† *Cockfield Cat.*, 31. Whitton,† *Benbow MSS.* Isleworth!,† *K. & L.*, 189. **4.** Side of the river Brent, c. 1840 (HE); c. 1869,† *Farrar*, *T. & D.*, 226. Near North End, Hampstead, *Burn. Med. Bot.*, 105; c. 1912,† *Whiting Fl.*, Neasden!, introduced, *K. & L.*, 185. **5.** By the Brent, Perivale,† *Lees*, *T. & D.*, 226. Between Alperton and Perivale!,† *Benbow MSS.* Osterley Park!, *Druce Notes*; 1972. **6.** Lee Navigation Canal, Tottenham,† *Benbow MSS.* Trent Park, 1958, *A. Vaughan*; 1966, *T. G. Collett* and *Kennedy*. **7.** Banks of Lee from Clapton to Tottenham,† *Francis*, *Coop. Suppl.*, 12. Hackney Wick,† *T. & D.*, 224. Highgate Ponds, 1944, probably planted, *Fitter*. Buckingham Palace grounds!, probably planted, *Eastwood*; occasional, 1964, *Fl. B.P.G.* Waste ground, Phillimore Walk, Kensington, 1952,† *Sandwith*, *K. & L.*, 185.

L. ciliata L.
Alien. N. America. Garden escape. Naturalised by rivers, streams, ditches, on waste ground, etc. Rare.
First record: *Boniface*, 1948.
1. Pinner, 1967, *Kennedy*!. **3.** Hanworth, 1966. **4.** Harrow Weald, 1965; near Mill Hill, 1966, *Kennedy*!. Brent Reservoir, 1966. **5.** Chiswick, 1948, *Boniface*; 1950!,† *K. & L.*, 185. Brook by Oldfield Lane, Greenford, plentiful, 1962; much reduced by road-widening operations, 1964. Canal bank, Alperton, 1962→. **6.** Enfield, 1965→; north of Enfield Lock, 1966, *Kennedy*!.

ANAGALLIS L.

A. tenella (L.) L. Bog Pimpernel.
Formerly native in bogs, on wet heaths, etc., but now extinct.
First record: *Blackstone*, 1737. Last record: *Hodson* and *Ford*, 1873.

1. On Harefield Moor, abundantly, *Blackst. Fasc.*, 65. **4.** Hampstead Bog, *Blackst. Spec.*, 60; c. 1813, *Bliss, Park Hampst.*, 29. **6.** Enfield, *H. & F.*, 147. **7.** Ken Wood, *Hunter, Park Hampst.*, 29.

A. arvensis L. subsp. **arvensis** f. **arvensis** Scarlet Pimpernel, Shepherd's Weather-glass.
Lit. *Notes Roy. Bot. Gard. Edinb.* **28**: 173 (1968).
Native on cultivated and waste ground, etc., also as a garden weed. Common.
First record: *T. Johnson*, 1629.
7. Common in the district.

f. **carnea** (Schrank) Hyland.
var. **carnea** Schrank
2. Abundant on disturbed ground near Hampton, 1966, *Tyrrell*!. **7.** Eaton Square, S.W.1, 1958, *McClintock, L.N.* 39: 53.

Var. **verticillata** Diard.
1. Harefield, 1951, *Lousley* and *McClintock*!, *K. & L.*, 186. **3.** Hounslow Heath, 1965. **6.** Enfield, 1966, *Kennedy*!. **7.** Lawn in front of new flats, Shoreditch, 1966.

Subsp. **foemina** (Mill.) Schinz & Thell.
Anagallis foemina Mill., *A. caerulea* Schreb., non L.
1. Uxbridge Common, 1884–90, *Benbow* (BM). Uxbridge, 1891, *Benbow MSS.* Ickenham, one plant, 1962, *Davidge.* **2.** West Drayton, *Druce Notes.* Near Heathrow, 1945, *Welch*!, *L.N.* 26: 62. **3.** Isleworth, *H. Banks, K. & L.*, 186. **4.** Hampstead, 1872, *Beeby* (SLBI). Mill Hill, 1960, *Warmington.* **6.** Tottenham, c. 1875, '*J.W.*', *Sci. Goss.* 1881: 46. Highgate, 1911, *Coop. Cas.* **7.** Regents Park, 1954, *Holme, L.N.* 34: 5.

A. minima (L.) E. H. L. Krause Chaffweed, Bastard Pimpernel.
Centunculus minimus L.
Formerly native on damp sandy heaths, but long extinct.
First record: *W. Watson*, 1746. Last evidence: *J. Banks*, c. 1800.
2. Near Hampton Court, *Huds. Fl. Angl.* 2: 63. Between Colnbrook and Cranford Bridge, *Lightf. MSS.* **3.** Low marshy ground near the paper-mills on Hounslow Heath, *W. Watson, Blackst. Spec.*, 13; . . . and on the sides of the Roman Camp on Hounslow Heath, *Lightf. MSS.* On Ashford Common in tolerable plenty in moist depressed situations; passing from Ashford to Hounslow Heath in similar situations in greater plenty, *Curt. F.L.*; c. 1795, *Dicks. Hort. Sicc.*; Hounslow Heath gravel pits to the south-east of Whitton, *J. Banks* (BM).

SAMOLUS L.

S. valerandi L. Brookweed.

Formerly native in brackish ditches near the tidal Thames, but now extinct.

First record: *J. Martyn*, 1732. Last evidence: *Trimen*, 1866.

7. Isle of Dogs, *Mart. Tourn.* 2, 335; ditch close by the timber-dock, 1866, *Trimen* (BM). About Blackwall, *M. & G.*, 329. Between Poplar and the Isle of Dogs, *Jones, With. Bot. Arr.* 2, 1: 221.

BUDDLEJACEAE

BUDDLEJA L.

B. davidii Franch. 'Buddleia', Butterfly-bush.

B. variabilis Hemsl.

Alien. China. Planted in parks, gardens, etc., where it readily regenerates. Seedlings, saplings and mature bushes are now naturalised on waste ground, in chalkpits, gravel pits, etc., and established on old walls, railway banks, etc. Common.

First record (of regeneration): *Melville* and *R. L. Smith*, 1927.

7. Common in the district.

OLEACEAE

FRAXINUS L.

F. excelsior L. Ash.

Lit. *J. Ecol.* **49**: 739 (1961).

Native in woods, hedges, fields, etc. Very common, though often planted.

First record: *T. Johnson*, 1632.

7. Frequent in the district.

SYRINGA L.

S. vulgaris L. Lilac.

Alien. S.-E. Europe, etc. Garden outcast. Naturalised in woods, hedges, etc. Very rare.

First record: *D. H. Kent*, 1939.

2. Laleham Park!; gravel pit between Dawley and Hillingdon!, *K. & L.*, 186. **3.** Between Hounslow Heath and Cranford!,† *K. & L.*, 186. **4.** Brent Reservoir. **5.** Horsendon Lane, Perivale!,† *K. & L.*, 186.

LIGUSTRUM L.

L. vulgare L. Common Privet.

Native on wood borders, in thickets, scrub, etc., on calcareous and other dry soils. Local, and sometimes planted.

First certain record: *T. Johnson*, 1632.

1. Harefield!, *Blackst. Fasc.*, 52. Pinner chalkpits;† South Mimms!, *T. & D.*, 183. Garett Wood, Springwell.† Northwood, *Fitter*. Stanmore Common. Pinner Hill. Between Uxbridge and Denham. **2.** Hammonds, near Staines, *T. & D.*, 183. Between Chertsey Bridge and Laleham. Railway banks, Yeoveney to Poyle, frequent. Stanwell Moor. Laleham!, *Monckt. Fl.*, 46. Ashford. Harlington, *Bangerter*. **3.** By the Crane, Hounslow Heath!, *T. & D.*, 183. **4.** Hampstead Heath!, *Johns. Eric.* Between Finchley and Hendon, *Newbould, T. & D.*, 183. **5.** Near Hanwell!, *Warren, T. & D.*, 183. Greenford!, *Druce Notes.* Hanger Hill, Ealing. **6.** Tottenham, *Johns. Cat.* Edmonton, *T. & D.*, 183. Bush Hill Park, *Tremayne*. Whitewebbs Park, Enfield. **7.** Near Fulham,† *M. & G.*, 15. Near Kilburn† (G & R). Ken Wood.

L. ovalifolium Hassk. Japanese Privet.

Alien. Japan. Frequently used for hedging, and occurring as a naturalised outcast in hedges, thickets, on waste ground, etc. Common.

First record: *D. H. Kent*, 1939.

7. Hampstead. Holloway. Isle of Dogs.

APOCYNACEAE

VINCA L.

V. minor L. Lesser Periwinkle.

Alien. Europe, etc. Garden escape. Naturalised in woods, thickets, on hedgebanks, etc. Local.

First record: *Merrett*, 1666.

1. Little Grove Wood, near Breakspears,† *Blackst. Fasc.*, 19. Harefield!; Northwood!, *K. & L.*, 187. Between Pinner and Eastcote. Yiewsley, *M. & S.* Near Harrow Weald Common, *De Cresp. Fl.*, 77. The Grove, Stanmore Common!, *Welch*. South Mimms, *T. & D.*, 184. Potters Bar!, *Kennedy*. **2.** River bank between Richmond Bridge and Hampton Court!, *Newbould, T. & D.*, 184; river bank near Kingston Bridge, 1944→. **3.** Harrow Grove!, *Melv. Fl.*, 51. Islip Manor Park, Northolt!, *Payton*. Village green, Northolt, 1962→. Hayes!, *Tremayne, K. & L.*, 187; Hayes Park, plentiful, 1966!, *Royle*. Near Wood End. Hatton Cross, 1966, *Kennedy*!. **4.** On the west of Hampstead Heath, plentifully,† *Merr. Pin.*, 27. Hendon,

Pamplin; Harlesden Green;† Willesden,† *A. B. Cole, T. & D.*, 184. Bentley Priory Woods!, abundant, *Melv. Fl.*, 51; 1967. Near North End, Hampstead,† *Whiting Fl.* Mill Hill!, *E. M. Dale*; near Finchley!, *C. S. Nicholson*; borders of Bishop's Wood!, *K. & L.*, 187. Harrow!, *Harley Fl.*, 20. Edgware, *Horder*!. **5.** Near Ealing Common.† Greenford,† *Anderson*. Near Perivale Church, 1965→. Shrubbery, Hanger Lane, Ealing!, 1965, *Britton*; 1966. Old shrubbery, Acton, 1966. **6.** Near Tottenham, *M. & G.*, 308. Abundant in almost every copse about Enfield and Southgate, *Phyt. N.S.* 6: 301; Whitewebbs Park, Enfield!, *Battley* (LNHS); and elsewhere about Enfield Chase. Southgate!, 1902, *L. B. Hall, K. & L.*, 167; 1967, *Kennedy*!; Trent Park, 1966, *Kennedy*. Colney Hatch, common, *W. G. Smith, T. & D.*, 183. **7.** On the moatside at Jack Straw's Castle,† *Pet. Midd.* and *Doody MSS.* Near Lisson's Green (near Paddington),† *Doody MSS.* Belsize, Hampstead, towards Tottenham Court Road, *Alch. MSS.*; between Primrose Hill and Hampstead,† *M. & G.*, 308. Near Kilburn, 1817† (G & R). Near Islington,† *Spencer*, 318. Holloway, 1965, *D. E. G. Irvine*.
Plants with variegated leaves are recorded from **4.** Between Hampstead and Willesden,† *M. & G.*, 308.

V. major L. Greater Periwinkle.
Alien. S.-E. Europe, etc. Garden escape. Naturalised on hedgebanks, in thickets, etc. Very rare.
First record: *How*, 1650.
1. Harefield!, *Blackst. Fasc.*, 19. Near Uxbridge, *A. B. Cole, T. & D.*, 185. **2.** Hampton, *Newbould*; near Kingston Bridge, towards Hampton Court!, abundant, *T. & D.*, 185; 1968. Near Colnbrook, 1884, *Benbow* (BM). **3.** Between Isleworth and Hounslow† (DILL). Isleworth,† *Druce Notes*. Twickenham,† *T. & D.*, 185. Near Hounslow Heath!, *Welch*; Northolt!,† *K. & L.*, 187. **4.** Hendon, *Irv. MSS.* Harrow!, *Melv. Fl.*, 51. Mill Hill, *J. E. Cooper*. **5.** Horsendon Lane,† *Melv. Fl.*, 51. Near Ealing, 1946,† *Boucher*!, *K. & L.*, 187. **6.** Enfield!, *Cockfield Cat.*, 13; c. 1873, *H. & F.*, 150; 1965, *Kennedy*!. Muswell Hill, *W. H. Lloyd, Coop. Fl.*, 104; 1841† (HE). Trent Park, 1966, *Kennedy*. **7.** Near Kingsland,† *How Phyt.*, 129. Between Bromley and Blackwell,† *M. & G.*, 312. Between Hampstead and Edgware Road, 1809,† *Winch MSS.* Kentish Town Churchyard,† *M. & G.*, 312. Stoke Newington, 1869,† *W. G. Smith, T. & D.*, 185.

GENTIANACEAE

CENTAURIUM Hill

C. pulchellum (Sw.) Druce Dwarf Centaury.
Erythraea pulchella (Sw.) Fr.

Formerly native in moist sandy places but now extinct.
First, and only record: *Shepheard*, 1896.
2. Near Staines, *Shepheard, J.B. (London)* 34: 511.

C. erythraea Rafn Common Centaury.
C. minus Moench, *C. umbellatum* auct., *Erythraea centaurium* auct.
Native in dry grassy places, open woods, on heaths, etc.; established on railway banks, etc. Formerly locally common, but now rare and decreasing.
First record: *T. Johnson*, 1638.
1. Harefield!, *Blackst. Fasc.*, 17. Ruislip Woods!; Pinner, *Melv. Fl.*, 51; 1866,† *Trimen* (BM). Bayhurst Wood!; Northwood; railway banks between Uxbridge and West Drayton; Ruislip Reservoir; Duck's Hill; Mimmshall Wood!, *Benbow MSS.* Eastcote, 1898,† *E. M. Dale* (LNHS). Stanmore Common, 1909,† *C. S. Nicholson.* New Years Green, *Davidge* and *T. G. Collett*!, *K. & L.*, 188. Warren Gate,† *P. H. Cooke.* **2.** Gravel pit, East Bedfont!,† *H. R. Davies, K. & L.*, 188. **4.** In many places about Hampstead Heath, *Newt. MSS.*; 1849,† *W. Davies* (QMC). Mill Hill!, 1837, *Children* (BM); 1960. Stanmore, 1815† (G & R). Railway bank, Harrow,† *Melv. Fl.*, 51. Scratch Wood!, *Benbow MSS.* **5.** Floor of disused reservoir, Ealing, 1967→, *McLean, L.N.* 47: 10. **6.** Highgate, *Blackst. MSS.*; c. 1830,† *Irv. MSS.* Whitewebbs Park, Enfield, *Johns, K. & L.*, 188. **7.** West side of Hampstead Heath, near the Vale of Health,† *Park Hampst.*, 29. By Paddington Canal near Kensal Green Cemetery,† *Britten, T. & D.*, 186.
Plants with white flowers have been recorded from **1.** Ruislip Reservoir, *Hind, T. & D.*, 186. **6.** Tottenham,† *Johns. Cat.*

BLACKSTONIA Huds.

B. perfoliata (L.) Huds. Yellow-wort.
Chlora perfoliata (L.) L.
Native in grassy places on calcareous and basic clay soils. Very rare, and almost extinct.
First record: *Ord*, 1864.
1. Roadside near Pinner Hill, *Ord, Melv. Fl.*, 51; about six plants, 1878,† *W. A. Tooke, Trim. MSS.* In, and above the chalkpits, Harefield, 1868,† *Newbould, T. & D.*, 185. Chalk grassland above Springwell Lock!;† borders of Garett Wood, Springwell!,† *Benbow, J.B. (London)* 22: 37. Both habitats were destroyed by chalk quarrying about 1960, but a few plants were noted in the chalkpit on the site of Garett Wood, 1965!, *T. G. Collett* and *Kennedy*; 1967.

GENTIANA L.

G. pneumonanthe L. Marsh Gentian.
Lit. *J. Ecol.* **33**: 295 (1946).
Formerly native in wet places on heaths, but long extinct.
First record: *Dillenius*, c. 1730. Last certain evidence: *Dickson*, 1795.
2. Heath by Colnbrook, *Cockfield Cat.* Probably an error. **3.** Hounslow, towards Hampton, *Dill. MSS.*; Hounslow Heath, *Goodenough* (BM); c. 1795, *Dicks. Hort. Sicc.* **7.** Fulham, 1840, *Jameson* (K)—see *Kew Bull.* 1907: 70. Surely some confusion here, a most unlikely locality for the species.

GENTIANELLA Moench

Lit. *Watsonia* **4**: 169 (1959): *loc. cit.* **4**: 218 (1960): *loc. cit.* **4**: 290 (1961).

G. germanica (Willd.) Börner
Gentiana germanica Willd.
Native in calcareous grassland, etc. Very rare, and almost extinct.
First evidence: *Benbow*, 1892.
1. Garett Wood borders!,† and field adjoining, above Springwell Lock!,† *Benbow* (BM); both habitats were destroyed by chalk-quarrying operations c. 1950–55, but the species is attempting to colonise the chalk rubble; two small seedlings only seen on bare chalk, 1965, *Pickess*!.

G. amarella (L.) Börner Felwort, Autumn Gentian.
Gentiana amarella L.
Native on dry calcareous grassland and steep chalky slopes. Very rare, and almost extinct.
First certain evidence: *Newbould*, 1868.
1. In the old chalkpit near Harefield Mill,† *Blackst. Fasc.*, 32. The latter record could refer to *G. germanica*. A little north of Harefield, sparingly, 1868, *Newbould* (BM); second pit on the down from Jack's Lock towards the Copper Mills, Harefield, 1888;† side of steep pasture near Springwell, 1884; on the chalky banks of Garett Wood!;† slopes of an old chalkpit by Springwell Lock!, and on the downs around it!,† *Benbow MSS.* Eradicated from the chalky banks of Garett Wood and from calcareous grassland above Springwell Lock by chalk-quarrying operations, c. 1950–55; it still survives in small quantity on the steep slopes of the old chalkpit.

Gentianella amarella × germanica = G. × pamplinii (Druce) E. F. Warb.
Gentiana × pamplinii Druce
1. Harefield,† *Druce Fl.*, 230. Calcareous slopes above Springwell Lock, 1950,† *Brenan*!, *K. & L.*, 189.

MENYANTHACEAE

MENYANTHES L.

M. trifoliata L. Buckbean, Bogbean.

Lit. *J. Ecol.* **52**: 723 (1964).

Native in bogs, marshes, wet meadows, on pond verges, etc. Formerly local, but now very rare, and sometimes planted.

First record: *T. Johnson*, 1632.

1. Harefield Moor, in several places, *Blackst. Fasc.*, 103; c. 1869, *T. & D.*, 187; c. 1884;† meadows from Jack's Lock, Harefield to Drayton Ford;† moors below Springwell,† *Benbow MSS.* Yiewsley, 1910, *Green, Williams MSS.*; 1912–19,† *J. E. Cooper, K. & L.*, 189. Uxbridge Moor, 1839,† *Kingsley* (ME). Pond near Stanmore Common, 1949!, *Boniface, K. & L.*, 189; 1963. **2.** Marshes about Staines, in many of which it is the principal plant,† *Curt. F.L.* Staines Moor, *Whale Fl.*, 19; 1884,† *Benbow* (BM). West Drayton,† *Druce Notes.* **3.** Hounslow† (DILL). Stream in grounds of West Middlesex Hospital, Isleworth, probably planted, 1966, *Boniface.* **4.** Hampstead Heath, *Johns. Enum.*; 1780, *Pamplin MSS.*; 1838, *E. Ballard*; still abundant in the bog behind Jack Straw's Castle, 1860, *Trimen* (BM); 1864 (SY); still there, but dwindling, c. 1898, *Garlick W.F.*; no longer flowers, 1901, *Whiting Notes*; still there in 1912,† but failed to flower,† *Whiting Fl.*; not there in 1913, *P. H. Cooke, K. & L.*, 189. Golders Hill, 1921,† *Richards* (BM). Abundant around the lake in Grimsdyke grounds, Harrow Weald!, perhaps planted, *L.N.* 26: 63; 1962. Harrow Park lake, planted!, *Harley Fl.*, 20. **6.** Enfield, *H. & F.*, 149. Trent Park, perhaps planted, 1966, *T. G. Collett* and *Kennedy.* **7.** Banks of the Thames, Fulham,† *Faulkn. Hist.*, 13. Hammersmith,† *Crowe MSS.*

NYMPHOIDES Seguier

N. peltata (S. G. Gmel.) O. Kuntze Fringed Waterlily.

Limnanthemum nymphoides (L.) Hoffmanns. & Link, *L. peltatum* S. G. Gmel.

Native in the still waters of rivers and streams, ponds, lakes, etc. Locally plentiful near the Thames, rare and sometimes planted for ornamental purposes elsewhere.

First record: *Pena* and *l'Obel*, 1570.

2. Near Walton Bridge!, *Viscount Lewisham, Smith Fl. Brit.* 1: 226. Creeks of the Thames near Sunbury, in vast plenty, and most creeks of the river in that district, *J. Banks, B.G.*, 401. Sunbury, *Moxon*; between Hampton and Hampton Court!, *T. & D.*, 187. Hampton!, *Huds. Fl. Angl.* 2: 85. In the Thames just above Kingston Bridge!, abundantly, *Blackst. Spec.*, 60. Long Water, Hampton Court!; near Laleham,† *Benbow MSS.* Bushy Park!, *Green* (SLBI). Above Walton Bridge!, *Benbow, J.B. (London)* 23: 37. **3.** Thames at Richmond, 1841† (HE). Ornamental water, Twickenham Park, ? planted,† *T. & D.*, 187. **5.** Brentford!, abundant, *Robson Fl.*, 72; c. 1869, *Lees, T. & D.*, 187; Syon Park, probably planted, *T. & D.*, 187. Pond in Clayponds Hospital Grounds, South Ealing, planted? **7.** Frequent in the Thames about London,† *Pena & Lob. Stirp. Adv.*, 258. Hammersmith, 1811,† *J. F. Young* (BM). Pond in London Fields, Hackney,† *Woods, B.G.*, 401. Small pools near Hampstead,† *Button, Coop. Suppl.*, 11.

BORAGINACEAE

CYNOGLOSSUM L.

C. officinale L. Hound's-tongue.
Native on wood borders, banks, in grassy places, waste ground, etc., on dry soils. Very rare, and perhaps extinct.
First record: *Blackstone*, 1737.
1. Harefield, *Blackst. Fasc.*, 22; 1903,† *Green* (SLBI). Canal bank, Cowley, 1862, *Benbow* (BM); 1871, *Britten, Trim. MSS.*; eradicated, 1882,† *Benbow MSS.* Uxbridge gravel pits, 1838;† near Uxbridge Common,† *Kingsley* (ME). **2.** Hampton, *Newbould*; near Walton Bridge; Bushy Park; by the towing path in several places between Hampton Court and Kingston Bridge,† *T. & D.*, 189. **3.** Between Isleworth and Richmond, 1815† (G & R). **4.** Between Child's Hill and Hendon,† *Irv. MSS.* **7.** Ken Wood,† *Hunter, Park Hampst.*, 28. Perhaps an error. Bombed site, Cripplegate, *A. W. Jones, K. & L.*, 190; 1955,† *Jones Fl.*

C. germanicum Jacq. Green Hound's-tongue.
C. montanum auct.
Formerly native on hedgebanks, in thickets, etc., but long extinct.
First record: *Petiver*, 1695. Last evidence: *Buddle*, c. 1710.
7. In a hedge facing the road on Stamford Hill, between Newington and Tottenham, *Pet. Midd.* On Stroud Green, near the Boarded river, near Islington, *Buddle* (SLO).

SYMPHYTUM L.

Lit. *J. Linn. Soc. Bot.* **41**: 496 (1913): *L.N.* **33**: 55 (1954): *Watsonia* **3**: 280 (1956): *Proc. Bot. Soc. Brit. Isles* **4**: 446 (1962). (*Abstr.*): *loc. cit.* **7**: 432 (1968). (*Abstr.*)
S. officinale L. Comfrey.
Native by river- and streamsides, in ditches, on pond verges, damp

waste ground, etc., established by canals, etc. Formerly common, but now local and mainly confined to the vicinity of the Thames, Colne and Lee.

First record: *T. Johnson*, 1638.

1. Locally common in the Colne valley. Ruislip. Mimmshall brook, South Mimms. **2.** Locally plentiful by the Thames and Colne. **3.** Twickenham. Hounslow Heath. Canal side, Bull's bridge, Southall. **4.** Hampstead,† *P. H. Cooke* (LNHS). **5.** Ealing,† *Loydell* (D). Hanwell.† Chiswick. **6.** Locally common in the Lee valley. Colney Hatch. Finchley Common. **7.** Between Lee Bridge and Tottenham!, *Cherry*; Regents Park;† Zoological Gardens;† Hackney Wick marshes, abundant, *T. & D.*, 190. Near Newington,† *Ball. MSS.* By the lake, St James's Park, 1965, *Brewis, L.N.* 46: 67. Canal side, Limehouse, 1966. Canal side, Stepney, 1967.

S. asperum Lepech. × **officinale** = **S.** × **uplandicum** Nyman Blue, or Russian Comfrey.

S. peregrinum auct.

Alien of hortal origin. Formerly cultivated as a fodder crop and now naturalised by river- and streamsides, in ditches, fields, on waste ground, etc., and established on railway banks, canal banks, etc. Abundant throughout the Lee valley, common and increasing elsewhere.

First evidence: *J. E. Cooper*, 1912.

7. Regents Park, 1966, *Holme*; garden weed, Grosvenor Road, S.W.1, 1965!, *L.N.* 46: 32. Upper Clapton. Tottenham. Stepney. Hackney Wick. Hackney Marshes.

Backcrosses between *S.* × *uplandicum* and *S. officinale* are common in the Lee valley.

S. orientale L.

Alien. Asia Minor. Garden escape. Naturalised in grassy places, on waste ground, river banks, etc., and established on railway banks. Rare.

First evidence: *A. R. Pryor*, 1874.

2. Thames bank near Shepperton Lock, in quantity, 1967, *Kennedy*!. **3.** Railway bank, Isleworth!, 1874, *A. R. Pryor* (BM); plentiful, 1960!, *L.N.* 41: 20; 1974. **4.** Harrow-on-the-Hill Churchyard!, *Welch, K. & L.*, 191; 1973. Established about Harrow!, *Harley Fl.*, 20; still present in several localities, 1973. Railway bank, Cricklewood, 1965!, *L.N.* 45: 20; 1967. **5.** Perivale Churchyard!, *Welch, K. & L.*, 191; also plentiful on waste ground nearby, 1963–67.† Canal bank, Brentford, a single large plant, 1967!, *Boniface*. **6.** Winchmore Hill, 1922, *L. B. Hall* (BM).

S. tuberosum L. Tuberous Comfrey.

Introduced. Naturalised in thickets, etc. Very rare.

First record: *J. W. White*, 1869.

1. Stanmore Common!, *Green* (SLBI); 1966. **2.** Near Walton Bridge, 1877, *H.* and *J. Groves*, *Rep. Bot. Loc. Rec. Club*, 1877, 218. An error, the specimen (BM), on which the record was based, is *S. officinale*. **4.** Harrow, *Horton Rep.* Error, *S. orientale* intended. **6.** Near Barnet, 1869, *J. W. White*, *Sci. Goss.* 1869: 138.

PENTAGLOTTIS Tausch

P. sempervirens (L.) Tausch Alkanet.
Anchusa sempervirens L., *Caryolopha sempervirens* (L.) Fisch. & Trautv.
Alien. Europe, etc. Garden escape. Naturalised in grassy places, on hedgebanks, waste ground, by roadsides, etc., and established on railway banks. Rather common, and increasing.
First record: *Melvill*, 1864.
1. Uxbridge, 1884–85, *Benbow MSS.* Pinner, 1947!, *K. & L.*, 193. Harefield!, *Pickess*. Hillingdon, 1966. By Ickenham Church, 1966. Ruislip Churchyard, plentiful, 1966. **2.** Between Hampton Court and Kingston Bridge, 1903,† *Green* (SLBI). Laleham!, *L.N.* 26: 62. Near Staines. Near Staines Moor, 1966, *Boniface*!. East Bedfont, 1960. Bushy Park, 1965. Sunbury, 1967, *Kennedy*!. Harmondsworth, 1967. **3.** Sudbury Hill!; Hounslow Heath!, *L.N.* 27: 32. Feltham, 1946, *H. R. Davies*, *K. & L.*, 193. Near Feltham, in quantity, 1965.† Cranford, 1965→. Isleworth!, 1945, *Westrup*; 1966. Strawberry Vale, Twickenham, 1965. **4.** Garden weed, Harrow!, *Melv. Fl.*, 53; locally common about Harrow, 1969. Harrow Weald, *Knipe*. Hampstead, 1966. Mill Hill, 1960,† *Sennitt*. **5.** Chiswick, *Murray List.* Hanwell, 1962. Railway bank, Hanger Lane station, Ealing, abundant, 1965→. Acton, 1965. **6.** Enfield, 1965→. Highgate, 1966. **7.** East Heath, Hampstead, 1948–49!,† *H. C. Harris*, *K. & L.*, 193. Marylebone Churchyard, abundant, 1960→. Isle of Dogs, 1963. Fulham Palace grounds, 1966, *Brewis*.

ANCHUSA L.

A. arvensis (L.) Bieb. Field Alkanet, Bugloss.
Lycopsis arvensis L.
Native in fields, on cultivated and waste ground, etc., usually on light sandy or calcareous soils, probably an introduction on other soils. Very rare, invariably in small quantity, and often merely casual.
First record: *T. Johnson*, 1632.
1. Harefield, *Blackst. Fasc.*, 12; c. 1899, *E. M. Dale*; 1915, *J. E. Cooper*, *K. & L.*, 193. **2.** Near Teddington, a single plant, 1867, *Dyer*; Shepperton, 1886, *Benbow* (BM), Staines Moor, *Whale Fl.*, 18. Ashford, *Hasler*; Hampton Court Park!, *K. & L.*, 193; 1966. Bushy Park, 1966. **3.** Between Twickenham and Isleworth, *T. & D.*, 190; c.

1908, *Montg. MSS*. Gravel pit near Hounslow Heath, a single plant, 1949. **4.** Hampstead, *Irv. MSS*. Finchley, 1908, *Coop. Cas*. Harrow School Farm, two plants, *Horton Rep*. Wembley, a single plant, 1965. **5.** Sandy field, Chiswick, *Newbould, T. & D.*, 190; 1887, *Benbow* (BM). Turnham Green, 1872, *Britten, Trim. MSS*. Hanwell, one plant, 1951!, *K. & L.*, 193. **6.** Bush Hill Park, 1947, *Hanson, L.N.* 28: 33. Enfield, 1965, *Kennedy*. **7.** Upon the dry ditch banks about Piccadilly . . .,† *Johns. Ger.*, 799. Newington, 1838, *E. Ballard* (BM). Finchley Road, 1869, *Trim. MSS*. Bombed site, Crutched Friars, E.C., 1946,† *Lousley* (L). Plentiful on disturbed soil in front of new flats, Shadwell, 1966.

PULMONARIA L.

P. officinalis L. Lungwort.

Alien. Europe, etc. Garden escape. Naturalised on hedgebanks, in thickets, etc. Very rare.

First record: *Melvill*, 1864.

1. Uxbridge, 1884,† *Benbow* (BM). Near Potters Bar, a small colony, 1966, *Kennedy*. **3.** Harrow Grove,† *Melv. Fl.*, 52. **4.** Well naturalised near Kennet House, Harrow!, *Harley Fl.*, 21. **5.** Near Brentford, 1941.†

MYOSOTIS L.

Lit. *Rep. Bot. Soc. & E.C.* **10**: 338 (1933): *J.B. (London)* **80**: 127 (1942).

M. scorpioides L. Water Forget-me-not.

M. palustris auct.

Lit. *Nat. Camb.* **4**: 18 (1961).

Native by river- and streamsides, on pond verges, in ditches, marshes, etc. Common, except in heavily built-up areas where owing to the eradication of its habitats it is very rare or extinct.

First certain evidence: —, 1834.

7. Primrose Hill, 1834† (BM). Isle of Dogs, *Coop. Fl.*, 115. Lee Navigation Canal between Tottenham and Edmonton!, *Cherry*; Hackney Marshes;† Hornsey Wood;† East Heath, Hampstead; Eel Brook Meadow, Parsons Green,† *T. & D.*, 192. Beyond Hornsey Wood,† *Ball. MSS*. Ken Wood lake!; Highgate Ponds!, *Fitter*.

M. secunda A. Murr. Water Forget-me-not.

Lit. *Nat. Camb.* **4**: 18 (1961).

Formerly native in bogs and marshes on peaty soils but now extinct.

First, and only certain evidence: *T. Moore*, 1847.

1. Ruislip Reservoir, *Melv. Fl.*, 52. Error, *M. caespitosa* probably intended. **4.** Hampstead Heath, 1847, *T. Moore* (BM). Ditches near Pinner Drive, *Melv. Fl.*, 52. Error, the species here is *M. caespitosa*, *Benbow MSS*.

M. caespitosa K. F. Schultz Water Forget-me-not.
M. laxa Lehm. subsp. *caespitosa* (K. F. Schultz) Hyland.
Lit. *Nat. Camb.* **4**: 18 (1961).
Native on pond verges, in marshes, etc., especially on heaths, also by river- and streamsides. Local.
First certain evidence: *E. Forster*, 1792.
1. Stanmore Common!; Elstree Reservoir!; between South Mimms and Potters Bar!, *T. & D.*, 192. South Mimms, 1965, *J. G.* and *C. M. Dony*!. Ruislip Reservoir!; Swakeleys; near Uxbridge; Harefield!; Potters Bar!, *Benbow MSS*. **2.** Shortwood Common!, *Benbow MSS*. West Drayton, 1936,† *P. H. Cooke*. **3.** Hounslow Heath,† *T. & D.*, 192. **4.** Hampstead Heath,† *Irv. MSS*. Ditches near Pinner Drive, *Melv. Fl.*, 52 (as *M. repens*); c. 1887;† between Edgware and Whitchurch,† *Benbow MSS*. Harrow† (ME). Scratch Wood!, *K. & L.*, 194. **5.** Greenford,† *Melv. Fl.*, 52. Canal between Hanwell and Brentford,† *Benbow MSS*. **6.** Finchley Common!; Enfield Chase!, *Benbow MSS*. Trent Park, *T. G. Collett* and *Kennedy*. **7.** Hackney Marsh, 1792,† *E. Forster* (BM). Walham Green,† *Irv. Ill. Handb.*, 468. Reservoirs in Ken Wood grounds, *T. & D.*, 190.

M. sylvatica Hoffm. Wood Forget-me-not.
Native in woods and banks on calcareous and other light soils about Harefield, elsewhere in grassy places, woods, by streams, on waste ground, etc., as a garden escape. The native plant is very rare, the garden escape is common and increasing.
First record: *Blackstone*, 1737.
1. In Gutter's Dean† and the Old Park, sparingly!, *Blackst. Fasc.*, 62; Old Park Wood, Harefield, very scarce, 1963. Above the cement works, Harefield, 1946!, *Welch* (K), conf. *Wade*. Meadows by canal above Denham, 1892, *Wolley-Dod* (TUN). Pinner Wood,† *De Cresp. Fl.*, 43. Pinner, 1960. Hillingdon, 1973. Near Swakeleys, 1966. Ruislip, 1973. Harrow Weald Common, 1965, *Kennedy*!. Stanmore Common, 1966. Potters Bar, 1966, *Kennedy*. **2.** Staines, 1962. Dawley, 1966. Sunbury, 1967. Shepperton Lock, 1967, *Kennedy*!. Littleton, 1968. **3.** Hounslow Heath, in several places, 1965. Southall, 1973. Hayes Park, 1966, *Royle*!. Near Northolt, 1966. Greenford Park Cemetery, 1966. **4.** Stanmore; Whitchurch Common, 1966, *Tyrrell*!. Willesden, 1966. Hendon, 1966. West Heath, Hampstead, 1960. Edgware, 1967. Edgwarebury, 1967. Mill Hill, 1967. **5.** Greenford, 1859, *Hind*, *T. & D.*, 192. This was most likely the large wood form . . . of *M. arvensis*, *T. & D.*, 192. Hanwell, 1947!, *L.N.* 27: 33; 1968. Perivale Churchyard, abundant, 1965→. Ealing golf course, 1973. Near Park Royal, 1965. **6.** Botany Bay, Enfield Chase, 1966, *Kennedy*. Enfield; near Hadley Common, 1966, *Kennedy*!. **7.** Highgate, 1964.

Plants with pure white flowers were seen at **3.** Greenford Park Cemetery, 1966.

The garden escape flowers about a month earlier and has larger, often brighter blue corollas than the native plant. In most cases they consist of distinct races. Two closely allied species *M. alpestris* Schmidt and *M. dissitiflora* Bak., have numerous cultivars which are grown in gardens, and these are easily confused with cultivated forms of *M. sylvatica*. Some of the very large-flowered cultivars seen may be referable to hybrids between *M. dissitiflora* and *M. sylvatica*. *M. arvensis* var. *sylvatica* (Schlecht.) Druce is also sometimes confused with *M. sylvatica*. Many of the above records may, therefore, be in need of revision.

M. arvensis (L.) Hill Common Forget-me-not.

Native on cultivated ground, in woods, fields, ditches, etc. Formerly common, but now local, though often plentiful where it occurs.

First certain record: *Blackstone*, 1737.

1. Locally common. **2.** Locally frequent. **3.** Twickenham, *T. & D.*, 193. Wood End.† Roxeth.† Hayes, 1966. **4.** Harrow!, *Melv. Fl.*, 52. Hendon, *J. Morris*; between Finchley and Hendon!, *Newbould*; near Bishop's Wood, *T. & D.*, 193. Edgware!, *S. G.* and *J. Alletson*. Mill Hill. North-west Heath, Hampstead, 1949. Between Willesden and Harlesden,† *De Cresp. Fl.*, 43. Kenton, 1966. **5.** Hanwell,† *Warren*; Ealing!, *Newbould*; near Brentford!, *Hemsley*, *T. & D.*, 193. Bedford Park,† *Cockerell Fl.* Greenford. **6.** Edmonton; Enfield!, *T. & D.*, 193; Enfield Chase. Trent Park. **7.** Marylebone Fields, 1817† (G & R). East Heath, Hampstead, 1948, *H. C. Harris*.

Var. **sylvestris** (Schlecht.) Druce

var. *umbrosa* Bab.

1. Harefield!, *K. & L.*, 194. **3.** Hounslow Heath,† *T. & D.*, 193. **6.** Whetstone, *T. & D.*, 193.

M. discolor Pers. Yellow and Blue Forget-me-not.

M. versicolor Sm.

Native in dry grassy places, on banks, etc., chiefly on light soils. Formerly rather common, but now rare and decreasing.

First record: *Blackstone*, 1737.

1. Harefield!, *Blackst. Fasc.*, 62. Between Harefield and Ruislip!, *T. & D.*, 193. Uxbridge, 1883–84;† Cowley Peachey, 1884,† *Benbow* (BM). Ruislip!, *K. & L.*, 195. Swakeleys, 1885,† *Benbow MSS*. Near Denham, 1966. **2.** Between Staines and Hampton, *T. & D.*, 193. Staines!; Ashford; Charlton!, *Benbow MSS*. East Bedfont,† *Westrup*. Teddington, *De Cresp. Fl.*, 43. Hampton Court Park!, *K. & L.*, 195. Staines Moor, 1965, *Clement*. **3.** Hounslow Heath,† *T. & D.*, 193. Wood End. **4.** Stanmore Marsh (=Whitchurch Common), *T. &*

D., 193; c. 1900,† *C. S. Nicholson*, *K. & L.*, 195. Hampstead, 1863,† *Trimen* (BM). Scratch Wood area, abundant!, *Welch*, *L.N.* 28: 33; 1966. **5.** Northolt;† Alperton, abundant,† *Melv. Fl.*, 53. Near Chiswick,† *Fox*; near Hanwell,† *Warren*; Acton,† *Tucker*, *T. & D.*, 193. Brentford, 1968, *J. L. Gilbert*. **6.** Near Enfield, *Tucker*; Colney Hatch;† Whetstone, *T. & D.*, 193. **7.** Ken Wood grounds,† *T. & D.*, 193.

Var. **balbisiana** (Jord.) Wade
2. Thames bank between Hampton Court and Kingston Bridge, *C. E. Britton*, *J.B.* (*London*) 48: 331.

M. ramosissima Rochel Early Forget-me-not.
M. collina auct., *M. hispida* Schlecht.
Native on heaths, particularly on ant hills, grassy places, etc., on dry soils, also established on railway tracks and tops of old walls. Very rare.
First evidence: *Hardwicke*, 1853.
1. Harefield Churchyard, 1853† (HE). Near Ruislip, abundant, 1866, *Trimen*; south of Uxbridge Moor, 1885;† Hillingdon;† Uxbridge Common;† between Cowley and Hillingdon, 1890;† railway bank, Cowley Peachey, 1884,† *Benbow* (BM). Swakeleys,† *Benbow MSS.* Springwell!, *Welch*; South Mimms, *P. H. Cooke*, *K. & L.*, 194. Railway tracks between Denham and Uxbridge, 1952. Ant hills near Potters Bar, 1966, *Kennedy*. **2.** East Bedfont,† *Westrup*, *K. & L.*, 194. Perhaps an error. Staines, 1958, *Briggs*. **3.** Hounslow Heath, abundant,† *T. & D.*, 193. West of Hounslow Heath, very abundant,† *Benbow*, *J.B.* (*London*) 23: 338. Wood End!;† Cranford!,† *K. & L.*, 194. **4.** Whitchurch, 1909,† *P. H. Cooke* (LNHS). Near Scratch Wood. **5.** Greenford Church wall, very scarce,† *Melv. Fl.*, 53. Near Hanwell,† *Warren*; wall near Brentford, *T. & D.*, 193; Syon Park, 1955, *Graham*, *Hepper* and *Shaw*. Osterley Park walls,† *Montg. MSS.* Wall near Chiswick,† *Benbow MSS.*

LITHOSPERMUM L.

L. purpurocaeruleum L. Blue Gromwell.
Formerly naturalised as a garden escape in an old orchard but now extinct.
First record: *T. G. Collett*, 1950. Last record: *D. H. Kent*, 1965.
1. Old orchard, Pinner!, 1950, *T. G. Collett*, *L.N.* 30: 7; 1951, *T. G. Collett*!, *K. & L.*, 195. Eradicated prior to building operations, 1965.

L. officinale L. Common Gromwell.
Formerly native in bushy places, hedgebanks, etc., on calcareous and other dry soils in the north-west of the vice-county, and elsewhere on waste ground as an introduction, but now extinct.

First record: *T. Johnson*, 1638. Last record: *J. E. Cooper*, 1914.
1. Harefield, *Blackst. Fasc.*, 54. Springwell; Uxbridge, *Benbow MSS.* Yiewsley, 1910–14, *J. E. Cooper, K. & L.*, 195. **6.** Tottenham, *Johns. Cat.* Bank of Lee Navigation Canal, 1865, *Grugeon, T. & D.*, 191. **7.** Isle of Dogs, *Mart. Pl. Cant.*, 65. Hackney Marshes, 1914, *Coop. Cas.*

L. arvense L. Corn Gromwell, Bastard Alkanet.
Buglossoides arvensis (L.) I. M. Johnston
Formerly a colonist in cornfields, on cultivated ground, etc., but now very rare and sporadic on waste ground, etc.
First record: *T. Johnson*, 1638.
1. Harefield, abundant, *Blackst. Fasc.*, 53; 1920, *J. E. Cooper, K. & L.*, 195. Springwell, abundant, 1888, *Benbow MSS.* Yiewsley, 1910–12, *Coop. Cas.*; 1916, *J. E. Cooper, K. & L.*, 195. **2.** Colnbrook, abundant, *Benbow, J.B. (London)* 23: 37. West Drayton!, *Benbow MSS.* **3.** Harrow, on soil brought from elsewhere, *Farrar, T. & D.*, 191. Hounslow Heath, 1960, *Philcox* and *Townsend.* **4.** Hampstead, *Irv. MSS.* Cricklewood, imported with grain, 1869, *Warren, T. & D.*, 425. Hendon, 1912, *P. H. Cooke* (LNHS). Brent Reservoir, in small quantity, 1965, *Kennedy*!. Mill Hill, *Harrison.* **5.** Acton, *Newbould, T. & D.*, 191. Chiswick, *Benbow MSS.* Canal path, Brentford, 1947, *Sandwith*!, *L.N.* 27: 33. Canal path between Hanwell and Southall, a single plant, 1948!, *L.N.* 28: 33. Hanwell, 1951, *T. G. Collett*!. **6.** Tottenham, *Johns. Cat.* Enfield, *H. & F.*, 149. Between Ponders End and Edmonton, 1888, *Benbow* (BM). Near Enfield, *T. & D.*, 191. North Finchley, 1925, *J. E. Cooper, K. & L.*, 195. **7.** St John's Wood, 1815 (G & R). Chelsea, *Newbould, T. & D.*, 191. South Kensington, *Dyer List.* Hackney Marshes, 1913, *Coop. Cas.*; 1923, *J. E. Cooper, K. & L.*, 195. Hornsey, 1887, *Coop. Cas.* Newington, *Ball. MSS.* Bombed site, Upper Thames Street, E.C., 1945!, *Lousley* (L).

ECHIUM L.

E. vulgare L. Viper's Bugloss.
Formerly native in grassy places on calcareous soils at Harefield and Springwell, and on alluvial soils near the Thames, but now extinct, still present on waste ground, railway banks, etc., as an introduction. Very rare.
First record: *Morley*, 1677.
1. Harefield, *Blackst. Fasc.*, 24; 1839, *Kingsley* (ME); c. 1869, *T. & D.*, 191; 1884, *Benbow* (BM); 1900,† *E. M. Dale* (LNHS). Springwell, scarce, 1886;† borders of Garett Wood,† *Benbow MSS.* Uxbridge Moor, a single plant, 1886,† *Benbow, J.B. (London)* 25: 17. **2.** West Drayton,† *Lightf. MSS.* By the towing path near Hampton Court, *H. C. Watson* (W); c. 1869,† *T. & D.*, 191. **4.** Hampstead,† *Morley*

MSS. Spoil bank over railway tunnel, Scratch Wood, abundant!, *L.N.* 26: 62; considerably reduced by the bulldozing of the area in connection with the construction of the M1 Extension, 1965. Scratch Wood railway sidings, formerly abundant, but now much reduced by the construction of the M1 Extension. Mill Hill, 1903, *Coop. Cas.* This may refer to the Scratch Wood locality. Railway embankment, Golders Green, 1903,† *P. H. Cooke* (LNHS). Railway bank, Preston Road, near Harrow, 1955!, *L.N.* 35: 5. East Finchley, 1909,† *Coop. Cas.* **5.** Railway bank, Gunnersbury, 1872,† *Britten, Trim. MSS.* Chiswick, *Benbow MSS.*; 1958, *L. M. P. Small.* **6.** Muswell Hill, 1903;† Whetstone, 1906;† Finchley, 1909,† *Coop. Cas.* **7.** Haverstock Hill,† *Irv. MSS.* Chelsea, 1859,† *Britten, Bot. Chron.*, 59. South Kensington, 1870,† *Dyer* (BM). Isle of Dogs, 1887,† *Benbow, J.B. (London)* 25: 365. Hackney,† *Benbow MSS.* Lee Bridge,† *P. H. Cooke.* Regents Park,† *Webst. Reg.*, 101. Bombed site, Cripplegate, 1952–53, *Scholey, K. & L.*, 196; 1957,† *Jones Fl.* Tottenham, 1967, *Kennedy.*

CONVOLVULACEAE

CONVOLVULUS L.

C. arvensis L. Small Bindweed, Cornbine.
Native on cultivated and waste ground, in fields, by roadsides, etc. Very common.
First record: *T. Johnson*, 1638.
7. Frequent in the district.

CALYSTEGIA R.Br.

C. sepium (L.) R.Br. Bellbine, Larger Bindweed.
Native in marshes, ditches, on cultivated and waste ground, climbing over hedges, fences, etc. Common.
First record: *T. Johnson*, 1632.
7. Common in the district.

Forma **colorata** (Lange) Dörfler
Var. *americana* auct.
1. Cowley!; Northwood!, *L.N.* 28: 33. The latter record is an error, the plant seen being *C. pulchra.* **4.** Hendon, 1917, *J. E. Cooper.* Near Mill Hill. **6.** Church End, Finchley, 1919, *J. E. Cooper.* **7.** Clapton, 1871, *F. J. Hanbury* (BM). Fulham Palace grounds, 1966, *Brewis*, teste *Brummitt.*
This pink-flowered form of *C. sepium* is a rare plant of marshy places and ditches, and it is likely that some of the records given above are referable to *C. pulchra.*

C. sepium × silvatica = C. × lucana (Ten.) G. Don
Lit. *Watsonia* **5**: 88 (1961).
Common and widespread in hedges, etc., especially near the Thames, and often occurring in the absence of either parent.
7. Lisson Grove, N.W.1, 1965–66!, *L.N.* 46: 32; 1968. Fulham Palace grounds. Parsons Green. Islington. Hornsey. Hackney Marshes.
A form with corollas divided into five segments for most of their length was noted at **5.** Shepherds Bush, 1966.

C. pulchra Brummitt & Heywood
C. sylvestris (Willd.) Roem. & Schult. var. *pulchra* (Brummitt & Heywood) Scholz, *C. sylvestris* (Willd.) Roem. & Schult. f. *rosea* Hyland., *C. sylvatica* (Waldst.) Griseb. subsp. *pulchra* (Brummitt & Heywood) Rothm., *C. sepium* (L.) R.Br. subsp. *pulchra* (Brummitt & Heywood) Tutin, *C. dahurica* auct., non (Herbert) G. Don
Lit. *Proc. Bot. Soc. Brit. Isles* **3**: 384 (1960).
Alien of unknown origin. Garden escape. Naturalised in hedges, etc. Locally common in the vicinity of the Thames, rather rare elsewhere, though apparently increasing.
First evidence: *Dyer*, 1867.
1. Northwood, 1948!, *L.N.* 28: 33 (as *C. sepium* var. *americana*); 1966. Springwell, 1955,† *Lousley* and *Welch*!, *K. & L.*, 354 (as *C. dahurica*). Error, the specimen (L) on which the record was based is *C. fraterniflora* (Mack. & Bush) Brummitt, det. *Brummitt*. Between Hillingdon and Uxbridge, 1968. Uxbridge, 1973. **2.** East Bedfont!, 1963, *Clement* (BM), det. *Brummitt*; 1967. Near Kempton Park, 1966, *Tyrrell*!. Staines, 1966. **3.** Twickenham Park, 1867, *Dyer* (BM), det. *Brummitt* and *Heywood*. Near Twickenham!; Hounslow!, 1961, *Buckle*; 1966. St Margaret's, 1966, *Brummitt*. Railway bank by Richmond Bridge, 1966→. Isleworth, 1968. **4.** Railway bank, West Hampstead!, *K. & L.*, 196 (as *C. sylvestris* f. *rosea*). **5.** Ealing, 1965→. South Ealing, 1965→. Gunnersbury, 1965, *T. G. Collett* (BM), det. *Brummitt*. Chiswick, 1966. Gunnersbury Park, Brentford, 1967→. **6.** Palmers Green, 1964, *B. E. Bayliss* (BM), det. *Brummitt*. Enfield, 1966, *Kennedy*!. **7.** Grounds of Natural History Museum, South Kensington, 1951!, *Bangerter*, *K. & L.*, 196 (as *C. sylvestris* f. *rosea*). Isle of Dogs, 1964, *L. M. P. Small* (BM), det. *Brummitt*. Shadwell, 1966. Side of railway near Paddington, 1966!, *L.N.* 46: 32. Tottenham, 1966, *Wurzell*. Stoke Newington, 1967.
The hybrid *C. pulchra* × *silvatica* has not been recorded for Middlesex but is possibly present.

C. silvatica (Waldst.) Griseb. Larger Bindweed.
C. sylvestris (Willd.) Roem. & Schult., *C. inflata* auct., *C. sepium* (L.) R.Br. subsp. *silvatica* (Kit.) Maire, *Volvulus inflata* auct.

Lit. *Rep. Bot. Soc. & E.C.* **13**: 265 (1948): *Watsonia* **1**: 382 (1950): *loc. cit.* **2**: 118 (1951).
Alien. S.-E. Europe, etc. Formerly cultivated as a garden plant, and now widely naturalised in hedges, thickets, on waste ground, etc., as an escape or outcast. Common.
First record: *Britten*, 1873.
7. Common in the district.
Plants with corollas divided into five segments for most of their length (cf. var. *quinquepartita* Terracciano) were noted at **5.** Great West Road, Chiswick, 1965. Near Kew Bridge, 1973. For comments on the inheritance of the schizoflorous character see Stace (*Watsonia* **9**, 370 (1973)).

CUSCUTA L.

C. europaea L. Large Dodder.
Lit. *J. Ecol.* **36**: 356 (1948).
Native. Parasitic on *Urtica dioica* and *Humulus lupulus* in thickets, on river banks, etc. Very rare, and now confined to the immediate vicinity of the Thames.
First evidence: *J. J. Bennett*, 1820.
2. Various localities about Shepperton, *C. E. Britton, J.B.* (*London*) 53: 326. Near Staines!, *Sherrin* (SLBI). Penton Hook, 1917–19, *J. E. Cooper*, *K. & L.*, 197. By the Thames near Chertsey Bridge!, 1942, *Lousley* (L); 1944–53!, *K. & L.*, 197; 1954–60. Not seen in any of its Thames side localities since 1960 despite a careful search annually. **4.** Not uncommon about Hampstead, *Cochrane Fl.* Error, *C. epithymum* intended. **5.** By the Thames, Chiswick, 1841,† *McIvor* (CGE). Strand-on-the-Green,† *McIvor* (SY). **7.** Hyde Park, in the sunken ditch under the wall of Kensington Gardens, 1820 and 1821,† *J. J. Bennett* (BM).

C. epithymum (L.) L. Common Dodder.
C. trifolii Bab.
Formerly native. Parasitic on *Ulex* and *Erica* species and on *Calluna vulgaris*, etc., on heaths and commons, also parasitic on *Trifolium* species, particularly on *T. pratense* when grown as a fodder crop. Extinct as a native, but recently accidentally reintroduced, though unlikely to persist.
First record: *Parkinson*, 1640. Last evidence (as a native): *Benbow*, 1887.
1. Near Uxbridge, *Sibth. MSS.* Hillingdon, on *Ulex*, 1871, *Warren*, *Trim. MSS.* Harefield, among crops, *Blackst. Fasc.*, 21; in clover fields, 1884–85; clover fields, Springwell, in the greatest abundance, 1884–85, *Benbow* (BM). **3.** Twickenham Common, on *Ulex*, 1841,

Moxon (SY). Hounslow Heath, abundant, 1866, *Trimen*; sparingly on *Ulex*, 1887, *Benbow* (BM). **4.** Hampstead Heath, *Park. Theat.*, 9; on *Ulex*, c. 1680, *Newt. MSS.*; c. 1859, *White Hampst.*, 366. **5.** Hanwell, 1847, *McIvor, T. & D.*, 189. **6.** Enfield, *H. & F.*, 148. **7.** Buckingham Palace grounds, 1964, accidentally introduced with plants of *Calluna vulgaris*, *Fl. B.P.G.*; 1966, *McClintock*.

SOLANACEAE

LYCIUM L.

L. barbarum L. Duke of Argyll's Tea-plant.
L. halimifolium Mill., *L. chinense* auct., non Mill.
Lit. *Proc. Bot. Soc. Brit. Isles* **6**: 72 (1965). (*Abstr.*)
Alien. E. Asia. Garden escape. Naturalised in hedges, thickets, scrub, on waste ground, etc., established on railway banks, etc. Rather rare, and sometimes planted.
First evidence: *Newbould*, 1867.
1. Harefield!, 1867, *Newbould* (BM); 1968. Canal side between Uxbridge and Cowley!; near Cowley Peachey!; Yiewsley!, *K. & L.*, 199. New Years Green, *Lousley*!. Uxbridge Moor. Hillingdon. Bentley Heath, *J. G.* and *C. M. Dony*. **2.** West Drayton!, *P. H. Cooke*; between Dawley and Hayes!, *K. & L.*, 199. Near Stanwell. Near Ashford. Sunbury. Near Harlington. **3.** Abundant on waste ground south of Hounslow Heath, *Sandwith* and *Simpson*!, *L.N.* 27: 33; 1968. Canal bank, Bull's Bridge, Southall. Hayes. **4.** Willesden!, *F. Naylor*, *T. & D.*, 195; 1968. Canal bank, Harlesden. The Hyde, Hendon. **5.** Near Osterley Park!, *K. & L.*, 199. Canal bank, Brentford and Park Royal. Near Hanwell. **6.** Edmonton, *Tremayne*, *K. & L.*, 199. Near Barnet. Enfield. Enfield Chase, *Kennedy*. Near Enfield Wash, *Kennedy*!. **7.** Railway bank near Paddington, in quantity!, *L.N.* 39: 53; 1968. Railway side between Fulham and Chelsea, 1967. Stepney.
It is possible that some of the records given above are referable to *L. chinense* Mill.

ATROPA L.

A. bella-donna L. Deadly Nightshade, Dwale.
Lit. *J. Ecol.* **34**: 345 (1947).
Native in woods, thickets, on waste and disturbed ground, etc., on calcareous soils about Harefield, elsewhere on waste ground, thickets, etc., as an introduction. Occurs also as a garden weed. Rare.
First record: *Gerarde*, 1597.
1. Harefield!, *Blackst. Fasc.*, 95; 1968. Springwell, frequent!, *L.N.* 26:

63; scarce, 1968. Denham. Potters Bar, *Johns, K. & L.*, 199. **2.** Hampton Court!, *Green* (SLBI); 1967. River bank between Hampton Court and Kingston Bridge!, *C. E. Britton, J.B.* (*London*) 48: 231; 1967. West Drayton, 1957, *I. G. Johnson.* **3.** Grounds of West Middlesex Hospital, Isleworth!; stream bank near Mill Platt, Isleworth!,† *L.N.* 26: 63. Hounslow Heath!, *D. Bennett.* **4.** The Hale, near Edgware, 1834,† *J. F. Young* (BM). **5.** Thames bank, Chiswick!, *Bangerter, K. & L.*, 199; waste ground, Chiswick, *Murray*; garden weed, Chiswick Mall, 1958, *Holme.* Syon Park, frequent in thickets, shrubberies, on waste ground, etc.!, *K. & L.*, 199; much reduced in quantity, 1968; cultivated there for medicinal purposes during the First World War, *A. B. Jackson.* Acton Green,† *Benbow MSS.* Waste ground, Brentford, 1956–60;† railway bank, Brentford, 1966→. **6.** Highgate!, *Ger. Hb.*, 269; North Hill, Highgate, 1901, *L. B. Hall, K. & L.*, 199; grounds of Highgate Hospital!, 1947, *J. M. B. King*; 1960. Near Broomfield Park railway station, 1964, *Wurzell.* Enfield!, *H. & F.*, 149; weed in grounds of Myddleton House, Enfield, 1966, *Kennedy*!. **7.** Islington, *Mill. Bot. Off.*, 416; now (1793) totally expelled, *M. & G.*, 325; 1839,† *E. Ballard* (BM). Old gardens near Highgate, *Mart. Mill. Dict.* Barnsbury Park,† *Ball. MSS.* **7.** Crouch End, 1896, *Coop. Cas.* Building site behind the British Museum, Bloomsbury,† *Shenst. Fl. Lond.* Regents Park, *Webst. Reg.*, 102; three colonies, 1965, *Potter, L.N.* 46: 32. Chelsea, *H. R. Davies, L.N.* 24: 12. Garden weed, Campden Hill, W.8, 1946, *Wilmott* (BM). Bombed site, Eaton Terrace, S.W.1,† *R. L. Bennett.*

HYOSCYAMUS L.

H. niger L. Henbane.

Denizen on waste ground, in fields, by roadsides, etc., established on canal paths, etc. Rare, and often merely casual.

First record: *T. Johnson*, 1638.

1. Harefield, *Blackst. Fasc.*, 43; a plant by the canal, six feet high, with its top branches nearly three feet long, with 38 to 40 capsules, 1888, *Benbow MSS.* Cowley; Uxbridge Moor; near Uxbridge Common, *Benbow, J.B.* (*London*) 23: 38. Bushey Heath, *Ralph, Irv. Lond. Fl.*, 139. Pinner, 1863–65, *Hind, T. & D.*, 195; 1870, *Hind* (BM). Yiewsley, 1910, *Coop. Cas.* Canal bank north of Uxbridge, 1917, *Tremayne*; Ruislip, 1952, *Graham, K. & L.*, 201. **2.** Staines (=Shortwood) Common, 1841, *J. F. Young* (BM); 1962!, *Boniface.* Hampton Court!, *H. C. Watson* (W). West Drayton, 1907, *Worsdell* (K). Ashford County Hospital grounds, one plant, 1941, *Hasler, L.N.* 26: 76. Shepperton, one plant, 1946, *Westrup*; 1948, *Welch*; Perry Oaks Sewage Farm, one plant, 1950, *K. & L.*, 201. **3.** Hounslow Heath, 1947,

H. Banks!, *L.N.* 27: 33. **4.** Near Stonebridge, a single plant, *Benbow MSS.* Neasden, 1908, *Coop. Cas.* Finchley, 1909–10 and 1923, *J. E. Cooper*; near Scratch Wood!, 1946, *Simmonds, K. & L.*, 201; frequent on ground disturbed in connection with the extension of the M1 Motorway, 1965–66, *Kennedy*!; 1968. Hendon, 1929, *J. E. Cooper, K. & L.*, 201. Crown Street, Harrow, many plants appeared when an old wall collapsed,† *Harley Fl.*, 22. **5.** Acton Green, *Newbould, T. & D.*, 105; 1905, *Green* (SLBI). Sandy field, Chiswick, *Benbow MSS.* Canal bank, Brentford, 1944, *Welch* and *Lousley*!, *L.N.* 26: 63. Alperton, 1917, *J. E. Cooper*; 1918–21, *L. M. P. Small*; allotments, Castle Bar, Ealing, 1935–36, *Bull, K. & L.*, 201; Haven Green, Ealing, 1968, *McLean*. **6.** Tottenham!, *Johns. Cat.*, 1947. Enfield!, *H. & F.*, 149; weed in grounds of Myddleton House, Enfield, 1966, *Kennedy*!. **7.** Bow Common, *M. & G.*, 287. Vale of Health, Hampstead, *Bliss, Park Hampst.*, 29; Hampstead, *Burn. Med. Bot.* 1: 9. St John's Wood, 1815 (G & R). Parsons Green, 1856, '*J.A.*', *Phyt. N.S.* 2:168. Chelsea, 1860, *Britten, Bot. Chron.*, 58; 1944, *H. R. Davies, L.N.* 24: 12. South Kensington, *T. & D.*, 195. Fulham, *Corn. Surv.* Ball's Pond, *Ball. MSS.* Hackney Marshes, 1910 and 1912–14, *Coop. Cas.*; 1916–17, *J. E. Cooper*; Finsbury Park, 1951, *Scholey, K. & L.*, 201. Isle of Dogs, *Benbow MSS.* Near Tottenham Court, 1729, *P. Miller, Blackst. MSS.* Duck Island, St James's Park, 1955, *Teagle, L.N.* 39: 53. Hyde Park, 1962, *Allen Fl.* Shoreditch, 1965, *D. Bennett.* Regents Park, 1966, *Potter, L.N.* 46: 32. Whitechapel, several large plants, 1967, *Lousley, L.N.* 47: 10.

Var. **pallidus** (Waldst. & Kit. ex Willd.) Fr.

7. South Kensington, 1865, *T. & D.*, 195. Finsbury Park, 1952, *J. Bedford, K. & L.*, 201.

The seeds of *H. niger* will remain viable below the surface of the soil for very many years, and will germinate when the soil is disturbed, cf. its occurrence at Shortwood Common in 1841 and 1962. Odum (*Dansk Bot. Arkiv* **24**(**2**) (1965)) presents data to show that in Denmark buried seeds of the species have been known to remain viable for many centuries.

PHYSALIS L.

P. alkekengi L. Winter Cherry.

Alien. Europe, etc. Garden escape. Naturalised on waste ground, etc. Rare.

First evidence: *J. E. Cooper*, 1917.

2. Shepperton, 1946!, *K. & L.*, 198. **3.** Hounslow Heath. South Harrow, *W. R. Phillips, L.N.* 43: 22. Between Twickenham and Isleworth, 1965. Hounslow, 1966. **4.** Finchley, 1917, *J. E. Cooper*

(BM). Stonebridge Park, 1966. **5.** Chiswick!,† *K. & L.*, 198. Hanwell!,† *Welch*. By the canal between Alperton and Park Royal, 1962→. Canal bank, Alperton, 1962. Near Boston Manor railway station, 1966. Near Hammersmith Bridge!,† *K. & L.*, 198. **7.** Bombed site, Cripplegate area, E.C.,† *Jones Fl.* South Kensington, 1953,† *Bangerter, K. & L.*, 198. Bombed sites north-west of Victoria Park, Hackney, 1963, *Wurzell, L.N.* 43: 22. Hackney Wick, *L. M. P. Small.*

SOLANUM L.

S. dulcamara L. Bittersweet, Woody Nightshade.
Native in marshes, hedges, woods, on waste ground, etc., established on railway banks, etc. Very common.
First record: *T. Johnson*, 1638.

Var. **villosissimum** Desv.
var *tomentosum* Koch
2. Railway bank, West Drayton, 1947!, *L.N.* 27: 33. **6.** Woodside Park, 1922, *E. F.* and *H. Drabble* (BM). Edge of lake, Whitewebbs Park, Enfield, 1965, *Kennedy*!.
Plants with white flowers have been noted at **4.** Hampstead, 1949, *Bangerter* and *H. C. Harris*!, *K. & L.*, 198. **7.** Chelsea Hospital grounds, 1963, *Brewis, L.N.* 44: 24, and plants with yellow fruits were noted at **5.** Ealing,† *Masters, J.B.* (*London*) 14: 309 and *Jacks. Ann.*, 341.

S. nigrum L. subsp. **nigrum** Black Nightshade.
Native on cultivated and waste ground, etc., established on railway tracks, etc. Occurs also as a frequent weed of flower-beds in parks and gardens. Common, especially on light soils.
First record: *T. Johnson*, 1638.

Var. **atriplicifolium** (Desp.) G. Mey.
7. Hackney Marshes, *M. & S.*

Subsp. **schultesii** (Opiz) Wessely
1. Yiewsley, 1965. **3.** Hounslow Heath, 1964. **6.** North of Enfield Lock, 1967, *Kennedy*!.

S. pygmaeum Cav.
S. pseudocapsicum auct., non L.
Alien. S. America? Formerly established on a canal path, but now extinct.
First record: *D.* and *J. M. B. King* and *D. H. Kent*, 1945. Last record: *D. H. Kent*, c. 1970.
5. Canal path between Hanwell and Southall, 1945, *D.* and *J. M. B.*

King!; a large patch extending for about four yards, 1946–53!, *K. & L.*, 198; persisted and increased until 1969, despite the fact that it was cut down annually. Eradicated c. 1970. The species was not known to set fruit in the locality and the spread was entirely by vegetative means.

DATURA L.

D. stramonium L. Thorn-apple.

Alien. S. America? Occasionally naturalised on waste ground, but usually merely casual on waste and cultivated ground, rubbish-tips, etc. Occurs also as a garden weed. Rather rare, and very uncertain in occurrence from year to year. Warm summers appear to be favourable to the species. The seeds of *D. stramonium*, like those of *Hyoscyamus niger*, appear to remain viable below the surface of the soil for many years.

First certain record: *Francis*, 1837.

1. Uxbridge, 1839, *Kingsley* (ME); 1886–87, *Benbow MSS.* Harefield, 1902, *Green* (SLBI). New Years Green, 1954, 1960 and 1964, *T. G. Collett*!; 1966, Pinner, *Hind, T. & D.*, 195. Ruislip, 1949, *Graham, K. & L.*, 200. Yiewsley!, 1964, *Wurzell*; 1966. **2.** West Drayton!, *Druce Notes.* Near Staines, *Benbow MSS.* Hampton Court, 1940, *P. H. Cooke* (LNHS). East Bedfont, 1960, *T. G. Collett*!. Harmondsworth, 1967, *Kennedy*!. **3.** Twickenham!, *T. & D.*, 195; 1938, *Gerrans* (BM); 1948!, *L.N.* 28: 33; 1953, *Milne-Redhead.* Hounslow Heath, 1947–50!, *K. & L.*, 200; 1960, *Philcox* and *Townsend*; 1964. Northolt, established on waste ground!,† *K. & L.*, 200. Isleworth, 1966. Feltham, 1952, *Russell* and *Welch*!, *K. & L.*, 200. **4.** Harrow, 1947, *Reid, Horton Rep.* Mill Hill, *Harrison.* **5.** Chiswick, *Fox, T. & D.*, 196; 1958, *Murray List.* Acton, 1867, *Newbould* (SY). Brentford, 1960. Hanwell!; Greenford!; West Ealing, 1953!, *K. & L.*, 200; 1956 and 1963. Wormwood Scrubs, 1967, *L. M. P. Small*; 1968, *McLean.* **6.** Church End, Finchley, 1925–27, *J. E. Cooper, K. & L.*, 200. Weed in grounds of Myddleton House, Enfield, 1966, *Kennedy*!. Southgate, 1973, *Kennedy.* **7.** Walham Green, *Francis, Coop. Suppl.*, 12. Chelsea!, *Britten, Bot. Chron.*, 58; c. 1869, *T. & D.*, 196; 1942–43, *H. R. Davies, L.N.* 24: 12; 1963. Notting Hill, *A. B. Cole*; Kilburn, *Warren*; Pimlico; Green Park; South Kensington; East Heath, Hampstead; Victoria Street, S.W.1, *T. & D.*, 196; near Victoria railway station, 1951!; Ranelagh Gardens!, *L.N.* 39: 54. Isle of Dogs, 1866, *Newbould*; St John's Wood, *C. Andrews*; Upper Holloway, 1878, *French* (BM). Highbury, 1869, *W. G. Smith, Trim. MSS.* Hackney Marshes!, *M. & S.*; 1946!, *K. & L.*, 200. Bombed site, Cripplegate area, E.C., *Jones Fl.* Hornsey, 1947, *Scholey*; Hampstead Heath railway station, 1941, *J. Bedford, K. & L.*, 200. Hammersmith, 1961, *M. McC.*

Webster. Railway bank, White City, 1947!, *L.N.* 27: 33; 1960. Stamford Hill, 1963, *Wurzell*. Near Maida Vale, 1964. Tower Hill, 1965, *Brewis*, *L.N.* 46: 32. Churchyard of St George's-in-the-East, Shadwell, frequent, 1966.

Forma **tatula** (L.) A. Blytt
var. *tatula* (L.) Torrey, *Datura tatula* L.
1. Yiewsley, *Coop. Cas.* **2.** West Drayton, *Druce Notes*. Hampton Wick, *Country-side* 1919, 162. Hampton Court, 1922, *P. H. Cooke*. **5.** Hanwell!; Greenford!, *L.N.* 26: 63. **6.** Weed in grounds of Myddleton House, Enfield, 1966, *Kennedy*!. **7.** Hornsey, 1947, *Scholey*, *K. & L.*, 200. Hyde Park, 1961, *D. E. Allen*, *L.N.* 41: 20.

SCROPHULARIACEAE

VERBASCUM L.

V. thapsus L. Aaron's Rod, Great Mullein, High Taper.
Native in open woods, on banks, waste ground, heaths, etc., on dry soils, also established on wall-tops, railway banks, etc. Formerly widespread and common, but now local on calcareous soils in the north-west of the vice-county and on alluvial soils near the Thames; rather rare elsewhere.
First record: *Gerarde*, 1597.
1. Locally common. **2.** Locally common near the Thames. Ashford, *Hasler*. East Bedfont. West Bedfont, *Kennedy*!. West Drayton. **3.** Twickenham, one plant; Hounslow Heath!, abundant, *T. & D.*, 197. Isleworth. Railway banks, Southall. **4.** Hampstead!, *Irv. MSS.* Harrow!, *Hind Fl.* Stanmore, *T. & D.*, 197. Golders Hill, *Whiting Notes*. Scratch Wood, and district. Mill Hill. Stonebridge. Northwick Park, *Harley*. Wealdstone. Railway banks, Cricklewood and Willesden. **5.** Chiswick!, *T. & D.*, 197. Hanwell, 1937. Ealing. Acton. Brentford. **6.** Highgate, *Ger. Hb.*, 630. Edmonton, a single plant, *T. & D.*, 197. Enfield!, *H. & F.*, 150. **7.** Parsons Green, 1856,† *A. Irvine*, *Phyt. N.S.* 2: 168. Fulham, one plant; South Kensington,† *T. & D.*, 197. Regents Park, *Webst. Reg.*, 10; 1955, *Holme*, *L.N.* 39: 54. Bombed sites, Cripplegate, *Wrighton*, *L.N.* 29: 86; 1952–53;† Gresham Street area,† etc., *Jones Fl.* Buckingham Palace grounds, *Fl. B.P.G.* Upper Clapton; by canal between Lisson Grove and Regents Park, 1969, *L. M. P. Small*.

V. phlomoides L.
Alien. Europe, etc. Garden escape. Naturalised in grassy places, on waste ground, etc. Rare, and apparently decreasing.
First record: *Payton*, 1941.

1. New Years Green, 1960, *T. G. Collett*!; 1961→. Yiewsley, 1965. **2.** Shortwood Common, two plants, 1947!;† near Penton Hook Lock, 1947!,† *L.N.* 27: 33. West Drayton, 1951. Hampton Court, 1954. East Bedfont, 1965. **3.** Near Sudbury Hill, 1941!, *Payton*; 1942–44.† Near the canal, Northolt, abundant!, *L.N.* 27: 33; still present, but reduced in quantity, 1967. Hounslow Heath, 1947!, *L.N.* 27: 33; 1964. **4.** Stonebridge!, *K. & L.*, 201. Harrow, 1968. **5.** Hanwell, 1944!, *Welch*, *K. & L.*, 201; 1945–62.† Chiswick, 1945!,† *K. & L.*, 201. Perivale Wood, 1950. Shepherds Bush, 1966. **6.** Near Enfield Lock, 1966, *Kennedy*!. **7.** Bombed site, Fore Street, E.C., 1945,† *Lousley*, *K. & L.*, 201. Bombed site near St Paul's Cathedral, 1966. Hyde Park!; Kensington Gardens!, *D. E. Allen* and *McClintock*, *L.N.* 39: 54.

Var. **albiflora** (Rouy) Wilmott
3. Northolt!, *L.N.* 27: 33; 1966. **4.** Harrow, 1968. **5.** Hanwell!,† *K. & L.*, 201.

V. nigrum L. Dark Mullein.
Native in dry fields, on banks, by roadsides, etc., on calcareous and other dry soils in the north-west of the vice-county, and on alluvial soils near the Thames, elsewhere on waste ground, railway banks, etc., as an introduction. Rare, and decreasing.
First record: *Blackstone*, 1737.
1. Harefield!, *Blackst. Fasc.*, 105. Drayton Ford!; Garett Wood!;† gravel pit between Harefield and Ickenham!, *K. & L.*, 202. Down above Springwell Lock, a few plants, 1958, *Pickess*. Railway banks, Ruislip Gardens station, abundant, 1957!, *I. G. Johnson*; 1967. **2.** Between Hampton and Sunbury, *Wats. New Bot.*, 100. By the Thames under the wall of Hampton Court gardens!, *Wats. New Bot. Suppl.*; by the towing-path towards Kingston Bridge!, *T. & D.*, 197. Ashford, *Hasler*, *K. & L.*, 202. Teddington!, 1902, *Green*, *Williams MSS.*; Teddington Churchyard, a small colony, 1966. Between Teddington and Hampton Wick, two plants, 1965. Longford, *L. M. P. Small*. **3.** In a meadow by Richmond Bridge,† *M. & G.*, 299. Twickenham, one plant,† *T. & D.*, 197. Railway banks, South Ruislip station in small quantity, 1966→. **4.** Hampstead,† *Mart. Mill. Dict.* Near Hampstead,† *Cund. Guide*, 99. Railway bank, Harrow, a few plants, 1958.† **5.** Strand-on-the-Green,† *Mart. Mill. Dict.* Near Norwood, a single plant, 1886;† Chiswick, a single plant,† *Benbow*, *J.B.* (*London*) 25: 364. Near Alperton, 1917–21,† *L. M. P. Small*, *K. & L.*, 200. Syon Park, a single plant, 1955, *Sandwith*; several plants, 1966; near the canal, Brentford, a single plant, 1965!, *Boniface*; 1966→. **7.** Regents Park,† *Webst. Reg.*, 102. Bombed sites, Cripplegate, in two places, 1952,† *A. W. Jones*, *K. & L.*, 202. Fulham Palace grounds, 1966, *Brewis*.

V. blattaria L. Moth Mullein.
Alien. Europe, etc. Naturalised in grassy places, on waste ground, etc., established in gravel pits, etc. Very rare, and sometimes merely casual.
First record: *Doody*, c. 1690.
1. Harefield, 1735, *Blackst. Fasc.*, 11. Pinner, 1867,† *T. & D.*, 197. Disused railway, Uxbridge, 1968, *L. M. P. Small.* **2.** Fields near Hampton Court, *Doody MSS.* Teddington, *T. & D.*, 198. Gravel pit, East Bedfont, *Welch* and *J. G. Dony*!, *K. & L.*, 202; 1966, *Wurzell.* **3.** Northolt, 1947!, *L.N.* 27: 33. **4.** Hampstead Heath,† *M. & G.*, 301. **5.** Chiswick, 1945!, *Lousley, L.N.* 26: 63. **7.** East Heath, Hampstead,† *Irv. Lond. Fl.*, 128. South Kensington,† *T. & D.*, 198.

V. virgatum Stokes Twiggy, or Large-flowered Mullein.
Introduced. Naturalised in grassy places, on waste ground, etc. Very rare, and sometimes merely casual.
First record: *A. Irvine*, 1858.
1. Pinner,† *Melv. Fl.* 2: 76. Near Cowley Peachey, 1885,† *Benbow* (BM). Near Yiewsley, 1963. **3.** Hounslow, 1948–50; Hounslow Heath, 1948–53!, *Westrup, K. & L.*, 201; 1961. **4.** Golders Green, 1921,† *Richards, K. & L.*, 202. **7.** Chelsea College,† *Irv. Ill. Handb.*, 460. Near Hyde Park Corner, 1962!,† *Brewis, L.N.* 42: 12.

MISOPATES Raf.

M. orontium (L.) Raf. Weasel's Snout, Lesser Snapdragon.
Antirrhinum orontium L.
Colonist on cultivated and waste ground, etc. Very rare, and now confined to light soils in the south-west parts of the vice-county where it occurs in very small quantity.
First record: *Varenne*, 1827.
2. Near Strawberry Hill, plentifully, 1867, *Dyer* (BM). Teddington, *T. & D.*, 199; grounds of the National Physical Laboratory, 1949, *D. P. Young, K. & L.*, 205; Teddington Cemetery, a single plant, 1965, *Clement.* Tangley Park, a single plant,† *T. & D.*, 199. Between Shepperton and Walton Bridge, *Benbow, J.B. (London)* 25: 17. Stanwell, 1913, *Coop. Cas.* Charlton, *Benbow* (BM). Grounds of Ashford County Hospital, *Hasler, L.N.* 26: 76. **3.** Twickenham, one plant, *T. & D.*, 199; c. 1885,† *Benbow MSS.* Near Feltham, abundant, 1886,† *Benbow* (BM). **5.** Perivale,† *Brown Chron.*, 137. Perhaps an error. Ealing, one plant, 1971, *McLean.* Canal path, Hanwell, two plants, 1971, *Gilbert.* **7.** Marylebone Infirmary garden, 1827,† *Varenne, T. & D.*, 199. Chelsea College, 1861,† *Britten, Phyt. N.S.* 6: 340.

ANTIRRHINUM L.

A. majus L. Snapdragon.

Alien. S. Europe, etc. Garden escape. Established on old walls, in chalk pits, etc. Common.

First certain evidence: *Goodger* and *Rozea*, 1815.

7. Hampstead!, 1815 (G & R); 1968. Wall at Eel Brook Meadow, *T. & D.*, 198. Common on City of London bombed sites!, *Lousley Fl.* Bombed site, Snow Hill, W.C., 1950,† *Bangerter* and *Whittaker* (BM). Hyde Park, 1959, *D. E. Allen* and *McClintock*; 1961, *Brewis*, *L.N.* 41: 10. Waste ground near Euston station, 1965!, *L.N.* 46: 32. Holloway, 1965, *D. E. G. Irvine*. St Katherine's Dock, Stepney, *Brewis*. Hammersmith. Notting Hill. Kilburn. St John's Wood. Highgate. Finsbury Park. Hornsey. Hoxton. Dalston. Stoke Newington. South Tottenham. Bethnal Green. Upper Clapton. Tottenham. Bromley by Bow. Old Ford. Blackwall.

LINARIA Mill.

L. purpurea (L.) Mill. Purple Toadflax.

Alien. S. Europe, etc. Garden escape. Naturalised on waste ground, established on old walls, etc. Common.

First record: *A. Irvine*, c. 1830.

7. Frequent on City of London bombed sites!, and likely to thrive, *Lousley Fl.* Kensington Gardens, one plant, 1961, *D. E. Allen*, *L.N.* 41: 20. Fulham Palace grounds, *Welch*!. Earls Court. Brompton. St John's Wood. Hampstead. Clapton. Hackney. Shadwell. Bombed site car park, South Kensington, 1969, *E. Young*.

L. repens (L.) Mill. Pale Toadflax.

Introduced. Naturalised on waste ground, by roadsides, etc., established on railway tracks, etc. Very rare.

First record: *G. C. Druce*, 1902.

1. Rubbish-tip, New Years Green, 1961. Disused railway, Uxbridge, 1968, *L. M. P. Small*. **2.** Railway tracks near West Drayton, 1951!, *K. & L.*, 203; 1964. Staines, 1956, *Briggs*. **3.** Hounslow, 1950,† *H. Banks*, *K. & L.*, 203. **4.** Roadside, Highwood Hill, 1957. Weed in grounds of North London Collegiate School, Edgware, 1967!, *Horder*. **5.** Acton goods yard, 1902,† *Druce Notes*. Railway sidings, Greenford Green, 1947!,† *K. & L.*, 203. Site of disused railway, Brentford, in quantity, 1966. **6.** Myddleton House grounds, Enfield, abundant weed of flower-beds, etc., 1966, *Kennedy*!.

L. vulgaris Mill. Yellow Toadflax.

Native on hedgebanks, waste ground, in fields, by roadsides, etc., established on railway banks and tracks, etc. Common.

First record: *T. Johnson*, 1638.

7. Common in the district.

Peloric forms have been recorded from **3.** Baber Bridge, Hounslow Heath, *Westrup*, *K. & L.*, 203. **4.** Near Highgate, '*E.K.*', *Mag. Nat. Hist.* 1: 379.

CHAENORHINUM (DC.) Reichb.

C. minus (L.) Lange Small Toadflax.

Linaria minor (L.) Desf.

Native on cultivated, waste, and disturbed ground on calcareous soils at Harefield and South Mimms, and on alluvial soils near the Thames, elsewhere on railway tracks, waste ground, etc., as an introduction. Formerly rare, but now rather common and widespread, especially along the railway systems of the vice-county which it has colonised during the last half century. It still remains rare in the eastern parts of Middlesex, but will doubtless eventually become common via the railway system.

First record: *T. Johnson*, 1638.

1. Harefield!, *Blackst. Fasc.*, 53. Pinner!, *Hind Fl.* Uxbridge!; Springwell!; Drayton Ford, *Benbow MSS.* Potters Bar!, *P. H. Cooke*; Uxbridge Moor, very abundant about the cement works!, *K. & L.*, 203. Disused railway, Uxbridge to Denham!, *K. & L.*, 204; 1968. South Mimms. **2.** Towing-path between Kingston Bridge and Hampton Court!, *Bloxam*; Stanwell Moor!, *T. & D.*, 200. Colnbrook!; Staines!, *Benbow*, *J.B.* (*London*) 25: 17. West Drayton!, *Druce Notes.* Railway tracks, Yeoveney!, *Welch*, *L.N.* 27: 33; 1967. Near Poyle, 1966, *J. E. Smith*!. **3.** Near Hounslow, 1868, *Newbould* (BM). Hounslow Heath, 1961, *Philcox.* Feltham. South Ruislip, *Wrighton*, *L.N.* 27: 33. **4.** Railway tracks, Wembley Park to Stanmore, plentiful. Disused railway between Stanmore and Belmont, *Roebuck.* Scratch Wood railway sidings. Disused railway, Mill Hill to Hendon, abundant, 1966, *Kennedy*!. Railway tracks, Harrow to Rayners Lane. **5.** Hanwell, *Sandwith*!, *K. & L.*, 204. Railway tracks, Southall to Brentford and Ealing to Greenford, abundant. Acton goods yard. Railway tracks, North Acton. **6.** Tottenham, *Johns. Cat.* Muswell Hill, 1900, *Coop. Cas.* Railway tracks, Hadley Wood and Enfield Town. Railway sidings, Crew's Hill, Enfield, *Kennedy*!. **7.** Downshire Hill, Hampstead, 1870,† *Dyer* (BM).

KICKXIA Dumort.

K. spuria (L.) Dumort. Round-leaved Fluellen.

Linaria spuria (L.) Mill.

Native on cultivated and disturbed ground, etc. Very rare, and mainly confined to light soils.

First record: *Gerarde*, 1597.
1. Harefield!, *Blackst. Fasc.*, 25. Pinner,† *Melv. Fl.*, 57. Springwell, 1884;† Uxbridge, 1887,† *Benbow* (BM). Swakeleys, very abundant;† Ickenham;† South Mimms!, *Benbow MSS.*; field by Mimmshall Wood, 1966, *J. G.* and *C. M. Dony*!. Cowley, 1946!,† *K. & L.*, 204. **2.** Between Hampton and Sunbury, in garden ground,† *Newbould, T. & D.*, 200. Stanwell Moor!, abundant; Staines, *Benbow, J.B. (London)* 25: 17. Yeoveney;† Shepperton!; Walton Bridge,† *Benbow MSS.* Between West Drayton and Colnbrook,† *Druce Fl.*, 248. **3.** Near Roxeth,† *Melv. Fl.*, 57. **4.** Stanmore, 1817† (G & R). Sudbury;† Willesden Green,† *Farrar, T. & D.*, 200. **5.** Chiswick,† *Ger. Hb.*, 501. Greenford,† *Hind Fl.* Between Ealing and Hanwell, c. 1820,† *Sneyd, Trim. MSS.* **7.** Hyde Park, 1962!,† *D. E. Allen* and *McClintock, Allen Fl.* Introduced with soil brought from elsewhere.

K. elatine (L.) Dumort. Sharp-leaved Fluellen.
Linaria elatine (L.) Mill.
Native in similar habitats to the preceding species with which it sometimes grows. Very rare, and mainly confined to light soils.
First record: *Gerarde*, 1597.
1. Harefield!, *Blackst. Fasc.*, 25. Pinner,† *Melv. Fl.*, 57. Uxbridge;† Ickenham;† Northwood,† *Benbow, J.B. (London)* 25: 17. Ruislip Common, 1908,† *Green*; between New Years Green and Harefield, 1922. *Tremayne, K. & L.*, 204. New Years Green, 1966, *J. Ruisl. & Distr. N.H.S.* 15: 15. Field by Mimmshall Wood, 1966, *J. G.* and *C. M. Dony*!. **2.** Between Hampton and Sunbury, in garden ground,† *Newbould*; cornfields, Tangley Park,† *T. & D.*, 200. West Drayton;† Staines, *Benbow, J.B. (London)* 25: 17. Between West Drayton and Colnbrook,† *Druce Fl.*, 248. **4.** Stanmore (G & R); disused railway between Stanmore and Belmont, 1968, *Roebuck.* Hampstead,† *Varenne, T. & D.*, 200. Near Scratch Wood, 1959, *Knipe, L.N.* 39: 40; 1960, *Warmington.* **5.** Chiswick,† *Ger. Hb.*, 501. Between Hammersmith and Turnham Green, 1826† (HE). **6.** Tottenham,† *Johns. Cat.* **7.** Kensington Gardens, 1962!,† *Allen Fl.* Buckingham Palace grounds, 1963, *Fl. B.P.G.* Chelsea, 1968, *E. Young.* Probably introduced with soil brought from elsewhere in all three localities.

CYMBALARIA Hill

C. muralis Gaertn., Mey. & Scherb. Ivy-leaved Toadflax.
Linaria cymbalaria (L.) Mill.
Alien. S. Europe, etc. Naturalised on bare waste ground, established on old walls and other brickwork, etc. Common.
First record: *Dillenius*, 1724.

7. Abundantly on the walls of Chelsea Garden, and in neighbouring places!, *Ray Syn.* 3, *282. Frequent about London, *Huds. Fl. Angl.*, 271. Walls of the Thames, *Mart. Pl. Cant.*, 65. On the Temple wall,† *Curt. F.L.* Somerset House, 1802,† *Winch MSS.* On Battersea Bridge,† *J. Sowerby*; Blackwall, 1836, *J. F. Young* (BM). Kentish Town, 1841 (HE). Ken Wood!; Haverstock Hill; Eel Brook Meadow, *T. & D.*, 199. Islington, 1839, *E. Ballard* (BM). Stoke Newington, 1871, *F. J. Hanbury* (HY). Bombed site, Ebury Street, S.W.1,† *McClintock.* Bombed site, Bedford Way, W.C., 1950,† *Whittaker* (BM). Fulham Palace grounds, abundant, *Welch*!. Bedford Row by Theobalds Road, W.C.1, 1960, *Knipe, L.N.* 41: 20. Marylebone Road, N.W.1!, 1966, *Brewis, L.N.* 46: 32; 1967. St John's Wood. Hammersmith. Brompton. Earls Court. Hampstead. Hornsey. Clapton. Tottenham. Grosvenor Road, S.W.1, 1967. Dean's Yard, Westminster, *Gush, L.N.* 47: 9. Highbury, Dalston.

Plants with pure white flowers are very rare but have been recorded from **2.** Sunbury, 1947. **5.** St George's Cemetery, Hanwell, 1966. **6.** Botany Bay, Enfield Chase, 1965, *Kennedy*!. **7.** Chelsea Botanic Garden wall, 1834, *Pamplin, Trim. MSS.*

SCROPHULARIA L.

Lit. *Proc. Bot. Soc. Brit. Isles* **7**: 433 (1968). (*Abstr.*)

S. nodosa L. Knotted Figwort.

Native in damp woods, ditches, on hedgebanks, etc. Common, except in heavily built-up areas, where owing to the eradication of its habitats it is very rare or extinct.

First record: *Gerarde*, 1597.

7. In the greatest abundance in a wood as you go from London to Hornsey (=Hornsey Wood),† *Ger. Hb.*, 580. Kensal Green,† *A. B. Cole*; East Heath, Hampstead!, *T. & D.*, 201. Ken Wood!, *H. C. Harris.* Buckingham Palace grounds, one plant, 1962, *Fl. B.P.G.* Canal side near Paddington Basin, a small colony, 1966!, *L.N.* 46: 32; 1967. Kilburn. Highgate.

S. auriculata L. Water Figwort, Water Betony.

S. aquatica L. *pro parte*

Native by river- and streamsides, in ditches, wet woods, etc., established by canals, gravel pits, etc. Common, except in heavily built-up areas where it is rare or extinct.

First record: *Blackstone*, 1737.

7. . . . between Hornsey Wood . . . . and Highgate,† *Ball. MSS.* Lee near Hackney; Isle of Dogs, *T. & D.*, 202. Lee Navigation Canal, Upper Clapton and Tottenham. Hyde Park, a solitary large plant, in an ornamental bowl (since demolished), near the Dorchester Hotel, 1960,† *D. E. Allen, L.N.* 41: 20.

S. umbrosa Dumort.
S. alata Gilib., *S. aquatica* L. *pro parte*
Formerly native in ditches, but now long extinct.
First, and only evidence: *J. de C. Sowerby*, 1841.
7. Belsize Park, 1841, *J. de C. Sowerby* (BM).

MIMULUS L.

M. guttatus DC. Monkey-flower.
M. luteus auct. angl. *pro parte*, non L., *M. langsdorffii* Donn ex Greene
Alien. N. America. Garden escape. Naturalised by rivers, streams, lakes and ponds, in ditches, etc., and established by gravel pits, etc. Very rare and confined to the valleys of the Colne and Lee.
First evidence: *Green*, 1902.
1. Uxbridge,† *L. B. Hall, K. & L.*, 206. Wall of canal, Harefield!, *Welch, L.N.* 26: 63. Colne near Harefield and Cowley Peachey, 1955. **2.** Staines, 1902, *Green* (SLBI); 1908;† West Drayton!,† *P. H. Cooke, K. & L.*, 206. **4.** By stream and on damp ground, Bentley Priory, 1970, *Paterson*. **6.** Ponders End!, *J. E. Cooper* (BM). Ditches by the Lee Navigation Canal, and in adjacent gravel pits near Waltham Cross. Gravel pit north of Enfield Lock, 1966, *Kennedy*!; 1967.

M. moschatus Dougl. ex Lindl. Musk.
Alien. N. America. Garden escape. Naturalised by water, and established as a weed in flower-beds, etc. Very rare.
First record: *D. H. Kent*, 1948.
6. Weed of flower-beds, Myddleton House grounds, Enfield, 1966, *Kennedy*!. **7.** The Cascade, Ken Wood, 1948→. Weed of flower-bed, Buckingham Palace grounds, *Fl. B.P.G.* Weed of flower-borders, Kensington Gardens, *Allen Fl.* Weed in derelict greenhouses, Fulham Palace grounds, 1966, *Brewis*.

LIMOSELLA L.

L. aquatica L. Mudwort.
Formerly native on pond verges, in muddy places, etc., but now extinct.
First record: *Merrett*, 1666. Last evidence: *French*, 1875.
1. In the bogs on Harefield Common, *Blackstone* (BM). By the side of the Warren Pond near Breakspeares, plentifully, *Blackst. Fasc.*, 7. **3.** Hounslow Heath, *Merr. Pin.*, 95.; towards Hampton, *Doody*, *Ray Syn.* 2: 344; 1846, *J. Morris, T. & D.*, 203. **4.** Finchley (BO). Between Hampstead and Ken Wood, *Irv. MSS.* **5.** Turnham Green, 1798, *R. Brown, Trim. MSS.* Pond on Acton Green, 1867, *Newbould* (BM). **6.** Between Southgate and Hornsey, *Alch. MSS.* Hadley Common, towards Barnet, 1875, *French* (BM). **7.** Near Hornsey, *Pet. Midd.*

DIGITALIS L.

D. purpurea L. Foxglove.

Native in open woods, on heaths, etc., in the northern parts of the vice-county, elsewhere in woods, thickets, on banks, etc., as an escape from cultivation. Local.

First record: *T. Johnson*, 1629.

1. Locally common. **2.** Bushy Park!, *Newbould*; Tangley Park, abundant,† *T. & D.*, 198. Littleton, *Benbow MSS.* Ashford, *Hasler.* Hampton, *Tyrrell*!. Shepperton. Halliford. Staines. Sunbury. **3.** Near Harlington, a single plant;† Whitton Park Inclosure,† *Benbow MSS.* Hounslow Heath. Between Hounslow Heath and East Bedfont. **4.** Locally common. **5.** Perivale Wood!, introduced, *W. M. Webb.* Syon Park. Railway bank, Ealing. Hanwell. Acton. Chiswick House grounds!, *Murray*. **6.** Colney Hatch, *Coop. Fl.*, 107. Trent Park!, *A. B. Cole*; Hadley!, *Warren*; Enfield!, *T. & D.*, 198. Barnet Gate Wood; Enfield Chase!, *Benbow MSS.* Gravel pit north of Enfield Lock, *Kennedy*!. Highgate Woods. North Finchley. Whetstone. **7.** Near Parsons Green, 1862,† *T. & D.*, 198. Regents Park!, *Webst. Reg.*, 102. Ken Wood!; East Heath, Hampstead, *Fitter.* Hyde Park, 1961, *Brewis, L.N.* 41: 20. Bombed site, Cripplegate, *Wrighton, L.N.* 29: 88; in some abundance, *Jones Fl.* Holland House grounds, Kensington. Fulham Palace grounds, *Brewis.* Dorset Square, N.W.1, planted, but regenerating. Canal side between Marylebone and Primrose Hill, frequent.

VERONICA L.

V. beccabunga L. Brooklime.

Native in brooks, shallow streams, marshes, wet meadows, etc. Formerly common, but now local and decreasing.

First record: *T. Johnson*, 1638.

1. Harefield!, *Blackst. Fasc.*, 106. Uxbridge,† *Benbow* (BM). Drayton Ford. Springwell. Cowley. Ruislip. Stanmore Common, frequent. Mimmshall Wood. Potters Bar. Near Potters Bar, *Kennedy*!. **2.** Staines!, *Whale Fl.*, 31. Penton Hook. Hampton Court. Near Harlington,† *P. H. Cooke.* **3.** Near Hounslow Heath. Hayes Park, *Royle*!. **4.** Hampstead Heath, 1848, *T. Moore* (BM); c. 1912,† *Whiting Fl.* Hampstead;† Hendon, *P. H. Cooke* (LNHS). Mill Hill, *Sennitt.* Between Mill Hill and Totteridge, *Lousley*!. Harrow!, *Hind Fl.* Northwick Park,† *Harley* (BM). Bentley Priory, abundant. Edgware, *Horder.* **5.** Between Hanwell and Southall. Syon Park, *Welch*!. **6.** Tottenham,† *Johns. Cat.* Hadley Common. Southgate!. Highgate Woods, *C. S. Nicholson, Lond. Nat. MSS.* Trent Park!, *T. G. Collett* and *Kennedy*. Finchley Common. **7.** Stoke Newington,† *E. Ballard*; Green Lanes,† *Sewell* (BM). Near Bayswater;† Hampstead Lane† (G & R). East Heath, Hampstead, *T. & D.*, 206. Parliament Hill Fields!; Highgate Ponds!, *Fitter*. Crouch End,† *C. S. Nicholson* (LNHS). Tothill Fields, Westminster,† *Dill. MSS.*

V. anagallis-aquatica L. Water Speedwell.
Lit. *Watsonia* **1**: 349 (1950).
Native in ponds, streams, marshes, wet meadows, etc. Very rare.
First certain evidence: *H. C. Watson*, 1838.
1. Springwell, 1947, *Welch*; Uxbridge Moor, 1948!. Both teste *J. H. Burnett, K. & L.*, 208. Harefield, *Pickess*. **2.** Hampton Court, 1838, *H. C. Watson* (BM), teste *J. H. Burnett*. Poyle. Colnbrook. **4.** Hampstead Heath, 1839,† *E. Ballard* (BM), teste *J. H. Burnett*. **6.** River Lee, Edmonton, *F. W. Payne* (BM), teste *J. H. Burnett*. **7.** Tottenham, 1840,† *R. Pryor* (BM), teste *J. H. Burnett*.
All published records of the aggregate *V. anagallis-aquatica* are omitted unless vouched for by specimens.

V. catenata Pennell Water Speedwell.
V. aquatica Bernh., non S. F. Gray, *V. comosa* auct.
Lit. *Watsonia* **1**: 349 (1950).
Native in similar situations to the previous species with which it sometimes grows. Rare.
First certain evidence: *Warren*, 1870.
2. Shortwood Common!; Staines Moor!; Stanwell Moor!. All teste *J. H. Burnett, K. & L.*, 208. Near Penton Hook, 1956, *Welch*. Ponds and marshy fields between Chertsey Bridge and Shepperton Lock, abundant, 1966, *Kennedy*!. Poyle, 1958, *Pickess*. Bushy Park, 1963. Between West Drayton and Colnbrook, *Benbow* (BM), teste *J. H. Burnett*. **3.** River wall of the Thames between Isleworth and Twickenham, locally abundant, 1961→. **6.** Gravel pit, Enfield Lock, 1959, *A. Vaughan*. **7.** Shepherds Bush, 1870,† *Warren* (BM), teste *J. H. Burnett*.

V. scutellata L. Marsh Speedwell.
Native in bogs, marshes, on pond verges, wet heathy places, etc. Very rare.
First record: *Merrett*, 1666.
1. Harefield!, *Blackst. Fasc.*, 107. Ruislip Reservoir!; Harrow Weald Common,† *Melv. Fl.*, 54. Elstree Reservoir;† Stanmore Common!, *T. & D.*, 205. Pond on Pinner Hill;† pond on Duck's Hill Farm, Northwood;† Ickenham Green;† near Swakeleys,† *Benbow MSS.* Uxbridge!,† *Benbow* (BM). **2.** Shortwood Common!; Staines Moor!, *K. & L.*, 207. **3.** Hounslow Heath,† *T. & D.*, 205. **4.** Hampstead Heath (DILL); c. 1760, *Huds. Fl. Angl.*, 5; 1839, *E. Ballard* (BM); c. 1860,† *Britten, T. & D.*, 205. Stanmore Marsh (=Whitchurch Common), *T. & D.*, 205; c. 1910,† *P. H. Cooke, K.*

& L., 207. **6.** Finchley Common, 1841 (HE); 1901, *C. S. Nicholson*; 1917–20,† *J. E. Cooper, K. & L.*, 207. Hadley,† *G. Johnson, T. & D.*, 205. Near Potters Bar, *Benbow MSS.* **7.** Tothill Fields, Westminster,† *Merr. Pin.*, 7. Eel Brook Common,† *Pamplin, Trim. MSS.*

V. officinalis L. Common Speedwell.
Native on heaths, in dry open woods, dry grassy places, etc. Local.
First record: *T. Johnson*, 1629.
1. Frequent. **2.** Abundant by roadsides about Charlton, towards Staines and Hampton, *T. & D.*, 207. Golf course, Hampton Court Park. **3.** Hounslow Heath!, *T. & D.*, 207. **4.** Hampstead Heath, *Johns. Enum.*; c. 1869,† *T. & D.*, 207. Scratch Wood!; Mill Hill!; Barnet Gate!, *Benbow MSS.* Deacon's Hill!, *T. & D.*, 207. **5.** Horsendon Hill!, *T. & D.*, 207. **6.** Hadley!, *Warren*; near Enfield!, *R. Tucker*; Edmonton, *T. & D.*, 207. Winchmore Hill Wood;† Enfield Chase!, *Benbow MSS.* Southgate, *E. M. Dale, Lond. Nat. MSS.* **7.** Hammersmith, 1845,† *J. Morris*; East Heath, Hampstead,† *T. & D.*, 207.

V. montana L. Wood Speedwell.
Native in damp woods, on shady hedgebanks, etc. Locally plentiful in the north of the vice-county, rare elsewhere, and very rare in the extreme south-west.
First record: *Merrett*, 1666.
1. Locally common. **2.** Spout Wood, Stanwell, in small quantity, 1966, *J. E. Smith*!. **3.** Harrow Grove,† *Melv. Fl.*, 56. **4.** Hampstead Heath, *Merr. Pin.*, 6; c. 1700, *Dodsworth* (SLO); c. 1793,† *M. & G.*, 26. Bishop's Wood, plentiful, *Irv. MSS.* Harrow Park, abundant (ME); Grove Park and Vicarage garden, Harrow, *Harley Fl.*, 23. Scratch Wood, *Welch*!, *L.N.* 28: 33. **5.** Brentford,† *M'Creight* (BM). Hanwell,† *Druce Notes.* Wyke Lane, Osterley!, *Montg. MSS.*; Long Wood, Wyke Green, 1966→. **6.** Between Highgate and Hornsey,† *Newt. MSS.* Highgate Woods, 1896, *J. E. Cooper* (BM); 1902, *L. B. Hall*; Whitewebbs Park, Enfield!, *Johns, K. & L.*, 306; 1965. **7.** Belsize Park,† *M. & G.*, 26. Near West End, Hampstead, 1869,† *Warren* (BM).

V. chamaedrys L. Germander Speedwell, Bird's-eye.
Native in meadows, woods, on hedgebanks, cultivated and waste ground, etc. Common.
First certain record: *Blackstone*, 1737.
7. Common in the district.

V. longifolia L.
Alien. Europe, etc. Garden escape. Naturalised in grassy places, on waste ground, by roadsides, etc. Rather rare, but increasing.
First record: *D. H. Kent*, 1946.

1. Harrow Weald Common, 1965, *Kennedy*!. **2.** Ashford golf course, 1966. **3.** Hounslow Heath, 1960. **4.** Neasden, 1965→. Stonebridge, 1965→. Wembley Park, 1966. **5.** Hanwell, 1953!, *K. & L.*, 209. Perivale, 1965→. Park Royal, 1965. Acton, 1965. Ealing, 1966. South Ealing, 1966. **6.** Near Finchley, 1946!, *K. & L.*, 209. **7.** Haggerston, 1966.

Some of the above records may be referable to the closely allied species *V. spuria* L.

V. serpyllifolia L. subsp. **serpyllifolia** Thyme-leaved Speedwell.

Native in grassy places, on heaths, banks, etc., also as a weed of lawns. Common.

First record: *W. Turner*, 1548.

7. East Heath, Hampstead!, *T. & D.*, 207. Stoke Newington, *Ball. MSS.* Buckingham Palace grounds, *Fl. B.P.G.* Tottenham.

V. arvensis L. Wall Speedwell.

Native on cultivated ground, heaths, in grassy places, by roadsides, established on wall-tops, etc. Common.

First record: *T. Johnson*, 1638.

7. Tottenham, *Johns. Cat.* Near Vauxhall Bridge, 1817 (G & R). Hammersmith, 1845, *J. Morris*; Haverstock Hill; Parsons Green; Kensington Gore, *T. & D.*, 207–208. Weed in open frame, Hyde Park, 1961, *D. E. Allen*, *L.N.* 41: 20. Buckingham Palace grounds, *Fl. B.P.G.* Weed of flower beds, Zoological Gardens, Regents Park, 1966, *Brewis*, *L.N.* 46: 33. Brompton Cemetery. St John's Wood.

Var. **nana** Poir.

var. *eximia* Towns.

1. Harefield. **3.** Hounslow Heath, *Welch*!. **4.** Scratch Wood. **5.** Syon Park.

A dwarf early flowering plant which is sometimes confused with *V. verna* L.

Var. **polyanthos** Murr.

2. Laleham.

V. hederifolia L. *sensu lato* Ivy-leaved Speedwell.

Lit. *Proc. Bot. Soc. Brit. Isles* **7**: 435 (1968). (*Abstr.*)

Native on cultivated ground, hedgebanks, in fields, by roadsides, etc., also as a frequent weed of gardens, flower-beds and shrubberies in parks, etc. Common.

First record: *T. Johnson*, 1632.

7. Common in the district.

Fischer (1967, *Österr. Bot. Zeitschr.* **114**: 189) divides *V. hederifolia* into two species—**V. hederifolia** L. *sensu stricto*—10–60 cm. tall; leaves fleshy, with rather acute teeth; corolla 6–9 mm., bright blue, streaked with violet; pedicels 2–4 times as long as calyx. Usually confined to shady places and arable land. The other species **V. sublobata** M. Fischer, sp. nov. differs as follows—5–40 cm. tall; leaves thin in texture with 5–7 (9) small lobes; corolla 4–6 mm., pale purple-lilac to white; pedicels 3·5–7 times as long as calyx. All Middlesex material seen appears to come under the latter segregate. More recently some authors have referred *V. sublobata* to **V. hederifolia** subsp. **lucorum** Klett & Richter.

V. persica Poir. Buxbaum's Speedwell.
V. buxbaumii Ten., non Schmidt, *V. tournefortii* auct.
Alien. Europe, etc. Naturalised on cultivated and waste ground, by roadsides, etc., and as a garden weed. Common.
First evidence: *Balfour*, 1841.
7. Frequent in the district.

Var. **corrensiana** (Lehm.) C. E. Britton
3. Hayes, *Loydell*, teste *G. C. Druce, Rep. Bot. Soc. & E.C.* 8: 750.

Var. **aschersoniana** (Lehm.) C. E. Britton
3. Hayes, 1897, *Loydell*, teste *G. C. Druce, Rep. Bot. Soc. & E.C.* 8: 314.

V. polita Fr. Grey Speedwell.
V. didyma auct.
Lit. *J.B. (London)* **69**: 180 and 201 (1931).
Native on cultivated and waste ground, and as a weed of flower-beds in parks and gardens. Formerly local, but now common.
First certain evidence: *Twining*, 1844.
7. Regents Park, 1847, *T. Moore* (BM), conf. *Bangerter*. Camden Town; Parsons Green; South Kensington, *T. & D.*, 208. Hyde Park, one plant, 1961, *Allen Fl.* Kensal Green. Paddington Green!, *L.N.* 46: 33.

Var. **thellungiana** (Lehm.) Hayek & Hegi
3. Twickenham, 1844, *Twining* (BM).

V. agrestis L. Field Speedwell.
Lit. *J.B. (London)* **69**: 180 and 201 (1931).
Native on cultivated and waste ground, and as a weed of flower-beds in parks and gardens. Locally common, except in south-west Middlesex where it is rare.
First record: *Blackstone*, 1737.
1. Locally frequent. **2.** Hampton!, *T. & D.*, 208. Harmondsworth!, *K. & L.*, 208. Staines, 1958, *Briggs*. West Drayton; Laleham Park, *Kennedy*!. **3.** Isleworth, *A. B. Cole*; near Hounslow, a single plant, *T. & D.*, 208. **4.** Hampstead!, 1861 (SY). West Heath, Hampstead!; Harrow Weald!, *T. & D.*, 208. Harrow!, *Melv. Fl.*, 55. Mill Hill!, *Harrison*, teste *Sandwith*, *K. & L.*, 208. Rayners Lane. Near

Cricklewood. **5.** Acton Green; near Ealing!, *Newbould, T. & D.*, 208. Hanwell!; West Ealing!, *K. & L.*, 208. South Ealing. **6.** Hadley!, *K. & L.*, 208. Winchmore Hill, *A. Vaughan*. Southgate, 1965, *D. E. G. Irvine* (BM), teste *Bangerter*. Near East Barnet!, *Kennedy*, teste *Bangerter*. Enfield, locally plentiful. **7.** Near Brompton, 1773, *J. Banks* (BM). Marylebone!, *Varenne, T. & D.*, 208; 1958. Isle of Dogs; Fulham!, *Newbould*; Kensington Gardens!, *Warren, T. & D.*, 208. Hyde Park, 1958, *D. E. Allen, L.N.* 39: 54. Weed in florist's garden, Moorfields, E.C., *Jones Fl.* Haverstock Hill, 1867 (SY). Holloway, 1965, *D. E. G. Irvine*. Clapton. Highgate Cemetery, 1966. Earls Court, 1968. Buckingham Palace grounds, 1965, *McClintock*.

Var. **micrantha** Drabble
4. Finchley, 1912, *E. F.* and *H. Drabble* (BM).

Var. **garckiana** P. Fourn.
5. East Acton, *Loydell*, teste *G. C. Druce, Rep. Bot. Soc. & E.C.* 8: 314.

V. filiformis Sm. Slender Speedwell.
Lit. *Proc. Bot. Soc. Brit. Isles* **2**: 147 (1957): *loc. cit.* **4**: 384 (1962).
Alien. Caucasus. Garden escape. Naturalised in meadows, on banks of rivers and streams, and established on roadside verges, paths, in churchyards, etc., also as a serious pest of lawns and turf. Locally abundant near the Thames, rather rare elsewhere but apparently increasing.
First record: *H. Banks*, 1942.
1. Road bank, Northwood, 1962, *Graham, K. & L.*, 209; several large colonies in turf and roadside verges by the Rickmansworth Road, 1965→. Banks of the Colne, Harefield Moor, 1958!, *Donovan, L.N.* 38: 21; increasing 1967. Garden pest, Ruislip, *Wrighton*; canal path, Harefield!, *Welch, Proc. Bot. Soc. Brit. Isles* 2: 208; increasing and invading turf of adjacent meadows, 1965→. Garden pest, Eastcote!, 1960, *T. G. Collett, Proc. Bot. Soc. Brit. Isles* 4: 390; frequent in a number of localities about Eastcote, and particularly abundant as a pest of lawns and turf in Haydon Hall grounds, 1968. Pinner Churchyard, frequent in turf, 1966→. Roadside verges, Pinner Hill, 1967→. Pest of lawns, Billy Low's Lane, Potters Bar, 1967, *Kennedy*!. Swakeleys; Ickenham, *L. M. P. Small*. **2.** Locally abundant in turf and meadows by the Thames from Staines to Hampton, and in profusion at Penton Hook Lock, Laleham, Laleham Park, Shepperton Lock and Sunbury. Laleham Churchyard, abundant. Road verges between Staines and Laleham, abundant, 1966→. Pest of lawns near Hampton, 1966. Plentiful in turf, Harmondsworth Churchyard, 1965→. **3.** Weed of tennis courts, Hounslow, 1942, *H. Banks, Proc. Bot. Soc. Brit. Isles* 2: 208. Marble Hill Park, Twickenham, 1965, *L. M. P. Small*; 1965, *Gerrans* (BM); 1966→. River wall, Twickenham, 1966. Pest of lawns, Isleworth, 1968. **4.** Garden weed, Harrow, *Lockett*,

Rep. Bot. Soc. & E.C. 13: 305; Pinner Road Cemetery, Harrow, abundant, 1966, *Wurzell.* Abundant on road verges, Harrow Weald, 1961, *L. M. P. Small, Proc. Bot. Soc. Brit. Isles* 4: 390; in profusion, 1966; pest of lawn, Kenton, 1966, *Kennedy*!. In profusion on road verges near Sudbury Hill, 1966. Roadside verges and lawns, Edgwarebury Lane, Edgware, 1967. Mill Hill, *Harrison.* Near Totteridge. **5.** Pest of lawns, Syon Park!, 1966, *Boniface*; 1968. Pest of lawn near Alperton, 1966→. Pest of lawns, West Ealing, 1968→. Pest of lawn, Chiswick, 1968. **6.** Roadside, East Barnet, 1961, *J. F.* and *P. C. Hall, Proc. Bot. Soc. Brit. Isles* 4: 390. **7.** Grass verge by Hammersmith Flyover, 1973.

PEDICULARIS L.

P. palustris L. Red-rattle, Bog Lousewort.

Formerly native in bogs, marshes, wet fields, on damp heaths, etc., but now extinct.

First record: *J. Newton*, c. 1690. Last record: *J. E. Cooper*, 1920.

1. Harefield Moor, *Blackst. Fasc.*, 73; c. 1869, *T. & D.*, 203; 1884, *Benbow MSS.*; 1920, *J. E. Cooper, K. & L.*, 212. Near Yiewsley, *Newbould, T. & D.*, 303. Cowley; Springwell; Drayton Ford, abundant, *Benbow, J.B. (London)* 25: 17. Uxbridge Moor, *Druce Notes.* **2.** Between Hampton and Hampton Court, *Newbould*; Shortwood Common, *T. & D.*, 203. Thames side below Hampton, 1836, *H. C. Watson* (W). **4.** Hampstead Heath, 1837, *E. Ballard*; 1859 (small stunted specimens), *Trimen* (BM); frequent on the West Heath, *White Hampst.*, 346; *P. sylvatica* probably intended. **7.** In the ditches between Bow and Blackwall . . ., *J. Newton, Ray Syn.* 3, *384.

P. sylvatica L. Heath Lousewort.

Native on wet heaths, in bogs, marshes, etc. Very rare, and decreasing.

First record: *T. Johnson*, 1632.

1. Harefield, *Blackst. Fasc.*, 73; c. 1869, *T. & D.*, 204; 1887,† *Benbow MSS.* Ruislip Common, frequent!, *Melv. Fl.*, 56; 1967. Harrow Weald Common, abundant;† Stanmore Common!, *T. & D.*, 204. Northwood!, *Benbow MSS.* New Years Green, 1956, *Donovan.* **3.** Hounslow Heath,† *T. & D.*, 204. Between Hounslow and Feltham,† *Benbow MSS.* **4.** Hampstead Heath, *Johns. Enum.*; 1866 (SY); c. 1869, *T. & D.*, 204; formerly common, c. 1912,† *Whiting Fl.* Brockley Hill, *Rudge* (BM). **6.** Hadley!, *Warren, T. & D.*, 204. Hadley Green,† *Benbow MSS.* **7.** East Heath, Hampstead,† *T. & D.*, 204.

RHINANTHUS L.

R. minor L. subsp. **minor** Yellow-rattle.

R. crista-galli auct.

Lit. *Watsonia* **4**: 101 (1959).
Native in meadows and grassy places on dry soils, established on railway banks, etc. Formerly common, but now rare and decreasing.
First record: *T. Johnson*, 1638.
1. Harefield!, *Blackst. Fasc.*, 73. Harrow Weald Common;† Elstree;† South Mimms, *T. & D.*, 204. Swakeleys;† Uxbridge!, *Benbow* (BM); 1967. Pinner,† *Melv. Fl.*, 56. Haste Hill, Northwood!; Eastcote!; Ruislip!, *Tremayne*; between Uxbridge and Denham!, *K. & L.*, 212; 1966. **2.** Staines, towards Hampton;† between Hampton Court and Kingston Bridge,† *T. & D.*, 204. Near Hampton Wick, 1928,† *Tremayne*; between Poyle and Yeoveney!, *K. & L.*, 212; 1967. Near Staines Moor, 1966, *Boniface*!. **3.** Roxeth,† *Hind Fl.* **4.** Hampstead, *Morley MSS.*; 1865 (SY); c. 1901,† *Whiting Notes*. Harlesden Green,† *J. Morris, T. & D.*, 204. Harrow,† *Hind Fl.* Mill Hill!, *E. M. Dale* (LNHS). Finchley, *E. M. Dale, Lond. Nat. MSS.* Stanmore, 1908, *C. S. Nicholson* (LNHS). **5.** Near Brentford,† *Hemsley*; near Twyford,† *Newbould*; Hanwell,† *Warren, T. & D.*, 204. **6.** Tottenham,† *Johns. Cat.* Whetstone, *A. B. Cole*; Enfield!, *T. & D.*, 204. Southgate, *L. B. Hall, K. & L.*, 212.

MELAMPYRUM L.

M. arvense L. Field Cow-wheat.
Formerly native or colonist in cornfields, etc., but now extinct.
First, and only evidence: *A. W. Bennett*, c. 1870.
2. Teddington, *A. W. Bennett* (BM).

M. pratense L. Common Cow-wheat.
Lit. *Watsonia* **5**: 336 (1963).
Native in woods, on heaths, etc. Rare, and confined to the north of the vice-county.
First record: *Gerarde*, 1597.
1. Harefield!, *Blackst. Fasc.*, 59. Ruislip! and Pinner Woods!, *Melv. Fl.*, 56. Northwood. Harrow Weald Common and adjacent copses, 1965, *Kennedy*!. Stanmore Common. Mimmshall Wood, *Benbow MSS.* **4.** Hampstead Heath, *Ger. Hb.*, 65; c. 1869,† *T. & D.*, 203. Little Wood, Finchley, 1865† (SY). East Finchley, 1875, *French* (BM); 1883,† *J. E. Cooper*. The two latter records probably refer to the Little Wood locality. Woods between Edgware and Stanmore, scarce, *Benbow MSS.* **6.** Highgate!, *S. Palmer, Mag. Nat. Hist.* **2**: 266; Queen's Wood, Highgate, 1963, *L. Martin*. Winchmore Hill!, *K. & L.*, 212. Muswell Hill,† *Church*; Winchmore Hill Wood,† *T. & D.*, 203. **7.** Hornsey Wood† (G & R). Ken Wood, 1847,† *Mawer* (BM).

Var. **lanceolatum** Spen.
1. Harefield. Copse Wood, Northwood!, *Welch*. Park Wood, Ruislip!, *Wrighton*. Pinner Wood.

EUPHRASIA L.

Lit. *J. Linn. Soc. Bot.* **48**: 467 (1930): *J.B.* (*London*) **70**: 200 (1932): *loc. cit.* **71**: 83 (1933): *loc. cit.* **74**: 71 (1936): *loc. cit.* **78**: 11 and 89 (1940).

E. nemorosa (Pers.) Mart. Eyebright.
Native, semi-parasitic on the roots of grasses and other plants in meadows, open woods, on heaths, etc. Very rare, with a marked preference for calcareous soils.
First evidence: *Dillenius*, c. 1730.
1. Harefield Common;† Garett Wood;† Springwell!; Ruislip Common!; between Ruislip and Northwood; Northwood!, *Benbow* (BM), all det. *Pugsley*. Uxbridge!, *Druce Notes*. Near Denham. **4.** Hampstead Heath (DILL), det. *Pugsley*; 1910, *Pugsley* (BM); 1927,† *Moring* (HPD), det. *Pugsley*. **5.** Ealing Common, 1884,† *Benbow* (BM), det. *Pugsley*. **6.** Botany Bay, Enfield Chase, *Benbow* (BM), det. *Pugsley*. **7.** Crouch End,† *French* (BM), det. *Pugsley*. Bombed site, Ebury Street, S.W.1, 1946,† *McClintock*.

Var. **transiens** Pugsl.
1. Harefield, 1868, *Hind* (BM), det. *Pugsley*. Springwell!, *Lousley* (L), det. *Pugsley*. **5.** Between Ealing and Acton, 1866,† *Hind* (BM), det. *Pugsley*. Garden weed, West Ealing, 1967!,† *L.N.* 47: 9.

E. confusa Pugsl. forma **albida** Pugsl. Eyebright.
Formerly native, semi-parasitic on the roots of grasses and other plants in a meadow, but now extinct.
First, and only evidence: *Benbow*, 1887.
1. Pield Heath, 1887, *Benbow* (BM), det. *Pugsley*.

E. borealis (Towns.) Wettst. Eyebright.
E. brevipila auct. non Gremli
Formerly native, semi-parasitic on the roots of grasses and other plants on damp moorland, but now extinct.
First, and only evidence: *Benbow*, 1884.
1. Harefield Moor, 1884, *Benbow* (BM), det. *Pugsley*, as *E. brevipila*.

E. anglica Pugsl. Eyebright.
E. rostkoviana auct., non Hayne
Native, semi-parasitic on the roots of grasses and other plants on heaths, in fields and open woods, particularly on heavy soils. Very rare.
First evidence: *Dillenius*, c. 1730.
1. Pinner,† *Green* (SLBI), det. *Pugsley*. Copse Wood, Ruislip!, *Welch*,

det. *Pugsley, L.N.* 26: 63; 1966. Ruislip Common, 1966. **3.** Hounslow Heath,† *Pugsley*. **4.** Hampstead Heath† (DILL), det. *Pugsley*. **6.** Hadley Wood; Hadley Green,† *Benbow* (BM), det. *Pugsley*.
All published records of the aggregate *E. officinalis* L. have been omitted from the above account.

ODONTITES Ludw.

O. verna (Bellardi) Dumort. Red Bartsia.
Euphrasia odontites L., *Bartsia odontites* (L.) Huds., *O. rubra* Gilib.
First record: *T. Johnson*, 1633.

Subsp. **verna**
Introduced? Waste ground, etc. Very rare.
1. Near Uxbridge, 1964. **7.** Bombed site, Ebury Street, S.W.1, 1946,† *Lousley* and *McClintock* (L).

Subsp. **serotina** (Dumort.) Corbiére
Native on cultivated and waste ground, pond verges, in rough fields, by roadsides, etc. Formerly rather common, but now local and decreasing.
1. Harefield!, *Blackst. Fasc.*, 21. Elstree, 1825† (HE). Pinner,† *Hind Fl.* South Mimms!, *C. S. Nicholson*; Uxbridge!, *K. & L.*, 211. Haste Hill, Northwood!, *Wrighton, L.N.* 27: 33. **2.** Shortwood Common, sparingly!; Hampton, *T. & D.*, 205. Yeoveney!; Stanwell Moor!, *J. E. Cooper, K. & L.*, 211. West Bedfont, abundant in a gravel pit, 1965, *Kennedy*!. Harlington,† *P. H. Cooke, K. & L.*, 211. Upper Halliford, rare, 1966. **3.** Hounslow,† *T. & D.*, 205. Yeading!; West End, Northolt!, *L.N.* 27: 33. **4.** Hampstead Heath, *Johns. Ger.*, 91; c. 1713, *Pet. H.B.C.*; c. 1830,† *Irv. MSS.* Near Hendon, *Hampst.* Mill Hill, *Children* (BM). Near Harrow, *Hemsley*; Stanmore, *T. & D.*, 205. Edgware, *C. S. Nicholson, K. & L.*, 211. Whitchurch Common, 1954. **5.** Perivale,† *Lees*; Alperton,† *T. & D.*, 205. **6.** Enfield Chase!, *T. & D.*, 205. Crew's Hill, Enfield, 1965, *Kennedy*!. **7.** Islington,† *Preedy, Ball. MSS.*
Plants with white flowers are recorded from **1.** Harefield, 1920, *J. E. Cooper, K. & L.*, 211.
Some of the records given above may be referable to the subsp. *verna*.

OROBANCHACEAE

LATHRAEA L.

L. squamaria L. Toothwort.
Native, parasitic on the roots of *Ulmus* spp. and *Corylus avellana*, in hedgerows, copses, etc., chiefly on calcareous soils. Very rare.

First evidence: *Blackstone*, 1737.

1. In a shady lane near Harefield town, plentifully, 1737, *Blackstone* (BM); it grew abundantly on the banks of Springwell Lane over a distance of nearly a quarter of a mile until 1961 when most of the old elm trees were felled; although now mainly confined to the vicinity of Springwell Lock it appears to be increasing and spreading back along Springwell Lane, 1969→. Plantation close to Jack's Lock, Harefield!, *Benbow* (BM); 1950. Harefield (=Old Park) Wood, 1839, *Kingsley* (ME); c. 1861,† *Phyt. N.S.* 6: 150. **2.** Colnbrook, 1908,† *Druce Notes.* **4.** Not uncommon about Hampstead, *Cochrane Fl.* Error. Plants with straw-coloured flowers are recorded from **1.** Harefield, *Roffey*, *J.B. (London)* 48: 164.

OROBANCHE L.

O. rapum-genistae Thuill. Greater Broomrape.
O. major auct.

Formerly native, parasitic on the roots of *Ulex* spp. and *Cytisus scoparius*, on heaths, banks, etc., but now extinct.

First record: *Gerarde*, 1597. Last record: *Ford*, pre-1873.

1. Harefield, in a lane near the village, *De Cresp. Fl.*, 48. Error; possibly *Lathraea squamaria* intended. **2.** Cornfield between Egham and Staines, *Whale Fl.*, 21. Error; no doubt *O. minor* intended. **4.** Hampstead Heath, *Ger. Hb.*, 1132; c. 1730 (DILL); 1828, *Varenne*, *T. & D.*, 196; 1830, *Irv. MSS.* and '*G.A.*', *Nature Notes*, 1902. **6.** In great abundance for several years growing upon broom on a gravelly hill at Old Park, Enfield: now cut away to obtain ballast for the Great Northern Railway, *H. & F.*, 150.

O. elatior Sutton Tall Broomrape.
O. major auct.

Formerly native, parasitic on the roots of *Centaurea scabiosa*, in a chalk-pit, but now extinct.

First, and only evidence: *Green*, 1902.

1. Chalkpit, Harefield, 1902, *Green* (SLBI).

O. minor Sm. Lesser Broomrape.
O. apiculata Wallr.

Native, parasitic on the roots of *Trifolium* spp., *Matricaria* spp., etc., in fields, on waste ground, by roadsides, etc., and in particular among sown clover crops. Rare.

First record: *Macreight*, 1837.

1. Abundant in clover fields from Harefield to the county boundary near Springwell, 1884;† railway banks between Uxbridge and West Drayton, 1885,† *Benbow* (BM). Harefield, 1908, *Green*, *Williams MSS.* Springwell chalkpit, a single plant, 1945, *Welch*, *L.N.* 26: 63. Canal bank by Springwell Lock, 1961, *J. G. Dony* and *Souster*!. A large colony above an old chalkpit, South Mimms, 1966. **2.** Cornfield between Egham and Staines (as *O. major*), *Whale Fl.*, 21. Staines!, 1958, *Briggs*; a single plant, 1966, *J. E. Smith*!. Stanwell Moor, abundant in a clover field, 1951, *Lousley*!; railway bank, Poyle, 1950!, *K. & L.*, 214. Near Stanwell, 1966, *Boniface*. Between Shepperton and Sunbury, 1903, *Green* (SLBI). Harmondsworth, 1906, *Green* (LNHS). Grounds of Ashford County Hospital, *Hasler*, *K. & L.*, 214. **3.** Twickenham, *Macreight Man.*, 175. **5.** Canal path between Hanwell and Southall, a single plant, 1945, *D.* and *J. M. B. King*!, *L.N.* 26: 63; a single plant, 1964. Greenford, 1917, *J. E. Cooper*. Duke's Meadows, Chiswick, many plants on *Trifolium repens*, 1961, *Boniface*. **6.** Railway bank, Gordon Hill, Enfield, 1953, *Johns*. Railway side, Crew's Hill, Enfield, a single plant, 1965, *Kennedy*!. **7.** Regents Park,† *Webst. Reg.*, 101. Railway bank, Hampstead Heath station, 1949, *Boniface*, *K. & L.*, 214. Alley off Ebury Bridge Road, S.W.1, a single plant, 1963, *Codrington*, *L.N.* 43: 21.

O. hederae Duby Ivy Broomrape.

Native, parasitic on the roots of ivy, etc., in south-west England, etc. In Middlesex an introduction. Very rare.

First record as an established introduction: *Bull*, 1972.

1. A single plant occurred in June and July, 1868, on the root of a scarlet *Pelargonium* in a pot, at Pinner; another is now (Jan. 1869) growing under the same circumstances; *Hind*: Imported from elsewhere, *T. & D.*, 196. **7.** Highgate Cemetery, on ivy, 1972, *Bull*.

LENTIBULARIACEAE

UTRICULARIA L.

U. australis R.Br. (including *U. vulgaris* L.) Greater Bladderwort. *U. neglecta* Lehm., *U. major* auct.

Native, submerged in stagnant lakes and ponds, growing on wet mud, etc. Very rare, and invariably sterile. Flowers have been seen on a few occasions, and these appear to be associated with warm summers.

First record: *Alchorne*, c. 1750.

1. Pond, Little Common, Stanmore, 1969, *Le Gros*; sterile, but probably this species, *Lousley*. **2.** Between Staines and Wraysbury, 1881, *G. Nicholson* (BM); in several large ponds, 1883,† *G. Nicholson*, *J.B.* (*London*) 21: 85. Pool near Yeoveney!, 1886, *Benbow*, *J.B.* (*London*) 25: 18; c. 1945!; East Bedfont;† Shortwood Common!, *K. & L.*, 215. Hampton Court pond! (D) (as *U. vulgaris*); Long Water, Hampton Court, c. 1948, *Milne-Redhead* (K) (*U. australis*, teste *P. Taylor*); still locally plentiful, but not known to flower, 1968. **3.** Hounslow Heath!, *Huds. Fl. Angl.* 2: 9 (as *U. vulgaris*); hundreds of plants in flower on the mud of two small dried-up ponds during the very warm summer of 1947!, *L. G. Payne, L.N.* 27: 33; two plants in flower, 1951!, *K. & L.*, 215; a few plants in flower in three ponds, 1956; one pond eradicated by gravel-digging operations, 1959; still present in two ponds, but did not flower, 1963; a second pond destroyed by gravel-digging operations, 1964; still present in very small quantity in the remaining pond, but did not flower, 1965–67. **7.** Between Newington and Southgate,† *Alch. MSS.* Ditches near Hornsey,† *Huds. Fl. Angl.*, 8 (as *U. vulgaris*).

The Yeoveney plant was referable to *U. australis*. The plants from Hampton Court, Shortwood Common and Hounslow Heath were originally identified as *U. vulgaris*, but *P. Taylor* has shown that material from Hampton Court is referable to *U. australis*; he has also found that this species is much more frequent than *U. vulgaris*. It seems very probable, therefore, that the Shortwood Common and Hounslow Heath plants are also referable to *U. australis*. It is impossible to suggest to which of the two species the early records from between Newington and Southgate, and near Hornsey, should be referred.

U. minor L. Lesser Bladderwort.

Formerly native, submerged in shallow streams, ditches, ponds, also in bogs, on wet mud, etc., but now extinct.

First record: *Doody*, c. 1696. Last record: *J. Hill*, c. 1746.

1. In Uxbridge river (=Colne), *J. Hill, Blackst. Spec.*, 45. **3.** Hounslow Heath, *Doody MSS.*; c. 1724, *Dandridge, Ray Syn.* 3: 287.

VERBENACEAE

VERBENA L.

V. officinalis L. Vervain.

Native by river- and streamsides, on waste ground, by roadsides, etc. Locally common near the Thames and Colne, rare elsewhere.

First record: *T. Johnson*, 1638.

1. Locally common near the Colne between Harefield and Yiewsley. Pinner Wood;† Waxwell Lane [Pinner],† *Melv. Fl.*, 62. Pield Heath;† Ruislip; Eastcote!, *Benbow MSS.* **2.** Locally common near the Thames and Colne. Shortwood Common!; Hanworth Road,† *T. & D.*, 223. Between Stanwell and West Bedfont!; between Staines and Ashford!; Charlton!, *Benbow MSS.* East Bedfont!, *P. H. Cooke, K. & L.*, 215. **3.** Isleworth;† Twickenham,† *T. & D.*, 223. Whitton,† *Benbow MSS.* Near Hounslow Heath, *H. Banks*!. **4.** Hampstead,† *Irv. MSS.* Between Finchley and Hampstead, 1815† (G & R). Mill

Hill, 1857,† *Children* (BM). Harrow,† *Hind Fl.* **5.** Near Lampton;† between Perivale and Alperton,† *Lees, T. & D.*, 223. **6.** Tottenham,† *Johns. Cat.* Bombed site, Wood Green, 1946,† *Scholey, L.N.* 28: 33. Edmonton,† *T. & D.*, 223. **7.** Between Westminster and Chelsea,† *Pet. MSS.* South Kensington,† *T. & D.*, 223. Bombed site, Earls Court, 1952,† *Corke, K. & L.*, 215.

LABIATAE

MENTHA L.

Lit. *Proc. Bot. Soc. Brit. Isles* **6**: 369 (1967).

M. requienii Benth.

Alien. S. Europe, etc. Garden escape. Established in turf and on a gravel path. Very rare.

First record: *D. H. Kent*, 1962.

6. Established on a gravel path, Myddleton House grounds, Enfield, 1966, *Kennedy*!. **7.** Established in turf, Brompton Cemetery, having spread from an adjacent grave, 1962.

M. pulegium L. Penny-royal.

Formerly native in damp heathy places, on pond verges, etc., but now extinct.

First record: *W. Turner*, 1562. Last record: *Hind*, 1871.

1. Harefield Common, abundantly, *Blackst. Fasc.*, 82. Pond verge, Pinner Hill, *Hind, J.B. (London)* 9: 272. **3.** Hounslow Heath, *Turn. Hb.*, 2. **7.** On the common near London called Mile End . . . whence poor women bring plenty to sell in London markets . . ., *Ger. Hb.*, 564. Tothill Fields, Westminster, *Rand* (BM). East Heath, Hampstead, *Irv. MSS.*; frequent there, 1850, *Syme* (BM); recently extinct, *Syme, T. & D.*, 213.

M. arvensis L. Corn Mint.

Native on heaths, commons, by river- and streamsides, on cultivated and waste ground, in grassy places, etc. Common except in heavily built-up areas where it is rare or has been eradicated.

First evidence: *Buddle*, c. 1710.

7. Bombed sites, Cripplegate,† *Wrighton, L.N.* 29: 88. Kensington Gardens, 1958!, *D. E. Allen, L.N.* 39: 54. East side of lake, Buckingham Palace grounds, 1967, *McClintock*.

Var. **obtusifolia** Lej. & Court.

1. Near Harefield!; Northwood; Breakspeares, *Benbow* (BM). All teste *Graham*. **2.** Thames bank, Kingston Bridge, 1920, *Fraser* (K).

Var. **nummularia** Schreb.
1. Harefield (D), teste *Graham*. **2.** Thames bank opposite Surbiton, 1914; Thames bank opposite Kingston, 1917, *Fraser* (K). **7.** By the New river near Stoke Newington,† *Buddle* (SLO), teste *Graham*.

Forma **angustifolia** Fraser
1. Harefield (D), teste *Graham*. Pinner Hill!; Stanmore Common!. Both teste *Graham*, *L.N.* 28: 33. **4.** Brent Reservoir!, *Graham*, *L.N.* 28: 33.

Var. **cuneifolia** Lej.
2. Thames bank, Hampton Court, 1930, *Fraser* (K); 1931, *Fraser* (L).

Forma **hirtipes** Fraser
2. Cornfield, Upper Halliford, 1928, *Lousley* (L), det. *Fraser*.

M. arvensis × spicata = M. × gentilis L.
M. gracilis Sole, *M. cardiaca* (Gray) Bak.
Lit. *Watsonia* **1**: 276 (1950).
1. Pinner, *Hind Fl.* Harefield, 1953, *F. M. Day*, det. *Graham*, *K. & L.*, 219. Northwood, 1957, *Graham*, *L.N.* 37: 185. **2.** Hampton Hill, 1965, *Clement*. **3.** Hounslow Heath, 1961, *Buckle*. **4.** Brent Reservoir!, *Graham*, *K. & L.*, 219. Finchley, 1964, *Wurzell*. **5.** Turnham Green, 1949, *Boniface*, *K. & L.*, 219. Ealing, 1967, *McLean*. **7.** Stoke Newington,† *W. Sherard*, *Smith Fl. Brit.* 2: 622. Stroud Green,† *With. Arr.* 7, 3: 705. River Lee below Higham Hill, 1792,† *E. Forster* (BM), det. *Graham*. Burton's Court, Chelsea, 1947,† *Graham*; Eccleston Square, S.W.1, 1953,† *McClintock*, *K. & L.*, 219. Buckingham Palace grounds, *Fl. B.P.G.*

M. aquatica L. Water Mint.
M. hirsuta auct.
Lit. *Watsonia* **3**: 109 (1954).
Native by river- and streamsides, on pond verges, in ditches, marshes, etc. Common, except in heavily built-up areas.
First evidence: *Buddle*, c. 1710.
7. Near Marylebone, abundantly;† between Newington and Hornsey,† *Huds. Fl. Angl.*, 224. Marylebone Fields, 1815 (G & R). East Heath, Hampstead!, *T. & D.*, 211. Hackney Wick.
Plants with white flowers were noted at **2.** Thames bank opposite Long Ditton, 1919, *Fraser* (K). **5.** Canal bank, Hanwell, 1964.

Var. **hirsuta** (Huds.) Huds.
1. Ruislip Reservoir!; Stanmore Common, *Graham*, *K. & L.*, 217.

Var. **subglabra** Bak.
2. Thames bank between Staines and Chertsey, 1885; opposite Long Ditton, 1919, *Fraser* (K).

Var. **lobeliana** Briq.
5. Osterley Park, 1948, *Welch*, det. *Graham, K. & L.*, 217.

Var. **minor** Sole
2. Thames bank below Staines, 1885, *Fraser, Rep. Bot. Soc. & E.C.* 10: 614.

Var. **major** Sole
1. Colne at Harefield and Uxbridge Moor!, *Benbow* (BM), det. *Graham*. **2.** Colne at Stanwell Moor!, *Benbow* (BM), det. *Graham*. **6.** Southgate (DILL), det. *Graham*. **7.** Near Hornsey Wood, 1837,† *E. Ballard* (BM), det. *Graham*.

Var. **denticulata** (Strail) H. Braun
M. denticulata Strail, *M. aquatica* var. *lupulina* Briq.
1. Canal between Denham and Harefield Moor Lock, 1884, *Benbow* (BM), det. *Graham*. Stanmore Common, 1950, *Graham, K. & L.*, 219.

M. aquatica × **arvensis** = **M.** × **verticillata** L.
M. sativa L.
1. Common in the Colne valley. Northwood, *Benbow* (BM). Ruislip Reservoir!, *Graham, K. & L.*, 218. **2.** Locally common by the Thames between Staines and Teddington, and by the Colne between Staines and West Drayton. Ashford, *Benbow MSS*. **3.** By the Thames in scattered localities between Twickenham and Isleworth. **4.** Stanmore, 1827–30, *Varenne, T. & D.*, 211. Near Stanmore; by the Brent in many places,† *Benbow MSS*. Near Deacon's Hill, *R. A. Pryor, Trim. MSS*. **5.** Alperton,† *Lees, T. & D.*, 211. By the Brent in many places,† *Benbow MSS*. Brentford, *A. Wood, Trim. MSS*. **6.** North Finchley, 1925, *J. E. Cooper*. **7.** Near Islington,† *J. Hill* (BM), det. *Graham*. Near Bow, 1832,† *Ralfs* (SLBI). Hackney,† *E. Forster, E.B.*, 448.
Plants with white flowers are recorded from **1.** Harefield, *Benbow MSS*.

Var. **paludosa** (Sole) Druce
1. Ruislip Reservoir, 1949, *Graham, K. & L.*, 218. **2.** Teddington, 1950, *Welch*, det. *Graham, K. & L.*, 218.

Var. **adulterina** Briq.
1. Canal near Denham Lock, 1884, *Benbow* (BM), det. *Graham*.

Var. **acutifolia** Sm.
7. Near Hackney,† *Buddle* (SLO), det. *Graham*.

Var. **subspicata** (Weihe) Druce
2. Thames bank near Laleham, 1885, *Fraser* (K).

Var. **rivalis** (Sole) Briq.
1. Colne near Uxbridge; Uxbridge Common; Harefield, *Benbow* (BM), all det. *Graham*. **2.** Laleham, 1885, *Fraser* (K). **4.** Between Stanmore and Edgware, 1896, *Benbow* (BM), det. *Graham*.

Forma **calva** Still
1. Stanmore Common, 1928, *Lousley* (L).

M. aquatica × **arvensis** × **spicata** = **M.** × **smithiana** R. A. Grah.
M. rubra Sm., non Mill.
Lit. *Watsonia* **1**: 88 (1948).
1. Riverside a mile below Denham, *J. Hill* (BM), det. *Graham*. Harefield, 1942, *E. M. Day*, *Rep. Bot. Soc. & E.C.* 12: 500 (as *M. rubra*). Near Springwell!,† *Welch*, det. *Graham*, *K. & L.*, 219. **4.** Northwick Park golf course, 1947;† Brent Reservoir, 1946,† *Graham*, *K. & L.*, 214. Harrow, *Benbow*, *J.B.* (*London*) 25: 364 (as *M. rubra*). **5.** Hanwell.† Hanger Hill, Ealing, 1967, *McLean*. **6.** Enfield Chase, 1899, *C. S. Nicholson* (LNHS), det. *Graham*. Enfield, 1966, *Kennedy*. **7.** Bow, 1832,† *Ralfs* (SLBI).

Var. **raripila** (Briq.) R. A. Grah.
1. Northwood, 1884, *Benbow* (BM), det. *Graham*. **7.** Stoke Newington,† *Buddle* (SLO), det. *Graham*.

Var. **laevifolia** (Briq.) R. A. Grah.
1. Near Northwood, 1884, *Benbow* (BM), det. *Graham*.

M. aquatica × **spicata** = **M.** × **piperita** L. Peppermint.
Lit. *Watsonia* **2**: 30 (1951).

Var. **vulgaris** Sole
M. × *piperita* var. *druceana* Briq. ex Fraser
1. Harefield Moor, 1839, *Kingsley* (W), det. *Graham*. **3.** Near Feltham, 1886, *Benbow* (BM), det. *Graham*. **4.** Brent Reservoir!, *Graham*. **7.** Near Kensington, 1720,† *Rand*; Stamford Hill,† *E. Forster* (BM). Both det. *Graham*.

Var. **subcordata** Fraser
7. Kensington,† *Rand* (BM), det. *Graham*.

Var. **citrata** (Ehrh.) Briq.
M. citrata Ehrh.
1. Potters Bar, 1933, *Lousley* (L).

M. spicata L. Spearmint.
M. longifolia auct., non L., *M. sylvestris* L., *M. viridis* auct.
Alien of unknown origin. Long cultivated as a pot-herb, and now

naturalised in grassy places, on waste ground, by roadsides, etc., as an escape. Common.
First record: *J. Martyn*, 1732.
7. Common in the district.

M. spicata × **suaveolens** French Mint.
M. × *niliaca* auct., non Juss. ex Jacq., *M. nemorosa* auct., *M. alopecuroides* Hull
1. Ruislip, *D. P. Young, K. & L.*, 216. Yiewsley, *Wurzell.* Harefield Grove, *I. G. Johnson* and *Kennedy*!. **2.** Near Staines Moor, *Boniface*!. **3.** Hounslow Heath!, *Welch*, det. *Graham, K. & L.*, 216. **4.** West Heath, Hampstead, *Boniface, Rose* and *Wallace*!, *K. & L.*, 216. Canal path near Willesden Junction. Grounds of Willesden General Hospital. Stanmore. **5.** Chiswick. Ditch by railway between Hanwell and Southall. Railway bank, Southall, abundant. Hanger Hill, Ealing; North Ealing, 1967, *McLean*. Acton, 1966. **6.** Enfield, *Kennedy*!. Near Dollis Brook, near Totteridge, *Hinson*. Wood Green; Pymmes Park, Edmonton, *Kennedy*. **7.** Ken Wood, *H. C. Harris*!, *K. & L.*, 216. St Marylebone!, *L.N.* 39: 54. Near Swiss Cottage. St Katherine's Dock, Stepney!, *Fitter, L.N.* 43: 21. Building site, Hampstead Road, N.W.1, 1966!, *L.N.* 46: 33. Tottenham, *Kennedy*.

M. suaveolens Ehrh. Apple-scented Mint.
M. rotundifolia auct., non (L.) Huds.
Introduced. Formerly naturalised in grassy places, etc., but now extinct.
First record: *Buddle*, c. 1710. Last evidence: *J. Hill*, c. 1760.
1. Harefield, *Blackst. Fasc.*, 60; c. 1760, *J. Hill* (BM), det. *Graham* (as *M. rotundifolia*). **6.** Hornsey Churchyard, *Budd. MSS.*; c. 1732, *Mart. Tourn.* 2: 116; c. 1746, *Blackst. Spec.*, 53.

LYCOPUS L.

L. europaeus L. Gipsy-wort.
Native by river- and streamsides, in ditches, pond verges, etc. Common, especially by the Thames.
First record: *Blackstone*, 1737.
7. In most of the dirty ditches about London, *Jenk. Gen. Spec.*, 7. Thames banks and river walls from Hammersmith to the Isle of Dogs. Banks and walls of Paddington Canal from Kensal Green to Paddington Basin, and of the Regents Canal from Paddington to Limehouse. By Serpentine, Kensington Gardens! and Hyde Park!, *Allen*; Buckingham Palace grounds!, *L.N.* 39: 54. Millfield Lane, Highgate;† Eel Brook Meadow;† Hackney Marshes!, *T. & D.*, 213. Whitehall,† *Cottam*. Bombed site, Portland Place, W.1, 1948,† *Hasler*. Highgate Ponds!; Ken Wood!, *Fitter*. East Heath, Hampstead.

Var. **glabrescens** Schmid.
2. West Drayton, *Rep. Bot. Soc. & E.C.* 5: 298.

ORIGANUM L.

O. vulgare L. Marjoram.
Native on calcareous banks, etc., at Harefield, elsewhere on waste ground, railway banks, etc., as an introduction. Very rare.
First record: *Blackstone*, 1737.
1. Harefield!, frequent, *Blackst. Fasc.*, 70; plentiful on a chalky bank near Old Park Wood!, *I. G. Johnson*; 1969; persisted in small quantity in a large old chalkpit until 1965 when the floor of the pit was bulldozed prior to being converted to a rubbish-tip. Uxbridge, 1884, *Benbow* (BM); 1905,† *Green* (SLBI). Pinner,† *Hind Fl.*; 'Mr Hind now considers this to have been an error', *T. & D.*, 214. **3.** Hounslow Heath, 1948,† *Westrup*, *K. & L.*, 214. **4.** Railway bank near Mill Hill, 1920, *J. E. Cooper*; Scratch Wood, 1943, *Fitter*, *K. & L.*, 220. Harrow, *Harley Add.* **5.** Abundant on railway banks between Brentford and Isleworth, 1879,† *Britten*, *Trim. MSS.* Duke's Meadows, Chiswick!,† *Bangerter* (SLBI). **7.** Bombed sites, Cripplegate area, E.C.,† *Jones Fl.*

THYMUS L.

Lit. *New Phyt.* **53**: 470 (1954): *J. Ecol.* **43**: 365 (1955).
T. pulegioides L.
T. chamaedrys Fr., *T. ovatus* Mill., *T. glaber* Mill., *T. serpyllum* auct., non L.
Native on heaths, banks, etc., in dry grassy places, chiefly on calcareous and alluvial soils. Rare, and decreasing.
First certain evidence: *Trimen*, 1866.
1. Harefield!, 1868, *Newbould* (BM), det. *Pigott*; 1969. Springwell!, det. *Pigott*, *K. & L.*, 221; 1969. Between Harefield and Springwell, 1966→. Ickenham Green. **2.** Hampton Court Park!, *Welch*; Bushy Park!; East Bedfont Churchyard, 1945!;† Knowle Green!, all det. *Pigott*, *K. & L.*, 221; still present in the first two localities, 1969. **3.** Hounslow Heath, 1866,† *Trimen* (BM), det. *Pigott*. **4.** Established in stonework of terrace, North London Collegiate School, Edgware, 1967, *Horder*!. **5.** Syon Park!, det. *Pigott*, *K. & L.*, 221; 1966. **6.** Finchley Common,† *Benbow* (BM).
In view of the taxonomic and nomenclatural confusion in the British species of *Thymus* prior to *C. D. Pigott*'s revision of the genus all early printed records are omitted. *T. praecox* Opiz subsp. *britannicus* (Ronn.) Holub (*T. drucei* Ronn.) has not been recorded from the vice-county but may be present.

CALAMINTHA Mill.

C. sylvatica Bromf. subsp. **ascendens** (Jord.) P. W. Ball Common Calamint.

C. ascendens Jord., *C. officinalis* auct., *Clinopodium calamintha* auct., *Satureja ascendens* (Jord.) Maly

Native on dry banks on calcareous soils, elsewhere on waste ground, etc., as an introduction. Very rare, and perhaps extinct.

First record: *Blackstone*, 1737.

1. In Harefield, street going to the river,† *Blackst. Fasc.*, 13; old chalk-pits, Harefield!, *T. & D.*, 215; in recent years confined to a single small chalkpit where it was last seen in very small quantity in 1963. Roadside north of South Mimms,† *T. & D.*, 215. **7.** Crouch End, 1897,† *J. E. Cooper, Sci. Goss.*, 1898, 223.

ACINOS Mill.

A. arvensis (Lam.) Dandy Basil-thyme.

Calamintha acinos (L.) Clairv., *Clinopodium acinos* (L.) Kuntze, *A. thymoides* Moench, *Satureja acinos* (L.) Scheele

Native on cultivated ground, banks, etc., on calcareous and other dry soils. Very rare.

First evidence: *Dyer*, 1867.

1. Harefield!, *Newbould* (BM). Field below Garett Wood,† *Benbow, J.B.* (*London*) 23: 37. Railway side between Denham and Uxbridge, in small quantity, 1953, *T. G.* and *M. G. Collett*!, *K. & L.*, 221. Yiewsley, 1963, *Wurzell, L.N.* 43: 22. **2.** By towing-path, Hampton Court, 1867,† *Dyer* (BM).

CLINOPODIUM L.

C. vulgare L. subsp. **vulgare** Wild Basil.

Calamintha vulgaris (L.) Druce, non Clairv., *C. clinopodium* Benth.

Native on cultivated ground, banks, in fields, etc., on calcareous soils at Harefield and South Mimms, and on alluvial soils near the Thames, elsewhere on waste ground, railway banks, etc., as an introduction. Rare.

First record: *W. Turner*, 1548.

1. Harefield!, frequent, *Blackst. Fasc.*, 19. Harrow Weald Common,† *Melv. Fl.*, 61. Springwell!; railway banks, Cowley; Hillingdon;† Ickenham Green;† between Potters Bar and South Mimms!, *Benbow MSS.* By Mimmshall Wood, 1966, *J. G.* and *C. M. Dony*!. Warren Gate!, *K. & L.*, 221. Pinner,† *C. S. Nicholson, Williams MSS.* Northwood, 1915, *J. E. Cooper*; north of Ruislip Reservoir, 1921, *Tremayne, K. & L.*, 221. New Years Green, *I. G. Johnson*. **2.** Staines Moor!; Teddington, *T. & D.*, 215. Stanwell!, *P. H. Cooke, K. & L.*,

221; Stanwell Moor, locally frequent, 1966, *J. E. Smith*!. **3.** Isleworth,† *Ray Syn. MSS.* Between Twickenham and Whitton;† Duke's river near Chase Bridge,† *T. & D.*, 215. Gravel pit west of Hounslow Heath, 1947,† *H. Banks*!, *L.N.* 27: 33. **4.** Stanmore,† *Varenne*; Woodcock Hill [Kenton],† *T. & D.*, 215. **5.** Syon Park,† *Turn. Names.* **7.** Coal wharf, North London railway, Finchley Road, 1870,† *Trim. MSS.*

MELISSA L.

M. officinalis L. Balm.

Alien. Europe, etc. Garden escape. Naturalised on waste ground, by roadsides, etc., established on railway banks, etc. Rare.

First record: *A. Irvine*, 1856.

1. Harefield, *J. E. Cooper*. Near Potters Bar, 1965, *J. G.* and *C. M. Dony*!. **2.** Teddington!, *T. & D.*, 215. Laleham Park,† *Westrup*, *K. & L.*, 222. Staines, 1893, *Wolley-Dod* (CYN). Hampton Hill, *Clement*. **3.** Railway bank near Hounslow East station, abundant, *K. & L.*, 222; 1968. Hounslow Heath, 1957, *D. Bennett*. **5.** River wall, Strand-on-the-Green, 1887,† *Benbow* (BM). Gravel pit, Hanwell, c. 1932!,† *K. & L.*, 222. **6.** Whetstone, 1906, *J. E. Cooper*, *K. & L.*, 222. Bounds Green, 1963, *Brewis*. Botany Bay, Enfield Chase, 1965, *Kennedy*!. **7.** Between Little Chelsea and Parsons Green,† *A. Irvine*, *Phyt. N.S.* 2: 168. South Kensington,† *Dyer List*. Primrose Hill, 1950!,† *L.N.* 39: 54.

SALVIA L.

S. verticillata L.

Alien. S. Europe, etc. Naturalised in grassy places, on waste ground, etc., established on railway banks, etc. Very rare.

First evidence: *Ridley*, 1882.

1. Uxbridge, 1908,† *Green* (SLBI). Yiewsley, 1909,† *Coop. Cas.* **2.** Near West Drayton,† *Druce Notes*. **4.** Finchley, 1907–8,† *Coop. Cas.* East Finchley, 1921,† *Redgrove* (BM). Near Willesden Junction, 1965, *L.N.* 45: 20; 1967. Railway bank, Cricklewood, 1966. **5.** Chiswick, 1887,† *Benbow* (BM). Ealing,† *Druce Notes*. Railway banks near South Greenford!, *K. & L.*, 223; 1967. **6.** Muswell Hill, 1905,† *J. E. Cooper* (K). **7.** Exhibition grounds [site of the present Natural History Museum], 1882, *Ridley*; grounds of the Natural History Museum, South Kensington, 1904;† Fulham;† Isle of Dogs, 1887,† *Benbow* (BM).

S. horminoides Pourr. Wild Clary.

S. verbenaca auct. brit., *pro parte*

Native in dry fields, by roadsides, etc. Very rare, and now confined to the vicinity of the Thames.

First record: *Gerarde*, 1597.

1. Harefield,† *Blackst. Fasc.*, 43. **2.** Hampton Court!, *Ray Syn. MSS.*; 1968. Between Hampton and Hampton Court!, *Newbould*; between Sunbury and Walton Bridge;† by the towing-path between Hampton Court and Kingston Bridge, abundant!, *T. & D.*, 213; still plentiful, 1968. Near Staines, but not common!, *Whale Fl.*, 26; railway bank near Staines Moor, *Farendon*; gravel pit, Shepperton, abundant, 1947, *Collenette, K. & L.*, 213. **3.** Near Richmond Bridge, 1815† (G & R). Twickenham, c. 1800, *J. Banks* (BM); c. 1887,† *Benbow MSS.* **5.** Syon Park,† *Ray Syn. MSS.* Strand-on-the-Green,† *T. & D.*, 213. Chiswick,† *Benbow MSS.* **7.** In the fields of Holborn, near Grays Inn;† near Chelsea, *Ger. Hb.*, 628; c. 1793,† *M. & G.*, 35.

PRUNELLA L.

P. vulgaris L. Self-heal.

Native in meadows, pastures, woodland rides, etc., mainly on basic and neutral soils, also as a frequent weed of lawns. Common.

First record: *T. Johnson*, 1638.

7. Between Westminster and Chelsea, *Pet. MSS.* Marylebone fields, 1817 (G & R). Hackney Wick!; East Heath, Hampstead!, *T. & D.*, 217. Ken Wood!, *H. C. Harris*. Buckingham Palace grounds. Fulham Palace grounds, *Welch*!. Hyde Park!; Kensington Gardens!, *D. E. Allen, L.N.* 39: 54. Finsbury Square, E.C.2!, *L.N.* 46: 33. Holland House grounds, Kensington. Notting Hill Gate. Kilburn. Highgate. Hampstead. Euston. Highbury. Hornsey. Stoke Newington. Bethnal Green. Stepney. Hackney Marshes. Isle of Dogs.

Plants with white flowers are recorded from **1.** Harefield, *Blackst. Fasc.*, 82. North of Harefield, 1950. **7.** East Heath, Hampstead, *Irv. MSS.*

STACHYS L.

S. arvensis (L.) L. Field Woundwort.

Native or colonist in cornfields, on cultivated ground, etc., elsewhere on waste ground, rubbish-tips, etc., as an introduction. Very rare, and now usually merely casual.

First record: *Blackstone*, 1737.

1. Harefield, *Blackst. Fasc.*, 93; c. 1887; Uxbridge, *Benbow MSS.*; 1965, *Davidge*. Garden weed, Pinner, *Hind Fl.* **2.** West Drayton, *Newbould*; Tangley Park,† *T. & D.*, 221. Abundant between Colnbrook and Stanwell Moor, *Benbow MSS.*; Stanwell Moor, 1913, *J. E. Cooper* (BM). Grounds of Ashford County Hospital, a single plant, 1944, *Hasler, K. & L.*, 225. **3.** Twickenham, *T. & D.*, 221. Hounslow Heath, 1949, *Westrup*, *K. & L.*, 225. **4.** Field in Bishop's Wood, abundant, *T. & D.*, 221. Hendon, 1912, *P. H. Cooke* (LNHS). Arable field near Barnet Gate Wood, 1954, *Hinson*. **6.** Edmonton, *T. & D.*, 231. **7.** Waste ground near Marble Arch, 1962!,† *D. E. Allen* and *Potter*, *Allen Fl.*

S. palustris L. Marsh Woundwort.
Native by river- and streamsides, in ditches, on pond verges, etc. Common.
First record: *l'Obel*, 1600.
7. Near Hackney, *Lob. Stirp. Ill.*, 111; c. 1640; between Chelsea and Kensington,† *Park. Theat.*, 587. Isle of Dogs (DILL). Near Paddington Cemetery!, *Warren*, *T. & D.*, 221; canal near Kensal Green Cemetery. Lee Navigation Canal between Clapton and Tottenham. Canal bank near Paddington!, and at Regents Park!, *L.N.* 39: 54.

S. palustris × **sylvatica** = **S.** × **ambigua** Sm.
1. Cowley, 1884, *Benbow* (BM). Harefield!, *K. & L.*, 225. **4.** Golders Green, 1840† (HE). **5.** Canal bank near Alperton, 1947!,† *L.N.* 27: 33. **6.** Hadley Common!, *L.N.* 27: 33. Enfield Chase, 1966, *Kennedy*. **7.** East Heath, Hampstead, 1925, *Moring MSS.*

S. sylvatica L. Hedge Woundwort.
Native in woods, ditches, on shady hedgebanks, waste ground, by river- and streamsides, roadsides, etc. Common.
First record: *Johnson*, 1638.
7. Tottenham!, *Johns. Cat.* Kilburn!, *Warren*; Regents Park!; South Kensington, *T. & D.*, 220. Bombed site near Goldsmith Hall,† *Sladen* and *McClintock*, *Lousley Fl. City*. Fulham Palace grounds, *Welch*!. Kensington Gardens!; Eaton Square, S.W.1!, *L.N.* 39: 54. Buckingham Palace grounds!, *Fl. B.P.G.* Holland House grounds, Kensington. Brompton. Kensal Green. Hampstead. East Heath, Hampstead. Ken Wood. Kentish Town. Highgate. Highbury. Stoke Newington. Old Ford. Hackney Wick. Hackney Marshes. Wapping. Isle of Dogs.
A virescent form was recorded from **7.** East Heath, Hampstead, 1949!, *H. C. Harris*.

BETONICA L.

B. officinalis L. Betony.
Stachys betonica Benth., *S. officinalis* (L.) Trev.
Native in open woods, dry grassy places, on heaths, hedgebanks, etc. Formerly common, but now local.
First record: *Gerarde*, 1597.
1. Locally common. **2.** Stanwell Moor!, *T. & D.*, 220. Ashford; Littleton, *Benbow MSS.* East Bedfont,† *P. H. Cooke*. **3.** Hounslow

Heath!, *T. & D.*, 220. Hayes, *Benbow MSS.* Golf course near Dormers Wells, Southall, in very small quantity. **4.** Hampstead,† *Morley MSS.* Hampstead Heath;† Bishop's and Turner's Woods; Scratch Wood!; Deacon's Hill, *T. & D.*, 220. Wembley Park;† woods between Edgware and Stanmore, *Benbow MSS.* Hendon; near Mill Hill, *Kennedy*!. **5.** Horsendon Hill!, *T. & D.*, 220. Perivale Wood,† *Benbow MSS.* Hanwell, 1938.† **6.** Hadley!, *Warren*; Winchmore Hill Wood;† Colney Hatch, *T. & D.*, 220. Finchley Common;† Whetstone; near Potters Bar!; Enfield!; Enfield Chase!, *Benbow MSS.* Coldfall Wood, Highgate, *Cherry, Trim. MSS.* **7.** Hornsey Wood,† *Ball. MSS.* East Heath, Hampstead,† *T. & D.*, 220.

Plants with white flowers are recorded from **4.** Wood near Hampstead, *Ger. Hb.*, 578.

BALLOTA L.

B. nigra L. subsp. **foetida** Hayek Black Horehound.

Native on hedgebanks, waste ground, by roadsides, etc. Very common.

First record: *Merrett*, c. 1670.

7. Frequent in the district.

Plants with white flowers are recorded from **1.** Yiewsley, 1920, *J. E. Cooper*; south of Uxbridge, 1949!, *K. & L.*, 228. **2.** Near Stanwell; Stanwell Moor, 1965. **5.** Hanwell, 1944–53!,† *K. & L.*, 228. Chiswick, 1938, *Lousley* (L); 1947, *L. G. Payne*. South Acton, 1965. **6.** Near Enfield, 1965, *Kennedy*!. **7.** Near Chelsea, *Merr. MSS.* Chelsea, 1825, *Pamplin MSS.* Near Hammersmith, *Woodward, With. Arr.* 2: 616.

Var. **borealis** (Schweig.) Reichb.

2. West Drayton, *Druce Notes.*

LAMIASTRUM Heister ex Fabr.

L. galeobdolon (L.) Ehrendf. & Polatschek Yellow Archangel, Yellow Dead-nettle.

Galeobdolon luteum Huds., *Lamium galeobdolon* (L.) L.

Lit. *Watsonia* **8**: 277 (1971).

Native in woods, thickets, shady places, etc. Locally plentiful in the north of the vice-county, very rare elsewhere, and absent from the south-western districts.

First record: *Gerarde*, 1597.

1. Harefield!, *Blackst. Fasc.*, 49. Between Pinner and Harrow Weald Common; Garett Wood!;† Mimmshall Wood!, *Benbow MSS.* Pinner Wood, *De Cresp. Fl.*, 108. Park Wood, Ruislip!, *Welch, K. & L.*, 228. Near Denham!, *Davidge.* Near Potters Bar!, *Kennedy.* **4.** Hampstead Heath,† *Johns. Eric.* Near Hampstead Heath, *Garlick*

W.F. Bishop's and Turner's Woods; Deacon's Hill!, *T. & D.*, 218. Big Wood,† *Hampst.* Scratch Wood!, *Benbow MSS.* Mote Mount Park, Mill Hill. Kingsbury,† *Farrar, T. & D.*, 218. Finchley, *C. S. Nicholson, K. & L.*, 228. **5.** Horsendon Lane;† near Greenford!, *Melv. Fl.*, 60. Perivale Wood!, *K. & L.*, 228. **6.** Muswell Hill,† *Coop. Fl.*, 104. Colney Hatch, *J. Morris*; Alder's Wood† and hedges near Whetstone, abundant,† *T. & D.*, 218. By the Brent in many places between Whetstone and Hendon,† *Benbow MSS.* Highgate Woods!, *C. S. Nicholson, K. & L.*, 228. **7.** Near Hampstead, *Ger. Hb.*, 568. Belsize Park, 1870,† *Hartog, Trim. MSS.* Pond Street, Hampstead,† *Blackst. Spec.*, 43. Stoke Newington,† *Cockfield Cat.*, 16. Hampstead Churchyard;† west side of Primrose Hill,† *Merr. Pin.*, 69. Marylebone Fields, 1817† (G & R). Near Highgate;† Hornsey Wood,† *Ball. MSS.* Islington, 1837,† *E. Ballard* (BM). Still common in the hedges of the northern suburbs, Kentish Town,† Hornsey Wood,† Millfield Lane, Highgate,† etc., *T. & D.*, 219.

Plants with cream-coloured flowers are recorded from **1.** Harefield, *T. & D.*, 219, and with variegated leaves from **4.** Turner's Wood, *T. & D.*, 219. Recent studies on British populations by S. Wegmüller (*Watsonia* **8**: 277 (1971)) have revealed that the most frequent plant is subsp. **montanum** (Pers.) Ehrendf. & Polatschek, a tetraploid with narrow bracts and more or less serrate leaf margins. The diploid subsp. **galeobdolon** which has relatively broad bracts on the flowering shoots and crenate leaf margins, though common in Scandinavia, is apparently very rare in Britain and has been detected only in North Lincolnshire.

LAMIUM L.

L. amplexicaule L. Henbit.

Native on cultivated and waste ground, in fields, by roadsides, etc., chiefly on light dry soils, also as a weed of flower-beds in parks and gardens. Common, especially near the Thames.

First record: *Blackstone*, 1737.

7. Buckingham Palace grounds, *Codrington, Lousley* and *McClintock*!; Regents Park, *Lousley*; St James's Park; Kensington Gardens, *D. E. Allen*; Eaton Square, S.W.1, 1958!, *L.N.* 39: 55. Hyde Park, *D. E. Allen, L.N.* 41: 20. Weed of flower-bed, Lincoln's Inn Fields, W.C.2, 1965, *Burton*; Tower Hill, 1966!, *L.N.* 46: 33. Garden weed, Bayswater Road, W.2, 1962. Bishop's Walk, Fulham, 1965. Chelsea Hospital grounds, 1967. Hackney, 1965, *Renson*. Hammersmith, 1967. Weed of flower-beds, Shoreditch, Poplar and South Bromley, 1967.

L. hybridum Vill. Cut-leaved Dead-nettle.

L. incisum Willd., *L. dissectum* With.

Native on cultivated and waste ground by roadsides, etc., on light soils.

Very rare, or overlooked, and mainly confined to the south-west parts of the vice-county.
First record: *Merrett*, 1666.
1. Uxbridge Common; near Uxbridge!, *Benbow* (BM). **2.** Teddington, *Newbould*; near Strawberry Hill, *T. & D.*, 218. Grounds of Hampton Grammar School, 1950, *Westrup, K. & L.*, 227. River wall, Hampton Court, 1967, *Burton, L.N.* 47: 10. Staines, 1964. **3.** Near Hounslow, *T. & D.*, 218. **4.** Hampstead, *Irv. MSS.* Mill Hill, *Harrison.* **7.** In the King's new garden near Goring House [near Green Park],† *Merr. Pin.*, 69. St James's Fields, Westminster,† *Willisel, Ray Cat.*, 186. Holloway, 1965, *D. E. G. Irvine.*

Var. **decipiens** (Sond.) Rouy
1. Uxbridge, *Druce Notes.*

L. purpureum L. Red Dead-nettle.
Native on cultivated and waste ground, by roadsides, etc. Very common.
First record: *T. Johnson*, 1638.
7. Common in the district.
Plants with white flowers have been recorded from **1.** Uxbridge Common, *Benbow* (BM). **5.** Hanwell!, *K. & L.*, 227. Ealing, 1966, *McLean.* **6.** Edmonton, *T. & D.*, 218. Enfield, 1966, *Kennedy.*

L. album L. White Dead-nettle.
Native on hedgebanks, cultivated and waste ground, in ditches, by roadsides, etc. Very common.
First record: *T. Johnson*, 1638.
7. Frequent in the district.
Plants with buff-coloured flowers were seen at **5.** Chiswick, 1946.

L. maculatum L. Spotted Dead-nettle.
Alien. Europe, etc. Garden escape. Naturalised in grassy places, on waste ground, hedgebanks, etc. Rare.
First evidence: *Dillenius*, c. 1730.
1. Uxbridge, 1884; near Uxbridge, 1890, *Benbow* (BM). Ruislip Reservoir, *Firrell, J. Ruislip & Dist. N.H.S.* 12: 7. Near Potters Bar, 1965, *J. G.* and *C. M. Dony*!. **2.** Staines, 1966. East Bedfont, *L. M. P. Small.* **3.** Yeading, 1965. Southall, 1973. **4.** Kenton, two plants,† *Melv. Fl.*, 60. Brickfield by Hampstead Heath, 1887,† *Benbow* (BM). Barn Hill, Wembley Park, 1965. Harrow, *L. M. P. Small.* **5.** Hanwell, 1959!, *L.N.* 39: 40. Ealing, 1966, *McLean.* **6.** Arnos Grove, 1962; Finchley Common, 1966, *Wurzell.* Gravel pit north of Enfield Lock, 1967, *Kennedy*!. **7.** Hackney (DILL). Formerly gathered about Bayswater,† but suspected to be the outcast of some botanic garden, *J. E. Smith, E.B.*, 2550. South Tottenham, 1966, *Wurzell.* Finsbury Park, *L. M. P. Small.*

Var. **laevigatum** L.
3. Twickenham, *Druce Notes.*

GALEOPSIS L.

G. angustifolia Ehrh. ex Hoffm. Narrow-leaved Hemp-nettle.
G. ladanum auct. *pro parte*
Lit. *Watsonia* **5**: 143 (1962).
Native on cultivated and waste ground, etc. Very rare.
First record: *Blackstone*, 1737.
1. Harefield!, *Blackst. Fasc.*, 93. Yiewsley, 1912, *J. E. Cooper* (BM). **2.** West Drayton; Harmondsworth!, *Druce Notes.* **5.** Acton, 1912,† *Wedg. Cat.* **6.** Muswell Hill, 1900;† Finchley, 1910,† *Coop. Cas.* **7.** Coal wharf, North London Railway, Finchley Road, 1870,† *Trim. MSS.* South Kensington,† *Trimen, J.B. (London)* 13: 275.
It is probable that some of the records given above are referable to *G. ladanum* L.

Var. **calcarea** Briq.
1. Harefield, 1868, *Trimen*; near South Mimms, 1888, *Benbow* (BM).

G. tetrahit L. Common Hemp-nettle.
Native on cultivated and waste ground, heaths, in woods and thickets, ditches, etc. Common.
First certain record: *Hind*, 1861.
7. Side of Duckett's Canal, Hackney, *Cherry, T. & D.*, 219. Isle of Dogs, *Benbow* (BM). Finchley Road, *Trim. MSS.* Hackney Marshes, *Coop. Cas.* Ken Wood!, *H. C. Harris.* Bishop's Park, Fulham. Regents Park!, *L.N.* 39: 55. Hyde Park Corner, 1963,† *Brewis, L.N.* 44: 26. Wapping. Clapton. Hackney Wick. Brompton Cemetery, 1965.
Plants with white flowers are recorded from **1.** Garett Wood,† *Benbow MSS.* Springwell, 1966. Stanmore Common, 1965, *Kennedy*!. **5.** South Acton, 1965.

G. bifida Boenn.
G. tetrahit var. *bifida* (Boenn.) Lej. & Court.
Native in similar habitats to *G. tetrahit* with which it sometimes grows. The two species are frequently confused with each other. Common.
First certain record: *Melvill*, 1864.
7. Weed in garden of 70 Adelaide Road, N.W., 1867, *Syme* (BM). South Kensington,† *Trimen, J.B. (London)* 13: 275. Brompton, 1874, *Trim. MSS.* Isle of Dogs, *Benbow MSS.* Hyde Park!; Kensington Gardens!, *D. E. Allen*; Finchley Road, N.W.8, 1959, *H. C. Harris, L.N.* 39: 55. Buckingham Palace grounds, *Fl. B.P.G.* Stag Place, S.W.1, 1966, *McClintock, L.N.* 46: 12.

G. speciosa Mill. Large-flowered Hemp-nettle.
G. versicolor Curt.
Native or introduced on cultivated and waste ground, etc. Very rare, and sometimes only casual.
First record: *Hind*, 1861.
1. Near Uxbridge Common, in plenty, 1892,† *Benbow* (BM). Northwood, 1950, *Graham, L.N.* 30: 7. **4.** Cornfields, Harrow,† *Hind Fl.* **5.** Garden weed, Ealing, a single plant, 1963, *L. M. P. Small, L.N.* 42: 12. **6.** Enfield, 1880, *Find. MSS.* **7.** Waste ground, Gloucester Road, S.W.7,† *Warren, J.B. (London)* 12: 247. Hackney Marshes, 1924,† *J. E. Cooper* (BM). Hornsey, 1950, *Browning, L.N.* 30: 7.

NEPETA L.

N. cataria L. Cat-mint.
Native on dry hedgebanks, by waysides, in thickets, etc., chiefly on calcareous and alluvial soils. Very rare.
First record: *Blackstone*, 1737.
1. Harefield!;† Moorhall!,† *Blackst. Fasc.*, 60. Persisted in the latter locality until c. 1957. **2.** Bushy places by the towing path between Kingston Bridge and Hampton Court!, *Newbould, T. & D.*, 217; 1965, *Boniface.* **3.** St Margaret's, 1940.† **5.** Between Perivale and Alperton;† Perivale, *Lees, T. & D.*, 217; 1908,† *Eland, Williams MSS.*

GLECHOMA L.

G. hederacea L. Ground Ivy.
Nepeta hederacea (L.) Trev., *N. glechoma* Benth.
Native in woods, ditches, on banks, waste ground, etc. Common.
First record: *T. Johnson*, 1638.
7. Tottenham!, *Johns. Cat.* Paddington, *Varenne*; near Highgate; Isle of Dogs, *T. & D.*, 217. Highgate Ponds!, *Fitter.* Bishop's Park, Fulham, *Welch*!. Tower of London gardens, near Traitors' Gate, 1949. Tower Hill, 1966!, *L.N.* 46: 33. Lawn near the National Gallery, 1964, *McClintock, L.N.* 44: 26. Wellington Road, N.W.8!, *Holme, L.N.* 46: 33; 1967. Holland House grounds, Kensington. Hackney.
Plants with large pure-white flowers occur in **1.** Old Park Wood, Harefield.

MARRUBIUM L.

M. vulgare L. White Horehound.
Introduced in waste places, by roadsides, etc. Very rare, and perhaps extinct.

First record: *Petiver*, c. 1710.

1. Uxbridge Moor, abundantly,† *Blackst. Fasc.*, 58. Yiewsley, 1911,† *Coop. Cas.* **2.** Between West Drayton and Harlington, 1880, *Benbow* (BM). West Drayton, *Druce Notes.* Harlington, 1922, *J. E. Cooper, K. & L.*, 325. The latter three records probably refer to a single locality. **4.** Hampstead Heath, *Huds. Fl. Angl.*, 228; two plants, c. 1877,† *De Cresp. Fl.*, 40. **5.** Old gravel pit, Hanwell, c. 1932!,† *K. & L.*, 325. **6.** Finchley Common,† *Mart. Tourn.* 2: 107. Muswell Hill, 1905,† *Coop. Cas.* **7.** Between Westminster and Chelsea,† *Pet. MSS.* Chelsea† (DB). Highgate,† *Mart. Tourn.* 2: 107. Parsons Green,† *Britten, T. & D.*, 222. Hackney Marshes, 1912, *Coop. Cas.*; 1918,† *J. E. Cooper*; Primrose Hill, 1949!,† *K. & L.*, 225.

SCUTELLARIA L.

S. galericulata L. Skull-cap.

Native by river- and streamsides, on pond verges, etc., established by canals, etc. Common.

First record: *l'Obel*, 1576.

7. About London!, *Lob. Stirp. Observ.* Tothill Fields;† St James's Park, *Johns. Ger.*, 479; c. 1746;† sides of the New river between Islington and Newington!, *Blackst. Spec.*, 49. New river, Islington,† *Coop. Fl.*, 104. Lee between Tottenham and Clapton, *Francis, Coop. Suppl.*, 12. Lee Bridge, *Cherry, T. & D.*, 216. Isle of Dogs, 1815; Kensington Gardens and Harrow Road, 1817 (G & R). Paddington Canal near Kensal Green Cemetery!, *Britten, Phyt. N.S.* 7: 348. By Thames side nigh Westminster, *Lawson MSS.* Thames bank, Fulham, *Clarkson-Birch.* Ken Wood. Highgate Ponds!, *Fitter.* Canal banks, Paddington!, Marylebone! and Islington!, *L.N.* 39: 55. Regent's Canal, Stepney, Hoxton, Victoria Park and Hackney Wick. River wall of Thames at Hammersmith and Chelsea. Lee Navigation Canal, Upper Clapton to Tottenham.

S. minor Huds. Lesser Skull-cap.

Native in wet heathy places, open woods on acid soils, etc. Very rare.

First record: *Gerarde*, 1597.

1. Moist parts of Harefield Common!,† *Blackst. Fasc.*, 56. Harrow Weald Common!, *Hind Fl.* Near Uxbridge Common;† Bayhurst Wood!; Ruislip Reservoir, *Benbow MSS.* Stanmore Common!, *C. S. Nicholson*; Ruislip Common!; Copse Wood, Northwood!, *Welch, K. & L.*, 224. **3.** Hounslow Heath,† *Dill. MSS.* Headstone† (ME). **4.** Hampstead Heath, *Ger. Hb.*, 466; abundant, 1867, *T. & D.*, 216; c. 1912,† *Whiting Fl.* Near Golders Hill, rare,† *Whiting Notes.* Sparingly amongst loose stones, Pinner Drive,† *Melv. Fl.*, 62. **5.**

Turnham Green,† *S. Palmer*, *Mag. Nat. Hist.* 2: 261. **6.** Highgate, *Lob. Ill.*, 66 and 68. Hadley,† *G. Johnson*, *T. & D.*, 216. **7.** Hampstead Heath, *Ger. Hb.*, 466; sparingly on the East Heath,† *T. & D.*, 216. Banks of Lee between Clapton and Tottenham, *Francis*, *Coop. Suppl.*, 12. Error; *S. galericulata* probably intended.

TEUCRIUM L.

T. scorodonia L. Wood Sage.
Lit. *J. Ecol.* **56**: 901 (1968).
Native in woods, on heaths, banks, etc., on dry soils. Common, except in heavily built-up areas where it is very rare or extinct.
First record: *T. Johnson*, 1638.
7. Marylebone fields, 1815† (G & R). Islington,† *Ball. MSS.* East Heath, Hampstead!, *T. & D.*, 222. Ken Wood.

AJUGA L.

A. reptans L. Bugle.
Native in moist woods and meadows, on damp hedgebanks, etc. Locally common.
First record: *T. Johnson*, 1629.
1. Harefield!, *Blackst. Fasc.*, 12. Pinner, *T. & D.*, 222. Uxbridge. Near Denham. Springwell. Ruislip. Mimmshall Wood. Near Potters Bar; Bentley Heath!, *J. G.* and *C. M. Dony*. **2.** Staines!, *T. & D.*, 222. **3.** Isleworth Churchyard. Near Hounslow Heath. Wood End. Northolt. Hayes Park, *Royle*!. **4.** Hampstead Heath,† *Johns. Eric.* Big† and Little Woods,† *Hampst.* Between Finchley and Hendon!, *Newbould*; Stanmore!; Harrow!, *T. & D.*, 222. Near Elstree, abundant. Edgware!, *Horder*. Mill Hill. **5.** Horsendon Wood,† *Lees*, *T. & D.*, 222. Horsendon Hill. Ealing!, *McLean*. Brentford. Syon Park. Near Brentford!, *Hemsley*; Hanwell!,† *Warren*, *T. & D.*, 222. Osterley Park. Chiswick House grounds. **6.** Tottenham, *Johns. Cat.* Near Enfield!, *Tucker*; Colney Hatch!; Edmonton!, *T. & D.*, 222. Near Enfield Wash, *Kennedy*. **7.** Highbury† (WD). Belsize Park;† Millfield Lane, Highgate;† Kentish Town,† *T. & D.*, 222. Highgate Ponds!, *Fitter*. Buckingham Palace grounds, very rare, *Fl. B.P.G.*
Plants with white flowers are recorded from **4.** Harrow Park, *Farrar*, *Melv. Fl.*, 59, and with pink flowers from **4.** Stanmore, *Longman*, *Melv. Fl.*, 59. **5.** Osterley Park, *Masters*, *T. & D.*, 223. **6.** Highgate, *Dennes*, *Mag. Nat. Hist.* 8: 390.

PLANTAGINACEAE

PLANTAGO L.

P. major L. Great Plantain.
Lit. *J. Ecol.* **52**: 189 (1964).
Native in fields, on waste ground, by roadsides, etc., and established on lawns, paths, etc. Very common.
First record: *T. Johnson*, 1638.
7. Very common in the district.
A plant with a flowering spike 29½ inches long is recorded from **1.** Yiewsley, 1921, *J. E. Cooper*, *K. & L.*, 230, and a proliferous form was noted at **5.** West Ealing, 1968, *I. M. Kent*!.

P. media L. Hoary Plantain.
Lit. *J. Ecol.* **52**: 205 (1964).
Native in grassy places on neutral and basic soils. Locally plentiful in the Colne valley and near the Thames, rather rare elsewhere.
First record: *Rand*, 1724.
1. Near Harefield!, *Blackst. Fasc.*, 78. Harefield Moor, abundant. Pinner Wood, *Longman*, *Melv. Fl.*, 63. Between Yiewsley and Iver Bridge!, *Newbould*; Waxwell, Pinner, abundant,† *T. & D.*, 229. Springwell. Uxbridge. Uxbridge Moor, abundant. Ruislip Common. Bentley Heath, *J. G.* and *C. M. Dony*. **2.** Common. **3.** Hounslow Heath,† *Rand*, *Ray Syn.* 3: 314. Twickenham, *T. & D.*, 229. Hatton. **4.** Hampstead, *Irv. MSS.* Lawn near Harlesden, introduced with turves, 1966. **5.** Greenford,† *T. & D.*, 229. Perivale. Lawns, Clayponds Hospital grounds, South Ealing. Ealing. Near Brentford. **6.** Near Edmonton, *T. & D.*, 229. Enfield!, *Kennedy*. **7.** On grass-plots in the British Museum gardens in plenty,† *Forst. Midd.* South Kensington,† *S. A. Naylor*, *T. & D.*, 229. East Heath, Hampstead, *H. C. Harris*. Kensington Gardens, two plants, 1959!, *D. E. Allen*, *L.N.* 37: 200. Buckingham Palace grounds, 1961, *Fl. B.P.G.* Lawns, Grosvenor Road, S.W.1, introduced with turves, 1965!, *L.N.* 46: 33.

P. lanceolata L. Ribwort.
Lit. *J. Ecol.* **52**: 211 (1964).
Native in fields, on waste ground, by roadsides, etc., and established on lawns, etc. Very common.
First record: *Gerarde*, 1597.
7. Very common in the district.

Var. **dubia** (L.) Wahlenb.
P. eriophylla Decne.
5. Hanwell, 1947, *Lousley* (L).

Var. **capitata** Schum.
subsp. *sphaerostachya* (Mert. & Koch) Hayek, var. *sphaerostachya* Mert. & Koch, var. *lanuginosa* Boenn.
4. Finchley, *E. F.* and *H. Drabble*, *Rep. Bot. Soc. & E.C.* 8: 532.

Var. **microstachys** Wallr.
4. Finchley, *E. F.* and *H. Drabble*, *Rep. Bot. Soc. & E.C.* 8: 532.

Var. **timbali** (Jord.) Gaut.
P. timbali Jord.
1. Clover field, Harefield, 1960. **2.** Tangley Park,† *T. & D.*, 228. **3.** Whitton, *T. & D.*, 228. **5.** Hanwell.

P. coronopus L. Buck's-horn Plantain.
Lit. *J. Ecol.* **41**: 467 (1953).
Native on heaths, banks, waste ground, in bare places, etc., on sandy and gravelly soils, and established on paths, railway tracks, etc. Locally plentiful, especially in the south-western parts of the vice-county.
First record: *Gerarde*, 1597.
1. Uxbridge Common!, *Blackst. Fasc.*, 79. Harrow Weald Common, *Melv. Fl.*, 64. Harefield!, *Newbould*; Hillingdon, *Warren*; Stanmore Common, *T. & D.*, 228. **2.** By Walton Bridge; between Staines and Hampton!; towing path near Hampton Court!; Hampton, *T. & D.*, 228. Hampton Court Park, abundant!, *Welch*; Shortwood Common!, *K. & L.*, 230. Bushy Park. Shepperton. **3.** Hounslow Heath!; Twickenham, *T. & D.*, 228. **4.** Hampstead Heath!, 1681, *Newt. MSS.*; still locally abundant on the West and North-west Heaths. Parade ground, Harrow School, *Harley Fl.*, 25. **5.** Ealing Common,† *Lees*, *T. & D.*, 228. Chiswick!, *Benbow MSS.* Hanwell, 1903,† *Green* (SLBI). Syon Park!, *Tremayne*; railway tracks, Ealing!, *K. & L.*, 230. **6.** Abundant on paths, Myddleton House grounds, Enfield, *Kennedy*!. **7.** Tothill Fields, Westminster,† *Ger. Hb.*, 347. Hyde Park,† *M. & G.*, 189. Kensington Gardens, 1871,† *Warren* (BM). Waste ground, Gloucester Road, S.W.7, 1874,† *A. R. Pryor*, *Trim. MSS.* Hackney Marshes, 1909–10,† *Coop. Cas.*

LITTORELLA Berg.

L. uniflora (L.) Aschers. Shore-weed.
L. lacustris L.
Formerly native on sandy and gravelly verges of ponds and lakes, and in shallow water, but now extinct.
First record: *Petiver*, c. 1716. Last record: *D. H. Kent*, c. 1935.
1. Harefield Common, abundantly, *Blackst. Fasc.*, 78. Ruislip Reservoir, 1872, *J. A. Brewer* (BM); abundant on the east margin, 1893,

Benbow, J.B. (London) 31: 218; sparingly, c. 1935!, *K. & L.*, 231. **2.** Ashford Common, *Curt. F.L.* **3.** Hounslow Heath, *Pet. Gram. Conc.*, 218; towards Whitton, *Huds. Fl. Angl.*, 53; particularly in the ditch on the south side of Whitton Gardens, *J. Banks, B.G.*, 412.

CAMPANULACEAE

CAMPANULA L.

C. latifolia L. Large Campanula, Giant Bellflower, Throatwort.
Native in woods on heavy soils. Very rare.
First certain evidence: *Green*, 1909.
1. Margin of Mimmshall Wood, 1909, *Green* (SLBI); 1910, *P. H. Cooke* (LNHS); c. 1934, *Brittain*, 5; 100+plants, 1972, *Widgery*.

C. trachelium L. Bats-in-the-Belfry, Nettle-leaved Bellflower.
Native on wood borders, shady hedgebanks, in thickets, etc., particularly on calcareous and clay soils. Rare, and sometimes an escape from gardens.
First record: *Blackstone*, 1737.
1. Old Park Wood, Harefield!, abundantly, *Blackst. Fasc.*, 99; still occurs there but is now very scarce. North of Harefield!, *Newbould*; old chalkpits, Harefield, *T. & D.*, 180. Garett Wood!,† *Benbow MSS.* Wood by Ruislip Reservoir!, *Green.* **4.** Harrow-on-the-Hill Churchyard, 1958!, *Hockaday*, fide *Welch, Harley Add.*; 1968 probably introduced. **5.** Borders of Horsendon Wood!, *Lees, T. & D.*, 180; still present but very scarce, 1968. Ealing, 1968. **6.** Myddleton House grounds, Enfield, 1966, *Kennedy*!. Introduced. **7.** South Kensington,† *Trimen, J.B. (London)* 13: 275. Ken Wood, 1945, *Fitter, K. & L.*, 179. Belgrave Square, S.W.1, casual, 1945!;† canal path near Marylebone, casual, 1957!,† *L.N.* 39: 55. Regents Park, 1955,† *Holme*. Hampstead Heath, 1963, *J. R. Phillips*. Highgate Cemetery, 1972, *Bull*. Adventive in all these metropolitan localities.

C. rapunculoides L. Creeping Campanula or Bellflower.
Alien. Europe, etc. Garden escape. Naturalised in thickets, grassy places, etc., on waste ground, by roadsides, etc. Rather rare, but apparently increasing.
First record: *Buddle*, 1708.
1. Uxbridge. Cowley Peachey, *L. M. P. Small*. Potters Bar, 1969. **2.** Hampton Court, in several places!, *Welch, K. & L.*, 180; 1967. **4.** Hampstead, 1948!, *H. C. Harris*; Brent Reservoir, 1952, *Graham*!, *K. & L.*, 180; 1953→. Cricklewood, 1965. Near West Hampstead, 1966. Harrow, 1966, *Temple*!. East Finchley, 1966. **5.** Chiswick, 1949!, *Boniface*; Ealing, 1946, *Boucher*!, *K. & L.*, 180. Hanwell,

1962→. North Ealing, 1966→. South Acton, 1968. **6.** Enfield!, *H. & F.*, 148; 1967. **7.** A troublesome weed . . . in gardens at Hoxton,† *Budd. MSS.* Waste ground, Finchley Road, *Trim. MSS.* Regents Park!, *Webst. Reg.*, 101; 1957, *Holme, L.N.* 39: 55. Garden weed, Maida Vale, 1965!; Primrose Hill!, *L.N.* 46: 33.

Plants with white flowers have been noted at **4.** Cricklewood, 1966. Near West Hampstead, 1966. **6.** Enfield Chase, 1965. **7.** Primrose Hill!, *L.N.* 46: 33.

C. glomerata L. Clustered Bellflower.

Native in meadows and grassy places on alluvial soils by the Thames, established on railway banks, etc. Very rare. Strangely absent from calcareous soils.

First record: *Gerarde*, 1597.

1. Harefield, *De Cresp. Fl.*, 10. Error, possibly *C. trachelium* intended. Ruislip, 1960. *Pickess, Moxey List.* Introduced. **2.** Laleham, 1815† (G & R). Near Hampton Court,† *J. Banks* (BM). By the towing-path between Kingston Bridge and Hampton Court, *Warren, T. & D.*, 180; a single plant, 1887,† *Benbow, J.B. (London)* 25: 17. Teddington, 1867,† *Dyer* (BM). Near Staines, 1872 (HY). Between Staines Bridge and Laleham!,† *Benbow, J.B. (London)* 25: 17. Meadow nearly opposite Penton Hook Lock!,† *Benbow* (BM). Between Sunbury and Walton Bridge, a single plant,† *Benbow, J.B. (London)* 25: 17. Shepperton, *C. E. Britton, J.B. (London)* 55: 326. Near Walton-on-Thames,† *Willmott* (K). East Bedfont,† *Westrup.* Lawns at Strawberry Hill, 1820,† *Gibbs, Pamplin MSS.* By the railway between Colnbrook and Staines!, *Benbow MSS.*; 1965. **5.** In Syon meadow near Brentford,† *Ger. Hb.*, 365.

C. rotundifolia L. Harebell.

Native on heaths, in dry grassy places, etc. Locally plentiful, especially on dry soils.

First record: *Blackstone*, 1737.

1. Harefield Common!, *Blackst. Fasc.*, 13. Ruislip! and Harrow Weald Commons!, *Longman, Melv. Fl.*, 49. South Mimms!, *T. & D.*, 180. Uxbridge Common!; Hillingdon Heath!;† Stanmore Common!; near Elstree; Northwood!, *Benbow MSS.* **2.** Bushy Park!, *T. & D.*, 180. Hampton Court Park. Staines, *Whale Fl.*, 8. Between Staines and Ashford Common!; Littleton, *Benbow MSS.* East Bedfont. **3.** Hounslow Heath!, abundant; Twickenham, frequent, *T. & D.*, 180. Golf course, Dormers Wells, near Southall. **4.** West Heath, Hampstead!, *T. & D.*, 180. Hampstead Heath Extension, *Fitter.* Harrow Weald. Edgware. Mill Hill. Lawn, Wembley Hill, introduced with turves, 1966. **5.** Hanwell!, *Cherry, T. & D.*, 180. Osterley Park. **6.** Hadley!, *G. Johnson*; Bush Hill, Enfield!, *T. & D.*, 180. Enfield Churchyard, 1966. Whetstone!, *Benbow MSS.* Trent

Park, *T. G. Collett* and *Kennedy*. Finchley Common. **7.** East Heath, Hampstead!, *T. & D.*, 180. Ken Wood!, *Fitter*. Kensington Gardens, 1871,† *Warren* (BM). Fulham,† *Corn. Surv.* Regents Park,† *Webst. Reg.*, 101. Lawn in front of University of London building, Malet Street, W.C., introduced with turves, 1947!,† *L.N.* 27: 32.
Plants with white flowers are recorded from **4.** Hampstead Heath, *Trim. MSS.*; 1901. *Whiting Notes.*

C. portenschlagiana Roem. & Schult.
C. muralis auct.
Alien. S.-E. Europe, etc. Garden escape. Established on old walls. Very rare.
First record: *D. H. Kent*, 1960.
2. Old farmhouse walls near Hampton, 1966, *Hubbard.* **5.** Old wall, Hanwell, plentiful, 1967→. **7.** Old wall, Upper Mall, Hammersmith, 1960→.

C. rapunculus L. Rampion.
Alien. Europe, etc. Garden escape. Formerly naturalised in grassy places, on hedgebanks, etc., but now extinct.
First evidence: *E. Forster*, 1805. Last evidence: *Green*, 1908.
1. Harrow Weald Common, 1908, *Green* (SLBI). **2.** Bushy Park, 1840, *J. Harris* (W). **3.** Twickenham, *Twining* (W). **6.** Edge of Common field near Weir Hall, Edmonton; Enfield, 1805, *E. Forster* (BM); Enfield Churchyard, *T. F. Forster*, *B.G.*, 401.

LEGOUSIA Durande

L. hybrida (L.) Delarb. Venus's Looking-glass.
Specularia hybrida (L.) A. DC., *Campanula hybrida* L.
Formerly native on cultivated ground, etc., on calcareous and other dry soils, but now extinct.
First record: *T. Johnson*, 1633. Last evidence: *J. E. Cooper*, 1914.
1. Harefield, *Blackst. Fasc.*, 96; c. 1869, *T. & D.*, 181; 1907, *Green* (SLBI); 1914, *J. E. Cooper*; between Ruislip Reservoir and Northwood, 1884, *Benbow* (BM). Pinner, a single plant, *Hind, Melv. Fl.* 2: 65. Near Mimmshall, 1909, *Green, Williams MSS.* South Mimms, *Druce Notes.* **7.** Chelsea, *Johns. Ger.*, 440; c. 1858, *Irv. Ill. Handb.*, 498.

JASIONE L.

J. montana L. Sheep's-bit.
Formerly native on heaths, in dry grassy places, etc., on sandy and gravelly soils, but now extinct.
First record: *T. Johnson*, 1632. Last record: *D. H. Kent*, c. 1935.

1. Between Harefield and Rickmansworth, 1912, *P. H. Cooke* (LNHS). Perhaps in Herts. **3.** Hounslow Heath!, . . . and other places near, abundant, 1866, *T. & D.*, 179; Hounslow Heath, c. 1935. **4.** Hampstead, *Pet. H.B.C.*; Hampstead Heath, c. 1780, *Curt. F.L.*; c. 1830, *Irv. MSS.*; c. 1877, *De Cresp. Fl.*, 34; sparingly, 1901, *Whiting Notes*; becoming scarce, 1912, *Whiting Fl.* Wood at Stanmore, 1815 (G & R). **7.** In Hackney Common-field, *Forst. Midd.*

Forma **laevis** Pugsl.

3. Hounslow Heath, *Trimen* (BM), det. *Pugsley*.

Plants with white flowers are recorded from **4.** Hampstead Heath, *Merr. MSS.*

RUBIACEAE

SHERARDIA L.

S. arvensis L. Field Madder.

Native on cultivated and waste ground, in fields, by roadsides, etc., established on canal paths, etc. Formerly rather common, but now locally common near the Thames and rather rare elsewhere.

First record: *W. Turner*, 1548.

1. Harefield!, *Blackst. Fasc.*, 88. Ruislip!, *Melv. Fl.*, 39. **2.** Locally common near the Thames. Tangley Park,† *T. & D.*, 139. Ashford, casual, *Hasler.* **3.** Hounslow Heath!; Twickenham, *T. & D.*, 139. **4.** Hampstead, *Irv. MSS.* and *Coop. Fl.*, 102. Harrow, *Farrar*, *T. & D.*, 139. Finchley, *George*, *T. & D.*, 424. Brent Reservoir, casual, 1945!. **5.** A little from Syon,† *Turn. Names*; among the corn beside Syon,† *Turn. Hb.* 1: 39. Turnham Green;† near Kew Bridge,† *Newbould*; Horsendon Hill,† *T. & D.*, 139. Hanwell. Canal path between Hanwell and Southall. Perivale Park. **6.** Highgate, *Coop. Fl.*, 104. Trent Park, Southgate, *A. B. Cole*, *T. & D.*, 139. Woodside Park,† *C. S. Nicholson, Lond. Nat. MSS.* Enfield, *Kennedy*. **7.** Chelsea,† *Newbould*; roadside, Prince's Gate, South Kensington, 1867,† *T. & D.*, 139. East Heath, Hampstead, casual, 1949,† *H. C. Harris.* Eaton Square, S.W.1, 1956!, *L.N.* 39: 55. Buckingham Palace grounds, *Fl. B.P.G.*

Var. **mutica** Wirtg.

5. Acton, 1912, *Wedg. Cat.*

Plants with white flowers were gathered at **2.** Laleham, *Benbow* (BM).

CRUCIATA Mill.

C. laevipes Opiz Crosswort, Mugwort.

C. ciliata Opiz, *C. chersonensis* auct., *Galium cruciata* (L.) Scop.

Native in open woods, meadows, grassy places, on hedgebanks, commons, etc., rare, and mainly confined to calcareous and other light soils.

First record: *Gerarde*, 1597.

1. Harefield,† *Blackst. Fasc.*, 32. Lane between Potters Bar station and North Mimms Wood!, *Phyt. N.S.* 1: 407. Between South Mimms and Warren Gate!, *Benbow, J.B. (London)* 25: 16; 1967. Mimmshall Wood!, *Benbow MSS.*; 1967. South Mimms Churchyard, 1967. Long Lane, Hillingdon, one plant,† *Benbow, J.B. (London)* 23: 37. Ickenham,† *Benbow MSS.* **2.** Staines Moor!; between Staines and Hampton; between Sunbury and Walton Bridge!, *T. & D.*, 140; 1967. Abundant in a dry ditch by the towing-path between Hampton Court and Kingston Bridge,† *T. & D.*, 140. Yeoveney;† Ashford Common,† *Benbow, J.B. (London)* 25: 16. Laleham!; Charlton; Littleton!; Shepperton!; Halliford, *Benbow MSS.* **4.** Hampstead Heath,† *Johns. Enum.* Wood near Finchley,† *Irv. MSS.* **6.** Tottenham,† *Johns. Cat.* Near Hadley Common, *Benbow, J.B. (London)* 25: 16. **7.** Hampstead Churchyard and a pasture adjoining thereto, *Ger. Hb.*, 965; c. 1640, *Park. Theat.*, 566; c. 1666, *Merr. Pin.*, 31; c. 1760,† *W. Watson, Hill Fl. Brit.*, 513. Hornsey, 1863,† *French* (BM).

GALIUM L.

G. odoratum (L.) Scop. Sweet Woodruff.

Asperula odorata L.

Native in woods, on shady hedgebanks, etc. Rare, and mainly confined to damp calcareous and base-rich soils chiefly in the northern parts of the vice-county. Occurs also as a garden escape.

First record: *Blackstone*, 1737.

1. Harefield!, *Blackst. Fasc.*, 8. Ruislip, *Farrar, T. & D.*, 139. Garett Wood!;† Top wood, north of Harefield!; Stanmore Common!; Mimmshall Wood!, *Benbow MSS.* Near Denham. Near Elstree. **3.** Wood near Hounslow Heath!, 1965, *P. J. Edwards*; 1967. **4.** Bishop's Wood!, *Irv. MSS.* Stanmore; Scratch Wood!, *T. & D.*, 140. Hampstead,† *Pet. H.B.C.* Harrow, *Farrar, T. & D.*, 140; casual, *Harley Fl.*, 25. Woods about Edgware and Brockley Hill, *Benbow MSS.* Highwood Hill!, *Moring MSS.* Mill Hill, *Sennitt.* **6.** Colney Hatch, 1821, *J. J. Bennett* (BM); c. 1840† (HE). Forty Hill, Enfield,† *T. & D.*, 140; Whitewebbs Park, Enfield. Hadley Common!, *K. & L.*, 141. **7.** Ken Wood,† *Pulteney, Hill Fl. Brit.*, 74. Hornsey Wood, *Forst. MSS.*; 1837,† *Ball. MSS.*

G. album Mill. subsp. **album** Hedge Bedstraw.

G. mollugo auct., non L., *G. erectum* Huds.

Lit. *Proc. Bot. Soc. Brit. Isles* **7**: 438 (1968) (*Abstr.*)

Native on hedgebanks, in open woods, ditches, etc. Locally plentiful on calcareous soils in the north-west of the vice-county, and to a lesser extent on alluvial and gravel soils near the Thames, elsewhere by roadsides, on hedgebanks, railway banks, etc., as an introduction.

First record: *Blackstone*, 1737.

1. Locally plentiful, especially on calcareous soils about Harefield. **2.** Tangley Park,† *T. & D.*, 141. Penton Hook!, *C. S. Nicholson, K. & L.*, 139; 1968. East Bedfont,† *Westrup, K. & L.*, 139. Hampton Hill, 1965, *Clement*. Hampton Water Works, 1967, *Kennedy*!. **3.** Hayes, *P. H. Cooke, K. & L.*, 139. Hounslow Heath, 1949!,† *L.N.* 29: 12. Near Hatch End, scarce, 1966. **4.** Harrow,† *Hind Fl.* Harrow Weald, *T. & D.*, 141. Edgware, 1871, *F. J. Hanbury* (HY). Stanmore Churchyard, 1957. Near Brent Reservoir, *Warren, T. & D.*, 424. Brent Reservoir, 1948, *Meikle, K. & L.*, 139. Deacon's Hill!, *K. & L.*, 139. Hampstead,† *Irv. MSS.* **5.** Greenford!, *Coop. Fl.*, 108. Chiswick,† *Newbould*; by the Brent near Hanwell,† *Warren, T. & D.*, 141. Perivale!; near Park Royal!,† *L.N.* 27: 32. Railway banks between Southall and Brentford, frequent. West Ealing goods yard. Ealing and Old Brentford Cemetery, South Ealing. **6.** Roadside near Botany Bay, Enfield Chase, a single patch, 1965; field near Enfield Wash, scarce, 1966, *Kennedy*!. **7.** Chelsea† (BM). Near Kilburn,† *Warren, T. & D.*, 141. Clapton Marshes,† *W. H. Griffin*; Highgate Cemetery, *Fitter, K. & L.*, 139.

G. verum L. Lady's Bedstraw.

Native in dry fields, on heaths, hedgebanks, etc., established on lawns, railway banks, etc. Common.

First record: *T. Johnson*, 1638.

7. Hyde Park; Kensington Gardens!, *Warren Fl.* Ken Wood!, *Fitter*. Fulham Palace grounds, *Welch*!. Buckingham Palace grounds, *Fl. B.P.G.* Brompton Cemetery. Holland House grounds, Kensington. Notting Hill. Kilburn. Islington. Upper Clapton. East Heath, Hampstead. Highgate Cemetery. Lower Thames Street, E.C., 1967, *Lousley, L.N.* 47: 10.

G. saxatile L. Heath Bedstraw.

G. harcynicum Weigel, *G. hercynicum* auct.

Native on heaths, in open woods, etc. Locally abundant, mainly on acid soils.

First record: *Merrett*, 1666.

1. Ruislip!, *Melv. Fl.*, 39. Harrow Weald Common!; Stanmore Common!, *T. & D.*, 142. Uxbridge Common!; Hillingdon Heath;† near Swakeleys; Pinner Hill!; Eastcote, *Benbow MSS.* **2.** Roadside from Staines to Hampton, abundant; Bushy Park!, *T. & D.*, 142. Staines!, *Benbow MSS.* West Drayton,† *L. B. Hall, Lond. Nat. MSS.* **3.** Hounslow Heath!, *Merr. Pin.*, 9. **4.** Hampstead Heath!, *Pet. H.B.C.* Deacon's Hill, *T. & D.*, 142. Scratch Wood!, *Benbow MSS.* Mill Hill, *Sennitt*. Harrow Weald. Bentley Priory. Brockley Hill,

Kennedy!. **5.** Syon Park. Perivale Wood,† *Shenst. Fl.* **6.** Highgate Wood, *C. S. Nicholson* (LNHS). Winchmore Hill Wood;† Enfield Chase!, *Benbow MSS*. Trent Park. Finchley Common. **7.** Between Hyde Park and Kensington Gardens, a single tuft, 1868,† *Warren*; East Heath, Hampstead!, *T. & D.*, 142. Ken Wood!, *Fitter*. Buckingham Palace grounds, *Fl. B.P.G.*

G. palustre L. Marsh Bedstraw.
Native in marshes, ditches, by streamsides, etc. Rather common.
First record: *T. Johnson*, 1638.

Subsp. **elongatum** (C. Presl) Lange
G. elongatum C. Presl
1. Colne near Yiewsley!, *Newbould, T. & D.*, 142. Near Cowley, *Benbow* (BM). **2.** Walton Bridge, 1862† (W). Staines!, *Newbould*; Bushy Park!, *T. & D.*, 142. Hampton; Hampton Court! (W). **3.** Roxeth,† *Melv. Fl.*, 39. By the Crane in many places, as at Hospital Bridge, *T. & D.*, 142.

Subsp. **palustre**
var. *witheringii* Sm.
1. Harefield Moor, frequent. Harrow Weald Common; near South Mimms!, *T. & D.*, 142. Hillingdon,† *Trim. MSS*. Between Uxbridge Common and Swakeleys;† Ickenham Green!,† *Benbow* (BM). Uxbridge. Denham. Potters Bar. **2.** Near Staines; Staines Moor!; Sunbury!, *T. & D.*, 143. Laleham. Yeoveney. **3.** Hounslow Heath,† *Lightfoot* (K). **4.** Hampstead Heath, *Coop. Fl.*, 102. Harrow!, *Hind Fl.* Between Finchley and Mill Hill, *Newbould, Trim. MSS*. Brockley Hill, *Kennedy*!. Edgware!, *Horder*. **5.** Greenford!, *Coop. Fl.*, 102. Between Acton and Turnham Green,† *Newbould, T. & D.*, 143. Osterley Park. Near Boston Manor.† **6.** Tottenham,† *Johns. Cat.* Trent Park!, *T. G. Collett* and *Kennedy*. Enfield, *Kennedy*!.
It is possible that some of the old records given may be referable to subsp. *elongatum*.

G. uliginosum L. Fen, or Bog Bedstraw.
Native in bogs and marshes. Very rare, and confined to acid soils.
First evidence: *Buddle*, c. 1700.
1. Near Harefield,† *Newbould, T. & D.*, 142. Bog by Moor Hall;† moors below Harefield and Springwell; Uxbridge Common;† near Potters Bar, *Benbow MSS*. Bog in field by Harrow Weald Common, 1940!,† *K. & L.*, 140; bog drained, c. 1944. Ruislip Common, 1967, *J. A. Moore*. **3.** Roxeth,† *Hind Fl.* **4.** Bogs on Hampstead Heath, *Buddle* (SLO); c. 1710 (DB); 1815† (G & R). Marshy field below Hampstead Heath, 1881,† *De Crespigny* (BM). **5.** Between Acton and Turnham Green,† *Newbould, T. & D.*, 142. **6.** Finchley, 1827–

30, *Varenne, T. & D.*, 142; Finchley Common, very scarce, 1946!,† *K. & L.*, 140.

G. aparine L. Goosegrass, Cleavers.
Native on hedgebanks, cultivated and waste ground, in woods, thickets, ditches, etc., established on railway banks, in gravel pits, etc. Very common.
First record: *T. Johnson*, 1638.
7. Common in the district.

Var. **pseudo-vaillantii** Ar. Benn.
1. Uxbridge, 1933, *Lady Davy* (BM).

G. parisiense L. subsp. **anglicum** (Huds.) Clapham Wall Bedstraw.
G. anglicum Huds.
Formerly native in stony places, established on old walls, etc., but long extinct.
First, and only record: *J. Sherard*, 1690.
7. . . . at Hackney, on a wall, *J. Sherard, Ray Syn.*, 237.

CAPRIFOLIACEAE

SAMBUCUS L.

S. ebulus L. Danewort, Dwarf Elder.
Alien. Europe, etc. Naturalised by roadsides, on waste ground, etc. Very rare.
First record: *Gerarde*, 1597.
1. Meadow near Breakspears;† Uxbridge Moor, plentifully,† *Blackst. Fasc.*, 24. Uxbridge Churchyard,† *M. & G.*, 464. **3.** Marsh Farm, Twickenham, near the railway!, *T. & D.*, 137; still plentiful, 1968. **5.** Brentford,† *Ger. Hb.*, 1238. Heston, *Masters*; Chiswick,† *F. Naylor*; near Greenford,† *Lees, T. & D.*, 137. Osterley Park!, perhaps planted, *K. & L.*, 137. Near Wyke Green, in two places, 1963→. **6.** Enfield Chase, *Phyt. N.S.* 6: 301. **7.** Plentifully in the lane at Kilburn Abbey,† *Ger. Hb.*, 1238. Wild in the Bishop of Ely's garden in Holborn,† *W. Watson, Hill Fl. Brit.*, 163. Regents Park,† *Webst. Reg.*, 102. Abundant by river Lee near Old Ford, 1957!, *L. M. P. Small*; 1967. Bombed site north-west of Victoria Park, Hackney, *Wurzell, L.N.* 43: 22.

S. nigra L. Elder.
Native in woods, hedges, on waste ground, by river- and streamsides, etc., established on railway banks, etc. Very common.
First record: *T. Johnson*, 1632.
7. Very common in the district.

Var. **laciniata** L.

1. Hillingdon Heath, *L. M. P. Small.* **2.** Near Shepperton. **3.** Between Twickenham and Isleworth, *Dyer* (BM). Hanworth, *Welch, L.N.* 26: 61. **4.** Harrow, 1966, *Temple*!. **5.** Syon Park!, *Tremayne*. Thames side, Chiswick!,† *K. & L.*, 137. **7.** Zoological Gardens, Regents Park, 1902, *Wright* (BM).

Probably always planted.

Var. **viridis** Ait.

var. *leucocarpa* Koch

1. Near Ruislip, a single tree!, 1942, *Batko*, fide *Sandwith, Rep. Bot. Soc. & E.C.* 12: 727; 1946, *Batko*!. Yiewsley, a single tree, 1920–23, *J. E. Cooper, K. & L.*, 197. **5.** Hedge near Perivale Church,† *Lees, T. & D.*, 138. Hedges of Drayton House, Ealing,† *Jacks. Ann.*, 341. Near the canal, Hanwell, in several places!, *K. & L.*, 197. **6.** Hadley, a single tree, 1923–24; Church End, Finchley, 1918–23, *J. E. Cooper, K. & L.*, 197. Whetstone Stray, 1960, *J. G. Dony*!; 1966.

Possibly planted in some, if not all, of its localities.

VIBURNUM L.

V. lantana L. Wayfaring Tree.

Native in hedges and thickets on calcareous soils at Harefield, and on alluvial soils near the Thames, elsewhere probably an introduction. Very rare.

First record: *T. Johnson*, 1638.

1. Harefield!, *Blackst. Fasc.*, 108; 1927, *Tremayne*; 1952!, *K. & L.*, 138; not recently seen, and perhaps extinct. Near Denham.† Ruislip, 1963, *Pickess, Moxey List.* **2.** Hammonds, near Staines, *T. & D.*, 138. Staines Moor!, *Benbow MSS.* **3.** By the Crane, Hounslow Heath, 1946, *Westrup, K. & L.*, 138. **4.** Hendon, 1823, *Maurice*, fide *Varenne, T. & D.*, 138. Hampstead, *Irv. MSS.*; Hampstead Heath,† *Whiting Notes.* Harrow Park, *Melv. Fl.*, 36; 1907,† *Sherrin* (SLBI). Between Harrow and Kingsbury,† *Foley*, 32. **6.** Tottenham,† *Johns. Cat.* Trent Park, 1966, *T. G. Collett* and *Kennedy*. **7.** Hornsey Wood,† *Ball. MSS.*

V. opulus L. Guelder Rose.

Native in damp woods, thickets, by river- and streamsides, etc., established on railway banks, etc. Locally common in the Colne valley and near the Thames, rather rare elsewhere.

First record: *Petiver*, 1695.

1. Harefield!, *Blackst. Fasc.*, 108; 1968. Between Yiewsley and Iver Bridge!, *Newbould, T. & D.*, 138. Cowley!; copse by Denham Lock!; Moorhall;† Ruislip!; Pinner!; Pinner Wood!, *Benbow MSS.* Swakeleys, 1927,† *Tremayne*; between Potters Bar and South Mimms!, *P. H. Cooke*, *K. & L.*, 138. Mimmshall Wood. Springwell. Ickenham, *L. James*. Uxbridge. Furzefield Wood, Potters Bar, *Kennedy*!. **2.** Staines!, *T. & D.*, 138. Littleton!, *Benbow MSS*. Colnbrook!, *Druce Notes*. Harmondsworth, *P. H. Cooke* (LNHS). Laleham Park, *Westrup*, *K. & L.*, 138. West Drayton, 1883,† *Pearce* (BM). Stanwell Moor. Railway side, Yeoveney to Poyle, frequent. Near Chertsey Bridge. Shepperton. Halliford. **3.** Wood End,† *Melv. Fl.*, 37. Marsh Farm, Twickenham;† between Twickenham and Worton, *T. & D.*, 138; 1908,† *Sprague* (K). Isleworth, 1969, *Sandford*. **4.** Bishop's Wood, *M. & G.*, 476; 1902, *C. S. Nicholson*; 1910, *J. E. Cooper*; Temple Fortune, 1910,† *P. H. Cooke*, *K. & L.*, 138. Harrow Park, *Melv. Fl.*, 37. Scratch Wood!, *Benbow MSS*. Hampstead Heath,† *M. & G.*, 476. North side of Turner's Wood, abundant, *Trim. MSS*. Between Harrow and Kingsbury,† *Foley*, 32. **5.** Perivale Wood!, *Benbow MSS*. Long Wood, Wyke Green, 1933 !,† *K. & L.*, 138. **6.** The Alder's Wood, Whetstone,† *T. & D.*, 138. Winchmore Hill Wood;† Coldfall Wood, Highgate!, *Benbow MSS*. Highgate Wood, 1900, *J. E. Cooper*, *K. & L.*, 138. Queen's Wood, Highgate, 1963, *N. H. Martin*. **7.** In a wood against the Boarded river,† *Pet. Midd.* Hornsey Wood, *M. & G.*, 471; c. 1837,† *Ball. MSS*. Near Kentish Town,† *M. & G.*, 471. Crouch End fields,† *C. S. Nicholson*, *K. & L.*, 138.

SYMPHORICARPOS Duham.

S. rivularis Suksd. Snowberry.

S. albus auct., *S. racemosus* auct.

Alien. N. America. Garden escape, or outcast. Naturalised on hedgebanks, waste ground, in thickets, etc., almost always near human habitation, and sometimes planted. Common, especially in the north of the vice-county.

First record: *G. C. Druce*, 1910.

7. Fitzroy Park, Highgate, *Fitter*. Kensal Green. Holloway. Stoke Newington. Upper Clapton.

LONICERA L.

L. xylosteum L. Fly Honeysuckle.

Introduced. Garden escape. Formerly naturalised in hedges, but now extinct.

First record: *A. Irvine*, c. 1830. Last evidence: *Benbow*, 1887.

1. Hillingdon, 1887, *Benbow* (BM). **3.** Two shrubs in Harrow Grove, *Melv. Fl.*, 38. **7.** In a hedge opposite Lower Nursery, Haverstock Hill, *Irv. MSS*.

L. periclymenum L. Common Honeysuckle, Woodbine.
Native in woods, thickets, on heaths, etc. Locally common in the north of the vice-county, rare elsewhere.
First record: *T. Johnson*, 1638.
1. Harefield!, *Blackst. Fasc.*, 75. Stanmore Common, locally abundant. Harrow Weald Common. Ruislip. Ickenham. Uxbridge. Near Denham. Northwood. South Mimms. Bentley Heath, *J. G.* and *C. M. Dony*!. Potters Bar, *Kennedy*!. Near Potters Bar. **2.** East Bedfont, *P. H. Cooke*. Stanwell. **3.** Hounslow Heath. **4.** Bishop's Wood, *M. & G.*, 343. Harrow, *Hind Fl.* Mill Hill. Scratch Wood. Whitchurch Common. Brockley Hill, *Kennedy*!. West Heath, Hampstead. East Finchley. **5.** Greenford, *Melv. Fl.*, 38. Perivale Wood. Horsendon Hill. Hanwell. **6.** Highgate Woods!, *Irv. MSS.* Near East Barnet. Hadley Common. Enfield Chase. **7.** Tottenham,† *Johns. Cat.* Hornsey Wood, *M. & G.*, 343; c. 1837, *Ball. MSS.* Ken Wood!, *H. C. Harris*. East Heath, Hampstead.

L. caprifolium L. Perfoliate Honeysuckle.
Alien. Europe, etc. Garden escape. Formerly naturalised in hedges but now extinct.
First, and only evidence: *E. Ford*, 1873.
6. Enfield, *H. & F.*, 149.

ADOXACEAE

ADOXA L.

A. moschatellina L. Moschatel, Townhall Clock.
Native, in woods, on shady hedgebanks, etc. Formerly local, but now rare and mainly confined to the north of the vice-county.
First record: *Merrett*, c. 1670.
1. Old Park Wood!, and many other places about Harefield, *Blackst. Fasc.*, 83; grove near Harefield Church, 1853 (HE); Harefield Grove, 1966, *I. G. Johnson*. Roadside between Uxbridge and West Drayton, 1855,† *Phyt. N.S.* 1: 62. Hillingdon;† Swakeleys!; Ickenham!; Moorhall;† near Bayhurst Wood!, *Benbow MSS.* Near Mimms Wash, 1946, *Welch, K. & L.*, 137. North of Bentley Heath, 1966, *Kennedy*. Furzefield Wood, Potters Bar, 1967, *Kennedy*!. **2.** Harmondsworth,† *Monckt. Fl.*, 36. Perhaps an error. **4.** Near Hampstead,† *Merr. MSS.* Hampstead Hill,† *Mart. Pl. Cant.*, 67. Hampstead Heath,† *Whiting Notes.* Kingsbury,† *A. B. Cole*; fields by Blackpot Farm [Kingsbury], near the Brent, 1865,† *Farrar*; near Hendon,† *W. Davies*, *T. & D.*, 136. Wembley Park,† *Benbow MSS.* Temple Fortune, 1914,† *J. E. Cooper* (BM). Harrow-on-the-Hill Churchyard!, *Knipe*; 1966. **5.** By the Brent, Hanwell,† *Benbow MSS.* **6.** Enfield, *Forst. Midd.*; c. 1873, *H. & F.*, 147. Near Edmonton,

Macreight Man., 108; c. 1869,† *T. & D.*, 136. Church End, Finchley, 1902,† *E. F.* and *H. Drabble* (BM). **7.** On the moatside as you enter into Jack Straw's Castle,† *Pet. Midd.* Hornsey Wood;† Tottenham,† *Blackst. Spec.*, 54. Highbury Barn,† *Forst. Midd.* Field near Ken Wood, *Dennes, Coop. Suppl.*, 11; abundant, 1866, *T. & D.*, 136.

VALERIANACEAE

VALERIANELLA Mill.

V. locusta (L.) Betcke Lamb's Lettuce, Corn Salad.
V. olitoria (L.) Poll.
Native on cultivated ground, in dry grassy places, etc. Locally frequent near the Thames, rather rare elsewhere, and sometimes only casual.
First certain record: *Blackstone*, 1737.
1. Harefield!, *Blackst. Fasc.*, 48. Pinner, *Hind, Trim. MSS.* Near Springwell, 1965. Near Denham, 1965. Near Potters Bar, *J. G.* and *C. M. Dony*. Potters Bar, *Widgery*. **2.** Between Hampton Court and Kingston Bridge!, *Newbould*; near Staines, *T. & D.*, 144. Staines, *Briggs*. West Drayton!,† *P. H. Cooke, K. & L.*, 142. Colnbrook, *P. H. Cooke*. Penton Hook Lock, *C. S. Nicholson* (LNHS). **3.** Roxeth,† *Melv. Fl.*, 40. Isleworth,† *T. & D.*, 144. **4.** Hampstead,† *Irv. MSS.* Kingsbury,† *Farrar, T. & D.*, 144. Willesden,† *De Cresp. Fl.*, 75. **5.** Hanwell,† *Newbould*; Turnham Green;† Chiswick!, *T. & D.*, 144. Syon Park, *A. B. Jackson*!, *K. & L.*, 142. Railway bank, Brentford, 1966. **7.** St John's Wood, 1815† (G & R). Brompton Cemetery,† *Britten, Phyt. N.S.* 6: 592.

V. carinata Lois.
Native on cultivated and waste ground, etc., chiefly on calcareous and other light soils. Very rare.
First evidence: *Trimen*, 1866.
1. Above the chalkpits, Harefield!; Springwell, *Benbow*; north-west of Harefield, 1886, *W. R. Linton* (BM). **2.** Railway bank between West Drayton and Uxbridge!,† very abundant, *Benbow, J.B. (London)* 25: 16. **5.** Railway bank, Kew Bridge station, very abundant, 1866,† *Trimen* (BM); a few specimens in other neighbouring places,† *T. & D.*, 145. Ealing,† *A. Wood, Trim. MSS.*

V. rimosa Bast.
Native in cornfields, on cultivated ground, established on railway banks, etc. Very rare, and perhaps extinct.
First evidence: *Kingsley*, 1839.
1. Harefield, 1839, *Kingsley* (ME); 1889; Springwell, 1889, *Benbow MSS.* Between Eastcote and Northwood,† *Benbow, J.B. (London)* 23: 339. Railway bank near Uxbridge, 1941.

V. dentata (L.) Poll.
Native in cornfields, on cultivated ground, etc. Very rare, and perhaps extinct.
First evidence: *Kingsley*, 1839.
1. Near Harefield, 1839, *Kingsley* (ME); 1864, *Hind* (BM). Between Harefield and Springwell!, *Benbow MSS.* Harefield, 1912, *J. E. Cooper* (BM). Ruislip, *Hind Fl.* Between Uxbridge and West Drayton, 1944!,† *K. & L.*, 143. **3.** Hounslow,† *T. & D.*, 145. **4.** Hendon, 1910, *P. H. Cooke* (LNHS).

Var. **mixta** (L.) Dufresne
1. Ruislip, *De Crespigny, Williams MSS.* **2.** Teddington, *Henfrey, Williams MSS.* **4.** Mill Hill, *Harrison*, fide *Sandwith.*

VALERIANA L.

V. officinalis L. Valerian.
Lit. *Watsonia* **2**: 145 (1952): *Proc. Bot. Soc. Brit. Isles* **1**: 202 (1954). (*Abstr.*)
Native by river- and streamsides, in marshes, on pond verges, etc., established by canals, etc. Formerly common, but now local and decreasing.
First record: *T. Johnson*, 1632.
1. In scattered localities in the Colne valley from Springwell to Yiewsley. Ruislip!, *Melv. Fl.*, 40. Stanmore Common, *Payton*!. South Mimms, *Benbow MSS.* **2.** Locally common by the Thames between Staines and Hampton Wick. Poyle!, *Druce Notes.* Staines Moor. Colnbrook. West Drayton. **3.** Near Hatton;† riverside between Twickenham and Richmond Bridge, *T. & D.*, 143. Harlington,† *P. H. Cooke.* **4.** Hampstead Heath,† *Johns. Enum.* Near Finchley, *Newbould*; Edgware; Turner's Wood, *T. & D.*, 143. Near Mill Hill, *Benbow MSS.* **5.** Chiswick, *T. & D.*, 144. Between Brentford and Hanwell,† *A. Wood, Trim. MSS.* Thames side, Syon Park. **6.** Tottenham, *Johns. Cat.* Southgate, *A. B. Cole, T. & D.*, 144. Shady stream, Highgate Woods, 1904, *C. S. Nicholson* (LNHS). Near Enfield Lock, *Kennedy*!. **7.** Bishop's Walk, Fulham,† *Coop. Fl.*, 104. Isle of Dogs in great abundance;† Millbank;† Tothill Fields,† *M. & G.*, 49. Millfield Lane;† East Heath, Hampstead, *T. & D.*, 144.

Var. **secundo-dasycarpa** Kreyer
Lit. *Proc. Linn. Soc.* **155**: 102 (1944).
3. Twickenham (K), fide *Sprague.*

V. dioica L. Marsh Valerian.
Native in bogs, marshes, wet fields, etc. Very rare.
First record: *Blackstone*, 1737.

1. Moist meadows near Harefield, *Blackst. Fasc.*, 105. Harefield Moor!; Springwell;† meadows by the Colne, Uxbridge, *Benbow MSS.* Ruislip,† *Farrar*, fide *Hind*, *T. & D.*, 144. Denham Lock, 1912;† Yiewsley, 1912–19,† *J. E. Cooper*, *K. & L.*, 142. **2.** Teddington, 1792,† *Forst. MSS.* Hampton, 1841† (W). Between Hampton and Hampton Court, by the river,† *Newbould*, *T. & D.*, 44. Yeoveney;† Staines Moor;† Laleham,† *Benbow MSS.* **4.** Hampstead Heath, *Irv. MSS.* and *Coop. Fl.*, 105; very scarce, c. 1912,† *Whiting Fl.* **5.** Between Acton and Turnham Green,† *Newbould*, *T. & D.*, 144. **6.** Near Trent Park, Southgate, *A. B. Cole*; Edmonton;† Alder's Wood† and meadows adjacent, Whetstone,† *T. & D.*, 144. Enfield, *H. & F.*, 150. **7.** Ken Wood,† *Hunter*, *Park Hampst.* Banks of the Thames, Fulham,† *Faulkn. Hist.*, 22. Error. Hackney Marsh,† *Warner*, *Coop. Fl.*, 105.

CENTRANTHUS DC.

C. ruber (L.) DC. Red Valerian.

Alien. Europe, etc. Garden escape. Naturalised on waste ground and established on old walls, etc. Rare.

First record: *Crowe*, c. 1780.

1. Old walls, Manor Farm, Ickenham, c. 1884,† *Benbow MSS.* Old walls, Uxbridge!, 1862, *Benbow* (BM); 1947→!; old walls, Northwood, 1940→!, *K. & L.*, 142. Old wall near Stanmore Common, 1965→. **2.** Old walls by West Drayton Church!, *L.N.* 26: 61; 1947→. Old wall, Laleham, 1964→. Old walls, Sunbury!, *L.N.* 26: 61; 1947→, also on the river wall of the Thames. **3.** Headstone Moat,† *Melv. Fl.*, 97. Old walls, Isleworth!, *L.N.* 26: 61; 1947→. **4.** Old walls, London Road, Harrow!, *Harley Fl.*, 26. **5.** Roadbank, Horsendon Hill, casual, 1945!;† Ealing Common, casual, 1946!,† *L.N.* 26: 61. Brickwork by railway siding, West Ealing, 1964→. **6.** Old walls, Enfield, 1965, *Bangerter* and *Kennedy*!. **7.** Old walls, Bishopsgate,† *Crowe MSS.* Bombed site, Gresham Street area, E.C.,† *Jones Fl.*

White-flowered plants often grow with the more frequent red-flowered plant.

C. calcitrapa (L.) DC.

Alien. Europe, etc. Garden escape. Formerly established on old walls but now long extinct.

First evidence: *E. Forster*, c. 1790. Last record: *Pamplin*, 1825.

6. Walls belonging to the Palace at Enfield, *E. Forster* (BM). **7.** Walls belonging to Chelsea Hospital, *Dickson . . . Caley . . .* reports it completely naturalised, *With. Bot. Arr.* 4, 1: 63; c. 1800, *A. Lambert*; 1815, *Rozea* (BM); 1825, *Pamplin MSS.*

DIPSACACEAE

DIPSACUS L.

D. fullonum L. subsp. **fullonum** Wild Teasel.
D. sylvestris Huds.
Native by river- and streamsides, in woods, copses, rough fields, on waste ground, etc., established on railway banks, etc. Common, especially on heavy soils, though sometimes planted for ornamental purposes.
First record: *T. Johnson*, 1638.
7. Tottenham!, *Johns. Cat.* Kilburn, 1815 (G & R). Isle of Dogs!, 1844, *R. Pryor* (BM); 1967. Finchley Road, 1845, *J. Morris*, *T. & D.*, 145; c. 1870, *Trim. MSS.* Between Kentish Town and Hampstead, *T. & D.*, 145. Roadside near, and beyond Hornsey Wood, *Ball. MSS.* Ken Wood, *Whiting Fl.* Canal bank, Westbourne Park. Bishop's Park, Fulham, *Welch*!. St Peter's Churchyard, Eaton Square, S.W.1, *McClintock*, *L.N.* 39: 55. Kensington Gardens, probably planted!, *L.N.* 41: 20. Site of Harrow Road–Edgware Road Flyover, Lisson Street, N.W.1, 1966!,† *L.N.* 46: 33. Railway bank, West Brompton, 1967. Regents Park, 1966. Hornsey. Wapping. Upper Clapton. Stepney. Hackney Wick. Hackney Marshes. Poplar. Bromley-by-Bow.

Subsp. **sativus** (L.) Thell. Fuller's Teasel.
D. sativus (L.) Honck.
1. Uxbridge!, *Druce Notes*; 1968, *L. M. P. Small.* Yiewsley, 1908, 1910–11 and 1913–14, *Coop. Cas.*; 1915–16, *J. E. Cooper*; 1916, *A. D. Webster* (BM). **2.** West Drayton, 1916, *Wedg. Cat.*; 1917, *B. Reynolds* (BM). East Bedfont, 1965. **3.** Hounslow Heath, 1966, *Wurzell.* Between Hayes and West Drayton, 1967.

D. pilosus L. Small Teasel.
Native in thickets, damp woods, etc. Very rare.
First record: *T. Johnson*, 1638.
1. Moor Hall;† Harefield,† *Blackst. Fasc.*, 24. Springwell, one plant, 1944!,† *L.N.* 26: 61. Thicket by the canal near Uxbridge!, c. 1870, *A. Wood*, *Trim. MSS.*; still present in this locality which is north of Denham Lock. Uxbridge Common, a single plant, 1887;† Cowley, *Benbow MSS.*; 1915,† *J. E. Cooper* (BM). **4.** Between Highgate and Finchley,† *Mart. Tourn.* 1: 230. Near Golders Green,† *Irv. MSS.* Between Bishop's Wood and Finchley, 1865,† *Grugeon* (BM). **6.** Tottenham,† *Johns. Cat.* Near Edmonton,† *Forst. Midd.* **7.** Between Fulham and Hammersmith,† *Merr. Pin.*, 33. Fulham,† *Pluk. MSS.*, *Pet. Midd.* and *Mart. Pl. Cant.*, 66. Hornsey Churchyard,† *M. & G.*, 154.

KNAUTIA L.

K. arvensis (L.) Coult. Field Scabious.
Scabiosa arvensis L.
Native in dry grassy places, etc., and established on railway banks. Locally plentiful, especially on light soils.
First record: *T. Johnson*, 1638.
1. Harefield!, *Blackst. Fasc.*, 91. Pinner, *T. & D.*, 146. Cowley; Hillingdon; Pield Heath; Breakspeares; Springwell!; South Mimms!, *Benbow MSS.* Near Uxbridge!; Northwood!; Potters Bar!, *K. & L.*, 144. Near Denham. Ruislip Common. **2.** Locally common, especially near the Thames. **3.** Twickenham, *T. & D.*, 146. Near Ashford; near Hanworth!; Feltham!, *Benbow MSS.* Cranford!, *C. S. Nicholson*; Hounslow Heath!; near Hayes!, *K. & L.*, 144. Hatch End!, *Bull.* Northolt. Sudbury. Southall. **4.** Stanmore, 1827–30, *Varenne, T. & D.*, 146. Hampstead,† *Irv. MSS.* Railway bank, Harrow!, *Melv. Fl.*, 41. Railway bank, Neasden. Brent Reservoir. Mill Hill. **5.** Greenford,† *Melv. Fl.*, 41. Hanger Hill, Ealing, 1945!,† *Boucher, K. & L.*, 144. Hanwell. Syon Park. **6.** Tottenham, *Johns. Cat.* Oakleigh Park; Hadley, *J. G. Dony*!. Enfield. Enfield Chase. **7.** Isle of Dogs,† *Benbow MSS.* Bombed site, Queen Victoria Street area,† *Jones Fl.* Bank of moat, Tower of London, 1967.

SCABIOSA L.

S. columbaria L. Small Scabious.
Formerly native in alluvial meadows by the Thames, and on the adjacent towing-path, but now extinct.
First evidence: *Dyer*, 1867. Last record: *D. H. Kent*, c. 1939.
2. By the towing-path between Hampton Court and Kingston Bridge!, 1867, *Dyer* (BM); 1923, *Spreadbury, E.K.* and *W. K. Robinson, Country-side* 1923, 137; c. 1939!, *K. & L.*, 143. Field near Teddington Church, with *Campanula glomerata*, *T. & D.*, 147. **7.** Hornsey, casual, 1872, *French* (BM).

S. atropurpurea L.
S. maritima L.
Alien. S. Europe, etc. Garden escape. Naturalised on waste ground, established on railway banks, etc. Very rare.
First record: *Bull*, 1963.
1. Yiewsley, 1963, *Wurzell.* **3.** Railway bank between Hatch End and Carpenders Park, 1963!, *Bull*; 1967.

SUCCISA Haller

S. pratensis Moench Devil's-bit Scabious.
Scabiosa succisa L.
Lit. *J. Ecol.* **43**: 709 (1955).
Native in meadows, marshes, damp woods, etc. Common, particularly on heavy soils.
First record: *Gerarde*, 1597.
7. Near Kentish Town,† *Blackst. MSS.* East Heath, Hampstead, *T. & D.*, 147. Ken Wood, 1837, *E. Ballard* (BM); c. 1886, *Wharton Fl.* Highgate Ponds!, *Fitter.* Tower of London Gardens, 1949!, *L.N.* 39: 55.

COMPOSITAE

HELIANTHUS L.

H. annuus L. Common Sunflower.
Alien. Central America, etc. Garden escape, also introduced with cage bird-seed. Naturalised on waste ground, established on rubbish-tips, etc. Common.
First record: *Melville* and *R. L. Smith*, 1927.
7. Kensington Gardens, one plant, 1961, *D. E. Allen, L.N.* 41: 21. Hyde Park, one plant by subway to Marble Arch, 1962!, *Allen Fl.* St John's Wood. Regents Park. Hornsey. Tower Hill. Upper Clapton. Tottenham.
Heiser, *et al.* (*Mem. Torrey Bot. Club* **22**(**3**) (1969)) refer the giant monocephalic plant grown for its seed to var. **macrocarpus** (DC.) Ckll. The plants normally cultivated for ornamental purposes are, however, referred to subsp. **annuus**.

H. annuus × **decapetalus** L. = **H.** × **multiflorus** L.
7. Bombed sites between Leather Lane and Hatton Garden, W.C.;† bombed site on north side of St Paul's Cathedral, 1950,† *Whittaker* (BM). Regents Park, 1966, *Holme, L.N.* 46: 33 (as *H. decapetalus*). Isle of Dogs. Tottenham. Victoria Park. Limehouse. Bromley-by-Bow.
Heiser, *et al.* (*loc. cit.*) point out that this sterile triploid hybrid which spreads solely by means of rhizomes is known only as a cultivated ornamental plant in which both single-flowered and double-flowered forms occur. It is quite frequent on waste ground, etc., as an outcast from gardens.

H. tuberosus L. Jerusalem Artichoke.
Alien. N. America. Escape, or outcast, from cultivation which is naturalised on waste ground, etc. Rare or overlooked.
First record: *G. C. Druce*, 1910.

1. Yiewsley!, *M. & S.* Uxbridge. Eastcote. **2.** West Drayton, *Druce Notes.* East Bedfont, 1965. **3.** Northolt. **5.** Greenford. North Acton.† **6.** North of Enfield Lock, *Kennedy*!.

H. rigidus (Cass.) Desf. subsp. **rigidus** Perennial Sunflower.
Alien. North America. Garden escape. Naturalised on waste ground, established on rubbish-tips, railway banks, etc. Common.
First record: *D. H. Kent*, 1950.
7. Hyde Park Corner, 1963, *Brewis.* Notting Hill. Hampstead. Kentish Town. Isle of Dogs. Bombed site, Holborn, 1966. Shoreditch. Hoxton. Dalston. Wapping. Tottenham. Stepney. Old Ford. Hackney Wick.
H. rigidus is an extremely variable species, and the hybrid *H. rigidus* × *tuberosus* (*H.* × *laetiflorus* Pers., *H. scaberrimus* Ell.) is frequently cultivated. It is likely therefore that some of the records given above may be in need of revision.

BIDENS L.

B. cernua L. Nodding Bur-Marigold.
Native on pond verges, in marshes, by river- and streamsides, etc. Local.
First record: *T. Johnson*, 1638.
1. Harefield!, *Blackst. Fasc.*, 28. Between Yiewsley and Iver Bridge!; Stanmore Common!, *T. & D.*, 152. Uxbridge!, *Newbould, T. & D.*, 152. Ickenham. Near Ruislip. Wrotham Park; west of Potters Bar, *J. G.* and *C. M. Dony*!. **2.** West Drayton!; Bushy Park!; Hampton Court!, *Newbould, T. & D.*, 152. East Bedfont.† Ashford. Staines. Shepperton. Walton Bridge. Teddington. **3.** Smallbury Green,† *Ray Syn. MSS.* New Richmond Bridge!; behind Kneller Park;† behind Whitton Dean,† *T. & D.*, 152. Feltham. Cranford. **4.** Harrow,† *Hind Fl.* Mill Hill, *Children* (BM). Near Brent Reservoir, *Farrar*; West Heath, Hampstead, *T. & D.*, 152. Finchley. Mill Hill. Harrow Weald. Edgware. **5.** Between Acton and Ealing;† Twyford,† *Newbould*; Alperton,† *T. & D.*, 152. **6.** Tottenham, *Johns Cat.* Southgate, *Brocas, T. & D.*, 152. Trent Park, *Castell* and *Kennedy*. Edmonton, *T. & D.*, 152. Whetstone. Whitewebbs Park, Enfield, *Johns.* Crew's Hill golf course, Enfield!, *Kennedy.* **7.** Near Hornsey Wood,† *Curt. F.L.* Bayswater† (G & R). Copenhagen Fields† (HE). Green Lanes, Newington, 1846,† *R. Pryor* (BM). Ditch between Kensington Gardens and Hyde Park,† *T. & D.*, 152. Viaduct pond, East Heath, Hampstead!, *Fitter.*

B. tripartita L. Tripartite Bur-Marigold.
Native on pond verges, in marshes, by river- and streamsides, etc. Common.

First record: *T. Johnson*, 1632.
7. Marylebone fields, 1815† (G & R). Hornsey Wood† (HE). Upper Clapton!, *Cherry*; Shepherds Bush, abundant;† Eel Brook Meadow, Parsons Green;† East Heath, Hampstead!; Hackney Wick!; side of canal, Camden Town!, *T. & D.*, 152. Islington!; Newington, *Ball. MSS.* Northumberland Avenue, 1878,† *Trim. MSS.* Kensington Gardens, *J.B.* (*London*) 13: 336. Canal side near Kensal Green Cemetery. Highgate Ponds!, *Fitter*. Regent's Canal from Marylebone to Kings Cross and Islington!, *L.N.* 39: 55. River wall, Chelsea. Regent's Canal, Shoreditch and Bethnal Green. Lee Navigation Canal, from Clapton to Tottenham. Side of Lee, Hackney Marshes.

Var. **integra** Koch
3. Twickenham, *T. & D.*, 152. **4.** Willesden!, *Warren* (BM). **5.** Hanwell, *Welch*!. **7.** Islington, '*J.L.G.*' (BM).

GALINSOGA Ruiz & Pav.

G. parviflora Cav. 'Gallant Soldier', Kew Weed.
Alien. S. America. An escape from the Royal Botanic Gardens, Kew. Naturalised on cultivated and waste ground, by roadsides, etc. A frequent pest of flower-beds in parks and gardens, etc., and a common weed of nurseries. Seedlings are often seen in the soil of pot-plants purchased from florists, and entangled with bedding plants, etc., supplied by nurseries. These are undoubtedly important factors in the spread of the species which is now locally abundant, especially on light soils, near the Thames, in Central London and in the western and eastern suburbs; less frequent, but increasing in the northern parts of the vice-county.
First record: *Britten*, 1862.
7. Very common in the district.

G. ciliata (Raf.) Blake 'Shaggy Gallant Soldier'.
G. quadriradiata auct., *G. aristulata* Bicknell
Lit. *Rep. Bot. Soc. & E.C.* **12**: 93 (1939): *Lond. Nat.* **24**: 11 (1945): *Watsonia* **1**: 238 (1950).
Alien. Central and S. America. Naturalised on cultivated and waste ground, by roadsides, etc. A common weed of flower-beds in parks and gardens. Locally plentiful in Central London and the East End, especially on heavy soils, common and increasing in other parts of the vice-county, though as yet less frequent in the extreme north-western and south-western districts.
First evidence: *Pierce*, 1909.
7. Common in the district.

SENECIO L.

S. jacobaea L. Common Ragwort.
Lit. *J. Ecol.* **45**: 617 (1957).
Native in fields, on waste ground, etc., established on railway banks, etc. Common, but apparently decreasing.
First record: *T. Johnson*, 1638.
7. Between Westminster and Chelsea!, *Pet. MSS.*; Chelsea Hospital grounds, 1959!, *L.N.* 39: 56. Marylebone, 1817 (G & R); Regents Park, one plant, 1956, *Holme*, *L.N.* 39: 56. East Heath, Hampstead!, *T. & D.*, 161. Ken Wood, *Fitter*. Buckingham Palace grounds, *Codrington, Lousley* and *McClintock*!; Kensington Gardens, one plant, 1958!, *D. E. Allen*, *L.N.* 39: 56. Hyde Park, 1960–61, *D. E. Allen*, *L.N.* 41: 21. Finsbury Square, E.C.2, 1966!, *L.N.* 46: 33. Bishop's Park, Fulham, *Welch*!. Isle of Dogs. Highgate.

Var. **floxulosus** DC.
var. *discoideus* Wimm. & Grab.
3. Near Hounslow Heath, 1947, *H. Banks*!, *L.N.* 27: 32.

S. aquaticus Hill Marsh Ragwort.
Native by river- and streamsides, in marshes, wet meadows, on moors, etc., established by canals, etc. Common near the Thames and Colne, less frequent elsewhere.
First record: *Blackstone*, 1737.
1. Harefield!, *Blackst. Fasc.*, 45. Elstree Reservoir!, *T. & D.*, 161. Uxbridge!, *Druce Notes*. Pinner,† *Hind*, *Trim. MSS.* Mimmshall Brook, South Mimms. **2.** Frequent near the Thames and Colne. Shortwood Common!, *T. & D.*, 161. Gravel pits at East Bedfont and West Bedfont. **3.** Common by the Thames. Hounslow Heath!, *T. & D.*, 161. Hayes,† *Trim. MSS.* **4.** Harrow,† *Melv. Fl.*, 48. Near Finchley,† *Newbould*; Stanmore;† West Heath, Hampstead!, *T. & D.*, 161. Edgware, *E. M. Dale* (LNHS). Finchley, *Pugsley* (BM). **5.** Frequent by the Thames. Canal side at Greenford Green and Alperton, scarce. Hanwell, 1953. **6.** Edmonton; Colney Hatch, *T. & D.*, 161. Hadley Common. Whitewebbs Park, Enfield!, *Johns.* **7.** Newington;† Hornsey Wood,† *Ball. MSS.* Between Lee Bridge and Bow, *Cherry*; Eel Brook Common;† East Heath, Hampstead, *T. & D.*, 161. Isle of Dogs, *Cowper*, 107. Hammersmith Marshes,† *Coop. Fl. MSS.* Clapton Marshes,† *R. W. Robbins*, *Lond. Nat. MSS.* Highgate Ponds!, *Fitter*.

Var. **discoideus** Druce
5. Canal bank, Greenford Green, 1947!, *L.N.* 27: 32.

S. aquaticus × jacobaea = S. × ostenfeldii Druce
2. Staines Moor!, *L.N.* 45: 20.

S. erucifolius L. Hoary Ragwort.
Native on field borders, in grassy places, by roadsides, on waste ground, etc., established on railway banks, etc. Common, especially on calcareous and clay soils, though less frequent in heavily built-up areas.
First record: *Petiver*, 1713.
7. Near West End Lane, *Trim. MSS.* Kensal Green Cemetery. Fulham. Brompton Cemetery. Kilburn. Hornsey. Tottenham. Old Ford.

S. squalidus L. Oxford Ragwort.
Lit. *Proc. Bot. Soc. Brit. Isles* **3**: 375 (1963): *loc. cit.* **5**: 210 (1964).
Alien. S. Europe. An escape from the Oxford Botanic Garden. Now naturalised on waste ground, river- and streamsides, by roadsides, etc., and established on wall-tops, railway banks and tracks, on canal paths, etc. Very common.
First record: *Trimen* and *Dyer*, 1867.
7. Very common in the district.
This native of Sicily and southern Italy was first recorded from Britain on walls at Oxford in 1794, and had undoubtedly 'escaped' from the Botanic Garden where it was cultivated. Trimen and Dyer discovered a single plant on a newly made road at Twickenham in 1867, but this was probably of casual occurrence, for it was not until the end of the nineteenth century that the plant spread rapidly from Oxford along the railway systems into many parts of England and Wales. By about 1890 large colonies had formed about the railway at Reading, Berks. and Slough, Bucks., and it was seen in Middlesex on the railway at Uxbridge, West Drayton, Southall and Acton in 1904. These were the first authentic records of the species as an established adventive in the county. By the outbreak of the First World War it had extended its range to the White City and Hackney Marshes, and since that time has spread to all parts of the vice-county.
The var. **incisus** Guss. is common, and the forma **subinteger** Druce, with entire leaves, is often encountered on rich waste ground. Plants with flowers lacking ray-florets were seen at **5.** Alperton, 1965, *McLean*. South Ealing, 1968.

S. squalidus × viscosus = S. × londinensis Lousley
Lit. *Rep. Bot. Soc. & E.C.* **12**: 869 (1946).
1. Uxbridge, 1947!, *K. & L.*, 160. Cowley, *L. M. P. Small.* **2.** Between East Bedfont and Stanwell, 1965. **3.** Hounslow Heath, 1951, *T. G. Collett*!. **4.** West Heath, Hampstead, 1950, *Bangerter* and *Morton*. **7.** Bombed sites, Bush Lane!;† Cannon Street!;† Sergeant's Inn† and Aldersgate Street, E.C.,† *Lousley, Rep. Bot. Soc. & E.C.* 12: 880. Chelsea, 1946, *Graham*; bombed site, Ravenscourt Park, 1946,† *Sandwith*; bombed site, Piccadilly, W.1.!,† *Batko*; Stamford Hill!; Hackney!, *K. & L.*, 160. Pithead Mews, W.1., 1947,† *A. H. G. Alston* (BM). Car park, Marsham Street, S.W.1,

1948,† *Chandler, L.N.* 28: 29. Grounds of the Natural History Museum, South Kensington, 1949,† *Bangerter*; bombed site, Bute Street, South Kensington,† *J. B. Evans, K. & L.*, 160. Bombed sites, Suffolk Lane, E.C.† and The Temple,† *Lousley* (L). Rotten Row, Hyde Park, 1949!,† *Watsonia* 1: 298. Bombed site by St Paul's Cathedral, Paternoster Row, E.C., 1947,† *Lousley*. Paddington. Marylebone, 1949. Hackney Wick, *L. M. P. Small*.

S. squalidus × vulgaris = **S. × baxteri** Druce
3. Hounslow Heath, 1948, *Welch, L.N.* 28: 33. **4.** Hendon, 1962, *Brewis*. **7.** Sergeant's Inn,† *Sandwith* (K). Hyde Park, one plant, 1960, *D. E. Allen*!, *L.N.* 41: 21.

S. sylvaticus L. Heath Groundsel.
Native on heaths, in grassy places, open woods, etc. Local, and chiefly confined to sandy soils.
First record: *Petiver*, 1713.
1. Harefield!; Harrow Weald Common!; Stanmore Common, *T. & D.*, 160. Near Uxbridge, towards Harefield, *Phyt. N.S.* 1: 62. Ruislip!, *Melv. Fl.*, 48. Hillingdon!, *Trim. MSS.* Yiewsley, *J. E. Cooper* (BM). Haste Hill, Northwood!, *Wrighton*. South Mimms. Bentley Heath. **2.** Roadsides about Staines, Charlton!, etc., to Hampton, abundant, *T. & D.*, 160. Stanwell, *J. E. Cooper*. West Drayton, *Druce Notes*. **3.** Hounslow Heath!, *T. & D.*, 160. **4.** Hampstead!, *Pet. H.B.C.* West Heath, Hampstead!; Child's Hill,† *T. & D.*, 160. Edgware!, *J.* and *S. G. Alletson*. Brent Reservoir. Grimsdyke golf course, *Kennedy*!. **5.** Near Hanwell,† *Newbould, T. & D.*, 160. Wyke Green, abundant;† eradicated, 1963. Bedford Park, *Cockerell Fl.* Chiswick, *Murray List*. **6.** Enfield. Wood Green, *J. E. Cooper*. **7.** Near Hornsey,† *Dillenius, Ray Syn.* 3: 179. Hornsey, 1875,† *French* (BM). Marylebone fields, 1815† (G & R). Primrose Hill, 1830,† *Varenne*; near Hampstead, *T. & D.*, 160. East Heath, Hampstead. Parliament Hill fields. Ken Wood fields.

S. viscosus L. Viscid, or Stinking Groundsel.
Introduced. Naturalised on waste ground, by roadsides, etc., established on railway tracks, rubbish-tips, etc. Formerly very rare, but now common.
First record: *J. Hill*, 1761.
7. Common in the district.

S. vulgaris L. Groundsel.
First record: *T. Johnson*, 1638.
Native on cultivated and waste ground, etc. Very common.
7. Very common in the district.

Var. **hibernicus** Syme
Var. *radiatus* Rouy, non Koch

Lit. *Watsonia* **6**: 280 (1968).

2. River wall, Hampton Court, a single plant, 1947, *Welch*!, *L.N.* 27: 32. **4.** Northwick Park golf course, *Graham*. **6.** Near Edmonton, 1966, *Kennedy*!. **7.** Bombed site near Wood Street, E.C.,† *Sladen* and *McClintock*, fide *Lousley*. Regents Park, 1947, *A. H. G. Alston*. East Heath, Hampstead, 1949!, *K. & L.*, 161.

Crisp and Jones (1970, *Watsonia* **8**: 47) suggest that this 'variety' may be the result of back-crosses involving *S. squalidus* and *S. vulgaris*.

DORONICUM L.

D. pardalianches L. Great Leopard's-bane.

Alien. Europe. Garden escape. Naturalised in grassy places, on banks, etc. Very rare.

First evidence: *Newbould*, 1868.

3. Near Hounslow Heath, 1945!,† *K. & L.*, 158. **4.** Fairly established in a little copse by the stream, near the road between Finchley and Hendon, 1868,† *Newbould* (BM). **5.** Bank of lake at Syon Park!,† 1923, *A. B. Jackson* (SLBI); no doubt originally planted but naturalised and abundant until 1967 when the site was destroyed during the laying out of the new Garden Centre. A small colony in grass, railway yard, Kew Bridge station, 1963 →.

TUSSILAGO L.

T. farfara L. Coltsfoot.

Native on disturbed ground, by roadsides, river- and streamsides, etc., established on railway banks, etc. Very common.

First record: *T. Johnson*, 1638.

7. Frequent throughout the district.

PETASITES L.

P. hybridus (L.) Gaertn., Mey. & Scherb. Butterbur.

P. ovatus Hill, *P. officinalis* Moench, *P. vulgaris* Desf.

Native by river- and streamsides, on pond verges, etc., established by canals, etc. Locally frequent in the Colne valley, rather rare elsewhere.

First record: *Petiver*, c. 1710.

1. Locally frequent by the Colne and canal from Springwell to Cowley. Verge of pond, Stanmore Common, *Payton*!, *L.N.* 27: 32; 1967. **2.** Hampton Court!, *Welch*, *K. & L.*, 157. Near Bushy Park, 1965, *Clement*. **4.** Bentley Priory!, *Hind*, *Melv. Fl.*, 47. Turner's Wood, *J. J. Bennett* (CYN). Side of brook, Scratch Wood!, *K. & L.*, 157. **5.** Banks of the Thames between Putney and Hammersmith,† *Savage*, *T. & D.*, 147. Still occurs on the Surrey side of the river. Thames bank downstream from Chiswick Bridge!,† *Boniface*, *K. &*

L., 157. Perivale Wood!, introduced, *Groves Veg.* **6.** Banks of the Lee about Waltham Abbey; Edmonton; between Fortis Green and Hornsey, 1864,† *Savage, T. & D.*, 147. By the Lee near Ponders End, *Benbow MSS.* **7.** Between Westminster and Chelsea, *Pet. MSS.*; c. 1780,† *Curt. F.L.* About Chelsea Waterworks,† *Blackst. Spec.*, 41. Both male and female plants are common in the Colne valley, elsewhere the male plant is usually encountered.

P. japonicus (Sieb. & Zucc.) F. Schmidt
Alien. Sakhalin. Garden escape. Naturalised in wet places, damp woods, etc. Very rare.
First evidence: *Harley*, 1953.
4. Well established by the Park Lake, Harrow!; Grove Wood, Harrow, *Harley* (L).

P. fragrans (Vill.) C. Presl Winter Heliotrope.
Alien. Europe, etc. Garden escape. Naturalised on hedgebanks, waste ground, by roadsides, in grassy places, etc., established on railway banks, etc. Rather rare.
First record: –1835.
1. Cowley, 1884, *Benbow* (BM). Pinner, 1926; Eastcote!; Swakeleys, *Tremayne*; near Uxbridge!, *K. & L.*, 157. Stanmore Common!, *L.N.* 26: 62. Harefield, 1928, *B. T. Ward.* Near West Ruislip station, *Donovan.* **2.** West Drayton!, *Newbould, T. & D.*, 148. Between Poyle and Colnbrook!, *L.N.* 26: 62. Ashford, *Hasler*; Hampton Waterworks!, *Yeo, K. & L.*, 157. **3.** Thames bank opposite Richmond!, *K. & L.*, 157. Crane Park, Twickenham. Mount Park, Harrow!, *Horton Rep.* **4.** Hendon!, *Trim. MSS.* Harrow!, *Melv. Fl.*, 2. Harrow Weald, *L. M. P. Small.* Mill Hill, 1934, *Chapple* (BM). Willesden!,† *L.N.* 26: 62. Finchley!, *L.N.* 28: 33. **5.** Carville Hall Park, Brentford, 1942,† *Welch, L.N.* 26: 62. Osterley Park, *Welch*!. Ealing, 1966 →!, *T. G. Collett.* Railway bank near Chiswick, 1967. **6.** Edmonton, *T. & D.*, 148. Muswell Hill, *J. E. Cooper, K. & L.*, 157. **7.** Bayswater,† *Mag. Nat. Hist.* 8: 339. Kilburn, 1868,† *Warren, T. & D.*, 148. Highgate, 1905, *C. S. Nicholson* (LNHS). Railway bank, Clapton, abundant, 1928, *Tremayne, K. & L.*, 157. Railway banks near Stamford Hill!, 1962, *Wurzell, L.N.* 42: 12; 1966. Railway bank, Stoke Newington!, *Kennedy.* Ken Wood, *Johns. Nat.*, 75. Buckingham Palace grounds, *Fl. B.P.G.* Canal side between Marylebone and Primrose Hill, 1968.

INULA L.

I. helenium L. Elecampane.
Alien. Central Asia. Garden escape. Naturalised in meadows, by roadsides, etc., established on railway banks. Very rare.

First record: *Gerarde*, 1597.
1. Harefield,† *Blackstone* in litt. to *Richardson*, 18 December 1736. Near Breakspeares,† *Blackst. Fasc.*, 25. Northwood, 1884, *Benbow* (BM); 1909, *Green, Druce Notes*; 1914–15,† *J. E. Cooper, K. & L.*, 150. Between Little Heath and Potters Bar,† *Kemble, Webb & Colem. Suppl.* **2.** Colnbrook, *Ger. Hb.*, 649. **4.** By Hampstead Heath,† *Merr. MSS.* **6.** Enfield Chase, *Phyt. N.S.* 6: 301; c. 1873, *H. & F.*, 149. Railway bank and nearby roadside, Winchmore Hill!, *Welch, L.N.* 30: 7.

I. conyza DC. Ploughman's Spikenard.
I. squarrosa (L.) Bernh., non L.
Native in thickets, on dry slopes, waste ground, etc., chiefly on calcareous soils. Very rare.
First record: *Parkinson*, 1640.
1. Harefield!, *Blackst. Fasc.*, 9; now confined to a calcareous bank near Old Park Wood!, *I. G. Johnson*, and an old chalkpit, where it grows in very small quantity. Pinner Hill, a single plant, 1920,† *Tremayne, K. & L.*, 150. **2.** East Bedfont, 1945,† *Westrup, K. & L.*, 150. Perhaps an error. **4.** Near Hampstead,† *Park. Theat.*, 127. **5.** Between Perivale and Horsendon Hill,† *Lees, T. & D.*, 151. **7.** Bombed site, Gresham Street area, E.C.,† *Jones Fl.* Introduced.

PULICARIA Gaertn.

P. dysenterica (L.) Bernh. Common Fleabane.
Inula dysenterica L.
Native in marshes, ditches, on heaths, by river- and streamsides, roadsides, etc. Common, except in heavily built-up areas, where owing to the eradication of its habitats it is very rare or extinct.
First record: *T. Johnson*, 1633.
7. St James's Park;† Tothill Fields,† *Johns. Ger.*, 432. Between Westminster and Chelsea,† *Pet. MSS.* Isle of Dogs, *T. & D.*, 151. Hornsey Wood,† *Ball. MSS.* Finchley Road,† *Trim. MSS.* Fulham,† *Corn. Surv.* Green Park, *Allen Fl.* Buckingham Palace grounds, *Fl. B.P.G.*; 1967, *McClintock*. Clapton. Stoke Newington, 1972, *Kennedy*.

P. vulgaris Gaertn. Small Fleabane.
Inula pulicaria L., *Pulicaria prostrata* Aschers., *P. pulicaria* (L.) Karst.
Lit. *Ann. Bot.* **31**: 699 (1967).
Formerly native in moist sandy places, on pond margins, etc., inundated in the winter, on heaths and village greens, but now extinct.
First record: *l'Obel*, 1570. Last record: *Green*, 1908.
1. Harefield, *Blackst. Fasc.*, 20. **2.** Between Hampton and Sunbury, *H. C. Watson, Wats. New Bot.*, 100. Staines Moor, 1878, *H.* and *J.*

Groves (BM). Shortwood Common; Knowle Green, in plenty, *Benbow, J.B.* (*London*) 25: 16. Halliford Green, 1905 and 1908, *Green, Williams MSS.* **3.** Hounslow, *Lightfoot* (LT). Smallbury Green, *Ray Syn. MSS.* **4.** Hampstead Heath, *Johns. Enum.* Golders Green, *Irv. MSS.*; not there in 1852, *Syme, Phyt.* 4: 860. **5.** Ealing, *Pamplin, Wats. New Bot. Suppl.*, 588. Alperton, *Lees, T. & D.*, 151. Turnham Green, *T. Moore* (BM). **7.** Benard Greyn (= Green) [this was situated near Paddington – cf. *J.B.* (*London*) 67: 307 (1929)], *Lob. Adv. Nov.* Shepherds Bush to Hammersmith, *Hill Veg. Syst.*

FILAGO L.

F. vulgaris Lam. Common Cudweed.
F. germanica (L.) L., non Huds., *F. canescens* Jord.
Native on heaths, in dry fields, by roadsides, etc. Very rare, and mainly confined to sandy soils.
First record: – c. 1730.
1. Harefield!, *Blackst. Fasc.*, 33. Eastcote,† *Hind, Melv. Fl.*, 47. Near Harrow Weald Common, *T. & D.*, 158. Uxbridge Common, 1888,† *Benbow*; Ruislip, 1864,† *Hind* (BM). Between Cowley and Harefield,† *Benbow MSS.* Hillingdon,† *Trim. MSS.* **2.** Staines!; between Hampton Court and Kingston Bridge;† Fulwell;† Teddington,† *T. & D.*, 158. Between Staines and Laleham!;† between Staines and Ashford Common; between Hampton and Sunbury,† *Benbow MSS.* **3.** Smallbury Green,† *Ray Syn. MSS.* Hounslow Heath, scarce,† *T. & D.*, 158. **4.** Hampstead,† *Irv. MSS.* Stanmore,† *Varenne, T. & D.*, 158. **5.** Hanwell,† *Newbould*; Turnham Green,† *T. & D.*, 158. **6.** Trent Park, *A. B. Cole, T. & D.*, 158. Palmers Green, 1905,† *P. H. Cooke, K. & L.*, 149.

F. pyramidata L. Spathulate Cudweed.
F. spathulata C. Presl, *F. germanica* subsp. *spathulata* (C. Presl) Rouy
Formerly native in fields on sandy soils, by roadsides, etc., but now extinct.
First evidence: *H. C. Watson*, 1850. Last evidence: *Benbow*, 1891.
2. Thames towpath opposite Long Ditton Ferry, 1850, *H. C. Watson* (W). Near Staines, 1867, *Trimen*; between Staines and Laleham, 1891; roadside, Littleton, very abundant, 1891, *Benbow* (BM). Near Hampton Court, by the towing-path, *T. & D.*, 158. **3.** Twickenham, 1867, *Trimen* (BM); station destroyed, 1870, *Trim. MSS.*

F. minima (Sm.) Pers. Slender Cudweed.
Native on heaths, banks, in fields, bare places, etc., on sandy and gravelly soils. Very rare.
First record: *Petiver*, c. 1713.

1. Harrow Weald Common, *Melv. Fl.*, 47; 1866, *Trimen* (BM). Stanmore Common, 1827–30, *Varenne*, *T. & D.*, 158. Railway banks between Uxbridge and West Drayton!; Uxbridge Common,† *Benbow* (BM). Near Harefield!, *P. H. Cooke* (LHNS). Between Harefield and Ickenham!, *L.N.* 26: 61. Yiewsley, 1924; near Potters Bar, *J. E. Cooper*. **2.** Teddington, *T. & D.*, 158. Near Ashford Common; Fulwell, a single plant,† *Benbow*, *J.B. (London)* 25: 16. Hampton Court!; Bushy Park!, *K. & L.*, 149. **3.** Hounslow Heath, abundant, 1866, *Trimen* (BM). Near Hounslow Heath!, *K. & L.*, 149. **4.** Hampstead,† *Pet. H.B.C.* Hampstead Heath,† *Stocks* (BM).

GNAPHALIUM L.

G. sylvaticum L. Wood Cudweed.

Native in dry open woods, dry pastures, on heaths, waste ground, etc., with a marked preference for acid soils. Very rare, and perhaps extinct.

First record: *Pena* and *l'Obel*, 1570.

1. Harefield, *Blackst. Fasc.*, 28; c. 1869,† *Newbould*, *T. & D.*, 159. Harrow Weald Common, 1867, *Trimen* (BM); 1898,† *C. S. Nicholson* (LNHS). Stanmore Common, 1866,† *Trimen* (BM). Yiewsley,† *Druce Notes*. **4.** . . . on this side of the Thames in a dense wood three miles from London,† *Pena & Lob. Stirp. Adv.*, 202. In the dark woods of Hampstead, *Ger. Hb.*, 518; c. 1746,† *W. Watson*, *Blackst. Spec.*, 28. Hampstead,† *Wharton Fl.* **6.** Highgate Wood, *Johns. Eric.* Between Highgate and Muswell Hill,† *Watson*, *Blackst. Spec.*, 28. **7.** Ken Wood, *Pet. MSS.*; c. 1760,† *J. Hill* (BM). Near Paddington Cemetery, on soil brought from elsewhere, 1867,† *Warren*, *T. & D.*, 159. Air-raid shelter, Hyde Park, 1945,† *L. G. Payne* and *Lousley* (L). Introduced.

G. uliginosum L. Marsh Cudweed.

Native in heaths, damp waste and cultivated ground, by roadsides, in bare places, cart ruts, etc. Common, but mainly confined to acid soils.

First record: *T. Johnson*, 1638.

7. Tottenham, *Johns. Cat.* Marylebone, *J. Banks* (BM). Near Newington, *Ball. MSS.* Back of Adelaide Road, N.W.; East Heath, Hampstead!; Hackney Wick!, *T. & D.*, 159. Kensington Gardens, *Warren*, *J.B. (London)* 13: 336. Near Clapton, *Cherry*, *T. & D.*, 159. Ken Wood, *Fitter*. Buckingham Palace grounds, *Codrington*, *Lousley* and *McClintock*!; railway tracks near Paddington station!, *L.N.* 39: 56. Tower Hill, *Brewis*, *L.N.* 44: 26. Stag Place, S.W.1, 1966, *McClintock*, *L.N.* 46: 13. Brompton Cemetery. Holland House grounds, Kensington. Dalston.

SOLIDAGO L.

S. virgaurea L. Golden-rod.

Native in dry woods, on heaths, hedgebanks, etc., established on railway banks, etc. Very rare.

First record: *Gerarde*, 1597.

1. Harefield,† *Blackst. Fasc.*, 111. Pinner Wood, *Melv. Fl.*, 47. Ruislip Reservoir;† near Stanmore Common,† *Benbow MSS.* Harrow Weald Common,† *A. B. Jackson* and *Green*, *Williams MSS.* Park Wood, Ruislip!, *Welch*, *K. & L.*, 145. **3.** Hounslow Heath, *Ray Syn. MSS.*; c. 1869, *T. & D.*, 150; c. 1887,† *Benbow MSS.* **4.** Hampstead Wood,† *Ger. Hb.*, 348, and many later authors. Hampstead Heath, *Blackst. Spec.*, 105; c. 1760, *J. Hill* (BM); c. 1869, *T. & D.*, 150; 1883,† *J. E. Cooper*, *K. & L.*, 145. Bishop's Wood, *T. & D.*, 150; 1902, *C. S. Nicholson* (LNHS). **6.** Winchmore Hill Wood,† *T. & D.*, 150. Enfield, *H. & F.*, 150. Between Enfield and Trent Park, *Benbow*, *J.B. (London)* 25: 16. Highgate Wood!, *C. S. Nicholson* (LNHS). Coldfall Wood, Highgate, *E. M. Dale*; Queen's Wood, Highgate!, *K. & L.*, 145. Railway banks near Muswell Hill, 1961, *Bangerter* and *Kennedy*!. **7.** Ken Wood, 1837,† *E. Ballard* (BM). East Heath, Hampstead,† *T. & D.*, 150.

S. altissima L. Tall Golden-rod.

S. canadensis auct. eur., non L.

Alien. N. America. Garden escape. Naturalised by rivers and streams, on waste ground, etc., established on railway banks, rubbish-tips, etc. Common.

First record: *D. H. Kent*, 1938.

7. Common in the district.

S. gigantea Ait. subsp. **serotina** (O. Kuntze) McNeill

S. serotina Ait., non Retz. var. *gigantea* (Ait.) A. Gray

Alien. N. America. Garden escape. Naturalised by rivers and streams, on waste ground, etc., established on railway banks, rubbish-tips, etc. Common, though much less frequent than *S. altissima*.

First record: *D. H. Kent*, 1945.

7. Bombed site, Mile End Road, 1952, *Henson* (BM). Buckingham Palace grounds, 1961,† *Fl. B.P.G.* Kensal Green. Notting Hill. Isle of Dogs. Hoxton. Bethnal Green. Stepney. Tottenham.

ASTER L.

A. tripolium L. Sea Aster.

Lit. *J. Ecol.* **30**: 385 (1942).

Native on river walls, mud and waste ground by the tidal Thames, and naturalised in marshy fields. Very rare.

First evidence: *Trimen*, 1866.

3. Marshy field near Yeading, abundant, with *Puccinellia distans*, 1949!, *Boniface*, *L.N.* 29: 13; 1950–64; in 1965 the field was enclosed by a tall fence and the ground was disturbed, the plant may, however, have survived. Several colonies in a marshy field by the canal, Northolt, 1967 →. **7.** Isle of Dogs, three or four feeble plants in a ditch, 1866, *Trimen* (BM); waste ground near the South-West India Dock, *F. Naylor*, *J.B.* (*London*) 10: 371; Isle of Dogs, abundant, 1887, *Benbow*, *J.B.* (*London*) 25: 363; a few plants by the inlet to Millwall Dock, 1955; not seen on subsequent visits, but with the extensive dock installations on the 'island' it is likely that the species survives on waste ground and dock walls. Thames side, Fulham, *Corn. Surv.* Hackney Marshes, 1910–12,† *Coop. Cas.*

A. novi-belgii L. Michaelmas Daisy.

Alien. N. America. Garden escape. Naturalised by river- and stream-sides, on waste ground, hedgebanks, by roadsides, etc., and established on railway banks, rubbish-tips, etc. Extremely variable in height, habit, leaf shape and flower colour, probably due to the presence of many cultivars. Very common.

First certain record: *Wedgwood*, 1922.

7. Common in the district.

A. lanceolatus Willd. Michaelmas Daisy.

A. paniculatus auct., *A. lamarckianus* auct.

Alien. N. America. Garden escape. In similar habitats to the preceding species, and equally variable. Common.

First record: *D. H. Kent*, 1944.

7. Common in the district.

A. salignus Willd. Michaelmas Daisy.

Alien. N. America. Garden escape. Naturalised by river- and stream-sides, on waste ground, etc. Rare, or overlooked.

First record: *Trimen* and *Dyer*, 1867.

1. New Years Green, 1965. **2.** Side of the Thames between Strawberry Hill and Teddington, 1867, *T. & D.*, 149. Thames bank, Staines, 1962. **3.** Thames bank, Isleworth, and between Richmond Bridge and Twickenham, 1966. Hounslow Heath, 1966. **4.** Between Kenton and Wealdstone, 1966.

This 'species' may prove to be a variety or form of *A. lanceolatus*.

ERIGERON L.

E. acer L. Blue Fleabane.

Native in dry grassy places, on heaths, banks, by roadsides, etc., established on railway tracks, etc. Local, and chiefly confined to light soils.

First certain record: *Warren*, c. 1866.

1. Harefield!, *J. E. Cooper*; Springwell!, *K. & L.*, 148. Uxbridge!, *Benbow*, *J.B.* (*London*) 25: 16. Railway banks between Uxbridge and West Drayton, *Benbow MSS*. Disused railway between Denham and Uxbridge. Yiewsley!, *Coop. Cas.* **2.** West Drayton!; Staines!; Laleham!; Hampton; Sunbury, *Benbow*, *J.B.* (*London*) 25: 16. Shepperton. Near Chertsey Bridge. Between Penton Hook and Staines. Hampton Court. River wall between Hampton Court and Kingston Bridge. East Bedfont!, *K. & L.*, 148. West Bedfont. **3.** Feltham!, *Benbow*, *J.B.* (*London*) 25: 16. Harlington, *Coop. Cas.* Hounslow Heath!, *Welch*, *K. & L.*, 148. **5.** Ealing!, 1871, *E. C. White*, *Williams MSS.*; Ealing Common!,† *K. & L.*, 148. Between Greenford and Hanwell, 1960.† Chiswick!, *Welch*, *K. & L.*, 148. **6.** Between Hadley Wood and Potters Bar, *Benbow*, *J.B.* (*London*) 25: 16. North Finchley, 1915 and 1918, *J. E. Cooper*, *K. & L.*, 148. **7.** Parsons Green,† *Merr. Pin.*, 29. Perhaps an error; ? *Conyza canadensis* intended. Edgware Road, 1866–67, *Warren*; South Kensington, *T. & D.*, 149. Isle of Dogs, *Benbow MSS*. Bombed site, Cripplegate area, E.C., *Jones Fl.*

CONYZA Less.

C. canadensis (L.) Cronq. Canadian Fleabane.

Erigeron canadensis L.

Alien. N. America. Naturalised on heaths, commons, cultivated and waste ground, by roadsides, etc., and established on railway tracks, etc. Formerly rather rare, but now very common.

First record: *T. Robinson*, 1690.

7. Frequent in the district.

BELLIS L.

B. perennis L. Daisy.

Native in short grassland, established on lawns, etc. Very common.

First record: *T. Johnson*, 1638.

7. Very common in the district.

EUPATORIUM L.

E. cannabinum L. Hemp Agrimony.

Native in marshes, by river- and streamsides, in wet woods, etc., established by canals, etc. Locally plentiful, but decreasing.

First record: *T. Johnson*, 1638.

1. Locally frequent in the Colne valley from Springwell to Yiewsley. Pinner, a single plant,† *Hind*, *T. & D.*, 147. **2.** Locally frequent by the Colne and adjacent streams from Yiewsley to Staines. Locally common by the Thames from Staines to Hampton. Queen's river, Hampton!, *T. & D.*, 147. Longford river near East Bedfont. Harlington, *P. H. Cooke, K. & L.*, 144. **3.** Isleworth!, *Ray Syn. MSS.* Thames bank between Twickenham and Richmond Bridge!, *T. & D.*, 147. Feltham. **4.** Hampstead,† *Irv. MSS.* Kingsbury,† *Varenne, T. & D.*, 1947. Brent Reservoir. **5.** Thames bank, Chiswick, 1947, *Bangerter, K. & L.*, 144. Near Gunnersbury,† *Newbould, T. & D.*, 147. Thames bank near Brentford, 1966. Park Royal, 1965. **6.** The Alders, Whetstone;† Edmonton, *T. & D.*, 147. Enfield, *P. H. Cooke, K. & L.*, 144. **7.** Between Westminster and Chelsea,† *Pet. MSS.* Tottenham, *Johns. Cat.*

ANTHEMIS L.

A. tinctoria L. Yellow Chamomile.

Alien. Europe, etc. Garden escape. Naturalised in fields, on waste ground, by streams, etc., established on railway banks, rubbish-tips, etc. Rare, and sometimes merely casual.

First evidence: *Dyer*, 1869.

1. Uxbridge, *Druce Notes.* Yiewsley!, *J. E. Cooper* (BM). New Years Green, 1961. Roadside near Dyrham Park, South Mimms, 1965!, *J. G.* and *C. M. Dony.* **2.** West Drayton, 1914–17, *J. E. Cooper*; between Colnbrook and West Drayton, 1951!, *K. & L.*, 153; 1952 →. Field near Harmondsworth, abundant, and adjacent streamside in small quantity, 1965. **3.** Hounslow Heath, 1964, *Wurzell.* **4.** Stonebridge,† *Benbow, J.B. (London)* 25: 20. Golders Green, 1910;† Whitchurch, 1936,† *P. H. Cooke* (LNHS). Mill Hill, 1892,† *W. H. Hudson, Sci. Goss.* 28: 22. Near Finchley, 1910,† *J. E. Cooper* (BM). **5.** Alperton,† *Druce Notes.* Chiswick, 1942!,† *Sandwith, L.N.* 26: 61. **7.** Finchley Road, 1869,† *Dyer* (BM). Bombed site, St Olave's, Silver Street, E.C.!,† *Sladen* and *McClintock, Lousley Fl. City.*

A. arvensis L. Corn Chamomile.

Native on cultivated and waste ground on light soils, introduced and usually merely casual on heavy soils. Very rare.

First record: *A. Irvine*, 1838.

1. Uxbridge Common, 1886, *Benbow, J.B. (London)* 25: 16. Between Uxbridge and Cowley Peachey, *Benbow MSS.* Yiewsley, *Coop. Cas.* **2.** Near Penton Hook Lock; Fulwell; between Staines and Laleham!; Ashford Common, *Benbow MSS.* Teddington, *Irv. Lond. Fl.*, 155. Near Hampton, 1966, *Tyrrell.* **3.** Whitton, a single plant, 1867, *Dyer* (BM). Hatton, 1965. **4.** Finchley, 1901, *J. E. Cooper* (BM). **5.** Alperton, 1917; Greenford, 1917, *J. E. Cooper*; Hanwell, 1950, *Welch, K. & L.*, 154. **6.** Muswell Hill, 1904, *Coop. Cas.* **7.** Islington, *Ball. MSS.* Hackney Marshes, *Coop. Cas.* Hyde Park; Kensington Gardens, 1962!, *Allen Fl.*

A. cotula L. Stinking Mayweed.
Lit. *J. Ecol.* **59**: 623 (1971).
Native on cultivated and waste ground, by roadsides, etc. Common.
First record: *T. Johnson*, 1638.
7. Tottenham!, *Johns. Cat.* Between Westminster and Chelsea!, *Pet. MSS.* Marylebone!, *Varenne*; close to Kensington Palace, 1866, *Warren, T. & D.*, 154. Whitehall, *Cottam.* Bombed site, Basinghall Street,† *Sladen* and *McClintock, Lousley Fl. City.* Westbourne Park, 1948. Hackney Marshes, *J. E. Cooper.* East Heath, Hampstead!, *H. C. Harris.* Notting Hill. Kilburn. Tower Hill. Dalston. Stoke Newington. Stepney. Old Ford. Hackney Wick. Buckingham Palace grounds, 1963, *Fl. B.P.G.* Paddington!, *L.N.* 46: 33.

CHAMAEMELUM Mill.

C. nobile (L.) All. Chamomile.
Anthemis nobilis L.
Native on heaths, village greens, in grassy places, by roadsides, etc., chiefly on light soils. Formerly common, but now very rare.
First record: *W. Turner*, 1548.
1. Harefield Common,† *Blackst. Fasc.*, 17. Hillingdon,† *Warren*; between South Mimms and Potters Bar, *T. & D.*, 154. Uxbridge Common!;† Pield Heath,† *Benbow MSS.* Mimms Wash, 1922, *K. & L.*, 154. **2.** Teddington,† *Irv. Lond. Fl.*, 155. Towing-path, Hampton Court!, *Newbould*; Feltham Green;† Harlington,† *T. & D.*, 154. Bushy Park!, *Welch, K. & L.*, 154. Hampton Court Park. Staines, *Briggs.* **3.** Hounslow Heath!,† *Turn. Names.* Near Hatton;† Twickenham,† *T. & D.*, 154. Feltham;† Hanworth!,† *Benbow MSS.* **4.** Hampstead Heath, 1837, *E. Ballard*; 1861, *Trimen* (BM); 1863,† *Parsons* (CYN). Golders Green,† *Irv. MSS.* Mill Hill,† *Irv. Lond. Fl.*, 154. **5.** Brentford,† *Turn. Hb.*, 1, 47. Turnham Green;† Acton Green,† *Newbould*; Ealing Common,† *Lees, T. & D.*, 154. Ealing Green,† *Benbow MSS.* **6.** Muswell Hill, plentifully,† *W. Watson, Hill Fl. Brit.*, 435. Enfield Chase, *Burn. Med. Bot.*, 1, 38. Hadley!, *G. Johnson*; Enfield Green,† *T. & D.*, 154. Between Barnet and Potters Bar, *Benbow MSS.* Coldfall Wood, Highgate,† *Cherry, Trim. MSS.* By Hadley Wood, 1929, *Lousley* (L). North of Cockfosters, *Welch.* **7.** Benard Greyn (= Green) (near Paddington),† *Pena and Lob. Stirp. Adv.*, 145. Tottenham,† *Johns. Cat.* Tothill Fields,† *Pet. Midd.* Near Hyde Park,† *Merr. Pin.*, 25. Lincoln's Inn gardens, 1869;† East Heath, Hampstead,† *T. & D.*, 154. Kensington Gardens, on lawn turf south side of the Palace!, *Warren Fl.*; 1959!, *D. E. Allen*;

lawn, Cremorne Estate, Chelsea, 1959!, *L.N.* 39: 6. Buckingham Palace grounds, originally planted as a 'lawn' but now established in various parts of the grounds, *Codrington*, *Lousley* and *McClintock*!.

ACHILLEA L.

A. ptarmica L. Sneezewort.

Native in damp meadows, marshes, on waste ground, by river- and streamsides, etc. Double-flowered forms are sometimes seen as escapes from gardens. Locally common, especially on heavy soils.

First record: *Gerarde*, 1597.

1. Harefield!, *Blackst. Fasc.*, 82. Harefield Moor, frequent. Pinner!, *Melv. Fl.*, 49. Stanmore Common!; Elstree Reservoir!,† *T. & D.*, 153. Uxbridge, *Benbow* (BM). Pinner Hill. Between Denham and Ickenham. Haste Hill, Northwood!, *Wrighton*. Potters Bar; Bentley Heath; South Mimms, frequent, *J. G.* and *C. M. Dony*!. **2.** Near West Drayton, *K. M. Marks*!. **3.** Field behind Harrow Grove,† *Melv. Fl.*, 49. Hounslow Heath!;† between Twickenham and Worton;† Whitton, *T. & D.*, 153. Isleworth. Roxeth.† **4.** Hampstead,† *Pet. H.B.C.* Pinner Drive,† *Melv. Fl.*, 49. West Heath, Hampstead;† Edgwarebury!, *T. & D.*, 153. Edgware!, *E. M. Dale*; Golders Green, 1909,† *P. H. Cooke* (LNHS). Willesden!,† *Druce Notes*. Between Fortune Green and Child's Hill, abundant,† *Trim. MSS.* Brent Reservoir. Harrow. Northwick Park!, *Harley Fl.*, 38. Mill Hill. **5.** Twyford;† between Acton and Turnham Green,† *Newbould*, *T. & D.*, 153. Bedford Park,† *Cockerell Fl.* Hanwell. Brentham. Perivale!, *Montg. MSS.* Southall, 1882, *Wright* (BM). Osterley Park, *Welch*!. Chiswick. **6.** Tottenham, *Johns. Cat.* Near Highgate, *Blackst. Spec.*, 78. Hadley!, *G. Johnson*; Colney Hatch, *Newbould*, *T. & D.*, 153. Finchley Common. Trent Park, *T. G. Collett* and *Kennedy*. Enfield. Ponders End. **7.** Kentish Town,† *Ger. Hb.*, 484. East Heath, Hampstead, *T. & D.*, 452. Highgate Ponds!, *Fitter*. Regents Park!, *L.N.* 39: 56.

A. millefolium L. Yarrow, Milfoil.

Native in meadows, pastures, on waste ground, by roadsides, etc., established on lawns, railway banks, etc. Very common.

First record: *T. Johnson*, 1638.

7. Very common in the district.

TRIPLEUROSPERMUM Schultz Bip.

T. inodorum (L.) Schultz Bip. Scentless Mayweed.

T. maritimum (L.) Koch subsp. *inodorum* (L.) Hyland. ex Vaarama, *Matricaria inodora* L., *M. maritima* L. subsp. *inodora* (L.) Clapham

Lit. *Watsonia* **7**: 130 (1969).

Native on cultivated and waste ground, by roadsides, etc., established on railway banks, rubbish-tips, etc. Very common.
First record: *Morley*, c. 1677.
7. Very common in the district.
Plants with double flowers were gathered at **1.** Yiewsley, *J. E. Cooper* (BM).

MATRICARIA L.

M. recutita L. Wild Chamomile.
M. chamomilla auct.
Native on cultivated and waste ground, by roadsides, etc., established on railway banks, rubbish-tips, etc. Common.
First record: *T. Johnson*, 1632.
7. Common in the district.

M. matricarioides (Less.) Porter Pineapple Weed, Rayless Mayweed.
M. discoidea DC., *M. suaveolens* (Pursh) Buchen. non L.
Alien. N. America. Naturalised on bare waste ground, by roadsides, established on paths, in crevices in pavements, etc. Very common.
First evidence: *Stebbing*, 1896.
7. Common in the district.

CHRYSANTHEMUM L.

C. segetum L. Corn Marigold.
Alien. Mediterranean region. Formerly naturalised in cornfields, on cultivated ground, etc., but now sporadic on waste ground, rubbish-tips, by roadsides, etc. Very rare.
Lit. *J. Ecol.* **60**: 573 (1972).
First record: *T. Johnson*, 1638.
1. Harefield!, *Blackst. Fasc.*, 18; c. 1910, *Green*, *Williams MSS.* Harrow Weald Common, *T. & D.*, 156; 1902, *Green*, *Williams MSS.* Yiewsley, 1909–10, *Coop. Cas.* **2.** Teddington, *T. & D.*, 156. Halliford, *Benbow MSS.* Laleham, 1946!, *L.N.* 26: 61. Perry Oaks, 1958, *A. Vaughan.* **3.** Sipson; Whitton; near Isleworth, *Benbow MSS.* **4.** Hampstead, *Irv. MSS.* Mill Hill, 1909, *C. S. Nicholson* (LNHS); c. 1962, *Harrison.* Golders Green, 1908, *J. E. Cooper.* Hendon, 1960, *Warmington.* **5.** Ealing, 1902, *Green*, *Williams MSS.*; 1949, *L. M. P. Small*, *K. & L.*, 154. Near Greenford, 1945, *Welch*; Hanwell, 1946!, *L.N.* 26: 61. Chiswick, *Benbow MSS.* **6.** Tottenham, *Johns. Cat.* Edmonton, *T. & D.*, 156. Hadley, *Benbow MSS.* Enfield, 1967, *Kennedy.* **7.** St John's Wood, 1815 (G & R). Paddington, 1856, *Hind*; Burgess Hill, 1870, *Trim. MSS.* Hackney Marshes, 1912–20, *J. E. Cooper*, *K. & L.*, 154. Clapton, 1951, *Ing.* Near Ken Wood, 1962, *N. H. Martin.*

LEUCANTHEMUM Mill.

L. vulgare Lam. Marguerite, Moon-Daisy, Ox-eye Daisy.
Chrysanthemum leucanthemum L.
Lit. *Watsonia* **4**: 11 (1957): *J. Ecol.* **56**: 585 (1968).
Native in rich meadows, grassy places, etc., established on railway banks, etc. Common.
First record: *T. Johnson*, 1638.
7. Tottenham!, *Johns. Cat.* Between Westminster and Chelsea, *Pet. MSS.* Hyde Park!; Grosvenor Square, W.1, 1869,† *Warren*; Green Park, 1862;† South Kensington, *T. & D.*, 156. Buckingham Palace grounds, *Codrington, Lousley* and *McClintock*!; Kensington Gardens!, *D. E. Allen, L.N.* 39: 57. Fulham Cemetery. Brompton Cemetery. Holland House grounds, Kensington. Notting Hill. Kilburn. East Heath, Hampstead. Highgate. Isle of Dogs. Highbury. Hornsey. Dalston. Stoke Newington. Hackney Wick. Bethnal Green. Stepney. Tower Hill.
The Middlesex plants may be referable to *L. praecox* Horvatič ampl. Villard.

L. maximum (Ramond) DC. Shasta Daisy.
Chrysanthemum maximum Ramond
Alien. Pyrenees. Garden escape. Naturalised in grassy places, on waste ground, etc., established on railway banks, etc. Common.
First record: *D. H. Kent*, 1939.
7. Bombed site, St Bartholomew's Close, W.C., 1950,† *Whittaker* (BM). Regents Park. Kentish Town. Isle of Dogs. Holloway. Hornsey. Dalston.
The closely allied species *L. serotinum* (L.) DC. (*Chrysanthemum serotinum* L., *C. uliginosum* Pers.) from S.-E. Europe is also cultivated as a garden plant and sometimes occurs as an escape.

TANACETUM L.

T. parthenium (L.) Schultz Bip. Feverfew.
Matricaria parthenium L.
Alien. Caucasus, Asia Minor, etc. Formerly cultivated as a medicinal herb and also as a garden plant, and now naturalised in grassy places, on hedgebanks, waste ground, etc., and established on tops of walls, railway banks, etc. Common.
First record: *T. Johnson*, 1638.
7. Common in the district.
Double-flowered forms are quite frequent.

T. vulgare L. Tansy.
Chrysanthemum vulgare (L.) Bernh., *C. tanacetum* Karsch, non Vis.
Native or denizen on banks of rivers and streams, waste ground,

by roadsides, etc. Rather common, especially near the Thames and Lee.

First record: *T. Johnson*, 1638.

1. Harefield!, *Blackst. Fasc.*, 97. Denham Marsh, 1839. *J. F. Young* (BM). Between Harefield and Ickenham. Ruislip Common. Stanmore Common, *Payton*!. Yiewsley!, *J. E. Cooper* (BM). Springwell!, *I. G. Johnson*. South Mimms. **2.** Common by the Thames. Frequent about West Drayton and Dawley. **3.** Common by the Thames. Crane Park, Twickenham, abundant. Hounslow Heath. Yeading, frequent. Northolt. Roxeth. Hatch End. Near Southall. **4.** Hampstead, *Morley MSS*. Harrow!, *Melv. Fl.*, 46. Neasden. Brent Reservoir. Scratch Wood. **5.** Common by the Thames. By the Brent near Hanwell!, *Warren, T. & D.*, 157. Canal bank, Northolt to Greenford, frequent. Railway side, Southall to Brentford. South Ealing. **6.** Common by the Lee Navigation Canal from Waltham Abbey to Tottenham. **7.** Tottenham!, *Johns. Cat.* Lee Navigation Canal, Tottenham to Stamford Hill. Near Highgate Archway, *Irv. MSS*. River wall, Hammersmith, 1947. Fulham Palace grounds, 1966, *Brewis*. Bombed site, Bunhill Row area, E.C.,† *Jones Fl.* Bank of Regents Canal, Limehouse, 1967.

ARTEMISIA L.

A. vulgaris L. Mugwort.

Native on waste ground, by roadsides, etc. Very common.

First record: *T. Johnson*, 1638.

7. Frequent in the district.

A. verlotorum Lamotte Chinese Mugwort, Verlot's Mugwort.

Lit. *Watsonia* **1**: 209 (1950).

Alien. China. Naturalised by river- and streamsides, roadsides, on waste ground, etc., established on railway banks, rubbish-tips, etc. Common, especially near the Thames.

First evidence: *Green*, 1908.

7. Bombed site, Fetter Lane, E.C., 1950,† *Whittaker* (BM). Bombed site, Cripplegate area, E.C., *Jones Fl.* Chelsea!; Islington!, *L.N.* 39: 57. Holloway!; Paddington Green!, *L.N.* 41: 21. In several places between Paddington and Marylebone, 1969. Euston, 1965!; Hoxton, 1966!, *L.N.* 46: 33. Fulham, in several localities. Car park, East Smithfield, E.C.!, 1966, *Brewis*, *L.N.* 46: 33; 1967. Shoreditch. Isle of Dogs. Hackney. South Hackney. Limehouse. Bromley-by-Bow. Tottenham, *Kennedy*.

A. absinthium L. Wormwood, Absinthe.

Native or denizen on waste ground, by roadsides, etc., established on rubbish-tips, etc. Rather rare.

First record: *T. Johnson*, 1638.

1. Moorhall, *Blackst. Fasc.*, 1. Yiewsley!, *K. & L.*, 156. Near Ickenham, 1966. **2.** Hampton Court Park, 1940, *P. H. Cooke*, *K. & L.*, 156. **3.** Between Greenford and Yeading, *Welch*, *Lousley* and *Woodhead*!; Northolt!, *L.N.* 27: 32; 1967. Hounslow Heath!, *Philcox* and *Townsend*; 1967. **4.** Between Hampstead and Hendon, *Button*, *Coop. Suppl.*, 11. Hampstead, *Morley MSS.* Hampstead Heath, two plants, *De Cresp. Fl.*, 5. East Finchley, 1917, *J. E. Cooper* (BM). Kingsbury, 1952!, *Graham*, *K. & L.*, 156. Locally frequent on industrial waste ground by the North Circular Road between Stonebridge and Finchley. Scratch Wood railway sidings, 1957. **5.** Chiswick!, 1780, *Pamplin MSS.*; 1946, *K. & L.*, 156. North Hyde. New Southall!, *K. & L.*, 156. Brentford, 1966!, *Boniface*; 1967 →. **6.** Tottenham!, *Johns. Cat.* Church End, Finchley, 1917, *J. E. Cooper* (BM). Enfield, 1965, *Kennedy*. Near Pickett's Lock, Edmonton, 1966, *Kennedy*!. **7.** South Kensington, 1867,† *Dyer* (BM). Building site, Bloomsbury,† *Shenst. Fl. Lond.* Bombed site, Chelsea, 1943, *Sladen*, *L.N.* 24: 9; Royal Hospital grounds, 1969, *L. M. P. Small*. Bombed site, Cripplegate!,† *Wrighton*, *L.N.* 29: 87. Bombed sites, Upper Thames Street† and Beer Lane, E.C.,† *Welch*. Bombed site near the Tower of London, 1967. Bombed site, Marylebone,† *Holme*. Bank of Lee Navigation Canal, Stamford Hill!, *L.N.* 27: 32. Shoreditch, 1966. St Katherine's Dock, Stepney, 1966, *Brewis*. Wapping, 1967, *Clement*.

ECHINOPS L.

E. exaltatus Schrad.

E. strictus Fisch. ex Sims, *E. commutatus* Juratzka, *E. sphaerocephalus* auct., non L.

Lit. *Proc. Bot. Soc. Brit. Isles* **6**: 124 (1965): *loc. cit.* **7**: 243 (1968). (*Abstr.*)

Alien. Europe, etc. Garden escape. Naturalised on waste ground and established on railway banks. Rare.

First evidence: *F. Ballard*, 1931.

2. Railway side near Staines station, 1958!, *Briggs*; several large colonies, 1960 →. Near Hampton, a small colony, 1966 →, *Tyrrell*!. **3.** Hounslow Heath, 1961, *Buckle*. Isleworth, a small colony, 1966. **4.** Hendon golf links, 1931, *F. Ballard* (K). The Hyde, Hendon, a large colony, 1965 →. **5.** Acton, a small colony, 1965 →. Pitshanger Park, Ealing, a small colony, 1966.

The related species *E. banaticus* Roch. ex Schrad., *E. ritro* L. and *E. sphaerocephalus* L. are also cultivated and some of the records given above may be in need of revision.

CARLINA L.

C. vulgaris L. Carline Thistle.
Formerly native in calcareous grassland, and in a few places on other soils as an introduction, but now extinct.
First record: – c. 1730. Last evidence: *Green*, 1905.
1. Harefield, *Blackst. Fasc.*, 16. Ruislip Common, *Hind, Melv. Fl.*, 45. Gravel pits, Northwood, 1905, *Green* (SLBI). Near Mimmshall Wood, 1886, *Benbow MSS.* **5.** Hanwell Heath, *Ray Syn. MSS.* **6.** Finchley Common, 1845 (HE); a few plants, 1886, *Benbow, J.B. (London)* 25: 17; 1905, *C. S. Nicholson, K. & L.*, 162. **7.** Isle of Dogs, casual, 1887, *Benbow* (BM).

ARCTIUM L.

Lit. *Ann. Mag. Nat. Hist.* ser. 2 **17**: 369 (1856): *J.B. (London)* **51**: 113 (1913): *Watsonia* **2**: 312 (1953). (*Abstr.*)

A. lappa L. Great Burdock.
A. majus Bernh.
Native by stream- and riversides, on waste ground, by roadsides, etc. Local.
First certain evidence: *Buddle*, c. 1700.
1. Harefield!, *Blackst. Fasc.*, 9. Uxbridge; Swakeleys; Springwell!, *Benbow MSS.* Cowley!; Yiewsley!, *J. E. Cooper*; Ruislip!, *K. & L.*, 162. Pinner, *Hind, Trim. MSS.* **2.** Hampton, *Newbould*; between Sunbury and Walton Bridge!, *T. & D.*, 162. Shepperton!, *Benbow MSS.* Staines!, *Green* (BM). Penton Hook. West Drayton. **3.** Wood End,† *Hind. Fl.* Near Twickenham, *T. & D.*, 162. Southall!, *Wright* (BM). Cranford!, *K. & L.*, 162. **4.** Harrow!, *Hind Fl.* By the Brent near Kingsbury and Stonebridge!; Neasden;† Wembley Park,† *Benbow MSS.* Edgware!, *J. E. Cooper, K. & L.*, 162. Hendon!, *P. H. Cooke* (LNHS). Mill Hill, *Hinson.* **5.** By the Brent near Hanwell!; near Willesden Junction;† Shepherds Bush,† *Newbould, T. & D.*, 162. Chiswick;† Perivale!; Horsendon Hill!, *Benbow MSS.* Greenford!, *J. E. Cooper, K. & L.*, 162. **6.** Lee Valley in many places!, *Benbow.* Whitewebbs Park, Enfield, *Kennedy*!. Near Edmonton, *Kennedy.* **7.** Neat-houses, Chelsea,† *Buddle*; Ken Wood,† *Petiver* (SLO). Zoological Gardens, Regents Park,† *Shove Fl.* Perhaps an error.

A. minus Bernh. subsp. **minus** Lesser, or Small Burdock.
Native on waste ground, by roadsides, in woods, etc., established on railway banks, etc. Common.
First evidence: *Buddle* and *Petiver*, c. 1700.
7. Common in the district.

Subsp. **nemorosum** (Lejeune) Syme
A. nemorosum Lejeune, *A. vulgare* auct., *A. intermedium* auct.
Native in open woods, copses, thickets, waste ground, by roadsides, etc. Local.
1. Harefield!; north of Harefield, *Newbould, T. & D.*, 162. Uxbridge!; Ickenham!; Ruislip!, *Benbow MSS.* Eastcote. Near Potters Bar, *Kennedy*!. By Mimmshall Wood, *J. G.* and *C. M. Dony*!. **2.** West Drayton, *Newbould, T. & D.*, 162. East Bedfont. Harlington, *Bangerter.* **3.** Hanworth; Twickenham!, *T. & D.*, 162. Near Hounslow Heath, abundant!, *K. & L.*, 162. Hounslow Heath, scarce. Between Richmond and Twickenham. **4.** Harrow! (ME). Stanmore!, *T. & D.*, 163. Hampstead, *Blow, Bot. Loc. Rec. Club* 1875 *Rep.*, 164. **5.** Alperton;† Horsendon Hill, *T. & D.*, 163. Back Common, Chiswick. **6.** Enfield; Enfield Chase, *Kennedy*!. **7.** Chelsea,† *Buddle* and *Petiver* (SLO). Bird Sanctuary, Hyde Park!, *D. E. Allen, L.N.* 41: 21; 1968. Shoreditch.
The two subspecies sometimes grade into each other and are difficult to separate.

CARDUUS L.

C. tenuiflorus Curt. Slender Thistle.
Formerly a denizen on waste ground, etc., but now extinct.
First record: *T. Johnson*, 1638. Last record: *Melville*, 1929.
1. Yiewsley, *Melville, Rep. Bot. Soc. & E.C.* 8: 330. **2.** Yeoveney, *Benbow* (BM); casual, *Benbow, J.B. (London)* 25: 20. **7.** Tottenham, *Johns. Cat.* Frequent about London, *Budd. MSS.*; growing in the very suburbs, *Curt. F.L.* Bethnal Green, *B.G.*, 410. Islington, 1841 (HE). Sadlers Wells, 1838, *E. Ballard*; between Old Ford and Hackney Wick, 1887, *Benbow* (BM). Shepherds Bush, abundant; Holland Park, Kensington, abundant, *T. & D.*, 166. Hackney Marshes, *M. & S.* Newington, *Newbould*; Chelsea, *Britten*; between Lee Bridge and Hackney, *T. & D.*, 166.
C. tenuiflorus has been much confused with the adventive *C. pycnocephalus* L. and some of the above records are probably in need of revision.

C. nutans L. subsp. **nutans** Musk, or Nodding Thistle.
Native in meadows, on waste ground, etc. Locally frequent on alluvial soils near the Thames and on alluvial and calcareous soils in the Colne valley, very rare and possibly adventive elsewhere.
First record: *Blackstone*, 1737.
1. Harefield!, *Blackst. Fasc.*, 15. Between Yiewsley and Iver Bridge!, *Newbould, T. & D.*, 166. Yiewsley!, *J. E. Cooper, K. & L.*, 163. Uxbridge Moor, plentiful!, *L.N.* 27: 32. Springwell. **2.** Laleham, 1815 (G & R). Staines Moor!; Sipson;† between Sunbury and

Walton Bridge!; between Hampton Court and Kingston Bridge;† Teddington!, *T. & D.*, 166. Between Shepperton and Chertsey Bridge!, *Green, Williams MSS.* East Bedfont!; Harmondsworth, *P. H. Cooke*; Halliford Green!, *Welch, K. & L.*, 163. Between East Bedfont and Stanwell. **3.** Hounslow Heath!, abundant, *T. & D.*, 166; now very scarce. Near Richmond Bridge,† *T. & D.*, 166. Twickenham!, *K. & L.*, 163. **4.** West Heath, Hampstead, *M. A. Lawson, T. & D.*, 166. Finchley,† *J. E. Cooper, K. & L.*, 163. **5.** Chiswick, 1944!,† *Welch, K. & L.*, 162. Syon Park. **6.** Edmonton,† *T. & D.*, 166. **7.** Ball's Pond, 1838, *E. Ballard* (BM). Hornsey, casual, 1894,† *C. S. Nicholson, K. & L.*, 163. Kensington Gardens, one plant on a soil dump, 1961,† *D. E. Allen, L.N.* 41: 21.

C. crispus L. subsp. **occidentalis** Chass. & Arènes Welted Thistle.
C. acanthoides auct., non L.

Native by river- and streamsides, in damp grassy places, on waste ground, etc., established on canal banks, etc. Locally common near the Thames and Lee, and in the East End, rare elsewhere.

First record: *Blackstone*, 1737.

1. Harefield!, *Blackst. Fasc.*, 15. Garett Wood!;† Springwell!, *Benbow MSS.* Uxbridge!, *Druce Notes.* Near Harrow Weald Common,† *T. & D.*, 166. Near Denham. Ruislip. South Mimms. Potters Bar. **2.** Locally common near the Thames. Yeoveney!; Staines Moor!; Charlton!, *Benbow MSS.* Poyle. Colnbrook. **3.** Hounslow Heath,† *Cockfield Cat.*, 30. Near Hayes; near Twickenham!, *Benbow MSS.* Between Richmond Bridge and Twickenham. Hayes, *J. E. Cooper.* North Hyde!, *L.N.* 28: 33. Southall. **4.** Willesden,† *W. R. Linton* (BM). Finchley; Brent Reservoir, *J. E. Cooper.* Stonebridge, frequent. **5.** Chiswick!; canal between Hanwell and Brentford!,† *Benbow MSS.* Park Royal, 1909,† *Green* (SLBI). Hanwell.† **6.** By the Lee Navigation Canal, Ponders End to Waltham Abbey!, *Benbow, J.B.* (*London*) 25: 17. **7.** Very common in the environs of London, *Curt. F.L.* Kentish Town, *Irv. MSS.* Side of canal between Lee Bridge and Bow, *Cherry*; Isle of Dogs!, *Newbould, T. & D.*, 166. Between Hackney Wick and Old Ford!, *Benbow MSS.* Hackney Marshes, *Coop. Cas.* Frequent about Limehouse, Poplar, Bromley-by-Bow, Shadwell, Isle of Dogs, etc.

C. crispus L. subsp. **occidentalis** × **nutans**

2. Shepperton, with both parents, *C. E. Britton, J.B.* (*London*) 55: 326.

CIRSIUM Mill.

C. vulgare (Savi) Ten. Spear Thistle.
Carduus lanceolatus L., *Cirsium lanceolatum* (L.) Scop., non Hill

Native in fields, on waste ground, by roadsides, etc., established on railway banks, etc. Very common.
First record: *How*, c. 1656.
7. Very common in the district.
Plants with white flowers are recorded from **7.** St James's, *How MSS.*, and with flesh-coloured flowers from **3.** Hayes, *J. E. Cooper*, *K. & L.*, 163. **7.** Hackney Marshes, 1916, *J. E. Cooper*, *K. & L.*, 163.

C. palustre (L.) Scop. Marsh Thistle.
Carduus palustris L.
Native in marshes, wet fields, woods, on heaths, etc. Common, except in heavily built-up areas where it is rare or extinct.
First record: *T. Johnson*, 1638.
7. East Heath, Hampstead, *T. & D.*, 167. Ken Wood. Hackney Wick, *L. M. P. Small*. Plants with white flowers are common.

C. arvense (L.) Scop. Creeping Thistle.
Serratula arvensis L., *Carduus arvensis* (L.) Hill
Native on cultivated and waste ground, by roadsides, etc., established on canal and railway banks, etc. Very common.
First record: *T. Johnson*, 1638.
The var. **mite** Wimm. & Grab. and var. **setosum** (Willd.) C. A. Mey. are common, but probably of foreign origin.
Plants with white flowers are recorded from **1.** Yiewsley, 1912, *J. E. Cooper*. **4.** Edgware, 1921, *J. E. Cooper*. Harrow, 1966. **6.** Highgate; Muswell Hill, 1913, *J. E. Cooper*.

C. acaule Scop. Stemless, or Dwarf Thistle.
Carduus acaulos L., *Cirsium acaulon* auct.
Lit. *J. Ecol.* **56**: 597 (1968).
Native on closely grazed pastures, commons, in dry fields, etc. Very rare, and mainly confined to calcareous soils at Harefield and Springwell and to alluvial soils near the Thames.
First record: *Hind*, 1861.
1. Ruislip Common!, *Hind Fl.* Harefield!, *T. & D.*, 168. Northwood; Ruislip Moor;† Springwell!; between Ruislip and Northwood, *Benbow MSS.* **2.** Opposite Thames Ditton ferry,† *T. & D.*, 168. Meadow (now used as a timber-yard) by Chertsey Bridge!,† *Welch*, *L.N.* 27: 32; habitat destroyed by gravel-digging operations, c. 1963. Meadows between Chertsey Bridge and Shepperton, 1966, *Kennedy*!. Laleham,† *P. H. Cooke*, *K. & L.*, 163. Staines Moor!, *Briggs*. Knowle Green, scarce, *Dupree*!. **3.** Field by Hospital Bridge over the Crane, a few plants,† *T. & D.*, 168.

Var. **caulescens** (Pers.) DC.
1. Harefield!, *T. & D.*, 168. **2.** Near Ashford Common,† *Benbow MSS.*

C. dissectum (L.) Hill Meadow Thistle, Marsh Plume Thistle.
C. pratense (Huds.) Druce, non DC., *Carduus dissectus* L., *C. pratensis* Huds., *C. anglicus* Lam.
Formerly native in marshes and bogs on wet peat, on damp heaths, etc., but now extinct.
First record: *T. Johnson*, 1632. Last evidence: *Benbow*, 1884.
1. Stanmore Common, *Webb & Colem. Suppl.*, 12. Ruislip, *Hind, T. & D.*, 167. Harefield, 1884, *Benbow* (BM). **3.** Hounslow Heath, *Doody MSS.*; c. 1745, *Blackstone* (BM). **4.** Hampstead Heath, *Johns. Enum.*; c. 1730, *Dill. MSS.* Near Highgate, *Johns. Ger.*, 1184.

SILYBUM Adans.

S. marianum (L.) Gaertn. Milk-Thistle.
Carduus marianus L.
Alien. Europe, etc. Naturalised on waste ground, by roadsides, etc. Very rare, and often merely casual.
First record: *T. Johnson*, 1638.
1. Harefield, *Blackst. Fasc.*, 15; 1920, *P. H. Cooke, K. & L.*, 165. Pinner, 1865, *Hind, T. & D.*, 168. Yiewsley!, *Coop. Cas.* Near Springwell, 1951, *Lousley* and *McClintock*!. South Mimms, 1890, *Find. MSS.* **2.** West Drayton, *P. H. Cooke, K. & L.*, 165. Dawley, 1865; Ashford, casual, *Benbow MSS.* Near West Drayton, 1965!, *Wurzell.* **4.** Hampstead, *Irv. MSS.* Hendon, 1925, *J. E. Cooper* (BM). Kenton, a fine colony, 1949,† *Bull, K. & L.*, 165. East Finchley, *J. E. Cooper* (BM). **5.** Between Chiswick and Turnham Green, 1780, *Pamplin MSS.* Turnham Green, *T. & D.*, 168. Alperton, 1917, *J. E. Cooper*; Hanwell, casual, 1949!,† *K. & L.*, 165. **6.** Tottenham, *Johns. Cat.*; 1960, *Kennedy*. Highgate, casual, 1917,† *C. S. Nicholson* (LNHS). **7.** Between London and Hackney, *E. Forster* (BM). Islington; Tufnell Park, 1844 (HE). Hackney Marshes, *M. & S.* Regents Park, 1963, *Wurzell.*
Plants with green leaves are recorded from **7.** Near London, *How Phyt.*, 22. Islington,† *Merr. Pin.*, 21. Near Clerkenwell, *Horsnell, Ray Cat.*, 56; c. 1710,† *Bloss* (SLO). Banks of the New river,† *Pet. Midd.* Near New river Head,† *W. Watson, Hill Fl. Brit.*, 412. Near Shoreditch,† *J. Sherard, Ray Syn.* 3: 196. Marylebone Park,† *Mart. Pl. Cant.*, 73. Hackney, *Forst. Midd.*

ONOPORDUM L.

O. acanthium L. Scotch Thistle, Cotton Thistle.
Alien. Europe, etc. Naturalised on waste ground, by roadsides, etc. Very rare, and sometimes merely casual.
First record: *W. Turner*, 1562.

1. Harefield, *Blackst. Fasc.*, 14. **2.** Near Colnbrook, *Craig* (BM). Opposite Long Ditton, 1834† (W). Littleton, 1960, *Philcox*. Feltham Green, a single plant,† *T. & D.*, 165. **3.** Twickenham, one plant, 1870,† *Trim. MSS*. Yeading, 1953 !, *T. G. Collett*. Hounslow Heath, 1960, *Philcox* and *Townsend*. Roxeth, 1965 →, *Merison*. **4.** Hampstead, *Irv. MSS*. Finchley, 1910, *J. E. Cooper* (BM); 1963, *Brewis*. **5.** Syon Park,† *Turn. Hb.* 2, 146. Chiswick !, *Benbow*, *J.B.* (*London*) 25: 363. Greenford, 1944 !, *K. & L.*, 464. Hanwell, 1958. Garden weed, Ealing. Isleworth, several plants, 1972, *McLean*, one plant, 1972. **6.** Tottenham, *Johns. Cat.* Near Colney Hatch, one plant, *T. & D.*, 165. Muswell Hill, 1913, *J. E. Cooper* (BM). **7.** Most ditch banks about Town, very common,† *Petiver* (SLO). Islington, *Ball. MSS*. Kentish Town, 1844 (HE). Bombed site near The Minories, E.C., 1947 !,† *Welch*, *L.N.* 27: 32.

CENTAUREA L.

C. scabiosa L. Greater Knapweed.

Native in dry pastures, on banks, etc., on calcareous soils at Harefield, Springwell and South Mimms, and on alluvial soils near the Thames, elsewhere on waste ground, railway banks and tracks, by roadsides, etc., as an introduction. Locally common.

First record: *Blackstone*, 1737.

1. Common on calcareous soils at Harefield, Springwell and South Mimms. Northwood, *B. T. Ward*, *K. & L.*, 167. Yiewsley, *J. E. Cooper* (BM). Disused railway between Uxbridge and Denham. **2.** Locally common near the Thames. Between East Bedfont and Feltham !, *L.N.* 26: 62. Railway banks, Colnbrook to West Drayton. **3.** Hounslow Heath !, *L.N.* 26: 62. Railway banks, Hatch End !, *Bull*. **4.** Hampstead,† *Irv. MSS*. Finchley, *J. E. Cooper*; Stonebridge !, *K. & L.*, 167. Railway bank, Neasden, 1949 →. Bank by canal, Willesden Junction, 1962. Disused railway between Stanmore and Belmont, 1968, *Roebuck*. **5.** Canal path between Hanwell and Southall !, *L.N.* 26: 62; 1967. Railway bank, North Acton !,† *K. & L.*, 167. Railway side between Boston Manor and Brentford. **6.** Roadside near East Barnet, 1965, *Bangerter* !. **7.** Finchley Road, 1870,† *Trim. MSS*. Bombed sites, Bunhill Row,† Cripplegate† and Queen Victoria Street areas, E.C.,† *Jones Fl.* South Hackney, *L. M. P. Small*.

Plants with white flowers were gathered at **1.** Harefield, 1910, *J. E. Cooper* (BM).

C. cyanus L. Bluebottle, Cornflower.

Formerly a colonist in cornfields, on cultivated ground, etc. Now mainly as a casual on waste ground, etc., and usually of garden origin. Rare.

First record: *T. Johnson*, 1638.

1. Harefield!, *Blackst. Fasc.*, 21. Pinner, *Hind, Melv. Fl.*, 46. Uxbridge; Swakeleys; near Denham Lock; near Elstree, *Benbow MSS*. Yiewsley!, *M. & S.* Northwood, 1948!, *L.N.* 28: 33. New Years Green, 1954. Springwell, 1957; Ruislip, 1962, *I. G. Johnson*. Borders of Bayhurst Wood, 1965, *Jeffkins, J. Ruisl. & Distr. N.H.S.* 15: 16. **2.** Near Staines; Sunbury; Teddington; Strawberry Hill, *T. & D.*, 165. West Drayton!; near Colnbrook; between Sunbury and Hanworth, *Benbow MSS*. Staines, 1946!, *K. & L.*, 166. East Bedfont, *Westrup*. **3.** Twickenham, *T. & D.*, 165. Feltham; near Ashford; Hounslow; Hanworth, *Benbow MSS*. Hounslow Heath. **4.** Hampstead, *Irv. MSS*. Stonebridge, *Benbow MSS*. Willesden, *De Cresp. Fl.*, 108. East Finchley; Finchley, *Coop. Cas.* Wembley, 1965. Edgware, 1967, *Horder*!. **5.** Chiswick!; Hanwell; between Brentford and Hanwell!, *Benbow MSS*. Brentford, *A. Wood, Trim. MSS*. Greenford, 1938!, *K. & L.*, 166; 1965, *M. Collett*. Horsendon Hill, 1962. Boston Manor, 1966. **6.** Tottenham, *Johns. Cat.* Edmonton, *T. & D.*, 165. Near Ponders End, *Benbow MSS*. Enfield, 1965; north of Enfield Lock, 1966, *Kennedy*!. **7.** Chelsea, *Britten, Bot. Chron.*, 58. East Heath, Hampstead; Cremorne; Kensington Gore, *T. & D.*, 165. South Kensington, *Dyer List*. Isle of Dogs!, *Benbow MSS*. Kilburn, 1856, *Hind*; Finchley Road, 1869, *Trim. MSS*. Hyde Park, 1940, *Vivian, Wild Fl. Mag.* 1940. Hackney Marshes, 1912, *Coop. Cas.* Clapton, 1884 (HY). Crouch End, 1894, *C. S. Nicholson, K. & L.*, 166. Bombed site, Cripplegate area, E.C.,† *Jones Fl.* Holloway, 1965, *D. E. G. Irvine*. Paddington, 1965–66!, *L.N.* 46: 34; 1967.

C. jacea L. Brown-rayed Knapweed.
Lit. Marsden-Jones, E. M. & Turrill, W. B. *British Knapweeds* (1954).
Alien. Europe, etc. Formerly naturalised in meadows and on waste ground but now extinct.
First record: *Newbould*, 1862. Last record: *Benbow*, 1887.
4. Willesden, *De Cresp. Fl.*, 108. **5.** Acton, a stray plant or two, 1862, *Newbould, T. & D.*, 164. Near Chiswick, *Benbow, J.B. (London)* 25: 365. **7.** Site of the International Exhibition of 1862, South Kensington, 1872, *Dyer List*. Isle of Dogs, two plants, 1886, *Benbow, J.B. (London)* 25: 365; plentiful and well established, 1887, *Benbow MSS*.

Subsp. **angustifolia** Gugler var. **integrata** Gugler
3. Twickenham, in plenty in a meadow, 1867, *Dyer* (W), det. *Britton*; habitat destroyed, 1882, *Benbow MSS*.

C. jacea × **nigra** subsp. **nigra** = **C.** × **drucei** C. E. Britton
C. pratensis auct., *C. jungens* auct.
5. River Brent near Boston Manor station, *Jones & Turrill*, 62.

C. jacea × **nigra** subsp. **nemoralis** = **C.** × **moncktonii** C. E. Britton
C. surrejana C. E. Britton
6. Muswell Hill, *G. C. Druce, Rep. Bot. Soc. & E.C.* 6: 844.

C. nigra L. subsp. **nigra** Lesser Knapweed, Hardheads.
C. obscura Jord.
Lit. Marsden-Jones, E. M. & Turrill, W. B. *British Knapweeds* (1954).
Native in grassy places, on waste ground, established on railway banks, etc. Common, especially on heavy clay soils, absent from calcareous soils.
First record (of *C. nigra*): *T. Johnson*, 1638.
7. Fulham. Isle of Dogs. Regents Park. Finsbury Square, E.C.2!, *L.N.* 46: 34. Eaton Square, S.W.1!, *McClintock, L.N.* 39: 57. Primrose Hill. Kilburn. Hackney.

Subsp. **nemoralis** (Jord.) Gugler
C. nemoralis Jord.
In similar habitats to subsp. *nigra* but preferring lighter soils, though sometimes found on London Clay. Possibly common, but not always readily separable from subsp. *nigra*.
7. Chelsea Hospital grounds!, *Brewis, L.N.* 44: 26. Kensington Gardens!, *Allen Fl.* Buckingham Palace grounds, one plant, 1962, *Fl. B.P.G.*

C. calcitrapa L. Star Thistle.
Alien. Europe, etc. Formerly naturalised on waste ground, by roadsides, etc. Still found in such habitats but always casual. Very rare.
First record: *T. Johnson*, 1638.
1. Harefield, *Blackst. Fasc.*, 15. Near Uxbridge, casual, 1896, *Benbow* (BM). **3.** Hounslow Heath, 1790, *Forst. MSS.* **5.** Hanwell, 1954!, *D. Bennett.* **6.** Tottenham, *Johns. Cat.* Near Fortis Green, 1911, *Coop. Cas.* **7.** In the fields about London in many places as at Mile End Green and Finsbury fields, *Park. Theat.*, 990. Between London and Mile End, plentifully, *Merr. Pin.*, 21. Near Whitechapel, *Pet. Midd.* Bethnal Green, *Jones, With. Bot. Arr.* 7, 3, 961. Tyburn; Hammersmith, *Hill Veg. Syst.* Isle of Dogs, 1844, *E. Palmer* (W). Bombed site near Victoria Station, 1951!, *McClintock, K. & L.*, 167.

SERRATULA L.

S. tinctoria L. Saw-wort.
Native on wood margins, heaths, in thickets, etc., established on railway banks, etc. Very rare.
First record: *Gerarde*, 1597.
1. Harefield,† *Blackst. Fasc.*, 93. Elstree,† *T. & D.*, 163. Pole Hill, near Hillingdon, 1884,† *Benbow, J.B. (London)* 23: 37. Copse Wood,

Northwood!, *Welch, L.N.* 26: 62. Near Denham, 1955,† *Donovan, Ryall* and *Pickess*. Potters Bar, 1910, *P. H. Cooke* (LNHS); 1970, *Widgery*. West of Potters Bar, 1965, *J. G.* and *C. M. Dony*!. South Mimms, 1908, *J. E. Cooper, K. & L.*, 165. **3.** Roxeth;† Headstone,† *Hind, Melv. Fl.*, 45. Hounslow Heath!, *T. & D.*, 163; 1966. Railway banks between Hounslow and Feltham!, *Benbow MSS*. Near Feltham, 1965, *P. J. Edwards*. **4.** Hampstead Wood, *Ger. Hb.*, 576. Bishop's Wood, abundant, 1859, *Trimen* (BM); sparingly, 1920, *C. S. Nicholson* (LNHS). Hampstead Heath, *Johns. Enum.*; c. 1760, *J. Hill*; 1842, *R. Pryor* (BM); c. 1877,† *De Cresp. Fl.*, 67. Kingsbury;† Scratch Wood, *T. & D.*, 163. Near Mill Hill!, *J. E. Cooper*; Brookside Walk, Hendon,† *J. Bedford, K. & L.*, 165. **6.** Winchmore Hill Wood;† Whetstone,† *Benbow MSS*. Church End, Finchley, 1917,† *J. E. Cooper, K. & L.*, 165. **7.** Islington,† *Ger. Hb.*, 576. Marylebone fields, 1815† (G & R). Ken Wood,† *Irv. MSS*. Hornsey Wood, 1837, *E. Ballard*; c. 1890,† *R. Braithwaite* (BM).

Plants with white flowers are recorded from **1.** South Mimms, 1918, *J. E. Cooper, K. & L.*, 165. **4.** Hampstead Heath,† *Johns. Eric.* and *Park. Theat.*, 475. Hendon, *Irvine*; Kingsbury,† *T. & D.*, 164.

CICHORIUM L.

C. intybus L. Chicory, Wild Succory.

Native in fields, by roadsides, etc., on calcareous soils at Harefield and South Mimms, and on alluvial and gravel soils near the Thames; elsewhere on waste ground, by roadsides, etc., usually as the var. **sativum** Bisch., an escape from cultivation. Locally plentiful.

First record: *T. Johnson*, 1638.

1. Harefield!, *T. & D.*, 169. Springwell!; Uxbridge!; South Mimms!, *Benbow MSS*. Cowley!; Potters Bar!, *P. H. Cooke*; Yiewsley!, *J. E. Cooper, K. & L.*, 168. Ruislip. Northwood. **2.** Locally common, especially near the Thames. **3.** Meadow near Richmond Bridge,† *T. & D.*, 169. Northolt!, *J. E. Cooper*; Southall!, *K. & L.*, 168. Between Southall and Hayes, *Williams MSS*. Hayes; near Harlington, *J. E. Cooper, K. & L.*, 168. Railway banks between Osterley and Hounslow. Hounslow Heath, *Philcox*. Near Feltham. Hatton. Roxeth. **4.** Hampstead, *Morley MSS*. Stonebridge, *Benbow MSS*. East Finchley, *Coop. Cas.* Wealdstone!, *L.N.* 26: 62. Harrow Weald, 1960, *Knipe*. Near The Hyde, Hendon. Edgware, *Tyrrell*. **5.** Chiswick, *Benbow MSS*. Greenford!, *J. E. Cooper*. Hanwell!; railway banks, Hanwell to Southall, frequent!; Perivale!, *K. & L.*, 168. Railway bank, North Acton, 1961.† North Hyde. Railway bank, Ealing, 1957.† Lawn, West Ealing, 1969. **6.** Tottenham!, *Johns. Cat.* Muswell Hill, *Coop. Cas.* Bush Hill Park station, *Kennedy*. Abundant in a field, Forty Hill, Enfield, 1967, *Kennedy*!. **7.** Between Westminster and Chelsea,† *Pet. Midd.* South Kensington,† *Dyer*; railway side, Olympia, 1929, *Hales*; bombed site, Shadwell, 1952, *Henson* (BM). Bloomsbury,† *Shenst. Fl. Lond.* Isle of Dogs!, *Benbow MSS.*; in quantity, 1967. Hackney Marshes!, *Griffin*; Clapton, *P. H. Cooke*; bombed site, Fore Street, E.C.1, 1951,† *J. Bedford, K. & L.*, 168. Stamford Hill railway station, *Scholey*. Railway bank, White City, 1957 →. Bombed sites, Cripplegate and Queen Victoria Street areas, E.C., 1952–53, *Jones Fl.*

Some of the above records may be referable to *C. pumilum* Jacq., a native of the Mediterranean region, which occurs as a bird-seed introduction.

LAPSANA L.

L. communis L. Nipplewort.
Lampsana communis auct.
Native on waste ground, banks, by roadsides, etc., established on railway banks, etc. Very common.
First record: *T. Johnson*, 1638.
7. Common in the district.
A variable species worthy of further study. A very hairy form with large heads was seen in **3.** Shrubbery near Sudbury, 1951!, *T. G. Collett*, and a robust form with flower-heads twice the normal size was seen at **1.** Pinner, 1965.

ARNOSERIS Gaertn.

A. minima (L.) Schweigg. & Koerte Lamb's, or Swine's Succory.
Hyoseris minima L., *Arnoseris pusilla* Gaertn.
Formerly native in fields on sandy soils, but now long extinct.
First record: *Doody*, 1696. Last record: *W. Hudson*, 1778.
2. In Teddington Field, near Hampton Court, *Doody*, *Ray Syn.* 2: 344; c. 1746, *Blackst. Spec.*, 42; c. 1778, *Huds. Fl. Angl.* 2: 346.

HYPOCHOERIS L.

H. radicata L. Long-rooted Cat's Ear.
Native in fields, grassy waste places, on banks, etc., established on railway banks, lawns, etc. Very common.
First record: *T. Johnson*, 1638.
7. Very common in the district.

Var. **leiocephala** Regel
4. Highwood Hill, *E. F.* and *H. Drabble*, *Rep. Bot. Soc. & E.C.* 8: 531.

H. glabra L. Smooth Cat's Ear.
Formerly native in fields on sandy soils, but now extinct.
First, and only evidence: *J. Vaughan*, 1843.
2. Teddington, 1843, *J. Vaughan* (D).

LEONTODON L.

L. autumnalis L. Autumnal Hawkbit.
Apargia autumnalis (L.) Willd.
Native in meadows, short turf, on heaths, etc., established on lawns, etc. Very common.
First record: *T. Johnson*, 1638.
7. Very common in the district.
Plants with lemon-yellow coloured flowers are recorded from **6.** Muswell Hill, 1895; North Finchley, 1929, *J. E. Cooper, K. & L.*, 175.

L. hispidus L. Rough Hawkbit.
Apargia hispida (L.) Willd.
Native in meadows, grassy places, etc., established on railway banks, etc. Locally common, especially on calcareous and other dry soils.
First record: *T. Johnson*, 1638.
1. Harefield!, *Blackst. Fasc.*, 22. Harefield Moor, locally abundant. Springwell. Pinner!, *Hind, Melv. Fl.*, 42. Elstree, *T. & D.*, 142. Uxbridge. Eastcote. Potters Bar. Dyrham Park, South Mimms, *J. G.* and *C. M. Dony*. **2.** Locally common near the Thames. Stanwell Moor, *J. E. Smith*!. **3.** Twickenham, *T. & D.*, 170. Hayes, *P. H. Cooke*. **4.** Hampstead!, *Irv. MSS.* Hampstead Heath, *De Cresp. Fl.*, 36. Mill Hill, *Sennitt*. Stanmore!, *Varenne*; near Finchley!, *Newbould, T. & D.*, 170. East Finchley. Whitchurch Common. Harrow!, *Harley*. Oakleigh Park, *J. G. Dony*!. **5.** Alperton,† *Hind, Melv. Fl.*, 42. Hanwell,† *Warren*; Horsendon Hill!, *T. & D.*, 170. **6.** Tottenham, *Johns. Cat.* Highgate (G & R). Edmonton; Enfield Chase!, *T. & D.*, 170. Coldfall Wood, Highgate, *C. S. Nicholson* (LNHS). Muswell Hill!, *Bangerter* and *Raven*. Railway bank, Finchley Central Station. **7.** Grosvenor Square, W.1, 1869,† *Warren*; South Kensington,† *T. & D.*, 170. Bombed site, Ebury Street, S.W.1,† *McClintock*. Lawns in Fulham Palace grounds, *Welch*!.

L. taraxacoides (Vill.) Mérat Hairy Hawkbit.
Thrincia hirta Roth, *Leontodon leysseri* Beck, *L. hirtus* auct., non L., *Crepis nudicaulis* auct.
Native in dry grassy places on base-rich soils. Rather rare, or overlooked.
First record: *Blackstone*, c. 1736.

1. Harefield!, *Blackst. MSS.* Breakspeares, *Blackst. Fasc.*, 23. Harrow Weald Common, *T. & D.*, 170. Uxbridge Common!; Ickenham Green; near Bayhurst Wood!, *Benbow MSS.* Hillingdon, *Warren, Trim. MSS.* Near Ruislip Reservoir, *Welch, K. & L.*, 175. South Mimms, *J. G.* and *C. M. Dony*!. **2.** Near Teddington!, *T. & D.*, 170. Bushy Park!, *Benbow MSS.* East Bedfont. Staines Moor!, *Briggs.* Hampton Hill, *Clement.* **3.** Hounslow Heath!, *Newbould*; near Strawberry Hill, *T. & D.*, 170. Meadows near Richmond Bridge,† *Benbow MSS.* **4.** Hampstead Heath!, *Curt. F.L.* Stanmore, *Varenne*; near Brent Reservoir, *Farrar, T. & D.*, 170. Harrow, *Melv. Fl.*, 42. Whitchurch Common,† *C. S. Nicholson* (LNHS). Mill Hill! (BM). **5.** Acton Green,† *Newbould, T. & D.*, 170. Ealing Common!,† *L.N.* 27: 32. Chiswick, *Benbow MSS.* Meadow by Perivale Wood, *T. G. Collett*!. Near Greenford. Osterley. **6.** Enfield Chase!, *T. & D.*, 170. Highgate, *Hall, K. & L.*, 175. **7.** Primrose Hill, 1959!, *L.N.* 39: 57. Hyde Park, 1962!, *Allen Fl.* Eaton Square, S.W.1, 1958, *McClintock, L.N.* 39: 57. New turf in the middle of Hyde Park Corner, 1965,† *Brewis, L.N.* 46: 34.

PICRIS L.

P. echioides L. Bristly Ox-Tongue.

Helmintia echioides (L.) Gaertn.

Native on field borders, waste ground, banks, by roadsides, etc., established on railway banks, etc. Locally plentiful, especially on heavy soils.

First record: *W. Turner*, 1568.

1. Harefield!, *Blackst. Fasc.*, 41. Woodhall, Pinner, *T. & D.*, 172. Between Cowley and Uxbridge; between Uxbridge and Ickenham; Bayhurst Wood; near Harefield, *Benbow MSS.* Yiewsley!, *J. E. Cooper, K. & L.*, 169. Near Eastcote, *T. G. Collett*!. Eastcote, Ruislip Reservoir. Railway banks north of Potters Bar. **2.** Between Hampton Court and Kingston Bridge!, *Benbow MSS.* Ashford, *Hasler*; Staines!; Walton Bridge!, *K. & L.*, 169. East Bedfont. West Bedfont, *Kennedy*!. Near Harmondsworth, frequent. Between Uxbridge and West Drayton. **3.** Near Down Barns; Roxeth,† *Benbow MSS.* West End, Northolt, *P. H. Cooke, K. & L.*, 169. Hayes. Hounslow Heath. **4.** Harrow!, *Melv. Fl.*, 62. Edgware; between Sudbury and Alperton,† *T. & D.*, 172. Wembley Park; Edgwarebury, *Benbow MSS.* Willesden,† *De Cresp. Fl.*, 108. East Finchley; Finchley; Temple Fortune,† *Coop. Cas.* Hampstead. **5.** Between Syon and Brentford,† *Turn. Hb.*, 144. Shepherds Bush;† Twyford,† *Newbould*; Horsendon Hill!; abundant on railway between Acton and Willesden Junction,† *T. & D.*, 172. Between Greenford and Greenford Green!, *Benbow MSS.* Railway banks, Perivale to South Greenford, plentiful.

West Ealing, 1946.† Hanwell. Acton, *Druce Notes*. **6.** Near Finchley Common (HE). Edmonton!, *T. & D.*, 172. Lee Navigation Canal, Waltham Cross to Edmonton!, *Benbow MSS*. Noel Park, Wood Green, *Scholey, L.N.* 28: 33. **7.** Kilburn† (G & R). Between Kilburn and Kensal Green† (HE). Near Brompton Cemetery,† *Britten, T. & D.*, 173. Fulham, *Corn. Surv.* Isle of Dogs, *Benbow MSS.* Between Newington and Hornsey Wood,† *Ball. MSS.* Near Wormwood Scrubs, abundant, 1872,† *Warren, Trim. MSS.* Bombed sites, Bunhill Row,† Cripplegate!† and Gresham Street areas, E.C.,† *Jones Fl.* Hyde Park, one plant, 1961,† *Brewis, L.N.* 41: 21. Near Dorchester Hotel, W.1, 1965, *D. E. Allen, L.N.* 46: 34. Tottenham, *Kennedy*. Side of New river, Clissold Park, Stoke Newington, 1967. Victoria Park; South Bromley, *L. M. P. Small*.

P. hieracioides L. Hawkweed Ox-tongue.
Native in grassy places, by roadsides, on waste ground, established on railway banks, etc. Very rare, and mainly confined to calcareous and other dry soils.
First record: *Blackstone*, 1737.
1. Harefield!, *Blackst. Fasc.*, 41. Springwell!;† Garett Wood;† near Mimmshall Wood, *Benbow MSS*. Railway banks, Potters Bar. **2.** Near Teddington, *T. & D.*, 172. **3.** Hounslow Heath, a small colony, 1966. **4.** Hampstead Heath,† *Britten, T. & D.*, 172. Railway embankment, Northwick Park, *Harley Fl.*, 29. By M1, Scratch Wood, 1972, *Franklin*. **7.** Brompton Cemetery, 1941, *H. R. Davies*, det. *Welch, K. & L.*, 169.

TRAGOPOGON L.

T. pratensis L. subsp. **minor** (Mill.) C. J. Hartman Goat's-Beard, Jack-go-to-bed-at-noon.
T. minor Mill.
Native in meadows, on waste ground, by roadsides, etc., established on railway banks, etc. Very common.
First record: *W. Turner*, 1548.
7. Common in the district.

Subsp. **pratensis**
Denizen on waste ground, by roadsides, etc., established on railway banks, etc. Local near the Thames, rare elsewhere.
1. Near Northwood!, *K. & L.*, 178. **2.** Between Teddington and Strawberry Hill, *T. & D.*, 171. **3.** Twickenham!, *T. & D.*, 171. **5.** Chiswick, *T. & D.*, 171. Railway bank, Kew Bridge station!,† *Sandwith*. Gunnersbury Park, South Ealing to near Kew Bridge. Boston Manor, 1966. North Acton, 1949!,† *K. & L.*, 178. **6.**

Colney Hatch, *T. & D.*, 171. Bounds Green, frequent. Southgate. **7.** Millbank, S.W.1, 1836,† *J. F. Young* (BM). Copenhagen Fields railway station,† *T. & D.*, 171. Regents Park, 1955, *Holme*, *L.N.* 39: 57.

LACTUCA L.

L. serriola L. Prickly Lettuce.
L. scariola L.
Denizen on waste ground, by roadsides, etc., established on railway banks, rubbish-tips, etc. Formerly rare, but now common.
First certain evidence: *E. Ballard*, 1839.
7. Common in the district.
The var. **dubia** (Jord.) Rouy is common.

L. virosa L.
Native by river- and streamsides, on waste ground, by roadsides, etc., established in gravel pits, etc. Locally plentiful.
First record: *T. Johnson*, 1632.
1. Locally common. **2.** Locally frequent. **3.** Hatch End, *T. & D.*, 173. Hayes, *P. H. Cooke*; Yeading!, *K. & L.*, 176. Hounslow Heath!, *Philcox* and *Townsend.* **4.** Hampstead Heath, *Johns. Enum.*; c. 1830,† *Irv. MSS.* Mill Hill, *Coop. Cas.* Harrow, *Hind* (BM). Near Willesden Junction; between Willesden and Neasden!, *De Cresp. Fl.*, 35. Neasden. **5.** Acton;† Ealing;† Brentford!, *Goodenough*, *Smith Fl. Brit.*, 819. Between Acton and Willesden!,† *Newbould*; between Perivale and Horsendon Hill!;† Hanwell!,† *T. & D.*, 173. Horsendon Hill!, *Britten* (BM). Greenford!, *Wright MSS.* **6.** Enfield, *Druce Notes.* Alexandra Park, 1910, *J. E. Cooper* (BM). **7.** About London, *Merr. Pin.*, 68 and *Budd. MSS.* Near London, *J. Banks* (BM). Between Kensington and Knightsbridge,† *Merr. MSS.* Marylebone fields (G & R); Regents Park!,† *Shove Fl.* Near Stepney; between Blackwall and Woolwich, *Jones*, *With. Bot. Arr.* 3: 677. Kentish Town† (HE). Kilburn;† near Copenhagen House,† *Burn. Med. Bot.* 1, 12. Isle of Dogs. Bombed sites, Queen Victoria Street area, 1953,† *Jones Fl.* By the Thames, Wapping, 1966. St Katherine's Dock, Stepney, 1966, *Brewis*. Poplar. Limehouse.

Var. **integrifolia** Gray
1. Yiewsley!, *M. & S.* Springwell!, *K. & L.*, 176. **2.** Poyle!, *K. & L.*, 176. West Drayton.

L. saligna L. Least Lettuce.
Formerly native or denizen on waste ground, by roadsides, etc., but now long extinct.

First record: *T. Johnson*, 1638. Last evidence: *J. Banks*, 1800.
7. Tottenham, *Johns. Cat.* About Pancras Church, *Willisel, Ray Cat.* 2: 69; c. 1695, *Pet. Midd.*; c. 1800, *J. Banks* (BM).

MYCELIS Cass.

M. muralis (L.) Dumort. Wall, or Ivy-leaved Lettuce.
Lactuca muralis (L.) Gaertn., *Cicerbita muralis* (L.) Wallr.
Native on hedgebanks, in woods, shady places, by roadsides, etc., established on tops of walls, etc. Rare.
First record: *Petiver*, 1713.
1. Between Harefield and Ruislip!, *Blackst. Fasc.*, 95. Breakspeare House!, *J. Morris*; Pinner!; Pinner Hill!; Headstone,† *Hind, Melv. Fl.*, 42. Pinner Green. Railway bank between Uxbridge and West Drayton; Garett Wood!,† *Benbow MSS.* The Grove, Stanmore Common, *Green* (SLBI). Harefield!, *Kingsley* (ME). Ruislip, *I. G. Johnson.* **2.** Between Hampton Wick and Teddington, scarce, 1965. **4.** Hampstead!, *Pet. H.B.C.* Wall by 'The Spaniards', Hampstead Heath!, *Curt. F.L.*; still present in small quantity, 1967. North-west Heath, Hampstead!, *H. C. Harris, K. & L.*, 177; 1967. Harrow, 1958 →. Willesden Old Churchyard, 1965 →. East Finchley, 1966. **5.** Garden weed near Kew Bridge, 1967. Near Perivale, 1969, *L. M. P. Small.* **6.** Near Hornsey,† *Huds. Fl. Angl.*, 296. Highgate Wood, scarce, 1947!, *K. & L.*, 177. Wood Green, *L. M. P. Small.* **7.** Regents Park!, *L.N.* 39: 57; 1968. Basement near the B.B.C., Portland Place, W.1, a vigorous plant, 1956,† *Holme, L.N.* 39: 57.

SONCHUS L.

S. palustris L. Marsh Sow-Thistle.
Formerly native in marshes, on pond verges and by the tidal Thames, but now extinct.
First record: *Ray*, 1696. Last record: *Pamplin*, 1835.
5. Among reeds in the claypits of the large brickfields at Shepherds Bush, 1835, *Pamplin, Wats. New Bot. Suppl.* '*S. arvensis* was probably mistaken for *S. palustris*', *T. & D.*, 176. 'This I will not admit . . . the species being well known to me . . .', *Pamplin MSS.* **7.** Blackwall, *Ray Syn.* 2: 71. Isle of Dogs, *Budd. MSS.*; c. 1713, *Pet. H.B.C.*; c. 1800, *E. Forster, B.G.*, 410; not there in 1852, *Syme, Phyt.* 4: 860. Sparingly in the marshes about Blackwall and Poplar, *Curt. F.L.*

S. arvensis L. Field Sow-, or Milk-Thistle.
Native by river- and streamsides, on pond verges, cultivated and waste ground, by roadsides, etc., established on railway banks, etc. Common.

First record: *Blackstone*, 1737.
7. Common in the district.

S. oleraceus L. Milk- or Sow-Thistle.
Lit. *J. Ecol.* **36**: 203 (1948).
Native on cultivated and waste ground, by roadsides, etc., established on railway banks, tops of walls, rubbish-tips, etc. Very common.
First record: *Buddle*, c. 1710.
7. Very common in the district.

S. asper (L.) Hill Spiny Milk- or Sow-Thistle.
Lit. *J. Ecol.* **36**: 216 (1948).
Native on cultivated and waste ground, by roadsides, etc., established on railway banks, rubbish-tips, etc. Very common.
First record: *T. Johnson*, 1638.
7. Very common in the district.

CICERBITA Wallr.

C. macrophylla (Willd.) Wallr. Blue Sow-Thistle.
Lactuca macrophylla (Willd.) A. Gray, *Mulgedium macrophyllum* DC., *Sonchus macrophyllus* Willd.
Alien. Caucasus. Garden escape. Naturalised in grassy places, on waste ground, by roadsides, etc., established on railway banks, etc. Rare, but increasing.
First evidence: *J. E. Cooper*, 1924.
1. Eastcote!, 1962, *I. G. Johnson*; well naturalised in the grounds of Eastcote House, 1964 →. Northwood, three colonies, 1965 →. **2.** Hampton Court, 1944, *Welch*. Harmondsworth, *L. M. P. Small*. **4.** Temple Fortune, 1924–27,† *J. E. Cooper* (BM). Hampstead Heath, 1963, *J. R. Phillips*. Near The Hyde, Hendon, 1965 →. **5.** Railway bank and adjacent roadside, Ealing!, *L. M. P. Small*, *K. & L.*, 177; 1969. West Ealing, 1967, *L. M. P. Small*. Railway bank, Acton, 1956!,† *K. & L.*, 254. **6.** Enfield!, 1948, *A. K. Lloyd* (BM); near Oakwood station, 1965!, *Bangerter* and *Kennedy*. Bush Hill Park, Enfield, 1954, *Ayland*. **7.** Railway side near Paddington, 1952,† *Fitter*; Regents Park!, *Holme*, *L.N.* 39: 57. Derelict garden, Bayswater Road, W.2, 1965, *Gleadow*, *K. & L.*, 354.

HIERACIUM L.

Lit. *J. Linn. Soc. Bot.* **54** (1948).

Sect. VULGATA F. N. Williams.

H. pellucidum Laest.
Formerly established on old walls as an introduction, but now extinct.
6. Enfield, 1866, *Trimen* (BM), det. *Pugsley* (1946) and *Sell* and *West* (1966).

H. exotericum Jord. ex Bor.
Introduced. Naturalised in grassy places, on waste ground, etc. Very rare, or overlooked.
7. Grounds of Natural History Museum, South Kensington, 1949 and 1952, *Bangerter* (BM), det. *Sell* and *West*. Kensington Gardens, 1950!,† *K. & L.*, 171. Ranelagh Gardens, Chelsea, 1959. Regents Park, 1966, *Brewis*, *L.N.* 46: 34. Moorgate station, E.C., 1966, *Brewis*, det. *Andrews*.

H. vulgatum Fr.
Introduced. Naturalised in grassy places, on waste ground, etc., established on railway banks and tracks, old walls, etc. Very rare, but apparently increasing.
1. Railway banks and tracks, adjacent waste ground and walls, Uxbridge, 1953, *M. G.* and *T. G. Collett*!, det. *Sell* and *West*, *K. & L.*, 171; railway now disused, tracks taken up and area partly levelled but the plant persists, 1968. Disused railway between Denham and Uxbridge, 1965 →. **4.** Disused railway, Mill Hill, 1966, *Kennedy*!. **5.** Disused railway, Brentford, 1966. **6.** Disused railway, Muswell Hill, 1961!, *Bangerter* and *Raven*, det. *Sell* and *West*.

H. lepidulum (Stenstr.) Omang
Introduced. Naturalised in grassy places. Very rare, or overlooked.
7. Kensington Gardens!, *D. E. Allen*, det. *Sell* and *West*, *L.N.* 39: 57.

Var. **hamatophyllum** Dahlst.
7. Buckingham Palace grounds, 1956, *Codrington*, *Lousley* and *McClintock*!, det. *Sell* and *West*, *L.N.* 37: 185.

H. maculatum Sm. Spotted Hawkweed.
Introduced. Naturalised in grassy places, established on old walls, etc. Very rare.
3. Old wall, Twickenham, 1866,† *Trimen* (BM), det. *Pugsley*. **6.** Myddleton House grounds, Enfield, 1966, *Kennedy*!.

H. diaphanum Fr.
H. anglorum (A. Ley) Pugsl.
Introduced. Naturalised in grassy places, on waste ground, etc., and established on railway banks, etc. Very rare.
1. Uxbridge, *Loydell* (D), det. *Pugsley* (as *H. anglorum*). **5.** Kew Bridge station yard!, 1958, *Murray*, det. *Sell* and *West*, *L.N.* 38: 21; 1968.

H. cheriense Jord. ex Bor.
H. lachenalii C. C. Gmel. var. *pseudoporrigens* Pugsl., *H. porrigens* auct., *H. tunbridgense* Pugsl.
Native in woods, etc. Very rare.

6. Hadley Wood, 1950, *Bangerter*, *Morton* and *Whittaker* (BM), det. *Sell* and *West*.

H. strumosum (W. R. Linton) A. Ley
H. lachenalii auct. brit., non C. C. Gmel.
Probably introduced. Naturalised in woods, on waste ground, by roadsides, etc.; established on railway banks, etc. Common.
7. East Heath, Hampstead. Ken Wood. Bird Sanctuary, Hyde Park!, *D. E. Allen*, det. *Miles*, *L.N.* 41: 21.

Sect. TRIDENTATA F. N. Williams

H. trichocaulon (Dahlst.) Johans.
H. tridentatum sensu Pugsl., *pro parte*
Native on heaths, commons, waste ground, in woods, etc., established on railway banks, etc. Common.
7. Railway bank near Kensal Green Cemetery. Victoria Park, Hackney. Railway bank, Isle of Dogs.

H. calcaricola (F. J. Hanb.) Roffey
Introduced. Naturalised on waste ground. Very rare.
7. Bombed site, Ropemaker Street, E.C., 1962, *Lousley*, det. *Sell* and *West*, *L.N.* 42: 12.

Sect. UMBELLATA F. N. Williams

H. umbellatum L. Umbellate Hawkweed.
Native in woods, on heaths, waste ground, etc., and established on railway banks, etc. Very rare.
1. Harefield Common, abundantly,† *T. & D.*, 178. Hillingdon Heath;† Stanmore Common!, *Benbow*, *J.B.* (*London*) 25: 17. **3.** Hounslow Heath!, *Newbould* (BM). Railway banks between Hounslow and Feltham!, *Benbow*, *J.B.* (*London*) 25: 17. **4.** Near Mill Hill, *Kennedy*!. Highgate. **5.** Railway banks, Gunnersbury, 1951!,† *K. & L.*, 173. **6.** Copse between Enfield and Potters Bar, *Benbow*, *J.B.* (*London*) 25: 17. **7.** Ken Wood,† *Buddle* (SLO), Hornsey Wood, 1838,† *E. Ballard* (BM). Chesham Place, S.W.1, *Rönaason*, det. *Sell* and *West*, *L.N.* 42: 12. Waste ground, Isle of Dogs, 1966.

Var. **coronopifolium** Bernh.
3. Hounslow, *Druce Notes*.

Sect. SABAUDA F. N. Williams

H. perpropinquum (Zahn) Druce
H. bladonii Pugsl., *H. boreale* auct.
Native on heaths, in woods, etc., chiefly on light soils, elsewhere on railway banks, etc., as an introduction. Locally frequent.
1. Harefield!, *Green* (SLBI). Springwell. Near Uxbridge. Park Wood,

Ruislip!, *Wrighton*. Duck's Hill, Northwood!, *Benbow* (BM). Harrow Weald Common. Stanmore Common. Potters Bar. South Mimms. **3.** Hounslow Heath!, *Green* (SLBI). Whitton; Hanworth!, *Benbow* (BM). **4.** Bishop's Wood, *Buddle* (SLO), det. *Pugsley* (as *H. bladonii*). Harrow, *Temple*!. Railway sidings, Neasden. Cemetery near Kilburn. **5.** Wyke Green, 1946.† **6.** Railway bank, Highgate Wood. Hadley Common. Trent Park. North of Cockfosters. Enfield Chase, *Kennedy*!. **7.** Isle of Dogs, *Benbow* (BM). Ken Wood. Chelsea Hospital grounds, 1966 →.

H. virgultorum Jord.

Native in woods, on shady hedgebanks, etc. Very rare.

1. Copse Wood, Northwood!, *Welch*, det. *Pugsley*, *L.N.* 28: 33; 1968. Near Harefield!,† det. *Pugsley*, *L.N.* 28: 33.

H. rigens Jord.

Native on heaths, in shady places, etc., established on railway banks, etc. Very rare, or overlooked.

1. Harrow Weald Common, *Linton* (BM), det. *Pugsley*. Harefield!, *Graham*, det. *Sell* and *West*, *K. & L.*, 174. **5.** Railway bank, Ealing!, *T. G. Collett*, det. *Sell* and *West*. **7.** Finsbury Park, 1952, *J. Bedford*, det. *Sell* and *West*, *K. & L.*, 174. Bombed site, Shadwell, 1952, *Henson* (BM), det. *Sell* and *West*.

H. salticola (Sudre) Sell & West

H. sublactucaceum sensu Pugsl. *pro parte*

Probably introduced. Naturalised in shady places, on waste ground, etc., established on walls, etc. Possibly common but overlooked.

1. Northwood, 1949, *Graham*, det. *Sell* and *West*. Ruislip Reservoir, *Groves* (BM), det. *Sell* and *West*. New Years Green. Uxbridge!, 1958, *Matthews*, det. *Sell* and *West*; 1966. **2.** Railway bank, Staines Moor, 1963. **5.** Ealing, 1958!, det. *Sell* and *West*; 1966. **7.** Bombed site near Green Park, W.1, 1956,† *Crawford*, det. *Sell* and *West*. Bombed site, Shadwell, 1954, *Henson* (BM), det. *Sell* and *West*.

H. vagum Jord.

H. sublactucaceum Druce & Zahn *pro parte*, *H. calvatum* (F. J. Hanb.) Pugsl., *H. croceostylum* Pugsl., *H. subquercetorum* Pugsl.

Probably introduced. Naturalised in woods, on waste ground, etc., established on railway banks, etc.

7. Common on railway banks in the district.

Most published records of *Hieracium* have been omitted unless substantiated by specimens.

PILOSELLA Hill

Lit. *J. Linn. Soc. Bot.* **54** (1948).

P. officinarum C. H. & F. W. Schultz subsp. **officinarum** Mouse-ear Hawkweed.
Hieracium pilosella L.
Native on banks, heaths, in grassy places, etc., established on railway banks, etc. Common, particularly on dry soils.
First record: *T. Johnson*, 1638.
7. Tottenham, *Johns. Cat.* Near Bayswater, 1817 (G & R). Fulham Palace grounds, *Welch*!. East Heath, Hampstead. Ken Wood grounds.

Subsp. **concinnata** (F. J. Hanb.) Sell & West
Hieracium pilosella L. var. *concinnatum* F. J. Hanb.
1. Ruislip, *Pugsley*; Uxbridge, 1960, *Matthews* (BM), det. *Sell* and *West*.

P. lactucella (Wallr.) Sell & West subsp. **lactucella**
Hieracium lactucella Wallr.
Alien of unknown origin. Established as a garden weed. Very rare.
First record: *Studley*, 1964.
4. Harrow, 1964, *Studley*, conf. *Sell* and *West*, *L.N.* 44: 16; still present as a weed in a vegetable garden, 1965, *West*.

P. aurantiaca (L.) C. H. & F. W. Schultz subsp. **aurantiaca** Orange Hawkweed.
Hieracium aurantiacum L.
Alien. Europe, etc. Garden escape. Naturalised in grassy places, on waste ground, by roadsides, etc., established on railway banks, etc. Rather rare.
First evidence: *Trimen*, 1866.
1. Stanmore Common, 1866,† *Trimen* (BM), det. *Pugsley* (as *H. aurantiacum*). **6.** Disused railway, Muswell Hill, 1961!, *Bangerter* and *Raven*, det. *Sell* and *West*.

Subsp. **brunneocrocea** (Pugsl.) Sell & West
Hieracium brunneocroceum Pugsl.
1. Railway bank, Eastcote, abundant, 1947!, *L.N.* 27: 32; 1948 →. **2.** West Drayton!; railway side between Poyle and Colnbrook!, *K. & L.*, 174; 1967. Stanwell, 1955, *Grigg*. Staines!, *Briggs*. **3.** Wood End, one plant, 1942.† Railway bank by river near Richmond Bridge, plentiful, 1966 →. **4.** Kenton, 1952!, *K. & L.*, 174; 1966, *Kennedy*!. Mill Hill. Harrow!, *Harley Fl.*, 30. Waste ground, Wealdstone, a few plants, 1966. Railway bank, Oakleigh Park, 1960, *J. G. Dony*!; 1966. **5.** Greenford, '*L.B.*', *Country-side*, 1929, 373. Railway banks, Castle Bar, Ealing, abundant!, *Bull*, *K. & L.*, 174; 1968.

Grassy roadside, Syon Lane, Osterley, abundant, 1946!,† *L.N.* 26: 12. Habitat eradicated by road-widening operations. **6.** Enfield Chase; waste ground north of Enfield Lock, 1966, *Kennedy*!. Enfield Churchyard, 1966. **7.** Highgate Cemetery, 1973, *Bull.*

CREPIS L.

Lit. *Proc. Bot. Soc. Brit. Isles* **4**: 398 (1962).

C. foetida L. Stinking Hawk's-beard.

Formerly native in fields and waste places on sandy soils, but now extinct.

First record: *Doody*, 1695. Last evidence: *Tempere*, 1873.

2. Field near Teddington, 1873, *Tempere* (BM), conf. *J. B. Marshall.* **7.** Chelsea, *Doody*, *Pet. Midd.*

C. vesicaria L. subsp. **taraxacifolia** (Thuill.) Thell. Beaked Hawk's-beard.

C. taraxacifolia Thuill.

Alien. Europe, etc. Naturalised in fields, by river- and streamsides, roadsides, on cultivated and waste ground, etc., established on railway banks, rubbish-tips, etc. Very common.

First evidence: *Newbould*, 1866.

7. Very common in the district.

A plant with entire leaves was gathered at **1.** Yiewsley, 1910, *J. E. Cooper* (BM), and plants with pale-lemon-coloured flowers were noted at **1.** Yiewsley, 1916–21, *J. E. Cooper.*

C. setosa Haller f. Bristly Hawk's-beard.

Alien. Europe, etc. Naturalised on waste ground. Very rare.

First record: *Clement*, 1965.

5. About ten plants by a footpath, Boston Manor, 1965, *Clement*, *L.N.* 45: 21; 1966, *Clement*!; 1967 →.

C. biennis L. Rough Hawk's-beard.

Native in meadows, on cultivated and waste ground, by roadsides, etc., established on railway banks, etc. Very rare, and mainly confined to calcareous and other dry soils.

First certain evidence: *Benbow*, 1884.

1. Pinner, 1884–87; Harefield!, *Benbow*; Yiewsley, 1918, *J. E. Cooper* (BM), all conf. *J. B. Marshall.* Near Pinner; Pinner Hill, *Benbow*, *J.B. (London)* 22: 56. Breakspeares; between Uxbridge and Harefield, *Benbow MSS.* Railway banks near Potters Bar. **2.** West Drayton, *Montg. MSS.* **4.** Stanmore, *Druce Notes.* Harrow, *Horton Rep.* Error, *C. vesicaria* subsp. *taraxacifolia* intended. **5.** Chiswick, *Newbould*, *T. & D.*, 176. Error, the specimen (BM) on which the record was based is *C. vesicaria* subsp. *taraxacifolia.*

Var. **runcinata** Koch
1. Chalkpit, Harefield, and field above, abundant, *Pugsley*!, *K. & L.*, 170; the chalkpit was bulldozed and filled with rubbish in 1965–66 but the plant still grows in the field above.

C. capillaris (L.) Wallr. Smooth Hawk's-beard.
C. virens L.
Lit. *Proc. Bot. Soc. Brit. Isles* **5**: 325 (1964).
Native in fields, on heaths, cultivated and waste ground, by roadsides, etc., established on railway banks and rubbish-tips, etc. Very common.
First record: *T. Johnson*, 1638.
7. Common in the district.

Var. **glandulosa** Druce
Var. *anglica* Druce & Thell.
1. Uxbridge (D). **7.** Hackney, *G. C. Druce*, *Rep. Bot. Soc. & E.C.* 8: 744.

TARAXACUM Weber

Lit. Suppl. to *Watsonia* 9 (1972).

T. officinale Weber *sensu lato* Common Dandelion.
T. vulgare Schrank, *Leontodon taraxacum* L.
Native in meadows, grassy places, on cultivated and waste ground, by roadsides, etc., established on lawns, railway banks, etc.
First record: *T. Johnson*, 1638.
7. Abundant in the district.

T. brachyglossum (Dahlst.) Dahlst.
2. Hampton Court, *Todd*, *Rep. Bot. Soc. & E.C.* 6: 734. **3.** Southall, *G. C. Druce*, *Rep. Bot. Soc. & E.C.* 6: 734.

T. laevigatum (Willd.) DC. *sensu lato* Lesser Dandelion.
T. erythrospermum Andrz. ex Bess.
Native in fields, on heaths, waste ground, etc., established on railway banks, wall-tops, etc. Rather common, especially near the Thames, and chiefly confined to dry soils.
First record: *Petiver*, 1709.
1. Harefield. Between Ickenham and Harefield. Disused railway between Uxbridge and Denham. Ruislip. Potters Bar. **2.** Common near the Thames. Colnbrook. **3.** Hounslow Heath. Isleworth. Twickenham. Feltham. Near Southall. Hayes. **4.** Mill Hill. Near Edgwarebury. **5.** Chiswick. Syon Park. Brentford. South Ealing. Ealing. Hanwell. Near Hammersmith. **6.** Tottenham. Near Enfield. Southgate, *Kennedy*!. **7.** Islington Churchyard,† *Pet. Bot. Lond.* Isle of Dogs, *Newbould*; Grosvenor Square, W.,† *Warren*, *T. & D.*, 174. Hyde Park!, *D. E. Allen*, *L.N.* 39: 58. St John's Wood.

T. rubicundum Dahlst.

7. Kensington Gardens, *D. E. Allen*, *L.N.* 41: 21.

The British species of *Taraxacum* have recently been revised by A. J. Richards (The *Taraxacum* flora of the British Isles. Supplement to *Watsonia* **9**. Pp. 141 (1972)).
The following are cited for Middlesex:

Section ERYTHROSPERMA
T. brachyglossum (Dahlst.) Dahlst.
T. lacistophyllum (Dahlst.) Raunk.
T. fulvum Raunk.

Section SPECTABILIA
T. nordstedtii Dahlst.

Section VULGARIA
T. lacerabile Dahlst.
T. tenebricans (Dahlst.) Dahlst.
T. adsimile Dahlst.
T. hamatum Raunk.
T. privum Dahlst.
T. cophocentrum Dahlst.

MONOCOTYLEDONES

ALISMATACEAE

BALDELLIA Parl.

B. ranunculoides (L.) Parl. Lesser Water-Plantain.
Alisma ranunculoides L., *Echinodorus ranunculoides* (L.) Engelm.
Formerly native on pond and lake verges, in marshes, ditches, etc., but now extinct.
First record: *Blackstone*, 1737. Last record: *D. H. Kent*, 1947.
1. Harefield, *Blackst. Fasc.*, 78. **3.** Near Hounslow, *E. Forster*, *B.G.*, 404. **4.** Brent Reservoir, 1917, *J. E. Cooper*, *K. & L.*, 282. **6.** Finchley Common, 1795, *Rayer*, *E.B.*, 326; frequent there, *Woods*, *B.G.*, 404; still plentiful, 1886, *Benbow*, *J.B.* (*London*) 25: 18; 1917–28, *J. E. Cooper*; scarce, 1947!; Hadley Common, a single plant, 1947!, *K. & L.*, 282. **7.** Eel Brook Common, abundantly, 1830, *Pamplin*, *T. & D.*, 286; 1837, *J. J. Bennett* (CYN).

ALISMA L.

Lit. *Proc. Bot. Soc. Brit. Isles* 7: 442 (1968). (*Abstr.*)

A. plantago-aquatica L. Water-Plantain.
Native by river- and streamsides, on pond verges, in ditches, marshes, etc., established by canals, etc. Common.
First record: *T. Johnson*, 1638.
7. Dartmouth Park,† *Jewitt*; Eel Brook Meadow,† *T. & D.*, 286. About Copenhagen House, Hornsey Wood,† *Ball. MSS.* Highgate Ponds!, *Fitter.* Water tank, Great Turnstile Street, W.C., 1950,† *Whittaker* (BM). Canal side near Paddington, 1948!, *L.N.* 39: 58. Lake, Buckingham Palace grounds!, *Eastwood*; possibly extinct, *Fl. B.P.G.*

A. lanceolatum With.
Native in similar situations to the preceding species with which it sometimes grows. Locally plentiful.
First evidence: *Newbould*, 1868.
1. Frequent in the Colne and canal from Springwell to Cowley. Between Yiewsley and Iver Bridge!, *Newbould*; Elstree Reservoir,† *T. & D.*, 286. Ruislip Reservoir. **2.** Near Staines!; Sunbury; Queen's river, Hampton; Thames between Kingston Bridge and Hampton Court,† *T. & D.*, 286. Shortwood Common!, *Welch, K. & L.*, 282. **5.** Alperton;† Hanwell,† *Warren, T. & D.*, 286. **7.** Hornsey Wood ponds,† *T. & D.*, 286.

A. lanceolatum × plantago-aquatica
1. Springwell. Ruislip Reservoir!, *Graham.*

DAMASONIUM Mill.

D. alisma Mill. Thrumwort, Star-fruit.
Actinocarpus damasonium (L.) Sm., *Alisma damasonium* R.Br.
Formerly native in ponds on gravelly soils, but now extinct.
First record: *Goodyer*, c. 1618. Last certain record: *Benbow*, 1886.
1. Harefield Common, *Blackst. Spec.*, 75; 1839, *Kingsley* (ME). Uxbridge towards Denham, *Mart. Pl. Cant.*, 74. **3.** Hounslow Heath, *Goodyer, How MSS.*; plentiful, 1736, *Blackst. MSS.*; in many shallow pools by the roadside from Colnbrook to Hounslow, 1761, *Lightf. MSS.*; c. 1800, *A. B. Lambert*; Twickenham Common, 1842, *Twining* (K). **4.** Golders Green, *Irv. MSS.* **5.** Turnham Green, 1798, *R. Brown, Trim. MSS.* Acton Green, 1864, *Newbould, T. & D.*, 287. The latter two records probably refer to the same locality. **6.** Near Southgate, *Alch. MSS.* Finchley Common, *Woods, B.G.*, 404. Hadley Common, tolerably plentiful, 1855, *Coleman, Trim. MSS.*; a single plant, 1886, *Benbow, J.B. (London)* 25: 18. 'Said to grow on

Hadley Green . . . [1904–65], but I have looked for it here in vain', *J. G. Dony, Fl. Herts*, 99. **7.** Ditches between Holloway and Highgate, *Park. Theat.*, 1245. Near Newington Butts, *Fysher, Ray Syn.* 3: 273.

SAGITTARIA L.

S. sagittifolia L. Arrow-head.

Native in shallow water at the sides of rivers, streams and ponds, established in canals, etc. Formerly common, but now local and decreasing.

First record: *Pena* and *l'Obel*, 1570.

1. Harefield!, *Blackst. Fasc.*, 89. Uxbridge. Yiewsley. Ruislip Reservoir, *Pickess*. Springwell. **2.** Locally common, especially in and near the Thames. **3.** Twickenham!, *T. & D.*, 287; very abundant in both streams of the Crane, Crane Park. **4.** Brent at Willesden† (HE). Stonebridge;† Forty Farm [Kenton],† *Farrar, T. & D.*, 287. **5.** Canal at Greenford!, *Melv. Fl.*, 80. Alperton;† Wembley,† *A. B. Cole*; Brentford!, *Hemsley, T. & D.*, 287. Canal and Brent, Hanwell. **6.** Between Hornsey and Tottenham,† *Coop. Suppl.*, 11. Edmonton, *T. & D.*, 287. Enfield!, *H. & F.*, 150. **7.** In the Tower ditch,† *Pena & Lob. Stirp. Adv.*; *Ger. Hb.*, 337, and later authors. Before the Earl of Peterborough's House, above the horse ferry on the Westminster side,† *Baylis*, 93. Tothill Fields,† *Blackst. Spec.*, 86. Isle of Dogs, 1792, *E.B.*; 1815 (G & R); 1836, *J. F. Young* (K); c. 1853, *Cowper*, 107. Paddington Canal, 1862,† *Britten*; Temple Mills, 1868, *Cherry, T. & D.*, 287. Between Westminster and Chelsea,† *Pet. Midd.* Paddington Canal near Westbourne Park,† and at Maida Hill,† *Trim. MSS.* Abundant in the Regent's Canal near Cumberland Basin, Regents Park, 1893,† *A. W. Bennett, J.B.* (*London*) 31: 249. Clapton, 1837,† *E. Ballard* (BM). Islington, 1914,† *W. C. R. Watson.*

BUTOMACEAE

BUTOMUS L.

B. umbellatus L. Flowering Rush.

Native in ponds, lakes, ditches, streams, etc., established in canals, etc. Formerly common, but now local and decreasing.

First record: *Pena* and *l'Obel*, 1570.

1. Harefield!, *Blackst. Fasc.*, 33. Brook between Hillingdon and Uxbridge,† *J. Hill* (BM). Near Swakeleys;† canal side from Denham Lock to Harefield;† Uxbridge!, *Benbow MSS.* Elstree Reservoir!, *K. & L.*, 283. Springwell, 1965!, *Pickess*. **2.** Bushy Park!, 1802, *Winch MSS.*; 1972. Shortwood Common!, 1841 (HE); 1973. Stanwell!

and Staines Moors!, *T. & D.*, 288; 1968. Knowle Green;† Shepperton, *Benbow MSS.* Penton Hook; Hampton Court!, *J. E. Cooper, K. & L.*, 283; 1967. **3.** Roxeth,† *Melv. Fl.*, 80. Hounslow Heath!,† *T. & D.*, 288. Between Strawberry Hill and Twickenham, 1820,† *Pamplin MSS.* **4.** Willesden† (HE). Near Willesden Junction, 1867,† *Newbould* (SY). Between Finchley and Hendon,† *Newbould, T. & D.*, 288. Mill Hill, *E. M. Dale, K. & L.*, 283. Brent Reservoir!, *Lousley* (L); locally abundant, 1972. **5.** Canal near Greenford!,† *Melv. Fl.*, 80. Twyford,† *Newbould*; in the Brent, profusely,† *Lees, T. & D.*, 288. Perivale!;† canal between Hanwell and Brentford!;† near Alperton;† near Southall,† *Benbow MSS.* Near Ealing, 1817,† *F. J. Young* (BM). Syon Park!, *L. G.* and *R. M. Payne, K. & L.*, 283: 1973. **6.** Edmonton, *T. & D.*, 288. Enfield!, *H. & F.*, 148. By the Lee Navigation Canal in many places!, *Benbow MSS.* **7.** By the Tower of London,† *Pena & Lob. Stirp. Adv.*, 44. Paddington Canal, near 'The Mitre', 1815;† Bayswater Canal, Kensington Gardens, 1817† (G & R); Paddington Canal near Westbourne Park,† *Trim. MSS.*; Paddington Canal by Kensal Green Cemetery, 1862,† *Phyt. N.S.* 6: 348. Blackwall,† *Curt. F.L.* Isle of Dogs, very common,† *Smith MSS.* Banks of Thames, Fulham,† *Faulkner*, 22. Ken Wood!, *Hunter, Park Hampst.*, 30. Clapton, 1837,† *E. Ballard* (BM). In the Lee at Hackney,† and ditches at Hackney Wick,† *T. & D.*, 288. Regents Park,† *Webst. Reg.*, 102.

HYDROCHARITACEAE

HYDROCHARIS L.

H. morsus-ranae L. Frog-bit.

Native. Floating on ponds, ditches, slow streams, etc. Formerly rather common, but now very rare, and decreasing.

First record: *Pena* and *l'Obel*, 1570.

1. Harefield!, *Blackst. Fasc.*, 65. Uxbridge!,† *Lightf. MSS.* Colne near Iver Bridge!,† *Newbould, T. & D.*, 267. Cowley!,† *Benbow MSS.* Yiewsley. Swakeleys lake!,† *Tremayne*; Stanmore Common!;† between Harefield and Northwood!,† *K. & L.*, 261. Potters Bar. **2.** Staines (= Shortwood) Common!,† *Phyt. N.S.* 4: 263. Staines Moor ditches!,† abundant, *T. & D.*, 267; persisted in one place until c. 1957.† Between Sunbury and Hampton,† *Newbould*; Hampton Court!, 1862, *Bell, T. & D.*, 267; 1950. Stanwell Moor!;† near Hanworth,† *Benbow MSS.* Hampton Wick,† *E. K.* and *W. K. Robinson, Country-side*, 1919; 162. Pools near Walton Bridge!;† Colne at West Drayton!,† *L.N.* 27: 34. Canal backwater, West Drayton, 1956–64;† Dawley!,† *K. & L.*, 261. Colnbrook. **3.** Thames at Twickenham, abundant, 1866,† *Masters*; Marsh Farm, Twickenham;† Hounslow

Heath;† Crane by bridge on Hanworth Road,† *T. & D.*, 267. Old cuts, Southall!,† *Benbow MSS.* **4.** Forty Green [between Wembley Park and Kingsbury];† Wembley Park,† *A. B. Cole, T. & D.*, 267. Hampstead,† *Whiting Notes.* West of Harrow† (BM). Canal, Willesden,† *De Cresp. Fl.*, 108. **5.** Canal near Brentford,† *Hemsley*; pond near Acton, towards Ealing,† *Newbould, T. & D.*, 267. Canal between Hanwell and Brentford;† Heston,† *Benbow MSS.* Syon Park!, *L. G.* and *R. M. Payne, L.N.* 27: 34. **6.** Tottenham,† *Johns. Cat.* Brimsdown, 1957. Trent Park, *A. Vaughan*; 1966, *T. G. Collett* and *Kennedy*. **7.** By the Tower of London,† *Pena & Lob. Stirp. Adv.*, 258. Many ditches about London,† *Park. Theat.*, 1253. Near Blackwall, 1822,† *J. Taylor* (BM). Isle of Dogs, 1836, *Irv. Lond. Fl.*, 109; 1853,† *Syme* (SY). Most abundant in the many stagnant ditches intersecting the fields and market-gardens at Ranelagh; the Neat Houses;† Tothill Fields, etc., 1820–25,† *Pamplin, T. & D.*, 267. Near Stoke Newington,† *Winch, Wats. New Bot.*, 101. New river, Stoke Newington;† Finsbury Park lake,† *C. S. Nicholson, K. & L.*, 261. Canal, Regents Park,† *Webst. Reg.*, 100.

STRATIOTES L.

S. aloides L. Water Soldier.

Introduced. Naturalised in ponds and lakes. Very rare.

First record: *Ford*, 1873.

1. Abundant in two ponds, Little Common, Stanmore!, *L.N.* 27: 34; apparently eradicated when the ponds were cleaned out, c. 1963; locally abundant in a lake near the cricket ground, Stanmore Common, 1962 →. **6.** Enfield!, *H. & F.*, 150; abundant in the lake at Wildwood, Whitewebbs Park, 1965 →, *Kennedy*!, *L.N.* 45: 20.

ELODEA Michx.

E. canadensis Michx. Canadian Pondweed or Water-weed.

Anacharis canadensis (L. C. Rich.) Planch.

Alien. N. America. Naturalised in rivers, streams, ponds, ditches, etc., established in gravel pits, canals, etc. Common, though apparently slowly decreasing.

First record: *A. Evans*, 1854.

7. River Lee between Lee Bridge and Upper Clapton, 1854; ditches by the Thames near the Bishop's Palace, Fulham,† *A. Evans, Phyt. N.S.* 1: 96. Ditches and streams at Hackney, etc., abundant; Newington;† Hornsey Wood;† Hampstead Ponds!; Serpentine, Kensington Gardens, where it flowers profusely,† *T. & D.*, 267. Vale of Health and Viaduct Ponds, East Heath, Hampstead.

E. ernstae St John
E. callitrichoides auct., non (Rich.) Casp.
Lit. *Proc. Bot. Soc. Brit. Isles* **1**: 321 (1961).
Alien. Temperate S. America. Naturalised in streams, established in gravel pits, etc. Very rare.
First record: *Grigg*, 1950.
1. Gravel pits, Springwell, 1965, *Pickess*. **2.** Longford river, Stanwell, abundant!, *Grigg, L.N.* 30: 7; in flower December, 1961, *M. McC. Webster*; very scarce, 1964; not seen since. Longford river between Hounslow Heath and East Bedfont, in very small quantity, 1966. Shortwood Common, Staines, 1971, *Clement*, *Tuck* and *E. Young*. **3.** Longford river, Hanworth, sparingly, in 1961!, *L.N.* 41: 13. **5.** Floating in the Thames, Strand-on-the-Green, 1955!,† *K. & L.*, 13.

LAGAROSIPHON Harv.

L. major (Ridl.) Moss
Alien. S. Africa. Established in a gravel pit and in a pond. Very rare.
First record: *Kennedy*, 1965.
Lit. *Proc. Bot. Soc. Brit. Isles* **1**: 321 (1961).
2. Locally plentiful in a gravel pit, West Bedfont, 1965!, *Kennedy, L.N.* 45: 20; 1971. **5.** Pond in Clayponds Hospital grounds, South Ealing, 1968 →.

VALLISNERIA L.

V. spiralis L.
Lit. *Watsonia* **9**: 253 (1973).
Alien. Warmer parts of the world. Established in a canal. Very rare.
First record: *Harris* and *Lording*, 1971.
6. Sparingly in Lee Navigation Canal, south of Enfield Lock, 1971, *Harris* and *Lording*, *Watsonia* **9**: 253 (1973).

JUNCAGINACEAE

TRIGLOCHIN L.

T. palustris L. Marsh Arrow-grass.
Native in wet grassy places, marshes, on pond and lake verges, etc. Very rare.
First record: *Blackstone*, 1737.
1. Harefield, *Blackst. Fasc.*, 38; 1887, *Benbow*, *J.B.* (*London*) 25: 18; 1918, *Redgrove*; between Yiewsley and Iver Bridge,† *Newbould* (BM). Cowley;† Uxbridge Moor;† Ruislip;† Springwell,† *Benbow*,

J.B. (*London*) 25: 18. **2.** Bushy Park!, *J. Harris* (W); locally plentiful, 1972. Hampton Wick, *E. K.* and *W. K. Robinson*, *Country-side*, 1919, 162. Staines Moor, 1957. **5.** Greenford,† *Melv. Fl.*, 81. Syon Park, abundant around a lake!, *L. G.* and *R. M. Payne*, *L.N.* 27: 34: 1971. **6.** Finchley Common,† *Benbow*, *J.B.* (*London*) 25: 18. Gravel pit north of Enfield Lock, plentiful, 1966, *Kennedy*!; 1967. **7.** Eel Brook Common,† *Pamplin*, *T. & D.*, 289. Chelsea, 1901,† *Cotton* (K). Lee Marshes, Clapton, 1883,† *P. S. King* (LNHS). Isle of Dogs, 1840, *H. Newman* (BM); 1850,† *Syme* (SY).

APONOGETONACEAE

APONOGETON L.f.

A. distachyos L.f. Cape Pondweed.
Alien. S. Africa. Naturalised in ponds. Very rare, and perhaps extinct.
First record: *Whiting*, 1889.
1. Pond adjoining The Grove, Stanmore Common!, *K. & L.*, 287; not seen recently, and the pond has become very overgrown. **4.** Leg of Mutton Pond, Hampstead Heath,† *Whiting*, *J.B.* (*London*) 32: 16; since destroyed by the cleaning of the water, *Whiting Fl.*

POTAMOGETONACEAE

POTAMOGETON L.

Lit. Fryer, A. & Bennett, A. *The Potamogetons* (*Pond Weeds*) *of the British Isles*. London. 1898–1915.
P. natans L. Broad-leaved Pondweed.
Native. Floating on the surface of ponds, lakes, ditches, slow streams, etc., established in gravel pits, canals, etc. Common, except in heavily built-up areas, where owing to the eradication of its habitats, it is very rare or extinct.
First record: *Petiver*, c. 1710.
7. Between Westminster and Chelsea,† *Pet. MSS.* Ponds about London, *Petiver* (SLO). Isle of Dogs, 1817† (G & R). Ponds beside Hornsey Wood House† and Copenhagen House,† *Ball. MSS.* Lee Canal, *Cherry*, *T. & D.*, 294. Highgate Ponds!, *Fitter*.

P. polygonifolius Pourr. Bog Pondweed.
P. oblongus Viv., *P. anglicus* Hagstr.
Formerly native in ditches, ponds, bog-pools and streams on acid soils, but now extinct.
First evidence: *Dillenius*, c. 1730. Last evidence: *Benbow*, 1884.

1. Harrow Weald Common, *Melv. Fl.*, 83; c. 1869, *T. & D.*, 294. Near Ruislip, 1884; brook between Uxbridge and Harrow, 1884, *Benbow* (BM). **3.** Hounslow Heath (DILL). **4.** West Heath, Hampstead, *T. & D.*, 294.

P. lucens L. Shining Pondweed.
Native in rivers, streams, lakes, ponds, etc., established in gravel pits, canals, etc. Very rare.
First record: *Petiver*, 1695.
1. Swakeleys lake;† Yiewsley!;† Ruislip Reservoir, abundant!, *Benbow* (BM), all conf. *Dandy* and *Taylor*. Colne, Uxbridge,† *Trim. MSS.* Pond by Bayhurst Wood, near Harefield, 1940, *Philipson* (BM), conf. *Dandy* and *Taylor*. **2.** Thames about Teddington Lock,† *T. & D.*, 294. Staines Moor,† *Whale Fl.*, 23. Probably an error. **3.** Hounslow Heath,† *Budd. MSS.* Whitton,† *Francis, Coop. Suppl.*, 13. Duke's river near Isleworth;† Thames, Twickenham,† *T. & D.*, 294. **5.** Canal near Brentford, *Hemsley*, *T. & D.*, 294; 1887,† *Benbow* (BM), conf. *Dandy* and *Taylor*. **6.** Lee Navigation Canal in many places!, *Benbow MSS.*; locally plentiful in various places between Tottenham and Ponders End until c. 1950, but now very scarce. **7.** In many places in the Thames between Fulham and Hampton Court,† *Pet. Midd.* Chelsea,† *M. & G.*, 206. Lee Navigation† and Hackney Canals, abundant, *Cherry*, *T. & D.*, 294; Lee Navigation Canal, Tottenham!,† *De Crespigny*; Hackney Canal, 1881, *Benbow* (BM), both conf. *Dandy* and *Taylor*. Regent's Canal, Stepney, 1967!, *L.N.* 47: 9: 1973.

P. alpinus Balb. Reddish Pondweed.
P. rufescens Schrad.
Formerly native in ditches on acid soils, but now extinct.
First evidence: *De Crespigny*, 1880. Last evidence: *Benbow*, 1886.
1. Ditches by river Colne and Grand Union Canal near Harefield, 1880, *De Crespigny*; ditch in meadows between Harefield and Springwell, 1884; ditches and brooks between Harefield Moor and Springwell, 1886, *Benbow* (BM), all conf. *Dandy* and *Taylor*.

P. perfoliatus L. Perfoliate Pondweed.
Native in ponds, rivers and streams, etc., established in gravel pits, canals, etc. Formerly rather common, but now rare and decreasing.
First record: *T. Johnson*, 1638.
1. Harefield!,† *Blackst. Fasc.*, 80. Uxbridge Common,† *Benbow*; Yiewsley, *J. E. Cooper* (BM), both conf. *Dandy* and *Taylor*. Formerly frequent in the Colne from Harefield to Cowley but now mostly eradicated by the pollution of the waters. **2.** West Drayton!,† *Newbould*; Queen's river, Hampton;† Thames between Hampton and Kingston;† Colne, Staines Moor!, *T. & D.*, 295. Penton Hook!;

Longford river, Stanwell!,† *K. & L.*, 285. **3.** Thames at Twickenham, abundant;† Crane at Hospital Bridge† and Twickenham,† *T. & D.*, 295. **4.** Hampstead,† *Irv. MSS.* North End,† *M. & G.*, 205. Paddington Canal near Willesden Junction, 1881,† *De Crespigny* (BM), conf. *Dandy* and *Taylor.* **5.** Canal at Greenford! and Alperton!,† *Melv. Fl.*, 83. Brent near Ealing,† *T. & D.*, 295. **6.** Tottenham, *Johns. Cat.* Lee Navigation Canal, Edmonton!, *K. & L.*, 285. New river near Myddleton House, Enfield, 1943,† *Stearn* (BM), conf. *Dandy* and *Taylor.* **7.** New river,† *Pet. Midd.* Between Westminster and Chelsea,† *Pet. MSS.* Pond by Copenhagen House,† *M. & G.*, 205. Paddington Canal, 1818, *J. J. Bennett* (BM), conf. *Dandy* and *Taylor*; 1869,† *Warren*; Lee and Hackney Canals, abundant, *Cherry*; ornamental basins at head of Serpentine,† *Warren, T. & D.*, 295. Thames near Putney, 1817† (G & R). Fulham† (HE). Hackney Marsh, 1797,† *Salt*, fide *Newbould, Trim. MSS.* Regent's Canal, Islington!, *L.N.* 39: 58.

P. friesii Rupr. Flat-stalked Pondweed.
P. compressus auct. mult., *P. mucronatus* Schrad. ex Sond.
Native in streams, ponds, lakes, etc., established in canals, etc. Very rare.
First certain evidence: *De Crespigny*, 1871.
1. Canal between Denham and Harefield Moor Locks, 1884; canal near Denham Lock, 1884;† canal at Uxbridge, 1884,† *Benbow* (BM), all det. *Dandy* and *Taylor.* **2.** Bushy Park, 1897, *Pugsley* (BM); 1903, *Loydell* (D); 1937, *Pugsley* (BM), all conf. *Dandy* and *Taylor.* Long Water, Hampton Court, 1949!, det. *Dandy* and *Taylor, L.N.* 29: 12; 1963. **3.** Old cuts, Southall† and North Hyde,† *Benbow* (BM), det. *Dandy* and *Taylor.* **6.** Lee Navigation Canal near Tottenham, 1871,† *De Crespigny* (BM), det. *Dandy* and *Taylor.* **7.** Hampstead Ponds, 1949, *H. C. Harris*!, det. *Dandy* and *Taylor, K. & L.*, 285; 1964.

P. pusillus L. Lesser Pondweed.
P. panormitanus Biv., *P. rutilus* auct.
Lit. *J.B.* (*London*) **78**: 1 (1940).
Native in streams, ponds and lakes, etc., established in gravel pits, canals, etc. Rare, and confused with *P. berchtoldii.*
First certain evidence: *Lemman*, 1833.
1. Ruislip Reservoir, 1943, *Philipson*; Grand Union Canal near Broadwater Farm, Harefield, 1946; Colne at Uxbridge and Harefield, 1946, *G. Taylor* (BM), all det. *Dandy* and *Taylor.* **2.** Hampton Court, 1873, *B. D. Jackson* (CGE), det. *Dandy* and *Taylor.* Bushy Park, plentiful, 1967, *Clement*, det. *Dandy* and *Taylor.* Feltham Green, 1886,† *Benbow*; gravel pit south-east of East Bedfont, 1945!,† *Welch* (BM), both det. *Dandy* and *Taylor.* Wyrardisbury river and

Colne, Staines Moor!, det. *Dandy* and *Taylor*, *K. & L.*, 286. **3.** Grand Union Canal, Isleworth, 1939,† *Dandy* (BM). Gravel pit west of Hounslow Heath!,† det. *Dandy* and *Taylor*, *L.N.* 27: 34. **5.** Near Shepherds Bush, 1833,† *Lemman* (CGE), det. *Dandy* and *Taylor*. **7.** Round Pond!, Kensington Gardens, 1871, *Warren*; 1946, *G. Taylor* (BM); 1947!, det. *Dandy* and *Taylor*, *L.N.* 27: 34; still plentiful, 1964, but apparently varies in quantity from year to year.

P. obtusifolius Mert. & Koch Grassy Pondweed, Blunt-leaved Pondweed.
P. gramineus auct. mult.
Native in ponds and slow streams. Very rare.
First certain record: *D. H. Kent*, 1947.
1. Abundant in a pond on Little Common, Stanmore!, det. *Dandy* and *Taylor*, *L.N.* 27: 34; not seen recently and possibly eradicated when the pond was cleaned out, c. 1963. Colne north of Harefield, 1956!, *Milner*, *L.N.* 36: 15. **3.** Twickenham, *De Cresp. Fl.*, 53. Probably an error. **7.** Paddington, *Wilson* (CGE), fide *Bennett*, *J.B.* (*London*) 46: 119. An error, the locality referred to is in South Lancaster, fide *Billups*, *J.B.* (*London*) 46: 119. Round Pond, Kensington Gardens, *De Cresp. Fl.*, 53. Error, *P. pusillus* probably intended.

P. berchtoldii Fieb. Small Pondweed.
P. pusillus auct. mult., non L.
Lit. *J.B.* (*London*) **78**: 49 (1940).
Native in streams and ponds, established in canals, etc. Very rare, and confused with *P. pusillus*.
First evidence: *Buddle* and *Doody*, 1705.
1. Near Springwell Lock, 1884; Swakeleys lake, 1884;† canal between Denham and Moor Locks, 1884;† ditches about Harefield Moor, 1890; near Uxbridge, 1892, *Benbow*; Colne, Harefield, 1946–47, *G. Taylor* (BM), all det. *Dandy* and *Taylor*. Grand Union Canal near Yiewsley, 1934, *Lousley* (L), det. *Dandy* and *Taylor*. **2.** Bushy Park, 1886, *Benbow*; 1940, *A. H. G. Alston* and *Sandwith*; Wyrardisbury river, Staines, 1947!, *Welch*; Colne, Staines Moor, 1949! (BM), all det. *Dandy* and *Taylor*. **3.** Hounslow Heath, 1705, *Buddle* and *Doody* (SLO), det. *Dandy* and *Taylor*.

P. trichoides Cham. & Schlecht. Hair-like Pondweed.
Lit. *J.B.* (*London*) **76**: 166 (1938).
Native in ponds, ditches and streams, etc., established in canals, etc. Very rare.
First evidence: *Benbow*, 1884.
1. Pond, Eastcote, 1884 and 1893,† *Benbow*; Colne at Harefield, 1939 and 1946; canal, Harefield and Uxbridge, 1946, *G. Taylor* (BM), all det. *Dandy* and *Taylor*. **2.** Staines!, 1885, *Fraser* (K); Shortwood

Common!, 1886, *Benbow* (BM); 1947!, *Welch*, *L.N.* 27: 34; 1963, all det. *Dandy* and *Taylor*. Hampton Court Park!, *Pugsley*; Colnbrook, 1886, *Benbow* (BM), both det. *Dandy* and *Taylor*. **3.** Old cuts, Southall, 1884,† *Benbow* (BM), det. *Dandy* and *Taylor*. Canal backwater north of Southall, 1947!; det. *Dandy* and *Taylor*, *L.N.* 27: 34; 1955;† backwater filled in c. 1960. **5.** Canal near Brentford, 1966, *Boniface*.

P. compressus L. Grass-wrack Pondweed.
P. zosteraefolius Schumach.
Formerly native or denizen in a canal but now extinct. There is a possibility that it may be refound elsewhere.
First evidence: *A. Irvine*, c. 1850. Last evidence: *G. Nicholson*, 1875.
5. Canal near Brentford: *A. Irvine* (SY); 1852, *Syme* (W); 1875, *G. Nicholson* (BM), all det. *Dandy* and *Taylor*.

P. acutifolius Link Sharp-leaved Pondweed.
P. cuspidatus Schrad.
Native in a lake. Very rare.
First record: *W. Hudson*, 1778.
2. About Staines!, abundantly, *Huds. Fl. Angl.* 2: 76 (cf. *J.B.* (*London*) 46: 120); Staines, 1879, *H.* and *J. Groves*, *Rep. Bot. Exch. Club* 1880, 41; 1882, *H.* and *J. Groves*; 1885, *W. R. Linton* (BM); abundant in the lake on Shortwood Common, 1947!, *Welch*, *L.N.* 27: 34, all det. *Dandy* and *Taylor*; still present, but varies in quantity from year to year.

P. crispus L. Curled Pondweed.
Native in ponds, shallow streams, etc., established in gravel pits, canals, etc. Common, except in heavily built-up areas, where owing to the eradication of its habitats it is rare or extinct.
First certain record: *Blackstone*, 1737.
7. New river, *M. & G.*, 207. Marylebone,† *J. Banks*; Marylebone Park,† *Rudge*; Kensington Gardens,! *A. H. G. Alston*; Hyde Park!, *Eastwood*! (BM), all conf. *Dandy* and *Taylor*. Paddington Canal,† *Warren*; pond beyond Primrose Hill;† Hackney Canal, *T. & D.*, 295. Isle of Dogs,† *E. Forster* (BM), conf. *Dandy* and *Taylor*. Highgate Ponds, *Fitter*. Vale of Health Pond, East Heath, Hampstead. Near Brompton, 1860,† *Britten*, *J.B.* (*London*) 46: 120. Regent's Canal, Islington!, *L.N.* 39: 58. Regent's Canal, Limehouse. Clapton Pond, 1962!, *Wurzell*.
A form with uncrisped leaves, superficially resembling *P. obtusifolius*, occurs in **5.** Canal backwater, Hanwell, 1965 →, det. *Dandy* and *Taylor*.

P. pectinatus L. Fennel-leaved Pondweed.
Native in rivers, streams, etc., established in canals, etc. Common, especially in heavily polluted waters.

First evidence: *Dillenius*, c. 1730.

7. Common in the canals of Central London and the East End. Round Pond, Kensington Gardens!, *Warren*, *T. & D.*, 297. Hyde Park, 1949, *Eastwood*; river Lee!; lake in Regents Park, *E. Forster*; 1840,† *Robinson* (BM), all det. *Dandy* and *Taylor*. Hackney Marsh! (DILL), det. *Dandy* and *Taylor*: 1973. Isle of Dogs!, 1900, *Riddelsdell* (BM); 1967.

GROENLANDIA Gay

G. densa (L.) Fourr. Opposite-leaved Pondweed.

Potamogeton densus L., *P. serratus* L.

Native in clear streams, ditches, ponds and lakes, etc., established in gravel pits and canals, etc. Formerly locally common, but now very rare and decreasing.

First evidence: *Goodger* and *Rozea*, 1815.

1. Colne near Iver Bridge!, *Newbould*, *T. & D.*, 297. Colne and canal in many places, *Benbow MSS.* Uxbridge Common;† ditch near Ruislip Reservoir, 1884,† *Benbow* (BM), both conf. *Dandy* and *Taylor*. Mimmshall brook, Bridgefoot, Potters Bar, 1970, *Widgery*. **2.** Colne, Staines Moor!;† common by Walton Bridge;† Bushy Park!, *T. & D.*, 297. West Drayton, 1884,† *Benbow* (BM), conf. *Dandy* and *Taylor*. Wyrardisbury river, Staines!, *Welch*; Shortwood Common!; Longford river, Stanwell!,† *K. & L.*, 286. **3.** Near Hanworth,† *Twining* (W), conf. *Dandy* and *Taylor*. Rivulets on Hounslow Heath, near Bedfont† (K), det. *Dandy* and *Taylor*. **4.** Hampstead,† *Irv. MSS.* Stanmore,† *T. & D.*, 297. Near Edgware,† *Benbow MSS.* **5.** Canal at Greenford,† *Melv. Fl.*, 52. **6.** Whetstone;† Lee Navigation Canal and adjacent ditches,† *Benbow MSS.* **7.** Canal towards Hampstead Road, 1815† (G & R). Ditches, Hackney Wick;† Isle of Dogs,† *T. & D.*, 297. Fulham Palace moat, plentiful,† *De Cresp. Fl.*, 54. River Lee below Wick Lane,† *Braithwaite* (BM), conf. *Dandy* and *Taylor*. Highgate Ponds, 1945, *Fitter*.

ZANNICHELLIACEAE

ZANNICHELLIA L.

Z. palustris L. Horned Pondweed.

Native in rivers, streams, ponds, lakes, ditches, etc., established in canals, gravel pits, etc. Local, and decreasing.

First record: *Petiver*, 1695.

1. Pinner,† *Hind*, *Melv. Fl.*, 83. South Mimms, *T. & D.*, 207. Between Pinner and Pinner Green;† Eastcote;† Swakeleys,† *Benbow MSS.* Harrow Weald Common, 1881,† *Blow* (BM). Stanmore Common!,

K. & L., 287; 1966. Potters Bar, 1970. **2.** Staines!, *Coop. Fl.*, 115. Near Staines!; Shortwood Common!. *Benbow MSS*. Bushy Park!, *Benn. MSS*. East Bedfont!, *Rose, K. & L.*, 287. Pond, Hampton Court Park, 1966, *Clement*: 1973. **3.** Isleworth, 1824,† *J. E. Smith* (BM). Pond below 'The Spaniards', Hampstead,† *T. & D.*, 297. **5.** Acton;† near Horsendon Hill;† near Southall,† *Benbow MSS*. **6.** Near Highgate, *Davies, T. & D.*, 297; 1875,† *French* (BM). Hadley!, *Warren, T. & D.*, 297. Whetstone!, *L.N.* 28: 34. Stroud Green,† *Doody MSS*. New river, Enfield, abundant, 1965, *Bangerter* and *Kennedy*!. Gravel pit north of Enfield Lock, 1966, *Kennedy*!. Trent Park, 1966, *Kennedy*. **7.** Near Islington,† *Pet. Midd.* By St Pancras Church,† *Dillenius, Ray Syn.* 3: 136. Tothill Fields,† *Sowerby, Smith MSS*. Kilburn, 1968,† *Warren*; Isle of Dogs, abundant, *T. & D.*, 297; 1887,† *Benbow MSS*. Kentish Town, 1848,† *T. Moore* (BM). Kensington Gardens!, *Warren, Trim. MSS.*; abundant in the Round Pond, 1949!; also in lily ponds, *D. E. Allen, L.N.* 39: 58. Duckett's Canal, Hackney, 1872, *Trim. MSS*. Pond near Finchley Road railway station,† *De Cresp. Fl.*, 78.

LILIACEAE

CONVALLARIA L.

C. majalis L. Lily-of-the-Valley.

Native in dry woods, on shady banks, in plantations, etc. Very rare, and often non-flowering.

First record: *Gerarde*, 1597.

1. Stanmore Common, a large patch!, *Warmington, L.N.* 37: 185. Harrow Weald Common, six non-flowering plants, 1959, *Knipe, L.N.* 39: 40. Shady hedgebank east of Potters Bar, 1966, *Kennedy*. **4.** Hampstead Heath, *Ger. Hb.*, 302, and many later authors; very sparingly under bushes near the bog, 1850–55,† *Pamplin*; Bishop's Wood, a single root, 1864,† formerly abundant, *T. & D.*, 276. Turner's Wood,† *Bliss, Park Hampst.*, 29. **5.** Norwood, Middlesex, abundantly, *Martyn, Baxt.*, 1. Norwood, Surrey, probably intended, *T. & D.*, 277. Osterley Park!, *Welch*. Perhaps originally planted. **6.** Winchmore Hill Wood, *J. Mitchell, Coop. Fl.*, 119; a large patch, 1887, *Benbow, J.B. (London)* 25: 365; 1906,† *Sabine, K. & L.*, 273. The Alder's Copse, Whetstone, 1863;† we could not find it in 1867, *T. & D.*, 277. Enfield, *H. & F.*, 148. **7.** Highgate,† *Coop. Fl.*, 104. Lord Mansfield's Wood near 'The Spaniards' (= Ken Wood)!, *Curt. F.L.*; still plentiful but rarely flowers, 1973. Holland House grounds, Kensington, planted, but now naturalised.

POLYGONATUM Mill.

P. odoratum (Mill.) Druce Angular Solomon's Seal.
P. officinale All.
Introduced. Naturalised in woods, on shady hedgebanks, etc. Very rare.
First record: *Hunter*, 1814.
1. The Grove, Stanmore Common!, *K. & L.*, 273; still present, but rarely flowers, 1973. Road bank, Bushey Heath, 1950–62; site partly built over but two plants survive in road verge, 1967. **7.** Ken Wood,† *Hunter, Park Hampst.*, 29.

P. multiflorum (L.) All. Solomon's Seal.
Denizen or native in woods, thickets, on shady hedgebanks, etc. Very rare.
First record: *W. G. Smith, M. C. Cooke* and *Grugeon*, 1864.
1. Furzefield Wood, Potters Bar, garden escape, 1967. **2.** Near West Drayton, garden escape, 1884,† *Benbow* (BM). **4.** Bishop's Wood, a single plant, 1864,† *W. G. Smith, M. C. Cooke* and *Grugeon*, *T. & D.*, 277. Mill Hill, 1871,† *George*, *Trim. MSS.* Hampstead Heath, a single small plant,† *Whiting Fl.* Newland's Wood, Harrow, probably planted, *Horton Rep.* **6.** Enfield!, *H. & F.*, 148; a few plants in scrub, Botany Bay, Enfield Chase, 1965, *Kennedy*!, *L.N.* 45: 20. **7.** Millfield Lane, Highgate, 1849,† *W. Davies* (QMC).
Some of the records given above may be referable to the hybrid *P. multiflorum* × *odoratum* (*P.* × *hybridum* Brügger) which occurs as an escape from gardens.

P. multiflorum × odoratum = P. × hybridum Brügger
2. Spout Wood, Stanwell, *Clement*, *L.N.* 43: 22. **7.** Holloway, 1965, *D. E. G. Irvine*.

MAIANTHEMUM Weber

M. bifolium (L.) Schmidt May Lily.
M. convallaria Weber, *Unifolium bifolium* (L.) Greene, *Smilacina bifolia* (L.) J. A. & J. H. Schult.
Lit. *J.B.* (*London*) **52**: 202 (1913).
Formerly native or denizen in woods, but extinct by c. 1924, reintroduced in 1933. Very rare, and perhaps extinct.
First record: *Hunter*, 1813.
7. Ken Wood, *Hunter, Park Hampst.*, 29; 1829, *A. Irvine* (BM); one patch destroyed just prior to 1829, *A. Irvine, Phyt. N.S.* 4: 233; several patches under the shade of fir trees, 1835, *E. Edwards, Phyt.* 1: 579; 1839, *Kingsley* (ME); 1840 and 1845 (BO); 1861, *A. Irvine* (CYN); a patch of about twenty square yards on an eminence under the

shade of a very large beech in the enclosure of Ken Wood grounds near its S.-E. angle, *T. & D.*, 277; 1884, *Benbow MSS.*; plentiful, 1912, *A. B. Jackson* and *Watt*; a good patch in flower, 1917; scarce, and did not flower, 1922; a single plant, which did not flower, 1924,† *Redgrove*. Reintroduced in 1933, *Gilmour*, *J.B.* (*London*) 71: 168; naturalised and increasing, 1945, *Fitter*, *K. & L.*, 273; 1953; not seen since but may still be present.

ASPARAGUS L.

A. officinalis L. subsp. **officinalis** Asparagus.

Alien. Europe. Bird-sown from gardens and naturalised in meadows, grassy places, on waste ground, by roadsides, etc., established on railway banks, old walls, etc. Common.

First record: *R. A. Pryor*, 1874.

7. Walls in Fulham Palace grounds, frequent, *Welch*!. Bombed sites, Cripplegate† and Gresham Street areas,† *Jones Fl.* Trinity Square, E.C.3, 1966!, *L.N.* 46: 34. Brompton Cemetery, 1964. Holland House grounds, Kensington, 1964. Railway bank, White City, 1965. Kilburn, 1964. Highgate Cemetery, 1965. Upper Clapton, 1964. By canal between Lisson Grove and Primrose Hill, 1969, *L. M. P. Small.*

RUSCUS L.

R. aculeatus L. Butcher's Broom.

Native in dry woods, thickets, on hedgebanks, in the north, particularly the north-east, of the vice-county, elsewhere in similar habitats as an introduction. Very rare.

First record: *Gerarde*, 1597.

1. Near Breakspeares,† *Blackst. Fasc.*, 88. Potters Bar, 1908, *P. H. Cooke*, *K. & L.*, 272; 1965, *J. G.* and *C. M. Dony*. East of Potters Bar, 1966!, *Kennedy*. **2.** Bushy Park, 1948, *Boniface*, *K. & L.*, 272. **3.** Kneller Hall Park, 1852† (WD). **4.** Hampstead Heath, *Ger. Hb.*, 759, and many later authors; formerly abundant there, *Loud. Arb.*, 2519; near Leg of Mutton Pond, but almost exterminated when the plantation of firs was made near West Heath Road, *Whiting Notes*; one plant, 1902, *Whiting*, *K. & L.*, 272; a few plants lingered until about 1906,† *Whiting Fl.* Bishop's Wood, in small quantity, 1861,† not there in 1867, *T. & D.*, 578. Mill Hill, almost certainly introduced, *Sennitt.* **5.** Gravel pit, Hanwell, 1903,† *Green* (D). Small copse between Brentford and Ealing, 1950, *Farenden*, *L.N.* 30: 7. Syon Park, *Tremayne*, *K. & L.*, 272. Hanger Hill, Ealing, 1965, *McLean.* **6.** Near Finchley, *Varenne*, *T. & D.*, 278. Frith Manor Farm, Finchley, abundant, *Hampst. Sci. Soc.* 1920–22 *Rep.*: 48. Hadley Wood, *Mawer*,

Sci. Goss. 1883; 262; a fine clump by the road to Potters Bar, 1907, *P. H. Cooke, K. & L.*, 272. South Lodge, Enfield, 1929, *Hansen* (LNHS). Whitewebbs Park!, *Johns.*, *L.N.* 27: 34. Copse, Vicarage Farm, Forty Hill, and elsewhere about Enfield and Enfield Chase, 1965, *Kennedy*!. Trent Park, two clumps, 1959, *A. Vaughan.* **7.** Lane near West End,† *Burnett, T. & D.*, 278.

LILIUM L.

L. martagon L. Martagon Lily, Turk's Cap.
Alien. Europe, etc. Garden escape. Naturalised in woods, old pastures, etc. Rare.
First record:–1863.
4. Mill Hill, two or three plants, 1871,† *George, Trim. MSS.* **6.** Old pasture, Enfield Chase, *Phyt. N.S.* 6: 573. Whitewebbs Park!, 1962, *L.N.* 42: 12; 1964.

FRITILLARIA L.

F. meleagris L. Snake's Head, Fritillary.
Native or denizen in damp meadows, etc. Very rare.
First record: *Ashby*, pre-1696.
1. '... grows in a meadow by a wood side near Harefield, and has done so about forty years as a neighbouring gentleman informs me',† *Blackstone*, in litt. to *Richardson*, 11 December 1736. In Maud-fields near Ruislip Common, observed above forty years by Mr Ashby of Breakspears,† *Blackst. Fasc.*, 29. 'In a small field going from Uxbridge to Ruislip Common, a little beyond a farm called Blackenburgh, in Harefield parish',† *Lightf. MSS.* The latter three records probably refer to a single locality. In a meadow near Swakeleys, in Ickenham parish; between Swakeleys and Long Lane, abundantly, 1781, *Lightf. MSS.*; Swakeleys Park, 1884, very scarce and nearly extinct; after being abundant there for over fifty years they are almost all rooted up by collectors, *Benbow* (BM); reappeared plentifully in 1890, then disappeared again; a single plant, 1906,† *Benbow MSS.* Fields at Pinner, in one very abundant, *Melv. Fl.*, 78; this field is a little north of the church and adjacent to Moss Lane, the plant also occurs in three other meadows near and also in an orchard, *Hind, T. & D.*, 279; East End Farm, Pinner, 1903, *Green* (SLBI); now (1934) extinct at Pinner, *Druett*, 193; abundant in an old orchard, Moss Lane, 1950!, *T. G. Collett, L.N.* 30: 7; undoubtedly Hind's locality; trees grubbed up and ground disturbed prior to being built over, 1965.† Grounds of a derelict mansion, Eastcote, 1961, *Perkins, L.N.* 41: 14. **4.** Finchley, 1842,† *Hankey* (BM). Near Finchley, *Trim. MSS.* Near Mill Hill, about twenty plants in a damp hollow,

1938, *Pigott*; 1946, *Lousley*!, *L.N.* 26: 76. Near Harrow Park lake, one plant, c. 1949, probably planted, *Harley Fl.*, 31. **6.** Near Enfield, *Huds. Fl. Angl.* 2: 144; near Bury, Enfield, *With. Bot. Arr.* 2: 1, 346. Colney Hatch, a single plant,† *Church, T. & D.*, 279. Meadows near Barnet, 1953, *Lansbury*. **7.** Lee Marshes, 1884,† *P. S. King* (LNHS).

TULIPA L.

T. sylvestris L. Wild Tulip.
Alien. Europe, etc. Naturalised in pastures, plantations, woods, etc., where it rarely flowers. Very rare.
First record: *Woods*, 1805.
1. Harefield!, 1839, *Kingsley* (ME); four plants in flower, 1881, *F. J. Hanbury, J.B. (London)* 19: 175; in flower 1886 and 1887; abundant, but non-flowering, 1895, *Benbow MSS.*; 1928, *Spooner, K. & L.*, 276; 1936, *Lousley* (L); 1939–54!, *K. & L.*, 276; still plentiful in the grove behind Harefield Church, but has apparently not been recorded in flower for many years. Copse near Breakspeares, 1956; verge of lake, Bayhurst Wood, Harefield, abundant!, *Pickess, L.N.* 45: 20; 1966 →. Ruislip,† *James* (SY). **3.** Roxeth, 1871,† *Parr*, fide *Hind, J.B. (London)* 10: 266. **6.** Top of Muswell Hill, *Woods, B.G.*, 403; hundreds of plants there, 1855, *Pamplin* and *A. Irvine, Phyt. N.S.* 1: 39; 1860–61; *Pamplin, T. & D.*, 278; 1900,† *Pugsley*.

ORNITHOGALUM L.

O. umbellatum L. Star-of-Bethlehem.
Denizen in meadows, copses, etc. Very rare.
First record: *Collins*, c. 1730.
1. Harefield!, *De Cresp. Fl.*, 48; frequent in meadows and copses about Knightscote Farm, 1956!, *Pickess*. Harefield Moor, 1959. Established in many meadows near Uxbridge, *Benbow, J.B. (London)* 25: 18. Bank of the Colne near Denham, 1968, *A. Vaughan*. **2.** Teddington,† *Collins, Dill. MSS.* Thames bank near Hampton Court!, *Montg. MSS.*; 1967. Hampton Court Park, 1957.

O. nutans L. Drooping Star-of-Bethlehem.
Alien. Europe, etc. Garden escape. Naturalised in grassy places. Very rare.
First record: *Chambers*, pre-1859.
2. Hampton Court, 1923, *Spreadbury*; 1940, *Gibson, K. & L.*, 275. **3.** Twickenham,† *Chambers* (BM). **5.** Perivale Churchyard, 1965!, *T. G. Collett*; abundant, 1971.

SCILLA L.

S. autumnalis L. Autumnal Squill.
Native in dry grassy places near the Thames. Very rare.
First record: *Parkinson*, 1629.
2. Sparingly in one spot on the sloping bank of the Thames towing-path between Hampton Court and Ditton Ferry, 1868, *Newbould* (BM); 1887,† *Benbow MSS*. Hampton Court Park!, *Montg. MSS.*; 1973. On Kingston Bridge,† *Ray Syn.* 3: 372. Near Teddington, 1846,† *Stevens* (K). **3.** Near Hounslow Heath,† *Merr. Pin.*, 64. Near Twickenham, 1844,† *Twining* (BM). **7.** Thames side, Chelsea,† *Park. Par.*, 132.
Plants with white flowers are recorded from **2.** Hampton Court Park!, *K. & L.*, 275.

ENDYMION Dumort.

E. non-scriptus (L.) Garcke Bluebell, Wild Hyacinth.
Scilla non-scripta (L.) Hoffmanns. & Link, *S. nutans* Sm., *Endymion nutans* Dumort.
Lit. *J. Ecol.* **42**: 269 (1954).
Native in woods, copses, rarely in pastures, etc., established on railway banks, etc. Locally abundant, especially in the northern parts of the vice-county.
First record: *W. Turner*, 1548.
1. Frequent. **2.** Ashford golf course. **3.** Harrow Grove; Roxeth!,† *Melv. Fl.*, 78. Between Pinner and Harrow!, abundant, *T. & D.*, 281. Northolt. Hayes. Hayes Park, *Royle*!. Cranford Park. Hounslow Heath. **4.** Hampstead Heath!, *Johns. Eric.* Harrow!, *Hind Fl.* Bishop's! and Turner's Woods!, *T. & D.*, 281. Mill Hill. Scratch Wood. Whitchurch Common. Edgware. Finchley. Barn Hill, Wembley. **5.** Syon Park!, *Turn. Names.* Hanwell. Long Wood, Wyke Green. Osterley Park. Hanger Hill, Ealing. Perivale Wood. Horsendon Hill. Chiswick. **6.** Hadley!, *Warren*; Winchmore Hill Wood;† Whetstone!, *T. & D.*, 281. Enfield. Enfield Chase. Trent Park. Highgate Woods. **7.** Ken Wood!; Hornsey Wood,† *T. & D.*, 281. Buckingham Palace grounds. Holland House grounds, Kensington. Small wood by railway near Stepney.
Plants with white flowers are not uncommon, and plants with rose-coloured flowers are recorded from **5.** Osterley Park, *Masters*, *T. & D.*, 281.

E. hispanicus (Mill.) Chouard
Scilla hispanica Mill., *S. campanulata* Ait.
Alien. Europe. Garden escape. Naturalised in grassy places, woods,

thickets, on waste ground, etc., established on railway banks, etc. Common.

First record: *C. E. Britton*, 1920.

7. Buckingham Palace grounds!, *Fl. B.P.G.* Regents Park. East Heath, Hampstead. Finsbury Park. Tottenham. Bromley-by-Bow.

E. hispanicus × non-scriptus

3. Hayes Park, *Royle*!. **4.** West Heath, Hampstead. **5.** Syon Park. Osterley Park. **7.** Buckingham Palace grounds!, *Fl. B.P.G.*

Probably more frequent than the few records given above suggest.

MUSCARI Mill.

M. atlanticum Boiss. & Reut. Grape-Hyacinth.

M. racemosum auct.

Introduced. Garden escape. Naturalised in dry grassy places, and established on the river-wall of the Thames and on railway banks. Very rare.

First record: *D. Cooper*, 1837.

2. Hampton Court!, *Coop. Suppl.*, 12; growing in some quantity in the stone wall of the Thames bank opposite Thames Ditton!, 1887, *Benbow* (BM); about a dozen plants, 1946!, *Welch*, *L.N.* 26: 64; still present, 1968. Hampton Court Park, a small colony in turf, 1962. Hampton Wick, *E. K. Robinson*, *Country-side* 1921, 73. **5.** Railway bank, West Ealing, a large well-established colony, 1965!, *L.N.* 45: 20; 1969. **7.** Bombed site, Cripplegate area, E.C.,† *Jones Fl.*

ALLIUM L.

A. vineale L. Crow Garlic.

Lit. *J. Ecol.* **34**: 209 (1947).

Native on river banks, waste ground, in fields, etc., established on railway banks, etc. Locally plentiful near the Thames, rather rare elsewhere.

First record: *Gerarde*, 1597.

1. Harefield, abundant above an old chalkpit!, *K. & L.*, 274. Uxbridge Moor, *L. M. P. Small*. **2.** Frequent by the Thames. Staines Moor, *Briggs*. Near Stanwell Moor, *Boniface*. Near Hampton, *Tyrrell*! East Bedfont!, *Rose*, *K. & L.*, 274. **3.** Thames bank between Twickenham and Richmond Bridge, *Benbow*, *J.B.* (*London*) 25: 365. Gravel pit near Cranford, *Battley*, *K. & L.*, 274. Roxeth,† *Hind*, *Melv. Fl.*, 78. Northolt. **4.** Near Hampstead, 1842, *Stocks* (BM). Whitchurch Common, scarce, 1954!, *K. & L.*, 274. Willesden Old Churchyard, 1966. **5.** Near Ealing, 1867, *Newbould*, *T. & D.*, 280. Greenford, *Battley*, *K. & L.*, 274. Railway bank near Barnes Bridge!, *Murray*.

Side of Great Chertsey Road, Chiswick, 1965. Railway bank near Brentford!, 1965, *Boniface*; plentiful, 1966. Hanwell, 1965. Near Drayton Green, plentiful, 1968. Osterley, 1966. Heston, 1966. **6.** Waste ground, Arnos Grove, 1962, *Wurzell*. **7.** Near Islington,† *Ger. Hb.*, 142. Near Marylebone,† *Budd. MSS*. Near Pancras,† *Blackst. MSS*. East Heath, Hampstead!, *L. M. P. Small*.

The common Middlesex plant is the var. **compactum** (Thuill.) Bor., but the var. **bulbiferum** Syme is recorded from **2.** Thames bank between Hampton Court and Kingston Bridge!, *C. E. Britton, J.B. (London)* 48: 331; 1967.

A. oleraceum L. Field Garlic.

Formerly native on dry hedgebanks but now extinct.

First, and only evidence: *H. C. Watson*, 1856.

2. Near Sunbury, 1856, *H. C. Watson* (W).

A. triquetrum L. Triquetrous Garlic.

Alien. Mediterranean region. Formerly established on waste ground, but now extinct.

First, and only evidence: *Banker*, 1852.

7. Isle of Dogs, 1852, *Banker* (BM).

A. paradoxum (Bieb.) G. Don

Alien. Caucasus. Naturalised in shrubberies, thickets, etc. Very rare.

First record: *Perkins*, 1961.

1. Grounds of a derelict mansion, Eastcote, 1961, *Perkins, L.N.* 41: 14. **4.** Whitchurch Common, a small colony, 1966, *Tyrrell*!, *L.N.* 46: 13: 1970.

A. ursinum L. Ramsons.

Native in damp woods, ditches, damp grassy places, thickets, on hedgebanks, by streamsides, etc. Locally abundant in the north of the vice-county, rare elsewhere.

Lit. *J. Ecol.* **45**: 1003 (1957).

First record: *Gerarde*, 1597.

1. Harefield,† *Blackst. Fasc.*, 3. By Ruislip Reservoir,† *Benbow MSS*. Pinner,† *Melv. Fl.*, 18. Northwood, 1966, *J. A. Moore*. **3.** Northolt Churchyard,† *Payton, K. & L.*, 274. **4.** Hendon!, *Mart. Pl. Cant.*, 66; Brent Park, Hendon, in profusion. Cricklewood, 1867,† *Fox*; Golders Green,† *A. B. Cole*; between Finchley and Hendon!, *Newbould*; Bentley Priory, *T. & D.*, 280. Sudbury;† Harrow Weald; Harrow!, *Melv. Fl.*, 78. Near Kingsbury, *De Cresp. Fl.*, 3. Near Hampstead!, *Dicks. Hort. Sicc.* North-west Heath, Hampstead!, *Lake*. Mill Hill!, *Benbow MSS*. Between Totteridge and Finchley, abundant, *Lousley*!. Banks of Brent, Brent Reservoir, in small quantity, 1967. Edgwarebury, abundant. Edgware!, *Horder*. **5.** Ealing,† *Boucher, K. & L.*, 274. **6.** Near Finchley Common!, 1793, *Rayer*,

Smith MSS.; 1945. Highgate Wood; Alder's Copse, Whetstone;† hedges near Whetstone, very abundant!, *T. & D.*, 280. By the Brent from Whetstone to Finchley; Holders Hill and Hendon, in great quantity!, *Benbow MSS.* Hadley Wood!, *P. H. Cooke* (LNHS). Woodside Park!, *Benbow MSS.* **7.** Near Hampstead!, *Ger. Hb.*, 142. Kentish Town,† *Newt. MSS.* Marylebone Fields, 1817† (G & R). Highgate (HE). Between Swiss Cottage and Hampstead, 1846,† *J. Morris, T. & D.*, 281. Islington, 1837,† *E. Ballard* (BM). Hornsey Wood,† *Ball. MSS.*

COLCHICUM L.

C. autumnale L. Meadow Saffron, Naked Ladies, Autumn Crocus.
Lit. *J. Ecol.* **42**: 249 (1954).
Introduced. Naturalised in damp grassy places. Very rare.
First record: *J. Sherard*, 1724.
6. Southgate, *J. Sherard, Ray Syn.* 3: 373. Forty Hill, Enfield Chase!, *Cullen, Coop. Suppl.*, 116; two small patches, having the appearance of being planted, by the lake in Forty Hall grounds, 1967!, *Kennedy*; Enfield!, *H. & F.*, 148.

TRILLIACEAE

PARIS L.

P. quadrifolia L. Herb Paris.
Native in damp woods and copses. Very rare.
First record: *Merrett*, 1666.
1. In the Old Park!,† *Blackst. Fasc.*, 41; grew sparingly in Old Park Wood until c. 1950; Hanging Wood,† and elsewhere about Harefield!, *Blackst. Fasc.*, 41; copse above Jack's Lock, Harefield, 1890,† *Benbow* (BM); Scarlet Spring, near Harefield, abundant, 1955!; unable to find a single plant though the copse appears to be undisturbed, 1965, *Pickess*; likely to reappear in due course. Abundant in a copse near Pinner Wood,† *Hind, Melv. Fl.*, 77. In a wood about half a mile beyond Pinner railway station,† *Gunning, Sci. Goss.* 1883: 139; extinct at Pinner, *Druett*, 193. Park Wood, Ruislip!, two or three plants, 1947, *Philipson, K. & L.*, 276; 1950. **4.** In a wood near Hampstead, *Merr. Pin.*, 61; c. 1677, *Morley MSS.*; c. 1778,† *C. Miller, Huds. Fl. Angl.* 2: 172. Near Mill Hill, 1766,† *Coll. MSS.* Near Hendon,† *Alch. MSS.* **6.** Friern Barnet, 1863,† *Trimen* (BM). The Alder's Copse, near Whetstone, abundant, *T. & D.*, 266. **7.** Ken Wood,† *Hunter, Park Hampst.*, 30.

JUNCACEAE

JUNCUS L.

J. squarrosus L. Heath Rush, Goose Corn.
Lit. *J. Ecol.* **54**: 535 (1966).
Native on wet heaths, in bogs, etc. Very rare, and decreasing.
First record: *Ray*, 1670.
1. Harefield Common!,† abundantly, *Blackst. Fasc.*, 47. Harrow Weald Common!, abundant, *Melv. Fl.*, 79. Stanmore Common, *T. & D.*, 283; c. 1900,† *C. S. Nicholson* and *L. B. Hall*, *K. & L.*, 278. **2.** Gravel pits, East Bedfont,† *Westrup*, *K. & L.*, 278. Perhaps an error. **3.** Hounslow Heath! (DILL); 1973. Gravel pits near Hounslow, sparingly,† *T. & D.*, 283. **4.** Hampstead Heath, *Ray Cat.*, 179; c. 1730, *Dill. MSS.*; 1846 (HE); 1921–27,† *Richards*, *K. & L.*, 78. **7.** East Heath, Hampstead,† *Benbow MSS.*

J. tenuis Willd. Slender Rush.
J. macer Gray
Lit. *J. Ecol.* **31**: 51 (1943).
Alien. N. America. Naturalised on heaths, in grassy places, woodland rides, established on canal banks, etc. Rare, but increasing.
First evidence: *Lousley*, 1933.
1. Canal bank, Uxbridge, 1933, *Lousley* (L). Canal bank south of Uxbridge, 1960. Canal bank, Cowley, 1955!, *K. & L.*, 278. Canal bank by Springwell Lock, abundant, *Brenan*, *Sandwith* and *C. West*!, *L.N.* 28: 34; 1968 →. Copse Wood, Northwood!, *Sandwith*, *L.N.* 27: 34; 1966. Potters Bar, 1969, *L. M. P. Small*. **2.** Thames bank, Hampton Court!, *Boniface*, *K. & L.*, 278; 1970. **3.** Hounslow Heath, *Welch*!, *L.N.* 26: 64; 1973. **7.** Ken Wood!, *Bangerter* and *Morton*, *L.N.* 30: 7. By St Peter's Church, Hobart Place, S.W.1, 1954!,† *McClintock*, *K. & L.*, 278. Kensington Gardens, in several places!, *D. E. Allen*, *L.N.* 39: 58; 1972. Buckingham Palace grounds, *Fl. B.P.G.* Regents Park, 1965, *D. E. Allen* and *Potter*, *L.N.* 46: 34. Holloway, 1965, *D. E. G. Irvine*.

J. compressus Jacq. Round-fruited Rush.
Native in marshes, damp grassy places, by river- and streamsides, on pond and lake verges, etc., established by canals, gravel pits, etc. Locally plentiful on alluvial soils near the Thames, probably extinct elsewhere.
First record: *Varenne*, 1827.
2. Locally frequent by the Thames from Staines to Penton Hook Lock!; Laleham!; between Laleham and Shepperton!; near Staines!, *Benbow*, *J.B.* (*London*) 25: 18. Near Walton Bridge!, *Groves*, *Trim.*

MSS. Hampton Court!, *Green* (SLBI). Between Hampton Court and Kingston Bridge. Shortwood Common!, *Welch*; Stanwell Moor!, *L.N.* 27: 34. Poyle, *L. M. P. Small.* **4.** Stanmore, 1827,† *Varenne, T. & D.*, 284. Golders Green,† *Irv. MSS.* **5.** Shepherds Bush, 1870,† *Warren, J.B. (London)* 13: 276. **7.** Isle of Dogs, very sparingly, 1866,† *Newbould*; Green Lanes, Newington,† *T. & D.*, 284.

J. gerardii Lois. Mud Rush.

Introduced. Formerly established in marshy places, on damp waste ground, etc., but now extinct.

First evidence: *De Crespigny*, 1878. Last record: *J. E. Cooper*, 1914.

4. Marshy place by a pond between Fortune Green and Hampstead Heath, two or three plants, 1878, *De Crespigny* (BM). **7.** Hackney Marshes, 1913–14, *J. E. Cooper, K. & L.*, 278.

J. bufonius L. Toad Rush.

Native on heaths, damp waste ground, in wet places, marshes, ditches, woodland rides, on pond verges, etc., established on damp paths, tracks, in cart-ruts, gravel pits, etc. Common.

First record: *T. Johnson*, 1632.

7. Between Tottenham Court and Hampstead,† *Johns. Ger.*, 4. East Heath, Hampstead!; Eel Brook Meadow;† Hackney Wick!; Isle of Dogs, *T. & D.*, 284. Hyde Park!, casual in a flower-bed, *Warren Fl.*; c. 1955!; Kensington Gardens!, *D. E. Allen, L.N.* 39: 58. Ken Wood.

Var. **fascicularis** Koch

1. Hillingdon Heath, 1884, *Benbow* (BM). Stanmore Common. **2.** Laleham. Shortwood Common. East Bedfont. West Drayton. **3.** Hounslow Heath. Near Southall. **4.** Mill Hill. **6.** Hadley Common. Enfield Chase, *Kennedy*!.

J. inflexus L. Hard Rush.

J. glaucus Sibth.

Lit. *J. Ecol.* **29**: 369 (1941).

Native in marshes, by river- and streamsides, damp grassy places, etc., established in gravel pits, by canals, etc. Very common, particularly on heavy soils.

First record: *T. Johnson*, 1638.

7. Common in the district.

J. effusus L. Soft Rush.

J. communis auct.

Lit. *J. Ecol.* **29**: 375 (1941).

Native in marshes, wet fields, damp woods, by river- and streamsides, etc., established in gravel pits, by canals, etc. Very common.

First record: *T. Johnson*, 1638.

7. Common in the district.

J. effusus × inflexus = J. × diffusus Hoppe
Lit. *Kew Bull.* **1958**: 392 (1959): *Watsonia* **9**: 1 (1972).
1. Pinner, 1871, *Hind*; Uxbridge Common, *Benbow* (BM). **2.** Stanwell Moor!, *Dyer* (BM). East Bedfont!, *K. & L.*, 277. **5.** Marshy ground by the Brent near Greenford, abundant,† *Benbow MSS.* **6.** Finchley Common!, *Benbow*, *J.B.* (*London*) 25: 18. Enfield, 1966, *Kennedy*.

J. effusus × pallidus
2. East Bedfont, 1945, *J. G. Dony*!; 1964, *Wurzell*.

J. subuliflorus Drej. Compact Rush.
J. conglomeratus auct., *J. communis* E. Mey.
Lit. *J. Ecol.* **29**: 381 (1941).
Native in marshes, wet fields, damp woods, on heaths, etc. Rare, and chiefly found on acid soils.
First certain evidence: *D. Cooper*, 1837.
1. Harrow Weald Common!, *Melv. Fl.*, 79. Harefield!, *Newbould*; Stanmore Common!, *T. & D.*, 282. Springwell. Potters Bar, *J. G.* and *C. M. Dony*!. **2.** Staines, *Whale Fl.*, 16. Perhaps an error. East Bedfont. **3.** Headstone,† *Benbow MSS.* **4.** Stanmore, 1837, *D. Cooper* (BM). Hampstead Heath!, *Coop. Fl.*, 103; West Heath, 1968. Harrow, *Hind Fl.* Scratch Wood, *Benbow MSS.* Mill Hill, *Sennitt*. **5.** Horsendon Hill,† *Benbow MSS.* Osterley Park. Bedford Park,† *Cockerell Fl.* Perhaps an error. Wyke Green.† **6.** Edmonton, *T. & D.*, 282. Enfield!, *Kennedy*. Trent Park, *Bangerter* and *Kennedy*!. **7.** East Heath, Hampstead. Kensington Gardens, one plant in turf, 1960, *D. E. Allen*, *L.N.* 41: 21.
Forms of *J. effusus* with congested panicles are often mistaken for *J. subuliflorus*.

J. pallidus R.Br.
Alien. Australia. Established in gravel pits. Very rare, and almost extinct.
First record: *H. R. Davies*, 1945.
2. Gravel pit, East Bedfont, 1945!, *H. R. Davies*, *L.N.* 25: 13; 1946, *Welch* and *Rose*!; *Lousley*!; 1947, *Welch* and *J. G. Dony*!, *K. & L.*, 279; still present in small quantity, 1960, *G. W.* and *T. G. Collett*!; 1964; original gravel pit partially filled with rubble during 1965, but other pits excavated; a number of plants still survive on 'islets' in the pits, 1965, *Mason*; 1966, *Boniface*!; 1967.

J. acutiflorus Ehrh. ex Hoffm. Sharp-flowered Rush.
J. sylvaticus auct.
Native in wet shady places, marshes, on damp heaths, pond verges, etc. Common, chiefly on acid soils, but absent from heavily built-up areas.

First record: *T. Johnson*, 1632.
7. Hornsey Fields, 1815† (G & R). East Heath, Hampstead!; Eel Brook Meadow,† *T. & D.*, 283.

J. acutiflorus × articulatus = **J. × surrejanus** Druce
6. Hadley Common, 1961, *Benoit, J. G.* and *C. M. Dony, Fl. Herts*, 102.

J. articulatus L. Jointed Rush.
J. lampocarpus Ehrh. ex Hoffm.
Native in wet meadows, marshes, swampy woods, on pond and lake verges, etc. Local, and mostly confined to acid soils.
First evidence: *Lightfoot*, c. 1780.
1. Harrow Weald Common!, *Melv. Fl.*, 79. Harefield!, *Newbould, T. & D.*, 283. Moors and meadows north of Uxbridge!, *Benbow* (BM). Hillingdon,† *Trim. MSS.* Ruislip. Stanmore Common. **2.** Between Hampton and Hampton Court!, *Newbould*; Teddington, *T. & D.*, 283. Hampton Court Park!, *Welch*. Bushy Park. Walton Bridge. East Bedfont. Shortwood Common. **3.** Hounslow Heath! (LT). Between Twickenham and Richmond Bridge;† Worton,† *T. & D.*, 283. **4.** Willesden, 1842;† Hampstead! (HE); West Heath, Hampstead. Whitchurch Common!, *P. H. Cooke* (LNHS). **5.** Hanger Hill, Ealing.† Shepherds Bush, 1870,† *Warren, J.B. (London)* 13: 276. **6.** Edmonton,† *T. & D.*, 283. Hadley Common. Finchley Common. **7.** East Heath, Hampstead!, *T. & D.*, 283. Ken Wood!; Highgate Ponds!; Parliament Hill Fields!, *Fitter*. Buckingham Palace grounds, 1961, *Fl. B.P.G.*
A very large robust form was noted at **2.** Gravel pit, East Bedfont, 1966, *Boniface*!.

J. bulbosus L. Bulbous Rush.
J. supinus Moench
Native in bogs, wet woodland rides on acid soils, on damp heaths, etc. Rare, and decreasing.
First record: *Ray*, 1670.
1. Harefield Common!,† *Blackstone* (BM). Harrow Weald Common!, *Melv. Fl.*, 79. Stanmore Common!, *T. & D.*, 273. Duck's Hill Heath, Northwood; Ickenham Green!;† Mimmshall Wood, *Benbow MSS.* Wrotham Park. **2.** Staines Moor. River wall of Thames near Kingston Bridge!, *Welch*. Stanwell Moor. **3.** Hounslow Heath!, *Benbow MSS.* **4.** Hampstead Heath!, *Ray Cat.*, 150; still present in the bog on the West Heath, behind Jack Straw's Castle. **6.** Hadley Common!, *C. S. Nicholson, Lond. Nat. MSS.* **7.** East Heath, Hampstead; Eel Brook Common,† *T. & D.*, 283. Ken Wood.

Var. **uliginosus** (Fr.) Druce
2. Staines Moor, *Whale Fl.*, 15.

LUZULA DC.

L. pilosa (L.) Willd. Hairy Woodrush.
Native in woods, on heaths, shady hedgebanks, etc. Common, except in heavily built-up areas where owing to the eradication of its habitats it is very rare or extinct.
First certain record: *Petiver*, 1716.
7. Ken Wood!, *Pet. Gram. Conc.*, 226. Highgate Wood, 1844, *S. P. Woodward* (CYN).

L. forsteri (Sm.) DC. Forster's Woodrush.
Native in woods, on shady hedgebanks, etc. Very rare, and mainly confined to the north-western part of the vice-county.
First evidence: *Stevens*, 1846.
1. Old Park Wood, Harefield!, 1846, *Stevens* (K); 1966. Pinner, *Hind Fl.*; 1885, *Benbow*, *J.B.* (*London*) 23: 339. Duck's Hill, Northwood, 1891, *Benbow* (BM). Mad Bess! and Park Woods, Ruislip!; Copse Wood, Northwood!, *Welch*, *L.N.* 26: 64; still present in all three localities, 1970. **4.** Harrow Weald, *Melv. Fl.*, 80, Scratch Wood, *Welch*!, *K. & L.*, 279.

L. forsteri × pilosa = L. × borreri Bromf. ex Bab.
Lit. *Watsonia* **5**: 251 (1962).
1. Stanmore Common!, *L.N.* 28: 34. **4.** Scratch Wood, *Welch*!, *L.N.* 28: 34.

L. sylvatica (Huds.) Gaudin Greater Woodrush.
L. maxima (Reichard) DC.
Native in woods, damp shady places, etc. Rare, and mostly confined to acid soils.
First record: *Ray*, 1670.
1. Harefield,† *Blackst. Fasc.*, 37. Pinner Wood!, *Melv. Fl.*, 79. **4.** Ditch by Hampstead Wood, *Ray Cat.*, 149; Bishop's Wood!, *Curt. F.L.* Scratch Wood!; Hale, near Edgware,† *Benbow MSS.* Barnet Gate Wood!, *Benbow* (BM). Hendon Park!, *Boniface*, *L.N.* 29: 13. Grimsdyke grounds, Harrow Weald, *Welch*!, *L.N.* 28: 34. **5.** Syon Park, two clumps by the lake, 1966,† *Boniface*; eradicated during the construction of the Garden Centre, 1967–68. **6.** Between Highgate and Muswell Hill,† *E. Forster* (BM). Between Whetstone and Colney Hatch,† *T. & D.*, 234. **7.** Ken Wood!, *Pet. Gram. Conc.*, 227; still present, 1968. Hornsey Wood, *Blackst. Spec.*, 31; 1837,† *E. Ballard*; near Highgate Archway,† *E. Forster* (BM).

L. campestris (L.) DC. Field Woodrush.
Native in dry grassy places, established on lawns, etc. Common.
First record: *T. Johnson*, 1632.
7. Hyde Park, *Merr. Pin.*, 53. Kensington!, *Rudge* (BM). Ken Wood!,

Pet. Gram. Conc., 224. Fields between Highgate and Kentish Town, *T. & D.*, 285. Buckingham Palace grounds, *Codrington, Lousley* and *McClintock*!; Regents Park!, *Holme*; Kensington Gardens!, *L.N.* 39: 58. Fulham Palace grounds, *Welch*!.

L. multiflora (Retz.) Lejeune Many-headed or Heath Woodrush.
L. campestris subsp. *multiflora* (Retz.) Buchen.
Native on heaths, in open woods, etc. Locally plentiful, chiefly on acid soils.
First record: *Petiver*, 1716.
1. Between Harefield Common and Batchworth Heath,† *Blackst. Fasc.*, 37. Harefield Common;† Harrow Weald Common, abundant!; Stanmore Common!, *T. & D.*, 285. Ruislip Woods!; Pinner Wood!; Mimmshall Wood!, *Benbow MSS.* Ruislip Common. **3.** Hounslow Heath!, *Newbould*; near Hatton,† *T. & D.*, 285. **4.** Between Finchley and Hendon, *Newbould*; West Heath, Hampstead!, *T. & D.*, 285. North-west Heath, Hampstead. Scratch Wood; Barnet Gate Wood, *Benbow MSS.* Willesden Cemetery, 1965, introduced with turves. **5.** Near Hanwell,† *Warren, T. & D.*, 285. Horsendon Hill!, *Benbow MSS.* Syon Park. **6.** Hadley!, *Warren*; railway bank, Colney Hatch,† *T. & D.*, 285. Winchmore Hill Wood;† Enfield Chase!, *Benbow MSS.* **7.** Ken Wood!, *Pet. Gram. Conc.*, 225; 1963.

Var. **congesta** (Thuill.) Koch
1. Mimmshall Wood, *Sherrin* (SLBI). Stanmore Common. Harrow Weald Common. **5.** Syon Park.

AMARYLLIDACEAE

LEUCOJUM L.

L. aestivum L. Summer Snowflake.
Formerly native in wet places by the tidal Thames, and established elsewhere as a garden escape, but now extinct.
First record: *Curtis*, c. 1783. Last evidence: *Hardwicke*, 1845.
7. Isle of Dogs, *Curt. F.L.*; western side of the Isle of Dogs, 1801, *Borrer* (BO). Hornsey, 1845 (HE); from a garden?, *T. & D.*, 276.

NARCISSUS L.

Lit. *J. Roy. Hort. Soc.* **58**: 17 (1933).
N. pseudonarcissus L. Wild Daffodil.
Denizen or native in damp woods, fields, etc., also established on railway banks, in old orchards, etc. Very rare.
First record: *Blackstone*, 1737.

1. Orchard at Breakspears, plentifully, *Blackst. Fasc.*, 63; several places near Harefield, *Blackst. Spec.*, 58; grove near Harefield Church, 1853† (HE). In great plenty . . . between Pinner and Rickmansworth . . ., 1867,† *T. & D.*, 275. Ruislip!, in great profusion, 1884, *Benbow* (BM); Mad Bess Wood, Ruislip, very scarce!, *Hanson*; naturalised in several orchards at Ruislip!, *Wrighton, K. & L.*, 271. Mimmshall Wood,† *Brittain*, 5. **4.** Mill Hill, *R. A. Salisbury, Trans. Hort. Soc.* 1; 348; 1840† (HE).

N. hispanicus Gouan
N. major Curt.
Alien. Europe, etc. Garden escape or outcast. Naturalised in woods, meadows, etc., established on railway banks, in old orchards, etc. Very rare.
First record: *Cockfield*, 1813.
3. Roxeth,† probably not indigenous, *Hind, Melv. Fl.*, 77. Cranford!; Hayes, *Montg. MSS.* Railway banks, Hounslow. **4.** Behind 'The Spaniards', Hampstead,† no doubt escaped or planted, *T. & D.*, 275. **5.** In a wood near Ealing,† introduced, *Hemsley, T. & D.*, 275. Osterley Park!, *K. & L.*, 271. **6.** Hornsey,† *Cockfield Cat.*, 16. Edmonton, garden escape, *T. & D.*, 275. Enfield, *H. & F.*, 149. Copse near Muswell Hill, rare, 1945!,† *L.N.* 28: 34. Railway bank near Barnet. Coppetts Wood, Colney Hatch, a single plant, *Welch, K. & L.*, 271. **7.** Stamford Hill,† *Cockfield Cat.* Abundant in Ken Wood!, but probably planted, *T. & D.*, 275.
The modern double daffodils of gardens which sometimes occur as escapes probably consist of hybrids between *N. hispanicus* and other European species.

N. × incomparabilis Mill.
Alien of hortal origin. Garden escape. Formerly naturalised in meadows and established in old orchards, but now apparently merely casual. Very rare.
First evidence: *Buddle*, 1711.
1. Ruislip, one plant, *Firrell, J. Ruisl. & Distr. N.H.S.* 12: 7. **6.** 'I found this 1711 in some orchards and closes adjoining near Hornsey Church', *Buddle* (SLO).

N. majalis Curt. Pheasant's-eye Narcissus.
N. poeticus auct., *pro parte*
Alien. S. Europe, etc. Garden escape. Formerly naturalised in meadows, etc., but now apparently merely casual. Very rare.
First record: *Mrs Tooke*, 1864.
1. Meadow at Pinner Hill, *Mrs Tooke, Melv. Fl.*, 77. Field between Ruislip Reservoir and Harefield, 1866, *Griffith, T. & D.*, 275. Harefield, 1945, *K. & L.*, 271; 1955.

N. poeticus L. × **tazetta** L. = **N.** × **medioluteus** Mill. Primrose Peerless.

N. × *biflorus* Curt.

Alien of hortal origin. Garden escape. Naturalised in meadows, also established in old orchards where it was originally planted. Very rare.

First certain record: *Blackstone*, 1737.

1. Near Harefield in several places, *Blackst. Fasc.*, 64. Meadow south of Ruislip Reservoir in plenty, *T. & D.*, 275; 1884,† *Benbow* (BM). Old orchard, Ruislip!, *Green* (SLBI). Between Pinner and Rickmansworth, 1864,† *Hind* (BM). **4.** Harrow,† probably planted, *Melv. Fl.*, 77. **6.** Near Hornsey Church,† *J. Sherard*, *Ray Syn.* 3: 371; 'Mr Dillwyn could not find it there', *B.G.*, 403. Error, the plant intended was *N.* × *incomparabilis*.

IRIDACEAE

IRIS L.

I. foetidissima L. Gladdon, Stinking Iris.

Denizen in woods and thickets, on hedgebanks, etc. Very rare, and sometimes obviously introduced.

First record: *Parkinson*, 1640.

2. Bushy Park, *A. W. Jones*, *K. & L.*, 270. **5.** Pasture at Perivale, apparently native,† *Lees*, *T. & D.*, 274. **6.** Muswell Hill, in a hedge,† *Mart. Pl. Cant.*, 72. Enfield, *H. & F.*, 149. **7.** 'Near to Kentish Town . . . I do verily think it not natural in that place', *Park. Theat.*, 258; in a hedge near Kentish Town,† *Mart. Tourn.* 2: 45. On Jack Straw's Castle, and in a hedge near it,† *Pet. Midd.* and *Mill. Bot. Off.*, 421. Near Hornsey,† *Huds. Fl. Angl.*, 14, and many later authors. Lord Mansfield's Park (= Ken Wood),† *Coop. Fl.*, 102. Newington;† Islington,† *P. Miller*, *Blair Pharm.-Bot.*, 33.

I. pseudacorus L. Yellow Flag.

Native in marshes, wet fields, by river- and streamsides, on pond verges, etc., established by canals and gravel pits, etc. Sometimes planted for ornamental purposes. Common.

First record: *T. Johnson*, 1638.

7. Tottenham!, *Johns. Cat.* Ponds in Ken Wood grounds!; Isle of Dogs, *T. & D.*, 274. By the lake, Buckingham Palace grounds!, *Eastwood*. East Heath, Hampstead!; Highgate Ponds!, *Fitter*. Highbury. Lee Navigation Canal, Upper Clapton. River Lee, Hackney Marshes. Regent's Canal, Stepney, Old Ford and Hackney Wick. Canal side between Marylebone and Primrose Hill.

The common Middlesex plant appears to be the var. **acoriformis** (Bor.) Bak. The var. **pallidiflava** Sims is recorded from **7.** Ken Wood, *Syme*, *J.B.* (*London*) 6: 69.

I. germanica L.

Alien. Mediterranean region? Garden escape or outcast. Naturalised in ditches, on waste ground, established by gravel pits, on railway banks, etc. Very rare.

First record: *D. H. Kent*, 1945.

2. Gravel pit, West Bedfont, 1965, *Goom* and *Kennedy*!. **5.** Greenford, 1945. Hanwell, 1957. Ealing, 1965. **7.** Bombed site, Cripplegate, *Wrighton, L.N.* 28: 26.

CROCUS L.

C. purpureus Weston Purple Crocus.

C. vernus auct., *C. albiflorus* auct.

Alien. S. Europe, etc. Formerly naturalised in meadows and pastures but now extinct.

First record:–pre-1800. Last record: *Butcher*, c. 1925.

4. Near Mill Hill, 1874, *Whitwell MSS.* Hampstead, *Wharton Fl.* Probably an error. **6.** In the meadows near the church at Hornsey, in plenty, 1842, *Flower* (LINN); 1852 (WD); 1867, *Trimen* (BM); abundant in the turf of a large meadow a little south of Colney Hatch, but in Hornsey parish. Prof. *A. H. Church* informs us that it grew in this last locality in the recollection of persons living in the neighbourhood, at the end of the last century, and that before houses were built it grew sparingly between Hornsey and Wood Green, *T. & D.*, 274; 1887, *J. E. Cooper, K. & L.*, 270; abundant, 1900, *E. M. Dale* (LNHS); abundant, 1901, *C. S. Nicholson, K. & L.*, 270; abundant, c. 1925, *Butch. MSS.* The habitat, now enclosed within the grounds of Colney Hatch Hospital, was apparently destroyed by being covered with rubbish shortly before the outbreak of the Second World War.

CROCOSMIA Planch.

C. aurea Planch. × **pottsii** (Bak.) N.E.Br. = **C. × crocosmiflora** (Lemoine) N.E.Br. Montbretia.

Alien of hortal origin. Garden escape. Naturalised in ditches, shady places, on waste ground, etc., established in gravel pits, on rubbish-tips, etc. Rare.

First record: *D. H. Kent*, 1946.

1. New Years Green!, *K. & L.*, 270; 1967. Harefield, 1966, *I. G. Johnson* and *Kennedy*!. Springwell, 1966. Potters Bar, 1965, *C. M. Dony*. **3.** Hounslow Heath!, 1964, *Wurzell*; 1967. **4.** Hendon, 1950. Stonebridge, 1947. Rayners Lane, *L. M. P. Small*. **5.** Hanwell, 1946–54!, *K. & L.*, 270; 1955 →. Near Osterley, 1965. **7.** Near Marble Arch, 1962, *D. E. Allen, L.N.* 44: 27.

DIOSCOREACEAE

TAMUS L.

T. communis L. Black Bryony.
Lit. *J. Ecol.* **32**: 121 (1944).
Native in woods, thickets, climbing over hedges, etc. Common, except near the Thames, and in heavily built-up areas where it is rare or extinct.
First record: *T. Johnson*, 1638.
7. Tottenham, *Johns. Cat.* Hampstead, *Merr. Pin.*, 17. Marylebone, 1815 (G & R).

ORCHIDACEAE

Lit. Godfrey, M. J. *Monograph and Iconograph of Native British Orchidaceae.* Cambridge (1933): Summerhayes, V. S. *Wild Orchids of Britain.* London, edition 2 (1968).

EPIPACTIS Sw.

Lit. *Watsonia* **1**: 102 (1949): *loc. cit.* **2**: 253 (1951): *loc. cit.* **5**: 127 (1962): *loc. cit.* **6**: 388 (1967).

E. helleborine (L.) Crantz Broad-leaved Helleborine.
E. latifolia (L.) All., *E. media* auct.
Native in woods and thickets, on hedgebanks, by roadsides, etc. Rare, and mainly confined to the northern parts of the vice-county.
First certain evidence: *Benbow*, 1886.
1. Uxbridge Common, 1886, *Benbow* (BM), det. *D. P. Young*. Ickenham Park, 1909, *Green* (SLBI), det. *D. P. Young*; Ickenham, a long-established patch destroyed by building operations, 1953; a second patch located nearby is also threatened by building operations, 1954, *Davidge, K. & L.*, 263; the patch survives, but is now enclosed within a garden in Swakeleys Road, 1961; thickets by the river Pinn!, c. 40 plants, 1962, *Davidge, L.N.* 42: 112; still present, 1971; Swakeleys House grounds, 1956; roadside near Brackenbury Farm, near Harefield, a single plant, 1966, *I. G. Johnson*. Garett Wood, Springwell,† *Green* (SLBI), det. *D. P. Young*. Mimmshall Wood, *Benbow*, *J.B. (London)* 25: 365. Wood west of Potters Bar, c. 40 plants, 1965, *J. G.* and *C. M. Dony*! Shady roadbanks east of Potters Bar, about six isolated plants over a distance of about a mile, 1966, *Kennedy*!. **2.** By a lake, Bushy Park, four plants, 1965; Upper Lodge, Bushy Park, four plants under ancient lime trees, 1966; one plant, 1967, *Clement*. Ashford golf course, three plants, 1966. **4.** Scratch Wood!, 1909, *Green* (SLBI), det. *D. P. Young*; 1954!, *K. & L.*, 263. Under a tree,

by pavement, Templewood Avenue, Hampstead, 1960, *Sworder*, fide *D. P. Young*. **5.** Horsendon Lane, Perivale, one plant, 1973, *Beaumont*. **6.** Hadley Wood, 1886, *Pugsley* and *Seagrott* (BM); 1908, *Green* (D), both det. *D. P. Young*. Whitewebbs Park, Enfield, a single plant, 1947, *Johns*, *L.N.* 27: 34; 39 flowering spikes, 1969, *Greenwood*. Queen's Wood, Highgate, a single plant, 1956, *Bangerter*.

E. purpurata Sm. Violet Helleborine.
E. sessilifolia Peterm., *E. violacea* (Dur. Duq.) Bor., *E. media* auct.
Native in woods, thickets, on hedgebanks, etc. Very rare, and confined to the northern parts of the vice-county.
First certain record: *R. J. Knight*, 1864.
1. Pinner Wood, 1872, *W. A. Tooke* (BM), det. *D. P. Young*. Near Uxbridge, *Benbow*, *J.B. (London)* 25: 365. Harefield, 1888, *Wall* and *Rolfe* (K); 1909, *Green* (D), both det. *D. P. Young*. Mad Bess Wood, Ruislip, 1955, *Pickess*. **4.** Grounds of Parsonage, Harrow Weald, *R. J. Knight*, *Melv. Fl.*, 76; in some plenty on a raised mound formed about nineteen years ago by the soil removed in digging the foundation of the church, and planted as a shrubbery between Harrow Weald Church and the Rectory, 1866,† *Trimen* (BM), det. *D. P. Young*. Scratch Wood, a fine plant, 1960, *Warmington*, *L.N.* 40: 21. **6.** Whitewebbs Park, Enfield!, *Johns.*; 1950! and 1954!, *K. & L.*, 264; 1962!, *L.N.* 42: 12. Wooded enclosure, Myddleton House grounds, Enfield, a single plant, 1966, *Kennedy*!, *L.N.* 46: 13.

E. phyllanthes G. E. Sm. Pendulous-flowered Helleborine.
E. vectensis (T. & T. A. Stephenson) Brooke & Rose, *E. media* auct.
Lit. *Watsonia* **2**: 262 (1952): *loc. cit.* **5**: 136 (1962).
Native in open woods, copses, etc. on damp soils. Very rare, and so far known only from the north-west of the vice-county but likely to be found elsewhere.
First record: *Pickess*, 1965.
1. Small covert on Harefield Moor, c. six plants, 1966!, *Crooks* conf. *D. P. Young*, *L.N.* 46: 13.

Var. **degenera** D. P. Young
1. Open wood in an old chalkpit, Harefield, c. 60 plants, 1965!, *Pickess*, det. *D. P. Young*, *L.N.* 45: 20; 1966 →.
Most printed records of *Epipactis* have been omitted from the above account unless substantiated by specimens.

SPIRANTHES Rich.

S. spiralis (L.) Chevall. Autumn Lady's Tresses.
S. autumnalis Rich.
Formerly native in dry pastures, on banks, etc., established on lawns, etc., but now extinct.

First record: *W. Turner*, 1548. Last record: *D. H. Kent*, c. 1936.

1. Uxbridge, *Lightf. MSS.* Pinner Hill!, *Mrs Tooke, Melv. Fl.*, 76; c. 1936. **2.** On the lawn at Strawberry Hill, c. 1780, *Gibbs, Pamplin MSS.* **3.** Near Isleworth, *Ger. Hb.*, 168. **4.** Plentifully in a field ... at Mill Hill; field at Highwood Hill, abundantly, *M. Collinson, Coll. MSS.* **5.** Beside Syon, *Turn. Names.* Hanwell Heath ..., sparingly, *Goodenough, Curt. F.L.* **6.** Enfield Chase, *Doody MSS., Alch. MSS.* and *Mart. Pl. Cant.*, 65; c. 1873, *H. & F.*, 150. **7.** Upon a common heath by a village called Stepney, *J. Coles*; field by Islington ..., *Ger. Hb.*, 168.

LISTERA R.Br.

L. ovata (L.) R.Br. Twayblade.

Native in moist woods, pastures, shady places, etc., established in chalkpits, etc. Rare, and mostly confined to the north of the vice-county.

First record: *Gerarde*, 1597.

1. In Whiteheath Wood† and Scarlet Spring, and in meadows near the river, Harefield,† *Blackst. Fasc.*, 10. Copse near Harefield towards Uxbridge,† *A. B. Cole, T. & D.*, 272. Meadows between Denham and Harefield Moor locks;† Moor Hall;† Harefield Moor!; Garett Wood!,† *Benbow MSS.*; Harefield Grove, 1955, *Pickess*; old chalkpit, Harefield, 1962!,† *T. G. Collett*; chalkpit filled with rubbish 1965–67. Copse near Pinner Wood; Ruislip!, *T. & D.*, 272; meadows near Bayhurst Wood; Ruislip Woods!; near Uxbridge Common;† near Denham Lock!;† near Ickenham!, *Benbow MSS.* Near Northwood!, 1902, *Whiting, K. & L.*, 262; Copse Wood, Northwood, c. 200–300 plants, 1955, *Pickess*; a small colony under bracken near Shrubbs Corner, between Northwood and Harefield, 1965. **2.** Spout Wood, Stanwell, one plant, 1961, *Pickess.* **3.** Harrow Grove,† *Melv. Fl.*, 75. Wood near Hounslow Heath, one plant, 1965, *P. J. Edwards.* **4.** Hampstead Wood, *Ger. Hb.*, 326; Bishop's Wood, *T. & D.*, 272. Hampstead Heath,† *Johns. Enum.* Between Highgate and Hampstead,† *Park. Theat.*, 505. Near Hendon,† *Coop. Suppl.*, 11. Bentley Priory woods, *Melv. Fl.*, 75. Scratch Wood!, *T. & D.*, 272. Near Finchley,† *Newbould, Trim. MSS.* **5.** Syon Park, a single plant!,† *K. & L.*, 262; habitat destroyed by the construction of the Garden Centre, 1966–67. **6.** Highgate,† *Ger. Hb.*, 326. The Alder's Wood, Whetstone,† *T. & D.*, 272. Enfield, *A. Irvine, Sci. Goss.* 8, 279. **7.** Ken Wood!, 1945–47, *Soper*; six plants, 1949!, *H. C. Harris, L.N.* 29: 13; 1950–53!, *K. & L.*, 262; 1962: not seen since.

NEOTTIA Guett.

N. nidus-avis (L.) Rich. Bird's-nest Orchid.
Native in shady woods, on shady banks, etc., usually under beech. Very rare, and perhaps extinct.
First record: *Blackstone*, 1735.
1. In Whiteheath Wood on Harefield Common, but very rarely, 1735,† *Blackst. Fasc.*, 67. Garett Wood!,† *Benbow* (BM); habitat destroyed by chalk-quarrying operations. Under beeches on a bank west of Rickmansworth Road, Harefield, 25 yards on the Middlesex side of the county boundary, 1946, *Welch*, *L.N.* 26: 63; not seen since, but may survive in a dormant state. Pinner Hill, *W. A. Tooke*, *Melv. Fl.*, 2: 129. Mimmshall Wood, *Benbow*, *J.B.* (*London*) 25: 365. **3.** Copse in Headstone Lane, Pinner,† *Bourne*, fide *Hind*, *J.B.* (*London*) 9: 272. Harrow Grove, 1868,† *Farrar*, *T. & D.*, 272. **4.** On a common laurel in the grounds of Stanmore Cottage, c. 1830,† *Varenne*, *T. & D.*, 272. Scratch Wood, *Benbow* (BM).

COELOGLOSSUM Hartm.

C. viride (L.) Hartm. Frog Orchid.
Habenaria viridis (L.) R.Br.
Formerly native in meadows and pastures, but now extinct.
First record: *W. A. Tooke*, 1871. Last evidence: *J. E. Cooper*, 1913.
1. Pinner Hill, *W. A. Tooke*, fide *Hind*, *J.B.* (*London*) 9: 272. Between Harefield and Northwood, 1906, *Green*, *Druce Notes*. Near Harefield, 1913, *J. E. Cooper* (BM).

GYMNADENIA R.Br.

G. conopsea (L.) R.Br. subsp. **conopsea** Fragrant Orchid.
Habenaria conopsea (L.) Benth., non Reichb. f.
Native in dry grassy places, etc. Very rare, and confined to calcareous and sandy soils in the north-west of the vice-county.
First record: *Blackstone*, 1737.
1. In Harefield chalkpit, *Blackst. Fasc.*, 69; in abundance there,† *M. Collinson*, *Coll. MSS*. Sloping banks of Garett Wood!, 1889; a single plant, 1892, *Benbow* (BM); a single plant, 1946!,† *L.N.* 26: 64; habitat since destroyed by chalk-quarrying operations. Harefield, 1907, *Green* (SLBI). Northwood golf course, a single plant on sandy soil; grubbed up about ten days after its discovery, 1965, *J. A. Moore*, *L.N.* 46: 13.
Plants with white flowers are recorded from **1.** Harefield chalkpit, sparingly,† *Blackst. Fasc.*, 69.

PLATANTHERA Rich.

P. chlorantha (Custer) Reichb. Greater Butterfly Orchid.
Habenaria chlorantha (Custer) Bab., non Spreng., *H. virescens* Druce, non Spreng.
Formerly native in woods, on banks, etc., but now extinct.
First record: *Gerarde*, 1597. Last evidence: *Benbow*, 1890.
1. Harefield, not frequent, *Blackst. Fasc.*, 70; copse near Harefield Park; copse just above Jack's Lock, 1885–90, *Benbow* (BM). **4.** North end of Hampstead Heath, *Ger. Hb.*, 166. In plenty about Mill Hill, 1757–66, *M. Collinson, Coll. MSS.* **6.** Fields . . . at Highgate, *Ger. Hb.*, 166; Highgate Wood, 1857, *Jewitt, T. & D.*, 271. **7.** Ken Wood, *R. & P.*, 218.

P. bifolia (L.) Rich. Lesser Butterfly Orchid.
Formerly native on heaths, in open woods, etc., but now extinct.
First record: *Uvedale*, c. 1700. Last evidence: *Benbow*, 1887.
4. Hampstead Heath, *Budd. MSS.*, *Doody MSS.* and *Alch. MSS.* In Bishop's Wood, 'recently' [c. 1830], *Irv. MSS.* Scratch Wood, two plants, 1887, *Benbow* (BM). **6.** Near Enfield Chase, *Uvedale, Doody MSS.*; in 1756 in great abundance for more than two miles amongst the bushes on Enfield Chase, between Southgate and the lodge now (1760) in the possession of Mr Jalabert, *M. Collinson, Coll. MSS.* Highgate Wood, *Jewitt, T. & D.*, 270. **7.** 'My father saw this in a wood between Hampstead and Highgate, now the property of Lord Mansfield, and since enclosed by him with pales [= Ken Wood]', *M. Collinson, Coll. MSS.*; Ken Wood, *Alch. MSS.*

OPHRYS L.

O. apifera Huds. Bee Orchid.
Native on banks, in hilly pastures, etc. Very rare and mainly confined to calcareous soils.
First record: *Blackstone*, 1737.
1. Harefield chalkpit, sparingly!,† *Blackst. Fasc.*, 70; it was last seen there about 1955, but the habitat has now been bulldozed and filled with rubbish. Sloping banks of Garett Wood!,† *Benbow* (BM); habitat destroyed by chalk-quarrying operations, c. 1956. About a dozen plants in a meadow south of Harefield, *Benbow, J.B. (London)* 22: 279. Frequent on hilly pastures about Harefield and Springwell, *Benbow MSS.*; still grows sparingly in at least four localities near Harefield, 1946!, *L.N.* 26: 64; sandpit, Harefield Grove,† *I. G. Johnson*; habitat destroyed by the tipping of rubbish; now apparently very rare in the Harefield area, and not recently seen, 1963; four plants discovered on calcareous grassland in 1966 by *I. G. Johnson*:

four plants, 1970, *I. G. Johnson, L.N.* 52: 119. Old sandpits, New Years Green, 1953!,† *Haywood, L.N.* 33: 54; habitat destroyed by the tipping of rubbish, 1954, the species may, however, be discovered elsewhere in the area. Sandpit, Breakspeares House, 1957, *I. G. Johnson*. **4.** Mill Hill,† *Foley*, 12. Perhaps an error.

O. insectifera L. Fly Orchid.
O. muscifera Huds.
Formerly native on wood borders, shady slopes and thickets, etc., on calcareous soils, but now extinct.
First record: *Blackstone*, 1737. Last evidence: *Green*, 1907.
1. Harefield chalkpit . . ., *Blackst. Fasc.*, 68; 'I found this in a chalkpit at Harefield, but not in the one described by Mr Blackstone, where I have often searched for it without success, 1758', *P. Collinson, Dillwyn Hort. Coll.*, 36; in Harefield chalkpit (Blackstone's station), 1790; 'never in the course of twenty six or twenty seven years observed by me before', *M. Collinson, Coll. MSS.* In two copses of the road from Harefield to Uxbridge, near Moor Hall and West End, 1867–68, *A. B. Cole, T. & D.*, 271. Sloping bank of Garett Wood, abundant, 1898, *Benbow* (BM); 1907, *Green* (SLBI).

ORCHIS L.

O. militaris L. Soldier Orchid.
Formerly native on wood borders on calcareous soils, established in chalkpits, etc., but now extinct.
First record: *Blackstone*, 1737. Last evidence: *Pugsley* and *C. R. P. Andrews*, 1900.
1. In the chalkpit near the paper mill at Harefield, plentifully, *Blackst. Fasc.*, 67. Chalk bank above Gulch Well (= Garett Wood), c. 1800, *J. Banks*; sloping banks of Garett Wood, 1885, *Benbow* (BM); 'several plants on the Middlesex side of the wood, and some twenty on the Herts side, 1885, some unfortunately taken up by the roots by my sons and nephews; a single plant on the Middlesex side, and a single plant on the Herts side, 1889; no plants on the Middlesex side, but twelve on the Herts side, 1890; no plants on the Middlesex side, but eight on the Herts side, 1891; no plants on the Middlesex side, but four on the Herts side, 1892 and 1895; no plants seen on either side, 1896; no plants on the Middlesex side, a single plant on the Herts side, 1899', *Benbow MSS.*; [a single plant on the Middlesex side], 1900, *Pugsley* and *C. R. P. Andrews* (BM); 'no plants on the Middlesex side, but two on the Herts side, 1900 and 1901; no plants on the Middlesex side, a single plant on the Herts side, 1902', *Benbow MSS.* The Rev. J. Roffey searched the wood thoroughly in 1910 but found it very overgrown with brambles and failed to

locate any plants of *O. militaris*. I searched the area many times between 1940 and 1952, when it was cleared prior to chalk-quarrying operations. Although apparently suitable habitats still existed I failed to rediscover the plant, and feel that unrestricted collecting by nineteenth-century botanists contributed to the extinction of this rare and beautiful species in the vice-county. H. W. Pugsley assured me that he was quite certain that the specimen gathered in 1900 grew on the Middlesex side of the wood.

O. ustulata L. Dark-winged, Burnt, or Dwarf Orchid.
Formerly native in short turf on calcareous soils, but now long extinct.
First, and only record: *Blackstone*, 1737.
1. In Harefield chalkpit, sparingly, *Blackst. Fasc.*, 69. Michael and Peter Collinson searched the area for the plant on many occasions between 1757 and 1790 without success.

O. morio L. Green-winged, or Meadow Orchid.
Native in damp meadows and pastures, formerly rather rare, but now very rare.
First record: *Blackstone*, 1737.
1. Harefield!, *Blackst. Fasc.*, 68; near Harefield Church,† *A. B. Cole, T. & D.*, 268; near Moor Lock, Harefield,† *Benbow* (BM); meadow, Knightscote Farm, Harefield, a small colony, 1955!; not seen recently, *Pickess*, 1964; near Harefield, 1912, *J. E. Cooper* (BM); 1913, 1920 and 1923, *J. E. Cooper, K. & L.*, 265. Pinner, abundant,† *Melv. Fl.*, 76. Ruislip Moor, plentiful,† *T. & D.*, 268. Uxbridge, 1884,† *Benbow* (BM). Swakeleys;† Ickenham;† Ruislip;† Northwood!,† *Benbow MSS.* **2.** Between Colnbrook and Staines,† *Benbow MSS.* **3.** Between Harrow and Pinner, 1862,† *T. & D.*, 268. Roxeth,† *Melv. Fl.*, 76. Ickenham Marsh,† *Benbow MSS.* Yeading,† *T. Moore, Trim. MSS.* Near Hanworth, *Pugsley*. Hayes Park, *Royle*. **4.** Between Hendon and Hampstead,† *Button, Coop. Suppl.*, 11. Plentiful in a meadow near Hendon, *Irv. MSS.*; 1907,† *Moring MSS.* Near Mill Hill,† *M. Collinson, Coll. MSS.* Mill Hill,† *Rudge* (BM). Near Highgate Wood;† Stanmore, 1827–30,† *Varenne, T. & D.*, 268. Near Hampstead, 1852† (SY). Near Edgware, c. 1908, *Champ. List.* Harrow,† *Benbow MSS.* **5.** Osterley Park, 1882,† *Britten* (BM). **7.** Near Islington,† *R. & P.*, 218.
Plants with white flowers are recorded from **1.** Harefield, *Blackst. Fasc.*, 68. **3.** Near the Crane, near Feltham, a single plant, 1967 →, *R. K. Edwards*, fide *P. J. Edwards.* **4.** Mill Hill,† *M. Collinson, Coll. MSS.*

O. mascula (L.) L. Early Purple Orchid.
Native in woods and copses, etc. Very rare, and mainly confined to base-rich soils in the northern parts of the vice-county.
First record: *T. Johnson*, 1632.

1. Harefield!, frequent, *Blackst. Fasc.*, 67; near Harefield Church,† *A. B. Cole*; Old Park Wood,† *T. & D.*, 268; wood between the chalk-pits, Harefield!,† *Green* (SLBI); Scarlet Spring, near Harefield, 1955!; not seen since, 1965, *Pickess*; woods, Harefield Grove, 1965, *I. G. Johnson*. Garett Wood, 1855,† *Phyt. N.S.* 1: 62. Between Harefield and Springwell, 1884, *Benbow* (BM); north of Harefield!, *Welch, L.N.* 27: 34. Copse near Pinner Wood,† *Hind, Melv. Fl.*, 97. Pinner;† Ruislip,† *De Cresp. Fl.*, 47. Northwood, 1967, *J. A. Moore, L.N.* 47: 10. **2.** Spout Wood, Stanwell, one plant, 1961, *Clement*. **4.** Hampstead Heath, *Johns. Enum.*; 1878, *De Ves. MSS.* Many places near Mill Hill,† *M. Collinson, Coll. MSS.* Mill Hill,† *Foley*, 12. Side of wood behind 'The Spaniards', Hampstead, 1821,† *J. J. Bennett* (BM).

Plants with white flowers are recorded from **4.** Mill Hill,† *M. Collinson, Coll. MSS.*

DACTYLORHIZA Nevski

Lit. Vermeulen, P. *Studies on Dactylorchids.* Utrecht (1947): *Proc. Bot. Soc. Brit. Isles* **6**: 372 (1967).

D. incarnata (L.) Soó subsp. **incarnata** Early Marsh Orchid, Meadow Orchid.

Orchis incarnata L. sec. Vermeul., *O. latifolia* auct., *O. strictifolia* Opiz, *Dactylorchis incarnata* (L.) Vermeul.

Formerly native in marshes, wet meadows, etc., established in damp chalkpits, etc., but now extinct.

First certain evidence: *Benbow*, 1890. Last record: *D. H. Kent* and *Lousley*, 1946.

1. Harefield Moor, and moist meadows adjoining, 1890; meadows by the Colne, Drayton Ford, *Benbow* (BM). Yiewsley, 1910, *Green* (SLBI); 1913, *J. E. Cooper* (BM); 1914–24; near Denham Lock, 1920, *J. E. Cooper, K. & L.*, 265. Old chalkpit, Harefield, one plant, 1946, *Lousley*!, *L.N.* 26: 64. **2.** Marshy unclaimed ground at the edge of Staines Moor, in profusion, 1899, *Shepherd* (K). West Drayton, 1909, *Rolfe* (BM).

Subsp. **gemmana** (Pugsl.) P. D. Sell

Orchis incarnata L. var. *gemmana* Pugsl., *Dactylorchis incarnata* (L.) Vermeul. subsp. *gemmana* (Pugsl.) H.-Harrison f.

2. West Drayton, 1909, *R. Bedford* (BM).

D. incarnata × praetermissa = D. × wintoni (A. Camus) P. F. Hunt

2. West Drayton,† *Rep. Bot. Soc. & E.C.* 5: 400.

D. maculata (L.) Soó subsp. **ericetorum** (E. F. Linton) P. F. Hunt & Summerhayes Heath Spotted Orchid.

Orchis maculata auct., *O. ericetorum* E. F. Linton, *O. elodes* auct., *Dactylorchis maculata* (L.) Vermeul. subsp. *ericetorum* (E. F. Linton) Vermeul.

Native on damp heaths, in bogs, etc. Very rare, and almost extinct.

First certain evidence: *Melvill*, 1863.

1. Harrow Weald Common, 1863† (ME). Stanmore Common, abundant, *T. & D.*, 269; c. 1900,† *C. S. Nicholson, K. & L.*, 266. Near Harefield, 1913,† *J. E. Cooper* (BM). Ruislip Common, 1965, *Pickess.* **3.** Hounslow Heath, *T. & D.*, 269; 1887,† *Benbow* (BM).

In view of the confusion between *D. maculata* subsp. *ericetorum* and *D. fuchsii* most printed records have been omitted unless substantiated by specimens.

D. fuchsii (Druce) Soó subsp. **fuchsii** Common Spotted Orchid.

Orchis fuchsii Druce, *O. maculata* auct., *Dactylorchis fuchsii* (Druce) Vermeul.

Native in woods, damp meadows, thickets, on heaths, by roadsides, etc., established in damp chalkpits, etc. Rare, but most frequent on calcareous and other base-rich soils.

First record: *Gerarde*, 1597.

1. Harefield chalkpit, plentifully!,† *Blackst. Fasc.*, 68; survived until the habitat was bulldozed and filled with rubbish, 1965–66; woods, Harefield Grove, 1966, *I. G. Johnson*. Roadside between Harefield and Northwood, two plants, 1965. Ruislip!, *Benbow MSS.*; Copse Wood, Ruislip!; Poor's Field, Ruislip, one plant, 1963, *Pickess*. Old sandpits, New Years Green, 1953!,† *Haywood, L.N.* 33: 54; habitat destroyed by the tipping of rubbish, 1954; a second locality, 1960! →, *Pickess*. Near Denham!; Mimmshall Wood, abundant,† *Benbow MSS.* Stanmore Common, 25 plants, 1971, *Roebuck*. **2.** Littleton gravel pit, 1968 →, *J. A. W. Jones*, fide *Ettlinger*. **3.** Harrow Grove,† *Melv. Fl.*, 76. **4.** Hampstead Wood, *Ger. Hb.*, 170. Wood beyond 'The Spaniards', *Irv. MSS.* Stanmore, *Varenne*; Bentley Priory, *T. & D.*, 269; 1968, *Roebuck*. Scratch Wood!, *T. & D.*, 269. Harrow Park,† *Melv. Fl.*, 76. **5.** Wood near Brentford,† *Hemsley, T. & D.*, 269. Syon Park, one plant, 1961,† *Hunt*; habitat destroyed during the construction of the Garden Centre, 1966–67. Wyke Green, c. 1936.† Bed of disused Fox Reservoir, Ealing, seventeen plants, 1968, *McLean*. **6.** Tottenham,† *Johns. Cat.* Alder's Wood, Whetstone;† Edmonton,† *T. & D.*, 269. Enfield Chase, *Benbow MSS.* Coldfall Wood, Highgate, *E. M. Dale, K. & L.*, 266. **7.** East Heath, Hampstead, 1963, *Springett*; a single plant, 1964, *Lousley, L.N.* 44: 16. Ken Wood; meadow off Millfield Lane, Highgate, 1964; Highgate Ponds, a single plant, 1968, *Springett*.

A plant with all the flowers inverted was noted at **1.** Harefield, *Benbow, J.B. (London)* 28: 120.

D. fuchsii subsp. **fuchsii** × **praetermissa** = **D.** × **grandis** (Druce) P. F. Hunt

1. Abundant near Harefield, with both parents, 1944–46!, *L.N.* 26: 44; this hybrid occurred in great quantity on the floor of an old chalkpit at Harefield until c. 1948 when it began to decrease quite rapidly. A few plants survived until 1959,† and searches made between 1960 and 1965 failed to reveal any hybrids or specimens of *D. praetermissa*, though *D. fuchsii* subsp. *fuchsii* still occurred in small quantity. The habitat was destroyed by bulldozing and the subsequent tipping of rubbish in 1965–66. Sandpit near New Years Green!, 1964, *I. A. Moore*; abundant, 1965!, *Pickess*; 1966 →.

D. praetermissa (Druce) Soó Druce's Marsh Orchid.

Orchis praetermissa Druce, *O. latifolia* auct., *Dactylorchis praetermissa* (Druce) Vermeul.

Native in marshes, wet meadows, etc., established in damp chalkpits, sandpits, by canal sides, etc. Very rare, and chiefly found on calcareous and other base-rich soils.

First certain evidence: *E. Forster*, 1792.

1. Near Uxbridge!,† *G. C. Druce, Rep. Bot. Soc. & E.C.* 5: 165. Frequent about Harefield! and Harefield Moor, 1946!; a specimen collected in an old chalkpit in 1945 measured nearly a metre in height!, *L.N.* 26: 46; now very rare, and much reduced in quantity as a result of gravel-digging operations on Harefield Moor: in the old chalkpit it had become extinct by c. 1960. Canal side between Denham and Harefield Moor Locks, 1947.† Sandpit, New Years Green, 1965!, *Pickess*. Yiewsley, 1913, *J. E. Cooper* (BM); 1914–24,† *J. E. Cooper*. **2.** Swampy meadow near West Drayton,† *Dymes* (K). Marshy ground by the railway, Staines Moor, 1958, *Briggs*. Littleton gravel pit, c. twenty plants, 1968, *J. A. W. Jones*, fide *Ettlinger*. **6.** Rammey Marsh, Enfield, 1929,† *C. E. Britton* (K); this locality which has now been transferred to S. Essex, v.c. 18, has been drained and raised with rubble and clay. **7.** Marshes, Lee Bridge, 1792,† *E. Forster* (BM).

A plant with snow-white, unspotted flowers was gathered in **1.** Marshy field by canal, Harefield, 1938,† *Hubbard, Nelmes* and *Sandwith* (K).

ANACAMPTIS Rich.

A. pyramidalis (L.) Rich. Pyramidal Orchid.

Orchis pyramidalis L.

Native in grassy places on calcareous soils. Very rare, and varying in quantity from year to year.

First record: *Blackstone*, 1737.

1. Harefield chalkpit, plentifully, *Blackst. Fasc.*, 69; in abundance there, c. 1790,† *M. Collinson, Coll. MSS.* Down above Springwell Lock, 1885–89, *Benbow*; 1904,† *Roffey* (BM). Sparingly on the downs between Jack's Lock and Springwell!, *Benbow, J.B. (London)* 23: 339; 1889; not seen again until 1904, *Benbow MSS.*; twenty seven plants, 1946!, *Welch, L.N.* 26: 64; one plant, 1947, *Welch*; six plants, 1964; c. fifty plants, 1965, *Pickess*; plentifully, 1966, *Johnson*; plentiful, 1967–69; at least 100 plants, 1970, *I. G. Johnson, L.N.* 52: 119.

A white-flowered plant was seen at **1.** Harefield, 1946!, *Welch.*

ARACEAE

ACORUS L.

A. calamus L. Sweet Flag.

Alien. S. Asia, etc. Naturalised in shallow water at the margins of rivers, streams, ponds, lakes, etc., established in canals, gravel pits, etc. Common, though often sterile.

First record: *Doody*, 1695.

7. Fulham Palace moat,† *Doody, Pet. Midd.* Ken Wood!; Copenhagen Fields,† *Burn. Med. Bot.*, 32. Hackney Wick!, *T. & D.*, 291. Hackney Canal between Hackney Wick and Lee Bridge!, *Benbow MSS.* Hackney Marshes!, *M. & S.* Near Kentish Town, *Pamplin MSS.* Thames side, Chelsea,† '*D.E.*' (BM). Regent's Canal near St Pancras, 1959! →, *L.N.* 39: 58. By the lake, Buckingham Palace grounds, probably planted, *Codrington, Lousley* and *McClintock*!, *L.N.* 36: 15.

CALLA L.

C. palustris L.

Alien. Europe, etc. Naturalised in a pond. Very rare, and perhaps extinct.

First record: *Welch*, 1946.

1. Pond on Stanmore Common!, *Welch, L.N.* 26: 34; 1954.

ARUM L.

A. maculatum L. Lords-and-Ladies, Cuckoo-pint, Wild Arum.

Lit. Prime, C. *Lords and Ladies*. London (1960).

Native in woods, on shady hedgebanks, etc. Common, especially on base-rich soils, though rare or extinct in heavily built-up areas.

First record: *T. Johnson*, 1629.

7. Between Kentish Town and Hampstead,† *Johns. Eric.* Marylebone Fields, 1817† (G & R). Many places in the northern suburbs; Crouch

End; Hornsey, abundant, *T. & D.*, 291. Fulham Palace grounds, *Welch*!. Buckingham Palace grounds, *Codrington, Lousley* and *McClintock*!, *L.N.* 39: 58; not seen recently, 1963, *Fl. B.P.G.* Hurlingham, *L. M. P. Small.*

LEMNACEAE

SPIRODELA Schleid.

S. polyrhiza (L.) Schleid. Great Duckweed.
Lemna polyrhiza L.
Native. Floating on still waters of ponds, lakes, ditches, etc., and in the slower parts of streams. Local.
First record: *Varenne*, c. 1827.
1. Feeder to Elstree Reservoir;† near South Mimms!, *T. & D.*, 292. Harefield!, *Druce Notes*. Between Harefield and Springwell. Hillingdon,† *Trim. MSS.* Colne near Denham!, *K. & L.*, 281. Near Uxbridge. Ruislip Common, *J. A. Moore*. Potters Bar. **2.** Between Hampton and Sunbury, *Newbould*; Staines Moor!; Shortwood Common!, *T. & D.*, 292. Between West Drayton and Colnbrook, *Benbow MSS.* Poyle!, *Druce Notes*. Hampton Court Park!, *K. & L.*, 281. Bushy Park. **3.** Hounslow Heath. Isleworth, sparingly,† *T. & D.*, 292. Hayes!; Northolt!;† Southall!,† *K. & L.*, 281. Longford river, Feltham, 1971, *P. J. Edwards*. **4.** Hampstead Heath† (HE). Harrow,† *Hind Fl.* Pond between Finchley and Hendon,† *Newbould*; in the Brent,† *Farrar, T. & D.*, 292. Mill Hill, *Sennitt*. **5.** Ealing Common, abundant,† *Newbould, T. & D.*, 292. Canal, Brentford,† *Wright MSS.* Thames at Strand-on-the-Green, 1955!,† *K. & L.*, 281. **6.** Hadley Green!; Town Park, Enfield!, *L.N.* 29: 12. Pond on Crew's Hill golf course, Enfield!, *Castell* and *Kennedy*. **7.** Hornsey, 1827–30,† *Varenne*; Hampstead Ponds, abundant,† *W. G. Smith*; Serpentine,† *T. & D.*, 292. Bishop's Walk, Fulham† (HE). Duckett's Canal, Hackney, 1872, *Warren, Trim. MSS.* Static water tank by the Thames, Millbank, S.W.1,† *McClintock, Rep. Bot. Soc. & E.C.* 12: 761.

LEMNA L.

L. trisulca L. Ivy-leaved Duckweed.
Native. Floating, often below the surface of the water, in ponds, lakes, ditches, etc. Locally plentiful.
First record: *A. Irvine*, c. 1830.
1. Near Harefield!, *Newbould*; Pinner; feeder to Elstree Reservoir,† *T. & D.*, 291. Stanmore Common!, *P. H. Cooke, K. & L.*, 282. Near Potters Bar, *J. G.* and *C. M. Dony*!. **2.** Staines Moor!; Shortwood Common!; common by Walton Bridge;† Bushy Park, very abundant!, *T. & D.*, 291. West Drayton, *Druce Notes*. Hampton Court

Park!, *R. W. Robbins*; ditch in front of East Bedfont Church!,† *Welch*; Poyle!; Yeoveney!, *K. & L.*, 282. Near Colnbrook. **3.** Grove Pond, Harrow,† *Melv. Fl.*, 82. Twickenham, *T. & D.*, 291. Hanworth, *De Cresp. Fl.*, 117. **4.** Hendon, *Irv. MSS.* Harrow Park lake!, *Melv. Fl.*, 82. Golders Green, 1853† (HE). Edgware!, *Druce Notes.* Whitchurch Common, in flower, 1906, *L. B. Hall* and *C. S. Nicholson, K. & L.*, 282. Golders Hill Park. **5.** Near Kew Bridge,† *Newbould, T. & D.*, 291. Syon Park!, *L. G.* and *R. M. Payne*; near Southall!,† *K. & L.*, 282. Backwater of Brent below Hanwell Church. **6.** Near Whetstone, *T. & D.*, 291. Near Ponders End, *De Cresp. Fl.*, 110. Highgate, *French, Trim. MSS.* Town Park, Enfield!, *L.N.* 29: 12. Pond, Crew's Hill golf course, Enfield, 1965!, *Castell* and *Kennedy*. Hadley Green. Near East Barnet, 1965, *Bangerter* and *Kennedy*!. Trent Park, *Castell* and *Kennedy*. **7.** Tottenham, *W. G. Smith, T. & D.*, 291. Ken Wood, *Whiting Fl.* Viaduct Pond, East Heath, Hampstead, *Hampst.* Isle of Dogs,† *E. Newman, Trim. MSS.* Notting Hill Marshes, 1853† (SY). Clapton, *C. S. Nicholson, K. & L.*, 282.

L. minor L. Small Duckweed, Duck's-meat.
Lit. *Proc. Bot. Soc. Brit. Isles* **7**: 447 (1968). (*Abstr.*)
Native. Floating on still waters of lakes, ponds, ditches, streams, canals, etc. Common.
First certain record: *Irvine*, c. 1830.
7. Common in the district.
Plants were noted in flower at **4.** Whitchurch Common, 1904, *L. B. Hall* and *C. S. Nicholson, K. & L.*, 281.

L. gibba L. Gibbous Duckweed.
Lit. *Proc. Bot. Soc. Brit. Isles* **7**: 447 (1968). (*Abstr.*)
Native. Floating on still waters of ponds, lakes, ditches, streams, canals, etc. Locally abundant, but apparently varying in quantity from year to year.
First record: *Varenne*, 1830.
1. Common in the Colne valley. Eastcote, *Farrar*; Potters Bar, *T. & D.*, 292. Hillingdon,† *Trim. MSS.* Ruislip!, *Wrighton*. Stanmore Common. **2.** Common in the Colne valley. Shortwood Common!; Hampton, *T. & D.*, 292. Poyle!, *Druce Notes.* East Bedfont!, *Welch, K. & L.*, 232. Longford river, Stanwell!, *Lousley* (L). Shepperton. Hampton Court Park. Bushy Park. **3.** Roxeth,† *Hind Fl.* Crane on Hounslow Heath; Whitton, abundant; Twickenham; Isleworth, *T. & D.*, 292. Hayes Park. **4.** Near The Hyde, Hendon, *Irv. MSS.* Harrow Weald, *T. & D.*, 292. Golders Hill Park, *H. C. Harris*!, *K. & L.*, 292. **5.** Between Harrow and Greenford,† *T. & D.*, 292. Syon Park!, *L. G.* and *R. M. Payne, K. & L.*, 282. Alperton, *McLean*. Near Brentford. **6.** Town Park, Enfield!, *L.N.* 29: 12. Myddleton House

and Forty Hall grounds, Enfield, *Kennedy*!. Hadley Green!, *K. & L.*, 282. **7.** Serpentine, Kensington Gardens, 1819,† *J. F. Young* (BM). Hackney Wick, abundant, 1867, *T. & D.*, 292. Hackney Marshes, *Druce Notes*. Shepherds Well Fields, Hampstead, 1869,† *T. & D.*, 425. Notting Hill Marshes, 1852† (SY). Ditches at Clapton,† *C. S. Nicholson*, *K. & L.*, 282.

WOLFFIA Hork. ex Schleid.

W. arrhiza (L.) Hork. ex Wimm.
Lemna arrhiza L.
Native. Floating, sometimes below the surface of the water, in still waters of ponds and ditches, etc. Very rare, and confined to the vicinity of the Thames.
First evidence (and as a British plant): *Trimen*, 1866.
2. Piece of water which probably communicates with the Thames, but looks like a pond, near the railway bridge on Shortwood Common, 1866, *Trimen*; 1867, *Newbould* (BM); still abundant there, 1868, *T. & D.*, 293. Stagnant pool, Sunbury, 1869, *Watson*, *Trim. MSS.* Pond on Knowle Green, 1884, *Benbow* (BM). Hampton Court Park, *C. E. Britton*; 1931, *Redgrove* (BM); 1940, *Pugsley*. West Bedfont, *C. E. Britton*. **5.** Pond by canal between Hanwell and Brentford, 1878,† *G. Nicholson* (BM).
This species, the smallest British flowering plant, is easily overlooked as it sometimes occurs in association with large colonies of *Lemna* species. Although not recently seen in Middlesex, it probably still survives in the Staines and Hampton Court areas.

SPARGANIACEAE

SPARGANIUM L.

Lit. *Watsonia* **5**: 1 (1961): *J. Ecol.* **50**: 247 (1962).

S. erectum L. subsp. **erectum** Bur-reed.
S. ramosum Huds.
Native on mud or in shallow water in ponds, lakes, ditches, streams, etc., established in gravel pits, canals, etc. Common.
First record: *T. Johnson*, 1638.
7. Tottenham!, *Johns. Cat.* Isle of Dogs, abundant, *Curt. F.L.* Near Hornsey, 1815† (G & R). Stream near White House, Temple Mills, *Cherry*; Hackney Wick!; side of ponds in Ken Wood grounds!, *T. & D.*, 290. Ponds at Copenhagen House† and Hornsey Wood,† *Ball. MSS.* Notting Hill Marshes, 1852† (SY). Highgate Ponds!, *Fitter*.

Subsp. **neglectum** (Beeby) Schinz & Thell.
S. neglectum Beeby
1. Frequent in the Colne valley. Lake in Hillingdon Park, 1891,† *Benbow* (BM). Ruislip Reservoir!, *Welch.* Stanmore Common. **2.** Common in the Colne valley. **4.** Brent at Hendon,† *C. S. Nicholson, Lond. Nat. MSS.*

S. emersum Rehm. Unbranched Bur-reed.
S. simplex Huds., *pro parte*
Native in streams, ditches, ponds, lakes, etc., established in gravel pits, canals, etc., often submerged and sterile. Common, especially in highly polluted waters.
First evidence: *Roberts*, c. 1710.
7. About Bow, *Roberts* (SLO). Lee between Clapton and Tottenham, *Francis, Coop. Suppl.*, 12. Sparingly in the old cut of the New river, *Munby, Nat.* 1867, 181. Ditches about Hackney Wick, abundant, *Benbow MSS.* Serpentine, Kensington Gardens, 1815,† *J. F Young* (BM). Clapton, 1837, *E. Ballard* (BM).

TYPHACEAE

TYPHA L.

T. latifolia L. Great Reed-mace, 'Bulrush'.
Native in swamps, lakes, ponds, slow-flowing rivers and streams, etc., established in gravel pits, canals, etc. Common, though often planted for decorative purposes.
First record: *T. Johnson*, 1638.
7. Tottenham!, *Johns. Cat.* Isle of Dogs, *Blackst. MSS.* Blackwall (HE). Notting Hill Marshes, 1852,† *Syme* (BM). Ken Wood Ponds!, *Whiting Fl.* Hampstead Heath Extension!; by railway between Royal Oak and Old Oak Common, 1945,† *Fitter.* Mecklenburgh Square, W.C.1; known for some years in uncompleted foundation of a new building, 1960, *Cramp, L.N.* 41: 21. Lake, Buckingham Palace grounds; exterminated by 1960,† *Fl. B.P.G.* East Heath, Hampstead.
Some of the planted specimens may be referable to the hybrid *T. angustifolia* × *latifolia* (*T.* × *glauca* Godr.), cf. D. E. Allen (1966, *Proc. Bot. Soc. Brit. Isles* **6**: 234).

T. angustifolia L. Lesser Reed-mace.
Native in swamps, lakes, ponds, slow-moving streams, etc., established in gravel pits, canals, etc. Rare.
First record: *Clusius*, 1581.
1. Ruislip Reservoir!, *Melv. Fl.*, 81. Ruislip Common, 1967, *J. A. Moore.* Pinner Hill, 1908,† *Green, Druce Notes.* Yiewsley, *J. E. Cooper*,

K. & L., 280. **2.** Shepperton, *Newbould, T. & D.*, 289. West Drayton, *Druce Notes*. Sunbury, *De Cresp. Fl.*, 74. Near Chertsey Bridge, *Benbow MSS.* Dawley, *J. E. Cooper, K. & L.*, 280. Perry Oaks!; Staines Moor, 1958, *Briggs*. Pond by railway between Staines and Ashford, 1965. Hampton Court Park. **3.** Hounslow Heath,† *Mart. Pl. Cant.*, 71. Cranford, *De Cresp. Fl.*, 117. Northolt, *Montg. MSS.* **4.** Hampstead,† *Irv. MSS.* Stanmore, *Varenne, T. & D.*, 289. Grimsdyke grounds, Harrow Weald!, *K. & L.*, 280. **5.** Canal between Hanwell and Brentford!,† *Hemsley, T. & D.*, 289; persisted until 1955. Osterley Park, 1955. Chiswick House grounds!, *Fitter*. Floor of disused Fox Reservoir, Ealing, 1967, *McLean*. **7.** . . . plentifully, 1581, in a pit by Tyburn Churchyard . . .† *Clus. Rar. Pl.* Shepherds Bush,† *Newbould, T. & D.*, 289. Notting Hill Marshes, 1852† (SY). Ditches by railway between Clapton and Broxbourne,† *De Cresp. Fl.*, 74. Perhaps in Herts. Lake, Buckingham Palace grounds,† *Eastwood.*

CYPERACEAE

ERIOPHORUM L.

E. angustifolium Honck. Common Cotton-grass.
E. polystachion L., *pro parte*
Lit. *J. Ecol.* **42**: 612 (1954).
Native in bogs, on wet heaths, moors, etc. Very rare.
First record: *Gerarde*, 1597.
1. Harefield!, *Blackst. Fasc.*, 39; Harefield Moor!, and meadows adjoining;† meadows and moors below Springwell, abundant,† *Benbow, J.B. (London)* 23: 339; almost eradicated by gravel-digging operations but may still survive in small quantity on one or more of the remaining unworked areas of Harefield Moor. Boggy meadow near Uxbridge, *Benbow, J.B. (London)* 25: 19; c. 1910,† *Druce Notes.* **3.** Heathy ground . . . near Hatton, a single specimen, 1867,† *T. & D.*, 303. **4.** Bog at further end of Hampstead!, *Ger. Hb.*, 27, and numerous later authors; still occurs in some quantity on the West Heath, in the bog below Jack Straw's Castle. **6.** Highgate Park,† *Ger. Hb.*, 27.

Var. **vulgare** Koch
1. Wet meadows, Moor Hall, 1908,† *A. B. Jackson*, fide *Williams, J.B. (London)* 47: 324.

TRICHOPHORUM Pers.

T. cespitosum (L.) Hartm. subsp. **cespitosum** Deer-grass.
Scirpus cespitosus L.

Formerly native on acid moorland but long extinct.
First, and only record: *Blackstone*, 1737.
1. Harefield Moor, plentifully, *Blackst. Fasc.*, 91.

ELEOCHARIS R.Br.

Lit. *J. Ecol.* **37**: 192 (1949): *Proc. Bot. Soc. Brit. Isles* **6**: 384 (1967).

E. acicularis (L.) Roem. & Schult. Slender Spike-rush.
Scirpus acicularis L.
Native on wet, sandy and muddy margins of ponds and lakes, etc. Very rare.
First record: *Doody*, c. 1696.
1. Ruislip Reservoir, abundant!, *Hind*, *T. & D.*, 299. Elstree Reservoir, plentiful, *Webb & Colem.*, 311. Old Sand's Heath, Northwood, abundant, 1890,† *Benbow*; Uxbridge Moor, 1747,† *Blackstone* (BM). Floating in the canal, Uxbridge, 1880,† *Beeby*, *Benn. MSS*. Near Uxbridge,† *Druce Notes*. **2.** Abundant on the margins of the Queen's river, Bushy Park!, *Benbow*, *J.B.* (*London*) 25: 19; 1973. Hampton Court Park!, *Lousley* (L); 1973. Shortwood Common, *Lousley*, *Welch* and *Woodhead*!, *L.N.* 27: 34; 1973. **3.** Hounslow Heath, towards Hampton, *Doody*, *Ray Syn.* 2: 344; c. 1736,† *Blackst. MSS*. **5.** Heathy spot at Greenford, near the Brent,† *Lees*, *T. & D.*, 299.

E. multicaulis (Sm.) Sm. Many-stemmed Spike-rush.
Scirpus multicaulis Sm.
Native in bogs, wet heathy places, etc. Very rare, and almost extinct.
First certain evidence: *Benbow*, 1884.
1. Ruislip Reservoir, *Melv. Fl.*, 84. Probably an error, *T. & D.*, 299; *E. acicularis* probably intended. Harefield;† Springwell Lock, 1884,† *Benbow* (BM). Marsh near Walton Bridge!, 1957, *Lousley* (L); 1960.

E. palustris (L.) Roem. & Schult. Common Spike-rush.
Native on pond and lake margins, muddy and sandy verges of rivers and streams, in marshes, ditches, etc. Common, except in heavily built-up areas where it is rare or extinct.
First record: *Buddle*, c. 1705.
7. Thames near Peterborough House,† *Budd. MSS.* and *Pet. Gram. Conc.*, 208. Kensington, 1817† (G & R). Hackney Canal, *Cherry*; Eel Brook Meadow;† Isle of Dogs, *T. & D.*, 299. Parliament Hill Fields, *Fitter*. Highgate Ponds; East Heath, Hampstead, *H. C. Harris*!. Ken Wood. Verge of lake, Buckingham Palace grounds!, *Eastwood*.
The common Middlesex plant is the subsp. **vulgaris** Walters, but the subsp. **palustris** (subsp. *microcarpa* Walters) was gathered at **5.** Riverside, Chiswick, 1937, *Bangerter* (BM), conf. *Walters*.

SCIRPUS L.

S. maritimus L. Sea Club-rush.

Native in shallow water, on mud banks, etc., by the tidal Thames, elsewhere naturalised by ponds and lakes and established in gravel pits, etc., as an introduction. Sometimes planted for ornamental purposes. Rare.

First record: *Parkinson*, 1640.

1. Ruislip Reservoir, 1961, *Streeter, Moxey List.*; 1962 →, *Pickess.* **3.** Church Ferry, Isleworth,† *B. D. Jackson, Sci. Goss.*, 10: 283. By the Thames, Twickenham, a small patch, 1949, *Boniface, L.N.* 29: 13. **4.** Eastern end of Brent Reservoir!, 1957, *Fitter, L.N.* 37: 57; 1958 →. **5.** Abundant from the Mall, above Hammersmith Bridge to Strand-on-the-Green and Brentford Ferry,† *Benbow, J.B. (London)* 25: 365. Lot's Ait, Brentford, a small patch on a mud bank, 1962 →. Canal backwater, Hanwell, one clump, 1965 →. **6.** Lake in Trent Park, well established, 1958, *A. Vaughan*; 1966, *T. G. Collett* and *Kennedy*. Lake in Scout Park, Wood Green, *Scholey.* **7.** Pentiful in the low marshes beyond Ratcliffe,† *Park. Theat.*, 1265. In the river of Thames, *Ray Syn.*, 147, and later authors; Thames where the water is not salt, and on the edges of the creeks running from it, *Curt. F.L.* Near Blackwall, 1815† (G & R). Isle of Dogs, *Mart. Pl. Cant.*, 65; sparingly, 1866, *T. & D.*, 299. Plentifully in a ditch between Ranelagh and the old Chelsea Waterworks near the neat-houses, 1820,† *Pamplin, T. & D.*, 299. Abundant around the lake, Buckingham Palace grounds!, though no doubt originally planted, *Eastwood.*

Var. **macrostachys** Willd.

7. Hammersmith,† *Lady Davy, Rep. Bot. Soc. & E.C.* 7: 751.

S. sylvaticus L. Wood Club-rush.

Native in marshes, wet places in woods, by streamsides, etc. Formerly locally common, especially in the northern parts of the vice-county and along the course of the Brent, but now very rare.

First record: *T. Johnson*, 1632.

1. Uxbridge Moor,† *T. F. Forster, B.G.*, 399. **4.** Hampstead Heath,† *Johns. Enum.* Meadows between Hampstead Heath and Highgate,† *Merr. Pin.* 33 and 52 and *Pet. Midd.* Turner's Wood, *Irv. MSS.* Bishop's Wood, 1861, *Phyt. N.S.* 6: 50; c. 1869, *T. & D.*, 300; abundant, 1902, *C. S. Nicholson* (LNHS). Little Wood, Hampstead, 1865† (SY). Neasden,† *De Cresp. Fl.*, 65. By the Brent in many places about Stonebridge† and Kingsbury,† *Benbow MSS.* By the Brent near Hendon, 1912–27, *J. E. Cooper, K. & L.*, 288. **5.** Canal at Northolt,† *Hind, Melv. Fl.*, 84. Canal side, Alperton,† *Newbould*; canal near Brentford,† *Hemsley*; by the Brent, Hanwell,† *Warren*; by the Brent, Perivale, *Lees, T. & D.*, 300; 1887;† Greenford,†

Benbow MSS. **6.** Tottenham,† *Johns. Cat.* Pond near Highgate,† *Hill Fl. Brit.*, 33. Southgate, *Woods, B.G.*, 399. The Alder's, Whetstone;† Edmonton,† *T. & D.*, 300. Between Hadley Common and Barnet,† *Benbow MSS.* Whitewebbs Park, Enfield!, *L. M. P. Small, K. & L.*, 288. **7.** Between London and Kentish Town,† *Park. Theat.*, 1173. Near Kentish Town,† *Blackst. MSS.* Between Kentish Town and Highgate, 1840† (HE); between London and Highgate,† *Blackst. Spec.*, 16. Near Pancras Church,† *Newt. MSS.* Ken Wood!, *Redgrove, Rep. Bot. Soc. & E.C.* 6: 751; still grows in quantity in a wide ditch by the East Heath, Hampstead.

BLYSMUS Panz.

B. compressus (L.) Panz. ex Link   Broad Blysmus.
Scirpus planifolius Grimm, *S. caricis* Retz., *S. compressus* (L.) Pers., non Moench
Formerly native in marshy places on calcareous soils but long extinct.
First, and only evidence: *J. F. Young*, c. 1830.
1. Harefield, *J. F. Young* (BM).

SCHOENOPLECTUS (Reichb.) Palla

Lit. *J.B.* (*London*) **69**: 151 (1931).

S. triquetrus (L.) Palla   Triangular Scirpus or Bulrush.
Scirpus triquetrus L.
Formerly native on the muddy banks of the tidal Thames but now extinct.
First record: *How*, 1650.   Last record: *Benbow*, 1887.
5. Abundant from The Mall, above Hammersmith Bridge to Strand-on-the-Green and Brentford Ferry, *Benbow, J.B.* (*London*) 25: 365. **7.** By the Horse-ferry [Westminster], *How MSS.*; c. 1660, *J. Dale, Merr. Pin.*, 67. By the river Thames side, both above and below London, *Ray Cat.*, 179, *Ray Syn.*, 201, etc. Isle of Dogs, copiously; in the Thames by Limehouse, *Doody MSS.* In the Thames between Peterborough House and the horse-ferry, Westminster, *Pet. Midd.*, etc. Millbank, *E. Forster* (K). Banks of the Thames near Chelsea, *Spencer*, 318.

S. lacustris (L.) Palla   Bulrush.
Scirpus lacustris L.
Native in rivers, streams, lakes, ponds, etc., especially where there is abundant silt, established in gravel pits, canals, etc. Locally plentiful.
First record: *Blackstone*, 1737.
1. Harefield!, *Blackst. Fasc.*, 46. Canal between Denham and Moor Locks!;† Colne behind Denham Lock!, *Benbow MSS.* Colne between Iver Bridge and Cowley. Ruislip Reservoir!, *Hind, T. & D.*,

302. **2.** Thames in various places between Staines and Shepperton. Colne in various places between Staines and West Drayton. Thames near Strawberry Hill, *T. & D.*, 302. Gravel pit near Walton Bridge!, *Welch, K. & L.*, 289. Gravel pits, East and West Bedfont. **3.** Crane on Hounslow Heath, *T. & D.*, 302. **4.** Kingsbury, in the Brent,† *T. & D.*, 302. **5.** Brent near Hanwell!, *Warren*; Ealing,† *T. & D.*, 303. Northolt,† *Melv. Fl.*, 84. Brent between Alperton,† Greenford,† Perivale!† and Hanwell!; Thames from Hammersmith to Brentford, very abundant,† *Benbow MSS.* Lake in Gunnersbury Park, Acton!, *Green* (SLBI). **6.** River Lee and its backstreams!, *Benbow MSS.* **7.** River Lee at Temple Mills, *T. & D.*, 302. Bombed, waterlogged basement in the City, 1945,† *Lousley*. Buckingham Palace grounds!, *Eastwood*; probably originally planted.

S. lacustris × triquetrus = S. × carinatus (Sm.) Palla
Scirpus × carinatus Sm.
5. Opposite Barnes, if not higher up the river,† *Newbould, T. & D.*, 301. Abundant from The Mall, above Hammersmith Bridge to Strand-on-the-Green and Brentford Ferry,† *Benbow, J.B.* (*London*) 25: 365; Thames opposite Kew, 1898,† *Parsons* (CYN). **7.** In the Thames by Limehouse,† *Doody MSS.*, *Budd. MSS.*, etc. Thames near Millbank, Westminster,† *Boott* (BM). Almost certainly occurs about Hammersmith† and Fulham,† *T. & D.*, 301.

S. tabernaemontani (C. C. Gmel.) Palla Glaucous Bulrush.
Scirpus tabernaemontani C. C. Gmel.
Denizen or native in rivers and lakes. Very rare.
First record: *Doody*, c. 1705.
5. Side of lake, Syon Park, 1955!, *Sandwith, L.N.* 35: 5; 1967. **7.** In a pond of breach a little beyond Limehouse,† *Doody MSS.*

ISOLEPIS R.Br.

I. setacea (L.) R.Br. Bristle Scirpus, or Club-rush.
Scirpus setaceus L.
Native in damp heathy places, on bare sandy or gravelly margins of ponds, etc. Very rare.
First record: *Petiver*, 1695.
1. Harefield Moor,† *Blackst. Fasc.*, 47. Harefield Common!,† *T. & D.*, 302. Pinner,† *W. C. R. Watson, K. & L.*, 290. Ruislip Common, 1967, *J. A. Moore*. Hampton Court Park!, *Welch, K. & L.*, 290; 1968. **3.** Hounslow Heath, *Blackst. MSS.* **4.** About Highgate† and Hampstead,† *Pet. Midd.* Hampstead Heath, 1863,† *Parsons* (CYN). **6.** Highgate,† *Coop. Fl.*, 104. **7.** Beyond Stepney† (DILL). East Heath, Hampstead, *Irv. Lond. Fl.*, 89; 1866,† *Trimen* (BM).

ELEOGITON Link

E. fluitans (L.) Link Floating Scirpus, or Club-rush.
Scirpus fluitans L.
First record: *Blackstone*, 1737.
1. Harefield Common, plentifully,† *Blackst. Fasc.*, 38. Pond near Uxbridge Common,† *Benbow* (BM). **3.** Hounslow Heath, *Huds. Fl. Angl.* 2: 18; 1867,† *Trimen* (BM). **4.** Whitchurch Common, 1827–30, *Varenne, T. & D.*, 302; 1912,† *P. H. Cooke* (LNHS). **6.** Finchley Common!, *Irv. Lond. Fl.*, 89. **7.** Bombed site near Holborn, 1950,† *Whittaker* (BM).

CYPERUS L.

C. longus L. Galingale.
Introduced. Naturalised by lakes and ponds, in damp grassy places, etc., though sometimes planted for ornamental purposes. Very rare.
First record: *L. G.* and *R. M. Payne*, 1947.
1. Damp grassland, Haste Hill, Northwood!, 1962; increasing and spreading on to Northwood golf course, 1966, *J. A. Moore*. New Years Green, 1966, *Clement*. **2.** Long Water, Hampton Court, a large well-established patch!, 1954, *Welch, L.N.* 34: 6; 1967. **3.** Hounslow Heath, a very pale form, 1968, *L. M. P. Small*, conf. *S. S. Hooper*. **4.** Verge of lake, North London Collegiate School, Edgware!, 1966, *Horder*; 1967. **5.** Verge of lake, Syon Park, 1947!, *L. G.* and *R. M. Payne, L.N.* 27: 34; 1967. **7.** Verge of lake, Buckingham Palace grounds!, *Eastwood.*
Gunther, 296, refers to this species, with some doubt, John Dale's record of *Juncus caule Triangulati* . . . at the horse-ferry. Dale's plant was undoubtedly *Schoenoplectus triquetrus.*

C. fuscus L. Brown Cyperus.
Lit. *J.B.* (*London*) **9**: 148 and 212 (1871): *L.N.* **37**: 180 (1958).
Native in damp places besides ponds and lakes, particularly in areas submerged during the winter, wet marshy meadows, etc. Very rare, and most erratic both in occurrence and quantity from year to year.
First evidence: *Haworth*, 1819.
2. Wet muddy ground by a pond at Staines, plentiful, 1957!, *Dupree, L.N.* 37: 180; abundant, 1959!, *Boniface, L.N.* 39: 39; one plant 1960, *T. G. Collett*; apparently not seen between 1961 and 1963; reappeared in small quantity in 1964; apparently not seen between 1965 and 1969; 30+ plants, 1971; 600+ plants, 1972; abundant, 1973. **7.** Fulham fields (= Eel Brook Common or Meadow), 1819, *Haworth* (CYN); 1828 and 1830, *Borrer* (BO); 1831, *J. J. Bennett* (CYN); 1832, *Borrer* (BO); 1833, *Kingsley* (ME); 1850, *J. A. Brewer*; 1852 and 1853,

Syme (SY); a single plant, 1956, *A. Irvine, T. & D.*, 298; about a score of small plants, 1862, *Trimen* (BM); c. 1865;† 'the plant has probably been seen there every year since its discovery up to 1865, though in 1836 it was said to be "extinct" (*Hooker*, in Smith *Compendium*, 14), but has varied in abundance. The meadow was drained in 1864 or 1865, and its present dry state prevents damp loving plants from growing there; the spot will soon be built over', *T. & D.*, 298; 'I believe now destroyed by the ground being drained and built over', *Syme, E.B.* 10: 41.

C. rotundus L. Tiger Nut.
Alien. Tropics. Naturalised among grass. Very rare.
First record: *D. H. Kent*, 1959.
7. Primrose Hill, a large well-established patch, 1959!, *L.N.* 39: 40; increasing, 1961; not seen since as the area is regularly mown, but the species possibly survives in a dormant state.

CAREX L.

Lit. Jermy, A. C. and Tutin, T. G. *British Sedges*. London (1968).

C. laevigata Sm. Smooth Sedge.
C. helodes Link
Formerly native on damp heaths but long extinct.
First, and only record: *Varenne*, c. 1827.
1. Stanmore Common, c. 1827–30, *Varenne, T. & D.*, 309.

C. binervis Sm. Green-ribbed Sedge, Moor Sedge.
Native on heaths and commons. Very rare, and perhaps extinct.
First record: *A. Irvine*, c. 1830.
1. Ruislip Common, sparingly;† Harrow Weald Common, *Melv. Fl.*, 85; 1866,† *Trimen* (BM). Stanmore Common!, *T. & D.*, 309; a single clump, 1947. Harefield Common, very sparingly,† *Benbow, J.B.* (*London*) 23: 28. **4.** Hampstead Heath, *Irv. MSS.*; very sparingly, c. 1887, *Benbow MSS.*; 1903,† *L. B. Hall, K. & L.*, 296.

C. demissa Hornem.
C. tumidicarpa Anderss., *C. flava* auct., non L.
Native in damp grassy places, woodland rides, etc. Very rare, and mainly confined to acid soils.
First evidence: *Goodger* and *Rozea*, 1817.
1. Stanmore Common!, *Trimen* (BM). Harefield Common,† *T. & D.*, 309. Ruislip Common!; Harefield Moor, sparingly!; Duck's Hill Farm, Northwood,† *Benbow, J.B.* (*London*) 23: 38. Ickenham Green!,† det. *Nelmes, L.N.* 27: 34. **2.** Staines Moor!, *Whale Fl.*, 9. Near Staines, 1883, *H.* and *J. Groves* (BM). **4.** West Heath, Hampstead, *Irv. MSS.*; 1866,† *Trimen* (BM). **7.** Hyde Park, 1817† (G & R).

East Heath, Hampstead,† *T. & D.*, 304. Buckingham Palace grounds, *Fl. B.P.G.* Introduced.

C. sylvatica Huds. Wood Sedge.
Native in damp woods, etc. Locally plentiful, especially on heavy soils in the north of the vice-county.
First record: *Petiver*, 1695.
1. Locally common. **2.** Bedfont, *Whale Fl.*, 9. An error. **3.** Harrow Grove, *Horton Rep.* An error. **4.** Harrow Park!; Bentley Priory!, *Melv. Fl.*, 85. Between Finchley and Hendon, *Newbould*; Bishop's Wood!; Scratch Wood!, *T. & D.*, 309. Barnet Gate Wood!, *Benbow MSS.* Temple Fortune, 1914,† *J. E. Cooper.* **5.** Between Acton and Ealing, 1835,† *J. F. Young* (BM). Perivale Wood!, *Benbow MSS.* Horsendon Wood,† *Trim. MSS.* Long Wood, Wyke Green. **6.** Whetstone!; Enfield Chase!, *Benbow MSS.* Colney Hatch, 1841† (HE). Between Wood Green and Colney Hatch,† *Munby, Nat.* 1867, 181. Winchmore Hill Wood,† *T. & D.*, 310. Highgate Woods!, *J. E. Cooper.* Hadley. **7.** Plentifully in a wood against the Boarded river,† *Pet. Midd.*

C. pseudocyperus L. Cyperus Sedge.
Native. Verges of lakes, ponds and streams, etc., often in woods or partial shade, established in gravel pits, etc. Sometimes planted for decorative purposes. Locally plentiful.
First record: *T. Johnson*, 1638.
1. Near Uxbridge,† *Forst. MSS.* Near Harefield Moor, very sparingly,† *Benbow, J.B. (London)* 23: 38. Pond near Stanmore Common!, *Boniface, K. & L.*, 294. Dyrham Park, South Mimms, 1953, *Lansbury, K. & L.*, 294; 1965, *J. G.* and *C. M. Dony.* **2.** Locally common. **3.** Hounslow Heath,† *Rudge* (BM). Twickenham,† *T. & D.*, 310. Hayes, 1938,† *McClintock.* Feltham, 1966. **4.** Stanmore (G & R); Bentley Priory!, *Green* (SLBI). Whitchurch Common, 1917, *L. B. Hall*; 1936,† *P. H. Cooke*; Edgware!, *J. E. Cooper*; Scratch Wood!, *K. & L.*, 294. Near Edgware!, *De Cresp. Fl.*, 13. Kingsbury,† *Benbow, J.B. (London)* 25: 19. Finchley, scarce,† *Benbow MSS.* **5.** Canal between Hanwell and Brentford, *Benbow MSS.*; 1919,† *A. B. Jackson* and *Sprague* (SLBI). Osterley Park, frequent!, *Hubbard*; 1970. **6.** Tottenham,† *Johns. Cat.* Between Highgate and Muswell Hill,† *Irv. Lond. Fl.*, 92. Bounds Green, 1848 (HE); 1867,† *Munby, Nat.* 1867, 180. Edmonton, *T. & D.*, 310. Between Edmonton and Waltham Cross, *Dyer* (BM). Finchley Common, 1920;† Clay Hill, Enfield!, *J. E. Cooper*; Whitewebbs Park!, *Welch, K. & L.*, 294. Frith Manor, Finchley, 1925,† *Moring MSS.* Trent Park, Enfield, *A. Vaughan.* Cockfosters, *Mason.* Near East Barnet, *Bangerter* and *Kennedy*!. **7.** Between the Boarded river and Islington Road,† *Pet. Midd.* Stoke Newington, common,† *Woods*; Hackney,† *E. Forster*,

B.G., 411. Near Highgate Cemetery, 1842,† *E. Edwards, Phyt.* 1: 428. Beyond Hornsey Wood,† *Ball. MSS.* Notting Hill Marshes, 1854 (SY); 1856,† *Syme* (W). Green Lanes, 1859,† *E. Ballard* (BM).

C. rostrata Stokes Beaked, or Bottle Sedge.
C. ampullacea Gooden., *C. inflata* auct., non Huds.
Formerly native in wet boggy places but long extinct.
First, and only record: *E. Forster*, 1792.
2. Marshy meadow near Teddington, 1792, *Forst. MSS.*

C. vesicaria L. Bladder Sedge.
Native by ponds and streamsides, in damp woods, etc. Very rare.
First record: *J. E. Smith*, 1800.
1. Elstree Reservoir,† *Webb. & Colem.*, 322. Boggy ground near Ruislip Reservoir!, *Hind, Melv. Fl.*, 86; c. 1887, *Benbow MSS.*; 1909, *Green* (SLBI); 1945, *Welch*; 1949!, *K. & L.*, 295; 1960. Plentiful round the pond at Knightscote Farm, Harefield, 1956!, *Pickess*; 1962. Dyrham Park, South Mimms, 1965, *J. G.* and *C. M. Dony*. Potters Bar, 1970, *Widgery*. **4.** Brent near Willesden, 1842† (HE). By the Brent in various places to Alperton,† *Benbow MSS.* **5.** By the Brent near Hanwell,† *Hemsley* and *Warren, T. & D.*, 310. By the Brent from Vicar's Bridge to Perivale,† Greenford† and Hanwell, abundant;† pond near Greenford;† canal between Brentford and Hanwell,† *Benbow MSS.* **6.** Highgate Woods, 1905,† *C. S. Nicholson* (LNHS); now destroyed, *C. S. Nicholson, Trans. L.N.H.S.*, 1915, 42. **7.** Near the water works at Pimlico and elsewhere about the Thames,† *J. E. Smith, E.B.* 11: 779.

C. riparia Curt. Great Pond-sedge.
Native by sides of slow-flowing rivers and streams, on verges of lakes and ponds, etc., established by canals, in gravel pits, etc. Common.
First record: *Blackstone*, 1737.
7. Marylebone Fields, 1817† (G & R). Isle of Dogs!, *T. & D.*, 311. Lee Navigation Canal between Clapton and Tottenham. Hackney Marshes!, 1871, *F. W. Payne* (SLBI); 1967. East Heath, Hampstead.

Carex acutiformis Ehrh. Lesser Pond-sedge.
C. paludosa Gooden.
Native in similar situations to the previous species, with which it often grows. Common.
First record: *Rand*, 1716.
7. . . . in the ditches at the King's Arms against Whitehall,† *Rand, Pet. Gram. Conc.*, 159 and *Ray Syn.* 3: 418. By the Lee between Clapton and Tottenham!, *Francis, Coop. Suppl.*, 12. Hackney Canal about Hackney Wick!, Old Ford, etc., *Benbow MSS.* Clapton!, 1884, *F. J.*

Hanbury (SY). Kensal Green, 1864, *Parsons* (CYN). Lake, Buckingham Palace grounds!, planted, *Eastwood.*

C. pendula Huds. Pendulous Sedge.
Native in damp woods, shady places, marshes, on pond verges, etc. Locally common, especially on clay soils in the north of the vice-county, less common elsewhere, and rare in the vicinity of the Thames. Often planted by ponds and lakes for decorative purposes.
First record: *Merrett*, 1666.
1. Locally common, especially in the north-eastern parts of the district. **2.** Teddington, 1965, almost certainly planted. **3.** Harrow Grove!, *Melv. Fl.*, 85. Headstone,† *Benbow MSS.* **4.** Locally common, except in heavily built-up areas. **5.** Alperton,† *Melv. Fl.*, 85. Perivale Wood!, *Green* (SLBI). Syon Park, *Welch*!. **6.** Alder's Copse, Whetstone;† Colney Hatch, *T. & D.*, 307. Enfield!, *T. F. Forster*, *B.G.*, 411. Near Highgate, *Newt. MSS.* Highgate Woods, *C. S. Nicholson*, *Lond. Nat. MSS.* Hornsey, 1844,† *S. P. Woodward* (CYN). **7.** Primrose Hill,† *Merr. Pin.*, 25. Between Marylebone and Kilburn,† *Huds. Fl. Angl.* 2, 44. Marylebone Fields, 1827–30,† *Varenne*; Millfield Lane, Highgate, 1867,† *T. & D.*, 367. Newington, 1838,† *E. Ballard* (BM). Hornsey Wood,† *Ball. MSS.* Between Pancras and Kentish Town,† *M. & G.*, 81. Between Stoke Newington and Hornsey,† *Woods*, *B.G.*, 411. Tottenham Wood,† *Dill. MSS.* Between Clapton and Tottenham,† *Francis*, *Coop. Suppl.*, 12. Isle of Dogs,† *Coop. Fl.*, 115. Between Kentish Town and South End, Hampstead, 1841† (HE). East Heath, Hampstead, 1921, *Richards.* Ken Wood. Bombed site, St Olave's, Hart Street, E.C., 1949,† *Welch*!, *L.N.* 29: 14. Island in lake, Buckingham Palace grounds, 1967, *McClintock*, *L.N.* 47: 9.

C. strigosa Huds. Broad-leaved Wood Sedge.
Native in damp woods, especially by brooks, on damp shady hedgebanks, etc. Very rare.
First record: *Hind*, 1864.
1. Moss Lane, Pinner, *Hind.*, *Melv. Fl.*, 85; in some quantity on the hedgebanks on both sides of the lane, 1866, *Trimen*; 1884;† Old Park Wood, Harefield!, 1884, *Benbow* (BM); at least a dozen plants, in two localities, 1966, *Rose* and *P. A. Moxey*, *L.N.* 47: 7, 1967, *Boniface*; 1968 →.

C. pallescens L. Pale Sedge.
Native in damp woods, wet meadows, etc. Very rare.
First record: *Petiver*, 1695.
1. Park Wood, Ruislip!; Duck Wood, Northwood, *Benbow* (BM). Meadow near Bayhurst Wood, abundant, *Benbow*, *J.B.* (*London*) 23: 38. Pinner Wood, scarce, *Benbow MSS.* **4.** Bentley Priory, 1861,†

Hind; Bishop's Wood, 1860,† *Trimen* (BM). East Finchley, 1890,† *J. E. Cooper*, *K. & L.*, 297. **6.** Highgate Wood, 1839 (HE); several plants, 1905,† *C. S. Nicholson* (LNHS). **7.** Plentifully in a wood against the Boarded river,† *Pet. Midd.*

C. filiformis L. Downy Sedge.
C. tomentosa auct.
Native in damp meadows near the Thames. Very rare, and perhaps extinct.
First evidence: *I. A. Williams*, 1928.
2. Between Shepperton and Chertsey!, *I. A. Williams* (BM); 1944, *Lousley* (L); 1950!, *K. & L.*, 217; 1960; the area has now been much disturbed by gravel-digging operations but it is possible that the species survives on one or more of the 'islets' which remain.

C. panicea L. Carnation-grass.
Native in marshes, wet grassy places, on damp heaths, moors, etc. Very rare.
First evidence: *Rudge*, c. 1810.
1. Harefield Common;† Harrow Weald Common!, abundant; Stanmore Common!, *T. & D.*, 307; 1966. Pinner Wood,† *De Cresp. Fl.*, 118. Wet meadows below Ickenham Green, 1884,† *Benbow* (BM). Uxbridge,† *Druce Notes*. Eastcote, *De Cresp. Fl.*, 118. Ruislip Common!, *K. & L.*, 298. Harefield Moor. Northwood, 1967, *J. A. Moore*. **3.** Hounslow Heath, *Rudge* (BM); c. 1869,† *T. & D.*, 307. **4.** Hampstead Heath, 1860, *Trimen* (BM); West Heath, c. 1921, *Richards*, *K. & L.*, 298. Harrow, common,† *Hind Fl.*; football fields, Harrow, *Horton Rep.* Error. Near Scratch Wood. Whitchurch Common,† *De Cresp. Fl.*, 113. **6.** Whetstone; Edmonton,† *T. & D.*, 307. Hadley Green, abundant, 1949, *Payton*!, *K. & L.*, 298.

C. flacca Schreb. Glaucous Sedge, Carnation-grass.
C. glauca Scop., *C. diversicolor* Crantz, *pro parte*
Lit. *J. Ecol.* **44**: 281 (1956).
Native in dry pastures, marshes, woods, on heaths, by streamsides, etc. Locally frequent on calcareous soils about Harefield, on alluvial soils in the Colne valley and near the Thames and in woods on clay soils.
First evidence: *P. Miller*, pre-1730.
1. Locally common in the Colne valley. Stanmore Common!, *Varenne*; Harrow Weald Common; Pinner!; Elstree,† *T. & D.*, 308. Pinner Hill!, *De Cresp. Fl.*, 118. Ickenham Green!; New Years Green!, *K. & L.*, 297. Ruislip Common!, *L.N.* 27: 34. Dyrham Park, South Mimms, *J. G.* and *C. M. Dony*. **2.** Locally common near the Thames. Near Colnbrook!, *Druce Notes*. **4.** Finchley, 1842† (HE). Between Finchley and Hendon,† *Newbould*; Bishop's Wood,

T. & D., 308. Hampstead Heath,† *Irv. MSS.* Stanmore, *De Cresp. Fl.*, 12. Scratch Wood!, *Green* (SLBI). Harrow!, *Hind, Melv. Fl.*, 85; rare!, *Harley Fl.*, 33. Northwick Park, *Temple*!. Mill Hill, *Sennitt.* **5.** Near Brentford,† *Hemsley*; near Twyford,† *Newbould, T. & D.*, 308. **6.** Hadley, *Warren*; near Colney Hatch,† *T. & D.*, 308. Finchley Common, sparingly, 1901,† *C. S. Nicholson, K. & L.*, 297. **7.** On the banks of the New river,† *P. Miller* (DB).

C. hirta L. Hairy, or Hammer Sedge.
Native in rough grassy places, woods, on heaths, waste ground, by roadsides, etc., established on railway banks, canal paths, etc. Common.
First record: *l'Obel*, c. 1600.
7. Near Highgate, *Lob. Stirp. Ill.*, 52. East Heath, Hampstead!; Isle of Dogs, *T. & D.*, 310. Kentish Town, 1838, *E. Ballard* (BM). South Kensington, *Dyer List.* Hyde Park!, *Warren Fl.* Kensington Gardens!, *D. E. Allen*; Buckingham Palace grounds, *McClintock*; Primrose Hill!, *L.N.* 39: 59. St James's Park, *D. E. Allen, L.N.* 41: 21. Ken Wood. Hackney Wick, *L. M. P. Small.*

C. pilulifera L. Pill-headed Sedge.
Native on heaths, in open woods, etc. Very rare, and mainly confined to acid soils.
First record: *Doody*, c. 1688.
1. Harefield Common,† *Blackst. Fasc.*, 36. Harrow Weald Common!, *Melv. Fl.*, 86, now very scarce. Stanmore Common, sparingly!, *T. & D.*, 308; 1966. Northwood, 1967, *J. A. Moore.* Ruislip!; Pinner Wood, *Benbow MSS.* Near Eastcote;† Ruislip Common!, *Benbow* (BM). Mimmshall Wood!, *Green* (SLBI). **2.** Near Staines, a single plant,† *T. & D.*, 308. **3.** Hounslow Heath!,† *T. & D.*, 308. Gravel pits near Heston, a few plants, 1866,† *Trimen* (BM). **4.** Hampstead Heath!, *Doody, Ray Fasc.*, 10. Bishop's Wood, *Benbow MSS.* Grimsdyke grounds, Harrow Weald, *Welch*!, *L.N.* 28: 34. **6.** Highgate Wood, *Benbow MSS.* Queen's Wood, Highgate, 1902, *C. S. Nicholson* (LNHS). **7.** Ken Wood!, *H. C. Harris*!, *K. & L.*, 297. Building site, Westminster,† *Parsons, Proc. Croyd. N.H. & Sci. Soc.* 6: 163.

C. caryophyllea Latourr. Spring Sedge.
C. verna Chaix, non Lam., *C. praecox* auct., non Schreb.
Native in dry pastures, meadows, open woods, on heaths, etc. Locally plentiful on calcareous soils at Harefield, rare elsewhere.
First record: *T. Johnson*, 1632.
1. Harefield!, *Trimen* (BM). Between Harefield and Springwell. Ruislip Common!, *Melv. Fl.*, 86. Woodready, Pinner, *T. & D.*, 307. Roadside by Blackenbury Farm, near Harefield;† Ickenham Green, abundant;† near Pinner,† *Benbow MSS.* Uxbridge Common,†

Trim. MSS. Stanmore Common, 1921, *Richards*; field near Bayhurst Wood, 1928, *B. T. Ward, K. & L.*, 296. Between Ickenham and Harefield, 1966. **2.** Hampton Court Park!, *Welch, K. & L.*, 297; 1967. **3.** Gutteridge Wood, Northolt!, det. *Nelmes, L.N.* 27: 34. **4.** Hampstead Heath, *Johns. Enum.*; c. 1830,† *Irv. MSS.* **5.** Greenford,† *Melv. Fl.*, 86. Syon Park, *Welch*!, *K. & L.*, 297.

C. acuta L. Tufted Sedge.
C. gracilis Curt.
Native by stream- and riversides, in marshes, etc. Rare.
First certain record: *Blackstone*, 1737.
1. Harefield!, *Blackst. Fasc.*, 35. Uxbridge,† *Benbow* (BM). Yiewsley!, *K. & L.*, 298. Springwell. **2.** River bank between Hampton Court and Kingston Bridge!, *Newbould.* Staines Moor!, *T. & D.* 306. West Drayton!,† *K. & L.*, 298. **3.** Isleworth;† Twickenham,† *T. & D.*, 306. **4.** Mill Hill, *Sennitt.* **5.** Brentford!, *Hemsley*; canal, Alperton,† *Newbould*; by the Brent near Hanwell!,† *Warren, T. & D.*, 306. Syon Park. **7.** Lee between Tottenham and Clapton,† *Francis, Coop. Suppl.*, 12. By the lake, Buckingham Palace grounds, *Codrington, Lousley* and *McClintock*!, *L.N.* 36: 15.

C. nigra (L.) Reichard Common Sedge.
C. angustifolia Sm., *C. goodenowii* Gay, *C. vulgaris* Fr., *C. cespitosa* auct., *C. fusca* All.
Native in marshes, bogs, wet grassy places, on heaths, etc. Formerly locally common, but now rare and decreasing, and mostly confined to acid soils.
First record: *A. Irvine*, c. 1830.
1. Harefield!, *Newbould, T. & D.*, 306. Harrow Weald Common; Stanmore Common!, *Trimen* (BM); 1963. Ruislip Moor,† *De Cresp. Fl.*, 118. Ruislip Reservoir!; Harefield Moor!; Springwell!;† Pinner,† *Benbow MSS.* Ickenham Green!, *L.N.* 27: 34. Dyrham Park, South Mimms, *J. G.* and *C. M. Dony.* **2.** Staines Moor, abundant!, *L.N.* 27: 34; now much reduced in quantity. **3.** Whitton Park Inclosure, *T. & D.*, 306; 1910,† *Green* (SLBI). Hounslow Heath.† **4.** West Heath, Hampstead, *Irv. MSS.*; 1837, *E. Ballard*; 1860,† *Trimen* (BM). Harrow,† *Melv. Fl.*, 85. Whitchurch Common,† *Lond. Nat. MSS.* **5.** Syon Park, *Welch*!. **6.** Edmonton, *Trimen* (BM). Hadley Common. Finchley Common!, *L.N.* 27: 34. Hadley Green. **7.** East Heath, Hampstead!, *T. & D.*, 306. Ken Wood.

C. paniculata L. Panicled Sedge.
Native in swamps, marshes, bogs, on pond and lake verges, established by canal sides in gravel pits, etc. Formerly locally plentiful in the upper parts of the Colne valley but now much reduced by the eradication of its habitats by gravel-digging operations. Now very rare in all parts of the vice-county.

First record: *Doody*, c. 1700.

1. In small quantity in a few localities in the Colne valley between Springwell and Yiewsley. **2.** Canal bank between Uxbridge and West Drayton!,† *Warren, J.B.* (*London*) 11: 208. Although the species occurred in this locality Warren's record may have been an error as a specimen collected from here by him in 1873 (BM), and labelled '*C. paniculata*', is *C. appropinquata*. Thames bank near Walton Bridge, a single tussock, 1945!,† *L.N.* 26: 64. By the lake, Shortwood Common, a single small tussock, 1958, *T. G. Collett*!; increasing, 1973. **4.** Near Hampstead Heath, *Doody MSS.*; c. 1789, *Forst. Midd.*; c. 1830,† *Irv. MSS.* Hampstead Wood,† *Huds. Fl. Angl.*, 347. Between Bishop's Wood and Finchley,† *Irv. MSS.* Stonebridge,† *Farrar, T. & D.*, 305. By the Paddington Canal, Willesden, 1867,† *Newbould* (SY). **5.** Canal near Alperton, sparingly, 1867,† *Newbould, T. & D.*, 305. Canal between Greenford and Northolt, one tussock,† *Benbow MSS.* **6.** The Alder's Copse, Whetstone, abundant, 1867,† *Trimen* (BM). **7.** By the New Cut over Lee Bridge, probably accidental,† *E. Forster* (BM).

C. appropinquata Schumach.

C. paradoxa Willd., non J. F. Gmel.

Formerly native in marshes on calcareous humus-rich soils, established in gravel pits, on canal banks, etc., but now extinct.

First evidence: *Warren*, 1873. Last evidence: *Lousley*, 1936.

1. Abundant on the moors below Springwell Lock, 1885, *Benbow* (BM); 1900, *Pugsley*; near Springwell Lock, 1936, *Lousley* (L); habitats destroyed by gravel-digging operations shortly before the outbreak of the Second World War. Meadows between Denham Lock and Harefield Moor, *Benbow, J.B.* (*London*) 23: 339. Near Denham, *Druce Notes.* Uxbridge Moor, 1910, *Green* (SLBI). Canal side between Cowley and Yiewsley, *Benbow MSS.* **2.** Canal at West Drayton, 1873, *Warren* (BM); 1885, *Benbow, J.B.* (*London*) 23: 339.

C. appropinquata × paniculata = C. × solstitialis Figert

1. Springwell,† *Benbow* (BM).

C. otrubae Podp. False Fox-sedge.

C. vulpina auct. occid., non L., *C. vulpina* L. subsp. *nemorosa* (Rebentisch) Schinz & Keller

Native in marshes, ditches, on pond verges, by river- and streamsides, etc., established by canals in gravel pits, etc. Common, except in heavily built-up areas where it is very rare or extinct.

First record: *T. Johnson*, 1632.

7. Between Clapton and Tottenham!, *Francis, Coop. Suppl.*, 12. Near Lee Bridge, *Cherry*; Paddington Canal!, 1869, *Warren, T. & D.*, 304; near Kensal Green. Kentish Town, 1846 (HE). Kensington Gardens,† *Warren, J.B.* (*London*) 13: 336.

C. otrubae × remota = C. × pseudoaxillaris K. Richt.
C. axillaris Gooden., non L.
1. Lane below Uxbridge Common, abundant;† between Ickenham and Hillingdon,† *Benbow*, *J.B.* (*London*) 23: 38. Near Ruislip Common;† Ruislip,† *Benbow MSS*. Near Uxbridge, 1892,† *Wolley-Dod* (BM). Ickenham Park, 1909,† *Green* (SLBI). **2.** Canal side, West Drayton, 1873, *Warren* (BM); c. 1887,† *Benbow MSS*. Stanwell Moor, 1908, *Green* and *A. B. Jackson* (SLBI). **4.** Near Harrow Weald, 1866,† *Trimen*; abundant in a lane near Edgware, towards Edgwarebury, 1887,† *Benbow* (BM). **5.** Near Northolt,† *Benbow*, *J.B.* (*London*) 25: 365. **6.** Tottenham,† *Cockfield Cat.*, 16.

C. disticha Huds. Brown Sedge.
Native in marshes, wet grassy places, on moors, etc. Formerly local in the Colne valley and near the Thames, but now very rare.
First evidence: *E. Forster*, 1792.
1. Harefield!, *Newbould*, *T. & D.*, 303; Harefield Moor!; Moor Hall,† *Benbow MSS*. Drayton Ford to Springwell!,† *Benbow* (BM). Near Uxbridge,† *Benbow MSS*. Uxbridge,† *Druce Notes*. Ruislip Moor,† *T. & D.*, 303; Ruislip Reservoir,† *De Cresp. Fl.*, 12. **2.** Near Teddington, 1792,† *E. Forster* (BM). Staines Moor!, *Benbow MSS*. West of Shepperton, *Welch*; Yeoveney!, *K. & L.*, 300. Near Walton Bridge.† Near West Drayton,† *Benbow MSS*. West Drayton,† *Druce Notes*. **4.** Hampstead,† *Irv. Lond. Fl.*, 90. Mill Hill,† *Newbould* (BM). **5.** Near the canal, Brentford,† *Hemsley*, *T. & D.*, 303. Syon Park, *Welch*!, *K. & L.*, 300. **6.** Edmonton,† *T. & D.*, 303. Marsh north of Enfield Lock, 1967.

C. divisa Huds. Divided Sedge.
Formerly native in ditches and marshes, chiefly near the tidal Thames, but now extinct.
First record: *Dillwyn*, 1805. Last evidence: *Benbow*, 1887.
4. Marshy place by a pond between Fortune Green and Hampstead, two or three tufts, 1878, *De Crespigny* (BM). **7.** Isle of Dogs, plentifully, *Dillwyn*, *B.G.*, 411; c. 1835, *Coop. Fl.*, 115; southern end of Millwall Docks, 1887, *Benbow* (BM).

C. divulsa Stokes Grey Sedge.
Lit. *Rep. Bot. Soc. & E.C.* **13**: 102 (1947).
Native on hedgebanks, in ditches, open woods, by roadsides, etc. Locally plentiful, especially on light soils near the Thames.
First record: *Blackstone*, 1737.
1. Locally common. **2.** Locally common, especially near the Thames. **3.** Roxeth,† *Melv. Fl.*, 84. Hounslow Heath, 1846† (BM). **4.** Harrow Weald; Deacon's Hill!, *T. & D.*, 304. Kingsbury,† *De Cresp. Fl.*, 112. Between Hendon and Finchley, *Newbould*, *T. & D.*,

304. Canal banks near Willesden Junction!,† *K. & L.*, 300. **5.** Twyford,† *Newbould, T. & D.*, 304. Osterley Park, *Welch*!. **6.** Between Wood Green and Colney Hatch,† *Munby, Nat.* 1867, 171. Whetstone; Enfield, *T. & D.*, 304; near Enfield Wash, *Kennedy*. Hadley Common!, *C. S. Nicholson* (LNHS). **7.** Grounds of the Natural History Museum, South Kensington, 1906,† *Sherrin* (SLBI). Millfield Lane, Highgate, 1924,† *Tremayne, K. & L.*, 300. Bank of Thames, just east of Chelsea Bridge, 1965, *Burton, L.N.* 45: 20. St John's Churchyard, Marylebone, 1966, *Holme, L.N.* 46: 34.

C. spicata Huds. Spiked Sedge.
C. contigua Hoppe, *C. muricata* auct. angl., *pro parte*, non L.
Lit. *Rep. Bot. Soc. & E.C.* **13**: 104 (1947).
Native in rough grassy places, on hedgebanks, by roadsides, etc., established on railway banks, in cemeteries, etc. Common, especially on heavy soils.
First certain evidence: *Groult*, c. 1800.
7. Hyde Park, *Groult* (K). Kensington Gardens!, 1871, *Warren* (BM); 1961!, *McClintock*. Bombed site, St Olave's, Hart Street, E.C., 1945!,† *Lousley* (BM, K, L). Bombed site, Ebury Street, S.W.1,† *McClintock*. Regents Park!, *L.N.* 39: 59; bank by Zoological Gardens, 1966, *Brewis, L.N.* 46: 34. Thames bank near Hammersmith, 1947. Kensal Green Cemetery, abundant. Fulham Cemetery. Brompton Cemetery. East Heath, Hampstead. Stoke Newington, 1972, *Kennedy*.

C. muricata L. Prickly Sedge.
C. pairaei F. W. Schultz, *C. muricata* subsp. *pairaei* (F. W. Schultz) Aschers. & Graebn.
Lit. *Rep. Bot. Soc. & E.C.* **13**: 103 (1947).
Native on hedgebanks, in meadows, dry places, by roadsides, etc. Locally plentiful on sandy and gravelly soils near the Thames, rather rare elsewhere.
First certain record: *D. H. Kent*, 1943.
1. Between Harefield and Ickenham!; near Elstree!, *L.N.* 26: 64. Near Pield Heath!; Ruislip!, *K. & L.*, 299. Northwood. **2.** Hampton Court, frequent!, *Welch*; Stanwell Moor!; road banks between Stanwell and Staines, abundant, *Welch*!, *L.N.* 26: 64. Bushy Park, abundant!; near Sunbury!, *Welch, K. & L.*, 299. **3.** Hounslow Heath, *Hubbard*!; near Colham Green!, *K. & L.*, 300. **4.** Harrow, *Harley Fl.*, 33. **5.** River wall, Duke's Meadows, Chiswick!, *Welch*; between Strand-on-the-Green and Great Chertsey Road, Chiswick!;† Hanwell, 1943!,† *L.N.* 26: 64. Between Hanwell and Osterley Park. **6.** Near Whitewebbs Park, Enfield, *Kennedy*!. **7.** Bird Sanctuary, Hyde Park, 1959!, *D. E. Allen*, conf. *Jermy, L.N.* 39: 59. Bennett's Yard, Victoria, S.W.1, *Rönaasen, L.N.* 44: 27. Hyde Park, traffic island north of Apsley House, 1961–62!,† *Allen Fl.*

Subsp. **leersii** (F. W. Schultz) Aschers. & Graebn.
C. leersii F. W. Schultz, non Willd., *C. polyphylla* auct., non Kar. & Kir.
Lit. *Rep. Bot. Soc. & E.C.* **13**: 101 (1947).
1. Above Harefield chalkpits!, det. *Nelmes* (as *C. polyphylla*), *L.N.* 26: 64. Near Jack's Lock, Harefield, 1955. Springwell, 1961. Northwood, 1965.

C. echinata Murr. Star Sedge.
C. stellulata Gooden.
Native in bogs, on damp heaths, etc. Very rare, and perhaps extinct.
First record: *Petiver*, 1695.
1. Harrow Weald Common!, abundant, *Melv. Fl.*, 84. Stanmore Common!; South Mimms, *T. & D.*, 305. Ickenham Green, 1890,† *Benbow* (BM). Harefield Moor; Harefield Common, sparingly;† Duck's Hill, Northwood,† *Benbow*, *J.B.* (*London*) 23: 38. Between Ruislip and Harefield, 1927,† *Tremayne*, *K. & L.*, 298. **3.** Hounslow Heath, *Rudge* (BM); c. 1950,† *Westrup*, *K. & L.*, 298. **4.** Woods about Highgate† and Hampstead,† *Pet. Midd.* West Heath, Hampstead, 1860,† *Trimen*; Stanmore, 1837,† *D. Cooper* (BM). **7.** East Heath, Hampstead,† *T. & D.*, 305. Near Copenhagen House, *Ball. MSS.*

C. remota L. Remote-spiked Sedge.
Native in ditches, damp shady places, on pond verges, etc. Common, except in heavily built-up areas where it is rare or extinct.
First record: *Blackstone*, 1737.
7. Between Clapton and Tottenham, *Francis*, *Coop. Suppl.*, 12. Clapton, 1837, *E. Ballard* (BM). Ken Wood grounds, *Fitter*. Brompton Cemetery. Albert Gate, Hyde Park, 1962, *Brewis*, *L.N.* 44: 27.

C. ovalis Gooden. Oval Sedge.
C. leporina auct., non L.
Native on heaths and commons, in rough grassy places, marshes, etc. Locally plentiful and mainly confined to acid soils.
First record: *Petiver*, 1695.
1. Near Uxbridge!, *Benbow*; Stanmore Common!, *Trimen* (BM). Ruislip. Harefield!; Harrow Weald Common, *Green* (SLBI). Uxbridge Common. Ickenham Green. Springwell. Wrotham Park!; Dyrham Park, South Mimms, *J. G.* and *C. M. Dony*. **2.** Near Staines!, *Trimen* (BM). Staines Moor. Bushy Park. Hampton Court Park. Teddington. **3.** Near Hatton, *Trimen*; Headstone, 1918,† *Redgrave* (BM). Hounslow Heath, abundant. **4.** Hampstead Heath!, *De Cresp. Fl.*, 12; still on the West Heath. Scratch Wood, *Green* (SLBI). Bentley Priory. Brockley Hill, *Kennedy*!. Mill Hill, *Sennitt*. **5.** Osterley Park, sparingly, *Welch*!. Wyke Green.† Horsendon Hill, sparingly. **6.** Southgate, *L. B. Hall*, *Lond. Nat. MSS.*; Trent Park!,

Kennedy. Finchley Common. Near East Barnet, *Bangerter* and *Kennedy*!. Edmonton, *Kennedy*!. **7.** Plentiful . . . near the Boarded river,† *Pet. Midd.* Kensington Gardens!, *E.B.* Hyde Park!, *D. E. Allen*, *L.N.* 41: 21. East Heath, Hampstead!, *Trimen* (BM). Buckingham Palace grounds, one plant, *Fl. B.P.G.*

C. pulicaris L. Flea-sedge.
Native in bogs and wet places on heaths, wet meadows on acid soils, etc. Very rare, and perhaps extinct.
First record: *Blackstone*, 1737.
1. Harefield Moor, plentifully,† *Blackst. Fasc.*, 35. Harrow Weald Common, in several places, 1866,† *Trimen* (BM). Stanmore Common!, 1921, *Richards*, *K. & L.*, 301; 1946!, *L.N.* 27: 34. Wet meadow near Bayhurst Wood,† *Benbow*, *J.B.* (*London*) 23: 38. **4.** Hampstead Heath, *Huds. Fl. Angl.*, 347; c. 1837,† *Irv. Lond. Fl.*, 90.

C. dioica L.
Formerly native in wet meadows on base-rich soils, but long extinct.
First, and only record: *E. Forster*, 1792.
2. Marshy meadow near Teddington, 1792, *Forst. MSS.*

GRAMINEAE

Lit. Hubbard, C. E. *British Grasses*. Edition 2. Penguin Books. London (1968).

PHRAGMITES Adans.

P. australis (Cav.) Trin. ex Steud. Common Reed.
P. communis Trin., *P. phragmites* (L.) Karst., *P. vulgaris* (Lam.) Crép.
Lit. *J. Ecol.* **60**: 585 (1972).
Native in swamps and shallow water, etc. Locally abundant.
First record: *T. Johnson*, 1638.
1. Locally common in the Colne valley. Pinner,† *Melv. Fl.*, 89. Ruislip Reservoir. Stanmore Common. Near Potters Bar!, *J. G.* and *C. M. Dony*. **2.** Stanwell Moor!; Staines!, *T. & D.*, 317. Shepperton. East and West Bedfont. West Drayton. Longford. Harmondsworth. Colnbrook. **3.** Cranford. Southall. **4.** Hampstead!, *Irv. MSS.*; West Heath, Hampstead. Willesden,† *De Cresp. Fl.*, 109. Bentley Priory. **5.** Between Acton and Turnham Green;† Ealing Common;† near Shepherds Bush;† near Kew Bridge,† *Newbould*, *T. & D.*, 217. Hanwell. Brentford. Osterley Park. **6.** Tottenham!, *Johns. Cat.* Locally common in the Lee valley. Enfield Chase. **7.** Clapton, 1837,† *E. Ballard* (BM). Lee Canal, Tottenham!, *Cherry*;

Hackney Wick; Isle of Dogs, *T. & D.*, 317. Lake in Buckingham Palace grounds!, *Eastwood*. Bombed sites, Cripplegate† and Gresham Street areas, E.C., 1952–53,† *Jones Fl.*

MOLINIA Schrank

M. caerulea (L.) Moench Purple moor-grass.
Native on damp heaths and commons, in wet grassy places on acid soils, etc., established on railway banks, etc. Locally abundant.
First record: *Blackstone*, 1737.
1. Meadows near Harefield Moor,† *Blackst. Fasc.*, 39; Harefield Common!, *T. & D.*, 323. Harrow Weald Common!, *Melv. Fl.*, 89. Stanmore Common!, *C. S. Nicholson, K. & L.*, 311. Near Copse Wood, Northwood, scarce, *J. A. Moore*. **3.** Hounslow Heath!, *T. & D.*, 323. Railway banks towards Feltham!, *Benbow MSS.* **4.** West Heath, Hampstead!, *T. & D.*, 343. North-west Heath, Hampstead. **6.** Hornsey Churchyard,† *M. & G.*, 89. **7.** Near Pancras,† *M. & G.*, 89. East Heath, Hampstead. Kensington Gardens, casual,† *Warren Fl.* Hyde Park, shrubbery at edge of Knightsbridge, S.W.1, presumably introduced with peat, 1965, *Brewis, L.N.* 46: 34. Buckingham Palace grounds, introduced with peat blocks, 1964, *Fl. B.P.G.* Walton Street, S.W.3, 1970, *E. Young*.

Var. **major** Lej. & Court.
1. Watts Common, Harefield, *Green* (SLBI).

SIEGLINGIA Bernh.

S. decumbens (L.) Bernh. Heath or Moor Grass.
Triodia decumbens (L.) Beauv., *Danthonia decumbens* (L.) DC.
Native on heaths, commons, moors, in rough grassy places, etc. Locally plentiful, especially on damp base-rich soils.
First record: *Blackstone*, 1737.
1. Harefield Common!, *Blackst. Fasc.*, 34. Harrow Weald Common!, abundant, *Melv. Fl.*, 90. Stanmore Common!, *T. & D.*, 322. Uxbridge Common!; Harefield Moor!, *Benbow MSS.* Ruislip Common!, *Bishop*; north of Harefield, *Welch, K. & L.*, 310. **2.** Between Bushy Park and Teddington, *Newbould, T. & D.*, 322. Bushy Park, abundant!, *Benbow MSS.* Staines Moor!, *K. & L.*, 310. **3.** Hounslow Heath!, *T. & D.*, 322. **4.** Hampstead Heath!, *J. Morris, T. & D.*, 322; North-west Heath, Hampstead. Near Scratch Wood!, *K. & L.*, 310. **6.** Hadley Green, abundant. **7.** East Heath, Hampstead, abundant!, *T. & D.*, 322.

GLYCERIA R.Br.

Lit. *Watsonia* **3**: 291 (1956).

G. fluitans (L.) R.Br. Floating Sweet-grass, Flote-grass.
Poa fluitans (L.) Scop.
Native in stagnant ponds, ditches, swamps, shallow streams, etc. Common, except in heavily built-up areas where it is rare or extinct.
First certain evidence: *Hardwicke*, c. 1840.
7. West End, Hampstead (HE). East Heath, Hampstead, sparingly!, *T. & D.*, 325. Ken Wood grounds.

G. fluitans × plicata = G. × pedicellata Townsend
1. Eastcote, *T. & D.*, 326. Uxbridge!, *Benbow* (BM). Harefield!, *Druce Notes*. Cowley. **3.** Hounslow, *Newbould*, *T. & D.*, 326. Southall. **4.** West Heath, Hampstead, *T. & D.*, 326. **5.** Pond near Brent between Greenford and Southall, 1929,† *Hubbard* (K). **6.** Near Whetstone, *T. & D.*, 326. **7.** East Heath, Hampstead!, in plenty; ditches at Hackney Wick, *T. & D.*, 326.

G. plicata Fr. Plicate Sweet-grass.
Native in ponds, ditches, shallow streams, swamps, etc. Probably common but confused with *G. fluitans* and *G. declinata* with which it sometimes grows.
First certain record: *T. Moore*, 1845.
1. North of Harefield, *Newbould*, *T. & D.*, 325. Harefield!, *A. B. Jackson* (D). Frays Meadows, Uxbridge!, *Benbow* (BM). Springwell!; near Ruislip!; Yiewsley!, *K. & L.*, 314. Cowley. **2.** West Drayton!, *Druce Notes*. Poyle!; near Colnbrook!, *K. & L.*, 314. Staines Moor!, *Briggs*. **3.** Near Heathrow!, *K. & L.*, 314. Southall. **5.** Near Brentford. **6.** Highgate Woods, 1904, *C. S. Nicholson*, *K. & L.*, 314. **7.** Field near Hampstead, 1845, *T. Moore*, *Phyt.* 2: 500. East Heath, Hampstead!; Eel Brook Common, abundant, 1862;† Isle of Dogs, 1866,† *T. & D.*, 325. Hackney Wick, 1867,† *Trimen* (BM).

G. declinata Bréb. Glaucous Sweet-grass.
Native in similar habitats to the preceding species. Rare, or confused with *G. fluitans* and *G. plicata* with which it sometimes grows.
First evidence: *T. Moore*, 1844.
1. Ruislip Common!, *Sandwith*, *L.N.* 27: 35. Wrotham Park, 1965, *J. G.* and *C. M. Dony*!. Stanmore Common, 1966. **2.** Bushy Park!, *Welch*; West Drayton!, both det. *Hubbard*, *K. & L.*, 314. Staines Moor, 1966. Hampton Court Park, 1966. **3.** Pool near Hounslow Heath, 1948, *Hubbard*!, *L.N.* 28: 34; 1966. **4.** Near Hampstead, 1844, *T. Moore* (K). West Heath, Hampstead!, det. *Hubbard*, *K. & L.*, 314. Mill Hill, *Sennitt*. **5.** Near Northolt. **6.** Near Barnet, 1867, *Stratton* (D). Perhaps in Herts. **7.** North of London, 1844, *T. Moore* (CGE). East Heath, Hampstead.

G. maxima (Hartm.) Holmberg Reed Sweet-grass.
Poa aquatica L., *Glyceria aquatica* (L.) Wahlberg, non J. & C. Presl
Lit. *J. Ecol.* **34**: 310 (1947).
Native in rivers, streams, ponds, lakes, etc., established in canals, gravel pits, etc. Common, though sometimes planted for ornamental purposes.
First record: *T. Johnson*, 1638.
7. Tottenham, *Johns. Cat.* Serpentine!, 1813, *Herb. Devon. Inst. Exeter ex T. & D.*, 325; 1961!, *D. E. Allen*, *McClintock* and *Rönaasen*, *L.N.* 41: 22; 1968. By Thames at Chelsea, *J. Morris*; London Canal, *Cherry*; Waltham Green, abundant, 1862,† *T. & D.*, 325. Thames near Fulham, *Wats. New Bot.*, 103. East Heath, Hampstead!; Isle of Dogs!, *T. & D.*, 325. Highgate Ponds!; Hackney Marshes!, *Fitter*. By Paddington Canal, 1863, *Parsons* (CYN). River Lee, Hackney, 1889, *Masterman* (BM). By the lake, Buckingham Palace grounds, perhaps planted, *Codrington*, *Lousley* and *McClintock*!, *L.N.* 39: 59. Lee Navigation Canal, Upper Clapton. Regents Canal, Stepney, Victoria Park and Hackney Wick.
A proliferous form was noted at **1.** Harefield, 1924, *J. E. Cooper*.

CATABROSA Beauv.

C. aquatica (L.) Beauv. Water Whorl-grass.
Native in ditches, shallow parts of streams, on muddy margins or ponds and lakes, etc., established in shallow parts of canals, etc. Rare, and rapidly decreasing.
First certain record: *Merrett*, 1666.
1. Meadows near Cowley, abundant, *Benbow MSS.*; 1927,† *J. E. Cooper*, *K. & L.*, 312. By the Colne, Uxbridge Moor!;† Yiewsley, *Benbow MSS.* Above Denham Lock, *Green* (SLBI). Harefield!, *Benbow* (BM). Springwell!, *K. & L.*, 311. **2.** Stanwell Moor!; near Strawberry Hill,† *T. & D.*, 327. West Drayton,† *Benbow MSS.* West Bedfont,† *Rose* and *Welch*!; Poyle!; Yeoveney!; Staines Moor!,† *K. & L.*, 313. Knowle Green.† **3.** Isleworth,† *T. & D.*, 327. **4.** Hampstead,† *Irv. MSS.* Hampstead Heath,† *Hampst.* **5.** Brentford, *T. & D.*, 327; c. 1905,† *Montg. MSS.* Ditch by railway between Ealing and West Ealing;† ditch filled in 1968. **6.** Hornsey,† *Cockfield Cat.*, 13. Near Whetstone, *T. & D.*, 327. Finchley Common, 1917;† Ponders End, *J. E. Cooper*, *K. & L.*, 313. **7.** Knightsbridge, *Merr. Pin.*, 53; in standing water [c. 1730†], *Dill MSS.* Islington,† *Pet. Midd.* Chelsea;† St James's Park,† *M. & G.*, 90. Lee between Clapton and Tottenham,† *Francis*, *Coop. Suppl.*, 12.
C. aquatica appears to be a shade-loving species, and J. F. and P. C. Hall have drawn attention to the marked frequency of its occurrence under bridges over streams and canals.

FESTUCA L.

Lit. *Rep. Bot. Soc. & E.C.* **13**: 338 (1948).

F. pratensis Huds. Meadow Fescue.
F. elatior L., *pro parte*
Native in meadows and grassy places, established on lawns, railway banks, etc. Common, though frequently sown.
First certain evidence: *S. P. Woodward*, 1848.
7. Frequent in the district.

Var. **pseudo-loliacea** Hack.
2. Hampton Court, *C. S. Nicholson* (LNHS).

F. arundinacea Schreb. Tall Fescue.
F. elatior L., *pro parte*
Native by river- and streamsides, on pond and lake verges, in meadows, etc. Common.
First evidence: *Doody*, 1688.
7. Common in the district.

Subvar. **strictior** Hack.
3. Hounslow Heath, *Hubbard*!. **7.** Front of Natural History Museum, South Kensington,† *Wilmott* (BM).

F. gigantea (L.) Vill. Giant Fescue.
Bromus giganteus L.
Native in damp open woods, copses, ditches, shady places, on hedgebanks, by river- and streamsides, etc. Common.
First record: *Doody*, 1688.
7. Fulham near the Bishop's palace!, *Doody*, *Ray Hist.* 2: 1909; 1968. Sandy End, Fulham,† *T. & D.*, 329. Kentish Town, 1846 (BM). Regents Park!, *L.N.* 39: 59. Hampstead. East Heath, Hampstead. Isle of Dogs. Bromley-by-Bow.

F. heterophylla Lam.
Alien. Europe. Naturalised in woods, etc. Very rare.
First certain record: *J. R. Palmer*, 1966.
4. Hampstead Heath, *J. R. Palmer*, *L.N.* 46: 12.

F. rubra L. subsp. **rubra** Red, or Creeping Fescue.
Lit. *J. Linn. Soc. Bot.* **46**: 313 (1924).
Native on heaths, commons, in rough fields, meadows, etc., established on railway banks, etc. Common.
First certain record: *A. Irvine*, 1830.
7. Common in the district.
The common Middlesex plant is the var. **vulgaris** Gaudin.

Var. **barbata** (Schrank) K. Richt.
1. Harefield, *Sandwith* (K), det. *Hubbard.* **4.** West Heath, Hampstead, det. *Hubbard.*

Subsp. **commutata** Gaudin Chewing's Fescue.
Festuca fallax Thuill., *F. rubra* L. var *fallax* (Thuill.) Hack.
1. Harefield Moor, det. *Hubbard.* **4.** West Heath, Hampstead, det. *Hubbard.* North-west Heath, Hampstead. **7.** East Heath, Hampstead, det. *Hubbard.*

F. ovina L. Sheep's Fescue.
F. sulcata auct.
Lit. *J. Linn. Soc. Bot.* **47**: 29 (1925).
Native on heaths, commons, in meadows, grassy places, etc. Common.
First certain record: *Winch*, 1810.
7. Hyde Park, *Gray*; gardens of Lincoln's Inn, *T. & D.*, 329. East Heath, Hampstead!, *Benbow MSS.* Newington, *Ball. MSS.* Bombed site, Ebury Street, S.W.1,† *McClintock.* Regents Park!, *Holme*; Kensington Gardens, *D. E. Allen*, *L.N.* 39: 59. Holland House grounds, Kensington. Brompton Cemetery. Bombed sites, Cripplegate and Gresham Street areas, 1952–53,† *Jones Fl.*
F. ovina is frequently confused with *F. tenuifolia.*

F. tenuifolia Sibth. Fine-leaved Sheep's Fescue.
F. ovina L. subsp. *tenuifolia* (Sibth.) Peterm., *F. capillata* Lam.
Lit. *J. Linn. Soc. Bot.* **47**: 29 (1925).
Native in similar situations to *F. ovina* with which it sometimes grows, and with which it is often confused. Local, or overlooked.
First certain evidence: *Green*, 1908.
1. Northwood, *Green* (SLBI). Stanmore Common!; Ruislip Common!; Mimmshall Wood!, all det. *Hubbard, K. & L.*, 314. Harrow Weald Common. Near Denham. **2.** Hampton Court!, *Welch, K. & L.*, 317. Bushy Park. **3.** Hounslow Heath!, det. *Hubbard, K. & L.*, 317. **4.** West! and North-west Heaths, Hampstead!, det. *Hubbard, K. & L.*, 317. **5.** Syon Park!, *Green* (SLBI). Castle Bar, Ealing!, det. *Hubbard, K. & L.*, 317. Perivale. **6.** Crew's Hill golf course, Enfield; near Enfield Wash, *Kennedy*!. **7.** East Heath, Hampstead, det. *Hubbard.* Ken Wood grounds. Bombed site, Cripplegate, *Wrighton, L.N.* 29: 88. Regents Park, *Holme, L.N.* 39: 59. Kensington Gardens!, *D. E. Allen*, conf. *Lousley*, *Allen Fl.*

F. longifolia Thuill. Hard Fescue.
F. trachyphylla (Hack.) Krajina, non Hack., *F. duriuscula* auct., non L.
Alien. Europe, etc. Sown on roadside verges, railway banks, etc., and now established. Probably common but overlooked.
First record: *A. Irvine*, c. 1830.
1. Harrow Weald Common; Potters Bar, *T. & D.*, 329. **2.** Staines

Moor. Abundant about Charlton and from Staines to Hampton, *T. & D.*, 329. Towing-path between Hampton Court and Kingston Bridge, *Clement* and *Mason*. Between Poyle and Yeoveney, *Wrighton* !, *K. & L.*, 316. **3.** Near Isleworth; Roxeth, *T. & D.*, 329. Road banks, Western Avenue, Northolt, abundant, 1960. **4.** Hampstead, *Irv. MSS.* **5.** Shepherds Bush, *Newbould*, *T. & D.*, 329. Verges of Great Chertsey Road, Chiswick, abundant !, det. *Hubbard*, *K. & L.*, 316. Hanwell. **6.** Wood Green, *Munby*, *Gard. Chron.*, 1868, 499. Near Edmonton; Muswell Hill, *Cherry*, *T. & D.*, 329. **7.** Kentish Town (HE). Kensington Gardens, *J. Morris*; Chelsea; Brompton Cemetery !, *T. & D.*, 352. Bombed site, Jewin Street, E.C., 1952,† *Lousley* (L). Green Park, 1957 !, *L.N.* 39: 59. Tottenham, 1967, *Kennedy*.

FESTUCA × LOLIUM = × FESTULOLIUM Aschers. & Graebn.

F. pratensis × Lolium multiflorum = × **Festulolium braunii** (K. Richt.) A. Camus

5. Paddock by Perivale Wood, 1973, *Hubbard*.

F. pratensis × Lolium perenne = × **F. loliaceum** (Huds.) P. Fourn.
Festuca loliacea Huds., *F. pratensis* Huds. var. *loliacea* (Huds.) With.
Lit. *New Phyt.* **72**: 411 (1973).

7. In the privy garden, Whitehall, a single plant,† *M. & G.*, 113. Common about Stoke Newington, *Woods*, *B.G.*, 400. About London, *Curt. F.L.* Chelsea (DILL). Kentish Town, *Irv. MSS.*; 1844, *T. Moore*; Hackney Marshes, 1882, *Paulson* (BM). Haverstock Hill, 1867 (SY). Regents Park, *Holme*, *L.N.* 39: 59.

A common hybrid which occurs wherever the two parents grow together in quantity.

LOLIUM L.

L. perenne L. Perennial Rye-grass.
Lit. *J. Ecol.* **55**: 567 (1967).
Native in meadows, grassy places, on waste ground, by roadsides, etc., established on railway banks, etc. Very common.
First record: *T. Johnson*, 1638.
7. Frequent in the district.

Var. **tenue** (L.) Huds.
2. Staines, *Hubbard*, *Wedg. Cat.*

L. multiflorum Lam. Italian Rye-grass.
L. perenne L. subsp. *multiflorum* (Lam.) Husnot, *L. italicum* A. Braun
Alien. S. Europe. Naturalised on cultivated and waste ground, in

fields, by roadsides, etc., established on railway banks, rubbish-tips, etc. Very common.
First record: *Hind*, 1864.
7. Common in the district.

L. multiflorum × perenne = L. × hybridum Hausskn.
Frequent on cultivated and waste ground, by roadsides, etc., throughout the vice-county.

VULPIA C. C. Gmel.

Lit. *Proc. Bot. Soc. Brit. Isles* **6**: 386 (1967).

V. bromoides (L.) Gray Barren or Squirrel-tail Fescue.
Festuca bromoides L., *F. sciuroides* Roth
Native on dry heaths, banks, waste ground, by roadsides, etc., established on railway tracks, canal paths, rubbish-tips, etc. Common.
First evidence: *Dillenius*, c. 1730.
7. Fields near Marylebone, *J. Banks*; Tufnell Park, *French* (BM). Kensington Gardens, *Winch MSS.* East Heath, Hampstead, *T. & D.*, 328. Bombed site, Ebury Street, S.W.1,† *McClintock*. Fulham Palace grounds, *Welch*!. Hyde Park, 1961, *McClintock* and *Rönaasen*, conf. *Lousley*; *D. E. Allen*, det. *Melderis*, *L.N.* 41: 22.

V. myuros (L.) C. C. Gmel. Rat's tail-Fescue.
Festuca myuros L.
Native on heaths, waste ground, etc., on light soils, established about, railways, on old walls, etc. Rare in native habitats, but common and increasing on railway tracks, in goods yards, etc.
First certain record: *Merrett*, 1666.
1. Potters Bar; Hillingdon, *Benbow MSS.* Yiewsley!, *Druce Notes.* Harefield!, *L.N.* 28: 34. Site of disused railway between Uxbridge and Denham, abundant. **2.** Walton Bridge, *Newbould*, *T. & D.*, 328. West Drayton!; Laleham!; Sunbury!; Hampton, *Benbow MSS.* Dawley!, *J. E. Cooper* (BM). Hampton Court Park. Railway tracks, Staines to Colnbrook, and Staines to Ashford, abundant. Ashford goods yard. Harmondsworth. **3.** Headstone Farm,† *Melv. Fl.*, 91. Twickenham!; Isleworth, *T. & D.*, 328. Whitton, *Benbow MSS.* Hounslow Heath!, *Philcox* and *Townsend*. The Grove, Harrow, *Melv. Fl.*, 91. **4.** Hampstead, *Irv. MSS.* Stanmore, *T. & D.*, 328; c. 1900; Canon's Park, 1906, *C. S. Nicholson*; Finchley, 1916, *J. E. Cooper*, *K. & L.*, 317. Between 'The Spaniards' and Highgate, *Benbow MSS.* Cricklewood, *Warren*, *Trim. MSS.* Scratch Wood railway sidings. Railway sidings, Neasden and Wembley Hill. Disused railway Mill Hill to Hendon, abundant, *Kennedy*!. **5.** Brentford!, in plenty, *T. & D.*, 328; canal path!, *Boniface*; abundant on disused railway.

Kew Bridge goods yard, abundant. Acton!; Chiswick!, *Benbow MSS.* Railway sidings near Greenford, abundant!, *K. & L.*, 317. Railway tracks, Southall to Brentford, abundant. Railway platform, West Ealing, abundant. Gunnersbury, 1971. **6.** Edmonton, *T. & D.*, 328. Botany Bay, Enfield Chase, *Benbow MSS.* Railway tracks, Enfield Town. Derelict railway, Wood Green. Hadley, 1875, *French* (BM). **7.** On the walls by Piccadilly,† *Merr. Pin.*, 55. Dalston, *E. Forster*; between Highbury and Seven Sisters Road, 1876, *French*; Chelsea, 1835, *Broome* (BM). Parsons Green, *T. & D.*, 328. Hackney Marshes, 1913–14, *Coop. Cas.* St Katherine's Dock, Stepney, *Brewis.*

NARDURUS (Bluff, Nees & Schau.) Reichb.

N. maritimus (L.) Murb. Matgrass Fescue.
Festuca maritima L.
Lit. *Proc. Bot. Soc. Brit. Isles* **4**: 243 (1961).
Denizen on dry banks, established on railway tracks, etc. Very rare.
First record: *D. H. Kent*, 1951.
1. Railway tracks between Denham and Uxbridge, 1951!, det. *Hubbard, L.N.* 31: 12; 1960. Abundant over a small area of a railway bank near Denham, 1966!, *Wurzell*; 1969. The latter locality is about 300 yards north of my 1951 station, and the plant grows in a tangled mass with *Pilosella officinarum*, a few plants also occurred on the cinders and granite chippings of the permanent way of the adjacent disused railway. This is undoubtedly the main colony from which the plants noted in 1951 and 1960 had spread.

PUCCINELLIA Parl.

P. distans (L.) Parl. Reflexed Salt-marsh Grass, Reflexed Poa.
Poa distans L., *Glyceria distans* (L.) Wahlenb., *Sclerochloa distans* (L.) Bab.
Denizen in rough fields, marshes, on waste ground, etc., established on canal paths, etc. Very rare, and sometimes merely casual.
First record: *Curtis*, 1786.
2. Canal path, West Drayton, abundant, *Wurzell*!, *L.N.* 45: 20; 1969 →. Rubbish-tip near Shepperton, 1969, *Clement* and *Ryves.* **3.** Abundant with *Aster tripolium* in a marshy field near Yeading, 1949!, *Boniface* and *Rose*, *L.N.* 29: 13; 1950–54!, *K. & L.*, 315; 1955–63; the field is now enclosed by a tall fence and the ground has been disturbed; the grass, however, probably survives. **4.** Hampstead Heath, *Coop. Fl.*, 102. East Finchley, 1910,† *Coop. Cas.* **6.** Verge of gravel pit north of Enfield Lock, 1966, *Kennedy*!; 1967. **7.** Near Hampstead,† *Curt. F.L.* Isle of Dogs, a single plant, 1966,† *Newbould*

(BM). On the new soil of the Thames embankment opposite Somerset House, 1866,† *T. & D.*, 326. Casual about Hyde Park,† *Warren Fl.* Hackney Marshes, 1909–10 and 1912, *Coop. Cas.*; 1916–17 and 1919–20. *Cooper, K. & L.*, 315. Bombed site by Cripplegate Institute, E.C., 1945,† *Lousley* (L).

CATAPODIUM Link

C. rigidum (L.) C. E. Hubbard Hard Poa, Fern-grass.

Poa rigida L., *Festuca rigida* (L.) Rasp., non Roth, *Sclerochloa rigida* (L.) Link, *Scleropoa rigida* (L.) Griseb., *Demazeria rigida* (L.) Tutin

Native on dry banks, bare ground, etc., chiefly on calcareous soils, established on tops of old walls, stone embankment of the Thames, railway tracks, etc. Formerly locally common, but now rare and decreasing.

First record: *W. Curtis*, c. 1780.

1. Canal bridge near Cowley Peachey wharf!;† Harefield!; between Denham and Harefield, *Benbow MSS.* Springwell!, *Welch.* Yiewsley, *J. E. Cooper.* Site of disused railway between Uxbridge and Denham. **2.** Hampton Court walls!; stone facing of river near Hampton Court!, *T. & D.*, 326. River wall by Kingston Bridge, abundant. Old walls, West Drayton!,† *Benbow MSS.* Laleham!,† *Green* (SLBI). Harmondsworth, *P. H. Cooke.* **3.** Harrow Grove;† Headstone Farm,† *Melv. Fl.*, 90. Isleworth!, *Trimen* (BM); 1968. **4.** Harrow,† *Hemsley, T. & D.*, 326. Old wall, Whitchurch Common,† *C. S. Nicholson, Lond. Nat. MSS.* **5.** Wall of Chiswick House grounds, *T. & D.*, 326; c. 1887;† Acton,† *Benbow* (BM). Turnham Green,† *Benbow MSS.* Bedford Park,† *Cockerell Fl.* Old wall near Gunnersbury, 1874,† *R. A. Pryor, Trim. MSS.* **7.** On most of the walls about London,† *Curt. F.L.* Chelsea,† *J. Morris*; Parsons Green,† *T. & D.*, 326. Beyond Hornsey Wood,† *Ball. MSS.* Highbury, 1876,† *French* (BM). Hornsey, 1885,† *J. E. Cooper.* Bombed sites, Cripplegate, E.C., 1947, *Wrighton, L.N.* 27: 35; 1952–53;† bombed site, Gresham Street area, E.C.,† *Jones Fl.* Brickwork near Porchester Road, Paddington, 1969, *L. M. P. Small*, conf. *Lousley.*

POA L.

P. annua L. Annual Meadow-grass or Poa.

Lit. *Watsonia* **4**: 1 (1957): *Proc. Bot. Soc. Brit. Isles* **7**: 452 (1968). (*Abstr.*)

Native on cultivated and waste ground, by roadsides, in grassy places, marshes, etc., established on lawns, garden paths, railway banks and tracks, etc. Very common.

First record: *T. Johnson*, 1633.

7. Abundant throughout the district.

Var. **aquatica** Aschers.

1. Harefield. **4.** West Heath, Hampstead, 1949!, det. *Hubbard*; 1966.

A form of *P. annua* with the vegetative shoots somewhat thickened at the base and coated with old sheaths occurs in **2.** in plenty over a limited area, chiefly among old granite-setts, on the towing-path near Hampton Court Bridge. The form was first noted by *E. Clement* in 1966, and Dr C. E. Hubbard, who has seen the plant *in situ*, considers that it may be perennial.

P. nemoralis L. Wood Poa.

Native in woods, in shady places, etc., and sometimes in open situations where it often attains a much more robust appearance. Locally abundant in the northern parts of the vice-county, but local and rare near the Thames.

First record: *A. Irvine*, 1830.

1. Frequent. **2.** Fulwell. Bushy Park. Near Sunbury. **3.** Roxeth;† Headstone,† *Melv. Fl.*, 90. Between Ickenham and Northolt. Northolt. Hayes Park. **4.** Common, except in very heavily built-up areas. **5.** Ealing!, *Newbould*; Syon Park!, *T. & D.*, 324. Osterley Park, *Welch*!. Hanwell. Horsendon Hill. Near Perivale. Chiswick House grounds. **6.** Frequent, except in heavily built-up areas. **7.** Between Kentish Town and Highgate, 1847† (HE). Kensington Gardens!, *J. Morris*; Kilburn,† *Warren*, *T. & D.*, 324. Bishop's Park, Fulham; Fulham Palace grounds, *Welch*!. Ken Wood!, *R. W. Robbins*, *Lond. Nat. MSS.* Hyde Park!, *Welch*. Buckingham Palace grounds, *Codrington*, *Lousley* and *McClintock*!, *L.N.* 39: 59. St Katherine's Dock, Stepney, 1966, *Brewis*.

This species is sometimes sown in grass seed mixtures, cf. D. E. Allen (*Proc. Bot. Soc. Brit. Isles* **5**: 233 (1964).

P. compressa L. Flattened Meadow-grass or Poa.

Native on dry banks, waste ground, by roadsides, etc., established on tops of old walls, on the stone embankment of the Thames, etc. Local, but more frequent near the Thames than elsewhere.

First record: *W. Hudson*, 1762.

1. Cowley; Cowley Peachey; Uxbridge Moor, *Benbow MSS.* Ruislip Common, *Benbow* (BM). Near Uxbridge, *Sherrin* (SLBI). Yiewsley, *J. E. Cooper*, *K. & L.*, 324. Harefield. New Years Green!, *B. M. C. Morgan*. **2.** Teddington!, *T. & D.*, 324. River wall of the Thames near Kingston Bridge and at Penton Hook Lock. West Drayton!, *Benbow MSS.* Railway yard, Ashford. **3.** Isleworth!, abundant, *T. & D.*, 324. Twickenham!, *Benbow MSS.* Hounslow Heath!; Roxeth!,† *K. & L.*, 314. Sudbury Hill. **4.** Mill Hill!, *F. W. Payne* (SLBI). Hampstead!, *Irv. MSS.* East Finchley; Finchley, *J. E. Cooper*, *K. & L.*, 313. **5.** Turnham Green,† *Newbould*; Osterley Park wall!, *T. & D.*, 325. Acton. St George's Cemetery, Hanwell. Shepherds

Bush, 1870, *Warren, Trim. MSS.* Disused railway platform, Brentford, abundant. Near Kew Bridge. **6.** Near Potters Bar, *Benbow MSS.* Queen's Wood, Highgate, *L. B. Hall, Lond. Nat. MSS.* Myddleton House grounds, and elsewhere about Enfield, *Kennedy*!. **7.** Near Marylebone, *Huds. Fl. Angl.*, 33. West India Docks, *Cherry*; Well Walk, Hampstead;† Kensington Gore, 1867,† *T. & D.*, 325. Hackney Marshes,† *J. E. Cooper, K. & L.*, 314. Bombed site, Cripplegate, E.C., 1960.† St Katherine's Dock, Stepney, 1963, *Fitter, L.N.* 43: 21. Wall of Tower of London, 1966, *Brewis, L.N.* 46: 34. Tottenham, 1967, *Kennedy*.

Forma **polynoda** (Parn.) Neum.
P. polynoda Parn.
3. Isleworth, 1932, *Hubbard* (K).

P. pratensis L. Meadow-grass.
Native in meadows, grassy places, waste ground, etc., established on railway banks, tops of walls, etc. Common, but extremely variable, and often sown.
First record: *W. Hudson*, 1762.
7. Frequent in the district.

Var. **latifolia** Weihe
5. Canal bank between Brentford and Hanwell, 1947!, det. *Hubbard, L.N.* 27: 35.

P. angustifolia L. Narrow-leaved Meadow-grass.
P. pratensis L. subsp. *angustifolia* (L.) Gaudin, *P. strigosa* Hoffm.
Lit. *Watsonia* **4**: 447 (1959).
Native in similar habitats to the preceding species with which it often grows. Common and widely distributed though sometimes confused with narrow-leaved forms of *P. pratensis*.
First certain evidence: *J. Banks*, c. 1800.
7. Common in the district.

P. subcaerulea Sm. Spreading Meadow-grass.
P. irrigata Lindm., *P. pratensis* L. subsp. *irrigata* (Lindm.) Lindb. f., *P. pratensis* subsp. *subcaerulea* (Sm.) Tutin
Lit. *Watsonia* **5**: 163 (1962).
Native in damp grassy places, etc. Locally plentiful, especially near the Thames, but often confused with forms of *P. pratensis*.
First evidence: *Trimen*, 1860.
1. Uxbridge Common, *Benbow MSS.* Between Uxbridge and Denham. Near Swakeleys. **2.** Bushy Park, frequent!, *Clement*, conf. *Hubbard.* Hampton Court Park. Staines Moor, and elsewhere about Staines. Stanwell Moor. Laleham Park, frequent, *Kennedy*!. Shepper-

ton. Sunbury. Upper Halliford. **3.** Hounslow, *T. & D.*, 324. **4.** Hampstead Heath!, *Trimen* (BM); locally plentiful on the West Heath, especially between the bog behind Jack Straw's Castle and the Leg of Mutton Pond. Harrow, *Temple*!. **5.** Horsendon Hill, *Benbow MSS*. Ealing, *Newbould, T. & D.*, 324. Acton, *Druce Notes*. Syon Park. **6.** North of Little Heath, Potters Bar, *Kennedy*. Meadow by the New river, Myddleton House grounds, Enfield, *Kennedy*!. **7.** East Heath, Hampstead!, *Benbow MSS*. Ken Wood grounds!, *Druce Notes*. Flower-bed by lake, St James's Park, *Brewis, L.N.* 46: 34.

P. trivialis L. Rough Meadow-grass.
Native in meadows, grassy places, on heaths, waste ground, by roadsides, etc., established on railway banks, etc. Common.
First record: *Petiver*, 1695.
7. Common in the district.

P. palustris L. Swamp Meadow-grass.
P. serotina Ehrh. ex Hoffm.
Alien. Europe, etc. Originally imported as a forage grass and now naturalised in swamps, marshes, by streamsides, on pond verges, waste ground, etc., established on the river wall of the Thames, etc. Very rare, and sometimes merely casual.
First record: *Lady Davy*, 1921.
1. Uxbridge, *Lady Davy, Rep. Bot. Soc. & E.C.* 6: 403; 1924, *Butch. MSS*. **2.** River wall, Hampton Court, a fine colony, stretching two or three yards; a few plants on the river wall about 400 yards eastwards, 1966, *Clement*, conf. *Hubbard*. **3.** Old market garden, Isleworth, 1933,† *Hubbard* (K). **5.** Swamp by lake, Chiswick House grounds, 1955!, *Boniface, K. & L.*, 313; 1959. Ditch by railway between Ealing and West Ealing, 1960–67;† ditch filled in 1968. **7.** Air-raid shelter, Hyde Park, 1945,† *Sladen* (L); waste heaps near Marble Arch, 1961!,† *McClintock* and *Rönaasen*, conf. *Lousley, L.N.* 41: 22; Serpentine, Hyde Park, 1962, *Brewis, L.N.* 42: 12. Shadwell Basin, 1945, *Lousley*; bombed site, Upper Thames Street, E.C.4, 1954,† *Graham, Lousley* and *McClintock* (L). Regents Park!, 1966, *Holme, L.N.* 46: 35; 1967.

P. chaixii Vill. Broad-leaved Meadow-grass.
P. sylvatica Vill., non Poll., *P. sudetica* Haenke
Alien. Europe, etc. Formerly naturalised in shady places, but now extinct.
First evidence: *Syme*, 1852. Last record: *A. Irvine*, 1859.
7. Chelsea Hospital grounds, 1852 and 1853, *Syme* (SY); 1859, *A. Irvine, Phyt. N.S.* 3: 345.

DACTYLIS L.

D. glomerata L. Cock's-foot.
Lit. *J. Ecol.* **47**: 223 (1959).
Native in meadows, grassy places, on waste ground, by roadsides, etc., established on railway banks, etc. Very common, but often sown.
First record: *T. Johnson*, 1638.
7. Frequent in the district.
Viviparous forms are recorded from **3.** Northolt. **4.** Hendon, *P. H. Cooke.*

CYNOSURUS L.

C. cristatus L. Crested Dog's-tail.
Lit. *J. Ecol.* **47**: 511 (1959).
Native in meadows, grassy places, etc., established on lawns, railway banks, etc. Very common, and often sown.
First record: *Cargill*, c. 1600.
7. Common in the district.
A viviparous form was noted at **5.** West Acton, 1968, *McLean.*

BRIZA L.

B. media L. Common Quaking Grass, Totter Grass.
Native in old meadows and pastures, dry banks, etc. Local and decreasing and chiefly confined to calcareous soils at Harefield and South Mimms, and alluvial soils near the Thames.
First record: *T. Johnson*, 1632.
1. Harefield!, *Blackst. Fasc.*, 39. Cowley;† Frays Meadows, Uxbridge; Ruislip Common!; Springwell!; Stanmore Common!; South Mimms!, *Benbow MSS.* Between Pinner and Harrow;† Pinner!, *Hind, Trim. MSS.* Yiewsley, *J. E. Cooper*; between Ruislip and Harefield!, *Tremayne*; Northwood, *J. E. Cooper*; Warren Gate!, *K. & L.*, 312. Near Denham. **2.** Staines!; Laleham!; Colnbrook!, *Benbow MSS.* Near Littleton, *Welch, K. & L.*, 312. Knowle Green, *K. & L.*, 3. Roxeth,† *Melv. Fl.*, 90. **4.** Hampstead Heath,† *Johns. Enum.* Near Stanmore (G & R). Harrow, *Hind Fl.*; c. 1935, *Temple.* Kenton, *Melv. Fl.*, 90. Harrow Weald; near Scratch Wood, *Benbow MSS.* **5.** Greenford!; Horsendon Hill!, *Benbow MSS.* Near Perivale!, *A. Wood, Trim. MSS.* Ealing,† *M. Taylor*, teste *Welch.* Bedford Park,† *Cockerell Fl.* **6.** Meadows near Potters Bar, *Benbow MSS.* Finchley Common,† *C. S. Nicholson, K. & L.*, 312.

MELICA L.

M. uniflora Retz. Wood Melick.
M. nutans auct., non L.
Native in woods, on shady hedgebanks, etc. Locally abundant in the north of the vice-county, rare elsewhere and completely absent from the districts near the Thames.
First record: *Merrett*, 1666.
1. Frequent. **4.** Hampstead Wood [= Bishop's Wood]!, and many other woods in the district!, *Merr. Pin.*, 49. Turner's Wood!; Scratch Wood!, *T. & D.*, 323. Harrow Park, *Melv. Fl.*, 89. Hendon!, *De Cresp. Fl.*, 41. East Finchley. **6.** Colney Hatch (HE). Hadley!, *Warren*; near Whetstone; Winchmore Hill Wood,† *T. & D.*, 323. Highgate Wood!, *Benbow MSS.* Enfield. Trent Park. **7.** Hornsey Wood;† near Highgate, *Blackst. Spec.*, 29. Ken Wood!, *E. Ballard* (BM). Regents Park!, *Holme, L.N.* 39: 60; 1969.

BROMUS L.

B. erectus Huds. Upright Brome.
Zerna erecta (Huds.) S. F. Gray
Native in dry pastures, meadows, etc. Locally plentiful on calcareous soils at Harefield, and on alluvial soils near the Thames, introduced and very rare elsewhere.
First record: *Lightfoot*, c. 1780.
1. Near Uxbridge!, *Lightf. MSS.* Railway side between Uxbridge and Denham. Harefield!, *J. Banks* (BM). Springwell!, frequent, *Benbow MSS.* Yiewsley, *J. E. Cooper* (SLBI). Ickenham Green, *Wrighton.* **2.** Between Hampton Court and Kingston Bridge!; between Staines and Penton Hook Lock!, *Benbow, J.B.* (*London*) 23: 339. Sipson, 1910,† *J. E. Cooper, K. & L.*, 318. **4.** East Finchley, 1909,† *J. E. Cooper, K. & L.*, 318. **5.** Hanger Hill, Ealing, a small patch, 1962 →. **7.** On the bank of the New river, Pentonville, in some quantity, 1870,† brought in with soil no doubt, *Trim. MSS.* Hackney Marshes, 1914 and 1921,† *J. E. Cooper, K. & L.*, 318. Bombed site, Cripplegate area,† *Jones Fl.*

Var. **villosus** Leight.
1. Harefield.

B. ramosus Huds. Hairy Brome.
B. asper Murr., *Zerna ramosa* (Huds.) Lindm.
Native in open woods, on wood margins, hedgebanks, in shady ditches, etc. Common, except near the Thames where it is local.
First record: *W. Curtis*, c. 1780.
1. Frequent. **2.** Staines Moor!, *T. & D.*, 330. Stanwell Moor.

Penton Hook. Littleton. Poyle. **3.** Near Whitton; Hounslow!; near Twickenham, very abundant, *T. & D.*, 330. Hounslow Heath. Hayes Park. Northolt. Hatch End. **4.** Frequent, except in heavily built-up areas. **5.** Frequent, except in heavily built-up areas. **6.** Frequent, except in heavily built-up areas. **7.** Upper Clapton; near Hornsey,† *Cherry*; East Heath, Hampstead, *T. & D.*, 331. Kensington Gardens,† *Warren*, *Trim. MSS.* Fulham Palace grounds, *Welch*!. St John's Wood, *Holme*, *L.N.* 39: 60. Bombed site, Ave Maria Lane, E.C., 1948!,† *Lousley*, *L.N.* 28: 30.

B. inermis Leyss. Hungarian Brome.
Zerna inermis (Leyss.) Lindm.
Alien. Europe, etc. Naturalised on waste ground, sown on railway banks, etc. Very rare.
First evidence: *Tester*, 1936.
1. Railway banks, Potters Bar, 1960, *J. G. Dony*!; 1973. **4.** Railway banks, Oakleigh Park, abundant, 1960, *J. G. Dony*!; 1966. **6.** Railway bank, Enfield Highway!, 1936, *Tester* (K); in quantity on railway banks for at least 400 yards, 1971 →. **7.** Bombed site, Shadwell, a few plants, 1966; several large well-established colonies, 1967.

B. sterilis L. Barren Brome.
Anisantha sterilis (L.) Nevski
Native on cultivated and waste ground, by roadside, etc., established on railway banks, etc. Very common.
First record: *T. Johnson*, 1638.
7. Very common in the district.

B. diandrus Roth Great Brome.
B. gussonii Parl., *B. maximus* auct. angl., *pro parte*, *B. rigens* auct., *Anisantha gussonii* (Parl.) Nevski, *A. diandra* (Roth) Tutin
Alien. Europe, etc. Naturalised on cultivated and waste ground, etc. Very rare, and sometimes merely casual.
First evidence: *J. E. Cooper*, 1907.
1. Yiewsley, 1909, *J. E. Cooper* (BM); 1911, *J. E. Cooper* (K). **5.** Greenford, 1953!, *Welch, K. & L.*, 318; 1966. Perivale, 1966. Waste ground, Brentford, abundant, 1966!, *Boniface*, *L.N.* 46: 38; still plentiful, 1969: destroyed, 1971.† **6.** Muswell Hill, 1907, *J. E. Cooper* (BM). Myddleton House grounds, Enfield, an abundant weed of flower-beds, etc., 1966, *Kennedy*!, *L.N.* 46: 38.

B. hordeaceus L. subsp. **hordeaceus** Soft Brome, Lop-grass.
B. mollis L., *Serrafalcus mollis* (L.) Parl.
Lit.: *Watsonia* **6**: 327 (1968).
Native in meadows, grassy places, on cultivated and waste ground, by roadsides, etc., established on railway banks, etc. Very common.

First certain evidence: *A. Irvine*, c. 1830.
7. Common in the district.

B. hordeaceus L. subsp. **hordeaceus** × **lepidus** = **B.** × **pseudothominii** P. Smith
B. mollis L. var. *leiostachyus* Hartm. *pro parte*, *B. mollis* L. var. *glabratus* Hartm. *pro parte*, *B. gracilis* Krösche var. *micromollis* Krösche *pro parte*, *B. lepidus* Holmberg var. *micromollis* (Krösche) C. E. Hubbard *pro parte*, *B. thominii* sensu Tutin, non Hardouin.
Lit. *Watsonia* **6**: 327 (1968).
1. Harefield, *Benbow* (BM). Uxbridge. Pinner. Ruislip. Eastcote. **2.** Staines. Near East Bedfont. Shepperton. Laleham. Charlton. **3.** Hounslow, *T. & D.*, 332. Southall. Cranford. Hayes Park. **4.** West Heath, Hampstead!, det. *Hubbard* (as *B. mollis* var. *leiostachyus*), *K. & L.*, 320. North-west Heath, Hampstead. Brent Reservoir, *Kennedy*!. Neasden. Finchley. Edgware. Harrow. **5.** Twyford, *Newbould* (BM). Shepherds Bush, *T. & D.*, 332. Hanwell!; Chiswick!, both det. *Hubbard* (as *B. mollis* var. *leiostachyus*), *K. & L.*, 320. Between Hanwell and Southall. Osterley Park. Norwood Green. Heston. **6.** Clay Hill, Enfield; near Botany Bay, Enfield Chase, *Kennedy*!. Trent Park, *Kennedy*. **7.** Bombed site, Upper Thames Street, 1945,† *Lousley* (L). Regents Park. Hackney Marshes. Hornsey.

B. lepidus Holmberg Slender Brome.
B. gracilis Krösche, *B. britannicus* I. A. Williams
Lit. *J.B.* (*London*) **67**: 65 (1929): *Watsonia* **6**: 327 (1968).
Alien of unknown origin. Naturalised on cultivated and waste ground, etc., established (? sown) on roadside verges, etc. Common, but confused with slender forms of *B. mollis*, and with *B.* × *pseudothominii*.
First evidence: *French*, 1871.
1. Yiewsley!, *J. E. Cooper* (K). Between Brockley Hill and Elstree!, det. *Hubbard*, *L.N.* 26: 65. Near Harefield Moor. Harefield!, *I. G. Johnson*. Springwell. Denham. Ruislip. South Mimms. **2.** Staines!; Poyle!, both det. *Hubbard*, *L.N.* 26: 65. Yeoveney. East Bedfont. Bushy Park. Laleham Park, *Kennedy*!. Stanwell Moor. **3.** Near Osterley!, det. *Hubbard*, *L.N.* 26: 65. Near East Bedfont, *J. G. Dony* and *Souster*!; near Heathrow!, both det. *Hubbard*, *L.N.* 27: 35. Hounslow, *Hubbard* (K). Southall. **4.** Cricklewood. Mill Hill. Harrow. Kenton. Edgware. **5.** Hanwell!, det. *Hubbard*, *L.N.* 26: 65. Osterley Park. Perivale. Greenford. Near Northolt. Norwood Green. **6.** Wood Green. Near Enfield Wash, *Kennedy*!. North Finchley. Whetstone. **7.** Finsbury Park, 1871, *French* (BM). Bombed site, Upper Thames Street, near Dowgate Hill, 1945,† *Lousley*; Hackney!, *Melville* (L). Ken Wood fields!; Kensington Gardens!, *L.N.* 39: 60. Regents Park, *Holme*, *L.N.* 46: 25. Hackney Marshes. Holland House grounds, Kensington. Hammersmith.

B. racemosus L. Smooth Brome.
Serrafalcus racemosus (L.) Parl.
Native in wet meadows and pastures, by river- and streamsides, on damp waste ground, etc., established on railway banks, etc. Very rare, or overlooked, and sometimes confused with *B. commutatus*.
First evidence: *E. Forster*, c. 1800.
1. Potters Bar, *T. & D.*, 331. Banks of the Colne, 1884, *Fraser* (K). **2.** Shepperton!, *Newbould*; meadows by the Thames, Sunbury!, plentiful, *T. & D.*, 331. Near Laleham. Between Penton Hook Lock and Staines. **3.** Near Twickenham, *A. Wood, Trim. MSS.* **4.** Hampstead, *Irv. MSS.* Brent Reservoir; Cricklewood, *Warren*; Harrow Weald, *T. & D.*, 332. Hendon, *De Cresp. Fl.*, 8. **5.** Near Twyford;† near Ealing;† near Acton,† *Newbould*; Hanwell,† *Warren*; near Chiswick, *T. & D.*, 332. Syon Park!, *A. B. Jackson*. **6.** Whetstone, *T. & D.*, 332. Near Finchley, *Newbould, Trim. MSS.* **7.** Hackney;† marshes near Lee Bridge,† *E. Forster* (BM). Kensal Green Cemetery, *Newbould*; Kilburn, *T. & D.*, 332. West India Docks, Limehouse, plentiful,† *Knapp Gram. Brit.* 1, 82. Hyde Park, casual,† *Warren Fl.*

Var. **subsecalinus** Parn.
5. Syon Park, *A. B. Jackson*.

B. commutatus Schrad. Meadow Brome.
B. pratensis Ehrh. ex Hoffm., non Lam., *Serrafalcus commutatus* auct.
Native in meadows, grassy places, on cultivated and waste ground, by roadsides, etc. Rare, or overlooked, and sometimes confused with *B. racemosus*.
First evidence: *E. Forster*, c. 1800.
1. Uxbridge!; Denham!, *Druce Notes*. Harefield, *J. E. Cooper, K. & L.*, 319. Yiewsley, *J. E. Cooper* (BM). Springwell. Ruislip. **2.** West Drayton; Harmondsworth!, *Druce Notes*. East Bedfont, *Welch*; Shepperton!, *K. & L.*, 319. **3.** Roxeth,† *Hind*; near Whitton, *T. & D.*, 332. Hounslow Heath, *Hubbard*!, *K. & L.*, 319. **4.** Harrow, frequent, *Melv. Fl.*, 92. **5.** Near Twyford, 1867; Brentford, 1867, *Newbould* (BM). Hanwell!, *K. & L.*, 319. **7.** Banks of the Lee above Lee Bridge;† Clapton,† *E. Forster*; Hackney Marshes, 1909,† *J. E. Cooper* (BM). Adelaide Road, Hampstead, 1870† (SY).

B. carinatus Hook. & Arn. Californian Brome.
B. laciniatus auct., *Ceratochloa carinata* (Hook. & Arn.) Tutin
Alien. California. An escape from the Royal Botanic Gardens, Kew, which is now abundantly naturalised on the Surrey bank of the Thames from near Hammersmith Bridge to beyond Richmond. It spread into Middlesex during the Second World War and is rapidly extending its range in the vice-county. Naturalised on the banks of

the Thames and Brent, waste ground, in grassy places, by roadsides, etc., established in gravel pits, on rubbish-tips, etc. Locally abundant, especially near the Thames.

First record: *D. H. Kent*, 1945.

1. Rubbish-tip, Harefield, 1969, *Clement* and *Ryves*. **2.** Gravel pits, East Bedfont, 1965!, *Mason, L.N.* 45: 21; increasing, 1966, *Boniface*!; locally plentiful, 1969. Gravel pits and rubbish-tip, Harmondsworth, 1967 →. **3.** Hounslow Heath, 1964, *Wurzell, L.N.* 44: 17. Thames side, Twickenham, in quantity, 1965 →. Duck's Walk, Isleworth, a few plants on the river-wall of the Thames, 1966. **5.** A few plants on the river-wall between Kew Bridge and Brentford, 1945!, *L.N.* 27: 35; locally abundant by the Thames at Brentford, 1969. Chiswick!, *Welch, L.N.* 28: 35; now very abundant by the Thames between Chiswick Bridge and Barnes Bridge; Chiswick Mall, 1965 →. Strand-on-the-Green!; rubbish-tip, Greenford!, *K. & L.*, 319; now plentiful along the banks of the Brent between Hanwell and Greenford. Rubbish-tip, Hanwell, 1953!, *Welch, K. & L.*, 319. Bombed site, Ealing!, 1951, *L. M. P. Small*; abundant, 1960; site built on in 1962, but a few plants survived on adjacent waste ground; very scarce, 1965. Neglected garden, Gunnersbury!, *Hubbard, K. & L.*, 319; still plentiful, 1973. Road verges, Great West Road, Brentford, 1965 →. Disturbed ground between Brentford and Hanwell, 1965 →. Gunnersbury Park, Brentford to South Ealing, plentiful, 1960 →. **7.** Waste ground, Hammersmith Broadway, a few plants, 1965. Inner Circle, Regents Park!, 1966, *Brewis, L.N.* 46: 35; 1968.

BRACHYPODIUM Beauv.

B. sylvaticum (Huds.) Beauv. Slender or Wood False-brome.

Native in woods and shady places, etc. Common, except in heavily built-up areas where it is rare or extinct.

First record: *T. Johnson*, 1629.

7. Regents Park!, *Holme*; Kensington Gardens!; Hyde Park!, *L.N.* 39: 60. Fulham Palace grounds, *Welch*!.

B. pinnatum (L.) Beauv. Tor-grass, Heath False-brome.

Native in dry grassy places, established on railway banks, etc. Very rare.

First evidence: *Benbow*, 1892.

1. Duck's Hill Heath, between Ruislip and Northwood, plentiful, 1892, *Benbow* (BM); Copse Wood, Northwood, locally plentiful!, *Welch, L.N.* 26: 65; 1968. The two records refer to the same locality. Sparingly above an old chalkpit, Harefield, 1949!,† *K. & L.*, 321. **5.** Railway bank, North Acton, a very large form!, 1968, *McLean*, det. *Hubbard*; 1969.

AGROPYRON Gaertn.

A. caninum (L.) Beauv. Bearded Couch-grass.
Triticum caninum L., *Roegneria canina* (L.) Nevski
Native in woods, thickets, shady places, etc. Rare, and decreasing, and mostly confined to the northern parts of the vice-county.
First evidence: *Goodger* and *Rozea*, 1817.
1. Between Harefield and Rickmansworth!, *T. & D.*, 333. Near Harefield!, *Benbow*, *J.B.* (*London*) 23: 38. Harefield, 1966, *I. G. Johnson* and *Kennedy*!. Between Yiewsley and Iver!, *Welch*, *K. & L.*, 322. South Mimms!, *Trimen* (BM). **2.** Between Poyle and Colnbrook!, *L.N.* 26: 65. Near Horton, a few yards on the Middlesex side of the county boundary!,† *K. & L.*, 322. **4.** Willesden,† *De Cresp. Fl.*, 73. By the Brent, Finchley!, 1902, *L. B. Hall*; 1927, *J. E. Cooper*, *K. & L.*, 322; 1964. Hendon golf course, plentiful, 1966, *Kennedy*!. **5.** Canal bank, Brentford, 1945!,† *Welch*, *L.N.* 26: 65. Hanwell!, *K. & L.*, 322; 1968. **6.** Church End, Finchley, *J. E. Cooper*. Near Highgate, *Hyde*, *Trim. MSS*. Enfield Chase. **7.** Marylebone Fields, 1817 (G & R). Hackney Marshes, 1912,† *J. E. Cooper*, *K. & L.*, 322.

A. repens (L.) Beauv. Couch-grass, Twitch.
Triticum repens L., *Elytrigia repens* (L.) Nevski
Lit. *J. Ecol.* **51**: 783 (1963).
Native on cultivated and waste ground, in fields, by roadsides, etc. Very common.
First record: *Gerarde*, 1597.
7. Frequent in the district.
The common Middlesex plant appears to be the var. **arvense** Reichb.

HORDEUM L.

H. secalinum Schreb. Meadow Barley.
H. pratense Huds., *H. nodosum* auct.
Native in meadows, in grassy places, etc., established on railway banks, on lawns, etc. Common.
First evidence: *F. H. Ward*, 1838.
7. Isle of Dogs, *F. H. Ward* (BM); c. 1869, *T. & D.*, 333. Hyde Park!, *Trim. MSS*. Camden Town, 1847 (HE). Hackney Marshes, *Cherry*, *T. & D.*, 333; 1909, *Green* (SLBI). Bombed site, Jewry Street, E.C., 1945!,† *Welch*. Highgate Ponds!, *H. C. Harris*. Regents Park, *Holme*; Kensington Gardens!, *L.N.* 39: 60. Hammersmith. Brompton Cemetery. Notting Hill. Kilburn. East Heath, Hampstead. Ken Wood. Shoreditch. Bethnal Green. Stepney. Hackney Wick.

Var. **brevisubulatum** (Trin.) Thell.
Hordeum brevisubulatum (Trin.) Link
5. Greenford, 1917, *J. E. Cooper* ex *Brown, Rep. Bot. Soc. & E.C.* 6: 327.
This species is sometimes sown in grass-seed mixtures.

H. murinum L. subsp. **murinum** Wall Barley.
Lit. *J. Ecol.* **59**: 493 (1971).
Native on cultivated and waste ground, in fields, by roadsides, etc., established on railway banks, rubbish-tips, etc. Very common.
First record: *T. Johnson*, 1638.
7. Very common in the district.
A plant with a trifurcate spike is recorded from **5.** Railway embankment, Kew Bridge station, *J.B.* (*London*) 53: 338.

Subsp. **leporinum** (Link) Aschers. & Graebn.
Hordeum leporinum Link
7. Bombed site between the Tower of London and Billingsgate;† The Highway, Shadwell, *Henson*, det. *Melderis, L.N.* 32: 82.

HORDELYMUS (Jessen) Harz

H. europaeus (L.) Harz Wood Barley.
Hordeum europaeum (L.) All., *H. sylvaticum* Huds., *Elymus europaeus* L.
Native in woods, shady places, etc., chiefly on calcareous soils. Very rare, and perhaps extinct.
First evidence: *Lightfoot*, c. 1780.
1. Harefield,† *Lightfoot* (LT). Garett Wood, 1886–89,† *Benbow* (BM). Denham, *C. E. Marks, Rep. Bot. Soc. & E.C.* 11: 48. Spontaneously in a garden at Northwood, 1951,† *Graham, K. & L.*, 324.

KOELERIA Pers.

K. cristata (L.) Pers. Crested Hair-grass.
K. gracilis Pers.
Native in dry grassy places on light soils. Very rare, and mainly confined to the vicinity of the Thames.
First record: *H. C. Watson*, c. 1850.
1. Uxbridge Common!,† *Benbow, J.B.* (*London*) 23: 240. **2.** Between Staines and Twickenham!, *Wats. MSS.* Staines, *Trimen* (BM). Hampton Court!, *J. E. Cooper, K. & L.*, 311; 1968. **3.** Meadow near Richmond Bridge, *Newbould*,† *T. & D.*, 322. Hounslow Heath!,† *L.N.* 27: 35. **5.** Horsendon Hill!, abundant, *Benbow, J.B.* (*London*) 25: 365; now very scarce. Syon Park!, *A. B. Jackson, K. & L.*, 311. **7.** Kensington Gardens, three or four tufts, 1871,† *Warren* (BM).

TRISETUM Pers.

T. flavescens (L.) Beauv. Yellow Oat.
Avena flavescens L.
Native in meadows, pastures, commons, etc., established on railway banks, lawns, etc. Common, especially on dry soils.
First evidence: *Goodger* and *Rozea*, 1817.
7. Common in the district.

AVENA L.

Lit. *Proc. Bot. Soc. Brit. Isles* **1**: 520 (1955). (*Abstr.*)

A. fatua L. Common Wild Oat.
Alien. Europe, etc. Naturalised on cultivated and waste ground, by roadsides, etc., established on railway tracks, canal paths, rubbish-tips, etc. Common.
First record: *J. Newton*, c. 1680.
7. Common in the district.
The vars. **fatua** (*pilosissima* Gray), **pilosa** Syme and **glabrata** Peterm. are all common.

A. fatua × sativa
2. West Drayton, 1909, *Druce*, det. *Hackel*, *Rep. Bot. Soc. & E.C.* 4: 431. **5.** Canal bank, Hanwell, 1925, *Hubbard* (K).

A. ludoviciana Durieu Winter Wild Oat.
Alien. Europe. Naturalised on cultivated and waste ground, especially on heavy soils, by roadsides, etc., established on rubbish-tips, etc. Rather rare, but increasing, and sometimes confused with *A. fatua*.
First record: *D. H. Kent*, 1946.
1. New Years Green, 1964. **2.** East Bedfont, 1965. Frequent near Colnbrook!, *L.N.* 45: 20. Near Poyle, 1966, *J. E. Smith*!. **3.** Hounslow Heath!, 1960, *Philcox* and *Townsend*; 1966. Near Feltham, 1965. Near Hanworth, 1965. Northolt, 1966. **4.** Canal path near Willesden Junction, 1962. Near Rayners Lane, 1966. **5.** Hanwell, 1946!, conf. *Hubbard*, *L.N.* 26: 65. Greenford, 1954, *Welch*, det. *Melderis*, *K. & L.*, 309. Between Perivale and Ealing, 1966. **6.** Enfield, 1965; near Enfield Wash, 1966; Brimsdown, 1966, *Kennedy*!. Near Colney Hatch, 1966.

HELICTOTRICHON Bess.

H. pratense (L.) Pilger Meadow Oat-grass.
Avena pratensis L., *Avenochloa pratensis* (L. emend. Holub) Holub
Native in dry grassy places by the Thames. Very rare.
First evidence: *R. S. Hill*, 1843.

2. Home Park, Hampton Court!, 1843, *R. S. Hill* (D); 1943, *Welch, L.N.* 26: 65; 1953!, *K. & L.*, 310; 1954→. **3.** Meadow near the Thames by Richmond Bridge,† *Newbould, T. & D.*, 321.

H. pubescens (Huds.) Pilger Hairy Oat-grass.
Avena pubescens Huds., *Avenochloa pubescens* (Huds.) Holub
Native in grassy places on calcareous soils at Harefield and South Mimms, and on alluvial soils near the Thames, elsewhere in fields, on banks, waste ground, etc., as an introduction. Very rare.
First record: *Ray*, 1688.
1. Canal side opposite Denham!,† *Trimen*; near Hillingdon, 1871,† *Warren, Trim. MSS.* Springwell to Harefield!, abundant, *Benbow, J.B.* (*London*) 23: 339. Ruislip;† Ickenham;† railway banks between Uxbridge and West Drayton, *Benbow, J.B.* (*London*) 25: 19. Hillingdon;† Frays Meadows, Uxbridge;† Drayton Ford!,† *Benbow MSS.* Northwood, 1912,† *J. E. Cooper*; Cowley, 1918,† *Redgrove* (BM). South Mimms, 1965, *J. G.* and *C. M. Dony*. **2.** Banks of the Colne between West Drayton and Staines Moor, abundant,† *Benbow, J.B.* (*London*) 25: 19. Between Hampton Court and Kingston Bridge!, *Benbow* (BM). **3.** Twickenham, *Ray Hist.* 2, 1910; abundant in the meadows between Richmond Bridge and Twickenham, 1884–87,† *Benbow* (BM). **5.** Syon Park, *Welch*!, *L.N.* 26: 65; 1965. **6.** Old pasture, Finchley Common, 1890,† *J. E. Cooper* (BM).

ARRHENATHERUM Beauv.

A. elatius (L.) Beauv. ex J. & C. Presl False Oat-grass.
A. avenaceum Beauv.
Lit. *J. Ecol.* **50**: 235 (1962).
Native in rough grassy places, ditches, on hedgebanks, cultivated and waste ground, established on railway banks, etc. Very common.
First record: *Gerarde*, 1597.
7. Very common in the district.
The var. **bulbosum** (Willd.) Spenn. (*Arrhenatherum bulbosum* (Willd.) C. Presl, *A. tuberosum* (Gilib.) Schultz, *A. elatius* (L.) Beauv. ex J. & C. Presl subsp. *bulbosum* (Willd.) Hyland.) is not uncommon on cultivated ground.

HOLCUS L.

Lit. *Biol. Journ. Linn. Soc.* **64**: 183 (1971).

H. lanatus L. Yorkshire Fog.
Lit. *J. Ecol.* **49**: 431 (1961).
Native in meadows, woods, marshes, fields, on heaths, commons, cultivated and waste ground, etc., established on railway banks, etc. Very common.

First record: *T. Johnson*, 1638.
7. Frequent in the district.
The hybrid *H. lanatus* × *mollis* is as yet unrecorded from the vice-county but is probably present.

H. mollis L. Creeping Soft-grass.
Lit. *J. Ecol.* **44**: 272 (1956).
Native in open woods, often carpeting the ground, on heaths, in grassy places, etc. Common, especially on light or moderately acid soils.
First record: *Lightfoot*, c. 1780.
7. Hornsey, 1815† (G & R). Primrose Hill, 1845,† *J. Morris, T. & D.*, 320. Ken Wood!; Parliament Hill Fields!, *Fitter*. East Heath, Hampstead. Kensington Gardens!, *Warren, Trim. MSS*. Holland House grounds, Kensington!, *Sandwith*. Bombed site, St Olave's, Hart Street, E.C.,† *Lousley, K. & L.*, 308. Bombed site, Billiter Square, E.C.,† *Lousley Fl. City*. Regents Park!, *Holme*; Buckingham Palace grounds, *Codrington, Lousley* and *McClintock*!, *L.N.* 39: 60. Fulham Palace grounds, *Brewis*.
A glabrous form is recorded from **2.** Bushy Park!, *Hubbard*.

DESCHAMPSIA Beauv.

D. cespitosa (L.) Beauv. subsp. **cespitosa** Tufted hair-grass.
Aira cespitosa L.
Native in marshes, wet meadows, rough fields, woods, etc., established on railway banks, etc. Common, especially on heavy, badly drained soils.
First record: *A. Irvine*, 1830.
7. Kensington Gardens!, 1850, *J. Morris, T. & D.*, 320; one tussock, 1949!; Regents Park!; Primrose Hill!, *L.N.* 39: 60. Side of London Canal, Hackney, *Cherry*; East Heath, Hampstead!, *T. & D.*, 320. Ken Wood. Parliament Hill Fields. Bombed site, Stepney. Car park by National Gallery, 1962, *Brewis, L.N.* 46: 28. Buckingham Palace grounds, one tuf ton islet, 1963, *Fl. B.P.G.* Islington. Dalston. Tottenham. Upper Clapton.

Subsp. **parviflora** (Thuill.) Richter
D. cespitosa (L.) Beauv. var. *parviflora* (Thuill.) Dumort.
1. Uxbridge, *Benbow* (BM). Harefield Grove, 1966, *I. G. Johnson* and *Kennedy*!. **4.** Turner's Wood, *Fitter* (K), det. *Hubbard* (as var. *parviflora*). Willesden Green, casual, 1949.† Mill Hill, 1966, *Kennedy*!.

D. flexuosa (L.) Trin. Wavy Hair-grass.
Aira flexuosa L., *Avenella flexuosa* (L.) Drejer
Lit. *J. Ecol.* **42**: 225 (1954).

Native on heaths, in open woods and grassy places on acid soils, etc. Locally abundant.

First record: *A. Irvine*, 1830.

1. Harrow Weald Common, abundant!, *Melv. Fl.*, 88. Stanmore Common!; Elstree!, *T. & D.*, 320. Duck's Hill, abundant; Ruislip!; Northwood!; Pinner Hill!; Pinner Wood!, *Benbow MSS.* **2.** Between Staines, Ashford and Charlton, *Benbow, J.B. (London)* 25: 19. Staines Moor!, *K. & L.*, 308. **3.** Hounslow Heath!, *K. & L.*, 308. **4.** Bishop's Wood!, *Newbould, T. & D.*, 320. Harrow Weald!; Bentley Priory!; woods between Edgware and Stanmore!; Barnet Gate Wood!; West Heath, Hampstead!, *Benbow MSS.* Brockley Hill, *Kennedy*!. Mill Hill, *Sennitt.* **5.** Horsendon Hill!, *Benbow MSS.* Syon Park, *Welch*!; Osterley Park!, *Hubbard, K. & L.*, 308. Wyke Green.† Chiswick House grounds!, *Murray.* **6.** Winchmore Hill Wood,† *Benbow MSS.* Highgate Woods, *J. E. Cooper, K. & L.*, 308. Finchley Common. Crew's Hill golf course, Enfield, *Kennedy*!. Trent Park, *T. G. Collett* and *Kennedy.* **7.** East Heath, Hampstead!, *Irv. MSS.* Ken Wood!, *Fitter, K. & L.*, 308. Highgate Ponds, *Fitter.* Kensington Gardens, in old turf, *D. E. Allen*, det. *Melderis, L.N.* 41: 22.

AIRA L.

A. praecox L. Early Hair-grass.

Native on heaths, commons, dry sunny banks, in open woods, etc., established on railway tracks, gravel paths, etc. Local, and chiefly confined to light soils.

Lit. *J. Ecol.* **55**: 539 (1967).

First evidence: *Blackstone*, c. 1736.

1. Harefield Common!, *Blackstone* (BM). Harrow Weald Common!, *T. & D.*, 320. Uxbridge Common!; Hillingdon Heath!; Stanmore Common!, *Benbow MSS.* Copse Wood, Northwood, *Wrighton.* Ruislip Common. Gould's Green. Mimmshall Wood, *J. G.* and *C. M. Dony*!. **2.** Near Charlton, sparingly, *T. & D.*, 320. Staines Moor!; Bushy Park!, *Benbow MSS.* Hampton Court Park!, *J. E. Cooper.* **3.** Hounslow Heath!, *T. & D.*, 320. Whitton Park Inclosure,† *Benbow MSS.* Golf course near Dormer's Wells, Southall. **4.** Hampstead Heath (HE); 1864, *Parsons* (CYN). Whitchurch Common, *Lond. Nat. MSS.* Scratch Wood. Barn Hill, Wembley, *Kennedy*!. Harrow, *Temple*! **5.** Near Hanwell,† *Warren, T. & D.*, 321. Ealing Common,† *Benbow MSS.* Syon Park, *Welch*! Chiswick House grounds. Plentiful on paths, Greenford Park Cemetery. Railway tracks near Brentford. **6.** Finchley Common!, *J. E. Cooper.* Hadley Green. Crew's Hill golf course, Enfield, *Kennedy*!. **7.** Hyde Park, 1816,† *J. F. Young* (BM). East Heath, Hampstead, *T. & D.*, 321.

A. caryophyllea L. Silvery Hair-grass.

Native on heaths, commons, dry banks, in open woods, etc., established on railway tracks, etc. Very rare, and mainly found on light soils.

First evidence: *Dickson*, c. 1795.

1. Railway between Uxbridge and West Drayton, 1890, *Benbow* (BM). Between Uxbridge Common and Swakeleys,† *Benbow MSS.* Disused railway between Uxbridge and Denham!, *K. & L.*, 307. Park Wood, Ruislip, *Turrill* (K). Near Pinner, 1919,† *Bishop, K. & L.*, 307. Mimmshall Wood!, *L.N.* 28: 35; 1964. Road bank, Great North Road, South Mimms, 1966. **2.** Teddington, 1867, *Dyer* (BM). Stanwell Moor, 1908, *Green* (SLBI). Hampton Court Park!, *Welch, K. & L.*, 307; 1969. Gravel pit, West Bedfont, 1965, *Goom and Kennedy*!. **4.** Hampstead Heath,† *Irv. Lond. Fl.*, 96. Oakleigh Park, 1914, *R. W. Robbins, K. & L.*, 308. **6.** North Finchley, 1925, *J. E. Cooper, K. & L.*, 308. **7.** Hyde Park,† *Dicks. Hort. Sicc.* Crouch End,† *French, Trim. MSS.* Hackney Marshes, 1913,† *Coop. Cas.* Site of 'The Hour Glass', Upper Thames Street, E.C., 1945,† *Lousley* (BM).

CALAMAGROSTIS Adans.

C. epigejos (L.) Roth Wood Small-reed, Bushgrass.

Native in damp woods, thickets, ditches, on field borders, etc. Very rare, and mainly confined to heavy soils.

First record: *l'Obel*, c. 1600.

1. Shady lane near Harefield leading to Rickmansworth;† Old Park Wood, Harefield,† *De Cresp. Fl.*, 9. Northwood, 1908,† *Green* (SLBI). **2.** Canal plantation, Bushy Park, a fine patch, 1965, *Clement*. **4.** Edge of field between Totteridge and Mill Hill!, 1957, *Hinson*; 1961, *J. F.* and *P. C. Hall*; 1964. **5.** Bedford Park, *Cockerell Fl.* Probably an error. Floor of disused reservoir, Ealing, 1968, *Clement* and *McLean*. **6.** Highgate,† *Lob. Ill.*, 42. Tottenham,† *Johns. Cat.* **7.** St John's Wood [Highbury],† *Park. Theat.*, 1181 and *Merr. Pin.*, 46. Wood against the Boarded river,† *Pet. Midd.* Between Newington and Hornsey,† *Huds. Fl. Angl.*, 43. Between Kentish Town and Pancras,† *M. & G.*, 127. Between Kilburn and Primrose Hill,† *Irv. MSS.* Near Lee Bridge, below Clapton,† *Pamplin, Irv. Lond. Fl.*, 96. Near Hornsey Wood, 1838,† *E. Ballard* (BM). Bombed site, Ave Maria Lane, E.C., 1948!,† *Lousley* (L).

AGROSTIS L.

Lit. *J. Linn. Soc. Bot.* **51**: 73 (1937).

A. canina L. Velvet Bent-grass.
Native on heaths, commons, in meadows, damp grassy places, etc., mostly on acid soils. Common.
First certain evidence: *Benbow*, 1884.
7. East Heath, Hampstead. Ken Wood.

A. stricta J. F. Gmel. Brown Bent-grass.
A. canina subsp. *montana* (Hartm.) Hartm., *A. coarctata* Ehrh. ex Hoffm. subsp. *coarctata*
Native on heaths, commons, in meadows, etc., mostly on acid soils, but in drier situations than the previous species. Common.
First certain evidence: *A, B. Lambert*, c. 1800.
7. Between London and Hampstead, *A. B. Lambert* (K), East Heath, Hampstead. Regents Park.

A. tenuis Sibth. Common Bent-grass, Brown Top.
A. vulgaris With., *A. capillaris* auct. brit.
Native on heaths, commons, in fields, by roadsides, etc. Established on railway banks, etc. Very common, particularly on poor and acid soils.
First evidence: *Buddle*, c. 1710.
7. Common in the district.

A. castellana Boiss. & Reut.
Alien. Spain, etc. Introduced with grass-seed mixtures. Possibly common, but overlooked as *A. tenuis* from which it differs in having lemmas awned from the base, a 5-nerved lemma, a longer uppermost ligule, and tufts of hair on the callus of the spikelet.
First record: *Hubbard*, 1973.
5. Field by Perivale Wood, 1973, *Hubbard*.

A. gigantea Roth Common Bent-grass, Black Bent-grass, Red Top.
A. nigra With., *A. capillaris* var. *nigra* (With.) Druce
Native in open woods, meadows, on cultivated and waste ground, by roadsides, etc., established on railway banks, rubbish-tips, etc. Common, but confused with *A. stolonifera* and *A. tenuis*.
First certain record: *Lousley*, 1944.
7. Common in the district.
The common Middlesex plant is the var. **dispar** (Michx.) Philipson.

Var. **ramosa** (Gray) Philipson
1. Uxbridge. Ruislip. Harefield. South Mimms. **2.** Staines. **4.** Hendon. **5.** Hanwell!, det. *Hubbard*, *L.N.* 27: 35. Syon Park!, det. *Hubbard*, *K. & L.*, 305. Greenford. **6.** Enfield; Enfield Chase, *Kennedy*!.

A. stolonifera L. Creeping Bent, Fiorin.
A. alba auct.
Native on heaths, cultivated and waste ground, in open woods, by roadsides, etc., established on railway banks, rubbish-tips, etc. Very common.

First certain evidence: *E. Forster*, c. 1800.
The var. **palustris** (Huds.) Farw. (*A. palustris* Huds.) is common in marshes, by pond verges and streams, in wet fields, etc.

A. stolonifera × tenuis
3. Hounslow Heath. **4.** West Heath, Hampstead.
This hybrid is probably common and widespread.

APERA Adans.

A. spica-venti (L.) Beauv. Loose Silky-bent, Silky Apera.
Agrostis spica-venti L.
Introduced. Naturalised on cultivated and waste ground, by roadsides, etc. Locally common especially on light soils near the Thames, though sometimes merely casual.
First record: *Willisel*, 1670.
1. Harefield!, *Blackst. Fasc.*, 34. Near Uxbridge, *Lightf. MSS.* Uxbridge; Hillingdon!, *Benbow MSS.* Yiewsley!, *Coop. Cas.* Ickenham!, *Tremayne.* Near Denham. **2.** Locally common, especially near the Thames. **3.** Twickenham, *Borrer, B.G.*, 400. Roxeth, *Hind, Melv. Fl.*, 88. Hounslow!; Whitton; Isleworth, *T. & D.*, 318. Hayes!; Cranford; Southall; Feltham, *Benbow MSS.* **4.** East Finchley, *Coop. Cas.* Finchley, 1921, *J. E. Cooper.* West Heath, Hampstead, 1955, *Welch, K. & L.*, 307. Mill Hill, 1960, *Sennitt.* **5.** Wyke Green; Alperton, *T. & D.*, 318. Near Southall!; Heston; Norwood Green; Hanwell!; Chiswick, *Benbow MSS.* Brentford, *A. Wood, Trim. MSS.* Near Brentford, casual, 1962. East Acton, 1847, *Dennes* (K). Between Hanwell and Ealing, 1877, *Britten, Trim. MSS.*; one plant, 1965. Ealing!, *Boucher.* Horsendon Hill, casual, *K. & L.*, 307. Acton, casual, 1966. Plentiful by Kew Bridge, 1971. **6.** Edmonton, *Woods, B.G.*, 400; abundant, 1960, *Dyer* (BM). Tottenham, *Dillwyn, B.G.*, 400. Southgate, *L. B. Hall, K. & L.*, 307. Muswell Hill, *Coop. Cas.* Near East Barnet, 1965, *Bangerter*!. **7.** About London, *Willisel, Ray Cat.*, 137. Pancras Churchyard,† *M. & G.*, 79. Near Tottenham, *Irv. Lond. Fl.*, 94. Chelsea, *Newbould, T. & D.*, 318. Crouch End; Hackney Marshes, 1909, *Coop. Cas.*; 1924, *J. E. Cooper.* South Kensington, *Dyer List.* Fulham, *Burchell*; Hornsey Wood,† *E. Ballard* (BM). Finchley Road, 1869, *Trim. MSS.* Eaton Square, S.W.1, *McClintock, L.N.* 34: 5. Primrose Hill, casual!, *K. & L.*, 307. Hornsey, 1888, *Coop. Cas.* Clapton, *P. S. King* (LNHS). Hyde Park, casual, *Warren Fl.*; 1961, *D. E. Allen*, det. *McClintock, L.N.* 41: 22.

PHLEUM L.

P. bertolonii DC. Smaller Cat's-tail.
P. nodosum auct.

Native in meadows, grassy places, etc., established on lawns, railway banks, etc. Very common.
First certain evidence: *Benbow*, 1891.
7. Frequent in the district.

P. pratense L. Timothy, Cat's-tail.
P. nodosum auct.
Denizen or native in meadows, grassy places, on waste ground, etc., established on railway banks, etc. Common.
First certain evidence: *Dillenius*, c. 1740.
7. Common in the district.
This species is often sown for hay.

ALOPECURUS L.

A. myosuroides Huds. Slender Foxtail, Black Grass, Black Twitch.
A. agrestis L.
Lit. *J. Ecol.* **60**: 611 (1972).
Native on cultivated and waste ground, by roadsides, etc., established on rubbish-tips, etc. Common, though often merely casual.
First record: *Buddle*, c. 1700.
7. Near Paddington, *Buddle*, *Ray Syn.* 3: 397. Camden Town (HE). Bow, *Cherry*; Chelsea, *Newbould*; Isle of Dogs; Hackney Wick; Kentish Town; East Heath, Hampstead; Victoria Street, S.W.1; Thames embankment; St George's Road, S.W.; Kilburn, *T. & D.*, 316. Kensington Gardens, *Warren Fl.* St Olave's Churchyard, Silver Street, E.C.!,† *Sladen* and *McClintock*, *Lousley Fl.* Bombed site, Upper Thames Street!,† *Lousley*. Regents Park, *Holme*; Eaton Square, S.W.1, 1958, *McClintock*, *L.N.* 39: 60. Hyde Park, 1961, *D. E. Allen* and *McClintock*, *L.N.* 41: 22. Site of Knightsbridge Barracks, S.W.1, 1966; weed of flower-beds, Zoological Gardens, Regents Park, 1966, *Brewis*; Holborn, W.C., several plants around the base of a tree, 1966!, *L.N.* 46: 35. Brompton, 1966. Islington, Tottenham. Wapping, 1967. Hackney, 1967.

A. pratensis L. Meadow, or Common Foxtail.
Native in meadows, grassy places, etc., established on railway banks, lawns, etc. Very common.
First record: *Kalm*, 1748.
7. Very common in the area.
Proliferous forms are recorded from **4.** Finchley; Mill Hill, *J. E. Cooper*, *K. & L.*, 304. **6.** Highgate; Muswell Hill, *J. E. Cooper*, *K. & L.*, 304.

A. geniculatus L. Marsh Foxtail.
Native in marshes, wet fields, on muddy pond verges, etc., rarely on dry waste ground. Common.

First record: *Milne* and *Gordon*, 1793.
7. Hyde Park, *M. & G.*, 76. Near Brompton,† *J. Banks*; near Finchley Road, 1870,† *Dyer*; Holloway, 1873, *French* (BM). Paddington, 1817 (G & R). Copenhagen Fields† (HE). Homerton,† *Cherry*; Isle of Dogs, *Newbould*; Kensal Green, *Warren*; Hackney Wick; East Heath, Hampstead!, *T. & D.*, 315. Canonbury fields,† *Ball. MSS.* Ken Wood. Parliament Hill Fields. Verge of lake, Buckingham Palace grounds, *Codrington*, *Lousley* and *McClintock*!. Kensington Gardens, 1959, introduced with soil, *D. E. Allen*, *L.N.* 39: 60. Verge of New River, Finsbury Park, 1969.

A. geniculatus × **pratensis** = **A.** × **hybridus** Wimm.
1. Marshy field by canal near Harefield, 1937, *A. H. G. Alston* and *Sandwith*, *Rep. Bot. Soc. & E.C.* 11: 153; 1938, *Hubbard* (K).

A. aequalis Sobol. Orange, or Short-awn Foxtail.
A. fulvus Sm., *A. geniculatus* L. subsp. *fulvus* (Sm.) Hartm., *A. geniculatus* L. subsp. *aequalis* (Sobol.) Fiori
Native on muddy pond verges, in wet fields, ditches, etc., chiefly on heavy soils. Rare.
First evidence: *Woods*, 1805.
1. Margin of Elstree Reservoir, abundantly,† *Webb. & Colem.*, 324; still present on the Herts side of the reservoir. Ruislip Reservoir!, *Hind*, *Melv. Fl.*, 86. Near Eastcote; near Northwood; Warren Pond, Breakspeares,† *Benbow MSS.* **2.** Hampton Court Park!, *Countryside* 1922, 209; 1966, *Clement*, conf. *Hubbard*; 1969. Shortwood Common!, *L.N.* 26: 65. Laleham, 1966, *Kennedy*!. **3.** Abundant on the margins of ponds, Cranford,† *Benbow*, *J.B. (London)*, 25: 19. Hounslow Heath!, *Welch*, *L.N.* 26: 65; 1966. **4.** Edgware, *E. M. Dale* (LNHS). Brent Reservoir!, *Richards*. Finchley, *Woods* (BM). Near Scratch Wood. **5.** Osterley Park!, *Flippance* (K). Perivale!, *Druce Notes*; wet field by Perivale Wood, 1964. **6.** Gravel pit near Tottenham, *De Cresp. Fl.*, 3. **7.** Pond in Ken Wood grounds, 1866, *Trimen* (BM).

MILIUM L.

M. effusum L. Wood Millet.
Native in damp woods, shady places, etc., especially on heavy soils. Locally plentiful in the northern parts of the vice-county, rather rare elsewhere.
First evidence: *Petiver*, c. 1683.
1. Harefield!, *Blackst. Fasc.*, 38. Uxbridge; Swakeleys; Pinner Wood!; Garett Wood!,† *Benbow MSS.* Springwell. Ruislip Woods!, *K. & L.*, 305. **2.** By Kingston Bridge, 1966. **3.** Wood End,† *Hind*, *Melv. Fl.*, 88. Harrow Grove, *Horton Rep.* Error. Near Hatch End. **4.**

Bishop's Wood!, *Warren*; Turner's Wood!, *T. & D.*, 316. Barnet Gate, 1872, *J. A. H. Murray* (BM). Scratch Wood, abundant!, *Benbow MSS.* Harrow Park!, *Harley Fl.*, 36. Near Harrow-on-the-Hill Churchyard, *Temple*!. Barn Hill, Wembley Park. Stanmore. Bentley Priory, abundant. Brent Park, Hendon. East Finchley. Near Brondesbury,† *De Cresp. Fl.*, 108. **5.** Perivale Wood!, *Benbow MSS.* Osterley Park!, *Loydell* (D). Wood near Hanwell!, *Melville* (K). Long Wood, Wyke Green. Boston Manor Park, near Brentford. **6.** Colney Hatch (HE). Near Whetstone; Winchmore Hill Wood,† *T. & D.*, 316. Highgate Woods!, *C. S. Nicholson, K. & L.*, 305. Enfield. Trent Park. **7.** Belsize, Hampstead,† *Petiver* (SLO). Hornsey Wood,† *Forst. MSS.* Ken Wood, *Macreight Man.*, 260; c. 1869,† *T. & D.*, 316.

ANTHOXANTHUM L.

A. odoratum L. Sweet Vernal-grass.

Lit. *BSBI Abstr.* **1**: 29 (1971). (*Abstr.*)

Native in meadows, grassy places, open woodlands, marshes, on heaths, commons, etc., established on railway banks, lawns, etc. Very common.

First record: *Stillingfleet*, 1762.

7. Common in the district.

PHALARIS L.

P. arundinacea L. Reed-grass.

Digraphis arundinacea (L.) Trin., *Typhoides arundinacea* (L.) Moench, *Baldingera arundinacea* (L.) Dumort.

Native on the verges of rivers, streams, lakes, ponds, etc., in marshes, established by gravel pits, in canals, etc. Common.

First record: *l'Obel*, c. 1600.

7. Not far from London, 'via qua itur Ratteam', *Lob. Ill.*, 45; in the low moist grounds by Ratcliffe . . .,† *Park. Theat.*, 1273. By the Thames banks, *How Phyt.* 50 and *Pet. Gram. Conc.*, 70. Isle of Dogs (G & R). Serpentine, 1813, *Herb. Devon Inst. Exeter*, ex *T. & D.*, 314. Hackney Wick!, *T. & D.*, 314. South Kensington,† *Dyer List.* Canal near Kensal Green Cemetery. Ken Wood grounds. Lake, Buckingham Palace grounds!,† Eastwood; perhaps planted, *L.N.* 39: 60; not seen since 1956, *Fl. B.P.G.* Thames side, Hammersmith. Lee Navigation Canal, Upper Clapton to Tottenham. Lee, Hackney Marshes.

Var. **picta** L. 'Gardeners' Garters'.

1. Eastcote, 1966. **3.** Hounslow Heath, 1965!, *Clement, Mason* and *Wurzell.* Hatch End, 1966. **6.** Gravel pit north of Enfield Lock, 1967, *Kennedy*!.

This variety, a garden outcast, has a broad white stripe down the centre of the leaf blade.

P. canariensis L. Canary Grass.
Alien. N. Africa, etc. Introduced with cage-bird seed. Naturalised on waste ground, by roadsides, etc., established on rubbish-tips, etc. Common.
First record: *T. Johnson*, 1638.
7. Common in the district.

NARDUS L.

N. stricta L. Mat-grass.
Lit. *J. Ecol.* **48**: 255 (1960).
Native on heaths, commons, rough grassy places on acid soils. Locally plentiful.
First record: *T. Johnson*, 1632.
1. Harefield Common!, abundantly, *Blackst. Fasc.*, 39. Harrow Weald Common!, abundant, *Melv. Fl.*, 87. Stanmore Common!, *T. & D.*, 316. Uxbridge Common!, *Benbow MSS.* Ruislip Common!, *L. & K.*, 323. **3.** Hounslow Heath!, *Clements*, *Hill Fl. Brit.*, 35. **4.** Hampstead Heath!, *J. Morris* (BM); still locally plentiful on the West and North-west Heaths. Whitchurch Common!,† *L. B. Hall*; Scratch Wood!, *H. C. Harris*, *K. & L.*, 323. **5.** Hanwell Heath,† *Hill Fl. Brit.*, 35. Wall by Turnham Green station, *Cockerell Fl.* Error. Syon Park!, *Welch*, *L.N.* 26: 65. **6.** Hadley Green, abundant!, *K. & L.*, 323. Trent Park, *A. Vaughan*. **7.** East Heath, Hampstead!, *T. & D.*, 316.

SETARIA Beauv.

S. viridis (L.) Beauv. Green Bristle-grass.
Panicum viride L.
Alien. Europe, etc. Introduced with cage-bird seed, etc. Naturalised on waste ground, by roadsides, etc., established on rubbish-tips, etc. Common.
First record: *Merrett*, 1666.
7. Common in the district.

Var. **weinmanni** Roem. & Schult.
1. Yiewsley, *M. & S.*

Var. **brevisetum** Döll.
1. Yiewsley, *J. E. Cooper* and *G. C. Brown*, *Rep. Bot. Soc. & E.C.* 8: 590.

CASUAL SPECIES

The following taxa have been noted as occurring casually in the vice-county. They consist of garden escapes, bird-seed, grain and soya-bean aliens, etc. It is possible that some of them may eventually become naturalised or established.

ASPLENIACEAE

Asplenium viride Huds. Green Spleenwort. **6.** Wall near Arnos Grove, Southgate, *O. E. Walker, Moore* 2: 115.

RANUNCULACEAE

Nigella arvensis L. **1.** Uxbridge, *Wedgwood, Rep. Bot. Soc. & E.C.* 6: 370. **7.** Crouch End, 1901, *Coop. Cas.*

N. damascena L. Love-in-a-mist. **1.** Near South Mimms, 1965, *J. G.* and *C. M. Dony*!. Harefield, 1966, *I. G. Johnson* and *Kennedy*!. **3.** Hounslow Heath, 1962.

Aconitum napellus L. *sensu lato.* Monkshood. **1.** Harefield, 1847, *Stevens* (K). **3.** Plantation adjoining Hounslow Heath, 1862, *T. & D.*, 21. **5.** Ealing, 1965, *McLean.*

Consolida ambigua (L.) P. W. Ball & Heywood, *Delphinium ajacis* auct., *D. gayanum* auct. Larkspur. **1.** Hillingdon Heath, 1885, *Benbow* (BM).

C. orientalis (Gay) Schrödinger subsp. **orientalis**, *Delphinium orientale* J. Gay. Larkspur. Common on rubbish-tips, etc. **7.** Chelsea; Brompton, frequent, *Britten, T. & D.*, 21. Crouch End, 1897, *Coop. Cas.* Hackney Marshes, *M. & S.* Bombed site, Lupus Street, S.W.1, 1943, *McClintock, Rep. Bot. Soc. & E.C.* 12: 700. Bombed site, Newgate Street, E.C. 1943, *Wilmott* (BM). Bombed site, Basinghall Street, E.C., 1944, *Sladen* and *McClintock, Lousley Fl.* Eaton

Square, S.W.1, one plant, 1958, *McClintock*; Eccleston Square, S.W.1, 1958–59!, *L.N.* 39: 43. Holloway. Isle of Dogs.

Anemone × hybrida Paxton, *A. japonica* auct. **4.** Harrow, *Harley Fl.*, 2. **5.** Hanwell, 1942–44.

Adonis annua L., *A. autumnalis* L. Pheasant's Eye. **1.** Uxbridge, 1884, *Benbow* (BM). New Years Green, one plant, 1954, *T. G. Collett, L.N.* 34: 6. **4.** Hampstead, 1878, *de Vesian* (HPD). **5.** Acton, *W. Watson, Blackst. Spec.*, 23.

Thalictrum aquilegifolium L. **4.** Willesden Green, 1956–62. **7.** Canal path near Old Ford, 1957.

T. minus L. *sensu lato*. Lesser Meadow Rue. **1.** Harrow Weald Common, 1958, *Knipe, L.N.* 39: 40. **4.** West Heath, Hampstead, 1948–52!, *K. & L.*, 1. **7.** Holloway, 1965, *D. E. G. Irvine.*

PAEONIACEAE

Paeonia mascula (L.) Mill., *P. corallina* Retz., *P. officinale* auct. Peony. **3.** Field near Norwood Green, 1965.

PAPAVERACEAE

Papaver orientale L. Oriental Poppy. **7.** Holloway, 1965, *D.E.G. Irvine.*

Argemone mexicana L. **3.** Hounslow Heath, 1959, *Buckle.*

Meconopsis cambrica (L.) Vig. Welsh Poppy. **1.** Pinner, *Hind, Melv. Fl.*, 5; 1950, *Knipe*. Stanmore Common, three plants, 1947!, *L.N.* 27: 30. **4.** Sewage works, Finchley, two plants, 1964, *Wurzell.* Roadside, East Finchley, four plants, 1966. **6.** Whetstone Stray, 1966, *Kennedy*!.

Roemeria hybrida (L.) DC. **1.** Near Potters Bar, 1912, *J. E. Cooper* (BM).

Glaucium flavum Crantz, *G. luteum* Crantz. Yellow Horned-poppy, Sea Poppy. **2.** Between Staines and Twickenham, *Wats. MSS.* **5.** Hanwell, a single plant, 1946!, *L.N.* 26: 57. **7.** Site of the International Exhibition of 1862, South Kensington, 1865, *T. & D.*, 24; 1872, *Dyer Key.*

G. corniculatum (L.) Rudolph, *G. phoenicium* Crantz. Red Horned-poppy. **1.** Spontaneously in the Parsonage garden, Pinner, 1865,

Hind, *T. & D.*, 25. **5.** Railway yard, Kew Bridge, 1956, *Murray*, *L.N.* 36: 17. **7.** In Chelsea Garden it has from time immemorial come up every year as a weed, *E.B.*

Eschscholzia californica Cham., *E. douglasii* (Hook. & Arn.) Walp. Californian Poppy. **1.** Springwell, 1962, *I. G. Johnson*. **4.** Harrow, *Harley Fl.*, 3. **4.** Mill Hill, *Sennitt*. **5.** Chiswick, 1965. **6.** Near Barnet, 1965, *Bangerter* and *Kennedy*!. Near Enfield, 1966, *Kennedy*!. **7.** Upper Holloway, 1869, *Cherry*, *Trim. MSS.* Chelsea, *Britten*, *Nat.* 1864. Bombed site, Newgate Street, deliberately introduced, *Lousley Fl.* Bombed site, Cripplegate!, *Wrighton*, *L.N.* 29: 86. Bombed site, Theobalds Road, W.C., 1950, *Whittaker* (BM).

FUMARIACEAE

Corydalis ochroleuca Koch. **3.** Hounslow Heath, 1960, *Buckle*. **6.** Gravel pit north of Enfield Lock, 1967, *Kennedy*!.

Fumaria agraria Lag. **1.** Between Iver and West Drayton, 1903–4, *G. C. Druce* (D).

CRUCIFERAE

Brassica oleracea L. *sensu lato*. Cabbage. Common on rubbish-tips, waste ground, etc., as an outcast from cultivation or bird-seed introduction.

B. elongata Ehrh. **2.** West Drayton, *Druce Notes*. **4.** Near Finchley, 1911, *J. E. Cooper* (BM). **5.** Hanwell, 1957.

B. juncea (L.) Czern. & Coss. Common on rubbish-tips, etc., but often confused with forms of *Sinapis arvensis*.

Erucastrum gallicum (Willd.) O. E. Schulz, *Brassica gallica* (Willd.) Druce. **4.** Hendon, 1927, *J. E. Cooper*, *K. & L.*, 22. Finchley, 1926, *J. E. Cooper* (BM). **7.** Hackney Marshes, 1919, *J. E. Cooper* (BM).

Rhynchosinapis cheiranthos (Vill.) Dandy, *Brassica cheiranthos* Vill. **7.** Chelsea, 1852, *Irv. Ill. Handb.*, 704. Regents Park, 1957, *Holme*, *L.N.* 37: 185.

Carrichtera annua (L.) DC., *Vella annua* L. **6.** Roadside near Botany Bay, Enfield Chase, 1965, *Kennedy*!, *L.N.* 45: 20.

Diplotaxis erucoides (L.) DC. **7.** Chelsea, 1864, *Molesworth* (K). Hackney Marshes, *M. & S.*

Eruca vesicaria (L.) Cav. subsp. **sativa** (Mill.) Thell., *E. sativa* L. **2.** West Drayton, *Druce Notes*.

Raphanus sativus L. Cultivated Radish. Common on rubbish-tips, etc., as an outcast from cultivation.

Enarthrocarpus lyratus (Forsk.) DC. **7.** Hackney Marshes, *M. & S.*

Rapistrum perenne (L.) All. **1.** Yiewsley, 1910, *J. E. Cooper* (BM). **7.** Bombed site, Hammersmith, 1952!, *Sandwith*, *L.N.* 32: 81; 1953–57.

Conringia orientalis (L.) Dumort., *Brassica orientalis* L., *Erysimum orientale* (L.) Crantz, non Mill., *E. perfoliatum* Crantz. Hare's-ear Mustard. **1.** Yiewsley, 1918, *J. E. Cooper*. **3.** Roxeth, 1941. **4.** Hendon, 1910, *P. H. Cooke* (LNHS). East Finchley, 1907; Finchley 1909, *Coop. Cas.* **5.** Greenford, 1917, *J. E. Cooper*. **6.** Winchmore Hill, 1919, *L. B. Hall*, *Lond. Nat. MSS.* North Finchley, 1925, *J. E. Cooper*. **7.** Near Edgware Road station, 1870, *Newbould*, *Trim. MSS.* Hornsey, 1887; Crouch End, 1889, *J. E. Cooper* (BM). Hackney Marshes, 1913, *Coop. Cas.*

Lepidium sativum L. Garden Cress. A common feature of rubbish-tips, etc.

Var. **latifolium** DC. **5.** Chiswick, 1960, *M. McC. Webster*, det. *Townsend.*

L. virginicum L. **1.** Near Yiewsley, 1927, *M. & S.* **2.** West Drayton, 1914, *Coop. Cas.* **3.** Twickenham, 1931, *Lousley* (L). **5.** Greenford, 1917, *J. E. Cooper*. **7.** Hackney Marshes, 1914, *J. E. Cooper* (BM); 1915 and 1921, *J. E. Cooper*.

L. densiflorum Schrad. **7.** Hackney Marshes, 1921, *J. E. Cooper* ex *G. C. Brown*, *Rep. Bot. Soc. & E.C.* 7: 864.

L. neglectum Thell. **1.** Uxbridge Moor, 1947!, *L.N.* 27: 30. **5.** Hanwell, 1946, *Sandwith*!, *L.N.* 26: 58; 1946–48!, *K. & L.*, 24.

L. perfoliatum L. **1.** Uxbridge, 1894 and 1900, *Benbow* (BM). **4.** Mill Hill, 1962, *Harrison*, det. *Sandwith*, *L.N.* 42: 12. **7.** Hackney Marshes, 1923, *J. E. Cooper* (BM).

L. graminifolium L. **5.** Greenford, 1954!, *K. & L.*, 338.

L. bonariense L. **7.** Canal side between Lisson Grove and Regents Park, 1969, *L. M. P. Small* det. *Bangerter*.

Isatis tinctoria L. Woad. **6.** Tottenham, 1638, *Johns. Cat.*

Iberis amara L. Candytuft. **1.** Harefield, one plant, 1958, *Pickess.* **7.** Kilburn, 1816 (G & R). Near Highgate, *Irv. Lond. Fl.*, 162. The two old records may be referable to *I. umbellata* or some other cultivated species.

I. umbellata L. Candytuft. **1.** Near Denham!, *H. Banks*. **2.** Teddington, 1965. **3.** Hounslow. **4.** Stonebridge, 1947. Wembley, 1965. **5.** Hanwell, 1947. **6.** Near East Barnet, 1966. **7.** Bombed sites, Cripplegate and Gresham Street areas, *Jones Fl.* Holloway, 1965, *D. E. G. Irvine*. Stoke Newington, 1967, *Kennedy*.

Thlaspi perfoliatum L. Perfoliate Penny-cress. **2.** West Drayton, 1903, *Dymes* (D); 1904, *Lindsay* (MY); 1905, *Montg. MSS.*

Hornungia petraea (L.) Reichb., *Hutchinsia petraea* (L.) R.Br. Rock Hutchinsia. **1.** Chalkpit near Ruislip, *Gray* and *Girard* (CYN).

Cochlearia officinalis L. Scurvy-grass. **5.** Bedford Park, *Cockerell Fl.* Probably an error. **7.** Hampstead, *Morley MSS.* Error. Waste ground near the Tower of London, 1951, *Lawfield, K. & L.*, 17.

Myagrum perfoliatum L. **7.** Tollington Park, 1875, *French* (BM). Hackney Marshes, *M. & S.*; 1927, *J. E. Cooper* (BM).

Euclidium syriacum (L.) R.Br. **7.** Hackney Marshes, 1912, *J. E. Cooper* (BM).

Neslia paniculata (L.) Desv., *Vogelia paniculata* (L.) Hornem. **1.** Uxbridge, 1917 (WE). Yiewsley, 1913, *Coop. Cas.* **3.** Hayes, 1916, *J. E. Cooper, K. & L.*, 25. **4.** Hampstead, 1888, *Norton* (BM). Temple Fortune, 1914, *J. E. Cooper, K. & L.*, 25. East Finchley, 1907, *J. E. Cooper* (BM). Mill Hill, *Harrison*. **5.** Chiswick, 1942, *Gepp* (SLBI); 1943!, *L.N.* 26: 58. **6.** Church End, Finchley, 1922 and 1928, *J. E. Cooper, K. & L.*, 25. **7.** Between Crouch End, Hornsey and Highgate, 1875, *French*; Highgate, 1899, *J. E. Cooper* (BM). Hackney Marshes, *Coop. Cas.*; c. 1927, *M. & S.*

Lunaria annua L. Honesty. **1.** Pinner, *Ger. Hb.*, 378. Near Harefield, 1903, *Green* (SLBI). Harrow Weald Common, 1965, *Kennedy*!. Harefield Moor, 1965, *I. G. Johnson, L.N.* 45: 21. **2.** Staines, 1966. Sunbury, 1967, *Kennedy*!. **4.** Harrow!, *Ger. Hb.*, 378; 1966, *Tyrrell*!. **5.** Chiswick, *T. & D.*, 38. Wyke Green, 1933, *K. & L.*, 16. Brentford, 1965!, *L.N.* 45: 20. Ealing, *L. M. P. Small*. **6.** Botany Bay, Enfield Chase, 1966, *Kennedy*. **7.** Bombed site, Cripplegate, *L.N.* 29: 86. Bombed sites, Finsbury Square, Gresham Street and Victoria Street areas, *Jones Fl.* Holloway, 1965, *D. E. G. Irvine*.

Alyssum alyssoides (L.) L., *A. calycinum* L. **2.** Between Penton Hook and Laleham, 1888, *Benbow* (BM).

Lobularia maritima (L.) Desv., *Alyssum maritimum* (L.) Lam. Sweet Alison. A common feature of rubbish-tips, disturbed road verges, etc.

Berteroa incana (L.) DC., *Alyssum incanum* L. **1.** Between Uxbridge and Harefield, *De Cresp. Fl.*, 135. Harefield; Uxbridge Common; Ickenham Green, 1884, *Benbow* (BM). **2.** Runnymede Range, 1898, *Fraser* (K). West Drayton, 1942!, *K. & L.*, 16. Near Harlington, 1936, *P. H. Cooke* (BM). **3.** Sudbury, 1881, *De Crespigny, Rep. Bot. Soc. & E.C.* 1: 46. **4.** Finchley, 1900 and 1909–10, *J. E. Cooper*; East Finchley, 1921, *Redgrove* (BM). Temple Fortune, 1914, *Coop. Cas.* **5.** Bedford Park, *Cockerell Fl.* Chiswick, 1887, *Benbow* (BM). **6.** Muswell Hill, 1914, *Coop. Cas.* **7.** Building site behind British Museum, Bloomsbury, *Shenstone Fl. Lond.* Crouch End, 1897, *Coop. Cas.* Queen's Gate Gardens, S.W., 1911, *Maude* (BM).

Draba muralis L. Wall Whitlow Grass. **7.** Wall of Chelsea Garden, *J. E. Smith* (LINN). Perhaps planted there.

Arabis alpina L. **6.** Wall at Highgate, *Britten, Nat.* 1864.

A. hirsuta (L.) Scop. Hairy Rock-cress. **7.** Hackney Marshes, 1914, *J. E. Cooper* (BM).

A. caucasica Willd. **7.** Bombed sites, Cripplegate and Gresham Street areas, *Jones Fl.*

Aubrieta deltoidea (L.) DC. **4.** Brent Reservoir, 1945!, *K. & L.*, 16.

Matthiola incana (L.) R.Br. Evening Stock. **1.** Harefield, *I. G. Johnson* and *Kennedy*!. **7.** Bombed site, St Bartholomew's Close, E.C., 1950, *Whittaker* (BM). Bombed site, Cripplegate!, *Wrighton, L.N.* 29: 86. Bombed sites, Finsbury Square and near the Tower of London; high up on the wall of a ruined building near Aldersgate Street, E.C., *Jones Fl.*

M. longipetala (Vent.) DC. subsp. **bicornis** (Sibth. & Sm.) P. W. Ball, *M. bicornis* Sibth. & Sm. Night-scented Stock. **5.** Hanwell, 1953!; Greenford, 1954!, *K. & L.*, 336.

Malcolmia maritima (L.) R.Br. Virginian Stock. **1.** Yiewsley, *M. & S.* **2.** Near Staines, 1866 (SY). **5.** Twyford, *Druce Notes.*

Erysimum repandum L. **1.** Yiewsley, 1907, *J. E. Cooper* (BM). **5.** Between Acton and Turnham Green, 1870, *Britten, J.B. (London)* 8: 224.

E. perofskianum Fisch. & Mey. **1.** Yiewsley, 1963, *Wurzell.* **5.** Hanwell, 1950. **7.** Stoke Newington; Highbury, *W. G. Smith, Nat.* 1864.

Sisymbrium strictissimum L. **7.** Bombed site, Chelsea, 1945, *E. J. Salisbury* (K). Belgrave Square, S.W.1, 1945!, *Congreve, K. & L.*, 19. Bombed site, St Olave's, Hart Street, 1945, *Lousley* (BM).

Camelina sativa (L.) Crantz. Gold of Pleasure. Common on rubbish-tips, etc. **7.** Highgate, *B.G.*, 407. Kensington Gardens, 1834, *Denison, Mag. Nat. Hist.* 8: 389. Isle of Dogs, *Woods, B.G.*, 407; 1884, *Benbow MSS.* Fulham, 1887, *Benbow, J.B. (London)* 25: 19. Stoke Newington, *B.G.*, 407. Near Lee Bridge, 1783, *E. Forster* (BM). Building site behind the British Museum, Bloomsbury, *Shenst. Fl. Lond.* Crouch End, 1887, *Coop. Cas.* Hackney, 1962, *Rudiger, L.N.* 42: 12. Regents Park, 1966, *Holme, L.N.* 36: 28. East Heath, Hampstead; South Kensington, *T. & D.*, 41.

Descurainia sophia (L.) Webb ex Prantl, *Sisymbrium sophia* L. Flixweed. **1.** Near Uxbridge, 1883–88; north of Uxbridge, 1902, *Benbow* (BM). Uxbridge, 1965, *Davidge*. Yiewsley, *Coop. Cas.*; c. 1927, *M. & S.*, Harefield, *Blackst. Fasc.*, 64; c. 1885, *Benbow* (BM); 1930, *P. H. Cooke, K. & L.*, 19. Cowley Peachey, 1958. New Years Green, 1960, *T. G. Collett*!. Near Potters Bar, *Coop. Cas.* **2.** West Drayton, 1909, *J. E. Cooper* (SLBI). Dawley, 1948!, *L.N.* 28: 32. **4.** Near Cricklewood, 1869, *Warren, T. & D.*, 421. Mill Hill, *P. H. Cooke*. Finchley, 1910, *Coop. Cas.* **5.** Alperton, 1908, *Montg. MSS.* Hanwell, 1946!, *Welch, L.N.* 28: 32; 1947–53. Greenford!; near Sudbury Hill!, *L.N.* 27: 30. Bedford Park, *Cockerell Fl.* Acton, 1908, *Green* (SLBI). **6.** Tottenham, *Johns. Cat.* Winchmore Hill, 1919, *L. B. Hall, K. & L.*, 19. **7.** Near St James's Palace, *Moris. Hist.* 2: 219. Ball's Pond, Islington, *Ball. MSS.* Strand, W.C., *Johns. Nat.* Possibly *Sisymbrium altissimum* intended. Hackney Marshes, 1912, *Coop. Cas.* Bombed site, Bunhill Row, E.C., 1954, *Lawfield*.

CAPPARIDACEAE

Cleome spinosa L. Giant Spider Flower. **7.** Kensington Gardens, 1961!, *D. E. Allen, L.N.* 41: 17; 1962, *D. E. Allen*!.

RESEDACEAE

Reseda alba L. White Mignonette. **1.** Weed in Parsonage garden, Pinner, *Hind. T. & D.*, 44. **4.** Mill Hill, 1961, *Warmington*. **5.** Acton Green, *Druce Notes*. **7.** Chelsea, *T. Moore* (BM); c. 1830, *A. Irvine*; c. 1868, *Fox, T. & D.*, 44. Brompton, c. 1834, *T. & D.*, 44; 1945, *Sladen*; 1947, *Welch*. Chelsea Square, S.W., *Graham, Rep. Bot. Soc. & E.C.* 13: 285. South Kensington, *Dyer Key*; 1948, *J. B. Evans* and *Bangerter* (BM).

R. odorata L. Sweet Mignonette. **1.** Uxbridge, *Druce Notes*. Yiewsley, *M. & S.* **2.** Harmondsworth, 1967, *Lousley*!. **7.** Cremorne, *T. & D.*, 45. Chelsea; Parsons Green, *Britten, Nat.* 1864.

VIOLACEAE

Viola lutea Huds. subsp. **sudetica** (Willd.) W. Becker × **V. tricolor** = **V.** × **wittrockiana** Gams, *V. tricolor* L. var. *hortensis* auct. Garden Pansy. A common feature of rubbish-tips, also on disturbed road verges, waste ground, etc.

TAMARICACEAE

Tamarix gallica L. Tamarisk. **1.** New Years Green, 1960!, *T. G. Collett, L.N.* 41: 14; 1961.

CARYOPHYLLACEAE

Silene conoidea L. **7.** Hackney Marshes, 1912, *J. E. Cooper* (K).

S. dichotoma Ehrh. **2.** Staines, 1909, *H.* and *J. Groves* (BM). Penton Hook, 1953, *B. M. C. Morgan, K. & L.*, 339. **7.** Hackney Marshes, 1912, *J. E. Cooper* (BM); 1916, *J. E. Cooper* (K).

S. gallica L. subsp. **anglica** (L.) Löve & Löve, *S. anglica* L. Small-flowered Catchfly. **1.** Yiewsley, 1908, *J. E. Cooper* (BM). **2.** Tangley Park, a single plant, *T. & D.*, 50. West Drayton, a single plant, 1862, *Benbow MSS.* **4.** Finchley, 1909–10, *Coop. Cas.* Hendon, 1915, *J. E. Cooper, K. & L.*, 33. West Heath, Hampstead, 1917, *Redgrove.* **5.** Acton, *Druce Notes.* **6.** North Finchley, 1925, *J. E. Cooper, K. & L.*, 33. **7.** Hackney Marshes, 1913, *Coop. Cas.*; 1915, *J. E. Cooper, K. & L.*, 33; c. 1926, *M. & S.*

Subsp. **quinquevulnera** (L.) Löve & Löve, *S. quinquevulnera* L. **3.** Near Whitton Park, 1907 (MY).

Lychnis chalcedonica L. **1.** Near Harefield, 1902, *R. W. Robbins* (LNHS).

Agrostemma coronaria L., *Lychnis coronaria* (L.) Desr. **4.** Kenton, 1951, *Bull.* **7.** Bombed site, Red Lion Passage, W.C., 1950, *Whittaker* (BM). Bombed site, Cripplegate, *Wrighton, L.N.* 29: 86.

A. coeli-rosea L. **3.** Hounslow, 1964, *K. Marks, Wild Fl. Mag.* 343: 29.

Dianthus barbatus L. Sweet William. **7.** Bombed sites, Cripplegate, *A. W. Jones.*

D. plumarius L. Common Pink. **3.** Twickenham, *Dillenius*, *Ray Syn.* 3: 336.

Vaccaria pyramidata Medic., *Saponaria vaccaria* L., *S. segetalis* Neck. **1.** Pinner, a single plant, *Hind, Melv. Fl.*, 13. Uxbridge, 1905, *Green* (SLBI). Yiewsley, 1913–14, *Coop. Cas.*; 1923, *J. E. Cooper, K. & L.*, 32. **2.** West Drayton, *Druce Notes*. **3.** Hounslow Heath, 1961, *Philcox* and *Townsend*. **4.** East Finchley, 1907; Finchley, 1907–9, *Coop. Cas.* Willesden Green, 1966. Mill Hill, *Harrison*, det. *Sandwith*. **5.** Alperton, 1918, *Summerhayes* (K). Greenford, 1917, *J. E. Cooper* (BM); 1920, *J. E. Cooper, K. & L.*, 32. West Ealing, 1959!, *Proc. Bot. Soc. Brit. Isles* 3: 442; 1965. **6.** Near Alexandra Palace, *Coop. Cas.* North Finchley, 1925, *J. E. Cooper* (BM). **7.** Isle of Dogs, 1887, *Benbow* (BM). Crouch End; Hackney Marshes, 1912–13, *Coop. Cas.*; 1914–15 and 1925, *J. E. Cooper, K. & L.*, 32. Tollington Park, 1876, *French* (BM). Trinity Square, E.C., 1964, *Mason, L.N.* 44: 20. Hyde Park, 1965; Kensington Gardens, 1965, *Allen Fl.* Regents Park, 1966, *Holme, L.N.* 46: 39.

Gypsophila viscosa Murray. **1.** Denham, 1930, *P. H. Cooke, K. & L.*, 32.

Spergularia marina (L.) Griseb., *S. salina* J. & C. Presl. **5.** Hanwell, a single plant, 1950!, *K. & L.*, 39.

Polycarpon tetraphyllum (L.) L., *Mollugo triphylla* L. Four-leaved All-seed. **4.** Garden at Harrow, occurring by accident, *Hind Fl.*

Herniaria hirsuta L., *H. cinerea* DC. **1.** Uxbridge, 1907, *Green* (SLBI). **5.** Garden ground, Ealing Common, 1817, *Burchell* (K). **6.** Highgate, *Dicks. Hort. Sicc.*

PORTULACACEAE

Portulaca oleracea L. Purslane. **2.** Near Harmondsworth, 1967, *Kennedy* and *Lousley*!. **7.** Trinity Gardens, E.C., 1969, *Clement.* Flower-beds near Tower of London, 1969, *L.N.* 49: 19.

P. grandiflora Hook. **5.** Hanwell, 1955, *McClintock, K. & L.*, 340.

AIZOACEAE

Mesembryanthemum crystallinum L., *Cryophytum crystallinum* (L.) N.E.Br. **7.** Fulham, *Clarkson-Birch, Corn. Surv.*

Tetragonia tetragonioides (Pallas) O. Kuntze, *T. expansa* Murr. New Zealand Spinach. **1.** Harefield, 1966!, *I. G. Johnson.* **5.**

Hanwell, 1947!; Greenford, 1951–52!, *K. & L.*, 125. **6.** North of Enfield Lock, 1967, *Kennedy*!. **7.** Hackney Marshes, 1947!, *K. & L.*, 125.

AMARANTHACEAE

Amaranthus caudatus L. Love-lies-bleeding. **1.** Yiewsley!, *M. & S.* Harefield, 1966, *I. G. Johnson* and *Kennedy*!. **2.** West Drayton!, *Druce Notes.* **5.** Hanwell, 1946–53!, *K. & L.*, 232.

A. hybridus L. subsp. **hybridus,** *A. hypochondriacus* L., *A. chlorostachys* Willd. **1.** Yiewsley!, *M. & S.* **2.** Shepperton, 1969, *Clement* and *Ryves.* **5.** Hanwell!, *L.N.* 26: 63. Greenford, 1947!, *L.N.* 27: 33. Garden weed, Ealing, 1954!, *T. G. Collett.*

Var. **pseudoretroflexus** (Thell.) Thell. **7.** Bombed site, Eastcheap, E.C., 1944!, *Lousley* (L).

Subsp. **incurvatus** (Timeroy ex Gren. & Godr.) Brenan var. **incurvatus,** *A. patulus* Bertol., *A. incurvatus* Timeroy ex Gren. & Godr., *A. hybridus* L. subsp. *cruentus* (L.) Thell. var. *patulus* (Bertol.) Thell. **1.** Near Springwell Lock, 1946!, *Sandwith* (K), det. *Brenan*; 1947!, *Lousley* (L), det. *Brenan.* New Years Green, 1961, *J. G. Dony*!.

Var. **cruentus** Mansf., *A. cruentus* L., *sensu stricto*, *A. paniculatus* L., *A. speciosus* Sims, *A. paniculatus* L. var. *cruentus* (L.) Moq., *A. hybridus* L. var. *paniculatus* (L.) Uline & Bray, *A. hybridus* L. subsp. *cruentus* (L.) Thell. var. *paniculatus* (L.) Thell. **5.** Hanwell, 1947!, *Lousley* (L), det. *Brenan.*

A. hybridus L. subsp. **hybridus × retroflexus = A. × ozanoni** Thell. **2.** Shepperton, 1969, *Clement* and *Ryves*, det. *Brenan* as 'probably *A. hybridus* × *A. retroflexus*'.

A. quitensis Humboldt., Bonpl. & Kunth. **2.** West Drayton, 1916, *A. B.* and *M. Cobbe*, *Rep. Bot. Soc. & E.C.* 5: 50.

A. albus L. **1.** Yiewsley, 1923, *J. E. Cooper* (BM). **3.** Northolt, 1947!, *Welch*, *L.N.* 27: 41.

A. blitoides S. Wats. **1.** Yiewsley, 1929, *Melville* (L). **5.** Greenford, 1952!, det. *Brenan*, *K. & L.*, 233.

A. graecizans L. subsp. **graecizans,** *A. angustifolius* Lam., *nom. illegit.* **6.** Whetstone, 1940, *Cruttwell*, det. *Brenan*, *Rep. Bot. Soc. & E.C.* 12: 240.

A. lividus L. subsp. **lividus,** *A. ascendens* Lois, *A. blitum* L. var. *ascendens* (Lois.) DC. **2.** Weed in nursery, Hampton, 1963,

Townsend, L.N. 43: 22. **5.** Chiswick, 1876, *Sim* (K), det. *Brenan.* **6.** Tottenham, *Atkins* (BM), det. *Kloos.* **7.** Chelsea (DILL); 1948 and 1850, *T. Moore*; near Walham Green, 1828, *E. Forster* (BM), all det. *Kloos.* Fulham (LT); 1846, *Stevens* (K), both det. *Brenan.* Near Parsons Green, 1864, *Newbould* (BM), det. *Kloos.*

Subsp. **polygonoides** (Moq.) Probst, *A. ascendens* Lois. var. *polygonoides* Thell. ex E. H. L. Krause, *A. lividus* L. var. *polygonoides* (Moq.) Thell. **7.** About London, 1822, *Blake* (D), det. *Kloos.* Hackney Marshes, 1913, *J. E. Cooper* (BM, K), det. *Brenan.*

A. standleyanus Parodi ex Covas, *A. vulgatissimus* sensu Thell., non Spegazz. **2.** West Drayton, 1969, *Clement*, det. *Brenan.*

CHENOPODIACEAE

Chenopodium ambrosioides L. **1.** Yiewsley, *Coop. Cas.* **5.** Syon Park, 1932, *Meinertzhagen* (BM). **7.** Hackney Marshes, *M. & S.*

C. berlandieri Moq. subsp. **platyphyllum** (Issler) Ludw., *C. platyphyllum* Issler. **1.** Near Springwell Lock, 1945!, det. *Brenan, K. & L.*, 236.

C. foliosum (Moench) Aschers. **5.** Chiswick, 1876, *Sim* (K), det. *Brenan.*

C. hircinum Schrad. **2.** West Drayton, *A. B.* and *M. Cobbe, Rep. Bot. Soc. & E.C.* 4: 500. **4.** Hendon, 1963, *Brewis.* **5.** Hanwell, 1946, *Sandwith*!, det. *Brenan, L.N.* 28: 34.

C. lanceolatum Muhl. **2.** West Drayton, *Druce Fl.*, 283. **5.** Acton, *Rep. Bot. Soc. & E.C.* 3: 174.

C. multifidum L., *Roubieva multifida* (L.) Moq. **1.** Uxbridge, *Wedgwood, Rep. Bot. Soc. & E.C.* 6: 396.

C. pratericola Rydb., *C. leptophyllum* auct. **1.** Yiewsley, *M. & S.* **2.** West Drayton, *Druce Notes.*

C. probstii Aell. **2.** Near Harmondsworth, 1967, *Kennedy* and *Lousley*!, *L.N.* 47: 9. Shepperton, 1969, *Clement* and *Ryves.*

C. rugosum Aell. **5.** Hanwell, 1946, *Sandwith*! (K), det. *Aellen.*

Var. **acutidentatum** (Aell.) Aell. **1.** Near Springwell Lock, 1945, *Sandwith*! (K), det. *Aellen.*

C. urbicum L. Upright Goosefoot. **4.** Hampstead, *J. E. Cooper. Sci. Goss.* 1887, 189. Hendon, 1912, *C. S. Nicholson, K. & L.*, 234. Building site, Brondesbury Park, 1965, *L. M. P. Small*, det. at *Brit.*

Mus. **7.** Crouch End, *J. E. Cooper*, *Sci. Goss.* 1887, 189. Kingsway, W.C., *Parsons* (CYN). Some of these records may be referable to forms of *C. rubrum*.

Var. **intermedium** (Mert. & Koch) Koch. **2.** Near Staines, 1908, *Green* (SLBI). **4.** Near Hampstead, 1817 (G & R). **7.** Kentish Town, 1845–46 (HE).

Beta trigyna Waldst. & Kit. **2.** Near Harmondsworth, 1967.

B. vulgaris L. subsp. **vulgaris.** Beet. Common on rubbish-tips, etc., as an outcast, and on arable land, etc., as a relic of cultivation.

Subsp. **cicla** (L.) Arcangeli, *Beta cicla* L. **5.** Hanwell, 1955, *McClintock*, *K. & L.*, 356. Greenford, 1960. **7.** Bombed site, Cripplegate. Hyde Park Corner, 1962, *McClintock*.

Spinacia oleracea L. Spinach. **1.** Yiewsley, *M. & S.* **3.** Hounslow Heath, 1961, *Buckle*. **5.** Hanwell!; Brentford!, *K. & L.*, 237. **7.** Hyde Park Corner, 1962, *McClintock*, *L.N.* 44: 21.

Atriplex littoralis L. Shore, or Sea Orache. **2.** Between West Drayton and Iver, 1892, *G. C. Druce*, *Rep. Bot. Soc. E.C.* 2: 60. **3.** Twickenham, *Newbould*, *Trim. MSS.* Roxeth, *Hind. Fl.* **5.** Hanwell, 1951!, *K. & L.*, 237. Chiswick, *Druce Notes.* **7.** Ditches between Old Street and Islington, *Petiver* (SLO). On sea-gravel used in making a new road near South End Green, Hampstead, *Dyer*, *Trim. MSS.*

A. hortensis L. Garden Orache. **1.** Yiewsley, *M. & S.* **2.** West Drayton, *Druce Notes.* **3.** Sudbury, 1881, *De Crespigny* (BM). **4.** Near Temple Fortune, 1915 and 1926, *J. E. Cooper* (BM). **5.** Greenford, 1954, *Welch*, *L.N.* 34: 6. Hanwell, 1950!, *K. & L.*, 237.

Var. **rubra** (Crantz) Roth. **1.** New Years Green, 1960, *T. G. Collett*!. Pinner, 1961, *Roseway*. **3.** Hounslow Heath, 1961, *Philcox* and *Townsend*; 1965. **5.** Hanwell, 1946!, *K. & L.*, 237. Greenford, 1954, *Welch*, *L.N.* 34: 6. Acton, *Druce Notes.* **7.** Bombed site, Noble Street, E.C., 1945, *Lousley* (L).

A. tatarica L. **1.** Yiewsley, *M. & S.* **2.** Between West Drayton and Iver, *Druce* (D). **7.** Hackney Marshes, *M. & S.*

A. rosea L. **3.** Hounslow Heath, 1965, *Wurzell*, det. *Lousley*, *L.N.* 45: 21.

Axyris amaranthoides L. 1. Uxbridge, *B. Reynolds*, *Rep. Bot. Soc. & E.C.* 5: 677. Yiewsley, 1924, *J. E. Cooper*, *K. & L.*, 237. **2.** West Drayton, *Druce Fl.*, 285. **5.** Alperton, 1916, *Rake*, *Dunn MSS.*

Kochia scoparia (L.) Schrad. Burning Bush, Summer Cypress. **1.** Yiewsley, 1963, *Wurzell, L.N.* 43: 22. **5.** Hanwell, 1960!; Greenford, 1951! and 1954!, *K. & L.*, 238.

Suaeda maritima (L.) Dumort., *Dondia maritima* (L.) Druce. Herbaceous Seablite. **1.** Yiewsley, 1917, *J. E. Cooper* (BM). **7.** Hackney Marshes, 1909, *J. E. Cooper* (BM).

Salsola kali L. subsp. **ruthenica** (Iljin) Soó, *S. pestifer* A. Nels., *S. tragus* auct. Russian Thistle. **3.** Hounslow Heath, 1946, *L. G. Payne* and *H. Banks, L.N.* 26: 63. **5.** Greenford, 1945!, *Welch, L.N.* 26: 63. Northolt, 1947, *Welch, Lousley* and *Woodhead*!, *L.N.* 27: 33. Hanwell, 1953. **7.** Bombed site, Lower Thames Street, E.C., 1941; bombed site, Mark Lane, E.C., 1944!, *Lousley Fl.* Bombed site, Kensington, 1947!, *L.N.* 27: 33.

Salicornia ramosissima Woods. Glasswort. **5.** Rubbish-tip, Greenford, c. 20 plants, 1954!, *L.N.* 34: 6.

NYCTAGINACEAE

Mirabilis jalapa L. Miracle of Peru. **3.** Hounslow Heath, 1961, *Buckle.*

PHYTOLACCACEAE

Phytolacca americana L. Pokeweed. **4.** Wembley Park, a single plant, 1966. Headstone, *L. M. P. Small.* **5.** Norwood Green Churchyard, 1969, *L. M. P. Small.* **7.** Fulham Palace grounds, a single plant, 1955, *Welch*!, *K. & L.*, 356.

MALVACEAE

Malva pusilla Sm., *M. rotundifolia* L. **1.** Uxbridge, 1890, *Benbow* (BM). North of Uxbridge, 1909, *C. S. Nicholson* (LNHS). Yiewsley, 1909–12, *Coop. Cas.* **2.** West Drayton, *Druce Notes.* **4.** Near Stonebridge, 1886, *Benbow* (BM). **5.** Chiswick, 1887, *Benbow* (BM). Acton, *Druce Notes.* **6.** Enfield, 1965, *Kennedy*!, *L.N.* 35: 20. **7.** Bombed site, Leather Lane, E.C., 1950, *Bangerter* and *Whittaker* (BM). Eaton Square, S.W.1, a single plant, 1958, *McClintock, L.N.* 39: 47.

M. parviflora L. **1.** Yiewsley, 1912, *J. E. Cooper* (BM). **3.** Hounslow Heath, 1961, *Buckle.* **4.** Hendon, 1927, *J. E. Cooper, K. & L.*, 44. **5.** Alperton, 1867, *Trimen* (BM). West Ealing, 1944!, *K. & L.*, 44. **7.** Hackney Marshes, 1912, *Coop. Cas.*; 1916, *Cooper, K. & L.*, 44.

Kensington Gardens, a single plant, 1961, *D. E. Allen*, conf. *Lousley*, *Allen Fl.*

Var. **microcarpa** (Desf.) Fiori & Paol. **4.** Finchley, 1910, *J. E. Cooper* (K). **7.** Hackney Marshes, 1926, *J. E. Cooper* (K); c. 1927, *M. & S.*

M. oxyloba Boiss. **4.** Fortis Green, 1911, *J. E. Cooper* (K).

M. verticillata L. **7.** South Kensington, 1872, *Dyer Key.*

M. crispa (L.) L. **5.** Acton Green, *Druce Notes.* **7.** In great abundance on the site of the 1862 Exhibition, South Kensington, *T. & D.*, 61.

Malope trifida Cav. **5.** West Ealing, 1959!, *Proc. Bot. Soc. Brit. Isles* 3: 442.

Lavatera trimestris L. **5.** Hanwell, 1967.

L. punctata All. **5.** Acton, *Druce Notes.*

Alcea rosea L., *Althaea rosea* (L.) Cav. Hollyhock. Common on rubbish-tips, railway banks, etc.

Kitaibelia vitifolia Willd. **7.** South Kensington, 1870, *Dyer* (BM).

Sida spinosa L. **1.** Springwell, 1945!, *L.N.* 26: 59; 1946–47!, *K. & L.*, 44.

Sidalcea canadensis A. Gray. **4.** Wembley Park, 1927 (BM).

S. spicata Greene. **4.** Near Colindale, 1966.

S. candida A. Gray × **malviflora** (DC.) A. Gray. **6.** Near Trent Park, 1966, *Kennedy.*

Wissadula spiciflora (DC.) Druce, *W. spicata* J. & C. Presl. **7.** South Kensington, 1867, *Dyer* (BM).

Abutilon theophrasti Medic., *A. avicennae* Gaertn. **1.** Near Springwell Lock, a single plant, 1945, *Welch*; frequent, 1946!, *L.N.* 26: 59; 1947–48 and 1951, *K. & L.*, 44. New Years Green, 1968 and 1969, *Clement* and *Ryves.* **2.** Near Harlington, 1968, *Clement* and *Ryves.* **5.** Hanwell, 1946!, *L.N.* 26: 59. **6.** Southgate, 1947, *Markham* (K). **7.** Love Lane, Tottenham, c. 12 plants, 1967, *Kennedy*!.

Hibiscus trionum L. Flower-of-the-hour. **1.** Near Springwell Lock, one plant, 1945!, *L.N.* 26: 59. New Years Green, one plant 1960, *T. G. Collett*!. **2.** Near Harmondsworth, 1967. **3.** Feltham, one plant, 1952, *Welch* (BM). Hounslow Heath, 1960 and 1964,

Wurzell. **4.** Fortis Green, 1909, *Coop. Cas.* **5.** Acton Wells, *Druce Notes.* **7.** In great profusion on the site of the Exhibition of 1862, South Kensington, 1867, *T. & D.*, 67.

LINACEAE

Linum grandiflorum Desf. **7.** Bombed sites, Cripplegate area, *Jones Fl.*

GERANIACEAE

Geranium nodosum L. **2.** Dawley, 1924, *Champneys.*

G. tuberosum L. **2.** Hampton Court, 1922; 1926, *R. W. Robbins* (LNHS); 1928, *Tremayne, K. & L.*, 49.

Erodium ciconium (L.) L'Hérit. **7.** Near Pimlico, *A. Irvine, Phyt. N.S.* 3: 335.

E. chium (L.) Willd. subsp. **littoreum** (Léman) Ball., *E. littoreum* Léman. **7.** Pimlico, *A. Irvine, Phyt. N.S.* 3: 335.

E. moschatum (L.) L'Hérit. Musk Storksbill. **1.** Harefield, 1946, *Stevens* (K). **2.** Laleham, 1915, (G & R).

Tropaeolum peregrinum L. *sensu lato.* Canary Creeper. **1.** South Mimms, 1965!, *C. M. Dony.* Harefield, 1966, *I. G. Johnson* and *Kennedy*!; 1967, *Clement.* **3.** Hounslow Heath, 1951, *Lousley* and *McClintock*!, *K. & L.*, 50. **4.** Brockley Hill, 1966, *Kennedy*!. **5.** Hanwell, 1953!, *L. M. P. Small.* Greenford, 1954. **7.** Chelsea, 1957, *Graham, L.N.* 39: 47.

T. majus L. 'Nasturtium'. **1.** Yiewsley!, *M. & S.* Harefield, 1966, *I. G. Johnson* and *Kennedy*!. **2.** Harlington, 1966. **5.** Hanwell, 1946!, *K. & L.*, 50; numerous plants near the Brent, 1966. **7.** Bombed site, Cripplegate, *Wrighton.* Canal side between Marylebone and Primrose Hill, 1968.

BALSAMINACEAE

Impatiens noli-tangere L. Touch-me-not. **1.** Pinner, *Hind, T. & D.*, 71.

RUTACEAE

Citrus aurantium L. Orange. **1.** Yiewsley. **2.** Rubbish-tip between West Drayton and Hayes, abundant, 1967. **3.** Rubbish-tip, Hounslow Heath, 1951, 1955, 1960→. **4.** Finchley, 1964, *Wurzell.* **5.** Rubbish-tips, Hanwell, 1945–51, and Greenford, 1946–50.

VITACEAE

Parthenocissus tricuspidata (Sieb. & Zucc.) Planch., *Vitis thunbergii* Druce, non Sieb. & Zucc. **1.** Yiewsley, *M. & S.*

PAPILIONACEAE

Lupinus angustifolius L. **1.** Harefield, 1922, *J. E. Cooper* (K).

Genista hispanica L. **7.** Bombed site, Parker Street, W.C.1, 1950, *Whittaker* (BM).

Ononis natrix L. **3.** Hounslow Heath, 1965, *Wurzell*, det. *Lousley*, *L.N.* 45: 20.

O. baetica Clemente, *O. salzmanniana* Boiss. & Reut. **5.** West Ealing, 1959!, *Proc. Bot. Soc. Brit. Isles* 3: 442.

Trigonella foenum-graecum L. **1.** Yiewsley, 1908, *J. E. Cooper* (BM); 1913, *J. E. Cooper* (K). **7.** Hackney Marshes, 1915, *J. E. Cooper* (BM); 1918, *J. E. Cooper* (K). South Kensington, 1919, *Wernham* (BM). Chelsea Embankment, 1955, *D. P. Young*, *L.N.* 35: 5.

T. procumbens (Bess.) Reichb., *T. besserana* Ser., *T. coerulea* (L.) Ser. subsp. *procumbens* (Bess.) Thell. **2.** West Drayton, *Druce Notes.* **5.** Acton, *Druce Notes.* **7.** Hackney Marshes, 1912, *J. E. Cooper* (BM).

Medicago minima (L.) Bartal. Small Medick. **4.** Neasden, a single plant, 1966. **5.** West Ealing, 1959!, *Proc. Bot. Soc. Brit. Isles* 3: 442. **6.** Near Hadley Wood, 1959.

M. laciniata (L.) Mill. subsp. **laciniata.** **5.** Alperton, 1917, *J. E. Cooper* (K).

Melilotus messanensis (L.) All., *M. sicula* (Turra) B. D. Jackson. **7.** Chelsea, *Britten*, *T. & D.*, 78.

M. sulcata Desf. **3.** Hounslow Heath, 1966, *Wurzell*, det. *Lousley.* **4.** Finchley, 1963, *Brewis.* **7.** South Kensington, *T. & D.*, 78.

M. caerulea (L.) Ser. **7.** South Kensington, *T. & D.*, 78.

Trifolium ochroleucon Huds. Sulphur Clover. **3.** Southall, 1903, *Druce Notes.* **5.** East Acton, *Druce Notes.* **6.** Muswell Hill, 1907, *Coop. Cas.* Error, the plant seen was *T. albidum*, *J. E. Cooper*, *J.B.* (*London*) 53: 69. Sparingly about Barnet, *Curt. F.L.* Perhaps in Herts.

T. incarnatum L. subsp. **incarnatum.** Crimson Clover. **1.** Springwell!; Harefield!, *Benbow* (BM). Denham; Uxbridge, *Benbow MSS.* **2.** Near Staines, 1866 (SY). Ashford Ford, *Benbow MSS.* Near Ashford, 1940–41, *Hasler*, *K. & L.*, 61. **3.** Whitton, 1866, *Trimen* (BM). **4.** Golders Green, 1921, *Richards*, *K. & L.*, 61; 1965, *J. L. Gilbert.* Hendon, 1941, *Bedford.* Finchley, 1914, *J. E. Cooper* (BM). **5.** Between Hanwell and Southall!, *L.N.* 26: 60. Near Chiswick, *Newbould*, *T. & D.*, 79. Bedford Park, *Cockerell Fl.* **6.** Highgate Wood, 1940, *Tremayne.* **7.** South Kensington, *F. Naylor*, *T. & D.*, 79.

Var. **stramineum** Presl. **1.** Springwell, 1884, *Benbow* (BM).

T. resupinatum L. Reversed Trefoil. **1.** Uxbridge, 1886; Ruislip, 1895, *Benbow* (BM). Yiewsley, *J. E. Cooper.* **2.** Near West Drayton, 1886, *Benbow* (BM). Dawley, 1919, *J. E. Cooper* (K). **3.** Twickenham, 1886, *Benbow* (BM). **4.** Near Hampstead, 1841, *Stocks* (BM). **5.** Alperton, *Druce Notes.* Hanwell, 1955, *Newey.* **6.** Muswell Hill, 1902, *J. E. Cooper* (BM). Southgate, 1903, *Findon MSS.* **7.** South Kensington, 1865, *Dyer* (BM); 1867, *F. J. Hanbury* (HY).

T. aureum Poll., *T. agrarium* L., *nom. ambig.* **2.** Staines Moor, 1964–65, *Briggs.* **4.** Finchley, 1929, *J. E. Cooper* (BM).

T. lappaceum L. **1.** Yiewsley, *J. E. Cooper* (BM). **4.** East Finchley, 1907, *Coop. Cas.* Finchley, 1910, *J. E. Cooper* (BM).

T. constantinopolitanum Ser., *T. alexandrium* auct. **3.** Twickenham, *Montgomery*, *Dunn Fl.*, 63. **5.** Acton, *Loydell* (D).

Var. **phleioides** Boiss. **2.** Ashford, 1906, *Shepheard*, *Rep. Bot. Soc. & E.C.* 4: 407.

T. retusum L., *T. parviflorum* Ehrh. **7.** Hackney Marshes, 1913, *J. E. Cooper* (K).

T. albidum Retz. **4.** Finchley, 1907, *J. E. Cooper*, *J.B.* (*London*) 53: 69. **6.** Muswell Hill, 1907, *J. E. Cooper*, *J.B.* (*London*) 53: 69.

Psoralea americana L. **3.** Hounslow, *K. Marks* and *Norman*, *Wild Fl. Mag.* 343: 30. **5.** Hanwell, 1952, *D. H. Kent* (BM).

Scorpiurus muricatus L. **4.** Mill Hill, 1967, *Harrison*, det. *McClintock.*

Var. **subvillosus** (L.) Fiori, *S. subvillosa* L. **4.** Fortis Green, 1911, *J. E. Cooper* (BM).

Ornithopus compressus L. **7.** Clapton, 1912, *P. H. Cooke* (LNHS).

Arachis hypogea L. **2.** Shepperton, 1971, *Clement*, *Mason* and *McLean.*

Cicer arietinum L. **3.** Hounslow Heath, 1964, *K. Marks* and *Norman*, *Wild Fl. Mag.* 343: 30. **5.** Acton, 1901, *Druce Notes.*

Lens culinaris Medic., *L. esculenta* Moench. **2.** Near Harlington, 1968, *Clement* and *Ryves*, det. *Lousley* (as *Vicia lens*).

Vicia tenuissima (Bieb.) Schinz & Thell., *V. gracilis* Lois., non Soland. Slender Tare. **5.** Garden weed, Ealing, 1949, *T. G. Collett*, *K. & L.*, 69.

V. villosa Roth subsp. **villosa.** **1.** Yiewsley, 1908, 1913 and 1914, *J. E. Cooper* (BM); 1920, *J. E. Cooper*, *K. & L.*, 68; c. 1926, *M. & S.* **2.** West Drayton!, *Druce Notes.* Near Harlington, 1968, *Clement* and *Ryves.* **3.** Southall, 1948!, *L.N.* 28: 32; 1955 and 1966. **4.** Stanmore, *Druce Notes*, East Finchley, 1909; Finchley, 1915, *J. E. Cooper* (BM). **5.** Greenford, 1920, *J. E. Cooper*, *K. & L.*, 68. Acton, 1908, *Loydell* (D). Hanwell, 1946!, *L.N.* 26: 60. Chiswick, 1943–46!, *K. & L.*, 68. **7.** Hackney Marshes, 1909, *J. E. Cooper* (BM). Ken Wood fields, 1948, *H. C. Harris* and *Lousley*!, *L.N.* 28: 32.

Subsp. **varia** (Host) Corbière, *V. dasycarpa* auct., ? an Tenore, *V. varia* Host. **1.** Yiewsley, 1913, *J. E. Cooper* (K). Harefield, 1958, *Graham*, *Harley* and *Lewis.* **2.** West Drayton, 1903, *G. C. Druce* (D). **3.** Hounslow Heath, 1948, *Welch*; 1960, *Philcox* and *Townsend.* **4.** Hendon, 1912, *P. H. Cooke* (LNHS). Edgwarebury, 1957. **5.** Acton, 1907, *Green* (SLBI). Hanwell, 1954, *G. W.* and *T. G. Collett.* **6.** Near Pickett's Lock, Edmonton, 1966, *Kennedy*!. **7.** Hackney Marshes, *M. & S.* Ken Wood fields, 1948!, *H. C. Harris*, *L.N.* 28: 32.

Subsp. **pseudocracca** (Bertol.) P. W. Ball, *V. pseudocracca* Bertol. **1.** Yiewsley, 1911–13, *Coop. Cas.*; c. 1926, *M. & S.* **4.** Temple Fortune, 1914, *Coop. Cas.* Finchley, 1909, *J. E. Cooper* (BM). **7.** Crouch End, 1897, *J. E. Cooper*, *J.B.* (*London*) 53: 59. Hackney Marshes, 1912–13, *Coop. Cas.*

V. benghalensis L., *V. atropurpurea* Desf. **1.** Yiewsley, *M. & S.* **2.** East Bedfont, 1960, *T. G. Collett*!. **5.** North Acton, 1960. **7.** Hackney Marshes, 1915, *J. E. Cooper*, *K. & L.*, 69.

V. ervilia (L.) Willd. **5.** Acton, 1907, *Druce Notes.*

V. pannonica Crantz subsp. **pannonica.** **2.** West Drayton, *Druce Notes.* **3.** Hounslow Heath, 1964, *Wurzell, Clement* and *Mason,* teste *Lousley, L.N.* 44: 17. **5.** Hanwell, 1953!, *K. & L.*, 344.

Subsp. **striata** (Bieb.) Nyman, *V. pannonica* Crantz subsp. *purpurascens* (DC.) Arcangeli, *V. pannonica* Crantz var. *purpurascens* (DC.) Boiss., *V. striata* Bieb. **5.** Acton, 1907, *Green* (SLBI).

V. peregrina L. **7.** Hackney Marshes, 1912, *Coop. Cas.*; 1918, *J. E. Cooper* (K); 1926, *M. & S.*

V. lutea L. subsp. **lutea.** Yellow Vetch. **2.** West Drayton, 1890, *Druce Notes.* **4.** Near Mill Hill, 1947, *H. C. Harris, L.N.* 28: 59. **5.** Hanwell, 1953!, *K. & L.*, 343. Chiswick, 1962, *Murray, L.N.* 43: 22. **7.** Hackney Marshes, 1915, *J. E. Cooper.*

Subsp. **vestita** (Boiss.) Rouy, *V. vestita* Boiss. **7.** Hackney Marshes 1912, *J. E. Cooper* (K).

V. hybrida L. **1.** Yiewsley, 1912 and 1923, *J. E. Cooper* (BM). **4.** Temple Fortune, 1915, *J. E. Cooper* (BM).

V. narbonensis L. Italian Vetch. **1.** South Mimms, 1911, *P. H. Cooke* (LNHS).

V. faba L. Broad Bean. **5.** Hanwell. Greenford.

Lathyrus hirsutus L. **1.** Mimms Hall, 1909, *Green* (SLBI). Uxbridge, *Druce Notes*; 1912, *Wedg. Cat.* **2.** Staines Moor, 1957, *Missen* and *Briggs.* **4.** East Finchley, 1909, *J. E. Cooper* (BM). **5.** Hanwell, 1953. **7.** Crouch End, 1897; Hackney Marshes, 1912–13, *J. E. Cooper* (BM).

L. tuberosus L. Tuberous Pea. **5.** Chiswick, 1933, *Bangerter* (SLBI); 1934–37, *Bangerter*; Hanwell, 1950, *E. A. Bennett, K. & L.*, 70. Syon Park, 1960!, *L.N.* 40: 21.

L. tingitanus L. **6.** Near Enfield Lock, 1966, *Kennedy*!

L. inconspicuus L. **7.** Hackney Marshes, 1912, *J. E. Cooper* (K); 1915, *J. E. Cooper* (BM).

L. cicera L. **7.** Hackney Marshes, 1912, *J. E. Cooper* (K).

L. odoratus L. Sweet Pea. **1.** Yiewsley!, *M. & S.* Harefield, 1966!, *I. G. Johnson.* **3.** Hounslow Heath. **5.** Hanwell, 1951.

Glycine max (L.) Merr., *G. soja* Sieb. & Zucc. Soya Bean. **1.** Springwell, 1945!, *K. & L.*, 72. **2.** West Drayton, 1969, *Clement* and *Ryves.*

Pisum sativum L., *P. arvense* L., *P. sativum* L. subsp. *arvense* (L.) Aschers. & Graebn. Garden Pea. **1.** Yiewsley!, *M. & S.* **2.** West Drayton!, *Druce Notes.* **5.** Hanwell. Greenford. **6.** Finchley Common. **7.** Hackney Marshes.

Phaseolus vulgaris L. Common Bean. A common feature of rubbish-tips, etc.

P. coccineus L. Scarlet Runner. **3.** Hounslow Heath, 1961, *Buckle*; 1964, *Wurzell.* **7.** Kensington Gardens, 1961, *D. E. Allen.*

Cassia fistulosa L. **1.** Yiewsley, 1934, *Lousley* (L).

ROSACEAE

Rubus phoenicolasius Maxim. Wine Berry. **5.** Osterley, 1959.

Potentilla norvegica L. **1.** Uxbridge, 1893, *Benbow* (BM); 1912, *P. H. Cooke* (LNHS). **2.** Between Staines and Laleham, *Shepheard, Trim. MSS.* **3.** Sudbury, 1881, *De Crespigny* (MANCH). Northolt, 1918, *Wernham* (BM). Hounslow Heath, 1961, *Philcox* and *Townsend.* **4.** Willesden, 1901, *Druce Notes.* Kingsbury, 1910, *P. H. Cooke* (LNHS). Hendon, 1916, *J. E. Cooper* (BM); 1945, *J. Bedford, K. & L.*, 102. East Finchley, 1908, *J. E. Cooper* (BM); 1923, *J. E. Cooper* (K). Belsize Park, 1921, *Richards, K. & L.*, 102. **5.** Acton, 1893, *Vallance* (K). Hanwell, *Loydell* (D). **6.** Church End, Finchley, 1918–19 and 1923, *J. E. Cooper, K. & L.*, 102. Muswell Hill, 1902, *J. E. Cooper* (BM). Winchmore Hill, 1905, *P. H. Cooke* (LNHS). Southgate, *Hall*; 1959, *A. Vaughan.* **7.** St James's Park, 1917, *Lester-Garland* (K). Crouch End, 1897, *J. E. Cooper* (BM). Hackney Marshes, 1919 and 1927, *J. E. Cooper, K. & L.*, 102. Clapton, 1909, *P. H. Cooke* (LNHS). Chelsea, *Graham, Rep. Bot. Soc. & E.C.* 13: 390. Fulham Palace grounds, 1966, *Brewis.*

P. intermedia L. **1.** Uxbridge, *Benbow* (BM). **2.** West Drayton, 1902 and 1904, *Dymes*; 1907, *Worsdell* (K). **3.** Twickenham, *Lousley.* **4.** Cricklewood, 1880, *Ridley*; Finchley, 1909 and 1933, *E. F.* and *H. Drabble* (BM); 1942, *Lousley* (L). **5.** Ealing, 1905, *Loydell* (D). **7.** Behind the British Museum, Bloomsbury, *Shenst. Fl. Lond.* Near Berkeley Square, W.1, c. 1942, *A. H. G. Alston, K. & L.*, 103.

P. hirta L. **1.** Yiewsley, 1908–9, *Coop. Cas.* **4.** Finchley, 1910, *Coop. Cas.* **5.** Acton, 1905, *Green* (SLBI). **6.** Muswell Hill, 1910, *Coop. Cas.* **7.** Parsons Green, 1856, *A. Irvine, Phyt. N.S.* 2: 168. Isle of Dogs, 1887, *Benbow*; Kensington Gardens, 1944, *Bostock* (BM).

P. inclinata Vill., *P. canescens* Bess. **1.** Uxbridge, 1909, *Loydell* (BM). Yiewsley, 1920, *J. E. Cooper* (K). **2.** Hampton, 1915, *J. E. Cooper, K. & L.*, 103. **4.** Finchley, 1915, *J. E. Cooper* (K); 1917–19, *K. & L.*, 103. **6.** Muswell Hill, 1903, *J. E. Cooper, K. & L.*, 103.

P. alba L. **7.** Chelsea, 1852, T. Moore (BM).

Duchesnea indica (Andr.) Focke, *Fragaria indica* Andr., *Potentilla indica* (Andr.) T. Wolf. **3.** Near Southall, 1946, *L. G. Payne, K. & L.*, 101.

Fragaria moschata Duchesne, *F. elatior* Ehrh. Hautbois Strawberry. **1.** Pinner, *Hind, T. & D.*, 94. **4.** Harrow (ME). Harrow Weald, *T. & D.*, 94. Near Totteridge, *L. M. P. Small.* **7.** Holloway, 1965, *D. E. G. Irvine.*

Alchemilla glabra Neygenf. **7.** Regents Park, 1963, *Wurzell, L.N.* 43: 22.

Prunus armeniaca L. Apricot. **7.** Bombed site, High Holborn, 1950, *Whittaker* (BM).

P. persica (L.) Batsch. Peach. Seedlings and saplings are a common feature of rubbish-tips.

P. dulcis (Mill.) D. A. Webb, *P. amygdalus* auct., *Amygdalus communis* L. Almond. **1.** Yiewsley!, *K. & L.*, 344. **5.** Hanwell!, *K. & L.*, 344. **7.** Hyde Park, seedlings, *Allen Fl.*

P. cornuta (Wall.) Steud. Himalayan Bird-cherry. **5.** Thames bank, Chiswick, 1951!, *A. H. G. Alston.*

Cotoneaster simonsii Bak. **2.** Between Kingston Bridge and Hampton Court, *C. E. Britton, Rep. Bot. Soc. & E.C.* 6: 727.

C. lindleyi Steud. **3.** Hounslow Heath, 1959, *D. Bennett, L.N.* 39: 39.

Crataegus punctata Jacq. **4.** Hampstead Heath, 1951, *Coates, K. & L.*, 114.

C. crus-galli L. **1.** Uxbridge, 1912, *P. H. Cooke* (LNHS).

CRASSULACEAE

Sedum forsteranum Sm. subsp. **elegans** (Lejeune) E. F. Warb., *S. rupestre* auct. **5.** Ealing, 1946!; Hanwell, 1946!, *K. & L.*, 118. **7.** Holloway, 1965, *D. E. G. Irvine.*

S. rosea (L.) Scop. **3.** Hounslow Heath, 1971, *Gilbert.*

S. maximum (L.) Hoffm., *S. telephium* L. subsp. *maximum* (L.) Krock. **1.** Harefield, 1962. New Years Green, 1964. **2.** Near Shepperton, 1966.

Sempervivum tectorum L. House Leek. **1.** Ruislip, 1926, *Tremayne*; 1945. Yiewsley, 1926, *Tremayne*. Cowley, 1950. **2.** Sunbury, *T. & D.*, 117. **3.** Twickenham; Whitton, *T. & D.*, 117. Feltham, *Tremayne*. **4.** Hampstead, *Irv. MSS.* Hendon, 1926, *Tremayne*. Harrow, *Hind. Fl.* Near Wembley Park, *Benbow MSS.* Golders Green, 1928, *Moring MSS.* **6.** Southgate; Colney Hatch, *Tremayne*. Edmonton, *T. & D.*, 117. Enfield, 1965, *Kennedy*!. **7.** Tottenham, *Johns. Cat.* Kentish Town (HE). Probably always planted.

Crassula tillaea L.-Garland, *Tillaea muscosa* L. **4.** Near Kingsbury, *Wharton*, fide *Farrar*, *T. & D.*, 115.

Umbilicus rupestris (Salisb.) Dandy, *U. pendulinus* DC., *Cotyledon umbilicus-veneris* auct. Wall Pennywort. **7.** Upon Westminster Abbey over the door that leadeth from Chaucer his tomb to the old palace, *Ger. Hb.*, 424; in this place it is not now to be found, *Johns. Ger.*, 529. In an old gravel-pit near Highbury Barn, *Mart. Pl. Cant.*, 71. Perhaps introduced with stone blocks.

SAXIFRAGACEAE

Saxifraga stolonifera Meerb. Strawberry Saxifrage. **7.** Buckingham Palace grounds, *Fl. B.P.G.*

Heuchera sanguinea Engelm. **2.** Between West Drayton and Hayes, 1967.

GROSSULARIACEAE

Ribes alpinum L. Mountain Currant. **4.** Between Highgate and Hampstead, 1838, *E. Ballard* (BM).

R. sanguineum Pursh. Flowering Currant. **5.** Near the Brent, Ealing, a single bush, 1965.

LYTHRACEAE

Lythrum junceum Banks & Soland. **3.** Hounslow Heath, 1964, *Wurzell*, det. *Lousley*. **4.** Mill Hill, *Harrison*, det. *Sandwith*.

ONAGRACEAE

Clarkia purpurea L. subsp. **viminea** (Dougl.) Lewis & Lewis. **3.** Hounslow Heath, 1960, *Philcox* and *Townsend.*

Fuchsia magellanica Lam. var. **macrostema** (Ruiz & Pav.) Munz, *F. riccartoni* auct., *F. gracilis* Lindl. **4.** Hendon, 1963, *Wurzell.*

UMBELLIFERAE

Eryngium planum L. **3.** Hounslow Heath, 1951, *Lousley* and *McClintock*!; 1952, *K. & L.*, 126. **7.** Side of the old Marylebone Workhouse, 1966, *Brewis.*

Anthriscus cerefolium (L.) Hoffm., *Chaerophyllum cerefolium* (L.) Crantz. Garden Chervil. **2.** Laleham Park, one patch, 1966, *Kennedy*!. **4.** Hampstead, *Irv. MSS.* **7.** Clapton, *E. Forster* (BM). Kentish Town, 1850 (HE).

Caucalis platycarpos L., *C. daucoides* L., *C. lappula* Grande. **1.** Uxbridge, *Druce Notes*. Harefield, 1906, *Coop. Cas.* Potters Bar, 1914, *J. E. Cooper* (BM). **3.** Hayes, 1916, *P. H. Cooke* (LNHS). **4.** Mill Hill, *Harrison.* **5.** West Ealing, 1961. **6.** Hadley Common, 1905, *P. H. Cooke* (LNHS). Muswell Hill, 1906, *J. E. Cooper* (BM). **7.** Hackney Marshes, 1912, *J. E. Cooper* (BM).

Coriandrum sativum L. Coriander. **1.** Yiewsley, 1915, *J. E. Cooper* (BM); c. 1926, *M. & S.* Harefield, 1955, *Pickess*; 1967. **2.** Shepperton, 1945!, *K. & L.*, 135. West Drayton, *Druce Notes.* East Bedfont, 1965, *Mason.* Near Harmondsworth, 1967. **3.** Isleworth, 1945, *Westrup, K. & L.*, 135. Hounslow Heath, 1960, *Philcox* and *Townsend*; 1964, *Wurzell*; 1965–66. **4.** Hendon, 1915, *J. E. Cooper, K. & L.*, 135. **5.** Hanwell, 1946!, *L.N.* 26: 61; 1948–50!, *K. & L.*, 135; 1954. Ealing, 1949, *L. M. P. Small.* Acton, *Druce Notes.* **7.** Chelsea, *Britten* (BM). Bombed sites, St Dunstan's-in-the-East, E.C., 1948, *Lousley, K. & L.*, 135. Hackney Marshes, 1913, *J. E. Cooper* (BM); 1915–26, *J. E. Cooper, K. & L.*, 135. Abundant in a derelict garden near Paddington, 1965!; Tower of London gardens, about twenty small colonies, 1966!, *L.N.* 46: 31. St Katharine's Dock, Stepney, 1966, *Brewis.*

Bifora testiculata (L.) Roth. **6.** Winchmore Hill, *L. B. Hall, Rep. Bot. Soc. & E.C.* 9: 355.

Bupleurum lancifolium Hornem., *B. protractum* Hoffmg., *B. subovatum* Lamarck. **1.** Yiewsley!, 1923, *J. E. Cooper, K. & L.*, 127;

1962. **2.** Between Hampton and Sunbury!, *Newbould*, *T. & D.*, 126; 1966, *Tyrrell*!. Staines, *Frost*. **3.** Hounslow, 1955, *Westrup*, *K. & L.*, 349. Hounslow Heath!, *Philcox* and *Townsend*; 1962→. Feltham, 1965, *P. J. Edwards*. **4.** Mill Hill, *Harrison*, det. *Sandwith*. **5.** West Ealing, 1959!, *Proc. Bot. Soc. Brit. Isles* 3: 442; 1960 and 1963. Hanger Hill, Ealing, 1968, *McLean*. **6.** Between Muswell Hill and Hornsey, *T. & D.*, 126. **7.** Hackney Marshes, 1913, *J. E. Cooper* (BM); 1918, 1920, 1923, 1925 and 1926, *J. E. Cooper*, *K. & L.*, 127. Hackney, 1962, *Rudiger*, *L.N.* 42: 12. Tottenham, 1960, *Kennedy*. Hyde Park!; Kensington Gardens!, *Allen Fl.*

B. fontanesii Guss. ex Caruel, *B. odontites* auct. **7.** Fulham, 1887, *Benbow* (BM).

Cuminum cyminum L. **7.** Hackney Marshes, 1917, *J. E. Cooper* (BM); 1918, *J. E. Cooper*, *K. & L.*, 135.

Ammi majus L. **1.** New Years Green, 1969, *Bot. Soc. Brit. Isles Field Meeting*. **7.** West India Dock, *Cherry*, *T. & D.*, 126.

Trachyspermum ammi (L.) Sprague, *T. copticum* (L.) Link. **1.** New Years Green, 1968, *Clement* and *Ryves*. **2.** Between West Drayton and Hayes, 1967!, *L.N.H.S. Excursion*, det. *Lousley*. Near Harlington, 1968!, *L.N.H.S. Excursion*. West Drayton, 1969, *Clement* and *Ryves*.

Pimpinella anisum L. Aniseed. **5.** Hanwell, 1950! and 1953!, *K. & L.*, 349.

Capnophyllum peregrinum (L.) Lange, *Tordylium peregrinum* L. **1.** Yiewsley, 1924, *J. E. Cooper*, *K. & L.*, 135.

Anethum graveolens L., *Peucedanum graveolens* (L.) Druce. Dill. **2.** Near West Drayton, 1965. Between West Drayton and Hayes, 1967. **3.** Hounslow Heath, 1954 and 1964. **5.** Hanwell, 1946!, *L.N.* 26: 61; 1947!; Greenford, 1947! and 1951!, *K. & L.*, 134. **7.** Bombed site, Great Tower Street, E.C., 1944, *Lousley Fl.*; 1945–46!, *Lousley*, *K. & L.*, 134. Bombed site near Holborn Circus, 1946, *Lousley* (L). Zoological Gardens, Regents Park, 1966, *Holme*, *L.N.* 46: 31.

CUCURBITACEAE

Cucumis melo L., *Melo sativus* Sageret ex Roem. Melon. **1.** Yiewsley, *M. & S.*

C. sativus L. Cucumber. A common feature of rubbish-tips.

Cucurbita pepo L. Vegetable Marrow. A common feature of rubbish-tips.

C. maxima Duchesne. Pumpkin. **1.** Yiewsley, *M. & S.*

Citrullus lanatus (Thunb.) Mansf., *C. vulgaris* Schrad., *Colocynthis citrullus* (L.) O. Kuntze. Water Melon. **1.** Harefield, *Clement* and *Ryves.* **3.** Hounslow Heath, 1959, *D. Bennett, L.N.* 39: 39.

EUPHORBIACEAE

Ricinus communis L. Castor-oil Plant. **1.** Yiewsley!, *M. & S.* **3.** Hounslow Heath, in quantity!, 1953, *Welch*; 1954, *T. G. Collett* and *Lousley*!, *K. & L.*, 249. **5.** Hanwell, 1950!, *K. & L.*, 249.

POLYGONACEAE

Polygonum arenarium Waldst. & Kit. subsp. **pulchellum** (Lois.) D. A. Webb & A. O. Chater, *P. pulchellum* Lois. **1.** Yiewsley, 1929, *Melville* (L). Harefield, 1966, *Johnson* and *Kennedy*!; 1969, *Clement* and *Ryves.* **2.** Between West Drayton and Hayes, 1967!, *L.N.H.S. Excursion*, det. *Lousley, L.N.* 47: 9. **5.** Greenford, 1954!, *Welch, K. & L.*, 241.

P. senegalense Meisn. **1.** New Years Green, 1967, *Clement* and *Lousley.*

P. orientale L. **7.** Burgess Hill, 1870, *Dyer* (K).

Var. **glabratum** Hook. **1.** Harefield, 1953!, *Graham*, det. *Brenan, K. & L.*, 356.

P. pensylvanicum L. **1.** New Years Green, 1967, *Clement* and *Lousley, L.N.* 47: 9. **2.** West Drayton, 1969, *Clement* and *Ryves.* **7.** South End Green, Hampstead, 1870, *Trim. MSS.*

Var. **laevigatum** Fern. **1.** Springwell, 1945–46!, *L.N.* 26: 76. **5.** Hanwell, 1946!, *K. & L.*, 341.

Fagopyrum esculentum Moench, *F. sagittatum* Gilib. Buckwheat. A common feature of rubbish-tips, etc.

F. tataricum (L.) Gaertn. **1.** Yiewsley, *M. & S.*

Rumex alpinus L. **7.** South Kensington, *Dyer Key.*

R. pseudonatranus Borbás, *R. fennicus* (Murb.) Murb. **7.** Hackney Marshes, 1910, *J. E. Cooper* (BM), det. *Rechinger* (as *R. fennicus*).

R. obovatus Danser, *R. paraguayensis* Thell., non Parodi. **5.** Greenford, 1920, *J. E. Cooper* (BM), det. *Rechinger.* **6.** North Finchley,

J. E. Cooper (BM), det. *Rechinger*. **7.** Hackney Marshes, 1912, *J. E. Cooper* (BM), det. *Rechinger*.

R. altissimus Woods. **4.** Temple Fortune, 1910, *J. E. Cooper* (BM), det. *Rechinger*.

URTICACEAE

Urtica pilulifera L. Roman Nettle. **2.** Poyle; West Drayton, *Druce Notes*. **5.** Acton Green, 1900, *Loydell* (K); 1901 (D). **7.** Haverstock Hill, *Irv. MSS*. Near Highgate, 1864 (SY). Regents Park, *Webster Reg*.

CANNABIACEAE

Cannabis sativa L. Hemp. A common feature of rubbish-tips, also seen on river and canal banks and adjacent waste ground.

ERICACEAE

Erica ciliaris L. Dorset Heath. **7.** Buckingham Palace grounds, introduced with peat blocks, *Fl. B.P.G.*; not seen since 1964, *McClintock*.

PLUMBAGINACEAE

Armeria maritima (Mill.) Willd., *Statice armeria* L., *S. maritima* Mill. Thrift, Sea Pink. **3.** Hounslow Heath, 1965. **7.** Bombed site, Cripplegate, 1948–49; bombed site off Cheapside, E.C., 1952, *A. W. Jones*, *K. & L.*, 183.

PRIMULACEAE

Lysimachia punctata L. **1.** Yiewsley, *M. & S.*

L. ephemerum L. **4.** Hampstead, 1923, *Garlick*.

BUDDLEJACEAE

Buddleja globosa Hope. **3.** Hounslow Heath, 1964, *K. Marks*, *Wild Fl. Mag.* 1965, 28.

POLEMONIACEAE

Polemonium caeruleum L. Jacob's Ladder. **1.** Pinner, *T. & D.*, 187. Uxbridge, 1885, *Benbow* (BM). **4.** Hampstead Heath, *M. & G.*, 304.

Phlox paniculata L. **3.** Roxeth, 1940!, *K. & L.*, 190. **5.** Hanwell, 1933!, *K. & L.*, 190.

BORAGINACEAE

Lappula squarrosa (Retz.) Dumort. subsp. **squarrosa,** *L. myosotis* Moench, *L. echinata* Gilib. **1.** Uxbridge, 1893, *Benbow* (BM). Yiewsley, 1910, *Coop. Cas.*; 1917, *Lester-Garland* (K); 1924, *J. E. Cooper, K. & L.*, 191. **2.** Near Staines, *Shepheard, J.B. (London)* 34: 511. West Drayton, *Druce Notes.* **4.** Brondesbury, 1880, *Ridley* (BM). Finchley, 1929, *J. E. Cooper, K. & L.*, 191. **7.** Hackney Marshes, 1913 and 1924, *J. E. Cooper* (BM).

Amsinckia menziesii (Lehm.) Nels. & Macbr. **2.** West Drayton, *Druce Notes.*

A. lycopsioides Lindl. **1.** Yiewsley, 1924, *J. E. Cooper* (K).

Symphytum asperum Lepech. **1.** Yiewsley, 1916, *J. E. Cooper.*

Borago officinalis L. Borage. **1.** Ickenham, 1889, *Benbow* (BM). Uxbridge, *Benbow MSS.* Yiewsley, 1913, *Coop. Cas.* **2.** Bushy Park, *T. & D.*, 189. West Drayton, *Druce Notes.* Between Staines and Laleham, 1947!, *K. & L.*, 192. Sunbury, 1886, *Fraser* (K). Near Hampton, plentiful, 1966, *Tyrrell.* **3.** Twickenham, *T. & D.*, 190. Whitton, 1886, *Benbow* (BM). Between Hanworth and Twickenham, *Benbow MSS.* Hounslow, 1954, *H. Banks.* Isleworth, 1966. **4.** Hampstead, *Wharton Fl.* East Finchley, 1908, *J. E. Cooper* (BM). Finchley, 1927, *J. E. Cooper, K. & L.*, 192. **5.** Ealing, *Boucher, K. & L.*, 192. Hanwell, 1954. **6.** Near Enfield Wash, 1966!, *Kennedy.* **7.** West End, Hampstead, *M. & G.*, 251. Kensington, *J. Banks* (BM). Kentish Town, 1846 (HE). Chelsea, *Britten*; East Heath, Hampstead, *T. & D.*, 190. Hornsey, 1896, *C. S. Nicholson, K. & L.*, 192. Highgate, 1945, *Fitter, K. & L.*, 192. Shadwell, 1954, *Henson* (BM).

Heliotropium peruviana L. **7.** Hyde Park, 1961, *Allen Fl.*

Anchusa officinalis L. **2.** West Drayton, *Druce Notes.* **4.** East Finchley, *Coop. Cas.* Hampstead, 1923, *Garlick.* **7.** Isle of Dogs, *Benbow MSS.*

A. undulata L. **7.** Isle of Dogs, 1889, *Benbow MSS.*

A. azurea Mill., *A. italica* Retz. **2.** West Drayton, *Druce Notes*. **6.** Muswell Hill, 1906, *J. E. Cooper* (BM). **7.** Crouch End, 1875, *French* (BM).

Myosotis dissitiflora Bak. **2.** Dawley, *Druce Notes*.

CONVOLVULACEAE

Convolvulus tricolor L. **5.** West Ealing, 1959!, *Proc. Bot. Soc. Brit. Isles* 3: 442.

Ipomoea batatas (L.) Lam., *Batatas edulis* Choisy. Sweet Potato. **1.** Rubbish-tip, Harefield, 1967, *J. G.* and *C. M. Dony*, det. *Verdcourt*, *L.N.* 47: 10.

I. purpurea Roth. **1.** Near Springwell Lock!, 1945, *Welch*; 1946!, *K. & L.*, 196. Yiewsley, 1963, *Wurzell*, *L.N.* 43: 22.

I. hederacea Jacq., *Pharbitis hederacea* (L.) Choisy. **1.** Near Springwell Lock, 1945–46, *Lousley*!, *L.N.* 26: 62; 1947–48!, *K. & L.*, 196.

Var. **integriuscula** Gray. **1.** Springwell, 1945!, *Lousley* (L).

I. lacunosa L. **1.** Near Springwell Lock, 1947!, *L.N.* 27: 33.

Calystegia fraterniflora (Mack. & Bush) Brummitt, *C. sepium* (L.) R.Br. var. *fraterniflora* (Mack. & Bush) Shinners. **1.** By Soya Foods Ltd, Springwell Lock, 1955!, *Lousley* (L), det. *Brummitt* (1966).

Cuscuta epilinum Weihe. Flax Dodder. **7.** Regents Park, on *Linum* species, 1867 (SY).

C. australis R.Br. var. **cesattiana** (Bertol.) Feinbrun. **6.** Enfield, on *Callistephus chinensis*, 1953, *Maude*, det. *Shillito*, *K. & L.*, 197. The plant seen may have been *C. campestris* Yuncker.

SOLANACEAE

Lycopersicon esculentum Mill. Tomato. Invariably occurs in quantity on sewage-farms, and to a lesser extent on rubbish-tips.

Nicandra physalodes (L.) Gaertn. Shoofly Plant. **1.** Springwell, 1948, *Brenan* and *Sandwith*!, *L.N.* 28: 33. New Years Green, 1961. Yiewsley, 1964–65!, *Wurzell*. Harefield, 1969, *Clement* and *Ryves*. Near Eastcote, 1969!. **2.** Frequent on rubbish-tip near Harmondsworth, 1967. **4.** Hendon, 1962, *Brewis*. **5.** Hanwell, 1946!, *L.N.* 26: 62. **6.** Whetstone, 1967. **7.** Parsons Green, 1860, *Britten*, *Trim.*

MSS. South Kensington, *T. & D.*, 195. By St Lawrence Jewry Church, E.C., 1967, *Lousley*.

Physalis ixocarpa Hornem. **3.** Hounslow Heath, 1959, *D. Bennett*, *L.N.* 39: 39.

P. angulata L. **1.** Springwell, 1946!, *K. & L.*, 198.

P. pubescens L. **1.** Yiewsley, 1927, *Melville* (K). **3.** Northolt, 1949, *Boniface*; Feltham, 1951!, *Russell* and *Welch*, *K. & L.*, 199.

P. peruviana L. **1.** Yiewsley, *M. & S.* Harefield, 1968, *Clement* and *Ryves*. **7.** Hackney Marshes, *M. & S.*

Solanum sarrachoides Sendtn., *S. nitidibaccatum* auct., non Bitter, *S. chenopodioides* auct. **1.** Springwell, 1945!, *Welch*; 1946–49!, *K. & L.*, 198.

S. cornutum Lam., *S. rostratum* Dunal. **2.** Between West Drayton and Hayes, 1967, *L.N.H.S. Excursion*, *L.N.* 47: 9. Shepperton, 1971, *Clement*, *E. Young* and *Tuck*. **4.** Near Stonebridge, 1886, *Benbow* (BM).

S. tuberosum L. Potato. Common on rubbish-tips, etc.

S. melongena L. var. **esculentum** Nees. Aubergine, Egg Plant. **7.** Between Tottenham Court Road and Rathbone Place, W.C., 1973, *McLean*.

S. ciliatum Lam. **1.** Yiewsley, *M. & S.*

S. capsicastrum Link ex Schauer. Winter Cherry. **7.** Bombed site, Leather Lane, E.C., 1950, *Whittaker* (BM).

Datura inermis Jacq. **5.** Acton Green, 1900, *Loydell*, *Rep. Bot. Soc. & E.C.* 3: 168.

D. ferox L. **5.** Greenford, 1954, *T. G. Collett* and *Lousley*!, *K. & L.*, 355.

Petunia axillaris (Lam.) Britton, E. E. Sterns & Poggenb. × **integrifolia** (Hook.) Schinz & Thell. = **P. × hybrida** (Hook.) Vilm. **1.** New Years Green, 1963. **5.** Hanwell, 1948. Greenford, 1952.

Nicotiana rustica L. **3.** Feltham, 1950, *Russell* and *Welch*!, *K. & L.*, 201. **5.** Chiswick, 1871, *F. Naylor*, *Trim. MSS*. Hanwell, 1951!, *K. & L.*, 201. Greenford, 1953, *Bangerter*, *T. G. Collett* and *Melderis*!, *K. & L.*, 201. **7.** South Kensington; Victoria Street, S.W.1, *T. & D.*, 195.

N. tabacum L. Tobacco. **1.** Yiewsley, *M. & S.* **5.** Hanwell!; Greenford!, *K. & L.*, 201. **7.** Isle of Dogs, 1945!, *K. & L.*, 201.

Hyde Park, 1961, *D. E. Allen*, *L.N.* 41: 20. Probably all errors. It appears unlikely that *N. tabacum* would grow in the open in south-east England, and the records given are probably referable to *N. alata*.

N. alata Link & Otto. **1.** New Years Green, 1954, *T. G. Collett*!; 1961, *Lousley*!. **2.** Near Harmondsworth, 1967. **3.** Hounslow Heath!, 1961, *Philcox* and *Townsend*; 1965, *Kennedy*!. **4.** Hendon, 1963, *Wurzell*. **5.** Hanwell, 1950. Greenford, 1944 and 1953. Near Chiswick, 1960. **7.** Isle of Dogs, 1945. Love Lane, Tottenham, 1967, *Kennedy*!.

Var. **grandiflora** Comes. **1.** Yiewsley, *M. & S.*

SCROPHULARIACEAE

Verbascum thapsiforme Schrad. **5.** Near Boston Manor, two plants, 1968, *J. L. Gilbert*.

V. lychnitis L. White Mullein. **2.** Bushy Park, 1950!, *K. & L.*, 202. **3.** Hounslow Heath, *Westrup*. **7.** South Kensington, *F. Naylor*, *T. & D.*, 197.

Alonsoa peduncularis Wettst. **4.** Finchley, 1909, *Coop. Cas.*

Linaria genistifolia (L.) Mill. subsp. **dalmatica** (L.) Maire & Petitmengin. **5.** Hanwell, 1946. Greenford, 1961.

L. incarnata (Vent.) Spreng., *L. bipartita* auct. eur., non (Vent.) Willd. **1.** New Years Green, 1961!, *McClintock*.

L. maroccana L.f. **5.** Hanwell, 1946, det. *Sandwith*. **7.** Regents Park, 1966, *Holme*, *L.N.* 46: 32.

Scrophularia vernalis L. Yellow Figwort. **7.** Garden weed, Long Acre, W.C., *Park. Theat.*, 261.

Nemesia strumosum Benth. **6.** Roadside, Botany Bay, Enfield, in quantity, 1965, *Kennedy*!, *L.N.* 45: 20.

Veronica peregrina L. American Speedwell. **4.** Mill Hill, *A. W. Bennett*; 1872, *Scott-White* (BM).

V. austriaca L. subsp. **austriaca,** *V. jacquinii* Baumg. **1.** Ruislip, 1937, *Moody* (L).

ACANTHACEAE

Acanthus mollis L. Bear's Breech. **5.** Brentford, 1967, *Payton*!.

A. spinosus L. **5.** Chiswick, 1946!, *K. & L.*, 215.

VERBENACEAE

Verbena patagonica Moldenke, *V. bonariensis* auct. **7.** Hyde Park, 1961, *Brewis*; *D. E. Allen* and *McClintock*; Kensington Gardens, 1961, *D. E. Allen*, *L.N.* 41 : 20. Buckingham Palace grounds, 1961, *Fl. B.P.G.*

V. hybrida Groenl. & Rumpl. **7.** Hyde Park, *D. E. Allen*, *L.N.* 41 : 20. Buckingham Palace grounds, 1961, *Fl. B.P.G.*

V. rigida Spreng. **5.** Hanwell, 1969, *J. L. Gilbert.*

LABIATAE

Elsholtzia ciliata (Thunb.) Hyland., *E. patrinii* (Lepech.) Garcke, *E. cristata* Willd. **7.** Chelsea; Parsons Green, 1856, *Irv. Ill. Handb.*, 437. Paddington, 1878, *Trimen* (BM).

Origanum majorana L. Sweet Marjoram. **3.** Potato field between Hounslow and Whitton, 1886, *Benbow* (BM).

Salvia officinalis L. Sage. **1.** Harefield, 1958. **3.** Between Hayes and Dawley, *Druce Notes.* **4.** Edgware, 1950!, *K. & L.*, 223. **5.** South Ealing, 1963.

S. sclarea L. **5.** Hanwell, 1953!, *K. & L.*, 223. Near Greenford, 1957. **7.** Finsbury Park, 1952, *J. Bedford*, *K. & L.*, 223.

S. viridis L. **5.** Hanwell, 1956.

S. reflexa Hornem. **1.** Yiewsley, 1966!, *Wurzell.* **2.** Near Harmondsworth, 1966–67. **4.** Mill Hill, *Harrison.* **5.** Hanwell, *Lousley* and *McClintock*!, *K. & L.*, 223; 1954, *T. G. Collett*!. **7.** Hyde Park Corner, 1962!; Hyde Park, 1962!, *Allen Fl.*

Nepeta mussinii Spreng. × **N. nepetella** L. = **N.** × **faasenii** Bergm. ex Stearn. Garden Catmint. **1.** Yiewsley, 1966. **5.** Hanwell, 1955.

N. nuda L. **1.** Uxbridge, *Loydell* (D).

Melittis melisophyllum L. Bastard Balm. **7.** Hornsey, 1952, *Browning*, *K. & L.*, 225.

Sideritis montana L. **4.** Harrow, 1885, *Winston* (BM).

S. romana L. **7.** Hackney Marshes, 1927, *J. E. Cooper*, *K. & L.*, 225.

Stachys annua (L.) L. **1.** Yiewsley, *Coop. Cas.* **2.** Near Staines, *Shepheard*, *J.B.* (*London*) 34: 511. **3.** Near Hounslow, 1887, *Benbow*

MSS. **4.** Near Stonebridge, *Benbow MSS.* Finchley, 1910, *Coop. Cas.* **5.** Chiswick, 1887, *Benbow, J.B. (London)* 25: 20. **7.** Crouch End, 1897, *Coop. Cas.* Isle of Dogs, *Benbow, J.B. (London)* 25: 20. Hackney Marshes, *Coop. Cas.*; c. 1926, *M. & S.* Near Marble Arch, 1962!, *D. E. Allen* and *Potter, Allen Fl.*

Lamium moluccellifolium Fr., *L. intermedium* Fr., *L. hybridum* auct. subsp. *intermedium* (Fr.) Gams. Intermediate Dead-nettle. **7.** Hackney Marshes, 1924, *J. E. Cooper* (BM).

Galeopsis ladanum L. **4.** Near Stonebridge, *Benbow, J.B. (London)* 25: 20. Hendon, 1913, *P. H. Cooke* (LNHS).

Leonurus cardiaca L. Motherwort. **4.** Hendon, 1907, *Coop. Cas.* **5.** Chiswick, *Benbow* (BM). Acton, 1900, *Green, Druce Notes.* Ealing, 1963!, *T. G. Collett, L.N.* 43: 22; 1964→.

PLANTAGINACEAE

Plantago indica L., *P. arenaria* Waldst. & Kit., *P. ramosa* (Gilib.) Aschers., *P. psyllium* auct., non L. **1.** Hillingdon, 1936, *P. H. Cooke* (LNHS). Yiewsley, 1934, *Lousley* (L). **2.** West Drayton, *Druce Notes.* Poyle, 1955, *Welch.* **3.** Whitton, *Montg. MSS.* **4.** Finchley, 1963, *Wurzell, L.N.* 43: 22. **5.** Acton Wells, *Druce Notes.* Greenford, 1920, *J. E. Cooper, K. & L.*, 229. **7.** Crouch End, 1897, *J. E. Cooper* (BM). Chelsea, *Mill. Gard. Dict.*; *Cullum Fl.*, 59. Hackney Marshes, 1912, *Coop. Cas.*; 1915, *J. E. Cooper, K. & L.*, 229. Suffolk Lane, E.C., 1963, *Holland, L.N.* 43: 22.

P. lagopus L. **2.** West Drayton, *Druce Notes.*

P. maritima L. Sea Plantain. **7.** Hyde Park, two plants, 1959, *D. E. Allen, L.N.* 39: 39.

CAMPANULACEAE

Wahlenbergia hederacea (L.) Reichb., *Campanula hederacea* L. Ivy-leaved Bellflower. **4.** Hampstead Heath, 1847, *H. Taylor* (BM). Probably deliberately introduced.

Campanula patula L. Spreading Bellflower. **4.** Willesden Green, 1945–48!, *K. & L.*, 180.

C. medium L. Canterbury Bell. **4.** Harrow, *Harley.* **5.** Greenford, 1951!, *K. & L.*, 180.

Legousia speculum-veneris (L.) Chaix, *Specularia speculum* (L.) DC., *Campanula speculum-veneris* L. **1.** Yiewsley, 1914, *J. E. Cooper* (BM).

LOBELIACEAE

Lobelia erinus L. **1.** Harefield, 1966, *I. G. Johnson* and *Kennedy*!.

RUBIACEAE

Asperula galoides Bieb. **2.** Hampton Court, 1871, *H. C. Watson*, *Rep. Bot. Loc. Rec. Club* 1874, 17; *loc. cit.* 1877, 258; *Rep. Bot. Soc. & E.C.* 4: 269.

A. arvensis L. **4.** East Finchley, 1908, *J. E. Cooper* (K). Harrow, 1960, *Knipe*. Mill Hill, 1962, *Harrison*, det. *Sandwith*. **5.** Ealing, 1958!, *Proc. Bot. Soc. Brit. Isles* 3: 442. **7.** Hackney Marshes, 1914, *J. E. Cooper* (BM); 1918, *J. E. Cooper, K. & L.*, 141.

Galium tricornutum Dandy, *G. tricorne* Stokes *pro parte*. Rough-fruited Bedstraw. **1.** Yiewsley, 1910 and 1913, *Coop. Cas.*; c. 1926, *M. & S.* **4.** Finchley, 1908, *Coop. Cas.* **5.** Hanwell, 1953. **7.** Stoke Newington, 1871, *F. J. Hanbury* (HY). Near Highgate, 1887; Hackney Marshes, 1912–13, *Coop. Cas.*; c. 1926, *M. & S.*

Rubia peregrina L. Wild Madder. **2.** West Drayton, 1916, *A. B.* and *M. Cobbe*, *Rep. Bot. Soc. & E.C.* 4: 489.

R. tinctorum L. **1.** Yiewsley, 1924, *J. E. Cooper* (K).

VALERIANACEAE

Valerianella eriocarpa Desv. **5.** Ealing, 1950, *L. M. P. Small* (BM).

DIPSACACEAE

Cephalaria syriaca (L.) Roem. & Schult., *Lepicephalus syriacus* Lag. **1.** Uxbridge, 1905, *Green, K. & L.*, 143. **2.** East Bedfont, 1965.

COMPOSITAE

Helianthus petiolaris Nutt. **1.** Yiewsley, 1956!, *K. & L.*, 351. **3.** Hounslow Heath, 1963.

Rudbeckia laciniata L. **2.** Between Perry Oaks and Stanwell, 1955–56, *Briggs, L.N.* 38: 21. **7.** Highgate, 1963, *Wurzell.*

R. bicolor Nutt. **7.** Kensington Gardens, 1961, *D. E. Allen* and *McClintock, L.N.* 41: 21.

R. serotina Nutt. **2.** Between West Drayton and Hayes, 1967.

Ambrosia artemisiifolia L., *A. elatior* L. Roman Wormwood, Ragweed. **1.** Springwell, 1945–46!, *L.N.* 26: 61; 1947–48!, *K. & L.*, 150. Harefield, 1947, *Sandwith*!, *L.N.* 27: 32. New Years Green, 1964, *T. G. Collett*!. **2.** West Drayton, *Druce Notes.* **3.** Near Hayes, 1967. **4.** Finchley, 1900, *J. E. Cooper* (BM). **5.** Hanwell, 1946, *J. G. Dony, Lousley* and *Woodhead*!, *L.N.* 26: 61. Greenford, 1947!, *L.N.* 27: 32. Gunnersbury Park, Acton, 1962!, *Boniface.* **6.** Muswell Hill, *Coop. Cas.* **7.** Crouch End, 1878, *French* (BM). Hackney Marshes, 1924, *J. E. Cooper.*

A. trifida L. Great Ragweed. **1.** Springwell, 1945–46!, *L.N.* 26: 61; 1947–48!, *K. & L.*, 150. **7.** Highgate, 1897; Hackney Marshes, 1912, *Coop. Cas.*

Var. **integrifolia** (Willd.) Torr. & Gray. **1.** Uxbridge, 1908, *Druce Notes.* Springwell, 1945!, *L.N.* 26: 61; 1946.

Xanthium strumarium L. Cocklebur, Woolgarie Bur. **1.** Uxbridge, 1912, *Wedgwood* (BM). **2.** Between [West] Drayton and Iver, *Ger. Hb.*, 665. Staines, *Lawson, Wilson Syn.*, 16. **5.** Acton Green, 1900, *Loydell* (BM). **7.** Newington, *Pet. Midd.*; c. 1740, *Blackst. MSS.* Islington, *W. Watson, Blackst. Spec.*, 6. Chelsea, 1866, *F. Naylor, J.B. (London)* 9: 371. Hackney Marshes, *M. & S.*

X. spinosum L. Spiny Cocklebur. **1.** Yiewsley, 1909, *Green* (BM). **2.** West Drayton, *Druce Notes.* Near West Drayton, a single plant, 1965!, *Clement, Mason* and *Wurzell.* **3.** Hounslow Heath, 1961, *D. Bennett.* **5.** Hanwell, two plants, 1953!, *Welch.* **7.** West India Dock, 1871, *F. Naylor, J.B. (London)* 9: 371. Hackney Marshes, 1913, *J. E. Cooper* (BM).

Guizotia abyssinica (L.f.) Cass. Common on rubbish-tips.

Sigesbeckia orientialis L. **5.** Kew Bridge, pre-1873, *A. Irvine* (K). Perhaps in Surrey.

Hemizonia pungens Torr. & Gray. **7.** Hackney Marshes, 1921, *J. E. Cooper* (BM).

Rhagadiolus stellatus (L.) Gaertn. **2.** Upper Lodge, Bushy Park, with bird seed, 1968, *Clement.*

Iva microcephala Nutt. **1.** Harefield, 1950, *Sayle, L.N.* 31: 12.

Coreopsis grandiflora Hogg. **2.** Between Laleham and Staines, 1930, *Tremayne*, *K. & L.*, 151. Near West Drayton, 1965, *Isherwood*!. **3.** Hounslow Heath, 1965.

Cosmos bipinnatus Cav. **1.** Harefield, 1961 and 1966. Yiewsley, 1955 and 1963. South Mimms, 1965, *J. G.* and *C. M. Dony*!. **2.** Dawley, 1948!, *K. & L.*, 151. **3.** Hounslow Heath, 1954→. **4.** Stonebridge. **5.** Ealing!; Hanwell!; Greenford!; North Acton!, *K. & L.*, 151. **7.** Bombed site, West Smithfield, E.C., 1945, *Lousley* (L). Bombed site, Fisher Street, W.C.1, 1950, *Whittaker* (BM).

Tagetes minuta L. **3.** Hounslow Heath, 1961, *Buckle*. **5.** Hanwell, 1952, *J. G. Dony*!, *K. & L.*, 351. **7.** Shoreditch, 1965, *Bennett*. Kensington Gardens, 1965, *Wurzell*. Hackney Marshes, *M. & S.*

Calendula officinalis L. Marigold. Common on rubbish-tips.

C. arvensis L. **7.** Hackney Marshes, 1915, *J. E. Cooper* (BM).

Pulicaria uliginosa Hoffmg. & Link., non Stev. **7.** Hackney Marshes, 1915, *J. E. Cooper* (BM).

Aster novae-angliae L. **1.** Yiewsley!, *M. & S.* **2.** Between Hampton Court and Kingston Bridge, 1947!, *K. & L.*, 147. **3.** Hounslow, 1960. **4.** Harrow Weald, 1946!, *K. & L.*, 147. **5.** Near Yeading!; Greenford!, *K. & L.*, 147. **6.** Finchley Common!, *K. & L.*, 147. **7.** Kensington Gardens, 1958, *Allen Fl.*

A. multiflorus Ait., *A. ericoides* Michx. **5.** Hanwell, 1963.

Callistephus chinensis (L.) Nees, *Callistemma hortense* Cass. China Aster. **1.** Harefield, 1966, *I. G. Johnson* and *Kennedy*!. **2.** Dawley, 1948!, *L.N.* 28: 33. East Bedfont, 1966, *Wurzell*. **3.** Hounslow Heath, 1965, *Kennedy*!.

Erigeron bonariensis L. **5.** Greenford, 1952!, *B. M. C. Morgan* (BM).

E. annuus (L.) Pers. **2.** Staines, 1902, *Riddelsdell* (BM).

Subsp. **septentrionalis** (Fern. & Wieg.) Wagenitz, *E. strigosus* Mühl. ex Willd. var. *septentrionalis* (Fern. & Wieg.) Fern., *E. septentrionalis* Fern. & Wieg. **5.** Chiswick, 1945–47!, *Welch*, det. *Sandwith*, *K. & L.*, 148.

Subsp. **strigosus** (Mühl. ex Willd.) Wagenitz, *E. strigosus* Mühl. ex Willd. **5.** Acton, 1908, *Druce Notes*.

Anthemis altissima L., *A. cota* L. **4.** Finchley, 1917, *J. E. Cooper* (BM). Mill Hill, *Harrison*, det. *Sandwith*.

A. mixta L. **7.** Hackney Marshes, 1916, *J. E. Cooper* and *G. C. Brown* (BM).

Achillea distans Waldst. & Kit., *A. tanacetifolia* All., non Mill. **1.** Yiewsley, 1910, *J. E. Cooper* (BM). **4.** Near Wembley, 1965.

A. aspleniifolia Vent. **3.** Twickenham, 1870, *Dyer* (BM).

A. filipendulina Lam. **5.** Ealing, 1934–37, *Bull., K. & L.*, 153.

Chrysanthemum morifolium Ramat., *C. sinense* Sabine. Garden Chrysanthemum. **1.** Yiewsley, *M. & S.*

C. coronarium L. **1.** Uxbridge, 1903, *A. B. Jackson* (SLBI).

C. uliginosum Pers., *C. serotinum* L. **4.** Stanmore, 1912, *Loydell* (D).

C. frutescens L. **5.** Acton Wells, *Druce Notes*.

Cotula coronopifolia L. **6.** Muswell Hill, 1913, *Coop. Cas.* **7.** Highbury, 1869, *P. Gray*; Hackney Marshes, 1912, *J. E. Cooper* (BM).

C. aurea Loefl. **5.** Acton, 1920, *Coules, Dunn MSS.*

C. dioica L. **6.** Highgate, 1961, *C. E. Marks, L.N.* 41: 14.

Artemisia biennis Willd. **1.** Uxbridge, 1922, *Lady Davy* and *Wedgwood, Rep. Bot. Soc. & E.C.* 6: 731. **4.** Hendon, 1925, *E. J. Salisbury, Dunn MSS.* **5.** Greenford, 1945!, *Welch, L.N.* 26: 61. Gunnersbury Park, Acton, 1965!, *Clement*, det. *McClintock, L.N.* 45: 20. **6.** Church End, Finchley, 1919, *J. E. Cooper* (BM). **7.** Hackney Marshes, 1913, *J. E. Cooper* (BM).

A. annua L. **1.** Yiewsley, 1923, *J. E. Cooper* (BM); 1927, *Melville* (L).

A. abrotanum L. Southernwood, Lad's Love, Old Man. **4.** Stonebridge, 1950. Kenton, 1951. **6.** Enfield Lock, 1965, *L. M. P. Small*, det. at *Brit. Mus. (Nat. Hist.)*.

A. pontica L. Roman Wormwood. **5.** Hanwell, 1957.

A. longifolia Nutt. **1.** Yiewsley, 1913, *Coop. Cas.*; 1914, 1917 and 1923, *J. E. Cooper.*

A. ludoviciana Nutt. **1.** Near Springwell, 1946!, *L. G. Payne, K. & L.*, 156.

A. scoparia Waldst. & Kit. **7.** South Kensington, 1867, *T. & D.*, 1870, *Dyer* (BM).

Echinops ritro L. Globe Thistle. **7.** Tottenham, 1963, *Wurzell, Lousley.*

Arctium tomentosum Mill. **1.** Yiewsley, 1916, *Wedg. Cat.*; 1919, *J. E. Cooper, K. & L.*, 162.

Carduus pycnocephalus L., *C. tenuiflorus* auct., non Curt. **1.** Yiewsley, 1913, *J. E. Cooper* (BM); 1914–23, *J. E. Cooper, K. & L.*, 163; 1929, *Melville* (K). **3.** Near Cranford, 1965. **4.** Temple Fortune, 1914, *Coop. Cas.* **6.** Church End, Finchley, 1923, *J. E. Cooper, K. & L.*, 163.

Cirsium eriophorum (L.) Scop., *Carduus eriophorus* L., *Cnicus eriophorus* (L.) Roth. Woolly-headed Thistle. **7.** Crouch End, 1896, *J. E. Cooper* (BM).

Cynara cardunculus L. Globe Artichoke. **7.** Hackney Marshes, *M. & S.* Regents Park, 1960·–63, *Wurzell.*

Centaurea paniculata L., *C. gallica* Gugler. **2.** West Drayton, 1916, *Wedg. Cat.*

C. montana L. **3.** Hounslow Heath, 1963, *Wurzell.* **7.** Holloway, 1965, *D. E. G. Irvine.*

C. solstitialis L. St Barnaby's Thistle. **1.** Near Uxbridge, 1883–85, *Benbow*; Harefield, 1912, *J. E. Cooper* (BM). Yiewsley, 1912, *Coop. Cas.* **3.** Hounslow Heath, 1954, *Welch.* **5.** Hanwell, 1953. **6.** Church End, Finchley, 1926, *Carrington* (K). **7.** Isle of Dogs, 1887, *Benbow* (BM). In a flower-tub, Mansell Street, S.W.1, 1965, *McClintock, L.N.* 45: 20.

C. amara L. **7.** South Kensington, 1871, *Dyer* (BM).

C. diffusa Lam. **1.** Uxbridge, 1921, *Wedgwood* (K).

C. adami Willd. **2.** West Drayton, *Druce Notes.*

C. melitensis L. **1.** Yiewsley, 1912, *J. E. Cooper* (K); 1924, *J. E. Cooper.* **2.** West Drayton, *Druce Notes.* **4.** Temple Fortune, 1914, *Coop. Cas.* Mill Hill, *Harrison.* **5.** Hanwell, 1951, *T. G. Collett*!. Greenford, 1954–55. **7.** Hackney Marshes, 1912–13, *Coop. Cas.* 1915, 1921, 1923–24 and 1927, *J. E. Cooper.*

C. pallescens Delisle. **4.** Fortis Green, 1911, *J. E. Cooper* (K).

C. moschata L. Sweet Sultan. **5.** Hanwell, 1946, *Sandwith*!; Greenford, 1951, *T. G. Collett*!, *K. & L.*, 168.

C. diluta Ait.f. Formerly a very rare casual, but now a common bird-seed introduction frequently encountered on rubbish-tips.

Carthamus tinctorius L. **1.** Yiewsley!, 1908–9, *Coop. Cas.*; 1961. Harefield, 1966, *I. G. Johnson* and *Kennedy*!. **2.** West Drayton!, *Druce Notes.* Near Hayes, 1967. Near Harlington, 1968!, *Clement* and

Ryves. **3.** Hounslow Heath, 1964, *Wurzell, Mason* and *Clement, L.N.* 44: 17. **4.** East Finchley, 1906, *J. E. Cooper* (BM). **5.** Hanwell, 1953 !, *K. & L.*, 353; 1954–55. **7.** Hackney Marshes, *M. & S.* Near the British Museum, Bloomsbury, *Shenst. Fl. Lond.* Regents Park, 1966, *Holme*.

Var. **inermis** Schweinf. **1.** Yiewsley, *M. & S.*

Cichorium pumilum Jacq., *C. divaricatum* Schousb. **1.** New Years Green, 1957. **2.** Near Harlington, 1968 !, *D. Turner*. **5.** Hanwell, 1960.

C. endiva L. Endive. **1.** Yiewsley, *M. & S.*

Scorzonera hispanica L. var. **latifolia** Koch. **3.** Twickenham, 1956, *L. M. P. Small*, det. *Melderis*.

Lactuca sativa L. Garden Lettuce. **5.** Hanwell. Greenford. Heston.

Tragopogon porrifolius L. Salsify. **1.** Uxbridge, 1889, *Benbow* (BM). **2.** West Drayton !, *A. B. Cole*, *T. & D.*, 172. **3.** Roxeth !, *Hind, Melv. Fl.*, 42. **4.** Harrow, *Hind, Melv. Fl.*, 42. **6.** Edmonton, *Hurlock, Blackst. Spec.*, 98. Barnet, *F. Naylor, T. & D.*, 172. Southgate, 1946–48, *Scholey, L.N.* 28: 37; 1949–51, *Scholey, K. & L.*, 178. **7.** Near Bow, *Pet. Midd.* Hammersmith, *D. Cooper* (BM). Walham Green, 1830, *Pamplin*; Finsbury, 1873, *Newbould*; Kensington, 1872, *Warren, Trim. MSS.* Belsize Park, 1921, *Richards*.

Crepis nicaeensis Balb. French Hawk's-beard. **1.** Pinner, 1886; Uxbridge, 1889, *Benbow* (BM), det. *J. B. Marshall*. **2.** Hampton Court, 1885, *Benbow* (BM), det. *J. B. Marshall*. **5.** Chiswick, 1958, *Murray, L.N.* 38: 21.

LILIACEAE

Hemerocallis fulva (L.) L. Day Lily. **3.** Hounslow Heath, 1960, *Philcox* and *Townsend*.

Alstroemeria aurantiaca L. Peruvian Lily. **7.** Bombed site, Cripplegate, *A. W. Jones, K. & L.*, 273.

Ornithogalum pyrenaicum L. Spiked Star-of-Bethlehem, Bath Asparagus. **2.** Strawberry Hill, *T. J. Woodward, B.G.* 403.

Hyacinthus orientalis L. Roman Hyacinth. **7.** Bombed site, Cripplegate, *A. W. Jones, K. & L.*, 275.

Allium porrum L. Leek. **5.** Hanwell, 1950.

A. schoenoprasum L., *A. sibiricum* L. Chives. **2.** Laleham Park, 1966!, *L.N.* 46: 12.

A. cepa L. Onion. Common on rubbish-tips.

COMMELINACEAE

Commelina nudiflora L. **7.** Hackney Marshes, *M. & S.*

Zebrina pendula Schnizl. Wandering Jew. **7.** Buckingham Palace grounds, *Fl. B.P.G.*

JUNCACEAE

Juncus aridicola L. Johnson. **2.** East Bedfont, 1946!, *Lousley* (L), det. *L. Johnson.*

J. usitatus L. Johnson. **2.** East Bedfont, 1946!, *Lousley* (L), det. *L. Johnson.*

J. continuus L. Johnson. **2.** East Bedfont, 1946!, *Lousley* (L), det. *L. Johnson.*

PALMACEAE

Phoenix dactylifera L. Date. Seedling date palms are a conspicuous feature of many rubbish-tips.

AMARYLLIDACEAE

Galanthus nivalis L. Snowdrop. **4.** Hampstead Heath, 1949, *A. W. Jones, K. & L.*, 271. **6.** Highgate, *P. Gray* (BM).

Narcissus serratus Haworth. **4.** Mill Hill, *E. Forster* (BM).

Vallota purpurea Herb., *V. speciosa* Th. Dur. & Schinz. Scarborough Lily. **6.** Finchley Common, one small clump, 1966.

IRIDACEAE

Sisyrinchium montanum Greene, *S. bermudiana* auct., non L., *S. angustifolium* auct., non Mill. Blue-eyed Grass. **4.** Brent Reservoir, 1965, *Kennedy*!. **7.** Shadwell Dock, 1945, *Lousley, K. & L.*, 270.

CYPERACEAE

Cyperus eragrostis Lam., *C. vegetus* Willd. **5.** Garden weed, Chiswick, 1966–67, *Veál*, *L.N.* 47: 9.

Carex vulpinoidea Michx. **5.** Brentford, 1903, *Druce Notes*. Acton, 1907, *Loydell* (BM).

GRAMINEAE

Vulpia ligustica (All.) Link. **5.** Chiswick, 1907, *Sherrin* (BM).

Lolium temulentum L. Darnel. Formerly as an impurity in grain and other crops, but now a common bird-seed introduction in gardens, on waste ground, rubbish-tips, etc. **7.** Islington, 1838, *E. Ballard* (BM). Camden Town, 1847 (HE). Near Ken Wood, *Irv. MSS.* Chelsea, 1861, *T. & D.*, 334. Hackney Marshes, *M. & S.* Bombed site, Cripplegate, E.C., *Fitter*. Bombed site, Mark Lane, E.C., 1948, *McClintock* and *Lousley* (L). Green Park, Piccadilly, W.1, 1960–61, *McClintock*, *L.N.* 41: 22. Plentiful in a neglected garden, Notting Hill, 1966.

Var. **arvense** Lilj. **1.** Harefield, *Benbow* (BM). Uxbridge, *Druce Notes*. **2.** Harmondsworth, 1967. West Drayton, *Druce Notes*. **5.** Acton, *Druce Notes*. **7.** Hackney, *E. Forster* (BM). Chelsea, 1955, *McClintock* (L).

L. rigidum Gaudin. **1.** Harefield, 1966, *I. G. Johnson* and *Kennedy*!. **2.** Upper Lodge, Bushy Park, 1967, *Clement*. Near Harmondsworth, 1967. **5.** West Ealing, 1959!, *Proc. Bot. Soc. Brit. Isles* 3: 442. **7.** Hyde Park, new traffic island by Dorchester Hotel, 1962!, *D. E. Allen*, det. *McClintock*, *Allen Fl.* New shrubbery bed at edge of Knightsbridge, S.W.1, 1965, *Brewis*, *L.N.* 46: 34.

Monerma cylindrica (Willd.) Coss. & Dur. **2.** Near West Drayton, 1965!, *Wurzell*, det. *Lousley*. **7.** Green Park, Piccadilly, W.1, 1961, *D. E. Allen* and *McClintock*, *L.N.* 41: 13.

Eragrostis parviflora (R.Br.) Trin. **7.** Shrubbery bed at edge of Knightsbridge, S.W.1, 1965, *Brewis*, *L.N.* 46: 35.

Cynosurus echinatus L. Rough Dog's-tail. **1.** Harefield, 1910, *J. E. Cooper* (BM); 1930, *P. H. Cooke*, *K. & L.*, 311. Northwood, 1936, *F. M. Day*. Yiewsley, 1927, *J. E. Cooper*, *K. & L.*, 311. **2.** West Drayton, 1920, *Dymes*; 1928, *G. C. Druce* (D). **3.** Hayes, 1923, *Tremayne*, *K. & L.*, 311. **4.** Fortis Green, 1909–10; East Finchley, 1910; Finchley, 1910–11, *Coop. Cas.*; 1916 and 1922, *J. E.*

Cooper. Hendon, 1917, *J. E. Cooper* (BM); 1927, *J. E. Cooper, K. & L.*, 311. Willesden, 1959. Wembley, 1924, *J. E. Cooper, K. & L.*, 311. **5.** Greenford, 1917, *J. E. Cooper, K. & L.*, 311. Chiswick, 1937, *Eliot, Rep. Bot. Soc. & E.C.* 11: 518. **6.** North Finchley, 1925, *J. E. Cooper, K. & L.*, 311. **7.** Hackney Marshes, 1909, *Coop. Cas.*; 1913 and 1916, *J. E. Cooper* (BM); 1918, 1921 and 1923, *J. E. Cooper*. Finchley Road, 1923, *Moring MSS*. Bombed site near Cannon Street, E.C., 1949, *I. A. Williams*; 1951, *I. A. Williams* and *Lousley*; 1953, *A. W. Jones, K. & L.*, 311.

Briza minor L. Small Totter-grass. **3.** Isleworth, 1866, *Trimen* (BM).

B. maxima L. Large Totter-grass. **1.** Yiewsley, *M. & S.* **5.** Hanwell, 1950!, *K. & L.*, 313. Strand-on-the-Green, 1968.

Bromus benekenii (Lange) Trim., *B. ramosus* Huds. subsp. *benekenii* (Lange) Schinz & Thell., *B. ramosus* Huds. var. *benekenii* (Lange) Druce, *B. asper* Murr. subsp. *benekenii* (Lange) Hack., *B. asper* Murr. var. *benekenii* (Lange) Syme, *Zerna benekenii* (Lange) Lindm. **7.** Kensington Gardens, *Trimen, J.B. (London)* 9: 270. Perhaps introduced with shrubs.

B. rigidus Roth, *B. maximus* auct., *Anisantha rigida* (Roth) Hyland. **1.** Yiewsley, *Coop. Cas.* **6.** Muswell Hill, 1907, *Coop. Cas.* Both records are erroneous, the specimens (BM) on which they were based being *B. diandrus*. **7.** Hackney Marshes, 1913, *Coop. Cas.* Probably an error.

B. tectorum L., *Anisantha tectorum* (L.) Nevski. Drooping Brome. **1.** Uxbridge, *Benbow*; Potters Bar, 1912, *J. E. Cooper* (BM), both det. *Melderis*. **2.** West Drayton, *Druce Notes*. **4.** Cricklewood, 1869, *Warren* (BM), det. *Melderis*. **6.** Enfield, 1965, *Kennedy*, det. *Melderis*. **7.** Hackney Marshes, 1912, *J. E. Cooper* (BM), det. *Melderis*.

B. madritensis L., *Anisantha madritensis* (L.) Nevski. Compact Brome. **1.** Yiewsley, 1922, *J. E. Cooper* (BM), det. *Melderis*.

B. interruptus (Hack.) Druce, *B. mollis* L. var. *interruptus* (Hack.) Druce. Interrupted Brome. **1.** Near Uxbridge, 1898, *G. C. Druce*; Uxbridge, 1907, *Loydell* (D). Yiewsley, 1912, *J. E. Cooper* (BM), det. *Melderis*. **3.** Southall, *Druce Notes*.

B. arvensis L., *Serrafalcus arvensis* (L.) Godr. Field Brome. **1.** Yiewsley, 1908, *J. E. Cooper* (BM). Uxbridge; Northwood, *Druce Notes*. Canal side near Uxbridge, 1968. **2.** Teddington, 1867, *Dyer* (BM); 1914, *Hinds* (K). West Drayton, *Druce Notes*. **3.** Twickenham, *Benbow MSS*. **4.** Canons Park, 1918, *C. S. Nicholson* (LNHS). East Finchley, *Coop. Cas.* **5.** Alperton; near Chiswick, *Benbow MSS*.

Acton, 1908, *Green* (SLBI). Bedford Park, 1888, *Fraser* (K). **6.** Muswell Hill, *Coop. Cas.* **7.** Newington, *Ball. MSS.* Primrose Hill, 1871, *Dyer*; Kensington, 1874, *Warren*; Hackney Downs, 1878, *Newbould, Trim. MSS.* South Kensington, *Trimen, J.B. (London)* 13: 275. Finchley Road, abundant on waste ground, 1869, *Dyer*; near Highbury, 1875, *French*; Hackney Marshes, 1914, *J. E. Cooper* (BM).

Bromus japonicus Thunb., *B. patulus* Mert. & Koch., *Serrafalcus japonicus* (Thunb.) Wilmott. **1.** Yiewsley, 1929, *Melville* (K). **2.** West Drayton, *Druce Notes*; 1928, *Melville* (K). **7.** Hyde Park, 1945, *Braybon* (K).

B. secalinus L., *Serrafalcus secalinus* (L.) Bab. Rye Brome. **1.** Uxbridge, 1884, *Benbow*; Harefield!, *J. E. Cooper* (BM). Near Uxbridge, *G. C. Druce, J.B. (London)* 36: 201. Denham!, *Druce Notes.* Yiewsley, 1914, *J. E. Cooper.* **2.** West Drayton, *Druce Notes.* Dawley, *J. E. Cooper, K. & L.*, 319. **3.** Hounslow Heath, 1965, *Mason.* **4.** Hampstead, *Kalm*, 49. Hendon, 1910, *P. H. Cooke* (LNHS). Finchley, 1908; Temple Fortune, 1914, *Coop. Cas.* **5.** Chiswick, 1887, *Benbow* (BM). Acton, *Druce Notes.* **6.** Finchley Common, 1841 (HE). North Finchley, 1913, *Coop. Cas.* **7.** Hampstead, *Merr. Pin.*, 49. Near Ken Wood, *Irv. MSS.* St Pancras, *M. & G.*, 116. South Kensington, 1875, *Dyer, J.B. (London)* 13: 275. Near Belsize Park, 1871, *Dyer, J.B. (London)* 9: 304. Hackney Marshes, 1909, *Coop. Cas.*; 1923, *J. E. Cooper.*

B. briziformis Fisch. & Mey. **1.** Yiewsley, 1913, *J. E. Cooper, K. & L.*, 319. **2.** West Drayton, *Druce Notes.* **5.** Greenford, 1917, *J. E. Cooper, K. & L.*, 319. **2.** West Drayton, *Druce Notes.* **5.** Greenford, 1917, *J. E. Cooper, K. & L.*, 319. **7.** Hackney Marshes, 1912, *J. E. Cooper* (BM).

B. scoparius L. **5.** Waste ground, Ealing Common, 1905, *Green* (SLBI).

B. squarrosus L. **1.** Uxbridge, *Benbow* (BM).

B. unioloides Humb., Bonpl. & Kunth, *B. willdenowii* auct., *Ceratochloa unioloides* (Willd.) Beauv. Rescue Grass. **1.** Yiewsley, 1916, *J. E. Cooper* (K). New Years Green, 1966, *Clement.* **4.** East Finchley, 1906, *J. E. Cooper* (BM), det. *Melderis.* Temple Fortune, 1915, *J. E. Cooper, K. & L.*, 318. Hendon, 1962, *Brewis.* **5.** Canal bank, Hanwell, 1925; Grove Park, 1927, *Hubbard* (K). **6.** Finchley, 1908 and 1910, *Coop. Cas.*; 1922; North Finchley, 1916, *J. E. Cooper, K. & L.*, 318. **7.** Hackney Marshes, 1914, *J. E. Cooper* (BM), det. *Melderis*; 1915–16 and 1920–21, *J. E. Cooper, K. & L.*, 318.

Brachypodium distachyon (L.) Beauv. **5.** West Ealing, 1959!, *Proc. Bot. Soc. Brit. Isles* 3: 442; 1962.

Agropyron desertorum (Fisch.) Schult. **7.** Hyde Park Corner, 1962!, *D. E. Allen*, det. *Melderis*, *Allen Fl.*

Secale cereale L. Rye. Common on rubbish-tips, weedy road verges, etc.

S. dalmaticum Vis. **1.** Uxbridge, 1917, *A. D. Webster* (BM).

Triticum aestivum L. Bread Wheat. Common on rubbish-tips.

T. turgidum L. Rivet. **7.** Hyde Park Corner, 1962, *Brewis*, *L.N.* 44: 28. Tottenham, 1967, *Kennedy*.

Aegilops cylindrica Host. **1.** Yiewsley, 1924, *J. E. Cooper*, *K. & L.*, 322. **4.** Fortis Green, 1909, *J. E. Cooper* (BM).

A. triaristata Willd. **4.** Neasden, 1965.

A. triuncalis L. **7.** Hackney Marshes, 1912, *Coop. Cas.*

Elymus canadensis L. **1.** Yiewsley, 1948!, det. *Hubbard*, *K. & L.*, 323; 1951, *Lousley* and *McClintock*! (L).

E. virginicus L. **5.** Hanwell, 1951, det. *Hubbard*.

Hordeum marinum Huds., *H. maritimum* Stokes ex With. Sea Barley, Squirrel-tail Grass. **1.** Uxbridge, 1887, *Benbow* (BM). **7.** Hackney Marshes, 1921, *J. E. Cooper*, *K. & L.*, 323.

H. hystrix Roth, *H. geniculatum* auct., *H. gussonianum* Parl. **1.** Yiewsley, 1926, *Melville* (K).

H. jubatum L. Fox-tail Barley. **2.** West Drayton!, *Druce Notes*. **3.** Feltham, 1951!, *David*, *K. & L.*, 323. **4.** Temple Fortune, 1924; Finchley, 1926, *J. E. Cooper*, *K. & L.*, 323. **5.** Near Boston Manor, 1968!, *Boniface*; plentiful and apparently sown, 1969. **7.** Hackney Marshes, 1909–10, *Coop. Cas.*; 1923 and 1927, *J. E. Cooper*, *K. & L.*, 323.

H. californicum Covas & Stebbins. **5.** Hanwell, 1953!, det. *Melderis*, *K. & L.*, 323.

H. vulgare L., *H. sativum* Pers. Barley. Common on rubbish-tips, weedy road verges, railway tracks, etc.

Var. **hexastichon** (L.) Aschers. **7.** Hackney Marshes, *M. & S.*

H. distichon L. Common on rubbish-tips, weedy road verges, railway tracks, etc.

Lophochloa phleoides (Vill.) Reichb., *Koeleria phleoides* (Vill.) Pers. **1.** Yiewsley, 1924, *J. E. Cooper* (BM), det. *Melderis* (as *Koeleria phleoides*).

Trisetum paniceum (L.) Pers. **7.** Hackney Marshes, 1913, *J. E. Cooper* (K).

Avena strigosa Schreb. Black, Bristle or Small Oat. **1.** Uxbridge, 1891, *Benbow* (BM). **4.** Hendon, 1963, *Brewis.*

Var. **glabrescens** (Marquand) Druce. **7.** Bombed site, Great Tower Street, E.C., 1946, *Brenan* and *Lousley, K. & L.*, 309.

A. byzantina C. Koch. Algerian Oat. **5.** Hanwell, 1949!, det. *Hubbard, K. & L.*, 309. **7.** East Heath, Hampstead, 1949!, det. *Hubbard, K. & L.*, 309.

A. sativa L. Common on rubbish-tips, etc.

Ammophila arenaria (L.) Link. Marram Grass. **1.** Northwood golf course, 1913, planted, *Tremayne* (LNHS).

Agrostis scabra Willd., *A. hyemalis* auct. **1.** Yiewsley, 1913, *Coop. Cas.* **2.** Shepperton, 1971, *Clement, Mason* and *McLean.* **4.** Finchley, 1910, *Coop. Cas.* **7.** Hackney Marshes, 1912, *Coop. Cas.* Buckingham Palace grounds, one large plant, 1967, *McClintock.*

Polypogon monspeliensis (L.) Desf. Annual Beard-grass. **1.** Pinner, 1971, *Hind, J.B. (London)* 9: 272. Near Uxbridge, 1885, *Benbow*; Yiewsley, 1921, *J. E. Cooper* (BM), both det. *Melderis.* **2.** West Drayton, 1905, *L. B. Hall*; 1969, *Clement* and *Ryves.* East Bedfont, 1964, *Wurzell, L.N.* 44: 17. **3.** Between Pinner and Harrow, *Farrar, T. & D.*, 319. **7.** Hackney Marshes, 1921, *J. E. Cooper* (BM), det. *Melderis.*

P. semiverticillatus (Forsk.) Hyland., *Agrostis semiverticillata* (Forsk.) C. Christ., *A. verticillata* Vill. **1.** Near Uxbridge, 1919, *Sherrin* and *L. B. Hall* (SLBI). **2.** West Drayton, 1969, *Clement* and *Ryves.* **5.** Acton, *Druce Notes.*

Lagurus ovatus L. Hare's-tail. **1.** Yiewsley, *M. & S.* **5.** Ealing, 1956, *Bence, K. & L.*, 359.

Phleum paniculatum Huds., *P. asperum* Jacq. **1.** Near Elstree, *Rudge* (BM). Perhaps in Herts.

P. subulatum (Savi) Aschers. & Graebn., *P. tenue* Schrad. **1.** Yiewsley, 1925, *J. E. Cooper.*

Anthoxanthum aristatum Boiss., *A. puelii* auct. **1.** Yiewsley, 1913, *Coop. Cas.* **7.** Hackney Marshes, 1909 and 1913, *Coop. Cas.*; 1924 and 1927, *J. E. Cooper, K. & L.*, 324. Holland House grounds, Kensington, *Fraser.*

Eleusine africana O'Byrne. **1.** New Years Green, 1967, *Clement*, teste *O'Byrne.*

Phalaris minor Retz. **1.** Yiewsley, 1924, *J. E. Cooper, K. & L.*, 303. New Years Green, 1961, *McClintock*!. Harefield, in quantity, 1966, *I. G. Johnson* and *Kennedy*!. **2.** West Drayton!, *Druce Notes*. East Bedfont, 1965. Near Harlington, 1968. **7.** Hackney Marshes, 1912, *Coop. Cas.*; 1918, 1921, 1925 and 1927, *J. E. Cooper, K. & L.*, 303.

P. paradoxa L. **1.** Uxbridge, 1908, *Green* (SLBI). Yiewsley, 1908 and 1912, *Coop. Cas.*; 1921, 1926 and 1927, *J. E. Cooper*. **4.** East Finchley, 1907, *Coop. Cas.*; 1918, 1921, 1925 and 1927, *J. E. Cooper*. **5.** Hanwell, 1954!, *L. M. P. Small*, det. *Melderis, K. & L.*, 304. **7.** Hackney, 1912, *E. G. Gilbert* (TUN). Hackney Marshes, 1921, 1926 and 1927, *J. E. Cooper, K. & L.*, 303.

Var. **praemorsa** Coss. & Dur. **1.** Yiewsley, 1924, *J. E. Cooper* (BM), det. *Melderis*. **5.** Hanwell, 1963. **7.** Hackney Marshes, 1924, *J. E. Cooper* (BM), det. *Melderis*. Bombed site near Holborn, 1950, *Bangerter* and *Whittaker* (BM), det. *Bor*. Disturbed ground, Paddington Green, abundant, 1967.

P. coerulescens Desf., *P. aquatica* auct. **2.** West Drayton, *Druce Notes*. **7.** Hackney Marshes, 1912, *J. E. Cooper* (BM), det. *Melderis*.

P. angusta Nees. **7.** Hackney Marshes, 1913, *Coop. Cas.*

Parapholis incurva (L.) C. E. Hubbard, *Lepturus incurvatus* (L.) Trin., *L. incurvus* (L.) Druce. **1.** Yiewsley, 1924, *J. E. Cooper, K. & L.*, 322.

P. pycnantha (Hack.) C. E. Hubbard, *Lepturus pycnantha* Hack. **2.** West Drayton, 1903, *G. C. Druce, Rep. Bot. Soc. & E.C.* 2: 33.

Cynodon dactylon (L.) Pers., *Capriola dactylon* (L.) Kuntze. Bermuda Grass. **2.** Hampton Court, 1925, *Bacon, Rep. Bot. Soc. & E.C.* 7: 904.

Echinochloa utilis Ohwi & Yabano, *E. frumantacea* auct., *Panicum frumantaceum* auct. Formerly a rare casual, but now a common feature of rubbish-tips.

E. colonum (L.) Link, *Panicum colonum* L. **2.** Near Harlington, 1968!, *Clement* and *Ryves*.

Digitaria sanguinalis (L.) Scop., *Panicum sanguinale* L. Crab-grass. **1.** Springwell, 1946!, conf. *Hubbard, L.N.* 26: 64. **2.** West Drayton, *Druce Notes*. Shepperton, 1969, *Clement* and *Ryves*. **5.** Bedford Park, 1901, *Sherrin* (SLBI). In quantity on a heap of ballast, West Acton station, 1968, *McLean*, det. *Bor*. **7.** Chelsea, 1849, *T. Moore, T. & D.*, 312; 1852, *T. Moore* (BM); 1967, *Cameron*, fide *M. McC. Webster*. Roadside, Holborn, 1967, *Brewis*.

D. ciliaris (Retz.) Koel., *D. adscendens* (Humb., Bonpl. & Kunth) Henrard, *D. marginata* Link. **1.** Springwell, 1945!, *Welch*, det.

Hubbard (as *D. adscendens*), *L.N.* 26: 64. **2.** West Drayton, 1969, *Clement* and *Ryves*. **3.** Hounslow Heath, 1964, *Wurzell, Mason* and *Clement*, det. *Hubbard* (as *D. adscendens*). **7.** Waste ground, Harrow Road by Edgware Road, W.2, 1969, *D. Turner, L.N.* 49: 19.

Setaria verticillata (L.) Beauv., *Panicum verticillatum* L. Rough Bristle-grass. **2.** West Drayton, *Druce Notes*. Upper Lodge, Bushy Park, 1967, *Clement*. **3.** Northolt, 1947!, *Welch*, conf. *Hubbard, L.N.* 27: 35. **5.** Chiswick, 1962, *Murray*, conf. at *Kew*. **7.** Chelsea, *J. Newton, Ray Hist.*, 2: 1263; weed in Chelsea Physic Garden, *Budd. MSS.*; 1849, *T. Moore* (BM); 1852, *T. Moore*; near Chelsea, 1852, *McIvor* (SY). Regents Park, 1860, *J. A. H. Murray* (BM). Bombed site, King William Street, E.C., 1950!, *Lousley* (L).

S. glauca (L.) Beauv., *S. lutescens* (Weigel) F. T. Hubbard, *S. pumila* (Poir.) Roem. & Schult., *Panicum glaucum* L., *P. lutescens* Weigel. Yellow Bristle-grass. **1.** Uxbridge, *Benbow MSS.* Springwell, 1945!, *Welch*, conf. *Hubbard, L.N.* 26: 65. Yiewsley, *Coop. Cas.* Harefield; New Years Green, 1966, *Clement*. **2.** Hampton Court, 1867, *T. & D.*, 313. Between West Drayton and Iver, 1922 (D). Staines, 1898, *Waddell, Rep. Wats. B.E.C.* 1698–99, 21. West Drayton, 1929, *M. Taylor*. East Bedfont, 1966. Near Harlington, 1968!, *Clement* and *Ryves*. **3.** Twickenham, *T. & D.*, 313. Sudbury, 1881, *De Crespigny* (BM). Feltham, 1951, *David*; 1952!, *Welch, K. & L.*, 302. Hounslow Heath, 1960, *Wurzell*. **4.** Finchley, East Finchley, *Coop. Cas.* **5.** Alperton, 1867, *Dyer*; Chiswick, 1887, *Benbow* (BM). Acton, 1902, *Wedg. Cat.* West Acton, 1968, *McLean*. Bedford Park, 1901, *Sherrin* (SLBI). Hanwell, 1952!, *Newey*. **6.** Muswell Hill; Highgate, *Coop. Cas.* **7.** Parsons Green, *T. & D.*, 313. Fulham, 1874, *H.* and *J. Groves* (BM). Kilburn, 1869, *Warren* (SY). Isle of Dogs, *Benbow MSS.* Crouch End; Hackney Marshes, *Coop. Cas.* Kensington Gardens, one plant, 1966, *Wurzell, L.N.* 46: 35.

S. italica (L.) Beauv., *Panicum italicum* L. Italian Bristle-grass. Common on rubbish-tips.

Brachiaria marlothii (Hack.) Stent. **7.** Spontaneously in a pot of cacti in an office at Paddington, 1965, *D. N. Turner*, det. *Lousley, L.N.* 45: 20.

Panicum miliaceum L. Millet. Frequent on rubbish-tips.

Forma **ramiflora** Junge. **1.** New Years Green, 1961. **5.** Hanwell, 1945!, det. *Hubbard, K. & L.*, 301.

P. capillare L. **1.** Yiewsley, 1966, *Wurzell*. **2.** Hampton Court, 1867, *T. & D.*, 313. **7.** Crouch End, 1897, *J. E. Cooper* (BM). Highgate, 1899, *Coop. Cas.*

Var. **occidentale** Rydb. **1.** Springwell, 1945, *Welch* and *Sandwith*!, det. *Hubbard, L.N.* 26: 64. Pinner, 1957, *I. G. Johnson*. **4.** Near Harlesden, 1887, *Benbow* (BM).

P. laevifolium Hack. **1.** Yiewsley, 1924, *J. E. Cooper* (BM); c. 1926, *M. & S.* New Years Green, 1966, *Clement*.

P. dichotomiflorum Michx. **1.** Springwell, 1945!, *Welch*, det. *Chase, K. & L.*, 301.

Zea mays L. Maize. **2.** West Drayton, *Druce Notes*. **5.** Hanwell, 1954!; Greenford, 1955!, *K. & L.*, 303. **7.** Hackney Marshes, *M. & S.*

Sorghum bicolor (L.) Moench, *S. vulgare* Pers., *S. saccharatum* (L.) Moench, *S. caffrorum* (Retz.) Beauv. **1.** Yiewsley, 1926, *J. E. Cooper* (BM); 1927, *J. E. Cooper* (K). **2.** West Drayton, *Druce Notes*. **5.** Hanwell, 1955, *McClintock* (L.), det. *Hubbard* (as *S. caffrorum* (Retz.) Beauv.).

S. halepense (L.) Pers. **2.** Near Harlington, 1968!, *Clement* and *Ryves*. Shepperton, 1969, *Clement* and *Ryves*. **3.** Northolt, 1947!, *Welch*, det. *Hubbard, L.N.* 27: 35.

SPECIES RECORDED IN ERROR

LYCOPODIACEAE

Lycopodium annotinum L. **3.** Hounslow Heath, *Forst. Midd.* No doubt an error, *T. & D.*, 344.

SELAGINELLACEAE

Selaginella helvetica Link **7.** Thames side at the neathouses, Knightsbridge, *Merr. Pin.*

EQUISETACEAE

Equisetum hyemale L. **3.** Hounslow Heath, *Mart. Pl. Cant.*, 71, and *Forster, B.G.*, 413. Specimens of *E. fluviatile* and *E. palustre* have been probably taken for this, *T. & D.*, 336.

THELYPTERIDACEAE

Thelypteris phegopteris (L.) Slosson, *Dryopteris phegopteris* (L.) C.Chr., *Polypodium phegopteris* L., *Phegopteris polypodioides* Fée. **5.** Norwood, three miles from Brentford, *Beevis*, *Francis*, 17; but in the fifth edition it reads 'Norwood, Surrey, and near Brentford'. Not a likely locality, *T. & D.*, 337.

OPHIOGLOSSACEAE

Botrychium lunaria (L.) Sw. **5.** . . . particularly near Acton, *R. & P.*, 218.

RANUNCULACEAE

Pulsatilla vulgaris Mill., *Anemone pulsatilla* L. **4.** West Heath, Hampstead, *White*, 363. A colour form of *Anemone nemorosa* was probably seen.

FUMARIACEAE

Fumaria capreolata L., *F. speciosa* Jord. **3.** Near Harrow, *Hind*, *Melv. Fl.*, 6. Whitton, *T. & D.*, 26. **4.** Kingsbury, *T. & D.*, 26. These records possibly refer to *F. muralis* subsp. *boraei* or luxuriant forms of *F. officinalis*. *F. capreolata* is unlikely to have occurred in Middlesex.

F. vaillantii Lois. Cited for Middlesex in *Druce C.F.* The record is possibly based on a specimen labelled *F.* '*vaillantii*' . . . Chiswick, 1887, *Benbow* (BM). This was determined by Pugsley as an abnormal plant of *F. officinalis*.

CRUCIFERAE

Rhynchosinapis monensis (L.) Dandy, *Sisymbrium monense* L., *Brassicella monensis* (L.) O. E. Schulz. Cited for Middlesex, with some doubt, in *Druce C.F.* I have been unable to trace the source of the record, and the species is most unlikely to have ever occurred in the vice-county.

CARYOPHYLLACEAE

Silene nutans L. **7.** Hackney Marshes, *Coop. Cas.* The specimen (BM), on which this record was based is the adventive *S. dichotoma.*

Holosteum umbellatum L. **1.** . . . '*Holosteum minimum tetrapetalon sive Alsine tetrapetalos caryophylloides*' . . . Harefield, *Blackstone* in litt. to *Richardson*, 18 December 1736. This record is referred, with some doubt, to *H. umbellatum* by Dawson Turner in *Richardson's Correspondence*. The species intended by Blackstone was *Moenchia erecta*.

Bufonia tenuifolia L., *Buffonia annua* auct. **3.** Hounslow Heath, *Doody MSS*. Presumably copied by Dillenius in *Ray Syn.* 3: 346, and by Hudson in *Huds. Fl. Angl.*, II; 'we found it . . . on Hounslow Heath, where Mr. Doody observed it', *M. & G.*, 199. Smith, *E.B.*, 1313, suspected that *Bupleurum tenuissimum* was mistaken for it, and Watson in *T. & D.*, 52, suggested that *Moenchia erecta* was the species seen. *Trimen* and *Dyer*, *loc. cit.*, 52, were, however, of the opinion that Doody had written his note of the locality against the wrong species, and that Milne and Gordon may have mistaken a late grown form of *Juncus bufonius* for it. *B. tenuifolia* is a south European species, and is unlikely to have occurred at Hounslow Heath even as a casual introduction, though there are correctly named, but regrettably unlocalised, specimens in DILL and LT.

CHENOPODIACEAE

Chenopodium botryodes Sm. Cited for Middlesex, with some doubt, in *Druce C.F.* I have been unable to trace the source of this record which was possibly based on a form of *C. rubrum*.

GERANIACEAE

Geranium sylvaticum L. Regents Park, *Webst. Reg.* Perhaps cultivated there.

PAPILIONACEAE

Astragalus glycyphyllos L. **1.** Woods about Harefield (old record), *De Cresp. Fl.*, 6. I have been unable to trace the source of this 'old record', and had such a conspicuous species occurred at Harefield I consider that it is unlikely that it would have been overlooked by Blackstone.

Vicia tenuifolia Roth. Cited for Middlesex in *Druce C.F.* There are no Middlesex specimens in D, and I have been unable to trace any other record of the occurrence of this adventive in the vice-county.

V. sylvatica L. **4.** Hampstead, *White*, 369. **7.** Regents Park, *Webst. Reg.* Possibly *V. sepium* was seen. *V. sylvatica* is unlikely to have occurred in Middlesex.

ROSACEAE

Sorbus domestica L., *Pyrus domestica* (L.) Ehrh. **7.** Hampstead, *Merr. Pin.*, 115. Ken Wood, *Hunter, Park Hampst.*, 30. Probably *S. aucuparia* was the species intended, *T. & D.*, 106.

CRASSULACEAE

Sedum sexangulare L. **4.** Hampstead Heath, *Coop. Fl.*, 102. Probably *S. acre* was intended, *T. & D.*, 116.

UMBELLIFERAE

Bupleurum falcatum L. Merrett's locality for *B. tenuissimum* was erroneously referred to *B. falcatum* in *E.B. Suppl.*, 2763.

Bunium bulbocastanum L. **7.** 'The Kensington Gardens plant was recorded by Smith, *Fl. Brit.* 1, 301, as *B. bulbocastanum* L. on the authority of Mr. W. Wood. In *B.G.*, 402, however, is a note stating that the latter botanist was then "of opinion the plants he gathered were only large specimens of *Conopodium majus*". See also Smith, *Engl. Fl.* 2: 55. Martyn's Hornsey plant has been often referred to *B. bulbocastanum* (see *Wats. Cyb. Brit.* 1: 436), and we were at first inclined to consider it that species, though all the places at present known to produce it are on a chalk soil, in consequence of some evident care shown in the record. But, from a note by Thomas Martyn in vol. 1 (article *Bunium*) of his edition of Miller's *Gardeners Dictionary*, it seems evident that the plant found by his father was a young and luxuriant example of *Conopodium majus* with the leaf-segments somewhat broader than usual, and not the true *B. bulbocastanum* L., which, at that time, though well known on the Continent, was not understood by British botanists', *T. & D.*, 123–124.

EUPHORBIACEAE

Euphorbia hyberna L. **1.** Harefield, *Cockfield Cat.*, 30. No doubt *E. platyphyllos* meant, *T. & D.*, 248. Dillenius, in *Ray Syn.* 3: 312, erroneously cites Doody's locality for *E. platyphyllos* under *E. hyberna*. This error was copied by Withering, Smith, and other authors.

BETULACEAE

Betula nana L. Recorded from **4.** Hampstead Heath, 1966, on the authority of *L. P. Turnbull*, 'London Day by Day', *Daily Telegraph* 29 August 1967. If it occurred on the heath it was deliberately planted as a hoax, as at the present time the species is not known to occur south of Teesdale.

SALICACEAE

Salix hastata L. **6.** Hollick Wood, Colney Hatch, *Woods, B.G.*, 413. Not at all likely to occur in England either wild or cultivated, *T. & D.*, 260.

ERICACEAE

Erica vagans L. **4.** Hampstead Heath, rare, '*G.A.*', *Nat. Not.* 1902, 230. Most unlikely to have occurred in the vice-county unless planted.

PRIMULACEAE

Primula elatior (L.) Hill. **1.** Wood End, Ruislip; Pinner, *Hind, Melv. Fl.* 2, addenda. No doubt *P. veris* × *vulgaris* intended.

GENTIANACEAE

Gentianella campestris (L.) Börner, *Gentiana campestris* L. Blackstone's record of '*Gentiana pratensis flora lanuginosa*' (= *G. amarella*) at Harefield (*Blackst. Fasc.*, 32) was erroneously referred to *G. campestris* in *B.G.*, 402. Watson (*Top. Bot.*, 278) recorded the species for Middlesex, with some doubt, on the basis of this record (*cf.* letter from Watson to Trimen, October 1873).

SCROPHULARIACEAE

Linaria pelisseriana (L.) Mill. Cited for Middlesex, with some doubt, in *Druce C.F.* I have been unable to trace the source of this record.

Sibthorpia europaea L. 'Mr. Mackay of Totteridge, Herts, showed us growing specimens which he had been informed were collected amongst the grass on the lawn at Ken Wood House (**7**). It is very improbable that it grows there unless cultivated', *T. & D.*, 205.

Veronica spicata L. **4.** Hampstead Heath, *Johns. Eric.*; near Leg of Mutton Pond, *White*, 365. Perhaps the adventive *V. longifolia* or large forms of *V. serpyllifolia* were seen. *V. spicata* is unlikely to have occurred in Middlesex, except as a garden outcast.

LABIATAE

Calamintha nepeta (L.) Savi, *Satureja nepeta* (L.) Scheele. **1.** Harefield (old record), *De Cresp. Fl.*, 9. No doubt *C. sylvatica* subsp. *ascendens* intended.

Salvia pratensis L. **4.** Hampstead Heath, *Coop. Fl.*, 101. Probably an exotic *Salvia* species escaped from a garden was seen.

CAMPANULACEAE

Phyteuma tenerum R. Schulz, *P. orbiculare* auct. angl., non L. **4.** In the hedges of a lane betwixt Kingsbury and Harrow; and in the meadows between Harrow and Pinner, *M. & G.*, 331. Perhaps *Jasione* was intended, *T. & D.*, 179. *P. tenerum* is unlikely to have occurred in Middlesex, but the localities given are equally unlikely to have yielded *Jasione montana.*

COMPOSITAE

Cirsium heterophyllum (L.) Hill, *Carduus heterophyllus* L. **3.** Hounslow Heath, *Doody*, *Ray Syn.* 3: 193; copied by many later authors. No doubt *C. dissectum* intended.

Taraxacum palustre (Lyons) Symons, *T. paludosum* (Scop.) Crépin. Cited for Middlesex in *Druce C.F.* I have been unable to trace the source of this record, and the species is unlikely to have occurred in the vice-county.

POTAMOGETONACEAE

Potamogeton lucens × natans = P. × fluitans Roth, '*P. crassifolius* Fryer. A plant which may be this is in *Lamb. Herb.* at Kew. It was gathered at (**3**) Hounslow', *Bab. Man.* 6: 362. Error, *Syme, E.B.* 9: 63.

LILIACEAE

Polygonatum verticillatum (L.) All. **7.** Ken Wood, *Hunter, Park Hampst.*, 29. Perhaps . . . in cultivation there, *T. & D.*, 277.

Scilla verna Huds. **7.** Ken Wood, *Hunter, Park Hampst.*, 29. An error, or cultivated?, *T. & D.*, 279.

JUNCACEAE

Juncus subnodulosus Schrank, *J. obtusiflorus* Ehrh. ex Hoffm. **4.** Hampstead Heath, *De Cresp. Fl.*, 35. A most improbable habitat for the species.

ORCHIDACEAE

Cephalanthera damasonium (Mill.) Druce, *C. latifolia* Janchen, *C. grandiflora* Gray, *C. pallens* Rich. **1.** Old Park Wood, Harefield, *I. G. Johnson, L.N.* 44: 16. Recorded in error, *I. G. Johnson*.

Epipactis palustris (L.) Crantz. **1.** Pinner Wood, *T. & D.*, 273. Error, the specimen (BM) on which the record was based is *E. helleborine*. **6.** Queens Wood, Highgate, *L. B. Hall, L.N.* 15: 82. Error, the specimen (*Hb. N. D. Simpson*) on which the record was based is *E. helleborine*, fide *Simpson*.

Herminium monorchis (L.) R.Br. **6.** Enfield, *Mart. Pl. Cant.*, 66, and *Mart. Mill. Dict.*, also apparently copied by *Robinson*, 272 and *H. & F.*, 150. It seems probable that Martyn accidentally entered this for *Spiranthes spiralis*, *T. & D.*, 271.

Orchis purpurea Huds. **1.** In the chalk-pit near the Paper-mill at Harefield, *Blackst. Fasc.*, 67; since Blackstone's time it has been gathered frequently in Middlesex, *Bicheno, Trans. Linn. Soc.* 12: 30. No doubt *O. militaris* intended.

Aceras anthropophorum (L.) Ait.f. **1.** Old chalkpits near Harefield (old record), *De Cresp. Fl.*, 1. I have been unable to trace the source of this 'old record'.

TYPHACEAE

Typha minima L. **3.** Hounslow Heath, *Dandridge, Ray Syn.* 3: 436. An error for *T. angustifolia*, *T. & D.*, 287.

CYPERACEAE

Eriophorum latifolium Hoppe, *E. paniculatum* Druce. **4.** Hampstead Heath, 1874; lost, 1895, *White*, 364. Probably a broad-leaved form of *E. angustifolium* intended.

Eleocharis uniglumis (Link) Schult. **1.** Uxbridge Moor, *E. Forster* (BM); cited with a query in *T. & D.*, 299. The specimen is a small form of *E. palustris*. **7.** East Heath, Hampstead, *T. & D.*, 299. Recorded with doubt by *Trimen* and *Dyer*, and likely to have been a small form of *E. palustris*.

Carex distans L. **4.** Hampstead Heath, *Doody*, *Ray Fasc.*, 10. An error for *C. binervis*, *T. & D.*, 309. **7.** Banks of the Lee between Clapton and Tottenham, *Coop. Suppl.*, 12. *Trimen* and *Dyer* suggest that *C. binervis* was seen but the habitat is a most unlikely one for that species.

C. hostiana DC., *C. fulva* auct., *C. hornschuchiana* Hoppe. **1.** Field near Harrow Weald Common, *Melv. Fl.*, 85. Probably an error of name as '*C. fulva*' from Harrow in *Herb. Harrow* is *C. panicea*, *T. & D.*, 309.

C. diandra Schrank, *C. teretiuscula* Gooden. **4.** Side of Paddington Canal near Willesden Junction, *De Cresp. Fl.*, 13. A most improbable locality for this local species.

C. curta Gooden., *C. canescens* auct., non L. **4.** Lanes in a hollow near Neasden, *De Cresp. Fl.*, 12. An unlikely locality for this species of bogs and acid fens.

GRAMINEAE

Melica nutans L. **6.** Copse near Hornsey, *M. & G.*, 86. No doubt *M. uniflora* intended, *T. & D.*, 323.

Spartina maritima (Curt.) Fern., *S. stricta* (Ait.) Roth. **7.** One plant on a wall near Whitehall Stairs, *M. & G.*, 83. Some other grass mistaken for it, *T. & D.*, 313.

Stipa pennata L. agg. **7.** Ken Wood, *Hunter*, *Park Hampst.*, 28. Cultivated in the garden there?, *T. & D.*, 316.

Poa bulbosa L. **4.** Hampstead Heath, *M. & G.*, 103 and *Coop. Fl.* 102; probably bulbous states of *P. pratensis* were intended, *T. & D.*, 323. **7.** Fulham, *M. & G.*, 103; probably bulbous states of *P. pratensis* were intended, *T. & D.*, 323.

ADDENDA AND CORRIGENDA

p. 21, line 7 from bottom. After BALLARD insert (1820–97)

pp. 31–49, additional **ABBREVIATIONS OF BOOKS, MSS. AND PAPERS CITED:**

Perring Suppl. *Critical Supplement to the Atlas of the British Flora.* Edited by F. H. Perring. London (1968).

Richards MSS. MSS. notebooks containing records of Middlesex bryophytes, c. 1922–27. Compiled by P. W. Richards.

Tranter Introd. Pl. Introduced plants in the Ruislip district. By G. G. Tranter, *J. Ruisl. & Distr. N.H.S.* 21: 37–40 (1973).

pp. 56–67, additions to **LIST OF RECORDERS, ETC.**

Beaumont, A. →
Cribb, P. J. →
Jasper, R. W. →
Murray, Sir J. A. H., fl. 1860–72
Roberts, Mrs W. →
Tranter, G. G. →
Warren, Mrs A. →
Wilkinson, L. K. →

p. 99, **S. crassicladum** Warnst.

Formerly native in bogs, but now extinct. **1.** Harrow Weald Common, 1922, *Richards MSS.*

S. fallax Klinggr.

Formerly native in bogs, but now extinct. **4.** West Heath, Hampstead, one tuft, 1927, *Richards MSS.*, det. *Sherrin.*

p. 138, **O. vulgatum** L. Adder's Tongue.

5. Perivale Wood area, 1,000+plants, 1974, *P. J. Edwards.*

p. 140, **P. cretica** L. Ribbon Fern.

5. Basement, Ealing, 1970! →, *McLean.* **7.** Near St Ann's Court, W.1, high up on a west-facing wall where a drainpipe leaks, 1973. This is c. 100 yards from the Dean Street locality, *McLean.*

p. 145, **P. setiferum** (Forsk.) Woynar Soft Shield-fern.

5. Ealing, 1973; introduced, *McLean.* **7.** Regents Park, 1973, probably introduced, *Widgery.*

p. 148, **A. filiculoides** Lam. Water Fern.

1. Colne, Springwell to Uxbridge, 1970, *I. G. Johnson, L.N.* 52: 119. Fray's river, Hillingdon, *Tranter Introd. Pl.*

p. 162, **C. submersum** L. Unarmed Horn-wort.

1. South of Potters Bar, *Widgery*.
p. 170, **H. incana** (L.) Lagr.-Foss. Hoary Mustard.
7. Danvers Street, S.W.3, *E. Young*, *L.N.* 52: 120.
p. 177, **B. orientalis** L.
3. Railway bank, Southall, still plentiful, 1974. **5.** Canal path, Greenford, 1969, *L. M. P. Small*, *L.N.* 49: 19.
p. 193, **V. hirta** L. Hairy Violet.
1. Potters Bar, *Widgery*.
p. 204, **S. vulgaris** var. **commutata** (Guss.) Coode & Cullen
5. Brentford, 1974.
A. githago L. Corn Cockle.
2. Sunbury, one plant in a garden, 1971, *Cribb*.
p. 230, **G. pratense** L. Meadow Cranesbill.
1. Potters Bar, *Widgery*.
p. 234, **E. cicutarium** (L.) L'Hérit. subsp. **cicutarium**. Common Storksbill.
1. Potters Bar, *Widgery*.
p. 242, **R. catharticus** L. Common Buckthorn.
3. By river Crane, Hanworth, one tree, 1956, *Cribb*.
p. 282, **P. recta** L.
5. Still near Acton Town, 1974.
p. 307, after line 9 from bottom insert Lit. *J. Ecol.* **62**: 279 (1974).
p. 309, after line 9 from bottom insert Lit. *J. Ecol.* **62**: 279 (1974).
p. 311, **E. roseum** Schreb. Small-flowered Willow-herb.
2. Bushy Park, *Clement*.
p. 312, **E. adenocaulon** × **hirsutum** = **E.** × **novae-civitatis** Smejkal
6. Near Alexandra Palace, 1973, *Wurzell*, *L.N.* 53: 84.
p. 351, **R. cristatus** DC.
6. By Lee Navigation Canal near Tottenham Lock, 1973, *Lousley*, *L.N.* 53: 84.
R. cristatus × **obtusifolius**
6. By Lee Navigation Canal near Tottenham Lock, 1973, *Lousley*, *L.N.* 53: 84.
p. 372, **C. vulgaris** (L.) Hull Ling, Heather.
3. Waste ground, Hanworth, 1960, *Cribb*.
p. 391, **M. ramosissima** Rochel Early Forget-me-not.
1. Potters Bar, *Widgery*.
p. 392, **E. vulgare** L. Viper's Bugloss.
3. Abundant on old railway marshalling yards, Feltham, 1972 →, *Cribb*.
p. 403, **V. blattaria** L. Moth Mullein.
3. Hounslow, 1973, *McLean*.
p. 404, **L. repens** (L.) Mill. Pale Toadflax.
1. Rubbish-tip, Harefield, one plant, 1972, *Cribb*.

p. 412, **V. spicata** L. subsp. **hybrida** (L.) E. F. Warb.
V. hybrida L.
Introduced. Garden escape. Naturalised in grassy waste places, etc. Very rare.
First record: *J. L. Gilbert*, 1973.
5. Hanwell; near Osterley, 1973, *J. L. Gilbert*.
p. 414, **V. filiformis** Sm. Slender Speedwell.
1. Ickenham Green, *Tranter Introd. Pl.*
p. 420, **O. minor** Sm. Lesser Broomrape.
7. Bishop's Park, Fulham, 1969, Mrs *W. Roberts*, *L.N.* 49: 19. Victoria Embankment Gardens, on *Nicotiana* sp., 1970, *Jasper*, *L.N.* 52: 119.
p. 430, **S. horminoides** Pourr. Wild Clary.
5. Near the Thames, Brentford, a small patch, 1974.
p. 436, **G. speciosa** Mill. Large-flowered Hemp-nettle.
6. Allotment ground, Edmonton, 1974, Mrs *A. Warren*.
p. 493, **C. macrophylla** (Willd.) Wallr. Blue Sow-Thistle.
4. Mill Hill, 1970, *L. K. Wilkinson*, *L.N.* 50: 115.
p. 498, **P. praealta** (Vill. ex Gochnat) C. H. & F. W. Schultz subsp. **praealta.**
Hieracium praealtum Vill. ex Gochnat
Alien. Europe. Naturalised in grassy places, etc. Very rare.
1. Elstree, *Perring Suppl.*
p. 500, **T. lingulatum** Markl.
7. West Kensington, 1974, *Brewis*, det. *A. J. Richards*.
T. cordatum Palmgr.
7. West Kensington, 1974, *Brewis*, det. *A. J. Richards*.
p. 513, **P. multiflorum** (L.) All. Solomon's Seal.
7. Gardens of Hurlingham Court, 1970, *D. S. Warren*.
p. 530, **E. helleborine** (L.) Crantz Broad-leaved Helleborine.
1. Harrow Weald Common, a single large plant, 1974, *J. R. Phillips*.
p. 581, line 22, after (1963). Add *loc. cit.* **61**: 353 (1973).
p. 586, **D. flexuosa** (L.) Trin. Wavy Hair-grass.
5. Ealing, 1969, *McLean*.
p. 611, **L. culinaris** Medic. **2.** Shepperton, 1971, *McLean*.
p. 638, **E. crus-galli** (L.) Beauv., *Panicum crus-galli* L. A common feature of rubbish-tips.

INDEX TO GENERA AND SPECIES

Names of genera are printed in small roman capitals, species are printed in small roman type and synonyms are in italics.

THE RAY SOCIETY

ITS HISTORY AND AIMS

The Ray Society was founded in 1844 by a group of British naturalists which included Thomas Bell, J. S. Bowerbank, Edward Forbes, William Jardine, George Johnston, Edwin Lankester, and Richard Owen. Its purpose, as they stated then, was 'the promotion of Natural History by the printing of original works in Zoology and Botany; of new editions of works of established merit; of rare Tracts and MSS; and of translations and reprints of foreign works; which are generally inaccessible'. The publication of learned books on natural history, with special but not exclusive relevance to the British fauna and flora, remains the object of the Society. It commemorates, with singular aptness, the great English naturalist John Ray (1627–1705). The founders evidently and rightly considered that Ray's breadth of interests, his learning and his eminence in all he touched provided ideals for emulation by the Society.

The publishing activities of the Ray Society began in 1845 with the issue of *Reports on the Progress of Zoology and Botany*, the first part of *A Monograph of the British Nudibranchiate Mollusca* by Alder and Hancock and a volume *On the Alternation of Generations* by Steenstrup, followed in 1846 by *Memorials of John Ray*, *Outlines of the Geography of Plants* by Meyen and *The Organization of Trilobites* by Burmeister. The diversity of subjects thus early established as coming within the scope of the Society has been worthily maintained in subsequent volumes, of which many have become classics. High-quality illustrations associated with authoritative text give most of them a lasting value. Although these publications are mainly systematic monographs on the fauna and flora of the British Isles, they possess a more general appeal and utility on account of the interest of many authors in morphology, ecology and history as well as in taxonomy, and also on account of the wide distribution within the northern hemisphere of many of the organisms concerned. Those pertaining to Zoology include accounts of Protozoa, Porifera, Polyzoa, Coelenterata, Platyhelminthes, Annelida, Trilobita, Crustacea, Arachnida, Insecta (Homoptera, Neuroptera, Odonata, Lepidoptera, aquatic Coleoptera), Mollusca, Urochordata, Amphibia, Reptilia, Birds, and Mammals. Botanical volumes deal with Desmids, Charophyta, planktonic Diatoms, Lichens, Cup Fungi, Knapweeds and Bladder Campions, as well as teratology; they include a translation into English of Linnaeus's *Critica botanica* and a facsimile of Linnaeus's *Species Plantarum* (1753) accompanied by a lengthy exposition of Linnaean method valuable both to zoologists and botanists, and facsimiles of William Turner's *The Names of Herbes* (1548), and John Ray's *Synopsis methodicum Stirpium Britannicarum* (1724). Biographical volumes have dealt with Ray, Hofmeister, and Pennant. Some of these important works would never have been written, let alone published with so many illustrations, had not the Ray Society undertaken their issue.

By providing workers in many fields of biology with standard reference books the Ray Society in turn merits their support. Membership of the Society is open to any person willing by subscription to promote its work of publishing scholarly contributions to British natural history, which are issued as funds and manuscripts permit. Members are entitled to purchase one copy of every new work published by the Society during their membership at a special concessionary price; they have also the privilege of purchasing, at a reduced rate, further copies of each work published by the Society.

The Society's business is conducted by a Council of botanists and zoologists; this is elected at the Annual General Meeting and consists of a President, six Vice-Presidents, an Honorary Treasurer, an Honorary Secretary, an Honorary Publications Secretary, an Honorary Foreign Secretary and twelve Councillors.